T0186746

CREEP AND FRACTURE OF
ENGINEERING MATERIALS AND STRUCTURES

CREEP AND FRACTURE OF
ENGINEERING MATERIALS AND STRUCTURES

Edited by J D Parker
Department of Materials Engineering
University of Wales Swansea

Proceedings of the 9th International Conference
held at University of Wales Swansea,
1st April – 4th April 2001

Book 0769

Published in 2001 by

The Institute of Materials
1 Carlton House Terrace
London
SW1Y 5DB

c. 2001 THE INSTITUTE OF MATERIALS

ISBN 1-86125-144-0

Printed in Wales at Gomer Press, Llandysul, Ceredigion SA44 4QL

PREFACE

The Ninth International Conference on 'Creep and Fracture of Engineering Materials and Structures' was held at the University of Wales, Swansea, from 1st - 4th April 2001.

The present Proceedings, containing the 75 papers presented, maintains the established tradition of providing an authoritative state of the art assessment of the major research programmes undertaken at key expert centres around the world.

This conference series has served as a key forum for scientists and engineers concerned with high temperature deformation and fracture to interact and exchange ideas. Professor Brian (George) Wilshire was the driving force behind initiating this conference series. It is fitting therefore, that the Ninth Conference will commemorate Brian's contribution to the creep field, made over a distinguished career spanning forty years.

Whilst the proceedings provide a lasting record of the importance of creep processes to key industries, the success of a conference is dependent on creating an environment where both formal and informal discussions can take place in a relaxed atmosphere. The advantages offered by the Clyne Castle site have been complimented by the personal commitment and excellence of all involved with this event. Special note should be made of the following contributions:-

- Session chairmen, authors and delegates for guaranteeing the scientific quality.

- The Right Worshipful the Lord Mayor of the City and County of Swansea, Councillor John Davies, and the City and County of Swansea Authority for their hospitality in providing the Civic Reception at the Maritime and Industrial Museum.

- Sponsorship provided by the Welsh Development Agency and ERA Technology Ltd., Leatherhead, Surrey.

- The organisational skills and administrative talents of Conference Secretariat, in particular, Dr Gavin Stratford, Dr Belinda Hulm and Ms Sophie Davies.

- Gomer Press, Printers and Book Manufacturers, Llandysul, Ceredigion, for extending the deadline for papers to the very last minute.

My personal thanks go to all those who have made this conference a success.

J D Parker
University of Wales Swansea
April 2001

CONTENTS

CREEP OF CERAMICS

CREEP OF MAGNESIUM

CREEP OF ALUMINIUM I

CREEP OF ALUMINIUM II

CREEP OF ALUMINIUM III

CREEP OF INTERMETALLICS

CREEP OF COMPOSITES

SUPERALLOYS I

SUPERALLOYS II

SUPERALLOYS III

ENGINEERING APPLICATIONS I

LOW ALLOY STEELS I

LOW ALLOY STEELS II

CREEP FUNDAMENTALS I

Key Microstructural Features of Diffusional Creep

K.R. McNee, H. Jones and G.W. Greenwood

Department of Engineering Materials, University of Sheffield,
Mappin Street, Sheffield, S1 3JD, UK

Abstract

Examination of creep data alone cannot always provide clear evidence for the identification of the operative deformation mechanism. Such uncertainty has led to a prolonged debate on the role of diffusional flow processes during deformation under low stress at elevated temperatures. More definitive information can be obtained from associated microstructural observations since clear distinctions can then be drawn between the different effects arising from different modes of deformation. On this basis low stress creep tests have been carried out on OFHC copper and on a precipitate strengthened magnesium alloy (ZR55). Evidence is presented here involving the displacement of surface marker lines, changes in grain boundary profiles and redistribution of precipitates with respect to the orientation of the applied stress. Observation of these microstructural features has assisted in defining situations where creep occurs by directional diffusion in these materials. The critical characteristics of these features, produced during the accumulation of strain, such that they may be interpreted as evidence for diffusional flow, rather than for any alternative deformation mechanism, are indicated.

1 Introduction

At low stresses and high temperatures, crystalline materials undergo time dependant plastic deformation that has some specific characteristics. For many years, it was assumed that the mechanism for plastic flow under these conditions was adequately described by theories based on diffusional creep (i.e. plastic deformation taking place as a result of the preferential flow of atoms between grain boundaries directed towards grain boundaries under tension). These theories were first proposed by Nabarro [1] and Herring [2] and extended by Coble [3]. In recent times, however, the interpretation of data using these theories has been disputed [4-6]. Distinction between operative mechanisms has traditionally been made on the basis of mechanical creep test data. However, in practice the distinction between different low stress mechanisms on such a basis is not clear. Experimental results rarely match fully the quantitative predictions of theoretical models and pose problems such as the existence of a threshold stress or the effect of grain growth. Any reliance purely on mechanical creep data in order to identify unambiguously the rate controlling mechanism is therefore not satisfactory.

Microstructural changes during creep testing should provide more compelling evidence of the operative mechanism. Scratch displacements at grain boundaries do not usually provide definitive evidence for the operative mechanism as diffusional creep. Instead, as noted by Bilde-Sorenson and Thorsen [7], it is important to show either deposition of material at grain boundaries perpendicular to an applied tensile stress and depletion of material at grain boundaries parallel to an applied tensile stress. This has been achieved in pure (99.95%) copper [8,9] and a copper-2wt%nickel alloy [7,10] by two different techniques, both of which allow the motion of grains to be measured three dimensionally. Creep generated precipitate free zones, first reported for magnesium alloys [11], were thought to provide evidence of deposition of material at grain boundaries perpendicular to the applied tensile stress and were therefore attributed by most workers at that time to diffusional creep [12-18]. More recently it has been claimed by Ruano et al [19] and Wolfenstine et al [20] that the precipitate free zones formed during low stress creep are not consistent with diffusional creep.

The available evidence for diffusional creep in terms of deposition and depletion of material at grain boundaries has been sought together with examination of precipitate distributions after creep. The extent to which these features can be interpreted as evidence for the occurrence of diffusional creep is discussed.

2   Experimental Procedure

Creep tests have been carried out on both OFHC copper (grain size d~55μm) and magnesium-0.5wt%Zr (ZR55) bar (d~1000μm). In addition some ZR55 bar was hot rolled at 773K to give 6mm thick plate (d~30μm) from which flat testpieces were machined. After machining the testpieces were polished by hand to a 1μm finish using diamond paste. The testpieces were then positioned, within an enclosed chamber, in a constant load stress rupture rig for testing. Tests were carried out in a flowing atmosphere of argon-5% hydrogen. The testpiece was heated to the test temperature and held for 24 hours prior to the application of the tensile stress. A thermocouple at the centre of the gauge length was used for the furnace controller, whilst an additional two thermocouples at the top and bottom of the gauge length were connected to the data logging equipment. The temperature along the gauge length was held within 3K of the set point and fluctuations in temperature during testing were less than 1K. A transducer at the top of the load arm also connected to the data logging equipment measured the extension of the testpiece. The temperature and stress dependence of creep rate ($\dot{\varepsilon}$) were determined by making step changes in temperature (T) or stress (σ) during testing. The surfaces of the testpieces were examined after testing using optical microscopy, scanning electron microscopy (SEM) and atomic force microscopy (AFM) which allowed precise measurement of the relative positions of grains in three dimensions. In addition some sections of the ZR55 testpieces were mounted in bakelite and polished to a 1μm finish using diamond paste and etched in a solution containing 2.5cc HCl, 10cc HNO_3, 25g malic acid and 467.5cc ethanol. These etched samples were examined optically and also using energy dispersive spectroscopy (EDS) in an SEM.

3   Results

3.1   Mechanical creep data

For copper a transition from diffusional creep (stress exponent n~1 in $\dot{\varepsilon} \propto \sigma^n$, with little or no threshold stress) to power law creep (n~3) at T=753K for the applicable grain size of ~55μm was found at ~4MPa (Figure 1). The activation energy for the low stress region was 99±5kJ mol[-1], consistent with control by grain boundary diffusion. The stress exponent was found to be 2 for the ZR55 bar (Figure 2). The creep rates determined at 6.3MPa and 9.5MPa were considerably higher with an apparent stress exponent of ~17±6. The transition from a low stress exponent to a high stress exponent occurred between 4.5MPa and 6.3MPa. The activation energy was determined as 175 ± 10 kJ mol[-1]. The stress exponent for the plate material between 1.3 and 4MPa was found to be ~4 and the creep rates were found to be much higher than those associated with the larger grain sized bar material.

3.2   Microstructural evidence

Examination of the surface of both OFHC copper (Figures 3 to 5) and ZR55 specimens (Figures 6 to 8) after creep testing at low stresses revealed displacement of scratches at grain boundaries and widened grain boundary grooves on grain boundaries transverse to the applied stress, which, while not seen consistently over all such boundaries, were found in areas associated with scratch displacements.   These features were clearly detectable using conventional microscopy techniques and have been further quantified using AFM.

The precipitate distribution after creep testing of ZR55 revealed no evidence of precipitate free zones or precipitate pile up in the negligibly stressed threaded region of the testpiece (Figure 9a).   Within the gauge length (Figures 9b-d) precipitate free zones can be seen predominantly on boundaries transverse to the applied stress and precipitate pile up can be observed predominantly on boundaries parallel to the applied stress. No precipitate free zones were detected within the testpieces from the high stress tests (6.3 and 9.5MPa).   The transverse sections (Figures 9c and d) provide more information on the location and shape of the precipitate free zones and indicate a clear orientation difference in the distribution and shape of these zones.

Figure 10 shows energy dispersive spectroscopy (EDS) results for a ZR55 testpiece strained to 2% at 723K and 1.7MPa.   Results for a) the bulk grain and b) grain boundaries transverse and c) parallel to the applied stress are shown.   The main peak corresponds to magnesium and the peaks for silicon (thought to arise from contamination during sample preparation) and zirconium are marked.   No zirconium was detected in the precipitate free zone while considerably more zirconium was found in the longitudinal boundaries than in the bulk of the grains.

4   Discussion

The temperature and stress dependence found in this work at low stresses are consistent with expectation for pure copper when a Coble creep is dominating the deformation [21, 22].   The creep rates however were an order of magnitude faster than predicted by Coble creep theory, which may be regarded as sufficient reason to question the direct applicability of diffusional creep theory.   However in the ZR55 material both the stress dependence and temperature dependence of creep rate in the present work were found to be inconsistent with both the Coble and Nabarro-Herring formulations.   Stress exponents of 2 and 4 were found rather than the Newtonian behaviour predicted and the activation energy of $178\pm10kJmol^{-1}$ for the bar material is higher than the reported value of $134kJmol^{-1}$ for lattice diffusion in pure magnesium [23].   It is possible that these inconsistencies, in part, could be addressed by incorporation of a temperature dependent threshold stress.   The observed creep rates for both the bar and plate material are typically an order of magnitude higher those predicted for Nabarro-Herring creep and Coble creep.   However it is clear that under similar test conditions the creep rate of the smaller grain size plate material was considerably higher than that of the bar material consistent with the operation of diffusional creep.

The microstructural evidence appears to be more consistent, with a definite similarity between the results for copper and ZR55.   Scratch displacements provide clear evidence that under the test conditions both OFHC copper and the ZR55 plate deform by diffusional creep and not by

Harper-Dorn creep, although grain boundary sliding accommodated by slip remains a possible alternative. Indeed on the basis of the scratch displacements any mechanism not involving a large element of grain boundary sliding is not valid.

The enlargement of surface grain boundary grooves on boundaries transverse to the applied tensile stress seen in Figure 3 can readily be explained in the context of diffusional creep theory and has already been shown to be an effect of diffusional creep for uranium dioxide [24]. In addition these widened grooves can be associated with scratch displacements. The fact that these features are strongly orientated with respect to the applied stress together with their clear association with scratch displacements, provides a firm indication of the operation of a diffusional creep process.

Demonstration of deposition and depletion of material at the relevant grain boundaries remains the most clear and conclusive indication of the operation of diffusional creep. It is argued later that the formation of precipitate free zones during creep provides such evidence. Scratch displacements can also provide such conclusive evidence when grain motions by which the displacements can be generated are considered in three dimensions.

4.1   Deposition at Grain Boundaries Transverse to the Applied Stress

Figures 4 and 7 for copper and ZR55 respectively both clearly show the deposition of material at grain boundaries transverse to the applied stress. In Figure 4 the scratch displacement, measured by comparison of the section across the grain boundary between A and B (Figure 4b) and through grain C (Figure 4d), is ~1.7µm and the surface displacement between adjacent grains is negligible (~100nm). This indicates that a thickness of 1.7µm of material has been deposited at the boundary between grains A and B. Similarly in Figure 7 a thickness of 2.8µm of material has been deposited at the boundary between grains A and B. Of additional interest in Figures 4 and 7 are the apparent grain boundary groove widths and depths. It can be seen by examination of Figures 4a and 7a that the boundary grooves transverse to the stress, sectioned in Figure 4c and 7c, are both deeper and wider than for the boundaries parallel to the applied stress.

4.2   Depletion at Grain Boundaries Parallel to the Applied Stress

Figures 5 and 8 again shows scratch displacements at grain boundaries and in these depletion of material at grain boundaries parallel to the applied stress is evident. It is clear from Figure 5a that grain C has moved upwards (i.e. in a direction normal to the applied stress) relative to grains A and B. This shift in grain positions can be quantified by comparison between scratch positions in Figures 5b and c indicating that grain C has moved towards grains A and B by 1.86µm in a direction transverse to the applied stress. The only feasible explanation for this displacement is therefore removal of material, by diffusional flow, from the grain boundaries parallel to the applied stress between grain C and grains A and B, leading to accommodation by grain boundary sliding. In Figure 8 the situation is similar with a 2.3 to 3.3µm thickness of material being removed from the boundary between grains A and C.

4.3   Formation of Precipitate Free Zones and Precipitate Pile Up Regions at Grain
       Boundaries

The microstructural evidence from the surface of ZR55 testpieces is entirely consistent with
the idea that the deformation is dominated by diffusional flow. Examination of the precipitate
free zone(PFZ) widths over a more extensive area, from a test on ZR55 bar at 1.7MPa and
723K and 733K, indicate that they are of a size sufficient to account for 90±7% of the
measured steady state strain. However it has been argued [25-27] that the observation of
PFZs after creep testing is not necessarily an indication of diffusional creep and other
mechanisms for their formation have been suggested. These alternative mechanisms are now
briefly considered against the evidence in the present work.

It has been suggested that PFZs claimed to be evidence for diffusional creep may be formed
by mechanical working prior to creep testing [25] or by annealing and are not produced as a
consequence of the creep strain. However PFZs have not formed in the unstressed region of
the low stress testpiece shown in Figure 9a. A distinct difference in appearance between the
sections taken transverse (Figure 9b) and parallel (Figures 9c and d) to the applied stress is
evident confirming the strong orientation of PFZs. No indication of PFZs or precipitate pile
up was found in the high stress tests. It is clear therefore that the PFZs found in this work
arise from the low stress creep of this material.

Dragging of precipitates by a migrating grain boundary [26] would lead to the formation of
PFZs but no precipitate pile up occurs adjacent to the PFZs. In fact accumulation of
precipitates occurs on the longitudinal boundaries which is clearly visible in Figure 9b and
has been detected using EDS (Figure 10). Dissolution of the precipitates by the migrating
grain boundary [15,16, 27] may also produce PFZs; however, the PFZs at issue, in addition to
being free of precipitates, do not contain zirconium in solution. The mechanism also provides
no means for the accumulation of precipitates on grain boundaries parallel to the applied
stress

In a recent paper by Wadsworth et al [27] the concurrent formation of PFZs and precipitate
pile up at grain boundaries transverse and longitudinal to the applied stress, respectively, has
been proposed using a mechanism based on dissolution of precipitates by migrating grain
boundaries. It is proposed that upon precipitate dissolution the zirconium, and presumably the
hydrogen, diffuse to the longitudinal boundaries where the hydride reforms giving rise to an
apparent precipitate pile up.   Whilst this proposed mechanism would produce the
microstructures observed in this work, it is not considered plausible for the following reasons

- Dissolution of ZrH_2 in this way is thermodynamically unlikely [28]
- No driving force exists for the zirconium to diffuse from a region denuded of
  zirconium to a region rich in zirconium.
- The accumulation of zirconium hydride in the grain boundaries parallel to the applied
  stress would lead to the generation of strain perpendicular to the applied stress, i.e. the
  mechanism operating in isolation would give rise to a negative strain rate.

It is clear that neither annealing nor grain boundary migration during creep can produce the
precipitate distributions after low stress creep seen in the present work.

5 Conclusions

1. Mechanical creep data alone cannot provide unambiguous evidence for the operation of diffusional creep particularly in alloyed material. The constitutive equations for diffusional creep are not strictly adhered to during creep deformation.

2. Scratch displacements together with widened grooves at grain boundaries provide a strong indication that the deformation mechanism under the conditions studied is dominated by diffusional creep. It is difficult to account for the concurrent formation of these features under any mechanism other than diffusional creep.

3. Atomic Force Microscopy produces clear evidence of deposition of material at grain boundaries predominantly transverse to the applied stress and depletion of material at grain boundaries predominantly parallel to the applied stress. These features can only be attributed to the operation of a diffusional flow mechanism.

4. Precipitate free zones at boundaries under tension together with precipitate pile up at boundaries parallel to the applied stress cannot be produced by any known mechanism other than diffusional creep.

Acknowledgement.

This work is supported by a UK EPSRC Research Grant Award (GR/N00296).

References

1. F. R. N. Nabarro, Report of a Conference on the Strength of Solids (Bristol), The Physical Soc., London, p. 75 (1948).
2. C. Herring, J. Appl. Phys., 21, 437 (1950).
3. R. L Coble, J. Appl. Phys., 34, 1679 (1963).
4. O.A. Ruano, J. Wadsworth, and O.D. Sherby, Acta Metall., 36, 1117 (1988).
5. O.A. Ruano, O.D. Sherby, J.Wadsworth, and J. Wolfenstine, Mater. Sci. Eng. A, A211, 66 (1996).
6. O.D. Sherby, O.A. Ruano, and J. Wadsworth, in Creep Behaviour of Advanced Materials for the 21st Century, ed. R.S. Mishra, A.K. Mukherjee, and K.L. Murty, p397, TMS, Warrendale, Pa (1999).
7. J.B. Bilde-Sorenson and P.A. Thorsen, idem, p441.
8. K.R. McNee, H. Jones and G.W. Greenwood, idem, p481.
9. K.R. McNee, H. Jones and G.W. Greenwood, Scripta Mater., in press.
10. P.A. Thorsen and J.B. Bilde-Sorenson, Mater. Sci. Eng., A265, 140 (1999).
11. R.L. Squires, R.T. Weiner, and M. Phillips, J. Nucl Mater., 8, 77 (1963).
12. J.H. Day, J. Nucl Mater., 11, 249 (1964).
13. B.W. Pickles, J. Inst. Met., 95, 333 (1967).
14. W.A. Backofen, G.S. Murty, and S.W. Zehr, Trans. Met. Soc. AIME, 242, 329 (1968).
15. R.B. Jones, in: Quantitative Relationship between Properties and Microstructure, ed. D.G. Brandon and A. Rosen, p43, Israel U.P., Haifa (1969).

9

16. A. Karim, D.L. Holt, and W.A. Backofen, Trans Met Soc-AIME, 245, 1131 (1969).
17. A. Karim and W.A. Backofen, Met. Trans., 3, 709 (1972).
18. G.F, Hines, V.J. Haddrell, and R.B. Jones, in: Physical Metallurgy of Reactor Fuel Elements, ed. J.E. Harris and E. Sykes, p72, The Metals Society, London (1975).
19. O.A. Ruano, J. Wadsworth, J. Wolfenstine, and O.D. Sherby, Mater. Sci. Eng. A, A165, 133 (1993).
20. J. Wolfenstine, O.A. Ruano, J. Wadsworth, and O.D. Sherby, Scripta Metall., 30, 383-386 (1993).
21. B. Burton and G.W. Greenwood, Met. Sci. J., 4, 215 (1970).
22. B. Burton and G.W. Greenwood, Acta Metall., 18, 1237 (1970).
23. B. Burton, and G.L. Reynolds, Mater. Sci. Eng. A, A191, 135 (1995).
24. P.G. Shewmon and F.N. Rhines, Trans. Metall Soc. AIME, 250, 1021 (1956),.
25. B. Wilshire, in: Creep Behaviour of Advanced Materials for the 21st Century, ed. R.S. Mishra, A.K. Mukherjee, and K.L. Murty, p451, TMS, Warrendale, PA (1999).
26. L.E. Raraty, J. Nucl. Mater., 20, 344 (1966).
27. W. Vickers and P. Greenfield, J. Nucl. Mater., 24, 249 (1967).
28. J. Wadsworth, O.A. Ruano, and O.D. Sherby, in: Creep Behaviour of Advanced Materials for the 21st Century, ed. R.S. Mishra, A.K. Mukherjee and K.L. Murty, p425, TMS, Warrendale, PA (1999).
29. F. R. N. Nabarro, idem, p391.

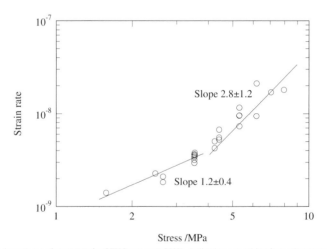

Figure 1 - Stress dependence of creep rate for OFHC copper at 753K. Indicating a transition from a low stress region (slope ~ 1) to a high stress region (slope ~ 3).

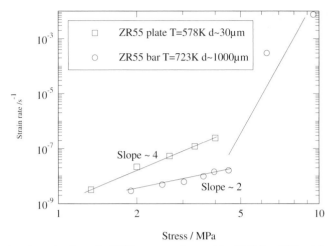

Figure 2 - Stress dependence of strain rate for ZR55 bar at stresses between 1.9 and 9.5MPa at 723K and plate at stresses between 1.33 and 4MPa at 578K giving rise to an apparent stress exponent at stresses below 5MPa of ~2 for the bar material and ~4 for the plate material and a stress exponent of ~17±6 at stresses above 4.5MPa.

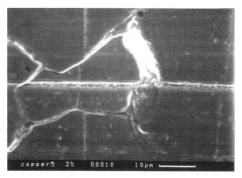

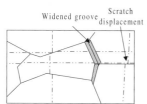

Figure 3 - SEM micrograph of the surface of an OFHC copper specimen strained 1% at 723K and 3.5MPa, then 0.7% at 753K and stresses between 5.3MPa and 8.0MPa and then 1% at 753K and 3.5MPa. Scratch displacements associated with a widened grain boundary groove can be seen. Stress axis horizontal.

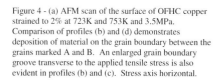

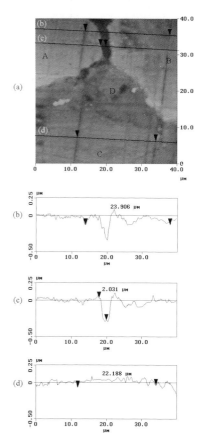

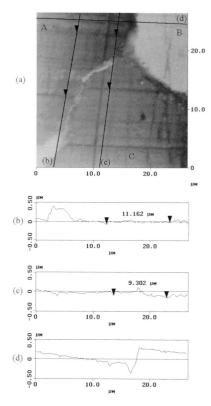

Figure 4 - (a) AFM scan of the surface of OFHC copper strained to 2% at 723K and 753K and 3.5MPa. Comparison of profiles (b) and (d) demonstrates deposition of material on the grain boundary between the grains marked A and B. An enlarged grain boundary groove transverse to the applied tensile stress is also evident in profiles (b) and (c). Stress axis horizontal.

Figure 5 - (a) AFM scan of the surface of OFHC copper strained 2% at 723K and 753K and 3.5MPa. Comparison of profiles (b) and (c) demonstrates a depletion of material at the grain boundary between grains A and C. An enlarged grain boundary groove transverse to the applied tensile stress is also evident in profile (d). Stress axis horizontal.

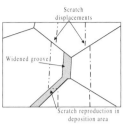

Figure 6 - SEM micrograph of the surface of a ZR55 plate specimen strained 7% at 623K and 2MPa. Scratch displacements associated with a widened grain boundary groove can be seen. Stress axis horizontal.

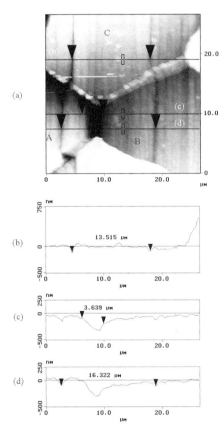

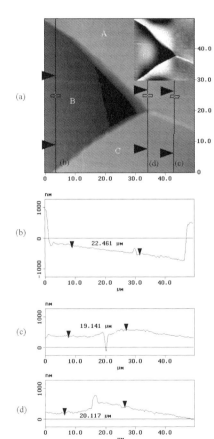

Figure 7 - (a) AFM scan of the surface of a ZR55 plate specimen strained 7% at 623K and 2MPa. Comparison of profiles (b) and (d) demonstrates deposition of material on the grain boundary between the grains marked A and B. An enlarged grain boundary groove transverse to the applied tensile stress is also evident in profiles (c) and (d). Stress axis horizontal.

Figure 8 - (a) AFM scan of the surface of a ZR55 plate specimen strained 10% at 598K and 2-5MPa. Comparison of profile (b) profiles (c) and (d) demonstrates a depletion of material at the grain boundary between grains A and C. Some rotation of grain C is also evident. Stress axis horizontal

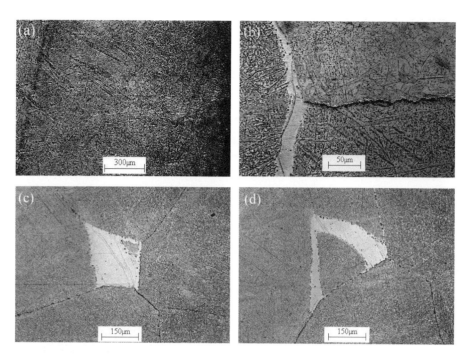

Figure 9 - (a) Optical micrograph of unstressed region of thread taken from longitudinal section of ZR55 bar strained by 2% at 723K and 733K at 1.7MPa. Stress axis horizontal. (b)-(d) - Optical micrographs taken from ZR55 bar strained by 3.5% in secondary creep at temperatures between 723K and 763K and a stress of 1.7MPa. In (b) Denuded zones can be seen on the grain boundaries transverse to the applied stress and precipitate pile up on the boundaries parallel to the applied stress. Stress axis horizontal. In (c) and (d) sections have been taken transverse to the applied stress. Large denuded areas are seen indicating regions near grain boundaries transverse to the applied stress.

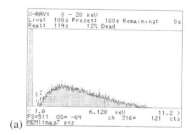

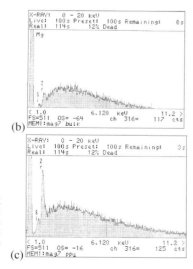

Figure 10 - Spot EDA profiles on ZR55 bar strained by 2% at 723K and 733K at 1.7MPa for (a) a denuded zone associated with a grain boundary transverse to the applied stress, (b) a bulk grain region showing the presence of zirconium and (c) a grain boundary parallel to the applied stress showing high zirconium content due to piled up precipitates.

Σ=3 GRAIN BOUNDARIES IN Cu-Ni SUBJECTED TO DIFFUSIONAL CREEP

Thomas Nørbygaard and Jørgen B. Bilde-Sørensen
Materials Research Department, Risø National Laboratory,
DK-4000 Roskilde, Denmark.

ABSTRACT

In earlier work (Mater. Sci. Eng. **A265** (1999) 140) the behaviour of individual boundaries during diffusional creep was studied by measuring the deformation of the boundaries on the surface of a grid-covered sample in a scanning electron microscope. It was found that the boundaries deform by a combination of grain boundary sliding, grain boundary migration and deposition or removal of material at the boundary. For boundaries that can be described within the framework of the coincident site lattice (CSL) concept this behaviour can be understood in terms of the movement of grain boundary dislocations.

It was also found that Σ = 3 boundaries with a misorientation close to an exact lattice CSL misorientation were inactive during deformation. This cannot be understood directly on basis of the contents of GBDs. Therefore selected Σ = 3 boundaries in Cu-2wt%Ni deformed in the difffusional creep regime have been studied by transmission electron microscopy. The boundaries were found to contain ledges, and dislocation tangles were found to emanate from some of these ledges. The interior of the grains was generally found to contain only few lattice dislocations. It is suggested that Σ = 3 boundaries close to an exact CSL misorientation are inactive because the ledges act as obstacles to the movement of GBDs.

1. INTRODUCTION

The mechanism for diffusional creep was proposed by Nabarro [1] in 1948. He pointed out that the equilibrium vacancy concentration in the vicinity of a grain boundary under stress is given by $C = C_o \cdot \exp(\sigma\Omega/kT)$ where C_o is the thermal equilibrium concentration of vacancies, Ω the atomic volume, k the Boltzmann constant and T the temperature. The stress, σ, is positive for tensile stresses and negative for compressive stresses. A concentration gradient will therefore be set up in the individual grains of a polycrystalline sample subjected to a stress, and this will result in a flow of vacancies from grain boundaries in tension to boundaries in compression (corresponding to a flow of atoms in the opposite directrion). Under the assumptions that the boundaries are perfect sinks and sources for vacancies and that the rate-controlling process is diffusion through the lattice, it can be shown that the rate equation is of the form

$$d\varepsilon/dt = B\sigma\Omega D/d^2kT \tag{1}$$

where B is a numerical constant around 12, D the lattice self-diffusion coefficient and d the grain size. At lower temperatures the grain boundaries may provide an easier diffusion path than the bulk lattice. Coble [2] showed that the rate equation for diffusional creep by boundary diffusion is of the form

$$d\varepsilon/dt = B'\sigma\Omega wD_g/d^3kT \tag{2}$$

where B' is a numerical constant around 50, w the grain boundary width and D_g the grain boundary self diffusion coefficient.

As mentioned, all boundaries were assumed to be perfect sinks and sources in these early models. Since then the understanding of grain boundary structures has increased considerably and it is clear that some of the phenomena observed during diffusional creep can only be understood in terms of the grain boundary structure.

In the following we shall use the coincident site lattice (CSL) model for our discussion of boundary structures [3]. The structural unit model provides an equivalent description, but the CSL model is chosen for the present purpose because it yields a more immediate illustration of some of the points. The basic concept of the CSL model is that some misorientations leads to a high degree of coincidence of lattice points on either side of the boundary. The fraction of coincident lattice sites is denoted by $1/\Sigma$. Such boundaries are often observed to have low energy. Small deviations from the CSL misorientation can be accommodated by secondary grain boundary dislocations (GBDs). These secondary GBDs have Burgers vectors that are translation vectors of the DSC lattice (the DSC lattice has translation vectors that preserve the grain boundary structure). The maximum deviation that can be accommodated by secondary GBDs, v_m, will in the following be given by the Palumbo-Aust criterion [4] $v_m=15°\cdot\Sigma^{-5/6}$. The boundaries that cannot be described within the framework of the CSL model are termed general boundaries. These will not be discussed here, but it should be mentioned that recent research has shown [5] that even in general boundaries there is a significant degree of order and that structural units found in special boundaries also appear in general boundaries.

In the context of diffusional creep it is important to note that it is generally assumed that vacancies can be absorbed or emitted only at grain boundary dislocations. This assumption has been experimentally corroborated by in-situ experiments in a transmission electron microscope [6]. A GBD climbs when it absorbs or emits a vacancy. Since the GBDs are confined to movements in the boundaries they will in the general case have to move by a combination of glide and climb. The glide of GBDs leads to grain boundary sliding. GBDs are associated with a step in the boundary [7]. This step moves with the GBD and this movement of the step leads to grain boundary migration. The implication of the CSL model is thus that deposition or removal of material at the boundary during diffusional creep is coupled to grain boundary sliding and grain boundary migration. This theoretically expected coupling of the diffusional creep process to grain boundary sliding and migration was recently demonstrated experimentally by Thorsen and Bilde-Sørensen [8,9] who by scanning electron microscopy measured the deformation at individual boundaries in Cu-2%Ni deformed in the diffusional creep regime. They thereby demonstrated that material had indeed been deposited on grain boundaries loaded in tension.

Thorsen and Bilde-Sørensen [10] also examined the activity of various CSL boundaries as observed in the scanning electron microscope. It was found that active $\Sigma = 3$ boundaries deviated more than 0.14 from the exact CSL misorientation on the Palumbo-Aust criterion, whereas the inactive boundaries deviated less than 0.21. A similar result was found for $\Sigma = 9$ boundaries. This indicates a threshold value for active boundaries with v/v_m in the range 0.14 to 0.21. It is obvious on the basis of the boundary model discussed above that a grain boundary can operate as an efficient source or sink for vacancies only if it contains a sufficient number of grain boundary dislocations. The question what in this context is a sufficient number has been discussed by Siegel, Chang and Baluffi [11]. They found that the sink efficiency is given by

$$\eta = \cfrac{1}{1 + \cfrac{\ell}{2\pi W} \ln(\cfrac{\ell}{2\pi r_o})} \tag{3}$$

where ℓ is the spacing of dislocation lines in the boundary, W is the distance of the boundary from a perfect planar source or sink and r_o the radius of the thin cylinder around the dislocation that acts as a perfect sink. This equation shows that the boundary should be an efficient sink as long as the spacing of GBDs is considerably smaller than W. It is disputable exactly which value to use for W. However, $v/v_m = 0.14$ corresponds to a dislocation spacing of appr. 14 nm, and any reasonable value of W will be considerably larger than this. In order to understand why the boundaries are inactive we have therefore examined $\Sigma = 3$ boundaries in Cu-2%Ni deformed in diffusional creep by transmission electron microscopy.

2. EXPERIMENTAL PROCEDURE

Cu-2%Ni was chosen as the material for the experiments in order to avoid the extensive grain growth during creep that was observed in preliminary experiments on pure Cu. After machining, the specimen was annealed in vacuum for 4 hours at 673 K and for 4 hours at 1073 K. It was then ground and polished and annealed for a second time. The specimen was finally given a combined etching and polishing treatment with a dispersion of 0.02 μm alumina in a mixture of 96 ml water + 2 ml 30% hydrogen peroxide + 2 ml 25% ammonia in order to obtain a smooth and clean surface. The grain size was measured to be 122 μm.

The specimen which had a reduced section with dimension 37.5x5x0.7 mm^3 was crept for 7 days with an applied stress of 1.14 MPa in a vacuum better than 10^{-5} torr to a total strain of 0.9 %. The strain was measured continually with a strain transducer. The creep rate before the onset of tertiary creep was $2 \cdot 10^{-9}$ sec^{-1}.

After creep, samples for transmission electron microscopy were prepared by jetpolishing a thin disc, cut from the sample, in a mixture of 200 ml H$_3$PO$_4$, 200 ml ethylene glycol and 400 ml water.

3. EXPERIMENTAL RESULTS AND DISCUSSION

Fig. 1 shows a near CSL $\Sigma = 3$ boundary. An obvious feature of this boundary is that it contains ledges. It is also noticed that the interior of the grains contains only few dislocations.

18

Fig. 1. Micrograph of a near CSL Σ=3 boundary exhibiting ledges inclined approximately 82° to the {111} facets.

500 nm

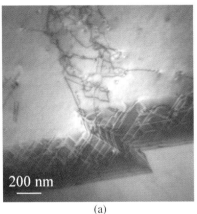

200 nm

(a)

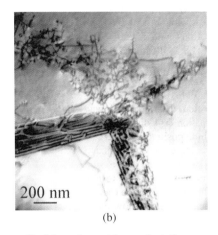

200 nm

(b)

Fig. 2. Dislocations emanating from ledges on a Σ= 3 boundary with an orientation close to the exact CSL misorientation. (a) θ = 59.94°, $v/v_m = 0.062$ (b) θ = 59.83°, $v/v_m = 0.040$.

This supports the view that the deformation has not taken place by a lattice dislocation mechanism. Another striking feature is seen in fig. 2 which shows that tangles of dislocations emanate from some of the ledges. In some cases the tangles lie in continuation of one of the planes forming the step in the boundary (fig. 2 (a)), in other cases in continuation of both planes (fig. 2 (b)).

Wolf et al. [12] performed an experimental and theoretical study of such ledges in Σ = 3, $[01\bar{1}]$ tilt boundaries in Cu. They found the existence of an energy minimum for a boundary with an inclination of 82° to the {111} boundary. This boundary exhibits a rhombohedral 9R structure formed in a 1 to 2 nm thick layer. The results were supported by high resolution electron microscopy.

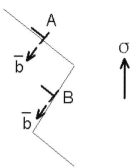

Fig. 3. Sketch showing two dislocations with identical Burgers vectors. Under the tensile stress indicated the dislocation marked A will move downwards to the right while the one marked B will move upwards to the right.

It seems highly likely that the presence of a rhombohedral structure at the ledges can act as an obstacle to the transfer of GBDs between the facets on the boundary. Even if the rhombohedral layer had not existed, there would be some configurations of the boundary for which it would be impossible to transfer secondary GBDs between the facets because dislocations with identical Burgers vector would move in opposite directions on the two facets This is illustrated in fig. 3. Two dislocations with the same Burgers vector are shown lying on two different facets of the boundary. The dislocation marked A could be a $\frac{a}{3}\langle 111\rangle$ dislocation lying in the coherent part of the boundary. Under the tensile stress indicated this dislocation must move downwards to the right in order to contribute positively to the deformation. The dislocation marked B, however, must move upwards to the right in order to contribute positively to the deformation. This sketch shows only one line orientation, but by considering these segments as parts of expanding loops on the two boundary segments, one can appreciate that the result applies to any line orientation: dislocations with the same Burgers vector will tend to move in opposite directions on the two boundary segments. It is therefore not possible to transfer GBDs from one boundary segment to another. The argument is only intended to show that ledges under certain circumstances can act as an obstacle to deformation. Transfer of this particular GBD could in principle occur if the ledge had turned the other way.

In this context it is interesting to note one of the results that Huang and Mishin [14] recently obtained in an experimental study of the ledges in $\Sigma = 3$ boundaries. They found that all the boundaries with ledges examined in their study had deviations of less than 2° from the exact CSL misorientation. A deviation of 2° corresponds to a value of $v/v_m = 0.33$ on the Palumbo-Aust criterion. The experimental threshold value for activity of $\Sigma = 3$ boundaries during diffusional creep is thus in quite good agreement with the value of v/v_m below which ledges are found. This supports the view that the threshold value for macroscopic activity of $\Sigma = 3$ boundaries during diffusional creep can be related to the presence of ledges on the boundaries.

The observation of dislocation tangles emanating from corners on the $\Sigma = 3$ boundaries also indicates that the lack of macroscopic activity during diffusional creep can be related to the presence of ledges. The tangles indicate that the boundaries have been active at the microscopic level during the deformation process. The dislocation tangles may arise from extrinsic grain boundary dislocations that are emitted from the boundary or they may be produced by sources in the neighbouring grain by stress concentrations from pile-ups of

GBDs. In both cases the appearance of dislocation tangles indicate that the boundary in principle has the ability to deform, but that the presence of a ledge impedes the deformation.

4. CONCLUSION

A number of $\Sigma = 3$ boundaries close to the exact CSL misorientation has been studied by transmission electron microscopy. It has been shown that these boundaries contain ledges and it is concluded that it is the presence of these ledges that make boundaries with $v/v_m < 0.14 - 0.21$ macroscopically inactive during diffusional creep. Dislocation tangles can be seen to emanate from the corners of some ledges, and this demonstrates that the boundaries are active at the microscopic level.

ACKNOWLEDGEMENT

We are grateful to Jan Larsen and Jørgen Lindbo for the experimental assistance. The present work was carried out within the Engineering Science Center for Structural Characterization and Modelling of Materials.

REFERENCES

1. Nabarro F. R. N., 1948, Report on a conference on the strength of solids, Bristol, 7-9 July 1947 (London: The Physical Society) pp. 75-90.
2. Coble, R. L., 1963, J. Appl. Phys., Vol. 34, pp. 1679-1682.
3. Hull, D., and Bacon, D. J., 1984, Introduction to Dislocations (Oxford: Pergamon Press) pp. 196-202.
4. Palumbo, G., and Aust, K. T., 1990, Acta metall. Vol. 38, pp. 2342-2352.
5. Farkas, D., 2000, J. Phys.: Condens. Matter. Vol. 12, pp. R497-R516.
6. King, A. H., and Smith, D. A., 1980, Philos. Mag., Vol. A42, pp. 495-512.
7. King, A. H., and Smith, D. A., 1980, Acta Cryst., Vol A36, pp. 335-343.
8. Thorsen, P. A., and Bilde-Sørensen, J. B., 1999, Mater. Sci. Eng., Vol A265, pp. 140-45
9. Bilde-Sørensen, J. B. and Thorsen, P. A., 1997, Boundaries and Interfaces in Materials; The David A. Smith Symposium (Warrendale: TMS), pp. 179-188.
10. Thorsen, P. A., and Bilde-Sørensen, J. B., 1999, Mater. Sci. Forum, Vol. 294-296, pp. 131-134.
11. Siegel, R.W., Chang, S. M., and Baluffi, R.W., 1980, Acta Metall., Vol. 28, pp. 249-257.
12. Wolf, U., Ernst, F., Muschik, T., Finnis, M., and Fischmeister, H. F., 1992, Philos. Mag., Vol A66, pp. 991-1016.
13. Huang, X., and Mishin, O., 2000, To be published in Acta Mater.

Creep of platinum alloys at extremely high temperatures

R. Völkl[1], U. Glatzel[1], T. Brömel[2], D. Freund[2], B. Fischer[2], D. Lupton[3]

[1] Friedrich Schiller University Jena, Loebdergraben 32, D-07743 Jena, Germany
[2] University of Applied Sciences Jena, Carl-Zeiss-Promenade 2, D-07745 Jena, Germany
[3] W.C. Heraeus GmbH & Co. KG, Engineered Materials Division, Heraeusstrasse 12-14,
63450 Hanau, Germany

Keywords: Creep, platinum, ODS, high temperature, fracture strain, internal oxidation

Abstract

Extremely high temperatures, mechanical stresses and severe corrosion conditions act in combination in equipment for the production of optical glasses. Common solid solution platinum base alloys such as Pt-10%Rh or Pt-5%Au often cannot be used because they can cause a detrimental discoloration of the glass. The materials selection is therefore limited to pure platinum or a dispersion hardened pure platinum matrix. However, compared to conventional alloys, dispersion hardened alloys are usually less ductile.

In the present investigation creep tests have been carried out on pure Pt (99.95 weight%) and the oxide dispersion hardened alloy Pt DPH in the temperature range between 1473 K and 1873 K. The specimens were heated directly by an alternating electric current in specially designed creep test facilities. Strain was measured with a video extensometer.

The measurements confirm much better creep rupture strength and lower stationary creep rates at high temperatures for the oxide dispersion hardened alloy Pt DPH compared to pure Pt. Activation energies and stress exponents are given. The hardening effect in Pt DPH is achieved by finely dispersed zirconium and yttrium oxides, which are formed during an internal oxidation heat treatment of the alloy in a compact state. Oxide particles can be found both inside grains and at grain boundaries. SEM and TEM investigations of the microstructure were performed in order to discuss the mechanical properties of the alloy at high temperatures.

1 Introduction

Platinum base alloys can be used at temperatures up to 2000 K. Despite their high prices, their exceptional chemical stability and high melting points make them interesting for structural applications in the glass industry and for some aerospace applications [1-5]. High-quality optical glasses and glass fibres require the use of platinum-tank furnaces, stirrers and feeders.

Pure platinum, like most engineering materials based on a metal matrix, requires a considerable improvement in strength to meet the design requirements for high temperature applications. Common strengthening mechanisms are cold working, solid solution hardening, fine grain hardening or precipitation hardening. However these conventional strengthening mechanisms become ineffective with increasing temperatures due to recrystallization, particle coarsening and dissolution. Several manufacturers have therefore developed oxide dispersion strengthened (ODS) platinum base alloys with improved high temperature properties for the most demanding applications [6-11]. In these alloys small amounts of stable oxide particles are finely distributed throughout the alloy matrix. The creep and stress-rupture strength at temperatures of 0.8-0.9 T_m are greatly increased due to reduced dislocation mobility and stabilization of the grain boundary structure even for long exposure times. Most ODS platinum base alloys are produced by complicated and expensive powder metallurgical processes (PM) [6-10]. The oxidation and corrosion resistance of the ODS materials is at least as good as that of the equivalent conventional solid solution alloys.

2   Experimental procedure

2.1   Materials

To produce the so-called platinum DPH materials, a new manufacturing route was developed by W. C. Heraeus in cooperation with the University of Applied Sciences Jena. In order to obtain Pt DPH, which is strengthened by zirconium and yttrium oxides, technical grade platinum is melted with additions of Pt-Zr and Pt-Y masteralloys in a vacuum induction furnace, homogenized and cast to an ingot under argon atmosphere. The ingot is then cold rolled to slabs, which are subjected to an extended annealing treatment in an oxidizing atmosphere. This treatment results in the essentially complete internal oxidation of the alloying elements Zr and Y within the compact platinum matrix. At present, the materials Pt DPH, Pt-10%Rh DPH and Pt-5%Au DPH are available.

2.2   Mechanical tests

The high prices of precious metals and the extremely high test temperatures prevent the use of ordinary creep test facilities. This is the reason why specially designed test units [12-15] are used at the University of Applied Sciences Jena for mechanical testing of platinum base alloys and refractory metal alloys at very high temperatures. The equipment permits tests at constant load either in air or under a protective gas atmosphere. A schematic diagram of the equipment is given in Fig. 1. The specimen is heated directly by an alternating electric current. Lower temperatures at the clamps than in the centre of the specimen allow the use of copper clamps. An infrared thermometer focussed on the centre of the sample monitors the temperature. The thermometer adjusts the heating current to maintain a constant temperature. The load is applied to the sample by means of calibrated weights. Due to thermal gradients near the clamps the deformation is limited to the hottest part of the specimen, i.e. the central zone. The temperature in a zone 15 mm to each side of the specimen centre is nearly constant (Fig. 1). The specimens have the form of strips with four small shoulders in the zone of constant temperature in order to measure strain (Fig. 2). The heated specimens are observed by a high-resolution camera, which is itself controlled by the program SuperCreep developed at the University of Applied Sciences for strain measurements by means of digital image analysis. SuperCreep continuously determines the distance between the shoulders with an accuracy better than 0.1 %.

The test equipment described has proven good reliability. The simplicity of the equipment offers the advantages of low costs and easily attainable high temperatures of up to 3300 K [15]. There is no need to use special materials for parts exposed to high temperatures. The free sight to the specimen makes it possible to measure the deformation by means of digital image processing, which allows not only longitudinal strain but also transverse strain and crack propagation to be measured.

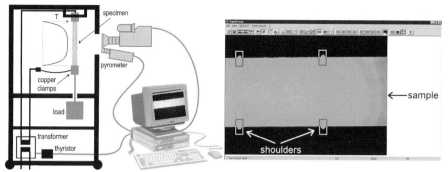

Fig. 1: Schematic diagram of the equipment used to measure stress rupture strength and creep properties of metals at temperatures up to 3300 K.

Fig. 2: Image of a tensile specimen with shoulders recorded by the image analysis software SuperCreep.

3 Results

3.1 High temperature mechanical properties

Figs. 3-5 show creep curves of the alloy Pt DPH and pure platinum at temperatures between 1473 and 1873 K. The creep curves are typical for the chosen temperature and stress range. The alloy Pt DPH has creep curves with a short primary creep, a long secondary creep and a well pronounced tertiary creep stage in the temperature and stress ranges investigated. No incubation period was observed for either pure platinum or Pt DPH similar to that found by Hamada et al [16]. The creep rates for pure platinum at 1873 K and 2.5 MPa (Fig. 5) increase steadily until fracture, i.e. the primary creep stage with a decreasing creep rate and the stationary creep stage are missing.

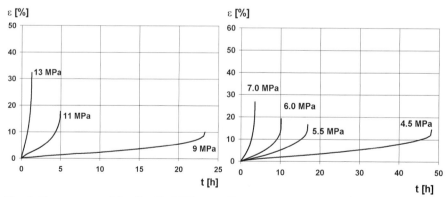

Fig. 3: Creep curves of the alloy Pt DPH at 1473 K and various loads.

Fig. 4: Creep curves of the alloy Pt DPH at 1723 K and various loads.

From the creep curves in Figs. 3-5 and Fig. 6 it is clearly seen that the fracture strains of the alloy Pt DPH increase with increasing load at temperatures between 1473 K and 1873 K. It is also interesting that an increase of the fracture strains with increasing temperature but constant load is observed for the alloy Pt DPH (Fig. 6). Although the fracture strains of

24

Pt DPH at 1873 K are not as high as those of pure platinum, the fracture strains of between 10 and 50 % (Fig. 6) observed on this material are exceptionally high compared with most other oxide dispersion strengthened platinum alloys [17].

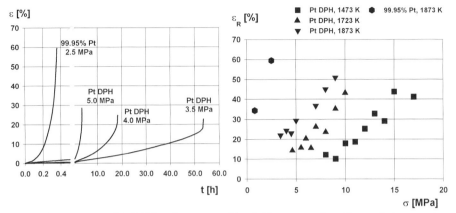

Fig. 5: Creep curves of Pt DPH and pure platinum at 1873 K and various tensile loads.

Fig. 6: Fracture strains of pure platinum and Pt DPH at temperatures between 1473 K and 1873 K.

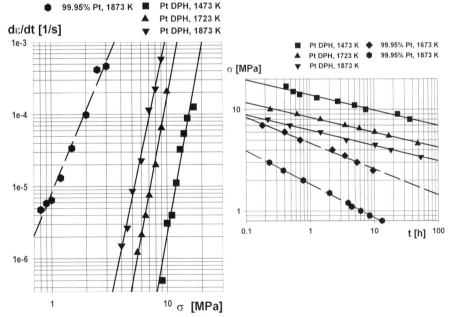

Fig. 7: Stationary creep rates of pure platinum at 1873 K and the alloy Pt DPH at temperatures between 1473 K and 1873 K.

Fig. 8: Stress rupture curves of the alloy Pt DPH and pure platinum between 1473 and 1873 K.

Fig. 7 shows double logarithmic Norton plots of the stationary creep rate of Pt DPH and pure platinum at temperatures between 1473 K and 1873 K. The stationary creep rates of Pt DPH are several orders of magnitude lower than for pure platinum at the same temperature and stress. The stationary creep rates of Pt DPH in the temperature and stress range considered can be fitted well with a Norton creep law:

$$\dot{\varepsilon} = 0.26 \cdot \sigma^{7.3} \cdot e^{-343\,kJ/RT} \left[\frac{1}{s}\right] \quad (\sigma \text{ in MPa})$$

The Norton exponent for Pt DPH is nearly twice as high as the exponent of 3.8 for pure platinum. The much higher creep strength of Pt DPH is also demonstrated by the creep rupture curves in Fig. 8.

3.2 Microstructure

In case of the alloy Pt DPH, zirconium and yttrium were added as oxide formers. Oxide particles can be recognized both at grain boundaries and inside the grains (Fig. 10). Unfortunately transmission electron microscopy (TEM) investigations have so far only been performed for the alloy Pt-10% Rh DPH. The alloy Pt-10% Rh DPH [18] is manufactured in the same way and shows high temperature behaviour comparable to that of the alloy Pt DPH. We can therefore assume that the TEM observations shown in Figs. 9-12 also give a first insight into the strengthening mechanisms of the alloy Pt DPH. Fig. 9 shows a very fine dispersion of small particles in the Pt/Rh matrix of the alloy Pt-10% Rh DPH. The mean particle size is less than 5 nm. Figs. 10 and 11 also reveal larger oxide particles at grain boundaries. Energy dispersive X-ray analysis (EDX) in the TEM gave a composition of approximately 65 at.% oxygen, 30 at.% zirconium and 5 at.% yttrium for the oxides at the grain boundaries (Fig. 12).

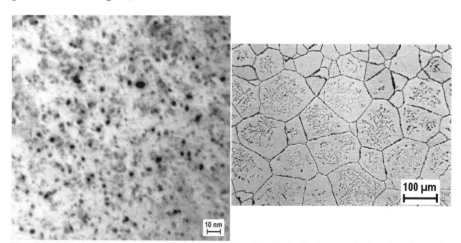

Fig. 9: TEM bright field image of the alloy Pt-10% Rh DPH showing a very fine dispersion of particles having diameters smaller than 10 nm.

Fig. 10: Optical micrograph showing the grain structure of the alloy Pt DPH.

26

 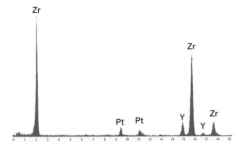

Fig. 11: Several TEM bright field images are connected in order to reveal a chain of oxide particles at a grain boundary in the alloy Pt-10% Rh DPH.

Fig. 12: EDX spectrum of the oxide particle marked with a cross in Fig. 11.

4 Discussion - Conclusions

Most platinum ODS materials are produced by powder metallurgical routes Disadvantageous mechanical properties have prevented the broader application of these platinum ODS materials. The powder metallurgical platinum ODS materials often obtain their high strength only at the price of poor ductility, i.e. these materials are brittle and tend to form cracks in high temperature service [17]. Furthermore these ODS materials cause processing problems, especially in welding. As far as they are weldable at all, the fusion welding process results in a considerable decrease of strength in the welding zone down to values of the conventional non-ODS materials. The DPH materials overcome these disadvantages of the powder metallurgically manufactured ODS materials. Creep tests at temperatures up to 1873 K confirm a considerable improvement in strength for Pt DPH compared to pure platinum with no great loss of ductility.

With the platinum DPH materials, a new manufacturing route has been introduced to form a fine dispersion of oxides in a metal matrix. The matrix alloy is melted with additions of masteralloys containing Zr and Y and cast to an ingot. Semi-finished slabs of the cold rolled ingot are then subjected to an extended heat treatment in order to achieve the complete internal oxidation of the alloying elements Zr and Y within the compact platinum matrix.

Two particle fractions are observed in the DPH materials. A very fine and homogeneous distribution of exceptionally small particles is found inside the grains. The mean particle size of less than 5 nm is smaller than in most ODS materials [19, 20]. These fine particles cause the pinning of dislocations and are therefore responsible for the substantial strengthening effect observed in the DPH materials. A second particle fraction with sizes ranging from tenths of a micron to a few microns can be found at the grain boundaries. EDX measurements indicate that this large particle fraction consists of ZrO_2-Y_2O_3 mixed oxides. Obviously, these bigger particles, which are visible in the optical microscope as well as in TEM images, inhibit the grain boundary mobility. Therefore, the material shows excellent grain stability and remains fine-grained at high annealing temperatures and long periods of exposure. Surprisingly, although many of these brittle oxide particles can be found at grain boundaries, the DPH materials show much better ductility than most platinum based ODS materials produced by powder metallurgy.

5 References

[1]   Lupton, D., May 1990, Noble and Refractory Metals for High Temperature Space Applications, Advanced Materials Journal, pp. 29-30

[2]   Whalen, M.V., 1988, Space Station Resistojets, Platinum Metals Review, Vol. 32, No. 1, pp. 2-10

[3]   Jansen, H.A., Thompson, F.A., 1992, Use of oxide dispersion strengthened platinum for the production of high-quality glass, Glastechnische Berichte, Vol. 65, No. 4, pp. 99-102

[4]   Thompson, F. A., July 1990, The Use of Platinum and its Alloys in the Glass Industry, GLASS, pp. 279-280

[5]   Stokes, J., 1987, Platinum in the Glass Industry, Platinum Metals Review, Vol. 31, No. 2, pp. 54-62

[6]   Johnson Matthey, GB, 1973, Patent specification 1 340 076

[7]   Owens-Corning Fiberglass Corporation, USA, 1981, Patent application WO 81 / 00977

[8]   Degussa AG, Germany, 1982, Patent application DE 30 30 751 A1

[9]   Johnson Matthey, GB, 1988, Patent specification DE 31 02 342 C2

[10] SCHOTT GLAS, Germany, 1995, Patent specification DE 44 17 495 C1

[11] Degussa AG, Germany, 1996, Patent specification DE 195 31 242 C1

[12] Fischer, B., Töpfer, H., Helmich, R., 1987, Patent specification DD WP G 0 1 N / 245 576 A3

[13] Fischer, B., Freund, D., Lupton, D., 1997, Gerät für Zeitstandversuche bei extrem hohen Prüftemperaturen, Proc. Werkstoffprüfung 1997, Verein Deutscher Eisenhüttenleute (VDEh), Bad Nauheim, Germany, December 4 to 5 1997

[14] Völkl, R., Freund, D., Fischer, B., Gohlke, D., 1998, Berührungslose Dehnungsaufnahme an widerstandsbeheizten Metallzugproben mit Hilfe digitaler Bildverarbeitung bei Prüftemperaturen bis 3000°C, Proc. Werkstoffprüfung, Verein Deutscher Eisenhüttenleute (VDEh), Bad Nauheim, Germany, pp. 211-218

[15] Fischer, B., Freund, D., Lupton, D., Stress-Rupture Strength of Rhenium at Very High Temperatures, Proc. International TMS Symposium on Rhenium and Rhenium Alloys, Orlando, Florida, USA, February 9 to 13 1997, pp. 311-320

[16] Hamada, T., Hitomi, S., Ikematsu, Y., Nasu, S., 1996, High-Temparature Creep of Pure Platinum, Materials Transactions, JIM, Vol. 37, No. 3, pp. 353-358

[17] Völkl, R., Freund, D., Fischer, B., Gohlke, D., 1999, Comparison of the Creep and Fracture Behavior of Non-Hardened and Oxide Dispersion Hardened Platinum Base Alloys at Temperatures Between 1200 °C and 1700 °C, Proc. CFEMS8, Key Engineering Materials, Vol. 171-174, ISBN 0-87849-842-7, pp.77-84

[18] Fischer, B., Behrends, A., Freund, D., Lupton, D., Merker, J., Dispersion Hardened Platinum Materials for extreme Conditions, Proc. TMS Annual Meeting, San Diego, USA, February 28 to March 4, 1999, ISBN 0-87339-420-8, pp. 321-331

[19] Heilmaier, M., Müller, F.E.H., April 1999, The role of Grain Structure in the Creep and Fatigue of Ni-Based Superalloy PM 1000, Journal of materials research, pp. 23-27

[20] Groza, J.R., Gibeling, J.C., 1993, Principles of particle selection for dispersion-strengthened copper, Materials Science & Engineering A, Vol. 171, pp. 115-125

Fatigue Behavior of Cu-SiO$_2$ Alloy at Elevated Temperatures

Hiromi Miura and Taku Sakai

Department of Mechanical Engineering and Intelligent Systems,
The University of Electro-Communications, Chofu, Tokyo 182-8585, Japan
tele/fax.: +81-(0)424-43-5409, e-mail: miura@mce.uec.ac.jp

Abstract

Copper polycrystal with dispersed sphere SiO$_2$ particles was cyclically deformed at temperatures between ambient temperature and 687K. For comparison, copper single crystal and bicrystal containing SiO$_2$ particles were also prepared to reveal effect of grain boundary on the fatigue behavior at the elevated temperatures. Though the polycrystal tended to show longer life than the single crystals at lower temperature, the life of the polycrystal became shorter than that of the single crystal especially under applied low stress amplitude at high-temperature region. At the critical stress amplitude where the life of the polycrystal becomes shorter than that of the single crystal, dominant cracking manner was supposed to change from nucleation in grain to on grain boundary. It is because that preferential crack nucleation at grain boundary and the propagation along the grain boundary took place more easily with increasing temperature and decreasing stress amplitude. All the bicrystals, whose grain boundaries could geometrically slide easily, fractured intergranularly when cyclically deformed at 673K. From these results, the observed loss of life or endurance at higher temperatures in the polycrystal should be caused by grain-boundary cracking induced by the occurrence of grain-boundary sliding.

1. Introduction

It is well known that second phase particles dispersed in metallic materials lower mobility of dislocations, and impede grain-boundary sliding (GBS) [1-3] and grain-boundary migration (GBM) [4] at high temperatures. Because such characters are expected to improve the high-temperature strength of metallic materials, numerous studies of dispersion-hardened alloys have been carried out at elevated temperatures. Excepting the effects above mentioned, some other superior natures of the dispersed particles to maintain strength of materials have been also revealed and known; dispersion of large amount of particles can suppress the occurrence of recrystallization which causes work softening [5]. However the studies using dispersion-hardened alloys at high temperatures were mainly about the creep and tensile strength, and so on. Even though the dispersion-hardened alloys are expected to be employed in fatigue circumstance at elevated temperatures, the research about their fatigue behavior is quite few as far as the authors know [6, 7].

At elevated temperatures, on contrary to the expected role of strengthening, such particles on grain boundary sometimes hasten grain-boundary cracking (GBC) during creep and static tensile deformation [2, 3]. It is because that grain-boundary particles inhibit GBS, hence stress concentration sites to promote the nucleation of cracks and voids are formed around them. It is reported that grain boundaries in single phase matrix seem to be the most preferential nucleation site for cracking even at ambient temperature [8]. Therefore, we naturally expect that the GBC can take place preferentially at grain boundaries and must be the most important life controlling factor in the dispersion-hardened alloys during high-temperature fatigue.

In the present study, fatigue behavior of Cu-SiO_2 polycrystal at temperatures between ambient temperature and 687K was investigated. Special emphasis was placed on the role of grain boundary and GBC in the fatigue behavior. For that purpose, Cu-SiO_2 single crystal and bicrystal were also prepared for comparison. This is the motivation of the present paper.

2. Experimental procedures

Polycrystal of a Cu-0.05mass% Si alloy of about 2mm in thickness was internally oxidized by the powder pack method with a mixture of Cu (1 part), Cu_2O (1 part) and Al_2O_3 (2 parts) at 1273K for 24h. By this treatment, we obtained Cu-0.41vol.%SiO_2 alloy polycrystal, where the initial grain size and mean particle radius in grain were 290mm and 57.2nm, respectively. The photograph of the polycrystal is shown in Fig. 1. After a degassing treatment at 1273K for 24h in a

Fig. 1 Initial equi-axised grain structure of the polycrystal sample.

graphite mold in vacuum to eliminate the excess oxygen, specimens of 11mm gage length and $6 \times 2mm^2$ cross section were cut by electric discharge machining. For comparison, specimens of Cu-SiO_2 single crystal and Cu-SiO_2 bicrystal containing [001] twist 48.7˚ boundary were also prepared by the Bridgman method followed by the same heat treatments above mentioned. The details of preparation of single crystal and bicrystal were written in elsewhere [2, 3]. Here after, the samples will be referred as polycrystal, single crystal and bicrystal. The loading direction in the single crystal was [001]. Such [001] single crystal is known as a typical multiple slip orientation, whose fatigue behavior is reported to be quite similar to that of polycrystals [9]. The bicrystal samples were machined to have boundaries inclined 45˚ against the loading direction so that GBS took place easily. The [001] twist 48.7˚ boundary is supposed to have an inherent nature of relatively easy sliding [10]. In order to observe the slip morphology, mirror-like surfaces were obtained by mechanical and electritical polishing before cyclic deformation. Fatigue tests were carried out in vacuum of about 10^{-3}Pa at the temperatures from 298K to 673K in a servo-hydraulic machine. Tensile-tensile cyclic loading was applied at 20Hz. Load ratio of R=0.1 was chosen to avoid buckling of the present thin and small samples. After the tests, the microstructure and fractography were observed by using scanning electron microscope (SEM).

3. Results and discussion

Figure 2 shows a series of the result of cumulative strain against cyclic number (ε-N curves)

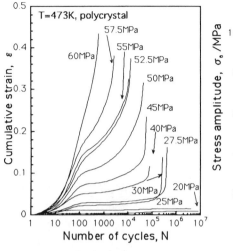

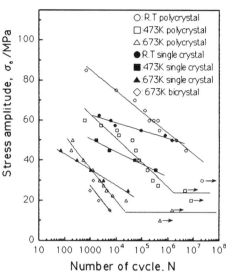

Fig. 2 Cumulative strain-cyclic stress amplitude curves for polycrystal cyclically deformed at 473K.

Fig. 3 S-N$_f$ diagram for the polycrystal, single crystal and bicrystal.

for the polycrystal at 473K. All the other curves also exhibited similar feature irrespective of temperature and crystal design, though they are not shown here. In all the curves in Fig. 2, the strain increases rapidly at lower cycles first, shows a cyclic-creep steady state or gradual increase at medium cycles, and then, finally shows abrupt increase at increasing tempo to rupture. However, in contrast, such increase in strain were not clearly observed at very lower stress amplitude region as σ_a = 20MPa and 25MPa. Their fatigue tests were stopped before rupture. The observed stagnation of strain increase should be because that the maximum applied stresses were almost the same with or lower than the yield stress of about 50MPa.

Figure 3 shows summarized results of the stress amplitude-life to rupture (S-N$_f$) curves for the all samples and temperatures tested. The bicrystal was cyclically deformed only at 673K, where GBS took place extensively, to investigate the effect of GBS on GBC during cyclic deformation. The lives of all the samples become shorter as the stress amplitude and temperature increase. Though some of the fatigue tests were interrupted in high-cycle fatigue region before the rupture, it is interesting to note that the fatigue limit in the polycrystal seemed to appear in the present Cu-SiO$_2$ alloy. What more interesting is that S-N$_f$ curves of the polycrystal and single crystal does not show same tendency of decrease in life, while the [001] single crystal is supposed to have similar fatigue behavior with that of polycrystal. The life of the polycrystal became shorter than that of the single crystal at certain critical stress amplitude. That critical stress amplitude becomes lower and appears at lower cycle as temperature increases. The longer life of the polycrystal at higher applied stress region would be because of higher yield stress and higher work-hardening rate due to existence of grain boundaries compared with that of the single crystal as understood by Hall-Petch relation.

32

intergranularly. However, initiated crack seems to propagate mostly transgranularly when cyclic deformed at room temperature, while that mainly propagates intergranularly at 673K. That is, intergranular cracking takes place more easily with increasing temperature. The wavy or serrated grain boundary seen in Fig. 5 (b) is due to GBM followed by dynamic recrystallization what was observed only near the fractured area in the samples deformed at 673K.

On the other hand, the occurrence of crack in single and bicrystal was quite different compared with that of polycrystal, as shown in Fig. 6. No crack nucleation could be observed on the surface of the single crystal when cyclically deformed to $N=10^3$ at a stress amplitude of 30MPa at 673K. Only multiply slipped lines are visible. Contrary to that, in the bicrystal, in which GBS is easiest, extensive crack along grain boundary

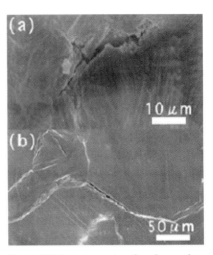

Fig. 5 SEM photographs of surfaces of the polycrystals cyclically deformed to (a) $N=10^5$ at a stress amplitude of 75MPa at room temperature and (b) $N=10^3$ at a stress amplitude of 25MPa

already took place when deformed under the same testing condition. These results indicate also that grain boundary cracking would take place more easily at higher temperature. Furthermore, it is interesting to see that about 7μm displacement of slip lines and unevenness formed on the grain boundary is taking place, which would be concerned with GBS. Miura et al. have shown experimentally and theoretically that GBS plays an important role in the occurrence of GBC in dispersion-hardened copper alloy during monotonic tensile test [2, 3]. If GBS occurs, grain-boundary particles act as inhibitors of GBS. This provides the stress concentration sites to induce the preferential nucleation of voids and cracks. It is reported that GBS can take place even at 450K [14].

GBS becomes more easily with increasing temperature. We believe that the occurrence of GBS at higher temperature causes or stimulates GBC even during cyclic deformation. Furthermore, GBS becomes more difficult as

Fig. 6 SEM photographs of the surfaces of (a) single crystal and (b) bicrystal cyclically deformed to $N=10^3$ at a stress amplitude of 30MPa at 673K.

strain rate increases. The strain rate increase corresponds to the increase in stress amplitude in the present stress-controlled cyclic deformation test. The temperature and strain-rate dependencies of GBS are supposed to cause the observed GBC and change of the critical stress amplitude in the present study.

Figure 7 exhibits grain-boundary fracture surfaces of the bicrystal. The elongated voids in Fig. 7 (a) suggests the occurrence of GBS during fatigue. On the other hand, mixed feature of striation and dimples can be seen in Fig. 7 (b). Grain-boundary particles, whose size is about one order of magnitude larger than that in grain, exist about half of the dimples. The relatively flat appearance of the serration with no SiO_2 particle indicates that crack propagated near the grain boundary, probably along particle free zone, but not exactly along the grain boundary. These features reveal that fracture is caused by combination of some cracking and fracture mechanisms; crack propagation to cause striation, elongated void formation due to GBS, ductile fracture to form dimples just before rupture, and etc. Of course, the manner of crack propagation would also change depending on fatigue conditions and also on grain-boundary character, although it can be not discussed in details here.

Schematic diagram of the critical stress amplitude can be drawn in Fig. 8. In the higher stress region than the critical stress amplitude, crack initiation in grain should be more dominant than that in grain. In contrast, cracking on grain boundary should be more dominant than that in grain in the lower stress region. However, the critical point must not be actually so sharp. It is because that some

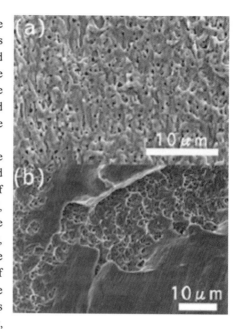

Fig. 7 Fractographs of the bicrystal cyclically deformed at 673K at stress amplitude of 30MPa with different magnifications.

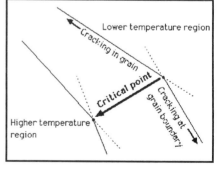

Number of cycle, N_f

Fig. 8 Schematic diagram of the critical stress amplitude and its temperature dependency.

of weak grain boundaries may sometimes trigger the cracking [2, 3] even in higher stress region and, still more, the crack propagation and fracture manners seem so complicated.

4. Summary

The current investigations yield the following results.

(1) Fatigue hardening behavior appeared at 473K in the $Cu-SiO_2$ alloy.

(2) Fatigue life became shorter with increasing temperature and stress amplitude.

(3) Grain-boundary cracking took place more easily as temperature increased and stress amplitude decreased.

(4) Critical stress amplitude, where dominant cracking manner changes from nucleation in grain to on grain boundary, was found, and the critical stress amplitude decreased with increasing temperature.

References

[1] Mori, T., 1985, Suppression of Grain Boundary Sliding by Second Phase Particles, Proceedings Eshelby Memorial Symposium, Cambridge University Press, pp. 277-291.

[2] Miura, H., Saijo, K., and Sakai, T., 1996, High-Temperature Deformation and Fracture Behavior of $Cu-SiO_2$ Bicrystals with [011] Twist Boundaries, Materials Transaction of JIM, Vol. 37, pp. 754-761.

[3] Miura, H., Sakai, T., Tada, N., Kato, M. and Mori, T., 1993, Temperature Dependence of Ductility and Fracture of $Cu-SiO_2$ Bicrystals with [001] Twist Boundaries, Acta Metallurgica Materialia, Vol. 41, pp.1207-1213.

[4] Manohar, A.P., Ferry, M. and Chandra, T., 1998, Five Decades of the Zener Equation, ISIJ International, Vol. 38, pp. 913-924.

[5] Blaz, L., Sakai, T. and Jonas, J.J., 1983, Effect of Initial Grain Size on Dynamic Recrystallization of Copper, Metal Science, Vol.17, pp. 609-616.

[6] Stobbs, M.W., Watt, F.D. and Brown, M.L., 1971, The Fatigue Hardening and Softening of Copper Containing Silica Particles, Philosophical Magazine, Vol. 23, pp. 1169-1184.

[7] Miura, H., Kotani, K. and Sakai, T., 2001, Temperature Dependency of Cyclic Deformation Behavior of $Cu-SiO_2$ Single Crystal, Acta Materialia, (in press).

[8] Hu, Y.M. and Wang, Z.G., 1998, Grain Boundary Effects on Fatigue Deformation and Cracking Behavior of Copper Bicrystals, International Journal of Fatigue, Vol. 20, pp. 463-469.

[9] Wang, Z.G., Li, W.X., Zhang, F.Z. and Li, X.S., 1999, Effects of Crystallographic Orientation and Grain Boundary on Cyclic Deformation in Copper Crystals, Proceedings of Fatigue '99, China, pp. 93-104.

[10] Monzen, R. and Suzuki, T., 1996, Philosophical Magazine Letters, Nanometer-Scale Grain-Boundary Sliding in Copper Bicrystals with [001] Twist Boundaries, Vol. 74, pp. 9-15.

[11] Suresh, S., 1998, Fatigue of Materials, Cambridge University Press, pp.152-154.

[12] Hatanaka, K., Kawabe, H., Tanaka, M. and Yamada, T., 1972, Microscopic Substructure in Face Centered Cubic Metals Fatigued at Elevated Temperatures, Proceedings of 15th Japan Congress on Materials Research, Kyoto, pp. 56-63.

[13] Sastry, L.M.S. and Ramaswami, B., 1973, Fatigue Deformation of Copper Single Crystals Containing Coherent Alumina Particles, Philosophical Magazine A, Vol. 28, pp. 945-952.

[14] Monzen, R., Sumi, Y., Kitagawa, K. and Mori, T., 1990, Nanometer Grain Boundary Sliding in Cu:[011] Symmetric Tilt Boundaries, Misorientation Dependence and Anisotropy, Acta metallurgica materialia, Vol. 38, pp. 2553-2560.

Acknowledgments

The authors acknowledge the material supplement from Nippon Mining and Metals Co. Ltd., Kurami Works, Japan and the help of Mr. Tozaki, K., The University of Electro-Communications.

CREEP FUNDAMENTALS II

EFFECT OF DYNAMIC STRAIN AGING ON DEFORMATION OF COMMERCIALLY PURE TITANIUM

F. Breutinger and W. Blum

Institut für Werkstoffwissenschaften, Lehrstuhl 1,
Martensstraße 5, D-91058 Erlangen, F. R. Germany

Abstract

The deformation resistance of commercially pure Ti-2 (0.5 at.-% O_{eq}, 50 μm recrystallized grain size) was investigated in uniaxial compression and tension both at constant stress σ and constant strain rate $\dot{\varepsilon}_{tot}$ in the temperature range between 423K and 723K (0.22 to 0.38 of melting point) in air. The testing conditions 723K and $8 \cdot 10^{-5}$/s define the value of temperature-normalized strain rate Z_D where the deformation characteristics change drastically: with Z_D increasing, the steady state relation between Z_D and σ shows a sharp kink with a change in stress exponent of strain rate from 3 to >100 and the evolution of deformation resistance changes from nonmonotonic (with a relative maximum) to monotonic. The observation of negative strain rate sensitivity of the flow stress points to dynamic strain aging and leads to interpretation of the observed effects in terms of transition from viscous motion of dislocations with a cloud of solute atoms to jerky, thermally activated motion of dislocations over fixed obstacles related with solutes. The effective stress model is used to model the observed pronounced transients from steady state deformation with jerky glide to steady state deformation with viscous glide.

1. Introduction

In technical applications of titanium the time-dependent deformation in the form of creep must be taken into account, since titanium is creeping even at temperatures T below ambient temperature and stresses distinctly below the yield stress (e.g. [1, 2]). The deformation behaviour of titanium is determined by the interaction between dislocations and interstitials. This can be seen e.g. from the strong influence of the interstitial content on the mechanical properties and the occurrence of dynamic strain aging, i.e. formation of solute atmospheres around dislocations during deformation, at intermediate T (e.g. [3, 4, 5]). In the present paper the deformation behaviour of titanium is quantified in the temperature range between 423K and 723K. It will be shown that dynamic strain aging has a strong effect on the deformation behaviour, including the steady state of deformation.

2. Experimental

Titanium of commercial purity (h.c.p. α-Ti-2; composition in wt.-%: O_2: 0.145, N_2: 0.0035, C: 0.01, H_2: 0.0015, Fe: 0.025) was obtained in the form of rods of 10 mm diameter produced by hot rolling followed by recrystallizing for 2 h at 948K in argon atmosphere. The initial microstructure consisted of equiaxed grains of 50 μm size (linear intercept). Loading was done parallel to the rod axis on cylindrical tensile specimens of 30 mm initial gage length at ambient temperature $l_{0,RT}$ and 4 mm diameter as well as on cylindrical compression specimens of initial aspect ratio (length over square root of cross section) $\kappa_0 \approx 1.3$ with $l_{0,RT}$ varying from

3.3 to 4.7 mm. Compression and tensile tests were done at $423K \leq T \leq 723K$ (0.2 to 0.4 of melting point) at constant stress σ (force per cross section at T; stress controlled test) as well as at constant true total (elastic plus plastic) strain rate $\dot{\varepsilon}_{tot}$ (strain rate controlled test) in air. The total strain ε_{tot} was measured by two symmetrically arranged rod-tube extensometers and determined as $\varepsilon_{tot} = |\ln(l/l_0)|$, where l and l_0 are the actual and initial length at T (using $1.0 \cdot 10^{-5}$ K^{-1} as linear thermal expansion coefficient). Elastic contributions of the specimen and creep machine were subtracted from ε_{tot} in order to get the plastic strain ε (see [6] for details).

Loading to the prescribed stress of creep was done either manually or by using a motor-driven table resulting in $\dot{\varepsilon}_{tot}$ varying between 10^{-3}/s and 10^{-2}/s during loading. Variations of strain rate due to deviations from constancy in stress were subsequently corrected by using a suitable stress exponent $n = d \log \dot{\varepsilon} / d \log \sigma$.

Friction between specimen and compression machine causes the stress σ acting on the dislocations to be reduced compared to the nominally applied stress σ_{nom} (force per average cross section at T) [7]: $\sigma = \sigma_{nom} \cdot [1\text{-}\exp(-\kappa/c)^p]$. $c = 0.25$ and $p = 1$ were obtained from fitting (see [7]. Friction becomes significant only for large ε exceeding 0.4.

3. Deformation Resistance

3.1 Evolution

The upper part of Fig. 1 displays the evolution of flow stress σ with ε at $\dot{\varepsilon}_{tot} = 10^{-3}$/s for different T. Circles mark the loading strain ε_{ld} in creep defined as that strain, where $\dot{\varepsilon}_{tot}$ has fallen to 10^{-3}/s just after loading. There is only a small difference between ε_{ld} and the corresponding strain for the same stress in the tests at constant $\dot{\varepsilon}_{tot} = 10^{-3}$/s indicating similar hardening in the loading phase of the creep tests and in work hardening at constant rate. The lower part of Fig. 1 shows the creep rate as function of strain. The factor $f_{723K} = \exp\{(Q/R) \cdot [(T/K)^{-1} -(723)^{-1}]\}$ ($R = 8.31$ kJ / mol K: gas constant; $Q = 242$ kJ/mol: activation energy for creep [8]) serves to approximately compensate deviations of test temperature from 723K. This normalization brings many of the curves measured at different T but similar σ into smooth connection.

The shape of the normalized creep rate-strain-curves has the following characteristics. For large normalized rates most of the curves have negative curvature, for small normalized rates there is a relative minimum in creep rate $\dot{\varepsilon}$. The varying curvature leads to unexpected variation of the creep time with temperature: The decrease of creep rate with strain starting from 10^{-3}/s is ε_{ld} is stronger for 523K than for 423K. More time is necessary for creep by a certain amount in addition to ε_{ld} at the higher one of the two temperatures.

Fig. 2 shows the influence of strain rate $\dot{\varepsilon}_{tot}$ on the σ-ε-curves at 723K. There is a remarkable change in the shape of the curves between 10^{-5}/s and 10^{-4}/s. Up to $\dot{\varepsilon}_{tot} = 10^{-5}$/s a relative maximum in flow stress is found, while for $\dot{\varepsilon}_{tot} \geq 10^{-4}$/s there is continuous strain hardening. At $\varepsilon \approx 0.02$ the flow stress increases significantly as $\dot{\varepsilon}_{tot}$ increases from $3 \cdot 10^{-6}$/s to 10^{-5}/s and then falls as $\dot{\varepsilon}_{tot}$ increases further.

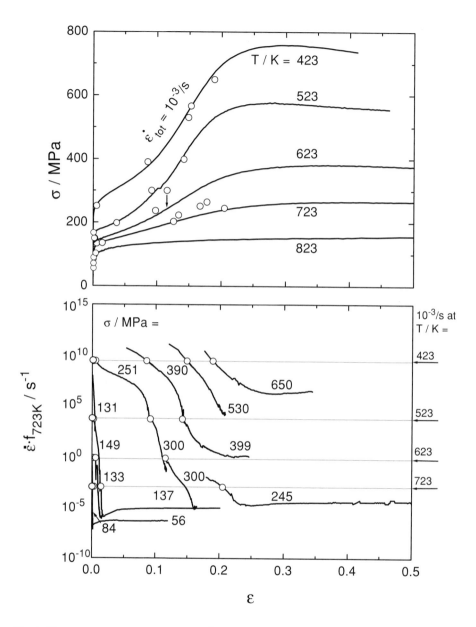

Fig. 1: Upper part: σ-ε-curves at $\dot{\varepsilon}_{tot} = 10^{-3}$/s for different T. Circles mark ε_{ld} as function of creep stress σ. Lower part: strain rate-strain-curves for different T and σ normalized for 723K. The intersections of the horizontal lines for $\dot{\varepsilon} = 10^{-3}$/s at given T with the strain rate-strain-curves define ε_{ld} (circles).

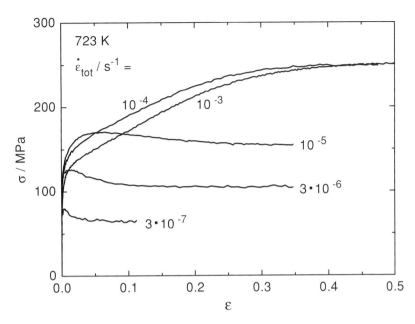

Fig. 2: σ-ε-curves at 723 K for different $\dot{\varepsilon}_{tot}$.

3.2 Steady State

The steady state of deformation is defined as the range where the deformation resistance remains constant in the course of deformation at constant conditions, i.e., where the strain rate-strain-curves of Fig. 1 and the stress-strain-curves of Figs. 1 and 2 become horizontal. Fig. 3 shows the steady state relation between T-normalized strain rate $Z_D = \dot{\varepsilon}\, k_B\, T\, /\, (D\, G\, b)$ (k_B: Boltzmann constant; $b = 2.95 \cdot 10^{-10}$m: length of Burgers vector [8], $D = 0.013$ m^2/s·exp(-242 kJ mol^{-1} / RT): diffusion coefficient of creep [8]) and normalized stress σ / G (shear modulus G / MPa = 49200-25.8·T/K [8]). The data measured at different T form a unique master curve. It lies at the right of the master curve for pure Al as a representative of pure materials [9], consistent with the solute hardening of Ti-2.

There is a sharp kink in the curve at $Z_D \approx 10^{-6}$ and $\sigma \approx 9 \cdot 10^{-3}\, G$, corresponding to $\sigma \approx 260$ MPa and $\dot{\varepsilon} \approx 8 \cdot 10^{-5}$/s at 723K. Below the kink the steady state data fall close to the three power law $Z_D = (\sigma /G)^3$. Above the kink the stress exponent takes extremely high values. The existence of a kink was confirmed in a test where deformation at a constant nominal stress of 275 MPa was carried to high compressive strain of more than 1.1 (Fig. 4a). Under these conditions the aspect ratio drops to 0.25 and the influence of friction becomes significant, leading to a continuous reduction of stress by up to 30%. The decrease of σ (Fig. 4a center) is accompanied by a decrease of $\dot{\varepsilon}$ (Fig. 4a top). The course of the Z_D-σ/G-variation of this test is shown in the insert of Fig. 3. There is good agreement with the steady state data points indicating that the deformation resistance in the test with gradually decreasing stress was close to steady state.

The slope of the curve goes through a maximum, separating regions with $n \approx 30$ at higher stresses and $n \approx 3$ at lower stresses. In other words: the steady state flow stress becomes insensitive to the strain rate. Fig. 4b confirms this by means of $\dot{\varepsilon}$ -changes between $\dot{\varepsilon} = 10^{-4}$/s ($Z_D = 2.7 \cdot 10^{-6}$) and $\dot{\varepsilon} = 10^{-3}$/s ($Z_D = 2.7 \cdot 10^{-5}$).

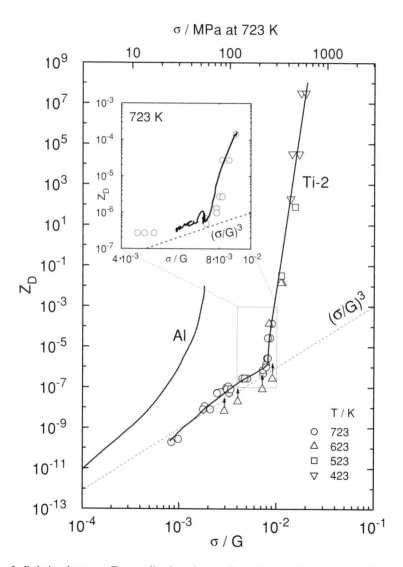

Fig. 3: Relation between T-normalized strain rate Z_D and normalized stress σ/G at steady state. The curve in the insert shows the Z_D-σ/G-relation of the test of Fig. 4a. For comparison the curve for pure Al (from [9]) is also shown. The dotted line marks the natural creep law.

a)

b)

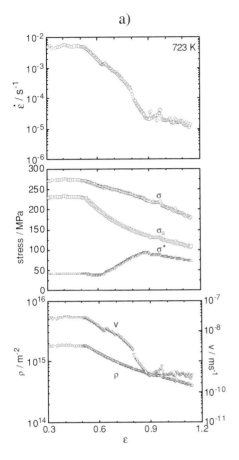

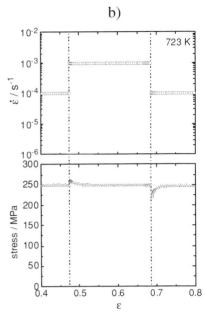

Fig. 4: Variation of $\dot{\varepsilon}$ and σ with ε at 723K for

a) test at approximately constant nominal stress σ_{nom} = 275 MPa; also shown: evolution of modelled σ_G, σ^*, ρ and v with ε,

b) test with strain rate changes between $\dot{\varepsilon}$ = 10^{-4}/s and $\dot{\varepsilon}$ = 10^{-3}/s at steady state.

4. Interpretation in terms of dynamic strain aging

The strain rate interval from 10^{-5}/s to 10^{-4}/s at 723K is special in the following respects: There is a maximum of initial work hardening resulting in negative strain rate sensitivity of the flow stress and the strain rate sensitivity of the steady state flow stress becomes negligible (stress exponent n diverges). It appears natural to relate these features to the interaction between dislocations and interstitial solute atoms. At high strain rates the dislocations move so fast that the obstacles formed by solute atoms are fixed compared to the dislocations. Dislocation motion occurs in a thermally activated jerky fashion. At low strain rates the mobility of solutes is high enough for them to enrich at the dislocations. A cloud of solutes forms at the dislocations which is dragged along with them. Thermally activated overcoming of the solutes is no longer necessary, the dislocations move viscously along with their clouds. The effective stress model has been shown in the past to be appropriate to describe the creep behaviour of solid solutions [10]. We will adopt the model here in order to find a semi-quantitative explanation of the observed deformation characteristics of Ti-2.

The constitutive equations of the model are as follows:

$$\dot{\varepsilon} \cdot M = b \cdot \rho \cdot v(\sigma^*, T) \tag{1}$$

$$\sigma^* = \sigma - \sigma_G \tag{2}$$

$$\sigma_G = \alpha \, b \, G M \sqrt{\rho} \tag{3}$$

$$\rho_{\infty,v} = (\sigma / Gb)^2 \qquad \rho_{\infty,j} = 2 \cdot \rho_{\infty,v} \tag{4}$$

Eq. (1) expresses $\dot{\varepsilon}$ by the product of density ρ and velocity v of dislocations; M is the Taylor factor. Eq. (2) defines the effective stress σ^* for dislocation glide as difference between the applied stress σ and the athermal stress component σ_G which in turn is related to ρ via eq. (3); the dislocation interaction constant is set as $\alpha = 0.2$. The steady state dislocation density ρ_∞ is assumed to vary in proportion to σ^2. However, for the same σ the steady state density $\rho_{\infty,j}$ in the case of jerky glide of dislocations is taken to be twice the steady state density $\rho_{\infty,v}$ in the case where dislocations move viscous (eq. (4)). This difference is motivated by the fact that dragging of solute clouds requires a relatively large effective stress. It is supported by the observation of the $\dot{\varepsilon}$ -minimum in the low stress $\dot{\varepsilon}$ -ε - curves of Fig. 1 indicating a decrease of ρ as the mode of dislocation motion changes from jerky to viscous [11].

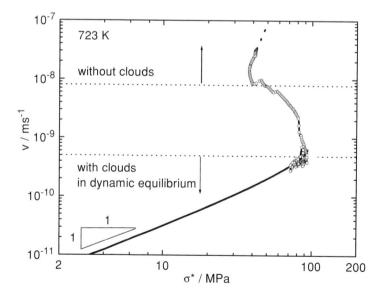

Fig. 5: Relation between v and σ^* at 723 K. Symbols refer to the test of Fig. 4a. Upper branch: jerky dislocation motion without clouds, lower branch: viscous dislocation motion with clouds in dynamic equilibrium.

46

The model was applied to the test shown in Fig. 4a in the following manner: The dislocation density is assumed to vary from the initial steady state value without clouds $\rho_{\infty,j}$ to the final steady state value with clouds $\rho_{\infty,v}$ according to

$$\rho\,(\Delta\varepsilon) = \rho_{\infty,j} + (\rho_{\infty,v} - \rho_{\infty,j})\cdot[1 - \exp(-\Delta\varepsilon/0.1)]. \qquad (5)$$

$\Delta\varepsilon$ is the difference between the current ε and that strain, where σ starts to decrease. This leads to the variation of σ_G, σ^* and v with ε given in Fig. 4a. The resulting variation of v with σ^* is shown in Fig. 5 (symbols). The variation is nonmonotonic and in qualitative agreement with calculations on the drag stress exerted on a moving dislocation by the solute atmosphere (e.g. [12]). It is interpreted in terms of dynamic strain aging as follows: As the effective stress

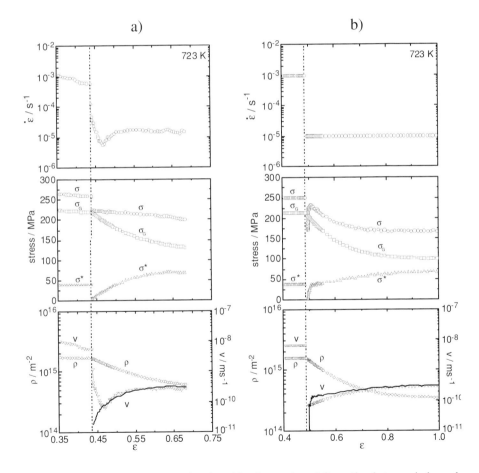

Fig. 6: Variation of $\dot{\varepsilon}$, σ, ρ, σ_G, σ^* and v with ε in transient deformation between jerky and viscous steady state deformation at 723K; a) stress reduction from 260 MPa to 225 MPa, b) strain rate reduction from 10^{-3}/s to 10^{-5}/s.

is decreasing continuously during deformation, a cloud of solute atoms starts to form at the dislocations. The configuration of the cloud is determined by the counteraction of capture of solutes by the dislocations and loss of solutes which are not fast enough to diffuse along with the dislocations. It is well known that the interaction between solute atoms and dislocations leads to a range of negative stress sensitivity of the dislocation velocity (and of $\dot{\varepsilon}$) between the ranges of jerky dislocation motion without clouds and viscous motion with clouds (e.g. [12]).

5. Transients with dynamic strain aging

For further study of the transition in mechanism of dislocation motion, tests with sudden changes in deformation conditions were performed. Fig. 6 shows two tests starting with jerky dislocation motion in the steady state at $\dot{\varepsilon} \approx 10^{-3}$/s and ending with viscous dislocation motion in the steady state at $\dot{\varepsilon} \approx 10^{-5}$/s. In both cases one observes a relative maximum in deformation resistance in the transient after the change. The explanation of the transient is based on the variation of ρ from the initial to the final steady state value according to eq. (4) and (5). Knowing ρ one derives from eq. (1) to (3) the variations of σ_G and σ^* as well as v depicted by symbols in Fig. 6. Alternatively, v is calculated from σ^* using the v-σ^*-relation of Fig. 5 (line in Fig. 6). The two v-ε-curves are in good agreement for the late stage of the transient. In the first part of the transient there is a strong discrepancy. This must be expected because the curve of Fig. 5 holds for clouds in dynamic equilibrium. After a sudden change, however, the cloud structure deviates from equilibrium and a certain strain is necessary for the clouds to approach their equilibrium state. In this strain interval the dislocations move faster than predicted from Fig. 5 because the clouds are still in the process of formation. This interpretation agrees with the findings derived from the transient characteristics for a solute hardened Ni-base alloy [10]. An additional contribution to the reduction in deformation resistance after the sudden change in deformation conditions comes through the strain associated with dislocation recovery. This strain contribution is not considered in the effective stress model. It yields one argument why deformation continues in forward direction in Fig. 4b even though the effective stress calculated from the present model is negative.

6. Concluding Remarks

It has been shown that the changes in deformation characteristics of commercially pure Ti can be associated with dynamic strain aging. The effective stress model of plastic deformation yields a semi-quantitative explanation of the pronounced transients accompanying the transition from fast steady state deformation with jerky motion of dislocations at high density to slow steady state deformation with viscous motion of dislocations at low density. It is also possible to explain the transients observed in the beginning of deformation for the transition from jerky to viscous dislocation motion as the dislocation density increases from the low initial value to the steady state value in terms of dynamic strain aging and the corresponding evolution of the dislocation density. However, this discussion is beyond the scope of the present paper.

It should be noted that the effective stress model used above implies that the dynamic recovery of the dislocation density which is necessary to reach the steady state of deformation is not a critical problem. This is best demonstrated by the strain rate change test of Fig. 4b. When the dislocations strip off their clouds, they become faster. However, the corresponding increase in dislocation generation rate is compensated by a corresponding increase in annihilation rate without measurable increase in steady state stress. In other words: the increase in dislocation density which must be expected for increase of the rate of dislocation annihilation produces an increase in athermal stress component which is virtually completely compensated by the reduction in effective stress due to loss of the clouds. This suggests that the dynamic recovery of the dislocations under the given conditions of deformation at low homologous temperature is controlled by the transport of dislocations to their sinks rather than by the annihilation of dislocation dipoles itself. Spontaneous annihilation of dislocations has been suggested to explain the behaviour of pure Al at $\sigma = 2 \cdot 10^{-3} G$ [13] and may well act as a significant mechanism of dynamic recovery at the position of the kink in Fig. 3 where $\sigma = 9 \cdot 10^{-3} G$.

Acknowledgements

Thanks are due to Audi AG for cooperation and the Bayerische Forschungsstiftung for financial support.

References

[1] Kiessel, W.R. and Sinnott, M.J., 1953, Creep Properties of Commercially Pure Titanium, Trans. AIME, 197, pp. 331-338.
[2] Kalkbrenner, E., 1969, Kriechen und elastische Nachwirkung von Reintitan, Metall, 23, pp. 424-431.
[3] Doner, M. and Conrad, H., 1973, Deformation Mechanisms in Commercial Ti-50A (0.5 at. pct O_{eq} at Intermediate and High Temperatures (0.3 - 0.6 T_M), Metallurgical Transactions, 4, pp. 2809-2817.
[4] Conrad, H., 1981, Effect of Interstitial Solutes on the Strength and Ductility of Titanium, Progr. Mater. Sci., 26, pp. 123-403.
[5] Santhanam, A.T., Garde, A.M. and Reed-Hill, R.E., 1972, The Significance of Dynamic Strain Aging in Titanium, Acta Metall., 20, pp. 215-220.
[6] Blum, W., Watzinger, B. and Zhang, P., 2000, Creep of Die-Cast Light-Weight Mg-Al-base Alloy AZ91hp, Advanced Engineering Materials, 2, 349.
[7] Weidinger, P., Blum, W., Hunsche, U. and Hampel, A., 1996, The Influence of Friction on Plastic Deformation in Compression Tests, phys. stat. sol. (a), 156, pp. 305-315.
[8] Frost, H.J. and Ashby, M.F., 1982, Deformation-Mechanism Maps, Pergamon Press, Oxford.
[9] Blum, W., 1991, Creep of Aluminium and Aluminium Alloys, in: Hot Deformation of Aluminum Alloys, ed. by Langdon, T. G. and Merchant, H. D. and Morris, J. G. and Zaidi, M. A., Warrendale, Pa, pp. 181-209.
[10] An, S.U., Wolf, H., Vogler, S. and Blum, W., 1990, Verification of the Effective Stress Model for Creep of NiCr22Co12Mo at 800°C, in: Proc. of the 4th Int. Conf. on Creep and Fracture of Engineering Materials and Structures, ed. by B. Wilshire and R.W. Evans, The Institute of Metals, London, pp. 81-95.
[11] Blum, W., 2000, Creep of crystalline Materials: experimental basis, mechanisms and models, Mater. Sci. Eng. A, in press.
[12] Yoshinaga, H. and Morozumi, S., 1971, A Portevin-Le Chatelier Effect expected from Solute Atmosphere Dragging, Phil. Mag., 23, p. 1351-1366.
[13] Blum, W. and Roters, F., Spontaneous dislocation annihilation explains the breakdown of the power law of steady state deformation, 2001, submitted to phys. stat. sol.

Primary Creep in Polycrystalline Alpha Titanium:
Coupled Observations and a Stochastic Cellular Automation Model

Neeraj Thirumalai, Michael F. Savage, Michael J. Mills and Glenn S. Daehn
Department of Materials Science and Engineering
2041 College Road, 477 Watts Hall
The Ohio State University, Columbus, OH 43210
Daehn.1@osu.edu
Fax: (614) 292-1537

1. ABSTRACT

The room temperature creep of titanium is rather remarkable. At stresses near the 0.2% yield stress extensive deformation can be observed that is often well approximated by a power law relationship between strain and time at moderate strains. Eventually at rather high strains a linear relation between strain and time often observed. This remarkable time dependence is largely due to the near-absence of traditional strain hardening mechanisms, instead the predominant mode of hardening appears to be related to a progressively larger fraction of the load being borne by hard oriented grains. This behavior is simulated by a number of different obstacles that are coupled by load shedding. This concept is carried out in a stochastic cellular automaton model that efficiently treats equilibrium and compatibility. The model captures many of the features experimentally observed in the plastic deformation of weakly-hardening alloys.

2. INTRODUCTION

One common conceptual method of modeling creep is by using state variables representing applied shear stress, τ temperature, T and one or more variables that represent the current structure of the material, denoted by $\hat{\tau}$. With this approach we can write the shear strain rate as:

$$\dot{\gamma} = f(\tau, T, \hat{\tau}) \tag{1}$$

And the structure of the material is expected to evolve with time. This can be represented by an equation of the form:

$$\frac{d\hat{\tau}}{dt} = f(\tau, T, \hat{\tau}) \tag{2}$$

In specific use of this approach, it is common to use the thermally-activated Arrhenius form to describe the plastic strain rate, $\dot{\gamma}$, using a stress dependent activation barrier ΔG^* with the equations:

$$\dot{\gamma} = A \exp\left[\frac{-\Delta G^*}{kT}\right] \quad \text{where} \quad \Delta G^* = \Delta F^* \left[1 - \left(\frac{\tau}{\hat{\tau}}\right)^p\right]^q \tag{3}$$

Here ΔF^* is the stress-free activation barrier, k is Boltzman's constant and A, p and q are fitting parameters.

In the specific cases of titanium alloys it has been noted that they can deform to large strains in room temperature creep at stresses below the nominal yield stress [1-5]. This has been attributed to the low rate of work hardening in these alloys [5,6]. This low rate of work hardening is in turn associated with slip being very planar, with a-type dislocations being most common. The planar slip keeps dislocations from becoming easily entangled nearly eliminating this classical mode of work hardening. The very planar slip can be attributed to the short-range ordering that takes place in Ti-rich Ti-Al alloys. [6,7]

The strength in α-Ti-Al alloys is very anisotropic at the crystalline level. For one particular Ti-Al alloy the critical shear stress required for *a*-type dislocation motion on the basal and prism planes was measured at about 200 MPa at room temperature [9]. However the *c+a* dislocations which are necessary to elongate crystals in the *c* direction require shear stresses of about 800 MPa to activate motion [9]. In this paper we propose a model of creep where the apparent hardening is based on the progressive redistribution of load from grains that are oriented for easy slip to those which have no easy slip systems available. It will be shown that this hardening by load redistribution gives the same near power-law form that is seen in experimental data. The model shows this transient can extend to several times the nominal elastic limit strain.

3. EXPERIMENTAL DETAILS

The Ti-6Al alloy was obtained as forged 12.7mm diameter rods from RMI Company, Niles, Ohio. The nominal composition of the alloy is shown in Table I. This is a single phase alloy (α-hcp phase) with equiaxed microstructure. Samples were annealed high in the α phase field at 900°C for 24h in sealed quartz tubes, evacuated and back-filled with high purity argon. They were then air cooled to room temperature (AC). All samples had an average grain size of approximately 75 μm.

Table I. *Nominal composition (in wt%) of the Ti-6Al alloy used in this study*

Material	Al	O	N	Fe	Ti
Ti-6Al	5.73	0.051	0.008	0.037	Bal

Creep samples were 4mm x 4mm x 12mm compression coupons, all faces of which were ground to 600 grit. They were tested in a dead load creep frame in a compression cage. The strains were measured using a strain gage mounted directly on one face of the sample. The approximate strain resolution of the strain gage was 5με. The strain and load data were acquired using a high-speed LabView based computerized data acquisition system.

For transmission electron microscopy studies (TEM), samples were prepared by electro-polishing. TEM investigations were performed using a Philips CM200 electron microscope (with a LaB_6 cathode) under an operating voltage of 200kV.

Because the material in its air-cooled state has a very low dislocation density, the AC material was first pre-strained by 5% at room temperature and stress-relieved. After the pre-strain, the samples were sealed and annealed at 900°C for 0.5h and air cooled to room temperature. Conventional mechanical testing showed this material to have an 0.2% offset yield stress of about 612 MPa. Accordingly, creep tests at room temperature were carried out at 80%, 90% and 100% of the 0.2% offset yield stress (490 MPa, 546 MPa and 612 MPa). Details of all procedures are available elsewhere [6].

The creep behavior of Ti-6Al at the three stress levels are shown in Figure 1. Creep behavior is of primary type, where the creep rate continually decreases with time. Steady state creep is not observed at these stress levels at the strains examined here. In the literature [2-6] creep data of both single and two-phase Ti have been represented by a power of the form $\varepsilon = At^a$ where ε is the creep strain, A and a are constants and t is the time [2-7]. The typical deformed structure is shown in Figure 2. Slip is very much confined to discrete planes, strongly limiting traditional work hardening.

Constant strain rate mechanical tests were also carried out on Ti-6Al single crystals. These tests were carried out with an alloy which was very similar but had a higher oxygen content (0.17 weight pct.). Compression tests were carried out on float-zone grown crystals that were oriented by the Laue Technique. Further details are available in the thesis of Savage [11] and the results of these experiments are shown in Figure 3 which shows the polycrystal to have significantly higher strength as well as a greater rate of strain hardening. Cleaner data was obtained in the polycrystalline sample which used a directly-mounted strain gauge as opposed to an LVDT to acquire strain data.

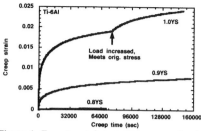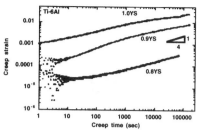

Figure 1. *Experimental strain-time curves developed in room temperature compressive creep of Ti-6Al. Experiments were run at 80%, 90% and 100% of the 0.2% offset yield stress. The same data is shown with linear axes in (a) and log-log axes in (b). In the test at the yield stress, the creep load was increased at about 2% plastic strain to match the true stress at the beginning of the experiment.*

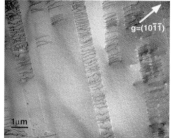

Figure 2. *TEM micrograph of deformed sample showing dislocation motion to be planar.*

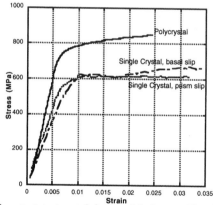

Figure 3: *Comparison of constant strain rate behavior of single crystal basal and prism slip with polycrystalline behavior in Ti-6Al. The data shows that in the polycrystals there is rapid strengthening in the small strain regime near the transition from elastic to plastic strains. This cannot be attributed to work-hardening of easy-slip grains as the strain levels are very small. A reasonable explanation is that load-transfer to hard-oriented grains occurs causing this rapid rise in the flow stress.*

52

4. MODELING

We consider a 'material' as being made up of a large number of small regions, each of which by virtue of its own orientation or intrinsic strength being somewhat different than adjoining regions. Each region is described an equation of the form of Equation 3 where local variations in local strength are represented by variations in ΔF^* and $\hat{\tau}$. In essence we consider one discrete plane through the material that is subjected to a remotely applied shear stress of value, τ_{ap}. The plane is discretized into a N x N array of material elements, each which is thought to be cubic with a side dimension of λ, as illustrated in Figure 4. Slip is also thought to be discretized into elements of magnitude, $\Delta\gamma$. This facilitates the simulation because we can take a stress, temperature, strain-rate relation, $\dot{\gamma}(\tau,\ T)$ and express it as a slip probability in that region by :

$$P_{slip}(\tau,\ T)\ =\ \frac{\Delta t}{\Delta\gamma}\ \dot{\gamma}(\tau,\ T)\ =\ A\ \frac{\Delta t}{\Delta\gamma}\ \exp\left[\frac{-\Delta G\ ^\star}{kT}\right] \qquad (4)$$

Here Δt represents the time interval over which probability is calculated and this is taken to be the reciprocal of the attempt frequency, or the atomic vibration period, ν, which is taken to be 10^{-12} second. The specific form taken for the probability law is the last term in the equation.

We note that as plasticity usually takes place by the passage of dislocations, $\Delta\gamma$ has a physical meaning as well as being a parameter that can be used to discretize the plasticity simulation. $\Delta\gamma$ is taken to be b/λ, which allows some study of size effects in this model. We note that for the conditions used here λ takes on sizes of 1μm or greater, the model approaches a continuum limit where results do not change with increasing λ. However, if λ is less than 1μm, results are size-dependent and the simulation takes on the character of a sandpile model [8].

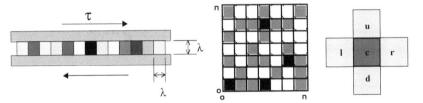

Figure 4. *Discetization scheme of a single slip plane being broken into a N x N matrix of elements, where each element communicates with the four near neighbors.*

The simulation is implemented as a stochastic cellular automaton model. The slip probability at each site is given by Equation 4 where the local value of stress is used instead of the remotely applied value. The local value varies from the remote value based on the difference between the number of slip events in the center element, n_c and the average of those of its four near neighbor elements, as:

$$\tau_{loc}\ =\ \tau_{rem}\ -\ \Delta\tau\left(n_c\ -\ n_{neigh.}^{av}\right) \qquad (5)$$

$\Delta\tau$ is the change in stress in a given element upon the passage of one slip event. It is related to the shear modulus, μ, as $\mu\Delta\gamma$. The key idea here is that as slip unloads one region it will increase the load on adjoining regions, keeping the stress averaged over the material constant. This stochastic cellular automaton approach is a computationally efficient method of roughly ensuring equilibrium and compatibility between neighboring elements.

The simulation operates by calculating the slip probability at each element in the period of one atomic vibration. Based on that a corresponding time increment is chosen and an element is randomly chosen to slip based on relative probabilities. By repeating this procedure the simulation progresses through time with strain on average increasing with the load generally increasing in 'hard' regions of the material, eventually approaching a steady-state. This method is described in detail elsewhere [10].

In order to roughly account for the relative variation in the hardness from one grain to the next, at each of the 1024 material sites in the simulation, a strength parameter, s, is randomly assigned based on a continuous distribution. This parameter modifies the stress free activation barrier to become $s\Delta F_o^*$ at each site, and the athermal bypass stress becomes $\sqrt{s}\tau_o$. The limits of the uniform distribution of s are 0.7 to 1.6. This and all the other parameters required in the simulation are shown in Table 2. These parameters were *not* specifically chosen or fit to match the experimental observations. The purpose was simply to demonstrate that a simple model based on many elements of varied strength can account for the creep behavior seen in polycrystalline α-Ti. Further refinements can certainly improve the fit to data. In accord with this, the slip probability is simply given as the exponential term in Equation 4.

The setup and results from the simulation are shown in Figure 5. The top left shows the distribution of the s parameter shown in gray scale in each case. The values nearest 1.6 are shown in white and those nearest 0.7 are black. The top right shows an output from the simulation -- the number of slip events in each of the material elements. Note that there is rough continuity, but there are local variations both locally and over longer lengths. The regions with the fewest slip steps correspond to the areas with the strongest sites, or clusters of strong sites. The creep curves that result from the simulation are shown in log-log form in the lower plots. The first shows the full range of the data from the simulations. This form containing slopes of unity at short and long time is characteristic of simulations of this type when structure values (ΔF^* and $\hat{\tau}$, here) held constant. This is discussed in detail elsewhere [10]. When only the experimentally accessible range of data is included (i.e., using the same axes and values shown in Figure 1(b)), it is apparent that the simulation and observation have very good qualitative agreement. The power-law exponent is similar to 0.25 in each case and in both cases there is some variation of the exponent. Also, the strain and strain rate values have similar absolute values to those measured. Lastly, the stress-strain shown in Figure 4 also shows that in single-crystal form when oriented for basal slip, this material exhibits virtually no hardening. This supports our contention that the power-law strain time behavior is largely a result of gradual load-shedding to stronger parts of the microstructure.

Table II. *Parameters used in the simulation*

Fundamental values		Derived values	
Parameter	Value	Parameter	Value
ΔF_o^*	325 kJ/mol	$\Delta\gamma$	3×10^{-4}
$\hat{\tau}_o$	500 MPa	$\Delta\tau$	10.5 MPa
p	1		
q	3/2		
μ	35 GPa		
b	0.3nm		
λ	1.0μm		
N	32		
s	Uniform distribution (0.7-1.6)		

5. CONCLUDING REMARKS

Here a model for creep based on a number of material elements that are varied in strength and coupled by load-shedding with their near neighbors is studied. It is shown that this model can develop extended power-law transients that are very similar to those seen in experiment. Such models can be quite important in attempts to infer the nature of local thermally activated events based on macroscopic measurements.

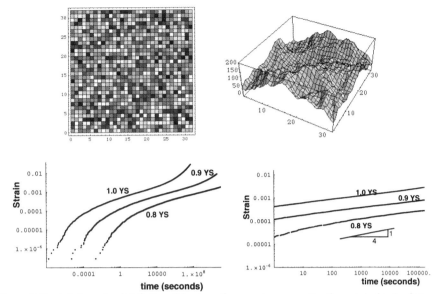

Figure 5. *Top left shows the distribution of the strength parameter, s, over the elements in the simulation. The top right figure shows the slip number of slip events in each element at the end of the simulation at the yield strength. The bottom figures show the simulated strain-time behavior. The full range of the model output is shown on the left and limits corresponding to Fig. 1(b) are shown on the right.*

6. ACKNOWLEDGMENTS

This work was supported by the Air Force Office of Scientific Research under grant F49620-98-1-039.

7. REFERENCES

1. H. K. Adenstedt, *Metal Progress*, 1949, **65**, 658.
2. A. W. Thompson and B. E. Odegard, *Metall. Trans.*, 1973, **4**, 899.
3. B. C. Odegard and A. W. Thompson, *Metall. Trans.*, 1974, **5**, 1207.
4. W. H. Miller, R. T. Chen, and E. A. Starke, *Metall. Trans.*, 1987, **18A**, 1451.
5. T. Neeraj, D.-H. Hou, G. S. Daehn, and M. J. Mills, *Acta Mater.*, 2000, **48**, 1225.
6. N. Thirumalai, Ph.D. Thesis, The Ohio State University, 2000.
7. M. J. Blackburn and J. C. Williams, *Trans. ASM*, 1969, **62**, 398.
8. P. Bak,, *How Nature Works*, Springer-Verlag, New York (1996).
9. N. E. Paton, R. G. Baggerly and J. C. Williams, Rockwell Science Center AFOSR Report SC526.FR, 1976.
10. G.S. Daehn, Modeling Thermally Activated Deformation with a Variety of Obstacles and its Application to Creep Transients, submitted to *Acta Mater* October 2000.
11. M. F. Savage, Ph. D. Dissertation, The Ohio State University, 2000.

Mechanisms and Effect of Microstructure on Creep of TiAl-Based Alloys

S. Karthikeyan[1], G. B. Viswanathan[1], Y-W. Kim[3], Vijay K. Vasudevan[2] and M. J. Mills[1]

[1] Department of Materials Science and Engineering, The Ohio State University, 477 Watts Hall, 2041 College Rd, Columbus, OH 43210

[2] Department of Materials Science and Engineering, University of Cincinnati, 497 Rhodes hall, PO Box 210012, Cincinnati, OH 45221

[3] Universal Energy Systems, Inc. 4401 Dayton-Xenia Rd., Dayton, OH 45432

Abstract

Transmission Electron microscopy studies on crept samples of an equiaxed Ti-48Al alloy deformed to strains near the minimum strain rate show a microstructure dominated by unit 1/2[110] type dislocations. These dislocations are pinned by superjogs. The jogged-screw model is adopted, where the rate controlling step is assumed to be the non-conservative dragging of a periodic array of jogs along the screw dislocations. The presence of tall jogs, and the existence of a stress-dependent upper bound on the jog height, have been incorporated into the model. These modifications lead to excellent agreement with experimental data for near-γ Ti-48Al. In contrast, lamellar structures show a highly inhomogeneous deformation behavior with dislocation activity in lamellae above a critical thickness, and negligible activity below this thickness. This limit is related to the minimum stress required to cause channeling of dislocations. The observation of jogged segments in the thicker lamellae suggests that a modification of the jogged-screw velocity law could be used by incorporating an effective stress approach. These modifications predict lower creep rates and higher stress exponents for lamellar structures, as observed experimentally

1. Introduction

High specific stiffness, good oxidation resistance up to 900°C, and high temperature strength (~600 MPa at 600°C) make ordered intermetallic compounds of the Ti-Al system promising as potential high temperature structural materials [1]. TiAl alloys have potential applications as high-pressure compressor stators, low-pressure turbine blades, transition duct supports, frames, seal supports and cases in the aerospace industry, and as turbocharger wheels and combustion valves in the automobile industry. The higher stiffness raises the vibrational frequencies, which is usually beneficial for structural components. γ-TiAl appears to be capable of substituting for superalloys in certain applications with substantial weight saving and minimal redesign [2]

The understanding of creep behavior in these materials is of utmost importance in view of many of these applications. Significant progress has been made recently in understanding the deformation behavior of γ-TiAl at low temperatures, including the yield strength anomaly [3]. However, despite the significant body of work done in the area of creep deformation, the present understanding of high temperature creep mechanisms, and the effect of alloying and microstructure, is very limited [2]. Clearly a difficulty is that creep is influenced by numerous microstructural parameters [4] and many of the investigations in γ-TiAl have been made on different structures. Widely-reported values of creep activation energies in equiaxed structures similar to those for self diffusion, and stress exponents in the range of 4-6, both suggest an underlying creep mechanism involving recovery of dislocations by climb [2, 3, 5, 6 , 7, 8, 9]. However, the absence of subgrain formation (in the minimum creep rate regime) suggests that the power law behavior is quite different from that of a pure metal. In this paper, we summarize a proposed model for creep in γ-TiAl which is based on a modification of the jogged screw model. The justification for this model is presented, along with a comparison of the model with measured creep data.

Two-phase TiAl alloys can have three broad categories of microstructures: equiaxed, duplex and fully lamellar. It is now established that lamellar microstructures, consisting of alternating γ-TiAl and α_2-Ti$_3$Al laths, have superior creep resistance when compared to the equiaxed and duplex microstructures. Several explanations have been cited for the superior creep resistance of the fully lamellar structures, such as the presence of interlocked lamellar boundaries [2, 6, 10], inhomogeneous deformation characteristics of lamellar structures [2, 11], composite behavior and reduction of the slip distance due to the presence of the interlamellar boundaries [2, 12, 13. 14].

It is imperative to understand the exact nature of deformation in the fully lamellar structures in order to sensibly employ strategies for further improvements in creep strength. Deformation modes including twinning [2, 8, 15, 16], dynamic recrystallization [7] , and interface sliding [5, 17, 18, 19] have been reported. It has been suggested previously that creep in the fully lamellar structures is controlled by the soft mode of deformation [2, 5, 17, 18]. Soft-mode deformation takes place by shear parallel to the lamellar interfaces. The mean free path of the mobile dislocations in the soft mode is related to the domain or colony size, which is two to three orders of magnitude larger than the lamellar spacing, making this slip mode relatively weak. Soft-mode deformation on the lamellar interfaces (i.e. interfacial sliding), has been proposed as a principal mode of creep deformation [5, 17, 18, 19]. However, if this were the case, then a finer lamellar spacing would result in more interfaces and hence a larger creep rate. Even though there is some evidence that there is an optimum lamellar spacing, and that too fine a spacing may cause weakening [20], a majority of work done in this area suggests that a finer lamellar structure yields lower creep rates, suggesting that other mechanisms may be dominant [6, 13, 14]. Twinning has also been frequently proposed as a creep mechanism[5, 15, 16, 17], though it is much more prevalent at seems at larger stresses and strain levels [5, 17]. In light of the large number of possible creep mechanisms, in this paper we attempt to determine the operative dislocation creep mechanisms through direct TEM characterization of the deformed microstructures. On the basis of these mechanisms, we will also attempt to rationalize the effect of microstructure on the creep performance.

2. Experimental Procedures

The creep results and microstructural evidence described in this paper were obtained from three different alloys: A Ti-48Al binary alloy, a "K5"[21] alloy and a Ti-48Al-2Cr-2Nb alloy. All three of these alloys were tested in a fully lamellar (FL) microstructure, while the binary Ti-48Al alloy was also tested in the equiaxed (EQ) near-γ microstructure. The binary alloy had a nominal composition of Ti-47.86 Al-0.116O-0.016N-0.041C-0.076H. Cylindrical blanks of the forged alloy were heat treated at 1473 K in the (α + γ) two-phase region just above the eutectoid temperature and followed by a stabilization treatment at 1173 K for 6 hrs. This heat treatment yields a near-gamma microstructure with γ grains are in the equiaxed morphology (grain size of 50 μm). These samples will be referred to as Ti-48Al(EQ) henceforth. Blanks of similar composition (Ti-48Al) were heat treated at 1658 K (above the α transus) to ensure a fully lamellar structure. These samples were also given a similar stabilization treatment. Different cooling rates were employed to get different lamellar spacings. These samples will be referred to as Ti-48Al(FL) henceforth. Cylindrical creep samples were prepared from these blanks and tested in tension at 768 and 815 °C. Ti-48Al-2Cr-2Nb was heat-treated similar to the binary lamellar alloy. The nominal composition of the K5 alloy is Ti-46.5Al-2.0Cr-3.0Nb-0.2W. These alloys were isothermally forged to 90%. Blanks of K5 were given a heat treatment involving solution treatment (in the single phase α phase field) at 1633 K for 5 minutes and furnace cooling to 1473 K, followed by air cooling to room temperature. The microstructure obtained thus is fully lamellar. An aging treatment of 1173 K for 24 hrs was given to the K5 samples. Parallelepiped samples were machined from the aged samples and creep tested in compression at 815 °C. Stress increment tests and monotonic tests were performed to obtain the stress dependence of the minimum creep rate.

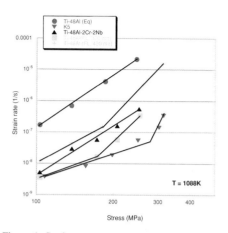

Figure 1. Strain rate versus stress curves at 1088K for equiaxed and fully lamellar Ti-48Al alloys, fully lamellar Ti-48Al-2Cr-2Al alloy and for the fully lamellar K5 alloy. Note the lower stress exponents at lower stresses and higher stress exponent at higher stress for the binary and K5 lamellar structures.

Thin foils for transmission electron microscopy observations were prepared from discs sectioned normal to the stress axis. The foils were thinned using a twin jet electro-polisher using a solution consisting of 65% ethanol, 30% butan-1-ol and 5% perchloric acid, at a voltage of 20V, current of 30 mA and temperature of –40°C. Observations on the microstructures were conducted on a Philips CM200 transmission electron microscope operated at 200 kV. Conventional TEM techniques were employed. Centered dark field images using α_2 reflections were used to image the α_2 laths and the measurements of the lath spacing were done by tilting the sample so as to get the lamellar interfaces edge-on. A line intercept method was used for measuring the lamellar thickness distribution.

3. Creep Results

The log strain rate versus log stress curves for the Ti-48Al (EQ and FL), Ti-48-2Cr-2Nb and K5 alloys are been shown in Figure 1. It is evident that the Ti-48Al(FL) is more creep resistant than the Ti-48Al(EQ) structure, while the K5 alloys containing Cr, Nb and W seem to have a much better creep response when compared to the binary alloy. As discussed later, it is important to recognize that the lamellar spacing is significantly finer for the K5 series alloys than for the binary case. For limited stress ranges, the creep response in γ-TiAl, both in the equiaxed and fully lamellar conditions, seems to obey a power law type Dorn equation relating the strain rate to the stress as

$$\dot{\varepsilon} = A \cdot \sigma^n \cdot \exp(-Q/RT) \qquad (1)$$

where A is a constant, Q is the creep activation energy, R is the gas constant and T, the temperature. In Figure 1 it can be seen that the equiaxed structures exhibit a stress exponent n of around 5. Similar results have been widely reported [2, 3, 5, 6, 7, 8, 9]. The stress exponents for fully lamellar structures however seem to be clearly lower than five (around 3) at lower stresses and significantly larger than five (~8) at higher stresses. Higher stress

58

exponent values for lamellar structures (at higher stresses) has been frequently reported [5, 11, 12]. The Ti-48Al-2Cr-2Nb(FL) alloy, however does not seem to exhibit two different stress exponents at lower and higher stresses. This alloy, despite being very similar in composition to the K5 alloy, has much higher creep rate and lower stress exponent, which seem to arise from the coarser lamellar spacing as discussed later.

The next section will present TEM observations of deformed microstructures in both equiaxed and lamellar forms, and relate these observations to possible creep mechanisms. An attempt has been made to understand the observed stress dependencies and their variation with microstructure on the basis of the modified jogged screw (MJS) model [9].

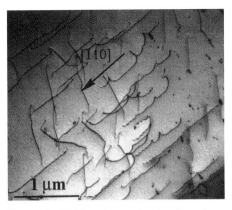

Figure 2. TEM micrograph showing the deformation structure in crept, equiaxed Ti-48Al. Screw segments are seen with numerous pinning points The segments on either side of the pinning points are significantly bowed [9].

4. Microstructural Observations and Discussion

4.1 Equiaxed Microstructure

The deformation microstructure in the equiaxed Ti-48Al alloy, at strains corresponding to the minimum creep rate, is dominated by $1/2[110]$ type dislocations. There is little tendency for subgrain formation as might be expected classically for an n~5 behavior. The dislocations tend to be elongated in the screw orientation and appear to be frequently pinned along their lengths, as can be seen in Figure 2. The segments on either side of the pinning points are seen to be bowed-out, forming local cusps along the length of the dislocations. Tilting experiments in the TEM have confirmed that these cusps are frequently associated with *tall* jogs on the screw dislocations, which are many times larger than the Burgers vector [9]. The observation of cusped screw segments in the equiaxed structures, and the general absence of subgrains, suggests that creep rate may be controlled by the non-conservative motion of jogs along the length of the screw dislocations. The jogs may be formed by the collision of two migrating kinks formed on two different glide planes [22]. Subsequent motion of the screw dislocation then requires the non-conservative dragging of these jogs (lateral glide motion is difficult). A balance between the work done by the applied force and the chemical force exerted by the vacancies produced by such non conservative motion of jogs gives rise to the velocity of such a jogged screw dislocations [23]:

$$v_s = (\frac{4\pi D_s}{h}) \cdot [\exp(\frac{\tau \Omega l}{hkT}) - 1]$$

(2)

where D_s is the self diffusion coefficient, h is the jog height, τ is the applied shear stress, Ω is the atomic volume, l is the jog spacing and k is the Boltzmann's constant. As these jogs grow taller, they could reach a critical height h_d, above which the oppositely signed, near edge segments attached to the top and bottom of the jog can bypass each other [24]:

$$h_d = (Gb)/(8\pi(1-v)\tau)$$

(3)

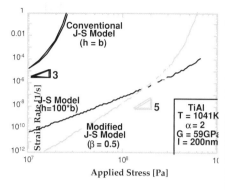

Figure 3. Predicted creep rate versus applied stress using the conventional and modified jogged-screw models, with model parameters appropriate for γ-TiAl. Included for comparison is experimental data from the work of Viswanathan [9].

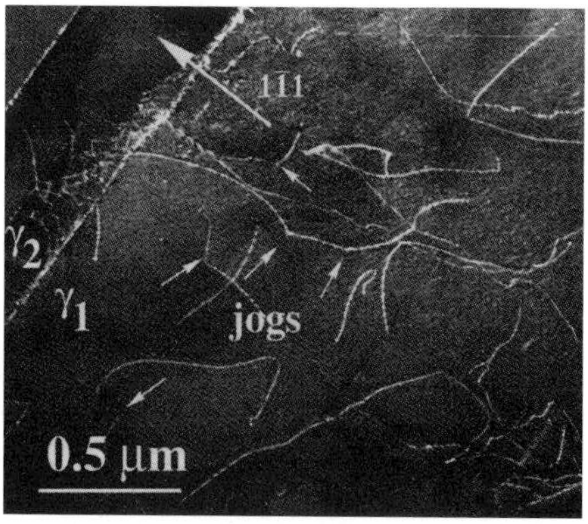

Figure 4. TEM micrograph showing the deformation structure in crept, lamellar Ti-48Al. Jogs are seen (indicated by arrows) in the thicker lath while the thinner lath has channeling dislocations

where G is the shear modulus, b is the Burgers vector and ν, Poisson's ratio. Jogs taller than h_d act as a source, rather than being dragged along with the screw dislocation. We presently make the assumption that the "characteristic" jog height is a constant fraction of h_d. Taking into account the stress dependence of h_d, dislocation density ρ (through Taylor's expression), the jogged-screw velocity law can be suitably modified to give an expression for the minimum creep rate. Assuming an equal number of vacancy producing and vacancy absorbing jogs, the strain rate can be expressed as[23]:

$$\dot{\gamma} = (\frac{\pi D_s}{\beta b h_d})(\frac{\tau}{\alpha G})^2 [\sinh(\frac{\tau \Omega l}{4 \beta h_d kT})]$$ (4)

where α is the Taylor factor which controls the dislocation density, and β is a parameter that characterizes the average jog height.

Figure 3 shows that there is excellent agreement between the MJS model and the experimental data for Ti-48Al (EQ). The "conventional" model assumes that the jogs originate from interaction of dislocations on different slip systems, which would produce jogs of unit height and thus predicts creep rates that are much too high and too stress-dependent. If it is assumed that the jogs are many Burgers vectors tall, the creep rates can be brought into closer agreement with experiment. However, the predicted stress dependence is close to 3, rather than the observed value of 5. Thus, the inclusion of a stress-dependent average jog height appears to be an important refinement offered in the MJS model[9].

4.2 Lamellar Microstructures

There is similarly ample microstructural evidence of 1/2[110] dislocation activity within the γ-laths of the lamellar structures. However, deformation is far more inhomogeneous, with smaller dislocation densities in the thinner laths and significantly larger densities in the thicker laths (Figure 4). Such inhomogeneity in deformation microstructure, and its possible role in increasing the stress exponents in lamellar structures, has been suggested earlier [11]. Lamellar interfaces of both types (γ/γ and γ/α2) appear to effectively constrain deformation to individual γ-laths, with little evidence for direct transmission under creep conditions. The thinner lamellae have hard-mode dislocations channeling through them, while the thicker lamellae have near-screw dislocations that are often cusped in configurations similar to those in equiaxed structures. There is little evidence of twinning in our samples (at minimum creep rate) and creep data supports the view that reducing the lamellar spacing reduces the creep rates. In light of this, it appears that dislocation activity within the laths may be very significant in controlling the creep response.

Channeling dislocations through a thin, capped film, must overcome a bowing stress, for which Nix has provided an expression [25]:

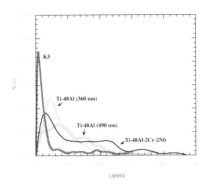

Figure 5. The distribution of lamellar spacings shows the fine lamellar spacing distribution of the K5 alloy and the relatively coarser lamellar spacings for the binary alloy and Ti-48Al-2Cr-2Nb.

$$\tau_b = (\frac{Gb}{4\pi\lambda}) \cdot \ln(\frac{\lambda}{b})$$

(5)

where λ is the lamellar spacing. If $\tau_{applied} < \tau_b$, then there is no effective stress. The fully lamellar structure has a distribution of lamellar thickness (see Figure 5), and so for a given applied stress, lamellae thinner than a critical cut off λ_c would not experience any effective stress and would not participate in the deformation process. This is the basic premise for an approach to modeling creep rates in fully lamellar structures. As the applied stress increases, the critical limit λ_c decreases and a greater volume fraction of the material is able to participate in the deformation process. This assumption seems to agree with microstructural observations of low dislocation activity in the thinner lamellae where dislocation multiplication appears to be limited.

To evaluate the creep rate it is required that the strain rate be averaged over the range of lamellar spacings:

$$\dot{\gamma} = (\int_{\lambda_{js}}^{\lambda_{max}} F(\lambda) \cdot v(\lambda) \cdot \rho(\lambda) \cdot b \cdot d\lambda)/(\int_{0}^{\lambda_{max}} F(\lambda) \cdot d\lambda)$$

(6)

where $F(\lambda)$ is the distribution of the volume fraction of gamma lamellae as a function of λ, $v(\lambda)$ is the velocity of dislocations under the effective stress and $\rho(\lambda)$ is the dislocation density as a function of the lamellar spacing. It should be noticed that this is an isostress approach which will only be valid (if at all) for small strains. Recent creep studies [4] indicate that creep rates in lamellar polycrystals are relatively close in magnitude to those for PST crystals oriented for "soft-mode" deformation. This correlation lends some support to the isostress assumption used here.

The observation of cusps in the wider lamellae strongly suggests that a suitable refinement of the MJS model may be constructed. In this refined model, the lower cutoff is assumed to be the minimum lamellar spacing (λ_{js}) required for jogs to be present. The model does not account for laths that are active (having a thickness greater than λ_c), but not thick enough for jogs to be present. However, this may not be a significant source of error since a large volume fraction of the material (up to 95% depending on the lamellar distribution of the material) is contained in lamellae which are wider than λ_{js} and thus most of the mobile dislocations will have jogs. We emphasize that this model assumes the rate-limiting process is the non-conservative motion of hard mode dislocations containing jogs. Strain contributions from soft mode dislocations are not considered. This assumption may

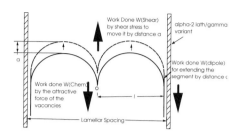

Figure 6. Schematic representation of the energy balance on the jogged dislocation within γ laths, under applied stress

be justified by considering that soft mode dislocations lying on trans-lamellar glide planes may not have large enough mean free paths to develop jogs sufficiently tall to make this process rate limiting. It is important to note that a smaller lamellar spacing leads to lower creep rates, suggesting a minimal role of interfacial sliding (which should increase the creep rates due to the presence of more lamellar interfaces in a finer lamellar material).

In order to proceed, we need to evaluate the terms inside the integral in Equation 6, namely $F(\lambda)$, $v(\lambda)$ and $\rho(\lambda)$, as discussed below.

(a) *Determination of F(λ):* TEM observations have been used to compile histograms of γ lath spacings (Figure 5). A log normal fit with a co-relation of more than 0.9 seems to be suitable for distribution $N(\lambda)$

$$N(\lambda) = a \cdot \exp[-(\ln(\tfrac{\lambda}{b}) - c)^2/(d^2)] \qquad (7)$$

where a, b, c and d are fitting parameters.

$$F(\lambda) = \lambda \cdot N(\lambda) \qquad (8)$$

(b) *Determination of v(λ):* As a jogged dislocation moves forward by the non-conservative dragging of the jogs, it lays out 60° segments along the bounding interfaces (both γ/γ and γ/α_2) of the lath. So at steady state, the work done by the applied shear stress in moving the dislocation forward by a, namely W_{Shear}, is balanced by the work done by the attractive forces due to the creation of vacancies (due to the non-conservative motion of jogs), W_{Chem} and the work done in extending the dislocation segments laid out along the interface, W_{dipole} (see Figure 6). The following expressions have been used for these work terms:

$$W(\lambda)_{chem} = n(\lambda) \cdot kT \ln(\tfrac{c}{c_0}) \qquad (9)$$

$$W(\lambda)_{dipole} = a \cdot E_{dipole}$$

$$E_{dipole} = \frac{G \cdot b^2}{2 \cdot \pi(1-v)} \cdot \ln(\tfrac{\lambda}{b}) + \frac{2E_{core}}{b} \qquad (10a, 10b)$$

$$W(\lambda)_{shear} = \tau \cdot b \cdot \lambda \cdot a \qquad (11)$$

where $n(\lambda)$ is the number of jogs present in a lamella of width λ, c and c_0 are the actual and equilibrium concentration of vacancies, E_{core} is the core energy of the dislocation (this value is assumed to be 2.2 eV [22]) and the other terms are as defined for equiaxed structures. At steady state we know that $W_{Shear} = W_{Chem} + W_{dipole}$. Using equations (9), (10) and (11), we can determine c/c_0. As in the case of equiaxed structures, h is equated to a constant fraction of h_d,

62

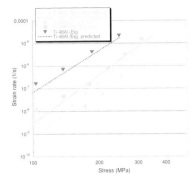

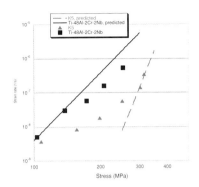

Figure 7: Comparison of minimum strain rate data for equiaxed and lamellar Ti-48Al (the latter having an average lamellar spacing of 360nm and 490 nm). The curves are those predicted by Eqn. 4 for equiaxed, and Eqn. 6 for lamellar structures.

Figure 8. Comparison of minimum strain rate data for K5 (FL) and lamellar Ti-48Al-2Cr-2Nb. The model predicts the effect of lamellar distribution well, as seen by the much lower creep rates (and higher stress exponents) for the K5 alloy.

and a is equated to Ω/bh, where Ω is the atomic volume. The velocity of the dislocation is then given by the drift velocity of the jogs [26]:

$$v(\lambda) = \frac{4 \cdot \pi \cdot D_s}{h} \cdot \{\sinh[\frac{\Omega \cdot \lambda}{h \cdot n(\lambda) \cdot k \cdot T} \cdot (\tau - \frac{E_{dipole}}{b \cdot \lambda})]\} \qquad (12)$$

(c) *Determination of $\rho(\lambda)$:* The dislocation density in the laths, $\rho(\lambda)$, may be modeled with the Taylor relation:

$$\rho(\lambda) = (\frac{\tau}{G \cdot \alpha})^2 \qquad (13)$$

where α may either be a constant, or may be a function of λ. At present we do not have sufficient data to determine this functional dependence, so in the present development we assume that ρ is independent of the lamellar spacing.

Preliminary observations in the deformed Ti-48Al-2Cr-2Nb (FL) samples indicate that the average jog spacing is of the order of 200nm, very similar to that observed in the Ti-48Al (EQ) structure. So a λ_{js} of 200nm is assumed. Using equations (8), (12) and (13) in equation (6), we can determine the strain rate. Figure 7. and 8. shows a comparison of the predicted creep rates and the experimental values. This model predicts lower creep rates and higher stress exponents for FL structures when compared to the EQ structure (Figure 7.). It also seems to reproduce the general trends associated with the lamellar spacing distribution. It predicts lower creep rates and higher stress exponents for the narrowest lamellar distribution (only at very high stresses in the case of K5) (Figure 8). The Ti-48Al-2Cr-2Al alloy does not exhibit two different stress exponents at lower and higher stresses and this could be because of the wide lamellar spacing, which makes its behavior similar to that of the EQ (Figure 8.). The absolute magnitude of the creep rates are not accurately predicted for all the microstructures. The probable source of this inaccuracy is the assumed invariance of ρ with λ. TEM results suggest in fact that narrow lamellae havelower dislocation densities than do the wider lamellae. Such a dependence would tend to produce higher stress exponents and significantly lower creep rates for very fine lamellar spacing distributions. Thus, this isostress approach may provide an explanation for both the reduced strain rates and higher stress exponents observed for the lamellar structures once the factor $\rho(\lambda)$ is more fully evaluated.

63

Note that this model does not provide an explanation for the significantly lower stress exponents observed at lower stresses. This regime has previously been attributed to interface dislocation motion [19]. Additional TEM study is necessary in order to determine whether an actual change in the principal deformation mechanisms is responsible for the markedly different stress dependence observed in this regime.

5. Conclusions

The creep behavior of an EQ microstructure of Ti-48Al has been investigated. Stress exponent values in the range of 5-6 were obtained stress increment tests at 1041K and 1088K. TEM analysis indicates that the deformation microstructure is dominated by 1/2<110] type dislocations, and that these dislocations are pinned by tall jogs of varying heights. The jogged-screw model has been adopted, where the rate controlling step is assumed to be the non conservative dragging of the jogs. The conventional jogged-screw model has been modified to include the presence of tall jogs having a stress-dependent average height. Incorporation of these changes into the MJS model yields excellent agreement with observed creep rates. Deformation in the FL microstructures is also dominated by 1/2<110] dislocations. However, the dislocation activity is inhomogeneous, with extensive activity in thicker lamellae and little or no activity in thinner lamellae. It is proposed that the critical lamellar thickness is related to the minimum stress required to cause channeling of dislocations: lamellae thinner than the cutoff thickness experience no effective stress. The observation of jogged segments in the thicker lamellae suggests that a refinement of the modified jogged-screw velocity law for the FL structures may be developed. A procedure to produce such a refined model has been discussed. To calculate the creep rate, it is required that the strain rate be averaged over the range of lamellar spacings. This isostress model predicts higher stress exponents and lower creep rates for finer lamellar distributions, in accord with observations. Further validation of the model requires additional TEM measurements for evaluating the dependence of dislocation density on lamellar spacing.

Acknowledgements

S.K., G.B.V. and M.J.M. acknowledge the support of the National Science Foundation under grant DMR- 9709029 with Bruce MacDonald as program manager. G.B.V. and V.K.V. are also thankful to the State of Ohio Edison Technology Center for financial support during the initial stages of this work under grant EMTEC/CT-29.

References:

[1] Maziasz, P. J., Ramanujan, R. V., Liu, C. T., Wright, J. L., "*Effects of B and W alloying additions on the formation and stability of lamellar structures in two-phase γ-TiAl.*" Intermetallics (1997), 5(2), 83-95.
[2] Beddoes, J., Wallace, W., Zhao, L., "*Current understanding of creep behavior of near γ-titanium aluminides.*". Int. Mater. Rev. (1995), 40(5), 197-217.
[3] Viguier, B., Hemker, K. J., Bonneville, J., Louchet, F., Martin, J. L., "*Modelling the flow stress anamoly in γ-TiAl, I. Experimental observations of dislocation mechanisms.*" Philosophical Magazine A (1995), 71, 1295-1312
[4] Parthasarathy, T. A., Subramanian, P. R., Mendiratta, M. G., Dimiduk, D. M., "*Phenomenological observations of lamellar orientation effects on the creep behavior of Ti-48 at.% Al PST crystals*", Acta Materialia (2000), 48(2), 541-551.

64

[5] Wang, J. N., Nieh, T. G., *"The role of ledges in creep of TiAl alloys with fine lamellar structures."*, Acta Materialia (1998), 46(6), 1887-1901.
[6] Parthasarathy, T. A., Mendiratta, M. G., Dimiduk, D. M., *"Observations on the creep behavior of fully-lamellar polycrystalline TiAl: identification of critical effects.*, Scripta Materialia (1997), 37(3), 315-321.
[7] Ishikawa, Y., Oikawa, H., *"Structure change of TiAl during creep in the intermediate stress range."*, Materials Transactions, Japan Institute of Metals (1994), 35, 336-345
[8] Lu, M., Hemker, K. J., *"Intermediate temperature creep properties of gamma TiAl."*, Acta Materialia (1997), 45(9), 3573-3585.
[9] Viswanathan, G. B., Vasudevan, V. K., Mills, M. J., *"Modification of jogged-screw model for creep of γ-TiAl."*, Acta Materialia (1999), 47(5), 1399-1411
[10] Wang, J. N., Schwartz, A. J., Nieh, T. G., Clemens, D., *"Reduction of primary creep in TiAl alloys by prestraining."*, Materials Science and Engineering A (1996), A206(1), 63-70.
[11] Zhang, W. J., Spigarelli, S., Cerri, E. Evangelista, E., Francesconi, L., *"Effect of heterogeneous deformation on the creep behavior of a near-fully lamellar TiAl-base alloy at 750 °C.*, Materials Science and Engineering A (1996), A211(1-2), 15-22.
[12] Morris, M. A., Lipe, T., *"I. Creep deformation of duplex lamellar TiAl alloys."*, Intermetallics (1997), 5(5), 329-337.
[13] Es-Souni, M., Bartels, A., Wagner, R., *"Creep behavior of near γ-TiAl base alloys: effects of microstructure and alloy composition."*, Materials Science and Engineering A (1995), A192/193 698-706.
[14] Loretto, M. H., Godfrey, A. B., Hu, D., Blenkinsop, P. A., Jones, I. P., Cheng, T. T., *"The influence of composition and processing on the structure and properties of TiAl-based alloys."* Intermetallics (1998), 6(7-8), 663-666.
[15] Skrotzki, B., Unal, M., Eggeler, G., *"On the role of mechanical twinning in creep of a near-γTiAl-alloy with duplex microstructure."*, Scripta Materialia (1998), 39(8), 1023-1029.
[16] Jin, Z., Bieler, T. R., *"An in-situ observation of mechanical twin nucleation and propagation in TiAl."*, Philosophical Magazine A (1995), 71(5), 925-47.
[17] Wang, J. G., Hsiung, L. M., Nieh, T. G., *"Formation of deformation twins in a crept lamellar TiAl alloy."* , Scripta Materialia (1998), 39(7), 957-962.
[18] Zhao, L., Tangri, K., *"Variation in the dislocation structure on lamellar titanium-aluminum (Ti3Al/TiAl) interfaces during deformation at different temperatures."*, Philosophical Magazine A (1992), 65(5), 1065-81
[19] Hsiung, L. M., Nieh, T. G.,*"Creep deformation of fully lamellar TiAl controlled by the viscous glide of interfacial dislocations."*, Intermetallics (1999), 7(7), 821-827.
[20] Yamamoto, R., Mizoguchi, K., Wegmann, G., Maruyama, K., *"Effects of discontinuous coarsening of lamellae on creep strength of fully lamellar TiAl alloys."*, Intermetallics (1998), 6(7-8), 699-702.
[21] Kim, Y-W., *"Effects of microstructure on the deformation and fracture of γ-TiAl alloys."*, Materials Science and Engineering A (1995), A192/193 519-33.
[22] Louchet, F., Viguier, B., *"Ordinary dislocations in γ-TiAl. Cusp unzipping, jog dragging and stress anomaly."*, Philosophical Magazine A (2000), 80(4), 765-779.
[23] Barrett, C. R., Nix, W. D., *"A model for steady state creep based on the motion of jogged dislocations."*, Acta Metallurgica (1965), 13(12), 1247-1258.
[24] Sriram, S., Dimiduk, D. M., Hazzledine, P. M., Vasudevan, V. K., *"The geometry and nature of pinning points of 1/2⟨110] unit dislocations in binary TiAl alloys."*, Philosophical Magazine A (1997), 76(5), 965-993.
[25] Nix, W. D., *"Yielding and strain hardening of thin metal films on substrates."*, Scripta Materialia (1998), 39(4-5), 545.
[26] Hirth, J. P., Lothe, J., *"Theory of Dislocations"*, John Wiley and Sons, Inc., 1982.

Some Thoughts on the Design of Dual Scale Particle Strengthened Materials for High Temperature Applications

J. Rösler, M. Bäker and C. Tiziani
Technical University Braunschweig, Institut for Materials, 38106 Braunschweig, Germany

Abstract

The creep behavior of dual scale particle strengthened materials containing particles of two different length scales, namely nanometer-size dispersoids in combination with "macroscopic" reinforcements, is analyzed theoretically. Provided certain design principles, related to the selected particle parameters, are obeyed, a synergistic strengthening effect and a creep strength level superior to today´s best particle strengthened high temperature materials is predicted. As example, dual scale particle strengthened copper is investigated and requirements on fiber properties are discussed using finite element analysis.

1. Introduction

Dual scale particle strengthening is a hardening concept that utilizes particles on two different length scales /1,2/, namely nanometer-size dispersoids in combination with "macroscopic" reinforcements having typical dimensions in the micrometer to millimeter range. Powder metallurgically processed aluminum composites /22-28/ belong to this material class as small amounts of Al_2O_3 dispersoids are always introduced during processing and it was pointed out by Park and co-workers /23,28/ that the additional hardening effect caused by the dispersoids must be considered in order to understand the unusually high stress sensitivity of the creep rate. Steady state creep equations for materials containing dispersoids and reinforcements in combination were proposed in /1,2,22,23/, taking both particle types into account. They were based on the threshold-stress concept in /22,23/, while in /1,2/ thermally activated dislocation detachment from dispersoid particles was considered.

Based on theoretical analysis, it was suggested in /1,2/ that creep strength levels far superior to today´s best particle strengthened high temperature materials can be achieved by dual scale particle strengthening, provided certain design guidelines, related to the selected particle parameters, are followed. In this article, we give a condensed theoretical analysis (chapter 2) following /1/ and discuss some design considerations on the example of dual scale particle strengthened copper (chapter 3). Noting that thermal and electrical conductivity is widely unaffected by the presence of dispersoids and reinforcements, dual scale particle strengthened copper is not only a good model material but is also believed to be of considerable practical significance for applications where high temperature strength is required in addition to excellent conductivity.

2. Theoretical Analysis

Before turning to dual scale particle strengthening in chapter 2.3, it is necessary to shortly review the individual strengthening contributions stemming from dispersion and reinforcement hardening in chapters 2.1 and 2.2.

66

2.1 Dispersion strengthening

It is today well established that dispersion strengthening at high temperatures results from an attractive interaction between dislocations and dispersoids which is caused by partial relaxation of the dislocation stress field at the incoherent particle/matrix interface /3-11/. Fig. 1 illustrates, as example, the relaxation of a shear stress field at a particle by diffusional matter transport. Key requirements for this mechanism to take place are (i) an incoherent interface between particle and matrix serving as source and sink for vacancies and (ii) a short diffusion distance, i.e. small particle dimensions, to ensure rapid stress relaxation at elevated temperatures. Both are met by dispersoid particles /3,4/ and, in consequence, a "pinning" effect on moving dislocations results. It is accounted for by the so-called relaxation factor k defined as the quotient between dislocation line energy at the particle T_p and in the matrix T_m ($k = T_p/T_m$) /7/. Hence, for $k < 1$ an attractive interaction results becoming stronger with diminishing k.

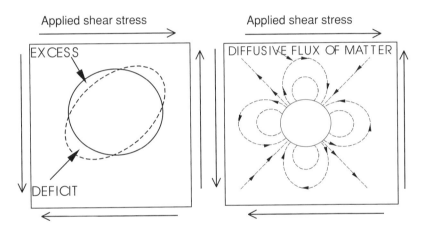

Fig. 1: On the left hand side, a spherical particle embedded in a metal matrix is shown (solid line) along with the shape change (dotted line) that is required to relax a shear stress field as indicated at the location of the particle. Provided the particle is non-deforming, matrix material has to be removed from the locations marked "excess" and deposited at locations of material "deficit" for stress relaxation to occur. This can be achieved by short range diffusive matter transport as indicated on the right hand side.

One interesting feature of dispersion strengthening is that the activation energy necessary for dislocation detachment from the departure side of the particle may become comparable to the thermal energy $k_B T$. Consequently, dislocation detachment must, in general, be treated as thermally activated process leading to the following rate equation for detachment controlled creep /10, 12/:

$$\dot{\varepsilon}_d = \dot{\varepsilon}_0 \cdot exp\left(-\frac{Gb^2 r[(1-k)(1-\sigma/\sigma_d)]^{3/2}}{k_B T}\right) \qquad (1)$$

where $\dot{\varepsilon}_0 = 6D_v\lambda\rho/b$ is a reference strain rate related to volume diffusivity D_v, dislocation density ρ, interparticle spacing 2λ and burgers vector b (G: shear modulus). The applied stress σ is normalized by the so-called athermal detachment stress σ_d that would be needed for dislocation detachment in the absence of thermal activation. It can be rewritten from /7/ for tensile loading as $\sigma_d = MGb / 2\lambda\sqrt{1-k^2}$ where M is the Taylor factor.

Inspection of eq. (1) shows that the activation energy (= nominator of the exponential expression) decreases with increasing k-value and decreasing dispersoid diameter $2r$ at given normalized stress σ/σ_d. Thus, thermal activation is particularly relevant when the interaction strength is moderate, say $k \geq 0.9$, and the particle size is small, say $2r < 50$nm. These conditions are met for a number of dispersion strengthened materials, e.g. ODS-superalloys /10,12/. Conversely, in situations of strong particle/dislocation interaction, relatively large dispersoid particles and/or moderate temperatures, eq. (1) leads to a stress dependence of the steady state creep rate that resembles a threshold stress behavior and is often described as such, e.g. /22-24/. In any case, it is noted that increasing the dispersoid volume fraction f_d beyond a few percent is a rather inefficient means to enhance the creep strength because it enters eq. (1) only via the athermal detachment stress ($\sigma_d \sim f_d^{1/2}$). This limits the strengthening potential of dispersoids which is unfortunate in view of the ever increasing demand for advanced high temperature materials. As will be shown below, dual scale particle strengthening is an avenue to overcome this shortcoming.

2.2 Reinforcement strengthening

In this paper, reinforcements are distinguished from dispersoids by their interparticle spacing 2λ which, for reinforcements, is too large to exhibit a significant back stress on moving dislocations. This condition is met for $2\lambda/b \gg 1000$, resulting in an Orowan shear stress significantly less than $G/1000$. Nevertheless, strengthening occurs due to load transfer from the softer matrix onto the mechanically stronger reinforcements /13-17/. As a result, the flow stress, driving matrix dislocations past obstacles, is reduced and the strengthening contribution can be accounted for replacing the externally applied stress σ by an effective stress σ/Λ. Hereby, $\Lambda > 1$ is the strengthening coefficient that depends on the reinforcement volume fraction f_r and aspect ratio l/R. Trends in Λ have been calculated in /17/ for aligned, axisymetric reinforcements using finite element analysis and approximated by an analytical equation in /18/ for the limiting case of perfect plasticity[1], leading to

$$\Lambda \approx 1 + 2(2 + \ell / R)f_r^{3/2} \qquad (2)$$

Note that the strengthening contribution is proportional to $f_r^{3/2}$ for high aspect ratio reinforcements. This is in contrast to dispersion hardened materials where a square root dependence between particle volume fraction and creep strength is found and is the basis for the concept of dual scale particle strengthening as outlined below.

There is an additional noteworthy discrepancy between the two particle hardening mechanisms. While dispersion strengthening at high temperatures relies on stress relaxation by diffusional matter transport at the particle (fig. 1), this mechanism is detrimental for reinforcements because load transfer strengthening requires that normal stress gradients along

---

[1] Note that perfect plasticity and power law creep with stress exponent $n \to \infty$ lead to the same solution for Λ.

the particle-matrix interface are maintained /18-21/. When these stress gradients are relaxed by short range diffusion at the particle, load transfer strengthening is lost. Consequently, reinforcements must either be coherent with the matrix, so that the particle/matrix interface cannot act as source and sink for vacancies, or sufficiently large so that stress relaxation by diffusional matter transport is insignificant. This is the reason why a particle cannot simultaneously exert an attractive interaction on moving dislocations and, at the same time, cause load transfer strengthening. Otherwise, high aspect ratio dispersoids would be the best choice for high temperature applications. There may, however, be an "intermediate" particle size range where neither reinforcement nor dispersion strengthening is effective because diffusional matter transport is already sufficiently fast to eliminate hardening by load transfer while the interparticle spacing is still too large to cause a significant back stress on moving dislocations. It is essential to avoid this particle size range.

2.3 Dual Scale Particle Strengthening

Let us consider a material containing „small" dispersoids and „large" reinforcements in the above sense. The dispersion strengthened matrix behaves under creep conditions according to eq. (1), provided dislocation detachment is the rate controlling mechanism. This is usually the case in high temperature / low strain rate loading situations. The effective stress "seen" by the dislocations is, however, reduced from σ to σ/Λ due to the presence of the reinforcements /1,2,22/. Consequently we have to rewrite eq. (1), replacing σ by σ/Λ, to come up with a constitutive equation for the steady state creep rate $\dot{\varepsilon}_{d,r}$ of dual scale particle strengthened materials under detachment controlled creep /1,2/:

$$\dot{\varepsilon}_{d,r} = \dot{\varepsilon}_0 \cdot exp\left(-\frac{Gb^2 r\left[(1-k)(1-\sigma/\Lambda\sigma_d)\right]^{3/2}}{k_B T}\right) \quad (3)$$

Considering furthermore a given dispersion strengthened material with fixed relaxation factor k and dispersoid radius r, the effect of particle parameters on the creep strength can be fully analyzed by inspection of the creep strength parameter $\Sigma = f_d^{1/2}(1+2(2+l/R)f_r^{3/2})$ which is proportional to $\Lambda\sigma_d/l$. Hereby, the approximation for Λ given in eq. (2) is used which seems to be reasonably accurate in view of stress exponents $n > 10$ normally observed for dispersion strengthened materials. Plotting Σ as a function of the particle volume fractions f_r, f_d (fig. 2) and requiring that the total particle content $f_d + f_r$ is constant, a number of interesting observations are made: at moderate particle volume fraction, represented by $f_d + f_r = 5\%$, dispersion hardening is most effective and there is nothing to be gained by dual scale particle strengthening[2]. However, at higher particle contents, represented by $f_d + f_r = 15\%$, a strength maximum is observed when a relatively small amount of dispersoid particles is combined with high aspect ratio reinforcements. This points to a synergistic effect when dual scale particle strengthening is utilized and appropriate particle parameters are selected. Detailed analysis /1/ reveals that $f_r / f_d = 3$ is ideal in many cases. The reason for the predicted synergism has been mentioned before: reinforcement strengthening is the more effective the higher the volume fraction is ($\Lambda \sim f_r^{3/2}$) while dispersion hardening profits only moderately from large particle contents ($\sigma_d \sim f_d^{1/2}$). Thus, it is most effective to combine a few percent dispersoid particles with a relatively high amount of reinforcements. Fig. 2 also shows that

---

[2] Please note again that detachment controlled creep is assumed here. Of course, this is unrealistic for $f_d \to 0$ and the prediction that Σ approaches zero for $f_d \to 0$ is an artefact of this assumption. For a more detailed analysis see /1/.

equiaxed reinforcements are fairly unattractive for strengthening purposes. Of course, they may be beneficial in altering other important engineering properties such as elastic modulus, coefficient of thermal expansion or abrasion resistance.

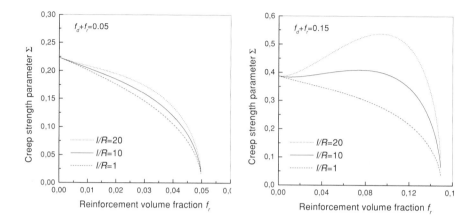

Fig. 2a,b: Calculated creep strength parameter Σ as a function of reinforcement volume fraction f_r and aspect ratio l/R at given total particle volume fraction $f_d+f_r = 5\%$ (a.) and 15% (b.) (from ref. /1/). Note the occurrence of a strength maximum in (b.) in case of high aspect ratio reinforcements.

3. Dual Scale Particle Strengthened Copper

Above, some fundamental aspects associated with dual scale particle strengthening were discussed under certain idealized assumptions. One idealization was that the reinforcements behave linear elastic but do not creep by themselves. Considering that the dispersion strengthened matrix may exhibit considerable creep strength, this is by no means guaranteed. Therefore, the effect of fiber properties on creep performance is analyzed in this paragraph on the example of dual scale particle strengthened copper. As mentioned above, this material class is not only a good model system but may also be of considerable practical significance. The dispersion hardened matrix may be manufactured by mechanical alloying using, for example, Al_2O_3 dispersoids /29/ or by internal oxidation of alloyed copper powder, e.g. Cu-Zr /30,31/. Tungsten, molybdenum, chromium and niobium are all potential fiber materials because of their low solubility in copper. Niobium is of interest due to its comparatively high ductility while chromium is appealing because of its good oxidation resistance. The drawback is, however, that their strength level is significantly inferior to molybdenum and tungsten fibers. Thus, it is of interest to understand which strength level is needed for dual scale particle strengthening to be effective.

To address the above question, finite element simulations using an axisymmetric unit cell model were performed applying a fixed strain rate $\dot{\varepsilon} = 10^{-6} \ s^{-1}$ at $T = 800°C$. For a detailed description of the finite element model see /1/. The strength of the dispersion hardened matrix at these loading conditions was assumed to be 16 MPa. For the tungsten and molybdenum fibers, elastic-plastic behavior with strain hardening was assumed and the yield strength at

800°C was set as 230 MPa and 150 MPa, respectively. As no information on strain-hardening behavior was available for chromium and niobium, perfect plasticity was assumed with a yield strength of 116 MPa and 55 MPa, respectively. Viscoplastic deformation of the fibers was not considered because of their high melting points (T/T_m < 0.5 is met for all four materials). Calculations were also performed assuming linear elastic fiber properties, giving an upper bound for the achievable strength level. The calculated stress-strain curves (fig. 3) show essentially identical results for tungsten and the elastic fiber, indicating that deformation of the tungsten fiber is predominantly elastic because of its high strength. In contrast, the yield strength of niobium and chromium is inadequate to utilize the full potential of dual scale particle strengthening under the loading conditions analyzed here.

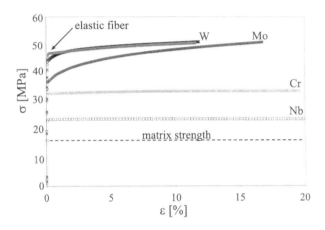

Fig. 3: Calculated stress-strain plot using a finite element unit cell model /1/ with fiber volume fraction f_r = 20%, aspect ratio l/R = 5 and T = 800°C, $\dot{\varepsilon}$ =10^{-6} s^{-1} as loading parameters. Shown are the dispersion strengthened copper matrix in comparison to dual scale particle strengthened copper containing the following fibers in addition to the dispersoid particles: niobium, chromium, molybdenum, tungsten, "elastic".

Inspecting the strain field in the finite element unit cell at an applied strain of 5% (fig. 4) gives further insight in the composite behavior. In case of the tungsten fiber, plastic deformation is essentially limited to the copper matrix as expected. Only the center of the fiber, which experiences the highest mechanical stress, shows a limited amount of plastic deformation. As the fiber strength decreases, the plastic zone in the fiber expands and the plastic strain is more equally distributed between matrix channel and fiber. In case of the niobium fiber, only the fiber ends remain elastic which readily explains the limited strengthening effect. Also, a shear band evolves in the copper matrix at an angle of 45° to the loading axis which is triggered by the sharp corner of the niobium fiber. In summary it is noted that high strength fibers are essential to make full use of dual scale particle strengthening whereby tungsten is an appropriate choice for copper alloys. For higher strength materials, it might be a considerable challenge to find appropriate fibers and it appears, therefore, that dual scale particle strengthening is best suited for relatively soft, low melting point metal matrices.

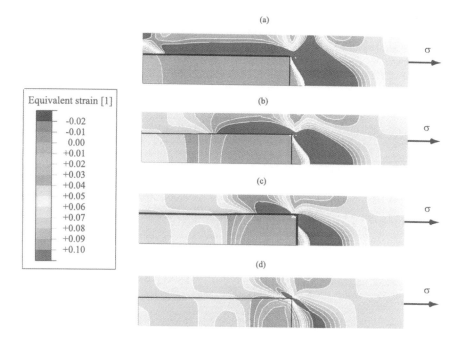

Fig. 4: Finite element unit cell model showing the equivalent total strain after a deformation of 5%. Fiber materials are tungsten (a.), molybdenum (b.), chromium (c.) and niobium (d.).

4. Summary

The creep behavior of dual scale particle strengthened materials, containing dispersoids and reinforcements in combination, has been investigated theoretically. A synergistic strengthening effect and a strength level surpassing todays best particle strengthened high temperature materials is predicted, provided (i) high aspect ratio reinforcements are selected, (ii) unloading of the reinforcements by diffusional matter transport is avoided and (iii) a volume fraction ratio between reinforcements and dispersoids of $f_r/f_d \approx 3$ is selected. Furthermore, potential fiber materials for dual scale particle strengthened copper were examined by finite element calculation. It was shown that tungsten is an appropriate fiber material whereas niobium and chromium exhibit insufficient strength levels for the loading conditions investigated here.

References

/1/   Rösler, J. and Bäker, M., Acta mater., **48**, 3553 (2000).
/2/   Rösler, J., in Werkstoffwoche 98, Vol. 6, R. Kopp, B. Beiss, K. Herfurth, D. Böhme, R. Bormann, E. Arzt and H. Riedel (eds.), Wiley-VCH Verlag, Weinheim, 1998, p. 509.
/3/   Srolovitz, D.J., Petkovic-Lutton, R.A. and Lutton, M.J., Acta metall., 1983, **31**, 2151.
/4/   Srolovitz, D.J., Lutton, M.J., Petkovic-Lutton, R. A., Barnett, D.M. and Nix, W.D., Acta metall., 1984, **32**, 1079.

72

/5/ Nardone, V.C. and Tien, J.K., kScripta metall., 1983, **17**, 467.
/6/ Schröder, J.H. and Arzt, E., Scripta metall., 1985, **19**, 1129.
/7/ Arzt, E. and Wilkinson, D.S., Acta metall., 1986, **34**, 1893.
/8/ Herrick, R.S., Weertman, J.R., Petkovic-Lutton, R.A. and Lutton, M.J., Scripta metall., 1988, **22**, 1879.
/9/ Rösler, J. and Arzt, E., Acta metall., 1988, **36**, 1043.
/10/ Rösler, J. and Arzt, E., Actal metall.mater., 1990, **38**, 671.
/11/ Rösler, J., Joos, R. and Arzt, E., Metall.Trans.A, 1992, **23**, 1521.
/12/ Rösler, J., VDI Fortschrittsbericht, Reihe 5, Nr. 154, 1988.
/13/ Christman, T., Needleman, A. and Suresh, S., Acta metall., 1989, **37**, 3029.
/14/ Tvergaard, V., Acta metall., 1990, **38**, 185.
/15/ Dragone, T.L. and Nix, W.D., Acta metall.mater., 1990, **38**, 1941.
/16/ Kelly, A. and Street, K.N., Proc. R.Soc.Lond. A, 1972, **328**, 283.
/17/ Bao, G., Hutchinson, J.W. and McMeeking, R.M., Acta metall., 1991, **39**, 1871.
/18/ Rösler, J., Bao, G. and Evans, A.G., Acta metall.mater., 1991, **39**, 2733.
/19/ Rösler, J. and Evans, A.G., Mater.Sci.Engng A, 1992, **153**, 438.
/20/ Nimmagadda, P.B.R. and Sofronis, P., Mech.Mater., 1996, **23**, 1.
/21/ Sofronis, P. and McMeeking. R.M., Mech.Mater., 1994, **18**, 55.
/22/ Li, Y. and Langdon, T.G., Metall.Trans. A, 1998, **29**, 2523.
/23/ Park, K.-T. and Mohamed, F.A., Metall.Mater.Trans. A, 1995, **26**, 3119.
/24/ Čadek, J., Oikawa, H. and Sustek, V., Mater.Sci.Engng, 1995, **A190**, 9.
/25/ González-Doncel, G. and Sherby, O.D., Acta metall.mater., 1993, **41**, 2797.
/26/ Greasley, A., Mater.Sci.Technol., 1995, **11**, 163.
/27/ Mishra, R.S. and Pandey, A.B., Metall.Trans. A, 1990, **21**, 2089.
/28/ Park, K.-T., Lavernia, E.J. and Mohamed, F.A., Acta metall.mater., 1990, **38**, 2149.
/29/ Broyles, S.E., Anderson, K.R., Groza, J.R. and Gibeling, J.C., Metall.Trans. A, 1996, **27**, 1217.
/30/ Nagorka, M.S., Levi, C.G. and Lucas, G.E., Met.Trans. A, 1995, **26**, 859.
/31/ Nagorka, M.S., Lucas, G.E. and Levi, C.G., Met.Trans. A, 1995, **26**, 873.

CREEP FUNDAMENTALS III

On Threshold Stress Concept

R.Kaibyshev, I.Kazakulov, O.Sitdikov, D.Malysheva

Institute for Metals Superplasticity Problems, Khalturina 39, Ufa 450001, Russia,
e-mail: rustam@anrb.ru, phone/fax: +7 (3472) 253856

Abstract
Threshold behavior of a Fe-3%Si steel and an 5083 aluminum alloy was considered in detail. It was shown that there existed a strong correlation between deformation behavior and threshold behavior. A temperature dependence of threshold stress is characterized by a value of the energy term, Q_o, representing activation energy to overcome an obstacle by a dislocation in equation:

$$\frac{\sigma_o}{E} = B_o \exp\left(\frac{Q_o}{R \cdot T}\right), \tag{1}$$

where B_o is a constant; σ_o is the threshold stress, E is the Young's modulus, R is the gas constant and T is the absolute temperature. Deformation mechanisms operating in the two materials under study were analyzed. It was demonstrated that a value of the energy term was constant in a temperature range where a process controlling deformation behavior did not change with temperature and strain rate variations. A transition from one type of deformation behavior to the other results in a change in Q_o value. It was concluded that a mechanism of interaction between a dislocation and an obstacle was dependent on ability of dislocations to climb. Deformation mechanisms operating in different temperature regions and their relation with mechanisms of dislocation bypass are discussed.

1. Introduction

Dispersion strengthened (DS) alloys are widely used for high-temperature applications due to their excellent creep resistance up to high homologous temperatures [1]. These materials exhibit threshold behavior, which is characterized by a continuous increase in the stress sensitivity with a strain rate decrease and high apparent activation energy [1-3]. Such behavior has been described in terms of threshold stress below which the strain rate is assumed to be negligible (Fig. 1). The steady-state strain rate, ε, of materials, where dispersoids impede the movement of dislocations, is generally represented by

$$\varepsilon = A \cdot \left(\frac{\sigma - \sigma_0}{E}\right)^n \cdot \exp\left(\frac{-Q_c}{R \cdot T}\right), \tag{2}$$

where σ is the applied steady-state stress, σ_0 is the threshold stress, E is the Young's modulus, n is the true stress exponent, Q_c is the true activation energy for plastic deformation, R is the gas constant, T is the absolute temperature and A is a constant. Plastic deformation of DS alloys is driven by effective stress, equal to σ-σ_0.

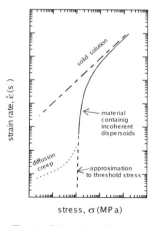

stress, σ (MPa)

Fig. 1. Schematic illustration of strain rate *vs* applied stress for a typical material exhibiting threshold behavior.

Several theoretical models were developed to explain the origin of threshold stress [4-8]. Almost all these models give the magnitude of athermal threshold stress. However, a strong temperature dependence of threshold stress was observed in numerous experimental studies [3, 9]. It has been proposed [3] to express the temperature dependence of normalized threshold stress, σ_0/E, by a relationship of the form (1). The activation energy, Q_0, is associated with a process by which mobile dislocations overcome the obstacles in their glide planes. At present this temperature dependence of threshold stress is not understood. It is a great contradiction in the theory of creep of DS alloys. Only a recent theoretical model [7] predicts an existence of weak temperature dependence by Eq. (1). Note that experimental data regarding to an examination of dependence (1) in wide temperature range are lacking. Eq. (1) was used to analyze threshold behavior of DS alloys with narrow temperature ranges, which do not exceed 100°C. At present there are no experimental data that can be used to provide a comparison between threshold behaviors of a material in the ranges of hot and warm deformation.

The aim of the present work is to report the threshold behavior of two materials containing incoherent dispersoids in a wide temperature range. Main efforts are focused to reveal a relationship between two different characteristic modes of deformation behavior being described by Eq. (2) at intermediate and high temperatures, respectively, and temperature dependence of threshold stress being described by Eq. (1).

2. Materials and Experimental Technique

Two materials were used as objects of the present study: a Fe-3%Si steel and a modified 5083 aluminum alloy. The Fe-3%Si steel was described in an earlier report [10]. The 5083 alloy has a chemical composition of Al-4.7Mg-1.6Mn-0.2Zr-0.18Cr-0.1Fe (in wt. %). It was fabricated at the Kaiser Center for Technology by direct chill casting and homogenization at 520°C for 10h. Second-phase particles were identified by the TEM analysis as Al_6Mn, Al_3Cr, and Al_3Zr. The Al_6Mn particles show a bimodal distribution and have a plate-like shape. The fine Al_6Mn particles were about 0.3μm in longitudinal and 0.1μm in transverse directions. They were precipitated from the supersaturated solid solution during homogenization annealing. The coarse Al_6Mn particles have a size in the range of 0.6 to 2μm. Al_3Cr and Al_3Zr particles are equiaxed, having sizes of 100 and 20nm, respectively. Coherent boundaries exhibit coherent stress fields distinguished by TEM as non-uniform diffraction contrast near dispersoids. No coherent boundaries were exhibited by the Al_3Cr and Al_3Zr dispersoids.

Details of experimental procedures were given earlier [10]. Notably, in the present study the compression tests with lubricant were carried out at initial strain rates ranging from 10^{-7} to 10^{-1} s^{-1} for both materials. For the 5083 alloy it is necessary to note that compression tests were carried out at temperatures ranging from 225 to 580°C.

3. Results
3.1 The shape of σ-ε curves

Typical true stress - true strain curves for the Fe-3%Si steel were presented in the earlier report [10]. It should be noted that the curves at initial strain rates lower than 10^{-5} s^{-1} exhibit a well-defined steady-state after initiation of plastic deformation.

For the 5083 aluminum alloy the typical true stress- true strain curves in the temperature range 250-550°C at different initial strain rates are presented in Fig. 2. All the curves exhibit a well-defined steady state.

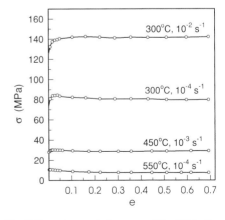

Fig. 2. True stress - strain curves of the 5083 alloy.

3.2 The variation of steady-state stress with strain rate

Fig. 3 shows the variation of strain rate with flow stress plotted on logarithmic axes for both materials. For the Fe-3%Si steel three different regions can be distinguished from the experimental data (Fig. 3a): high-stress region (I), moderate stress region (II) and low stress region (III). The datum points of high-stress region lie to the right of the broken line and may

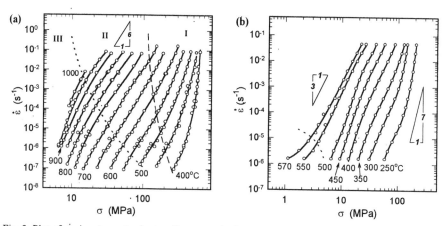

Fig. 3. Plot of strain rate *vs* steady state flow stress in double logarithmic scale.
(a) The Fe-3%Si steel. Dashed line shows a boundary between moderate stress and low stress regions of power-law creep. Broken line deviates the region of exponential creep from region of power-law.
(b) The 5083 alloy. Dashed line represents a boundary between diffusion creep and dislocation creep.

be equally described by the exponential law as the range of exponential creep:

$$\varepsilon = B \cdot \exp(\beta\sigma) \cdot \exp\left(\frac{-Q}{R \cdot T}\right), \tag{3}$$

where B is a constant and β is a coefficient. The power-law relationship

$$\varepsilon = A \cdot \left(\frac{\sigma}{E}\right)^{n_a} \cdot \exp\left(\frac{-Q}{R \cdot T}\right), \tag{4}$$

where n_a is the apparent stress exponent, is valid for datum points lying to the left from the broken line. Two types of strain rate ε, *vs* applied steady-state stress, σ, dependence can be distinguished in the power-law creep regime. The data for moderate stress region lying between broken and dashed lines can be adequately described by linear dependence. The value of $n_a=6\pm0.5$ was estimated from the slop of the plot and is constant with increasing temperature. No significant variation in n_a value with decreasing strain rate can be found in this region. The third set of data belonging to low stress region lies to the left from the dashed line. The n_a value was found to be variable and tended to increase with decreasing imposing strain rate. Strain rate exhibits a well-defined trend to decrease with decreasing stress.

For the 5083 alloy the experimental data can be described by Eq. (4) at all temperatures and strain rates examined (Fig. 3b). The stress exponent decreases from $n_a\sim7$ at the lowest temperature of 250°C to $n_a\sim3$ for the highest temperatures 550-570°C. The stress exponent values tend to increase with decreasing strain rate at temperatures ranging from 250 to 500°C. At higher temperatures the stress exponent exhibits an increasing trend with decreasing strain rate and attends ~1.7. It is apparent that transition to diffusion creep can take place at strain rates lower than 10^{-5} s^{-1} at pre-melting temperatures (The melting temperature is 572°C).

4. Discussion

As it has been mentioned above the increase in the stress exponent, n, with decreasing strain rate indicates that a material exhibits threshold behavior. Deformation behavior should be now examined in reference to the possibility of the existence of a threshold stress.

4.1 Estimation of threshold stresses
4.1.1 The Fe-3%Si steel

Careful inspection of Fig. 3a indicates that there are evidences for the existence of a threshold stress at temperatures ranging from 550 to 900°C. At lower temperatures there is no method to estimate the threshold stress in the range of exponential law (3). At higher temperatures 950-1000°C no stable trend in a change of the stress exponent was found. It can be caused by extensive dissolution of MnS dispersoids at these temperatures [11]. Therefore, threshold stress changes, randomly, with holding time and minor deviation in the testing temperature.

In estimating the threshold stress, σ_o, the experimental data were plotted as $\dot{\varepsilon}^{1/n}$ vs σ on a double linear scale at a single temperature [3] in the range 550-900°C. Fig. 4a shows a typical plot. To provide a validity of this graphic method in part of independence of σ_o value

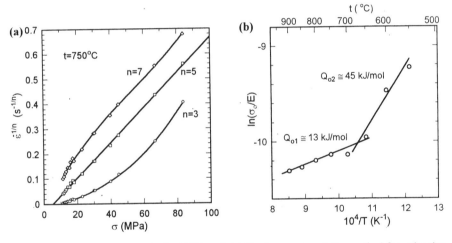

Fig. 4. Threshold behavior of the Fe-3%Si steel. (a) The linear extrapolation method for estimating the threshold stress at 750°C using values for n of 3, 7 the datum points exhibit curvature; 5, datum points provides the best linear fit. (b) Semi-logarithmic plot of the normalized threshold stress vs the reciprocal of absolute temperature.

from imposing strain rate the datum points were taken from stress region of the power-law. In addition, the experimental points at highest stresses were discarded. The stress exponent of 5 yields the best linear fit between $\dot{\varepsilon}^{1/n}$ and σ for datum points lying in the moderate and low stress regions and is the true stress exponent of the Fe-3%Si steel in the range 550-900°C. Calculated values of threshold stress are presented in Table 1. It is seen that absolute values of threshold stress in the steel are not so high. In the moderate stress region the σ_o value does not exceed 20pct of the applied stress. As a result, the linear dependence between strain rate and applied stress is observed in region II. In region III the relative value of threshold stress ranges from 15 to 50 pct of the applied stress value and effects strain rate vs applied stress dependence in the region of low stress.

Table 1. Threshold stresses of the Fe-3%Si steel at different temperatures.

t, °C	550	600	650	700	750	800	850	900
σ_o, MPa	16	12	7	5.5	5	4.5	4	3.7

The variation of normalized threshold stress, σ_o/E, with temperature in semi-logarithmic scale is shown in Fig. 4b. Two different temperature dependencies of threshold stress obeying Eq. (2) can be distinguished. It is seen that in the range 550-650°C the threshold stresses fit to a line for which the value of Q_o may be estimated as ~45kJ/mol. In the range 650-900°C the threshold stresses exhibit the linear temperature dependence with the Q_o value of ~13kJ/mol. Thus, a transition in temperature dependence of threshold stress is observed with increasing temperature with inflection point at t=650°C.

4.1.2 The 5083 alloy

Application of the above mentioned procedure for estimating threshold stresses in the 5083 alloys showed that there were three different temperature intervals distinguished by values of the true stress exponent, n. In the temperature range 250-300°C the stress exponent n of 6 yields the best linear fit (Fig. 5a). In the temperature range 350-500°C the stress exponent of 4 yields the best linear fit (Fig.5b). At higher temperatures small values of threshold stress can be obtained by discarding datum points at strain rates less than 8×10^{-4} s^{-1} (Fig. 3b).

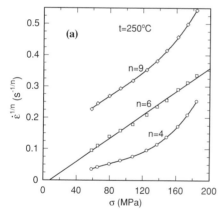

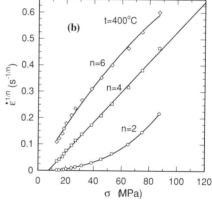

Fig. 5. Threshold stress in the 5083 alloy. The linear extrapolation method for estimating the threshold stress at (a) t=250°C, true stress exponent n=6 provides the best linear fit; (b) t=400°C, n=4; (c) t=570°C, n=3.

The stress exponent of 3 yields the best linear fit in the range 550-570°C (Fig. 5c). Calculated values of threshold stress and the true stress exponent values are summarized in Table 2 as a function of testing temperature.

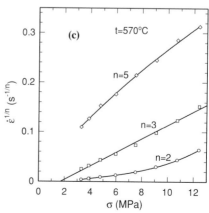

Table 2. Threshold stresses and n values of the 5083 alloy at different temperatures.

t, °C	250	300	350	400	450	500	550	570
n	6	6	4	4	4	4	3	3
σ_o, MPa	30.4	16.2	8.9	7.5	6.0	4.5	1.7	0.9

Fig. 6 shows the temperature dependence of the modulus-normalized threshold stress. By constructing this plot, the value of Q_o may be estimated as ~94, 17, 31 kJ/mol for ranges, 250-350°C, 350-500°C and 500-570°C, respectively. Notably, that Q_o values obtained in the range 400-570°C lie within the range values being available for the range of Al-based materials [9]. The Q_o value at the lowest temperature interval is essentially higher.

The both materials contain fine particles. Nanoscale incoherent dispersoids of MnS in the Fe-3%Si steel, Al_3Cr and Al_3Zr in the 5083 alloy represent effective obstacles to dislocation motion and can provide sources for the threshold stress. Slight temperature effect on the size of dispersoids and their volume fraction takes place at the examined temperatures in the both materials [11]. Thus, the observed transition in threshold behavior can be associated with a change in the mechanism of interaction between dislocations and dispersoids.

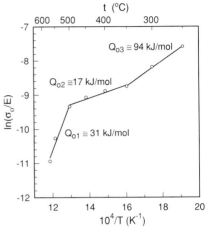

Fig. 6. Semi-logarithmic plot of the normalized threshold stress *vs* the reciprocal of absolute temperature for the 5083 alloy. Three different intervals of threshold behavior can be distinguished.

4.2 Activation energy

The true activation energy for plastic deformation, Q_c, in Eq. (1) was determined by the method described in earlier report [10] in the detail. For the Fe-3%Si steel Q_c values were plotted as a function of inverse temperature, $1/T$, in Fig. 7. It is seen, that the Q_c value is practically constant in the range 700-900°C and is equal to 290±30 kJ/mol. A decreasing trend resulting in 250±15 kJ/mol is observed in the range 550-700°C. The first value of Q_c is close to the value of the activation energy for lattice self-diffusion in pure α-iron above the Curie temperature (Q_l=281.5 kJ/mol) [13]. The minor decreasing trend in Q_c value can be explained in terms of reduction of lattice diffusion activation energy from a value of 281.5 kJ/mol above the Curie temperature to a value of 239 kJ/mol below it. The following decrease in Q_c value with

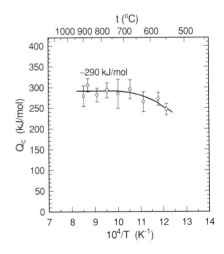

Fig. 7. The true activation energy for plastic deformation of the Fe-3%Si steel as a function of inverse of absolute temperature.

decreasing temperatures can be associated with transition from dislocation climb controlled by lattice diffusion to dislocation climb controlled by pipe-diffusion (Q_p=174 kJ/mol) [14].

For the 5083 alloy the true activation as a function of inverse temperature, 1/T, is presented in Fig. 8. In the range 250-300°C a value of Q_c is equal to 118±10 kJ/mol and exhibits the increasing trend with increasing temperature. At temperatures ranging from 350 to 500°C the value Q_c=140±8 kJ/mol is essentially close to the value of the activation energy for lattice self-diffusion in pure aluminum (Q_l=143.4 kJ/mol) [9]. With following temperature increase Q_c tends to decrease to a value of 114±15 kJ/mol. This value of Q_c is slightly lower than the activation energy for diffusion of Mg atoms in aluminum solid solution (Q_{Mg}=136 kJ/mol).

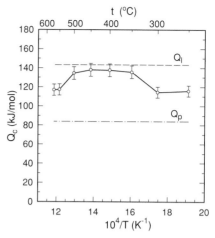

Fig. 8. The true activation energy for plastic deformation of the 5083 alloy as a function of inverse of absolute temperature.

The temperature dependence of the true activation energy obtained in the range 250-500°C can be interpreted in terms of transition from low temperature climb controlled by pipe-diffusion (Q_p=82 kJ/mol [16]) to high temperature dislocation climb controlled by lattice diffusion. It may be assumed that a decrease in Q_c values with increasing temperature in the range 550-570°C can be associated with transition to viscous dislocation glide.

4.3 An examination of normalized deformation data
4.3.1 The Fe-3%Si steel

Fig. 9a shows a double logarithmic plot of the normalized strain rate, $\varepsilon\, kT/(D_l Eb)$, against the normalized effective stress, $(\sigma-\sigma_0)/E$. No datum points from region I of exponential creep and from the region 900-1000°C are presented due to impossibility to calculate the threshold stress. It is seen that the Fe-3%Si steel exhibits two characteristic modes of deformation behavior. First, at the low normalized strain rate, $\varepsilon\, kT/(D_l Eb)<2\times10^{-6}$, there is a well-defined power-law relationship with the stress exponent ~4.8. Such deformation behavior is associated with high temperature dislocation climb controlled by lattice self-diffusion. This region locates below and slightly above the Sherby-Burke criterion ($\varepsilon\, kT/(D_l Eb)\cong2.5\times10^{-8}$) which represents the breakdown of power-law (3). Second, at the moderate normalized strain rate, $\varepsilon\, kT/(D_l Eb)>2\times10^{-6}$, the stress exponent increases up to 7. Such transition could be interpreted in terms of transition from high temperature climb controlled by lattice diffusion to low temperature climb controlled by pipe-diffusion along the dislocation cores. The classic relationship n=n+2 is observed at this transition from high to intermediate temperature. A transition from the power-law (2) to the exponential creep (3) takes place at the normalized strain rate ~10^{-3}. Therefore, the numerous experimental datum points in the power-law creep region lie above the Sherby-Burke criterion.

4.3.2 The 5083 alloy

Inspection of Fig. 9b shows that there are three different ranges for deformation behavior of the 5083 aluminum alloy. At the high normalized strain rate, $\varepsilon\, kT/(D_l Eb)>2.5\times10^{-8}$, there is a power-law relationship with the stress exponent, n, of ~6. At the normalized strain rate ranging from 2.5×10^{-8} to 5×10^{-13} the slope of straight line dependence $\dot{\varepsilon} kT/(D_l Eb)$ vs $(\sigma - \sigma_o)/E$ is ~4. It is seen that a classic transition from high temperature climb controlled by vacancy diffusion through the lattice to low temperature climb controlled by vacancy diffusion along dislocation cores, in accordance with the rule n=n+2 [16], takes place at $\dot{\varepsilon} kT/(D_l Eb)=2.5 \times 10^{-8}$. Notably the transition in the normalized strain rate matches Sherby-Burke criterion for the power law breakdown. However, no breakdown in the power-law is observed at this value of the normalized strain rate.

In the range of low normalized strain rates, $\dot{\varepsilon} kT/(D_l Eb) < 5 \times 10^{-13}$, the power-law relationship with n~3 is observed. This type of deformation behavior can be interpreted in terms of viscous dislocation glide [15]. Thus, the 5083 alloy exhibits three characteristic modes of deformation behavior. A temperature increase results in sequential transition from low rate to high rate dislocation climb and to viscous dislocation glide, finally. This interpretation of normalized deformation data is supported by the above-mentioned inspection of temperature dependence of the true activation energy.

4.4 Threshold behavior and deformation mechanisms

Inspection of deformation data shows that a strong correlation between characteristic modes of deformation behavior represented by Eq. (2) and temperature dependence of threshold stress (1) takes place in both materials examined. The point of transition from one region of power law to another in Fig. 9 and the inflection point for threshold behavior in Figs. 4b and 6 are essentially similar. It is obvious that transition from one rate-controlling mechanism of plastic deformation to another with temperature is accompanied by a change in the mechanism of interaction between lattice dislocations and incoherent particles. A precise origin of the mechanism of such interaction is not yet clear, but on the basis of the result presented it can be

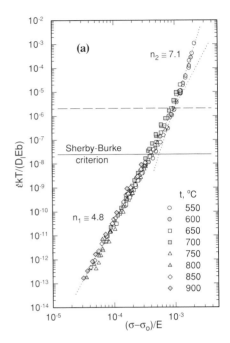

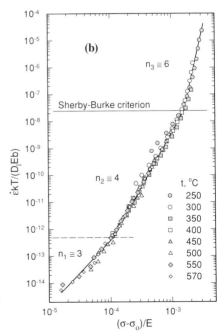

Fig. 9. Normalized strain rate vs normalized stress. (a) Fe-3%Si steel. (b) The 5083 alloy.

concluded that the mechanism of dislocation movement plays an important role in the process of overcoming the obstacle by lattice dislocations. At higher temperatures the general dislocation climb is a controlling mechanism of plastic deformation and dislocations overcome obstacles, easily. A decrease in the rate of diffusion with decreasing temperature impedes dislocation climb. At intermediate temperatures the deformation behavior is controlled by vacancy diffusion along dislocation cores. As a result, the ability of a lattice dislocation to overcome an obstacle is hindered, and a significant growth of activation energy for surmount particles by dislocation is observed. It is possible to presume that there is a connection between the mechanism of interaction of a dislocation with an obstacle and the rate-controlling mechanism of plastic deformation. Thus, threshold stress is a specific characteristic for a particular deformation mechanism of dislocation creep. It is obvious that mechanisms yielding threshold behavior of materials have a thermally activated origin. In addition, inspection of datum points in Fig.3b for the 5083 alloy shows that a transition from dislocation creep to diffusion creep results in disappearance of the threshold behavior.

There is a suggestion that may, in general, explain the correlation between the rate-controlling process and the mechanism of interaction between lattice dislocation and hard particles in case of dislocation creep. The mechanism of plastic deformation of the materials examined can be considered as an operation of two sequential processes. Fig. 10 represents a scheme of this mechanism based on the well-known second Weertman's model [17] as an example. Sources S_1 and S_2 lying in parallel crystallographic planes generate lattice dislocations under the applied stress in mutually opposite directions. Dislocations glide toward one another. To attend the encounter point the mobile dislocation should overcome a hard particle. The dislocation is assumed to

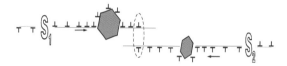

Fig. 10. A scheme of deformation process in a DS alloy. Dynamic recovery via the second Weertman's models is controlling mechanism of dislocation creep.

overcome the obstacle by means of local climb and following detachment. Dislocations with opposite Burgers vectors form a dislocation dipole in site of the encounter point. This dipole plays a role of barrier for dislocation glide and locks dislocation sources. Following climb of dipole's dislocation to one another results in their mutual annihilation and unlocking sources. The slower process in this scheme is overall glide and climb of lattice dislocations. This process controls deformation behavior. The process of surmount of particles by lattice dislocation is a rapid process. This process yields threshold behavior.

5. Conclusion

1. The Fe-3%Si steel and the 5083 aluminum alloy containing nanoscale incoherent dispersoids exhibit threshold behavior, like that in dispersion strengthened alloys. The apparent stress exponent is not constant and increases with decreasing strain rate in the range of low strain rates.

2. The analysis of the deformation data of the Fe-3%Si steel and the 5083 aluminum alloy reveals the presence of a threshold stress. Several types of the temperature dependence of threshold stress distinguishing by values of energy term, Q_o, were found in a wide temperature range for both materials examined.

3. By incorporating a threshold stress into the analysis, it was shown that the true stress exponent, n, decreases from n~n+2 at intermediate temperatures to n~n at high temperatures for the both materials. In the 5083 aluminum alloy the transition to the range of n=3 was found at pre-melting temperatures. Taking into account the temperature dependence of the

true activation energy for plastic deformation, Q_c, the deformation behavior of the both materials in the range of moderate and high normalized strain rates was interpreted in terms of transition from low temperature climb to high temperature climb. The deformation behavior of the 5083 aluminum alloy in the range of low normalized strain rate was interpreted in terms of transition from dislocation climb to viscous dislocation glide.

4. It was found that transitions in deformation behavior for the both alloys are correlated with transitions in the temperature dependence of threshold stress.

6. References

[1]. Lund, R.W., Nix, W.D., 1975, On High Creep Activation Energies for Dispersion Strengthened Metals, Metal.Trans., Vol. 6A, pp. 1329-1333.

[2]. Oliver, W.C., Nix, W.D., 1982, High Temperature Deformation of Oxide Dispersion Strengthened Al and Al-Mg Solid Solutions, Acta Metall., Vol. 30, pp. 1335-1347.

[3]. Mohamed, F.A, Park, K.T., Lavernia, E. J., 1992, Creep Behavior of Discontinios SiC-Al Composites, Mat.Sci.Eng., Vol. A150, pp. 21-35.

[4]. Rösler, J., Arzt, E., 1988, The Kinetics of Dislocation Climb over Hard Particles-I. Climb without Attractive Particle-dislocation Interaction, Acta Metall., Vol. 36, No. 4, pp. 1043-1051.

[5]. Arzt, E., Rösler, J., 1988, The Kinetics of Dislocation Climb over Hard Particles-II. Effects of an Attractive Particle-dislocation Interaction, Acta Metall., Vol. 36, No. 4, pp.1053-1060.

[6]. Rosler, J., Arzt, E., 1990, A new Model-based Creep Equation for Dispersion Strengthened Materials, Acta Metall.Mater., Vol. 38, pp. 671-683.

[7]. Pichler, A., Arzt, E., 1996, Creep of Dispersion Strengthened Alloys Controlled by Jog Nucleation, Acta Mater., Vol. 44, No. 7, pp. 2751-2758.

[8]. Dunand, D.C., Jansen, A.M., 1997, Creep of Metals Containing High Volume Fraction of Unshearable Dispersoids - Part I. Modeling the Effect of Dislocation Pile-ups Upon the Detachment Threshold Stress, Acta Mater., Vol. 45, No. 11, pp. 4569-4581.

[9]. Li, Y., Langdon, T.G., 1998, High Strain Rate Superplasticity in Metal Matrix Composites: The Role of Load Trasfer, Acta Mater., Vol. 46, No. 11, pp. 3937-3947.

[10]. Kaibyshev, R., Kazakulov, I., 2000, Deformation Behavior of Fe-3%Si Steel at High Temperatures, Key Engineering Materials, Trans Tech Publications, Swizerland, Vols. 171-174, pp. 213-218.

[11]. Sokolov, B.K., 1977, The Interaction between Grain Boundaries and Precipitations of Second Phase Particles, Phys.Met.Metall., Vol. 43, pp.1028-1035 (in russian).

[12]. Kim ,W.J., Yeon, J.H., Shin, D.H., Hong, S.H., 1999, Deformation Behavior of Powder-metallurgy Processed High-strain-rate Superplastic $20\%SiC_p/2124$ Al Composite in a Wide Range of Temperature, Mater.Sci.Eng., Vol. A269, pp. 142-151.

[13]. Landolt-Bornstein, 1990, Numerical Data and Functional Relationships in Science and Technology, New Series. Group III:Crystals and Solid State Physics. Diffusion in solid Metals and Alloys, Vol. 26, p. 49.

[14]. Frost, H.J., Ashby, M.F., 1982, Deformation-mechanism maps, Pergamon Press, p. 328.

[15]. Yavari, P.,Langdon, T., 1982, An Examination of the Breakdown in Creep by Viscos Glide in Solid Solution Alloys at High Stress Levels, Acta Metall, Vol. 30, pp. 2181-2196.

[16]. Luthy, H., Miller, A., Sherby, O., 1980, The Stress and Temperature Dependence of Steady State Flow at Intermediate Temperature for Pure Policrystallyne Al, Acta Metall. Vol.28, No.2, pp.169-182.

[17]. Chadek, J., 1994, Creep in Metalic Materials, Academia, Praque, p.302.

COMPUTATIONALLY EFFICIENT MODELS OF THE PARTICLE COARSENING PROCESS FOR USE IN MODELLING CREEP BEHAVIOUR

R. N. Stevens*† and C. K. L. Davies*‡

*Department of Materials, Queen Mary, University of London, LONDON E1 4NS, UK

*FAX 44(0) 20 8981 9804

†R.N.Stevens@qmw.ac.uk

‡C.K.L.Davies@qmw.ac.uk

Abstract: Models for the kinetics of the coarsening of embedded particles in a solid matrix driven by the reduction of interfacial surface energy are described. These are based on the chemical rate theory in which the growth rate of a particle of a particular size class is determined by replacing all the other particles in the system by a homogeneous medium in which solute atoms are uniformly created and destroyed. The particle and a concentric shell of normal matrix are embedded in the homogeneous medium and the diffusion equations solved. The relationship between the outer radius of the matrix shell and the particle radius is chosen in a way that is consistent with the volume fraction of the particulate phase in the material being modelled. The solution of the diffusion problem allows the asymptotic form of the particle size distribution to be determined in explicit analytical form, although numerical computation is required to find the parameters for use in the resulting expression. The same process yields an expression for the growth rate of the average particle size together with values of the numerical parameters involved in the expression. The fact that explicit expressions are involved make the results of the theory very readily available for comparison with experiment, and computationally efficient for incorporation in modelling the time dependence of mechanical properties at elevated temperature of such materials.

1 Introduction

Microstructures in which small particles are embedded in a continuous matrix are a common form for many high strength alloy systems, particularly those used at elevated temperatures. The interfacial energy in such a system provides a driving force for competitive particle growth which causes the size and inter-particle spacing to increase while the number of particles per volume decreases, a process usually called Ostwald ripening [1]. Such growth can seriously degrade the performance of the alloys and limit their useful life at elevated temperatures. Microstructural instability of this sort will tend to increase the creep rate and hence can be regarded as creep damage. Numerical models of creep behaviour will have to model the damage process, one component of this being the microstructural changes brought about by coarsening. Numerical modelling of the coarsening process is itself a computationally very expensive undertaking, and there is therefore considerable technological interest in the provision of theories which give analytical expressions for particle growth rates and particle size distributions. This paper provides such expressions for the case of small volume fractions of particles with zero misfit. It represents therefore only an initial step in the process.

The theory of coarsening was first worked out for the limiting case of zero volume fraction by Liftshitz and Slyozov [2] and by Wagner [3] (the LSW theory) following earlier work by Greenwood [4]. The theory predicts that at long times, the cube of the particle size is proportional to time and that the probability density of relative particle sizes (size divided by mean size) is time independent. Expressions for both of these relationships are provided by the theory. When the volume fraction approaches zero the surroundings of every particle of a give size becomes the same and therefore so does its diffusion field and growth rate. For a finite volume fraction however, even particles of the same size have different surroundings and thus different growth rates. In order to develop an equivalent of the LSW theory the mean growth rate for particles of a given size must be found. A number of theories, analytical and numerical, have been put forward accounting for the effect of particle volume fraction on coarsening. Reviews have been provided by Jayanth and Nash [5], Voorhees [6, 7] and Mullins and Viñals [8]. The aim of the work described herein is to develop the effective medium theory of Brailsford and Wynblatt [9].

2 The model

The solution to the diffusion problem for a finite volume fraction polydisperse system of spherical particles in which the solute concentration at the surface of each particle is fixed by the Gibbs-Thomson equation is very difficult. In the Brailsford and Wynblatt model the mean growth rate for particles of a given size class is found by replacing the surroundings of a representative particle by a homogeneous medium which averages the effect of the particle/matrix interfaces in emitting or absorbing particles. At every point in the effective medium, solute is absorbed and created. Corresponding to particles of radii

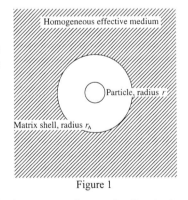

Figure 1

between r and $r + dr$ there is an emission rate $P(r, t)dr$ (moles per second per mole of system) and a sink rate $Dk_T^2 (r,t)c_\infty dr$ with the same units where t is time, D the diffusivity, c_∞ the mole fraction of solute and k^2 a parameter called the sink factor density. The total emission rate, $P_T(t)$ and the total sink factor, $k_T^2(t)$ are found by integrating Pdr and k^2dr over all particle sizes.

The model for solving the diffusion equation to find the growth rate of the representive particle is illustrated in Figure 1. A shell of matrix of radius r_A (the averaging sphere) surrounds the particle of radius r with the rest of space filled with the homogeneous medium. Steady state radial flow is assumed, with the flux divergence-less between r and r_A and having a divergence equal to the net total emission rate in the homogenous medium, while the solute concentration, c, and its gradient are continuous at $r = r_A$. At large r, c approaches c_∞.

With these boundary conditions the model yields, for the growth rate of a particle of size r

$$\frac{dr}{dt} = \frac{2\sigma V\, Dc_e}{RT} \frac{1}{r}\left[\frac{1}{r} - \frac{1}{r}\right] W(r, r_A, t) \qquad 1$$

where σ is the interfacial energy, V_m the molar volume, c_e the concentration in equilibrium with particles of infinite size, R the universal gas constant, T thermodynamic temperature, r^* the critical size of the particle which neither grows nor shrinks and W a dimensionless function of r, r_A and t. W becomes time-independent at long times when it is given by

$$W(r, r_A) = \frac{1 + k_T r_A}{1 + k_T(r_A - r)} \qquad 2$$

Also arising from the solution is the important equation

$$k_T^2 = 4\pi \int_0^\infty r W(r, t) f(r, t) dr \qquad 3$$

where $f(r, t)dr$ is the number of particles per volume with r between r and $r + dr$.

While these results are of interest they are not of the greatest importance in themselves. Expressions for the rate of change of the mean particle size and for the particle size distribution are of much more interest. The method for finding these expressions was set out by Liftshitz and Slyozov [1] and Wagner [2] and remains the same in the more complex models. It is based on recasting the problem in terms of dimensionless variables, y and t being respectively measures of particle size and time, and also a dimensionless sink factor, q and Y, a dimensionless form for r_A,

$$y = r/r^*, \qquad Y = r_A/r^*, \qquad \tau = \ln(r^*/r_0^*), \qquad q = k_T r^* \qquad 4$$

where r_0^* is the critical size at zero time. The quantities q and Y are dependent on volume fraction, ϕ. In terms of the dimensionless quantities, W becomes

$$W = \frac{1 + qY}{1 + q(Y - y)} \qquad 5$$

and equation (1) transforms to

$$U = \frac{dy}{d\tau} = \frac{\gamma(\tau) W(y, Y)(y - 1)}{y^2} - y \qquad 6$$

where $\gamma(\tau)$ is given by

$$\gamma(\tau) = \frac{2\sigma V_m Dc_e}{RTr^{*2}} \frac{dt}{dr} \qquad 7$$

The quantity dt/dr^* is closely related to the rate of change of the mean particle size, $\langle r \rangle$, and if γ can be found, equation (7) may be integrated to give the required relationship between $\langle r \rangle$ and t provided the relationship between r^* and $\langle r \rangle$ is known.

It can be shown that, at large τ (the asymptotic limit), $\gamma[1,2]$ and q [9] become independent of time, and U and dU/dt are zero at $y = y_c$, beyond which size no particles are found. In addition, the number of particles per volume with sizes between y and $y + dy$ can then be factorised into the product of a probability density, $p'(y)$ depending only on y and a number per volume, $n(t)$ depending only on t [1,2]. It can be shown that the normalised probability density is given by

$$p'(y) = -\frac{3}{U} \exp\left[3 \int_0^y \frac{dy'}{y'} \right] \qquad 8$$

where the pre-exponential constant of value 3 is that required for nomalisation independent of volume fraction, ϕ.

It is clear that no solution to the problem (for finite volume fraction, ϕ) can be obtained until Y and therefore also W are known as functions of y. It will also be necessary to have some method for finding q as a function of volume fraction.

3 Development of the model

In their original work Brailsford and Wynblatt [9] chose the simplest possible model, setting $Y = y$, giving $W = 1 + qy$. They suggested however that the form $Y = y/\phi^{1/3}$ was to be prefered since this would make the ratio of the volumes of the particle and the averaging sphere (of radius r_A) equal to the volume fraction, ϕ, for all particles and thus make the model self-consistent. They did not implement this suggestion. These proposals for the form of the Y function are illustrated in Figure 2 as the two full straight lines.

Voorhees [6] pointed out small particles are much less influenced by the environment than larger particles and suggested that the relative averaging sphere radius should be constant and equal to $\phi^{-1/3}$ from $y = 0$ to $y = 1$ and equal to $y/\phi^{-1/3}$ from $y = 1$ to $y = y_c$. This form is shown as the chain dotted line in Figure 2. The discontinuity in slope of the Voorhees suggestion makes it inconvenient for analysis and it is better to use a polynomial approximation such as the full curve.

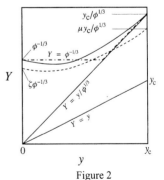

Figure 2

In the present work a more general definition of self-consistency has been used; this is that only the *mean value* of the ratio of the particle and averaging sphere volumes for all particles needs be equal to ϕ. The dotted curve illustrates a Y function of the form required to achieve this. Particles for which the Y curve lies above the line $Y = y/\phi^{-1/3}$ make a contribution to the mean less than ϕ while particles with Y lying below the line make a contribution greater than

ϕ. The parameters ζ and μ in Figure 2 can therefore be chosen to make the model self-consistent.

The following quadratic was used for the Y function:

$$Y = \phi^{-1/3}\left[\zeta + y\left(\mu - 2\zeta / y_c\right) + y^2 \zeta / y_c\right]$$

9

and this gives rise to W of the form

$$W = \frac{\lambda_0 + \left(\lambda_1 + q\right)y + \lambda_2 y^2}{\lambda_0 + \lambda_1 y + \lambda_2 y^2}$$

10

where

$$\lambda_0 = 1 + \zeta\alpha / \mu, \qquad \lambda_1 = \alpha\left[1 - 2\zeta /\left(\mu y_c\right)\right] - q, \qquad \lambda_2 = \alpha\zeta /\left(\mu y_c^2\right)$$

11

with

$$\alpha = \mu q / \phi^{1/3}$$

12

Putting equation (9) into equation (6) gives an expression for the growth rate, U

$$U = \frac{-\lambda_2 y^5 - \lambda_1 y^4 + \left(\gamma\lambda_2 - \lambda_0\right)y^3 + \gamma\left(\lambda_1 + q - \lambda_2\right)y^2 + \gamma\left[\lambda_0 - \lambda_1 - q\right]y - \gamma\lambda_0}{y^2\left(\lambda_0 + \lambda_1 y + \lambda_2 y^2\right)}$$

13

By applying the condition that both U and dU/dy are zero at $y = y_c$ we can find expressions for γ and y_c. These are

$$\gamma = \frac{y_c^3\left(1 + \beta y_c\right)}{\left(y_c - 1\right)\left(1 + \alpha y_c\right)}$$

14

and

$$-2\alpha\beta y_c^3 + \left(3\alpha\beta - 3\beta - \alpha\right)y_c^2 + \left(4\beta + 2\alpha - 2\right)y_c + 3 = 0$$

15

where

$$\beta = \alpha - q$$

16

We can now put equation (13) into equation (8) and find the distribution function. The form of the result depends on how many real roots the polynomial in the numerator of equation (13) has. The only case of interest here arises when in addition to the double root at y_c there is one real root, y_1 and a quadratic factor $y^2 + v'y + w'$ without real roots. The result is

$$p'(y) = \frac{3y^2\left(\lambda_0 + \lambda_1 + \lambda_2 y^2\right)\exp\left[\dfrac{-cy}{y_c - y}\right]\exp\left[K\tan^{-1}\left(\dfrac{y\sqrt{w' - v'^2/4}}{w' - v'y/2}\right)\right]}{\gamma\lambda_0\left(1 - y/y_c\right)^a\left(1 + y/y_1\right)^{b_1}\left(y^2/w' + v'y/w' + 1\right)^{b_2}}$$

17

Expressions are available for the parameters, a, b_1, b_2, c and K, but are not given here.

4 Obtaining the parameters and developing a self-consistent version

The parameters in equation (17) are not known explicitly in terms of volume fraction but only implicitly in terms of q. Putting equation (3) in terms of the dimensionless variables we find

$$q^2 = \frac{3\langle yW\rangle\phi}{\langle y^3\rangle} \qquad\qquad 18$$

where the angle brackets denote mean value over all particle sizes. This equation provides the basis for determining the q corresponding to a given ϕ. An estimate of q is made, p' calculated and from this the mean values in equation (18) found and thus a value for ϕ. An iterative procedure is then used to adjust q until the target volume fraction is achieved.

In order to make the results self-consistent ζ is fixed at 1 and a value of the parameter μ (where $1 > \mu > \phi^{1/3}$) is found which makes the mean ratio of particle and averaging sphere volumes equal to the target volume fraction. There are thus two nested iterative loops required, the inner loop finding q for the given ϕ, ζ and μ and the outer loop adjusting μ to make the result self-consistent. The process converges rapidly and efficiently.

One problem remains. The quantity $y = r/r^*$ is not measurable. We require to put the results in terms of the quantity $\rho = r/\langle r\rangle$ and find p where the probability of finding a particle with ρ between ρ and $d\rho$ is $p d r$. We define

$$p = p'\langle y\rangle, \qquad \rho = y/\langle y\rangle, \qquad \rho_c = y_c/\langle y\rangle \left.\right|$$
$$\rho_1 = y_1/\langle y\rangle, \qquad w = w'/\langle y\rangle^2, \qquad v = v/\langle y\rangle \left.\right\} \qquad 19$$

in terms of which equation (17) becomes

$$p(p) = \frac{3\rho\left(A_0 + A_1 + A_2\rho^2\right)\exp\left(\dfrac{-c\rho}{\rho_c - \rho}\right)\exp\left[K\tan^{-1}\left(\dfrac{\rho\sqrt{w - v^2/4}}{w - v\rho/2}\right)\right]}{\gamma\left(1 - \rho/\rho_c\right)^a\left(1 + \rho/\rho_1\right)^{b_1}\left(\rho^2 + v\rho/w + 1\right)^{b_2}} \qquad 20$$

where

$$A_0 = \langle y\rangle^3, \qquad A_1 = \langle y\rangle^4 \lambda_1/\lambda_0, \qquad A_2 = \langle y\rangle^5 \lambda_2/\lambda_0 \qquad 21$$

Now that the (time independent) value of γ is known for a given volume fraction, equation (7) can be integrated to give the relation between mean particle size and time

$$\langle r\rangle^3 - \langle r_0\rangle^3 = \frac{6D\sigma V_m c_e \langle y\rangle^3}{\gamma RT}t \qquad 22$$

where $\langle r_0\rangle$ is the mean value at $t = 0$. Table 1 gives values of all the parameters required for evaluating equation (20) as a function of volume fraction and the volume fraction dependent values of γ and $\langle y\rangle$ for use in equation (22). Linear interpolation to find values not in the table is reasonably accurate except near $\phi = 0$.

ϕ	q	A_0	$-A_1$	A_2	a	b_1	b_2	c
0	0	1	0	0	11/3	7/3	1	1
10^{-6}	.0016309	0.99981	0.065000	0.062265	3.6680	2.3328	0.99961	1.0015
10^{-4}	0.016401	0.99843	0.12995	0.11529	3.6708	2.3318	0.99868	1.0091
10^{-3}	0.052487	0.99555	0.18823	0.15037	3.6668	2.3312	1.0010	1.0203
0.005	0.11972	0.99067	0.25134	0.17691	3.6491	2.3303	1.0103	1.0338
0.01	0.17177	0.98704	0.28789	0.18829	3.6312	2.3292	1.0198	1.0414
0.02	0.24787	0.98177	0.33236	0.19918	3.6008	2.3271	1.0361	1.0502
0.04	0.36075	0.97385	0.38671	0.20893	3.5480	2.3229	1.0646	1.0600
0.06	0.45190	0.96723	0.42385	0.21363	3.4993	2.3188	1.0909	1.0658
0.08	0.53210	0.96119	0.45275	0.21620	3.4521	2.3148	1.1166	1.0698
0.10	0.60553	0.95544	0.47659	0.21658	3.4052	2.3106	1.1421	1.0725
0.12	0.67435	0.94983	0.49690	0.21816	3.3580	2.3065	1.1678	1.0744
0.14	0.73985	0.94427	0.51453	0.21813	3.3098	2.3023	1.1939	1.0756
0.16	0.80288	0.93869	0.53003	0.21763	3.2605	2.2980	1.2208	1.0762
0.18	0.86404	0.93303	0.54374	0.21761	3.2095	2.2937	1.2484	1.0763
0.20	0.92380	0.92725	0.55590	0.21542	3.1565	2.2892	1.2771	1.0759
0.22	0.98250	0.92130	0.56665	0.21377	3.1012	2.2846	1.3071	1.0750
0.24	1.0404	0.91513	0.57612	0.21178	3.0432	2.2800	1.3385	1.0737
0.26	1.0979	0.90870	0.58435	0.20943	2.9819	2.2751	1.3715	1.0719
0.28	1.1550	0.90193	0.59138	0.20669	2.9167	2.2700	1.4066	1.0697
0.30	1.2121	0.89477	0.59717	0.20354	2.8470	2.2467	1.4442	1.0669
0.32	1.2693	0.88710	0.60168	0.19991	2.7716	2.2591	1.4846	1.0636
0.34	1.3269	0.87881	0.60477	0.19573	2.6892	2.2532	1.5288	1.0597

ϕ	ρ_c	ρ_1	$-v$	w	K	γ	$\langle y \rangle$
0	3/2	3	1	∞	0	27/4	1
10^{-6}	1.5005	2.9943	1.0372	16.048	0.0036548	6.7364	0.99994
10^{-4}	1.5037	2.9592	1.0790	8.6033	0.030341	6.6366	0.99948
10^{-2}	1.5096	2.8948	1.1274	6.4871	0.082103	6.4351	0.99852
0.005	1.5186	2.7987	1.1822	5.3570	0.16242	6.1165	0.99688
0.01	1.5249	2.7343	1.2134	4.9250	0.21797	5.8963	0.99556
0.02	1.5338	2.6494	1.2504	4.5111	0.29332	5.6006	0.99389
0.04	1.5469	2.5369	1.2941	4.1016	0.39679	5.2020	0.99121
0.06	1.5576	2.4544	1.3233	3.8569	0.47526	4.9059	0.98896
0.08	1.5673	2.3863	1.3457	3.6773	0.54135	4.6605	0.98689
0.10	1.5766	2.3270	1.3641	3.5327	0.59974	4.4461	0.98492
0.12	1.5858	2.2737	1.3799	3.4099	0.65274	4.2532	0.98299
0.14	1.5949	2.2247	1.3938	3.3019	0.70166	4.0758	0.98107
0.16	1.6041	2.1790	1.4062	3.2045	0.74736	3.9104	0.97913
0.18	1.6137	2.1358	1.4175	3.1151	0.79037	3.7543	0.97716
0.20	1.6235	2.0945	1.4280	3.0318	0.83107	3.6056	0.97514
0.22	1.6338	2.0547	1.4377	2.9532	0.86970	3.4628	0.97305
0.24	1.6448	2.0162	1.4470	2.8783	0.90642	3.3248	0.97087
0.26	1.6565	1.9785	1.4558	2.8062	0.94131	3.1904	0.86859
0.28	1.6691	1.9414	1.4644	2.7363	0.97440	3.0588	0.96618
0.30	1.6828	1.9046	1.4728	2.6680	1.0056	2.9290	0.96361
0.32	1.6981	1.8678	1.4813	2.6005	1.0348	2.8001	0.96085
0.34	1.7153	1.8308	1.4901	2.5334	1.0618	2.6710	0.95785

Table 1 (a) and (b)

92

5 Results

The relative particle size distributions for various volume fractions are shown in figure (3). The value of μ cannot be less than $\phi^{1/3}$ (otherwise r_A would be smaller than r for some particles) and there is therefore an upper limit of ϕ (~ 0.445) for which the theory is applicable. The particle size distributions for a given volume fraction are shown in Figure 3.

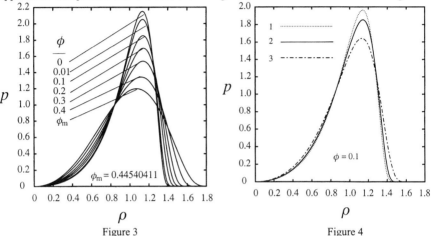

Figure 3   Figure 4

Comparision of the present results (full line 2) for a volume fraction of 0.1 with those of the original work of Brailsford and Wynblatt [9] (chain line 3) and the self-consistent scheme given by the line $Y = y/\phi^{1/3}$ in Figure (2) [10] (dotted line 2) is shown in Figure 4.

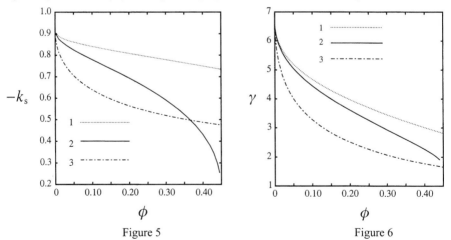

Figure 5   Figure 6

In Figure (5) the coefficient of skewness, k_s is plotted for the present work and the other two models identified as in Figure 4. Note that the skewness is negative and that $-k_s$ is plotted for convenience. Figure 6 plots the value of γ (used in equation (22)) for the three models in the same way.

Discussion and conclusions

Experimental observations [5] indicate that the effect of a finite volume fraction is to produce particle size distributions which are broader and less strongly skewed than that given by the LSW zero volume fraction theory while the inverse rate constant, γ is reduced. These effects are clearly seen in the current work. However, particle size distributions are notoriously difficult to determine experimentally with accuracy, and experiments based on a far greater number of particles than are typical of the experiments conducted so far are necessary if theories are to be critically tested.

Theories which do not take into account spatial correlation effects are unlikely to be valid at volume fractions above 0.3 [7]. The upper limit on the volume fraction in these theories is therefore not of great significance. It will be of significance, in the application of this theory in modelling creep behaviour since high volume fractions are commonly used in creep resistant alloys.

The only criterion the theory provides for choosing a form for the Y function (the relative size of the averaging sphere) is self-consistency. If self-consistency is defined as the mean value of the ratio of particle and averaging sphere volumes being equal to the volume fraction, then there is, in principle, an infinite number of ways this can be achieved of which the method used here is only one. This, however, gives the theory some flexibility which may be used to advantage. An alternative to analytical theories is provided by numerical simulation, which although computationally expensive, has reached a state of development in which good results can be expected. The flexibility of the effective medium theory may allow it to be fine-tuned to reproduce the results of numerical simulation in a much more accessible and computationally efficient form.

The mean particle size at any time can be found from equation (22) and the particle size distribution relative to the mean can be found from equation (20) using the values for the parameters given in Table 1. From these data, a mean inter-particle spacing can be calculated and estimates made of the distribution of inter-particle spacings. These are microstructural factors which numerical models of creep must use to predict long term creep behaviour. If such a modelling program stored the parameters given in figure (1) as data, the time required to calculate a complete distribution function and to find the necessary mean values would be of the order of milliseconds or less, whereas numerical models of the coarsening process of the kind produced by Akaiwa and Voorhees (11) would take many hours, adding very significantly to the computational work done by a creep modelling program. Theories of the type developed here can therefore make a significant contribution to the efficiency of a numerical model for the predicition of long term creep behaviour.

94

References

1   Ostwald, W. *Analytische Chemie*, (1901) 3rd Edition, p. 23, Engelman, Leipzig,

2   Lifshitzt, I. M. and Slyozov V. V. (1961) *The kinetics of precipitation from supersaturated solid solutions,* Journal of Physics and Chemistry of Solids, 19: 35 - 49

3   Wagner, C (1961) *Theorie der Alterung von Niederschlägen durch Unlösen (Ostwald-Reifung)* Zeitschrift für Elektrochemie, 65: 581 - 591

4   Greenwood, G. W. (1956) *The growth of dispersed precipitates in solutions* Acta Metallurgica 4: 243 - 248

5   Jayanth, C. S. and. Nash P, (1989) *Factors affecting particle coarsening kinetics and size distribution,* Journal of Materials Science, 24: 3041 - 3052

6   Voorhees, P. W. (1985) *The theory of Ostwald ripening* Journal of Statistical Physics 38: 231 - 252

7   Voorhees, P. W. (1992) *Ostwald ripening of two-phase mixtures,* Annual Reviews of Materials Science, 22: 197 - 215

8   Mullins, W. W. and Viñals, J. *Self-similarity and growth kinetics driven by surface free energy reduction* (1989) Acta Metallurgica 37: 991 - 997

9   Brailsford, A. D. and Wynblatt, P (1979) *The dependence of Ostwald ripening kinetics on particle volume fraction* Acta Metallurgica 27: 489 - 497

10   Stevens, R. N. and Davies, C. K. L. (1997) *I Recontre transfrontalière Matèriaux Biphasiques,* (A. Mateo, L. Lanes and M. Anglada, Eds), Barcelona. 21 - 26

11   Akaiwa, Norio and Voorhees, P. W. (1994) *Late-stage phase separation: Dynamics, spatial correlations, and structure functions* Physical Review E 49: 3860 - 3880

THE CREEP AND FRACTURE BEHAVIOUR OF OHFC COPPER UNDER BENDING

J D Parker

**Department of Materials Engineering,
University of Wales, Swansea SA2 8PP, UK**

ABSTRACT

High sensitivity creep tests on OFHC copper have shown that the mechanisms of deformation and fracture for miniature samples in bending are similar to those in uniaxial testing. Evaluation of the distribution of grain boundary cavities demonstrates that an agreement with analysis, the stress state established is equi-biaxial.

1.   INTRODUCTION

Traditionally creep and fracture performance is characterised by undertaking a series of uniaxial tests covering an appropriate range of conditions. Since the creep behaviour of most metals and alloys is highly dependent on stress and temperature, testing procedures must be closely controlled to minimise data scatter. Even using high sensitivity equipment to assess the behaviour of pure metals some variation in results is typically observed.   As a consequence of the microstructural complexities associated with these advanced materials, greater scatter has then been identified in programmes assessing the creep behaviour of engineering alloys

Improved reproducibility can normally be achieved by:

(i)   Increasing the gauge diameter since this will ensure that the testpiece includes representative structures, and

(ii)   Increasing the gauge length since for an extensometry system of a given accuracy greater displacement will occur.

In some situations creep data must be generated when the available material is limited. In these cases alternative testing configurations may offer the best option for generating data. Thus, for example, recently the need to monitor actual properties from components has been addressed by evaluation of the performance of small disc shaped specimens (1, 2, 3). Whilst tests for a range of commercial alloys have demonstrated that meaningful results can be achieved with this approach, the present research will permit details of the deformation and fracture behaviour to be established. Pure copper was selected for this investigation since:

(i)   background creep data were available for uniaxial loading.

(ii)   samples could be produced with a stable face centred cubic microstructure at appropriate grain sizes, and

(iii)   creep fracture of copper is typically the result of the nucleation and growth of cavities.

This paper demonstrates the comparability of mechanisms for creep in uniaxial loading and bending and indicates an approach for correlating data from the two loading regimes.

2. UNIAXIAL CREEP BEHAVIOUR

The creep and fracture behaviour of oxygen-free, high conductivity (OFHC) copper was evaluated. This involved reviewing published data (4, 5, 6, 7) and undertaking selected additional tests where necessary. Specific consideration was given to:

(i)   evaluation of strain:time behaviour and damage development, and
(ii)   characterisation of the stress and temperature dependence of deformation and rupture

To provide a comprehensive comparison with the results of the disc bend experiments, creep and rupture data were obtained for stresses in the range 10.3 to 83 MPa for temperatures between 335°C and 550°C.

Creep Curves and Damage Development

In general, the strain:time behaviour exhibited a decreasing rate during primary, a period where the creep rate remained approximately constant, and a region of increasing rate in tertiary leading to fracture.

Data revealed that the durations of primary, secondary and tertiary creep were approximately equal. The general shape of the creep curves was compared by normalising the strain:time data. This involved dividing the individual strain points by the failure strain and the individual time points by the rupture life. It was apparent that the curve shapes were similar for the full range of stress and temperature conditions considered.

The failure ductility observed was invariably less than 10%, with a decrease in failure strain to about 2% in the tests of the longest duration, Figure 1. In all cases fracture was a consequence of the nucleation, growth and link up of grain boundary cavities.

Stress and temperature dependence of deformation and rupture

The stress, σ, and temperature, T, dependence of the minimum creep rate, $\dot{\varepsilon}_{min}$, is typically described by the expression:

$$\dot{\varepsilon}_{min} = A\sigma^n \, exp\left(-\frac{Q_c}{RT}\right) \tag{1}$$

Where A and R are constraints, n is the stress exponent and Q_c the activation energy for creep.

The results obtained could be represented using equation 1 with data described by $n \cong 4.5$ and Q_c of 120 kJ/mol. These values suggest that creep is a consequence of the diffusion controlled generation and movement of dislocations (8). It should be noted that the activation energy for creep is about half that of self diffusion. This lower activation energy value has been explained on the basis that at temperatures below about 550°C, vacancy movement occurs by pipe diffusion along dislocations.

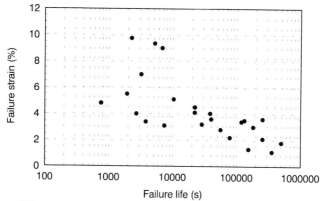

Figure 1. Creep failure ductility of OHFC copper for uniaxial tests

In a similar manner, the stress and temperature dependence of rupture were accurately described by the equation:

$$\frac{1}{t_f} = B\sigma^m \, exp\left(-\frac{Q_f}{RT} \right)$$ (2)

Values of the stress exponents, m and the activation energy for fracture, Q_f were similar to those noted for n and Q_c. This is consistent with the observation that fracture in the range 0.4 to 0.6 T_m, where T_m is the absolute melting point, is related to the deformation behaviour. Thus, as expected, it was noted that the measured minimum creep rate was proportional to the rupture life, i.e. the present data could be described by the Monkman-Grant relationship (9), Figure 2.

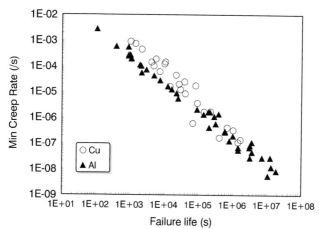

Figure 2. Relationship between minimum creep rate and time to failure for OHFC copper and super purity aluminium.

3. DISC BEND BEHAVIOUR

Disc bend testing was carried out using purpose-built, high-sensitivity equipment (1) in which samples, 9.5mm in diameter and 0.5mm thick, were loaded under the action of a hemispherical indentor. The specimen was located in a die which allowed clamping at the periphery of the test piece. The specimen thickness, punch diameter and die size were controlled so that the necessary clearance was available for deformation and fracture to take place without interference. The test temperature was controlled and monitored to better than ± 1°C, with detailed displacement:time readings taken for each test using a specialist extensometry system linked to a computer controlled ASL data acquisition unit.

The majority of tests were undertaken on material cold worked and annealed to produce an average grain size of about 20µm. Thus, the bulk of the testing, performed using loads in the range 23.3 to 115N at temperatures between 255°C and 413°C, was undertaken with approximately 25 grains across the section thickness. The annealing schedule was varied to give material with equiaxed average grain sizes of 30µm and 50µm. The effect of grain size on creep behaviour was therefore assessed by performing tests at 23.3N load and 372°C on specimens with approximately 10, 17 and 25 grains across the thickness.

Creep Curves and Damage Development

Testing was initially performed in air. At relatively high loads after the initial loading displacement, the creep rate decreased during primary until a region of approximately steady creep was established. Eventually the deformation rate accelerated during tertiary creep until fracture resulted. At lower applied loads for tests in air, the behaviour was different. Thus, initially the rate decreased as before. However, it was noted that the reduction in creep rate with time continued until the deformation rate became very slow. This phenomenon was investigated and it was apparent that, at low deformation rates, oxide formation was increasing the friction between punch and sample such that metal flow became very difficult. This problem was overcome by performing tests in an argon environment such that no significant oxidation of the test piece occurred.

Testing in argon allowed the displace:time characteristics to be established over the full range of stress and temperature conditions without the complication of surface oxide introducing friction. It was noted that the creep curves obtained were tertiary dominated. Thus, the majority of displacement occurred near to the end of the test. Moreover, assessment of creep rate:time behaviour showed that near fracture the creep rate was approximately 100 times the minimum value. These curves' shapes were markedly different to those measured during uniaxial loading.

To investigate the development of damage, repeat tests were performed with interruptions at selected life fractions between 50% and 80%. These samples were examined by optical metallography to measure the strain distribution and damage in the cross-sectional thickness of the specimen and by scanning electron microscopy to assess surface damage. Typical micrographs are shown in Figure 3.

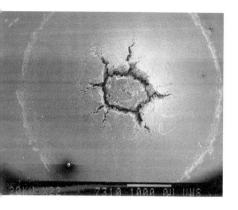
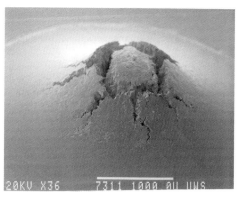
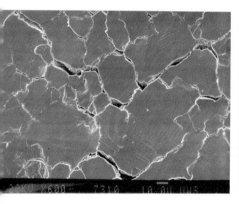
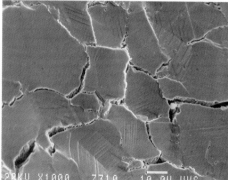

Figure 3. Scanning electron micrograph showing an OHFC copper sample interrupted at 80% of the apparent creep life illustrating the annular and radial cracking present and showing detail of the slip lines, grain boundary and associated cavitation damage.

It was found that:

(i)   the local strain developed was highlighted by the presence of slip lines

(ii)   the grain boundaries present were evident as a consequence of grain boundary sliding,

(iii)   cavities were present on grain boundaries, with the number and size of cavities and cracks increasing with life fraction,

(iv)   study of the orientation of damaged boundaries indicated that cavitation was present in approximately equal amounts on grain boundaries perpendicular to the hoop and radial stresses, and

(v)   at 80% life fraction macro-cracking through the sample thickness was present. This cracking was approximately circular, centred on the middle of the disc, with a diameter of approximately 1mm. In addition at this life fraction, radial macro-cracks had developed running perpendicular to the anular defect for about 0.5mm in length.

It is apparent that the failure in these samples was a consequence of the nucleation and growth of grain boundary cavities. The distribution of cavities and cracks indicates that the stresses present were maximum on the lower surface of the samples and were approximately equi-biaxial. The strain profiles obtained by metallographic evaluation indicated that although initially the indentor contacts the centre of the disc, maximum strain occurred between the tip and periphery of the indentor. Estimates of the reduction in thickness of the samples suggested that through thickness cracking occurred at strains of less than 10%.

Consideration of these observations provided an explanation for the anomalous creep curve shapes. Through thickness macro-cracks were present at 80% of the overall life measured, i.e. failure of the specimens had taken place before the end of the test was initially apparent. This behaviour can be explained by the fact that the diameter of the indentor was 2mm. Since the anulus of initial macro-cracking was 1 mm even when this crack had grown through thickness, the specimen was still capable of supporting the load. The final part of the original creep curves was thus recorded as the hole in the disc was widened to allow the indentor to pass through. The error introduced was thus a consequence of defining failure as the time when the specimen no longer supported the applied load. It was noted that the point of macro-crack penetration in these disc tests occurred when the creep rate had reached four times the minimum value. Indeed, this change in rate was very similar to that measured at the end of life of the uniaxial creep curves. Thus, it was possible to conduct further disc tests with the duration set by designating failure as having taken place when the instantaneous creep rate reached four times the minimum value.

Application of this 'effective failure' criterion allowed the deformation:time behaviour to be evaluated against the real point of failure. Curves developed on this basis showed evidence of primary, secondary and tertiary regions in a similar way to those for uniaxial loading, Figure 4. Furthermore, when the extension data were normalised against displacement and time to failure, the curve shape noted was insensitive to loading conditions. Repeat tests, under the same conditions, were reproducible to within a factor of two. Moreover, a similar level of reproducibility was noted in samples with different grain sizes. Thus, within sensible limits the results obtained were independent of grain size.

Stress and Temperature Dependence of Deformation and Fracture

For each creep curve, minimum creep rate and effective rupture life were noted. In a similar manner to the uniaxial data, the stress and temperature dependence of the minimum creep rate and rupture life could be described using equations 1 and 2 respectively. Analysis of the disc bend data revealed that values of the stress exponents n and m were 6 and 5.9 over the range of conditions studied. Moreover, values of Q_c and Q_f were calculated as about 100kJ/mol. The stress and temperature dependence of creep and fracture measured by the disc bend testing were in reasonable agreement to those established from uniaxial data.

Under all conditions, the rupture life was inversely proportional to the minimum creep rate, i.e. the Monkman-Grant relationship was obeyed. The product of minimum creep rate and time to rupture was found to be about 0.2, ie in good agreement with the uniaxial results, Figure 5.

4. DATA ANALYSIS

Traditional approaches for establishing creep and fracture behaviour are typically based on experiments performed using uniaxial loading. These approaches are frequently applied to characterise creep behaviour even though in most high temperature applications materials are subjected to a multiaxial stress state. It is apparent that the disc bend testing technique provides results, with the current arrangement and geometry, in an approximately equibiaxial loading condition. Evidence in the present programme has demonstrated that:

(i)   results are reliable and, within sensible limits, were not influenced by changes in grain size.
(ii)   the rate controlling processes of creep and fracture were similar to those identified using traditional approaches for appropriate stress and temperature regimes.
(iii)   the creep rate and time to fracture were related, with data for the disc bend testing agreeing with the relationship established for uniaxial results, and
(iv)   variations in failure ductility were consistent with established materials behaviour. Thus, the low ductility failures, reduction of thickness at fracture less than 10%, were a consequence of the nucleation and growth of grain boundary cavities.

It is therefore apparent that creep bending of small disc samples occurs with deformation and fracture mechanisms similar to those identified in uniaxial test programmes. Relationships should therefore exit between the data produced using the different methods. Initially, correlation of results has been examined on an isochronus basis, i.e. for a given test temperature the time to failure in a disc test at a given load has been compared to the stress to give the same rupture life under uniaxial conditions. This showed that a unique relationship described the behaviour indicating that the disc data can be correlated with uniaxial results through the definition of a reference stress, Figure 6. Thus, the stress linking uniaxial and multi axial creep data is a fraction of the appropriate limit behaviour of the structure.

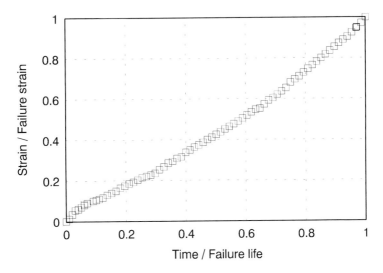

Figure 4. Normalized creep curve representation of OHFC copper disc bend tests. The data have been analysed using the actual rather than the apparent point of fracture (this occurred at 80% of the time when the specimen failed to sustain load.

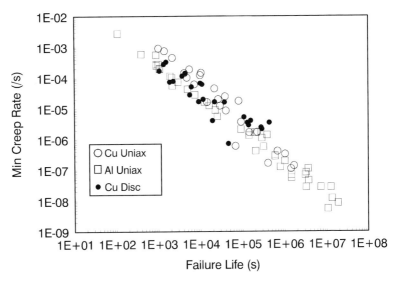

Figure 5. Comparison of uniaxial and disc bend creep data showing the relationship between minimum creep rate and time to fracture.

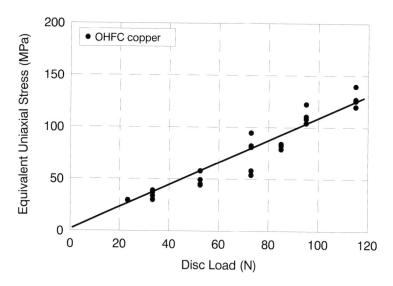

Figure 6. Correlation between uniaxial and disc bend creep data for OHFC copper

5. CONCLUDING REMARKS

The research programme was carried out with the view to assessing the capability of bend testing of miniature samples to measure creep and fracture behaviour. Results on OFHC copper produced using traditional approaches when compared with the bend test data, suggest that the small specimen technique does provide meaningful information. However, it is apparent that difficulties may be encountered with:

(i)   surface oxidation effects leading to significant levels of friction, and

(ii)   identification of the point of fracture in materials which fail with low overall ductilities.

Nevertheless, the present work has shown that these difficulties can be overcome so that this new approach can be used to monitor creep strength and ductility. Thus, this technology is of interest to engineering performance assessment of large scale plant operating under high energy conditions where structural integrity decisions involve knowledge of specific creep properties.

ACKNOWLEDGEMENT

Financial support for the present programme was provided by the Engineering and Physical Sciences Research Council, with the laboratory testing performed by Dr P Hollinshead and Dr G C Stratford.

104

REFERENCES

1. Parker, J.D., and James, J.D., Proc. Int. Conf. On Plant Life Management and Extension, pp2.1 – 2.10 (1992)
2. Parker, J.D., Stratford, G.C., Shaw, N., Spink, G. and Metcalfe,H., BALTICA IV, Plant Maintenance, pp 477-488 (1998)
3. Parker, J.D., Proc. Int. Symposium on Materials Aging and Life Management (Eds B. Raj, B Sankara Rao, T Jayakumar and R K Dayal), pp 1437 – 1444, (2000)
4. Feltham, P. and Meakin, J.D., Acta Met., $\underline{7}$, 614, (1959)
5. Davies, P.W. and Williams, K.R., J.I.M., $\underline{97}$, 337, (1969)
6. Davies, P.W., Nelmes, G., Williams, K.R. and Wilshire, B., Metal Sci J, $\underline{7}$, 87, (1973)
7. Parker, J.D. and Wilshire, B., Mat. Sci and Eng, $\underline{43}$, 271, (1980)
8. Barrett, C.R. and Sherby, O.D., Trans Met Soc AIME, $\underline{230}$, 1322, (1964)
9. Monkman, F.C. and Grant, N.J., Proc ASTM, $\underline{56}$, 593, (1956)

CREEP OF CERAMICS

The Effect of Lutetium Doping on the Transient Creep in Fine-grained Single-phase Al$_2$O$_3$

H. Yoshida, Y. Ikuhara* and T. Sakuma

Graduate School of Frontier Sciences, The University of Tokyo
* School of Engineering, The University of Tokyo
7-3-1 Hongo, Bunkyo-ku, Tokyo 113-8656, Japan
E-mail: yoshida@ceramic.mm.t.u-tokyo.ac.jp

Abstract

The creep deformation in fine-grained polycrystalline Al$_2$O$_3$ is highly suppressed by the addition of 0.1mol% LuO$_{1.5}$. The transient creep behavior in undoped and Lu-doped Al$_2$O$_3$ was examined at the testing temperature of 1250 - 1350ºC, and the data were analyzed in terms of the effect of stress and temperature on the extent of transient time and strain. The experimental data on the transient creep in the present materials shows a good agreement with the prediction from a time function of the transient and the steady-state creep associated with grain boundary sliding in fine-grained single phase materials. The difference in the transient creep between Lu-doped Al$_2$O$_3$ and undoped-one can be explained from the retardation of grain boundary diffusion due to the Lu^{3+} ions' segregation in the grain boundaries.

1. Introduction

In polycrystalline alumina, high-temperature creep deformation often occurs by diffusional flow[1-8]. The predominant deformation mechanism in Al$_2$O$_3$ with a grain size of less than about 10μm is grain boundary diffusion creep at temperatures of 1100 – 1400 ºC and the applied stress of less than 100MPa[2-5], and the grain boundary sliding contributes dominantly to high-temperature creep deformation in polycrystalline alumina with a grain size of less than about 2μm at 1400ºC[2,7].

The transient region in the creep curve has been observed in fine-grained Al$_2$O$_3$ as well as metallic materials[6,9]. In our earlier paper we have examined the transient creep associated with the grain boundary sliding of high-purity polycrystalline Al$_2$O$_3$ with an average grain size of 0.9 μm[10]. In order to explain the observed transient creep, a time function of creep strain is proposed from a two-dimensional model based on grain boundary sliding for fine-grained single-phase materials[10]. The observed transient behavior in high-purity, fine-grained alumina can be explained in terms of the time function at temperatures between 1150 and 1250ºC.

There is also a striking fact that the high-temperature creep resistance in fine-grained polycrystalline Al$_2$O$_3$ is highly improved by a small amount of dopant cation, such as ZrO$_2$ [11] or YO$_{1.5}$ or LuO$_{1.5}$ [12-14] even in the level of 0.1mol%. It is possible to explain the improved creep resistance in terms of the suppression of grain boundary diffusion of Al^{3+} ions by the segregation of dopant cations. However, the dopant effect has only been discussed in a steady-state creep region. The purpose of this paper is to examine the observed transient creep in fine-grained LuO$_{1.5}$–doped Al$_2$O$_3$ on the basis of the transient creep associated with the grain boundary sliding in fine-grained single-phase materials.

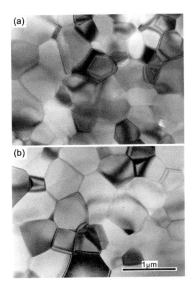

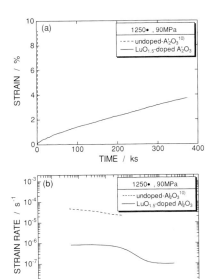

Figure 1 Conventional transmission electron micrographs of an as-sintered undoped, high-purity Al_2O_3 (a) and 0.1mol% $LuO_{1.5}$-doped Al_2O_3 (b).

Figure 2 Creep curves (a) and a plot of logarithmic strain rates against time (b) in fine-grained high-purity Al_2O_3 and 0.1mol% $LuO_{1.5}$-doped Al_2O_3 at 1250• under an applied stress of 90MPa.

2. Experimental Procedure

The materials used in this study is high-purity, undoped Al_2O_3 and Al_2O_3 with 0.1mol% of $LuO_{1.5}$. High-purity alumina powders with 99.99% purity (TM-DAR, Taimei Chemicals, Japan) and lutetium acetate (99.99%, Rare Metallic, Tokyo, Japan) were used for starting materials. The fabrication procedure is described in the earlier papers [12-14]. In order to obtain a relative density of more than 99% and an average grain size of about 1 µm, the green compacts of undoped and Lu-doped Al_2O_3 were sintered at 1300°C and 1400°C for 2 h in air, respectively. High-temperature creep experiments were carried out under uniaxial compression in air at a constant load using a lever-arm testing machine with a resistance-heated furnace (HCT-1000, Toshin Industry, Tokyo, Japan). The applied stress and temperature were in a range of 10 – 200MPa and 1150 – 1350°C, respectively.

3. Results and Discussion

Figure 1 shows a conventional transmission electron micrograph of an as-sintered high-purity, undoped Al_2O_3 (a) and 0.1mol% $LuO_{1.5}$-doped Al_2O_3 (b). The fabrication procedure is described in the earlier papers [12-14]. Fairly uniform and equiaxed grain structure with a grain size of about 1µm is obtained for the present materials. An average bulk density for the samples is more than 99% of the theoretical density. The microstructure of Lu-doped Al_2O_3 seems to be a single-phase material as well as the undoped one.

Figure 2 shows the comparison of creep curves (a) and a plot of logarithmic strain rate against time (b) in undoped-Al_2O_3 and 0.1mol% $LuO_{1.5}$-doped Al_2O_3 at 1250°C under an applied stress of 90MPa. Both creep strain and creep strain rate in polycrystalline Al_2O_3 are suppressed by Lu doping even in the level of 0.1mol%. The creep strain rate decreases with time, and becomes nearly a constant from a time of about 4ks in high-purity Al_2O_3 and 100ks in Lu-doped one. The total creep strain ε in undoped-Al_2O_3 at a time t is well described by the following equation [10];

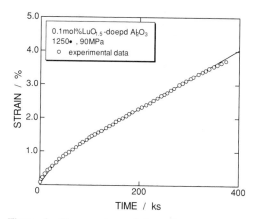

Figure 3 The experimental data of creep curve in $LuO_{1.5}$-doped Al_2O_3 under an applied stress of 90MPa at 1250•, together with the fitting line from Equation (1).

$$\varepsilon = a_1 t + a_2(1 - \exp(-a_3 t)) \tag{1}$$

where the first and second terms indicate the steady-state creep strain and transient creep strain, respectively. Eq. (1) has the same functional form as that originally proposed by Mcvetty[15], which is extensively applicable for creep curve fitting [16,17]. Figure 3 shows the experimental creep curves in Lu-doped Al_2O_3 together with the fitting curve from Eq. (1) for a specimen tested at 1250°C and the stress of 90 MPa. In the fitting line, a_1, a_2 and a_3 are estimated to be 8.80×10^{-8}, 5.22×10^{-3} and 3.47×10^{-6}, respectively. The values of a_1 and a_3 in Lu-doped Al_2O_3 are about two hundred times lower than those in undoped-Al_2O_3; a_1 and a_3 in undoped-one are 2.24×10^{-5} and 9.67×10^{-4}, respectively. However, the value of a_2 in Lu-doped one is rather close to that in undoped-one of 1.50×10^{-2}. Though the value of the parameters a_1 and a_3 in Lu-doped Al_2O_3 are different from those of undoped-one, the creep curve in both of the materials is well given by Eq. (1).

Figure 4 is an example of high-resolution electron micrograph (HREM) of a grain boundary (a) and a X-ray energy dispersive spectrum (EDS) taken from grain interior (b) and grain boundary (c) with a probe size less than 1nm in 0.1mol% Lu-doped Al_2O_3. No second phase or amorphous phase is seen even in the grain boundary, and the segregation of lutetium ions in the grain boundary is clearly detected in the EDS spectrum. The suppression of creep strain and strain rate in Al_2O_3 must be caused by the dopant segregation.

There are several models to describe the grain boundary sliding as a steady-state plastic flow in fine-grained single-phase polycrystalline materials on the basis of dislocation activity in grain boundaries [18-21]. One of the models is to assume that grain boundary sliding occurs by the glide of grain boundary dislocations accommodated by the climb of leading dislocations at the forefront of pile-ups. We showed that transient and steady-state creep in fine-grained single-phase materials can be derived from this model [10]. According to the analysis, the creep strain ε with time t is written as follows,

$$\varepsilon = \dot{\varepsilon}_s t + \varepsilon_T(1 - \exp(-\frac{t}{t_T})) \tag{2}$$

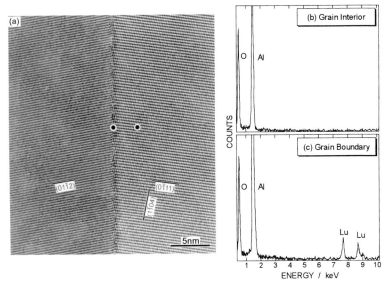

Figure 4 A high-resolution electron micrograph of a grain boundary in an as-sintered 0.1mol% $LuO_{1.5}$-doped Al_2O_3 (a) and the EDS spectra taken with a probe size of about 1nm from (b) the grain interior and (c) the grain boundary.

where $\dot{\varepsilon}_s$, ε_T and t_T are the steady-state strain rate, the total transient strain and the time extent of transient region, respectively. From Eq. (2), ε_T and t_T are approximately given as a function of stress and temperature as follows,

$$t_T = -\frac{T}{HA\theta D_{gb}\,\sigma} \quad \propto \quad \frac{T}{D_{gb}\,\sigma} \tag{3}$$

$$\varepsilon_T = \frac{\sigma}{H\theta^2} \propto \sigma \tag{4}$$

where H describes the work hardening rate determined by interaction between grain boundary dislocations, A is the material constant depending on shear modulus, Burgers' vector and grain size, θ is one scalar value depending on the grain size, D_{gb} is the grain boundary diffusivity. The values of H, A and θ are independent on temperature or stress. The transient creep time t_T is inversely proportional to the applied stress and D_{gb}, because the number of the grain boundary dislocations reaches a steady-state rapidly as the rate of dislocation climb is faster.

Figure 5 shows a log-log plot of transient time t_T against an applied stress σ in undoped [10] and Lu-doped Al_2O_3. The values of t_T and ε_T are determined from an iteration procedure in which the sum of square error of the datum points with respect to the average curve is minimized. The log t_T against log σ relationship in both of the materials is approximately single straight line with a slope of -1 at each temperature. It is noted that the t_T value in Lu-doped material is larger than that of undoped one, as typically shown in the data of 1250 °C and decreases with increasing applied stress and with increasing testing temperature.

Figure 6 shows a log-log plot of ε_T against the applied stress in undoped and Lu-doped Al_2O_3. The plot is a little bit scattered, but the value of ε_T in undoped and Lu-doped

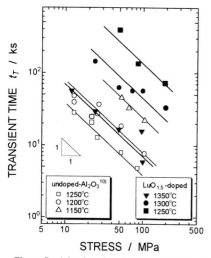

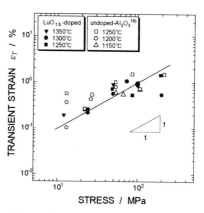

Figure 5 A log-log plot of time extent of transient creep against applied stress in LuO₁.₅-doped Al₂O₃. The previous data in high-purity Al₂O₃ [10] are also plotted for comparison.

Figure 6 A log-log plot of transient creep strain against applied stress in $LuO_{1.5}$-doped Al_2O_3, together with the previous data in undoped-Al_2O_3 [10].

Al_2O_3 can be approximated to be proportional to applied stress and is almost independent of temperature. It is noted that the strain extent of transient creep in Lu-oped material is nearly the same with undoped-one.

As shown in Figure 5 and 6, the stress and temperature dependence of the transient creep behavior in undoped and Lu-doped Al_2O_3 is consistent with our model of dislocation glide in grain boundaries. The difference in the t_T between both of the materials can be explained in terms of the difference in D_{gb} as follows. Figure 7 shows an Arrhenius plot of t_T at an applied stress of 53MPa against inverse temperature in undoped- and $LuO_{1.5}$-doped Al_2O_3. The activation energy for t_T in $LuO_{1.5}$-doped Al_2O_3 is estimated about 700kJ/mol, which is much larger than that of undoped-one, 340 kJ/mol [10]. The activation energy obtained from an Arrhenius plot has a tolerance of about 50kJ/mol, but the obtained activation energy in undoped-Al_2O_3 is nearly the same with that for Al^{3+} or O^{2-} ions grain boundary diffusion of 418 kJ/mol [4] or 460 kJ/mol [25], and is much lower than that for Al^{3+} or O^{2-} ions lattice diffusion; 510 kJ/mol [23] or 636 kJ/mol [24]. In contrast, the activation energy in Lu-doped Al_2O_3 is even larger than that of lattice diffusion in undoped-Al_2O_3. The difference in the activation energy in the transient time between Lu-doped and undoped-Al_2O_3 is nearly the same with that of steady-state creep strain rate; the activation energy for the steady-state creep rate is obtained to be 410kJ/mol for undoped- Al_2O_3 and 780 kJ/mol for Lu-doped one [12]. The activation energy of more than 700 kJ/mol in 0.1mol% $LuO_{1.5}$-doped Al_2O_3 is much larger than those for Al^{3+} or O^{2-} ions grain boundary diffusion or lattice diffusion. The large activation energy is not likely to be associated with sluggish lattice diffusion by the presence of dopant cation, because the solubility of lutetium ions in Al_2O_3 is extremely small [25], and they tend to segregate in Al_2O_3 grain boundaries as shown in Figure 4. The high activation energy in the Lu-doped Al_2O_3 is likely to result from the suppression of grain boundary diffusion by the segregation of dopant cation in grain boundaries [12-14].

On the other hand, ε_T depends not on the diffusivity, but on the shear modulus and

112

Burgers' vector of the grain boundary dislocations. Such values in Lu-doped Al_2O_3 have not been reported so far, but they may not be so different from that of undoped-Al_2O_3. It is thus possible to expect that the value of ε_T in Lu-doped Al_2O_3 is thus not so different from that in undoped- one. The present results suggest that both transient and steady-state creep behavior in Lu-doped Al_2O_3 can be explained in terms of the retardation in grain boundary diffusion due to the Lu ions segregation in grain boundaries.

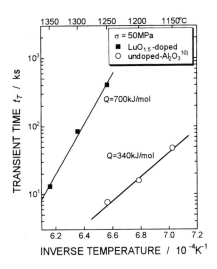

Figure 7 An Arrhenius plot of time extent of transient creep against inverse temperature in fine- grained Lu-doped Al_2O_3, together with the previous data in undoped-Al_2O_3 [10]. The slope of these lines represents the activation energies for time extent of transient creep.

4. Conclusions

The creep deformation in fine-grained polycrystalline Al_2O_3 is highly retarded by the grain boundary segregation of Lu ion. The creep deformation in Lu-doped Al_2O_3 at temperatures of 1150-1350 °C under applied stress of 10 – 200 MPa occurs by grain boundary sliding accommodated by grain boundary diffusion. The transient creep and the steady-state creep behavior in Lu-doped alumina can be described by the time function of the transient and the steady-state creep associated with grain boundary sliding as well as undoped-Al_2O_3, the stress dependence of t_T and ε_T estimated from the experimental creep curves is in accordance with the model. The change in the transient creep behavior of polycrystalline Al_2O_3 due to $LuO_{1.5}$-doping is explained in terms of the retardation of grain boundary diffusion due to the Lu ions segregation in grain boundaries.

Acknowledgement
The authors wish to express their gratitude to the Ministry of Education, Science and Culture, Japan, for the financial support by a Grant-in-Aid for Developmental Scientific Research (2)-10450254 for Fundamental Scientific Research. We also wish to express our thanks to Research Fellowships of the Japan Society for the Promotion of Science for Young Scientists for their financial aid.

References
[1] A.E. Paladino and R.L. Coble, *J. Am. Ceram. Soc.*, **46** (1963) 133-36.
[2] A.H. Heuer, R.M. Cannon and N.J. Tighe, ``Plastic Deformation of Fine-Grained Alumina, '' in *Ultrafine-Grain Ceramics*, edited by J.J. Burke, N.L. Reed and V. Weiss (Syracuse University Press, Syracuse, NY, 1970) pp. 339-65.
[3] T.G. Langdon and F.A. Mohamed, *J. Mater. Sci.*, **13** (1978) 473-82.
[4] R.M. Cannon, W.H. Rhodes and A.H. Heuer, *J. Am. Ceram. Soc.*, **63** (1980) 46-53.
[5] H.J. Frost and M.F. Ashby, p.98 in *Deformation - Mechanism Maps*. Pergamon, Oxford, U.K., (1982).
[6] A.H. Chokshi and J.R. Porter, *J. Mater. Sci.*, **21** (1986) 705-10.

113

[7] A.H. Chokshi, *J. Mater. Sci.*, **25** (1990) 3221-28.
[8] A.H. Chokshi and T.G. Langdon, *Mater. Sci. Tech.*, **25** (1991) 577-84.
[9] A.G. Robertson, D.S. Wilkinson and C.H. Cáceres, *J. Am. Ceram. Soc.*, **74** (1991) 915-922.
[10] H. Yoshida and T. Sakuma, *J. Mater. Sci.*, **33** (1998) 4879-4885.
[11] H. Yoshida, K. Okada, Y. Ikuhara and T. Sakuma, *Phil. Mag. Lett.*, **76** (1997) 9 - 14.
[12] H. Yoshida, Y. Ikuhara and T. Sakuma, *J. Mater. Res.*, **13** (1998) 2597-2601.
[13] H. Yoshida, Y. Ikuhara and T. Sakuma, *Phil. Mag. Lett.*, **79** (1999) 249-256.
[14] H. Yoshida, Y. Ikuhara and T. Sakuma, *J. Inorganic Mater.* **1** (1999) 229-234.
[15] P.G. Mcvetty, *Mech. Eng.*, **56** (1934) 149-154.
[16] A.J. Kennedy, *Process of Creep and Fracture in Metals* (John Wiley & Sons, Inc., New York, 1963) p.147.
[17] R.W. Evans and B. Wilshire, *Creep of Metals and Alloys* (The Institute of Metals, London, 1985) p. 274.
[18] T.G. Langdon, *Phil. Mag.*, **22** (1970) 689-700.
[19] A.K. Mukherjee, *Mater. Sci. Eng.*, **8** (1971) 83-89.
[20] R.C. Gifkins, *Metall.Trans.*, **A7** (1976) 1225-1232.
[21] A. Arieli and A.K. Mukherjee, *Mater.Sci.Eng.*, **45** (1980)61-70.
[22] Y. Oishi and W.D. Kingery, *J. Chem. Phys.*, **33** (1960) 480-486.
[23] M.L. Gall, B. Lesage and J. Bernardini, *Phil. Mag.*, **A70** (1994) 761-73.
[24] D. Prot and C. Monty, *Phil. Mag.*, **A73** (1996) 899-917.
[25] S.J. Schneider, R.S. Roth and J.L. Waring, *J. Research Natl. Bur. Standerds*, **65A** (1961) 345-374.

CHARACTERISTICS OF TENSILE CREEP
IN AN YTTRIA-STABILIZED TETRAGONAL ZIRCONIA

Siari S. Sosa and Terence G. Langdon

Departments of Aerospace & Mechanical Engineering and Materials Science
University of Southern California, Los Angeles, CA 90089-1453, U.S.A.

ABSTRACT

Experiments were conducted to evaluate the creep characteristics of a high purity 2.5 mol % yttria-stabilized zirconia (termed 2.5Y-TZP). Tensile creep tests were performed at elevated temperatures and measurements were taken to record the creep strain as a function of time. The results are compared with published creep data for 3Y-TZP under both tensile and compressive testing conditions. A comparison is made also with tensile creep data for low purity 3Y-TZP. It is shown there is a tendency for the creep rate to decrease continuously in tests performed at the lower stresses. Observations using scanning electron microscopy suggest this decrease is not related to the occurrence of grain growth or the development of internal cavitation. Detailed analysis shows also that the decreasing creep rate at low stresses has no influence on some of the basic parameters associated with high temperature creep, including the exponent of the inverse grain size where $p \approx 1$ and the apparent activation energy for creep where $Q_{app} \approx 570$ kJ mol^{-1}. However, there is a change in the value of the stress exponent from $n \approx 2$ in the early stages of the tests to $n \geq 3$ in the later stages.

1. INTRODUCTION

Although the deformation mechanisms occurring in high temperature creep are generally well characterized in pure metals and metallic alloys, less information is currently available concerning the creep of ceramic materials. It is now established that significant differences may exist between the creep of metals and ceramics, where these differences arise because of special features associated with some ceramic systems: for example, as a consequence of the occurrence of a restricted number of interpenetrating independent slip systems in ceramics so that there is a failure to satisfy the von Mises criterion for homogeneous deformation [1].

In addition, no firm conclusions have been reached concerning the flow processes associated with superplasticity in ceramic materials such as yttria-stabilized tetragonal zirconia (Y-TZP). In practice, a review of the literature shows that several different interpretations have been put forward to explain the deformation of these materials based on various sets of data obtained from a combination of creep testing in tension and compression and standard tensile and compressive tests under conditions of constant strain rate. Unlike metals, there is an additional problem in oxide ceramics because the nature of the ionic bonding imposes a condition of electrical balance and it follows that this characteristic should be incorporated into any detailed analysis of the mass transport processes. In this paper, tensile creep data are presented for a superplastic Y-TZP ceramic. From the results obtained in these experiments, it is suggested that the space charge layer formed at the grain boundaries in oxide ceramics may represent a possible source for the continuous decrease in strain rate that is often observed when testing Y-TZP ceramics in tensile creep at low stress levels.

2. EXPERIMENTAL MATERIALS AND PROCEDURES

The samples used in this investigation were fabricated from high purity commercial powders of 2.5 mol % yttria-stabilized tetragonal zirconia (termed 2.5Y-TZP). These powders were obtained from The Tosoh Corporation (Nanyo Plant, Shinnanyo, Yamaguchi-ken, Japan) and the reported chemical composition (in wt %) was 2.85% Y_2O_3, <0.005% Al_2O_3, <0.002% SiO_2, 0.004% Fe_2O_3 and 0.017% Na_2O with the balance as ZrO_2. A two-step sintering procedure was used to fabricate the samples for use in this investigation: details of this technique are given elsewhere [2]. Different sintering treatments were applied to the samples to obtain a range of ultrafine grain sizes. After sintering at 1723 K for periods of 1 and 2 h, the mean linear intercept grain sizes, L_o, were measured as 0.47 and 0.60 µm, respectively: sintering for 12 h at 1823 K gave a mean linear intercept grain size of 0.83 µm. These mean linear intercept grain sizes were determined by taking measurements in two orthogonal directions, as described earlier [3]. Specimens were machined in the form of flat tensile samples with gauge lengths of 5 mm. Inspection showed all of the samples contained an essentially uniform distribution of grain sizes and grain shapes with the grains having an equiaxed appearance.

Tensile creep tests were performed using a creep machine designed to maintain conditions of constant stress: the characteristics of this machine were described previously [4]. In each test, the load was imposed smoothly after reaching and stabilizing at the desired testing temperature. Tests were performed at different temperatures and stresses and the strains were recorded continuously using a linear displacement transducer. Microstructures were observed after testing, from positions both within the gauge length and in the grip section, using a scanning electron microscope (SEM). Selected samples were mechanically polished on silicon carbide paper and then finely polished with a diamond compound. Each sample was thermally etched at a temperature of approximately 1623 K for 1 h. These observations were performed using a voltage of 15 to 20 kV. During each observation, the samples were oriented with the tensile axis lying parallel to the horizontal axis within the field of view.

3. RESULTS OF CREEP TESTING

The creep curves for tests conducted on 2.5Y-TZP samples at 1673 K are presented in Fig. 1. These samples had an initial grain size of 0.60 µm. It is apparent that the samples tested at the larger stresses show an immediate steady-state creep rate whereas at lower stresses there is a continuous decrease in the creep rate that is followed, ultimately, by an essentially steady-state condition. This decrease in strain rate is most evident at the lowest applied stress of 5.6 MPa. The phenomenon of a decreasing creep rate during testing has been reported for several ceramic materials tested over a range of different conditions. For example, there are reports also for high purity 3Y-TZP/20% Al_2O_3 composites tested in tensile creep [4], for high purity 3Y-TZP tested in both tensile creep [5,6] and compressive creep [7], for low purity Y-TZP tested in tensile creep [8,9] and for magnesia-doped Al_2O_3 tested in tensile creep [10].

It is usually found that extensive cavitation develops in the later stages of creep and an example is shown in Fig. 2 for a specimen with a grain size of 0.26 µm tested at 1623 K with a stress of 5.6 MPa. This test was stopped after testing for 165 h to a true strain of 0.05 and inspection then showed the presence of several well-defined and angular cavities. Since the strain rate increases when cavities are formed, it is possible that a quasi-steady-state condition is ultimately established because of an approximate balance between the process leading to a decreasing strain rate and the increasing strain rate associated with the presence of cavities.

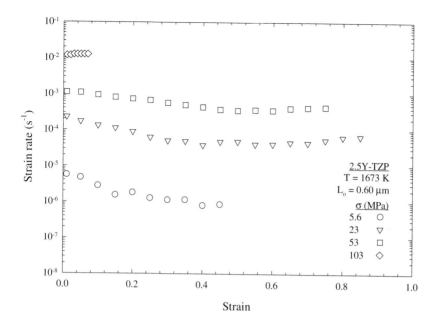

Fig. 1. Creep curves for 2.5Y-TZP samples with $L_o = 0.60$ μm tested at 1673 K.

Fig. 2. SEM image of a 2.5Y-TZP sample with $L_o = 0.26$ μm tested at 1623 K and 5.6 MPa. The test was stopped after 165 h at a true strain of $\varepsilon = 0.05$. The tensile axis is horizontal.

An alternative explanation for the decreasing strain rate at the lower stresses is based on the effect of dynamic grain growth during testing [5,6,11]. For example, the microstructure shown in Fig. 2 has a mean linear intercept grain size after testing of 0.53 μm and this is approximately double the initial size of 0.26 μm. Although it is true there is often a considerable grain growth in these long-term tests, it is also found there may be essentially negligible grain growth after relatively large decreases in the strain rate. For example, the microstructure shown in Fig. 3 was taken on a sample with an initial grain size of 0.83 μm tested at 1623 K and 5.6 MPa. This test was stopped after 48 h at a true strain of 0.04 and measurements from Fig. 3 showed the measured mean linear intercept grain size was 0.89 μm. This value of L_o is very close to the initial grain size of 0.83 μm but the strain rate decreased during the 48 h of the test from an initial value of 1×10^{-6} s^{-1} to a final value of 5×10^{-8} s^{-1}. Thus, this sample exhibited a very significant decrease in strain rate although there was relatively little change in the grain size. It is reasonable to conclude that the decreasing creep rate observed at low stresses is not due exclusively to any increase in grain size.

The decrease in creep rate leads to an apparent increase in the value of the stress exponent, n. This is illustrated in Fig. 4 which shows the strain rate plotted as a function of stress for tests conducted at 1623 K using samples with an initial grain size of 0.47 μm: the upper line shows the strain rates recorded at an early strain of 1%, designated $\varepsilon_{1\%}$, and the lower line shows the rates recorded at the ultimate strain, ε_{ult}. The values of n change from an initial value of ~2.1 which is consistent with deformation controlled by grain boundary sliding to a final value of ~3.4 for the rates measured prior to failure. It is evident from Fig. 4 that the strain rate decreases significantly, by up to two orders of magnitude, at the lowest stress used in these experiments.

It is possible that deformation may occur under these conditions by interface-controlled diffusion creep [12] and the decrease in the strain rate may be a consequence of the small number of grain boundary dislocations present in the material to act as discrete sources and sinks for vacancies. One of the consequences of deformation by this mechanism is the presence of a threshold stress. Previous studies have suggested the threshold stress may be a parameter exclusively affecting grain boundary sliding [13] or it may be interface-controlled [14]. In either case, the magnitude of the threshold stress is estimated to be fairly small and it is important to note that care must be taken in attempting to estimate the threshold stress using graphical methods [15].

The results from different creep tests performed at the same stress of 23 MPa are shown in Fig. 5 for samples with $L_o = 0.60$ μm tested at temperatures from 1573 to 1673 K. It is apparent from inspection of these curves that there is a similar decrease in strain rate at each temperature. This consistency is evident in Fig. 6 where the variation of strain rate is plotted against the inverse temperature for both the initial strain rate at a strain of 1% and the final or ultimate strain. The values of the apparent activation energy, Q_{app}, are very similar and estimated as ~560 and ~575 kJ mol^{-1} for the initial and final strains, respectively. These results are also consistent with the values of the apparent activation energies reported from compressive tests on high purity 3Y-TZP where $Q_{app} \approx 510 - 590$ kJ mol^{-1} [7] and low purity 2Y-TZP where $Q_{app} \approx 530$ kJ mol^{-1} [8].

A similar effect is observed in samples tested under the same conditions with different initial grain sizes, where the exponent of the inverse grain size, p, was measured as ~1.2 at both the initial and the final strain. This value of p is also consistent with results reported from compressive creep tests at low stresses in high purity 3Y-TZP where $p \approx 1.2 \pm 0.3$ [7].

The consistency in the values of Q_{app} and p estimated at both the low and high strains demonstrates that, unlike the effect on n shown in Fig. 4, these basic parameters of creep are not significantly influenced by the occurrence of decreasing strain rates at the lower stresses.

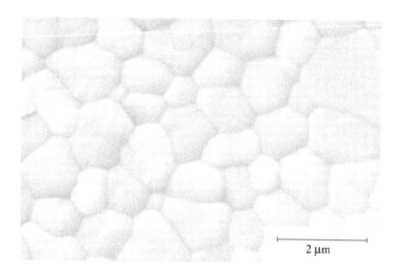

Fig. 3. SEM image of a 2.5Y-TZP sample with $L_o = 0.83$ μm tested at 1623 K and 5.6 MPa. The test was stopped after 48 h at a true strain of $\varepsilon = 0.04$. The tensile axis is horizontal.

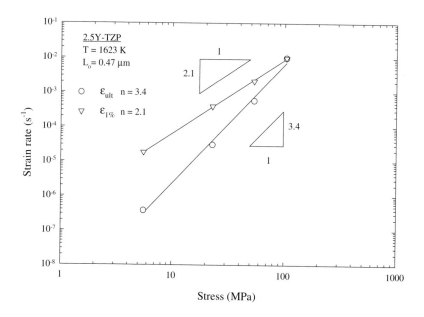

Fig. 4. Variation of strain rate with stress for 2.5Y-TZP tested at 1623 K

120

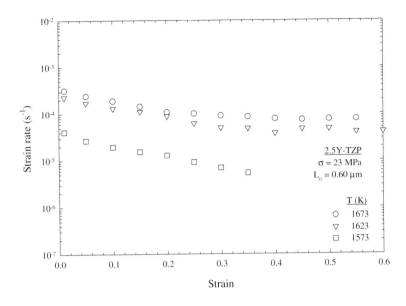

Fig. 5. Creep curves for 2.5Y-TZP samples with $L_o = 0.60$ μm tested with a stress of 23 MPa

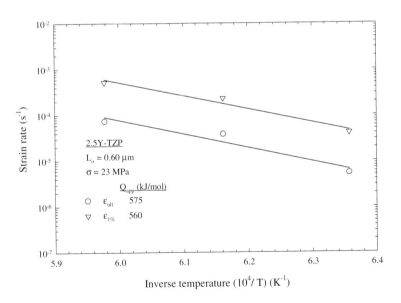

Fig. 6. Variation of strain rate with inverse temperature for 2.5Y-TZP samples tested with a stress of 23 MPa.

4. DISCUSSION

In Fig. 7, the results documented in Fig. 4 are compared with data reported from compressive tests on 3Y-TZP at the same testing temperature of 1623 K and with an initial grain size of 0.41 μm [7]. For 3Y-TZP in Fig. 7, the solid points relate to the final strain, ε_{ult}, and it is apparent these points are consistent with those obtained at the final strain in the present investigation. It is also evident there is an increase in the stress exponent at the lower stresses. The data in Fig. 7 are also compared with the predictions from a model of interface reaction-controlled diffusion creep [16] as developed in an earlier report [12]. For this model, the predicted steady-state strain rate is obtained from the relationship [12]

$$\dot{\varepsilon} = \frac{33.4\,D_{gb}\,G\,b}{k\,T}\left(\frac{\delta}{b}\right)\left(\frac{b}{d}\right)^3\left(\frac{\sigma}{G}\right)\left(\frac{N^2}{N^2 + \frac{1}{2}}\right) \qquad (1)$$

where D_{gb} is the coefficient for grain boundary diffusion, G is the shear modulus, b is the Burgers vector, k is Boltzmann's constant, T is the absolute temperature, δ is the grain boundary width, d is the spatial grain size (defined as $1.74 \times L_o$), σ is the applied stress and N is the number of grain boundary dislocations in a single grain boundary wall (defined as $\sigma d/2Gb_{gb}$ where b_{gb} is the Burgers vector for grain boundary dislocations). This prediction is given by the solid line for $L_o = 0.41$ μm and it is evident this line provides an almost perfect fit to the experimental data at the lower stresses although the measured creep rates are faster at the higher stresses. It should be noted that the fit to eq. (1) in Fig. 7 is better than reported in the earlier analysis [12] because of the incorporation of a conversion from a linear intercept grain size to a spatial grain size.

Figure 8 provides a direct comparison between data obtained on 3Y-TZP with $L_o = 0.41$ μm at a temperature of 1723 K in compressive creep [7] and results obtained from the tensile testing of low purity 3Y-TZP with $L_o = 0.29$ μm [9]: again the predicted line for diffusion creep is calculated from eq. (1) for a linear intercept grain size of 0.41 μm. Several conclusions may be reached from inspection of Fig. 8. First, the reduction in the strain rate when testing in compression creep is not as large as in tensile creep. Second, there is generally good agreement between the data for the low and high purity 3Y-TZP samples. Third, the results are consistent with the predictions of the model for interface reaction-controlled diffusion creep. Fourth, the low purity results lead to a stress exponent of ~2 which is similar to the value of n obtained at the higher strain rates in the high purity samples. This result suggests the decrease in strain rate at low stresses may be related to the presence, and possibly the level, of impurities within the material.

In zirconia ceramics, the role of diffusion creep may be affected as a result of the well-established observation that yttrium segregates to the grain boundaries [17]. Thus, when the structure associated with the grain boundaries has a different chemical composition from the matrix material, there is a loss of stoichiometry in ionic compounds and this defect structure requires electrical compensation. This electrical balance is provided by a space charge layer [18] that is produced through the presence of defects immediately adjacent to the grain boundary. These space charge clouds have been measured directly in several oxide ceramics such as TiO_2 [19], Al_2O_3 [20] and $SrTiO_3$ [21]. There are also reports of the electrical nature of the grain boundaries in Y-TZP ceramics [22,23] and these investigators have documented a decrease in the grain boundary resistivity after compression tests such that this decrease becomes more significant when the applied stress is higher or when the testing temperature is increased. Results of this type support the suggestion of a stress dependence associated with the grain boundary properties in these materials.

122

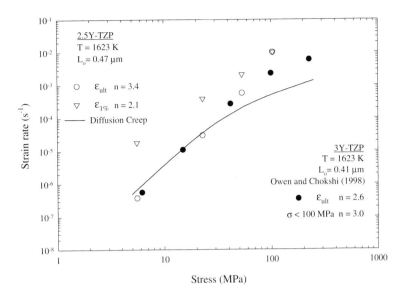

Fig.7. Variation of strain rate with stress for 2.5Y-TZP samples tested at 1623 K: data also shown for 3Y-TZP samples tested in compressive creep [7].

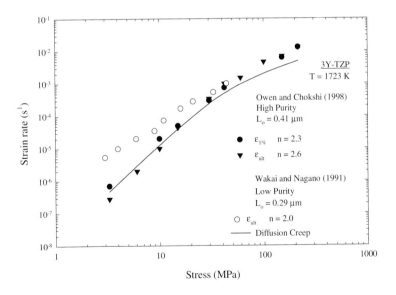

Fig. 8. Variation of strain rate with stress for low and high purity 3Y-TZP samples tested at 1723 K [7,9]

Although the precise influence of any space charge layer on the creep properties is not understood at the present time, it is apparent from these results that there is a reasonable consistency between different sets of experiments on yttria-stabilized zirconia ceramics including both high and low purity materials. Furthermore, there is generally good agreement with the predictions of an interface reaction-controlled model for diffusion creep.

5. SUMMARY AND CONCLUSIONS

5.1. Tensile creep tests were conducted on 2.5Y-TZP at elevated temperatures. The results show the measured strain rate decreases with increasing strain when testing at the lower stress levels.

5.2. This decrease cannot be attributed exclusively to the presence of grain growth during the test because it is observed also in tests where the extent of grain growth is essentially negligible. Also, it is not associated with the development of internal cavitation since this is expected to lead to an increasing strain rate.

5.3. It is shown that the exponent of the inverse grain size, p, and the apparent activation energy for creep, Q_{app}, are not affected by the decrease in strain rate at the lower stresses. In the present investigation, the exponent of the inverse grain size has a value of ~1.2 and the apparent activation energy is ~570 kJ mol^{-1}.

5.4. There is good agreement between the present results and earlier data obtained on high purity 3Y-TZP tested in compressive creep. The interface reaction-controlled model for diffusion creep is reasonably consistent with the experimental data for results obtained in both tension and compression.

ACKNOWLEDGEMENTS

This work was supported by the United States Department of Energy under Grant No. DE-FG03-92ER45472.

REFERENCES

[1] Chokshi, A.H. and Langdon, T.G., 1991, Characteristics of creep deformation in ceramics, Materials Science and Technology, Vol. 7, pp. 577-584.

[2] Zhou-Berbon, M., Sørensen, O.T. and Langdon, T.G., 1996, A simple technique for the preparation of tensile specimens of yttria-stabilized zirconia, Materials Letters, Vol. 27, p. 211-214.

[3] Sosa, S.S. and Langdon, T.G., 2001, Deformation characteristics of a 3Y-TZP/20%Al$_2$O$_3$ composite in tensile creep, Materials Science Forum, Vol. 357-369, pp. 135-140.

[4] Sosa, S.S. and Langdon, T.G., 2000, Creep behavior of a superplastic Y-TZP/Al$_2$O$_3$ composite, Materials Research Society Symposium Proceedings, Vol. 601, pp. 111-116.

[5] Morita, K. and Hiraga, K., 2000, High temperature deformation behavior of a fine-grained tetragonal zirconia, Scripta Materialia, Vol. 42, pp. 183-188.

124

[6]   Morita, K., Hiraga, K. and Sakka, Y., 2000, Creep deformation of a 3mol% Y_2O_3-stabilized tetragonal zirconia, Materials Research Society Symposium Proceedings, Vol. 601, pp. 93-98.

[7]   Owen, D.M. and Chokshi, A.H., 1998, The high temperature mechanical characteristics of superplastic 3 mol % yttria stabilized zirconia, `Acta Materialia, Vol. 46, pp. 667-679.

[8]   Wakai, F. and Nagano, T., 1988, The role of interface-controlled diffusion creep on superplasticity of yttria-stabilized tetragonal ZrO_2 polycrystals, Journal of Materials Science Letters, Vol. 7, pp. 607-609.

[9]   Wakai, F. and Nagano, T., 1991, Effects of solute ion and grain size on superplasticity of ZrO_2 polycrystals, Journal of Materials Science, Vol. 26, pp. 241-247.

[10]   Kottada, R.S. and Chokshi, A.H., 2000, The high temperature tensile and compressive deformation characteristics of magnesia doped alumina, Acta Materialia, Vol. 48, pp. 3905-3915.

[11]   Kim, B-N., Hiraga, K., Sakka, Y. and Ahn, B-W., 1999, A grain boundary diffusion model of dynamic grain growth during superplastic deformation, Acta Materialia, Vol. 47, pp. 3433-3439.

[12]   Berbon, M.Z. and Langdon, T.G., 1999, An examination of the flow process in superplastic yttria-stabilized zirconia, Acta Materialia, Vol. 47, pp. 2485-2495.

[13]   Jimenez-Melendo, M. and Dominguez-Rodriguez, A., 2000, High temperature mechanical characteristics of superplastic yttria-stabilized zirconia. An examination of the flow process, Acta Materialia, Vol. 48, pp. 3201-3210.

[14]   Raj, R., 1993, Model for interface reaction control in superplastic deformation of non-stoichiometric ceramics, Materials Science and Engineering, Vol. A166, pp. 89-95.

[15]   Berbon, M.Z., and Langdon, T.G., 1997, The variation of strain rate with stress in superplastic zirconia, Materials Science Forum, Vol. 243-245, pp. 357-362.

[16]   Artz, E., Ashby, M.F. and Verrall, R.A., 1983, Interface controlled diffusional creep, Acta Metallurgica, Vol. 31, pp. 1977-1989.

[17]   Primdahl, S., Thölen, A. and Langdon, T.G., 1995, Microstructural examination of a superplastic yttria-stabilized zirconia: implications for the superplasticity mechanism, Acta Metallurgica Materialia, Vol. 43, pp. 1211-1218.

[18]   Kingery, W.D., 1974, Plausible concepts necessary and sufficient for interpretation of ceramic grain-boundary phenomena: I, Grain boundary characteristics, structure and electrostatic potential, Journal of the American Ceramic Society, Vol. 57, pp. 1-8.

[19]   Ikeda, J.S., Chiang, Y-M., Garratt-Reed, A.J. and Vander Sande, J.B., 1993, Space charge segregation at grain boundaries in titanium dioxide: II, Model experiments, Journal of the American Ceramic Society, Vol. 76, pp. 2447-2459.

[20]   Li, C.W. and Kingery, W.D., 1984, Solute segregation at grain boundaries in polycrystalline Al_2O_3, Advances in Ceramics, Vol. 10, Ed. by W.D. Kingery, American Ceramic Society, Columbus, OH, pp. 368-378.

[21]   Johnson, K.D. and Dravid, V.P., 1999, Direct evidence for grain boundary potential barrier breakdown via in situ electron holography, Microscopy and Microanalysis, Vol. 5, pp.428-436.

[22]   Chen, C.S., Boutz, M.M.R., Boukamp, B.A., Winnubst, A.J.A., de Vries, K.J. and Burgraaf, A.J., 1993, The electrical characterization of grain boundaries in ultra-fine grained Y-TZP, Materials Science and Engineering, Vol. A168, pp. 231-234.

[23]   Boutz, M.M.R., Chen, C.S., Winnubst, L. and Burgraaf, A.J., 1994, Characterization of grain boundaries in superplastically deformed Y-TZ ceramics, Journal of the American Ceramic Society, Vol. 77, pp. 2632-2640.

High Temperature Creep/Fatigue of Single or Multi-phase Ceramics with an additional Glass/Liquid phase.

C.K.L. Davies*, D. Oddy*, M.J.Reece*, R.N. Stevens*, K.Williams*, G. Fantozzi**, C. Olagnon**, and R. Torrecillas***.

*Department of Materials, Queen Mary, University of London, Mile End Road, London, E1 4NS, UK. Fax no: 020 8981 9804 E-mail: c.k.l.davies@qmw.ac.uk

**Groupe d'Etude de Metallurgie Physique et de Physique des Materiaux, Instituit National des Sciences Appliquees de Lyon, Lyon, France. Fax no: +33 72 438528 E-mail: gemppm@insa-lyon.fr

***Instituto Nacional de Carbon, Consejo Superior de Investigaciones Cientificas, Oviedo, Spain. Fax no: +34 852 97662 E-mail: rtorre@muniellos.incar.csic.es

Abstract: This paper is concerned with mostly tensile creep, fatigue and fracture studies of Mullite and Zircon-Mullite-Zirconia (ZMZ) both containing a glass/liquid phase. The creep curves exhibited only a continuously decelerating primary stage leading to fracture in the case of Mullite, whereas this was frequently followed by a tertiary stage in the case of ZMZ. Primary creep strain initially arose from flow of the viscous glassy phase but as the grains interlocked further strain probably resulted from solution/precipitation. The accelerating tertiary creep was probably due to cavitation in the liquid. All specimens fractured as a result of single crack growth with cyclic fatigue lives being the longest due to viscous crack bridging and crack healing on unloading.

1   Introduction

The microstructures of many single and multi-phase ceramics contain a glassy/liquid phase which is distributed to varying extents at grain corners, edges and faces. The distribution and mechanical behaviour of this phase have a profound effect on the mechanical properties of ceramics particularly at elevated temperatures.   The static creep behaviour and damage development of such systems has been discussed extensively [1-3]. Liquid phase related creep strain can result from liquid flow [4], solution/precipitation [5] and cavitation in the liquid phase [6], the relative extents depending on the detailed microstructure, the test system and particularly the temperature. The liquid phase may assist crack growth during static loading by direct action at the crack tip [7] or as a result of growth into a cavitated liquid damage zone [8]. However, the liquid phase may 'retard' crack growth by crack bridging during cyclic loading [9].

The work reported here is part of a European collaborative study of the role of a liquid /glassy phase in creep, fatigue, and crack growth. [10-14] The current work reports the results for a Mullite and a Zircon (ZrSiO$_4$) - Mullite - Zirconia material tested at elevated temperatures statically and cyclically in tension with some tests carried out in four point bending for comparison purposes.

126

2 Experimental

2.1 Materials and Specimens

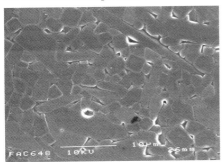

Fig.1 SEM Micrograph of Mullite

The Mullite specimens in the form of plates or rods were produced from a commercially available powder (Baikowski SA 193 CR, France) with an average particle size of 2.7μm. The powder was mixed with additives and then spray dried and cold isistatically pressed into plates or cylindrical rods at 200MPa. These were sintered at 1750°C for 5h and then annealed at 1450°C for another 5h, in order to attempt to reduce the amount of glassy phase. The sintered microstructure of the Mullite obtained (Fig.1) shows equiaxed grains of 4μm mean grain diameter and some elongated grains with an aspect ratio of about 4.The porosity was less than 1% and a silica rich glassy phase was present in pockets at triple points and spread to various extents along grain boundaries. Transmission electron microscopy and associated EDX analysis suggested that the maximum volume fraction of glass was no more than 2vol% with some grain faces being covered by a layer of 5-10mm thickness. The composition of the glass was approximately 90% SiO_2 with 10% Al_2O_3. Traces of CaO and Na_2O from the starting powder were probably present but were not detected.

The Zircon-Mullite-Zirconia (ZMZ) material was manufactured by reaction sintering 90% zircon powder (Zircosil 1 Cookson UK) of 1.2μm mean particle size with 10% α-alumina powder of 0.5μm mean particle size (HPA-0.5 Condea). The powders were first mixed by attrition milling for 30 minutes in isopropyl alcohol, dried and cold isistatically pressed at 200mPa into plates and rods. These were sintered at 1560°C for 24 hrs. The resulting sintered microstructure is shown in Figures 2a and 2b.

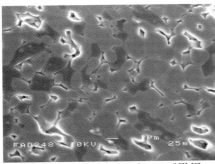

Fig. 2a Secondary electron image of ZMZ

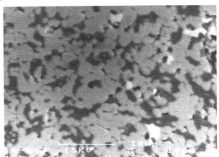

Fig. 2b Back scattered electron image of ZMZ

From these SEM micrographs and X-ray diffraction analysis the minimum zircon content was 70vol% of mean grain size 2.5μm, with maximum mullite and zirconia volume fractions of 22% and 8% respectively. TEM studies showed the zirconia to be present as bulk particles of the monoclinic phase and small tetragonal particles of the tetragonal phase in the mullite grains. The glassy phase covered most of the zircon grain faces with a thickness of 10-20nm, and was present in an overall volume fraction of up to 6 vol%. The glass contained up to 10% Al_2O_3 with about 1% ZrO_2.

The sintered rods of both materials were ground into 150cm long cylindrical specimens of 9.0mm diameter with reduced gauge lengths in the centre for tensile creep and fatigue testing. The creep specimens had a gauge length of 20mm and a gauge diameter of 5mm (Fig. 3). The fatigue specimens had double shoulders with a minimum reduced gauge diameter of 4.0mm of length 5mm (Fig. 3). The long specimens were heated in furnaces containing MoS_2 elements and the protruding ends were gripped in water cooled chucks. In creep tests the specimens were dead loaded via universal joints and in the fatigue tests loads were applied hydraulically, statically (R=1), and fully reversed, tension/compression (R=-1) and partially tension unloaded (R=0.1) at a frequency of 10Hz. In all cases the bending of the specimens was maintained to less than 5%. More extensive details of these testing methods can be found elsewhere [14].

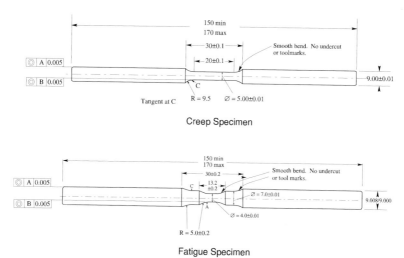

Fig. 3 Tensile Creep and Fatigue Test Specimens

The plates were ground into bend specimens of dimensions 4x3x40mm which were creep tested in four point bending with inner and outer spans equal to 18 and 36mm respectively. The tensile faces of the specimens were polished down to 3μm. The tensile strains were calculated from the measured deflections using the usual approximation [15].

3   Results and Discussion

3.1 Creep Curves

Creep curves for Mullite tested in tension or bending at 1400°C exhibit only a primary period of continuously decreasing creep rate leading to fracture. The same is true for the Zircon-Mullite-Zirconia (ZMZ) materials in some high stress tests. However, for this material tested in tension at low stresses the curves exhibit a primary stage, a minimum creep rate and an accelerating tertiary stage leading to fracture. In all cases the primary creep strain (ε_p) could be fitted to relationships of the forms (16);

$$\varepsilon_p = \beta \, t^{1/3} \tag{1}$$

and
$$\varepsilon_p = \theta_1(1\text{-}exp(\text{-}\theta_2 t)) \qquad (2)$$
Where β and θ_2 are rate parameters.

When full creep curves were observed the creep strain (ε_c) could be fitted with the full relationship of the form;

$$\varepsilon_c = \theta_1 (1\text{-}exp (\text{-}\theta_2 t)) + \theta_3(exp (\theta_4 t) \text{-}1) \qquad (3)$$

Where the terms θ_1 and θ_3 scale the primary and tertiary stages, respectively, while θ_2 and θ_4 are the respective rate parameters which define the primary creep decay rate and the tertiary rate of acceleration.

Equations (1-3) were fitted to all the creep curves and the resultant β and θ_i values were determined as functions of temperature and stress.

3.2 Primary/Exhaustion Creep

For both materials the magnitudes of the β and θ_2 rate parameters increase with increasing stress and temperature as illustrated in Fig. 4 and Fig. 5 with the tensile data for the ZMZ material.

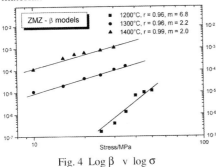

Fig. 4 Log β v log σ

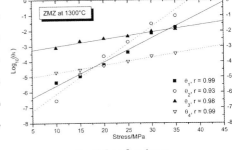

Fig. 5 Log θ_i v log σ

For Mullite at 1400°C the bending curves follow the shape and strain magnitude of tensile curves but exhibit extended creep lives. For ZMZ, the limited bending data shows continuously decelerating creep curves leading to fracture. The tensile and bending curves follow each other at low strains during primary but diverge when tertiary creep occurs in tension. Hence, the β and θ_2 values are similar at given stresses in tension and bending. If tensile specimens of either material are completely or partially unloaded, during

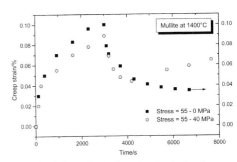

Fig. 6 Creep Strain from loading/unloading

primary, considerable reverse strain is observed (Fig.6), the extent of which increases with increasing magnitude of the previous forward primary strain, suggesting that creep is occurring by a viscoelastic process (17).

Creep models have been developed to describe creep by direct viscous flow of a grain boundary glassy phase (1,4) for hexagonal, square or cubic grains. The predicted results of the models differ by only a small constant factor. For a Newtonian viscous fluid the predicted initial creep rate $(\dot{\varepsilon})$ for a system of cube grains is given by

$$\dot{\varepsilon} = \frac{5}{54} \frac{\sigma f^3}{\eta_0}$$

(4)

Where η_0 is the glass viscosity, and f is the nominal volume fraction of glass defined as the volume fraction located at grain faces, $f=3H/L$, where H is the thickness of the glass layer and L is the face length (grain diameter).

In the present case values of glass viscosity (η_0) were calculated using the measured initial creep rates $(\dot{\varepsilon})$ and values for the nominal volume fraction of glass (f), determined from the measured micro-structural parameters.

For Mullite, a glass thickness H= 5nm - 10nm was assumed with a grain diameter L =4μm, giving values of f=3.75×10^3 - 6.0×10^3. When these values were inserted into equation 4, using data for the highest test stress, the resulting values for η_0 at 1400°C were 1.2×10^4 to 8.16×10^4 Pa·s. The values increase with decreasing stress as the initial creep rate had a power law stress dependence of n=3. This suggested that the glass behaved as a non-newtonian liquid with a stress dependence of the viscosity being similar to that observed for silica rich liquids in Si_3N_4 (1,3) and Al_2O_3 (18) systems. Values of the viscosities of silica rich melts have been calculated using the empirical relationships derived by Urbain et al (19), the results of which are shown in Fig. 7. From the figure it can be seen that the viscosity of a SiO_2-10% Al_2O_3 glass at 1400°C is about 7×10^4 Pa·s which corresponds well to the values calculated from the creep curves, particularly for a glass thickness of 10nm.

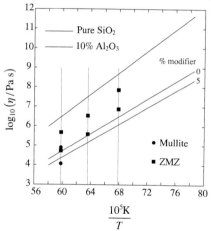

Glass viscosities (ref. 19) compared to those calculated from the initial creep rates.

Fig.7

For the ZMZ material a glass thickness H = 10-20nm was assumed with a zircon grain diameter L =2.5μm, giving values of f=1.2×10^{-2} - 2.4×10^{-2}. When these values were inserted into equation 7, again using data for the highest test stresses the resulting values of η_0 were as shown on Fig.7. These creep values are comparable with those for a SiO_2-10% Al_2O_3 glass layer of thickness 10nm.

The viscous flow model predicts the rate at which the creep rate declines and the limiting creep strain. In the case of mullite the initial primary creep curves and the resulting β and θ_2 parameters are of a form consistent with the viscous flow model. However, at later stages in primary the creep rate does not slow as rapidly as predicted by the model. The limiting predicted creep strains are larger than those observed at fracture.

For ZMZ the initial forms/shapes of the creep curves are consistent with a viscous flow model but at a latter stage the creep rate decays more slowly than predicted leading to a minimum creep rate.

For both materials it hence seems likely that while viscous flow of the liquid is an important initial strain producing process, it is likely to be superseded by a more rapid process as strain from viscous flow is exhausted.

The activation enthalpy for primary creep at 1400°C was measured for the Mullite by temperature cycling and was found to be 1140 kJ/mol. The temperature dependence of the initial creep rates for the ZMZ materials could be represented by an activation enthalpy of 600 kJ/mol, while that for the primary rate constants β and θ_2 was 750 kJ/mol. All of these values are very high when compared to the calculated temperature dependence of the viscosity of a SiO_2-10%Al_2O_3 glass (19) of 438 kJ/mol.

The stress dependence of the initial creep rates for Mullite yield an exponent of n=3 and for ZMZ a value of n=2.5. These values also represent the stress dependences of the primary rate parameters. These are clearly significantly larger than would be predicted for creep by viscous flow of a Newtonian liquid (n=1).

It is likely that for both materials creep commences by liquid flow and slows as grains interlock/contact. The initial distribution of grain sizes and glass layer thickness on grain faces will mean that the extent of grain contact increases gradually as a function of time from point to point in the system. In fact, for Mullite it is likely that some grain contact exists at time t=0. As grain contact occurs elastic back stresses σ_b, opposing creep, will build up in the grains and if not relieved would result in the cessation of creep. The effective stress (σ_{eff}) driving creep hence decreases with strain/time. This effective stress is given (20), by,

$$\sigma_{eff} = \sigma_a - \sigma_b \qquad (5)$$

where σ_a is the applied stress.

When specimens are partially unloaded the resultant effective stress of opposite sign would drive creep backwards, as observed in the present case (Fig.6), until the system settled to the new stress distribution and liquid flow re-occured.

In many metallic systems very high values of creep activation enthalpy and high n values are associated with the development of such large back stresses (20) and this may be the case for the present materials.

Even when grains come into contact further creep could occur by solution/precipitation processes (5,21). Normally this would yield a stress dependence n=1 for the creep rate, with an activation enthalpy associated with solution and diffusion through the glass phase. Even for systems with large heats of solution it is unlikely that such large creep activation enthalpies as measured here could result. It is hence also probable that this process is driven by an effective stress ($\sigma_a - \sigma_b$).

In the case of Mullite it is probable that solution/precipation gradually takes over from viscous flow but that the creep rate continues to slow right up to fracture. Even at fracture the measured creep rates (corrected for grain size differences) are still at least an order of magnitude faster than those reported for a glass free Mullite, presumably deforming by diffusion creep (22).

In the case of ZMZ solution precipitation will take over from viscous flow but at low stresses cavitation in the liquid phase (21), may become the important process leading to tertiary creep and fracture.

3.3 Minimum Creep Rates

In the case of Mullite at 1400°C in bending and tension the creep rates decrease continuously with time and the minimum occurs just prior to fracture. The stress dependence of this minimum yields n=3.5. This high apparent value could result from the fact that as fracture occurs at shorter times at higher stresses, it occurs at an earlier stage in primary creep resulting in anomalously high values of n. This explanation seems unlikely as the initial creep rates and the rate constants β and θ₂ all yield values of n of around 3. Published values of n for various Mullites mostly lie in the range 1-2 often with very high activitation enthalpies of 600-1000 kJ/mol. However, values of n as high a 4 have been reported for various Mullite composites. The high values of n and of the activation enthalpy reported here hence seem most likely to be a consequence of the presence of large elastic back stresses developed during creep.

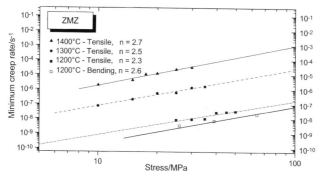

Fig. 8 Log minimum creep rate v log σ for ZMZ.

In the case of ZMZ at most temperatures and stresses the minimum creep rate in tension occurs at the transformation of primary to tertiary creep. However, in bending the minimum occurs prior to fracture with little tertiary creep being observed. The measured minimum creep rates as a function of applied stress are shown in Fig. 8 with the stress exponent being given by n = 2.5, and the creep activation enthalpy being 750 kJ/mol. The fact that the very different bending and tensile creep curves yield similar results for the minimum creep rate suggest that the major strain producing process is solution/precipitation with the possibility of cavitation in the liquid phase giving rise to tertiary creep in tension.

3.4 Time to Fracture

Fracture occurs by growth of a single crack which is easy to distinguish, on the final fracture surface (Fig.9a).

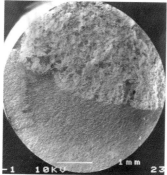

Fig. 9a SEM Micrograph of Mullite specimen fracture at low stress at 1400°C

132

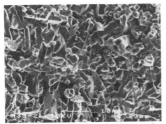

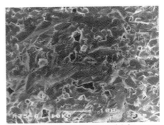

Fig. 9b Intergranular slow crack growth zone   Fig. 9c Transgranular final rapid fracture zone

As the slow crack growth process is intergranular through the glassy phase (Fig.9b), while the final catastrophic rapid fracture is clearly transgranular (Fig.9c).

There is little evidence of damage away from the main crack although this would be difficult to observe in the ZMZ material due to the large volume of glass. It is hence not precluded that cavitation in the glass could contribute to the tertiary creep strain in ZMZ but it is clear that final failure does not occur, even at low stresses, by linking up of creep cavity damage. The crack length at failure increases with decreasing stress as would be expected if a crack has to achieve a given critical stress intensity factor prior to catastrophic failure (Fig. 10).

The times to fracture for Mullite and ZMZ tested at 1400°C statically (R=1), fully reverse loaded, tension/compression (R=-1) and partially tension unloaded (R=0.1) are shown in Figs 11, 12.

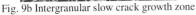

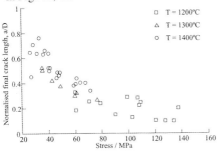

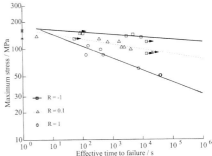

Fig. 10 Failure crack length versus stress for ZMZ   Fig. 11 Log maximum stress v log effective time to failure for Mullite at 1400°C

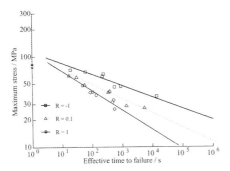

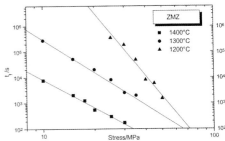

Fig. 12 Log maximum stress v log effective time   Fig. 13 log time to fracture versus log stress
to failure for ZMZ at 1400°C

The times to fracture for the creep specimens lie on the R=1 curve when allowance is made for the different stressed volumes of the creep and fatigue specimens. The creep lives in bending are longer than those in tension as a result of the much smaller maximum stressed volume and the polished nature of the tensile face in the bend specimens. The tensile creep times to fracture can be reasonably predicted using the crack velocity versus stress intensity factore data measured in double torsion (12). The time to fracture decreases with increasing temperature at a given applied σ. (Fig.13) but the fracture strain increases. The temperature dependence of θ_4, for ZMZ, is similar to that for the minimum creep rate suggesting that both maybe associated with the same process. However, the stress dependence of the tertiary θ_4 is much less than that for the primary θ_2, suggesting that the large elastic back stress developed in primary maybe partially released in tertiary.

From Fig. 11, 12 it is clear that cracks growing statically grow more rapidly than those partially cyclically un-loaded. It has been suggested (9) that this occurs when cracks are bridged by viscous glassy ligaments which in effect shield the crack reducing the effective stress intensity factor. This results in a longer time being required to grow the crack to a critical length and to a critical stress intensity. In the present case, fully reverse specimens (tension/compression) lasted the longest. It is possible that significant crack healing, either of the bridged crack, or of an equivalent cavity damage zone ahead of the crack tip, results in a reduction of the crack growth increment per cycle. The effect of temperature on this crack growth process would be expected to follow the temperature dependence of the viscosity of the glass and of its associated relaxation time (9).

4.0 Summary

It seems likely that in these materials creep takes place initially by liquid flow to be replaced by solution/precipitation as grains lock up. Tertiary creep occurs in the ZMZ material, maybe as a result of cavitation damage in the liquid phase. Failure always occurs as a result of single crack growth and the time to failure depends on the mode of cyclic loading.

5   References

1. Wilkinson, S.D. (1998) 'Creep mechanisms in multiphase ceramic materials'; Journal of the American Ceramic Society: Vol. 81; No2; pp.275-299.
2. Hynes, A. and Doremus, R. (1996) 'Theory of creep in ceramics'; Critical review in Solid State Materials Science: Vol. 21; pp.129-187.
3. Chan, K.S. and Page, R.A. (1993) 'Creep damage development in structural ceramics' (1993); Journal of the American Ceramic Society: Vol. 76; No 4; pp.803-825.
4. Dryden, J.R., Kucerovshy, D., Wilikinson, D.S. and Watt D.F. (1989) 'Creep deformation due to a viscous grain boundary phase'; Acta Metallurgica: Vol. 37; pp.2007-2015.
5. Pharr, G.M., and Ashby, M.F. (1983) 'On creep enhanced by a liquid phase'; Acta Metallurgica: Vol. 31; pp.129-138.
6. Luecke, W.E., Wiederhorn, S.M., Hocke, B.J., Krauser Jr, R.E., and Long, G.G. (1995) 'Cavitation contributes substantially to tensile creep in silicon nitride'; Journal of the American Ceramic Society: Vol. 78; pp.2085-2096.
7. Thouless, M.D. and Evans A.G., (1986) 'On creep rupture in materials containing an amorphous phase'; Acta Metallurgica: Vol.34; No. 1, pp.26-31.
8. Thouless, M.D.; Hsuch, C.H. and Evans, A.G., (1983) 'A damage model of creep crack growth in polycrystals'. Acta Metallurgica: Vol.31; No. 10; pp.1675-1687.

9. Ramamurty, U., (1996) 'Retardation of fatigue crack growth in ceramics by glassy ligaments: a rationalization'. Journal of the American Ceramic Society: Vol. 79; No.4; pp.945-952.

10. Torrecilas, R, Calderon, J.M., Moya, J., Reece, M.J., Davies, C.K.L., Olagnon, C. and Fantozzi, G., (1999) 'Suitability of mullite for high temperature applications'; Journal of the European Ceramic Society: Vol. 19; No. 13; pp.2519-2527.

11. Davies, C.K.L., Guiu, F., Li, M., Reece, M.J., and Torrecillas, R. (1998) 'Subcritical crack propagation under cyclic and static loading in mullite and mullite zirconia.' Journal of the European Ceramic Society: Vol.18; pp.221-227.

12. Rhanim, H., Olagnon, C., Fantozzi, G., Torrecillas, R., (1997) 'Crack propogation behaviour in Mullite at high temperature by the double-torsion technique'. Journal of the European Ceramic Society: Vol. 17; pp.85-89.

13. Davies, C.K.L., Guiu, F., Li, M., Oddy, D., Reece, M.J., Stevens, R. N., and Williams, K.M., (1996) 'High temperature creep and fatigue behaviour of mullite ceramics.' Advances in the characterisation of ceramics: Edited by R. Freer. Published by the Institute of Materials, London, U.K. pp.248-252.

14. Davies, C.K.L., Guiu, F., Reece, M.J., Stevens, R.N., Olagnon, C, Fontozzi, G., Torrecillas, R., (1998) 'Development of high temperature fatigue creep and thermal shock resistant zircon and Mullite-zirconia ceramics', Technical Report Brite-Euram 111 Contract BR2-CT94-0613 Project BE-8058: pp.27-33

15. Hollenberg, G.W., Terwillinger, G.R., and Gordon, R.S., (1971). 'Calculation of stresses and strains in four-point bending creep tests.' Journal of the American Ceramic Society: Vol. 54; No. 4; pp.196-199.

16. Evans, R.W. and Wilshire, B. (1985) 'The θ projection concept', creep of metals and alloys: Ch 6. The Institute of Metals, London, U.K. pp.197-256.

17. Woodford, D.A. (1998) 'Stress relaxation, creep recovery; and Newtonian viscous flow in silicon nitride.' Journal of the American Ceramic Society: Vol. 81, No. 9, pp.2327-2332.

18. Weiderhorn, S.M., Hockey, R.F., Krause Jr, R.F. (1986) 'Creep and fracture of a vitreous bonded aluminium oxide'. Journal of Materials Science: Vol. 21, pp.810-824.

19. Urbain, G., Cambier, F., Deletter, M. and Anseau, M.R. (1981) 'Viscosity of silicate melts.' Transactions of the Journal of the British Ceramic Society: Vol. 80, pp.139-141.

20. Davies, C.K.L., Poolay-Mootien, S., Stevens, R.N., (1992) 'Internal stress and unloading experiments in creep'. Journal of materials Science: Vol. 27, pp.6715-6724.

21. Wiederhorn, S.M. (2000) 'Particulate ceramic composites: their high-temperature creep behaviour.' Key Engineering Materials: Vol.175-176, pp.267-288.

22. Okamoto, Y., Fukudone, H., Hayashi, K., Nishikawa, T. (1990) 'Creep deformation of polycrystalline mullite.' Journal of the European Ceramic Society: Vol. 6, pp161-168.

ANAMALOUS GRAIN SIZE DEPENDENCE OF CREEP IN MOLYDISILICIDES

K. Sadananda and C.R. Feng
Code 6323, Materials Science and Technology Division
Naval Research Laboratory
Washington D.C. 20375

ABSTRACT

Creep deformation of $MoSi_2$ has been investigated under compression at temperatures at $1200^{\circ}C$. Compression specimens showed three regimes of creep; a Newtonian viscous regime at low stresses, power-law creep at intermediate stresses and breakdown of power-law at high stresses. The most unusual behavior observed under compression was the large grain size dependence. The grain size exponent is greater than four, which is unusually high. Furthermore, this large grain size dependence persists in the power-law creep regime. Most importantly these grain-size effects are history-dependent. Specimens that were deformed previously in the Newtonian viscous regime fail to show this grain size dependence in the power-law regime. This large grain-size dependence in power-law regime is accounted due to the deformation constraints introduced by crystal structure and diffusion kinetics.

1. Introduction

There is an increasing demand for materials that can withstand higher operating temperatures, than those currently achieved using the advanced nickel base superalloys. Intermetallics and ceramics are two available choices. Efforts on high temperature intermetallics are being pursued vigorously during the past decade. Among the intermetallics, aluminides and silicides are the two competing choices. The formation of alumina or silica provides a protective coating, stabilizing them for high temperature applications. Among the silicides, $MoSi_2$ received considerable interest due to its relatively good oxidation resistance and creep ductility [1-24]. There have been significant improvements in terms of processing, microstructural control and properties of $MoSi_2$. Several review papers [7,18,22]on $MoSi_2$ and its composites in terms of their processing and properties have been published. Grain size [23] found to have a significant effect on the creep properties of $MoSi_2$. Since these effects overshadow, to some extent, the effect of other variables, it is important to understand and recognize these effects. In this paper, we present results related to the anomalous effects of grain size.

2. Materials and Methods

The materials tested were obtained from Los Alamos National Laboratory (LANL). They were processed mainly through powder consolidation. Grain size was varied by changing the temperature of the hot press. To evaluate the grain size effects in $MoSi_2$, three types of tests were conducted; (a) incremental load tests at low stress range

136

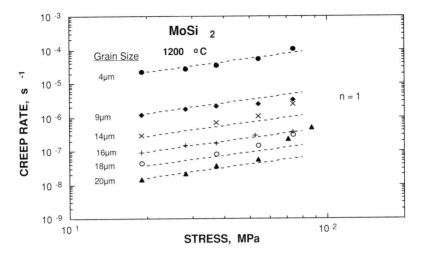

Fig. 1. Variation of creep rates with grain size in the Newtonian viscous region

in the Newtonian regime, (<75 MPa) (b) incremental load tests in the intermediate power-law creep range (>80 MPa), and (c) incremental load tests in the full range starting from Newtonian and ending into the power-law range. At each load, the tests were conducted till the strain rates reached their steady state values. Few step down load tests were also performed to evaluate the history effects. All tests were done at temperatures above 1000°C. The results are presented below.

3. Results and Discussion

Grain boundaries play a dominant role in the creep deformation. The creep rates as a function of stress follow a three-stage behavior[23]. At low stresses, stage I creep rates vary linearly with stress indicative of Newtonian viscous flow behavior, with stress exponent n = 1. With increasing stress, stage II occurs with n around 3-4 commonly referred to as power-law creep. At higher stresses, the deviations from power-law occur with stress exponents higher than 4.

Phenomenologically, the power-law relation ship is represented as

$$\varepsilon = A \, (1/d)^p \, (\sigma)^n \, \exp(-\Delta Q/RT) \qquad (1)$$

where A is a structure sensitive factor, p and n are the grain size and the stress exponents, respectively, ΔQ is the activation enthalpy, R and T have the usual meaning.

Fig. 1 shows the effect of grain size in the Newtonian creep regime. The grain sizes are in the range of 4 μm to 20 μm. Correspondingly, the creep rates change by three to four orders of magnitude in this narrow grain size range. At the larger grain size (20 μm), the power-law creep regime starts at around 60 MPa. This transition to the power-law occurs

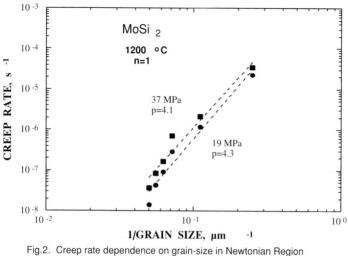

Fig.2. Creep rate dependence on grain-size in Newtonian Region

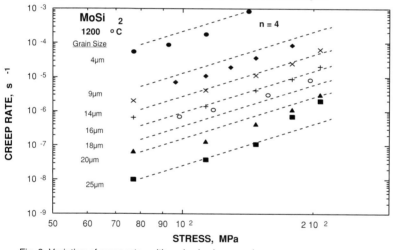

Fig. 3 Variation of creep rates with grain size in power-law creep

at higher stresses for finer grain size specimens. Fig. 2 shows the creep rates vary inversely with the grain size with an exponent, p equal to 4.2. It is inherently assumed in plotting Fig. 2 that all the contributing factors in Eq. 1, other than the grain size, are independent of grain size. That this may not be true is also a realistic possibility.

Fig. 3 shows the effect of grain size in the power-law creep regime. Normally in the power-law regime, grain size effects are rarely observed, since inter-dislocation spacing rather than grain size is the controlling factor. Fig. 3 shows that for the decrease in grain size from 20 μm to 4 μm, creep rates increased by three to four orders of magnitude. In

138

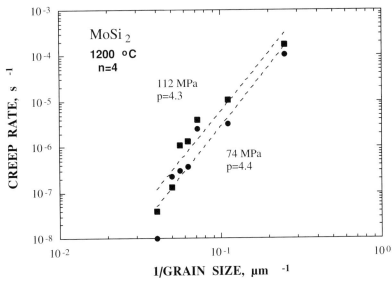

Fig. 4 Creep rate dependence on grain size in power-law creep

the larger grain size specimen, the break down of power-law occurs at a lower stress, 150 MPa. Fig. 4 shows the grain size exponent in the power-law creep range. The exponent is of the order of 4.3 to 4.4, much similar to the value in the Newtonian range. Such large grain size effects persisting even in the power-law creep range have rarely been observed before. One can dismiss that the results are erroneous, but our repeated tests confirmed the behavior. There is only one reported case[25] of such grain size effect in an intermetallic 50Ti-50Al (equi-atomic composition), with the grain size exponent of the same order, p = 4.2. Interestingly, in the TiAl case, when the composition was altered to 51Al, the grain size effects almost disappeared[26] with p value dropping from 4.2 to 0.7. It appears, therefore, that the observation of such large grain size effect in the power-law creep regime is not restricted to $MoSi_2$. The observed grain size effect at only 50:50 TiAl, and its disappearance with small change in the Al concentration require further analysis of the source for the grain size effect. Addition of 1% excess Al can offset the vacancy defect concentration and diffusion currents, and thus can affect the creep rates. In contrast to TiAl, $MoSi_2$ is a line compound. Therefore, defect tolerance (chemical vacancy solubility or off-site occupancy) is expected to be limited. The common feature of significant grain size effects in the stoichiometric compositions in both TiAl and $MoSi_2$ is worth noting.

The results of the third type of tests, wherein the loads increased in steps, starting from the Newtonian regime and ending up into the power-law creep regime, are shown in Fig. 5. The results differ from those in Fig. 3. When the specimens, that were pre-deformed in the Newtonian viscous regime, are loaded into power-law creep regime, the observed grain size dependence disappears. The data for all the grain sizes nearly fall on a single power-law line. The transition to the power-law for each grain specimen occurs at a different stress

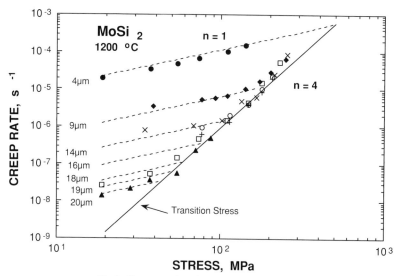

Fig.5 Absence of grain size effect in power-law creep for specimens crept in Newtonian region

level. Larger stresses are required to arrive at the power-law behavior for the smaller grain size materials. Examination of the results in Fig.3 and 5 indicates that the anomalous grain size dependence observed in the virgin samples in Fig. 3 disappears when they are pre-crept at lower stress regime. Whether this behavior is also true for the equiatomic TiAl remains to be known. Thus the grain size dependence in power-law creep arises from a process that is strain history dependent. We may note that this is the first time such history-dependent grain size effects in power-law creep has been observed. Lack of such observations could be partly due to lack of detailed investigations using materials that have restricted slip (stoichiometric compounds). Based on these results such behavior is expected in ceramics or intermetallics where low symmetric crystal structure restricts the number of slip systems available for deformation. For ceramics, the presence of viscous glassy phase at the grain boundaries contributes to most of their creep deformation, at low stresses[27-29]. In $MoSi_2$, a presence of viscous SiO_2 was reported, and it may well contribute to the creep in the Newtonian regime. But at high stresses, the material also deforms by power-law creep involving dislocation process. The observed strain-history dependence on the grain size effects should be inherently related to a dislocation process.

History effects on creep were further examined using two different grain-size specimens. In one specimen, the loads were applied in increasing steps; and in another, in decreasing steps, considering only the steady state strain rates for each load. Fig. 6 shows the results for a larger grain size specimen (19 µm). For incremental loads, the transition from Newtonian to power-law creep occurs around 48 MPa, with n = 4. On the other hand, for a specimen that was step-down loaded, the creep rates seem to follow Newtonian viscous behavior with n = 1 for the stress in the range of 220-150 MPa. With further decrease in loads from 150-70 MPa, the rates fall down rapidly with the slope much steeper than

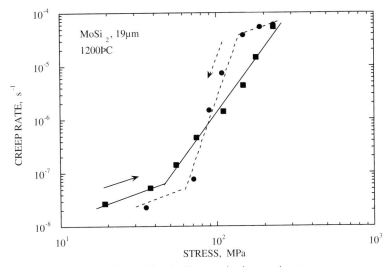

Fig. 6 History effect in 19 μm grain size specimen

n = 4. With an additional decrease in loads, Newtonian behavior is again seen with the transition from power-law occurring at a higher stress, 60 MPa. There is a definite hysteresis, with the decreasing load curve rapping around the increasing load curve. Observation of some Newtonian creep at high stresses in the decreasing load case is noteworthy. This could be due to the presence of some viscous silica at the boundaries. But as the silica is squeezed out under compressive creep, the deformation moves to the power-law creep. The material in the unloading case has a higher n value in the power-law creep, perhaps due to decreased dislocation mobility induced by work hardening at higher stresses. As the stress drops, the dislocations can get trapped initiating a slower rate deformation process. The history effects in a smaller grain size specimen, 9 μm, is shown in Fig.7. In contrast, the specimen under decreasing loads deforms predominantly by Newtonian flow. The creep curve shows a small step indicative of some superimposed power-law creep at some intermediate stress range. But these changes could be within the experimental scatter. The creep rates remain, for the most part, higher than those for increasing loads.

Thus, the transition from the Newtonian to the power-law creep depends on the grain size and load-history. We examine the grain size dependence of transition stress that initiates the power-law in increasing load tests, Fig. 8. The data is extracted from Fig. 5. The transition stress varies inversely with grain size with the exponent between 2 and 3 (slope close to 2 than 3). The dependence of transition stress on grain size is much larger than the conventional Hall-Petch type of relation. At small grain sizes (last data point in Fig. 8), there is a deviation from the curve. Assuming the exponent is close to 2, one can express the grain size dependence of the transition stress in the form

$$\sigma_T = \sigma_0 + (K_T/d)^2 \qquad (2)$$

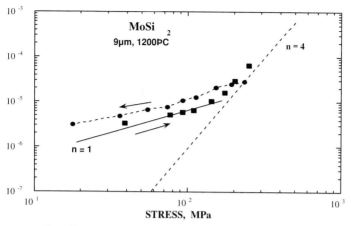

Fig. 7 History effects in smaller grain size specimen

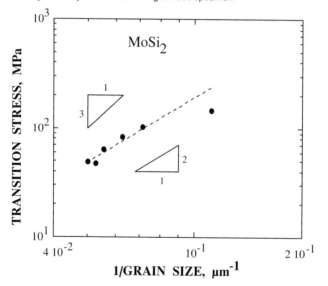

Fig. 8 Variation of transition stress on grain size

The second power relation on grain size is intriguing. It is possible that the observed high p value, 4.3, could be partly due to the incorporation of stress dependence of grain size. Normally for the Newtonian creep n=1, p is expected to be 2 for Naborro-Herring creep and 3 for Coble creep. It is conceivable that the normal grain size effects on creep are accentuated by the stress variation due to grain size. Thus p consists of two components - normal grain size dependence of 3 involving grain boundary diffusion and Coble creep, plus an additional grain size dependence due to the transition stress.

Phenomenologically, one can express the total strain rate as

$$\dot{\varepsilon} = A \, (1/d)^p \, (\sigma_a - \sigma_T)^n \, \exp(-\Delta Q/RT) \qquad (3)$$

which is a modification of Eq. 1, where σ_a is the applied stress and σ_T is the transition stress, which is a function of a grain size given by equation (3). This assumes that the transition stress is a pre-requisite to initiate the power-law creep, much like a threshold stress. This remains to be established. Here, we summarize the important anomalous effects of grain size on creep observed in our study of MoSi$_2$, requiring further confirmation and theoretical analysis. (a) Large grain size dependence was observed in both the Newtonian (n=1) and the power-law creep (n=4) regions. (b) The grain size exponent, p in both regions is round 4.3 much larger than that can be expected on the basis of the diffusion process involving Naborro-Herring (p=2) or Coble (p=3) creep. (c) The grain size dependence in the power-law region disappears if the specimens were creep-prestrained in the Newtonian region. (d) Transition from the Newtonian Creep to the power-law creep occurs at different stresses for different grain sizes. (e) The transition stress varies inversely with the grain size with a power of 2, indicative of some basic mechanism involved. (f) Creep rates under incremental loading differ from that under decremental loading, and this hysteresis varies with grain size.

There are no known current mechanisms that can fully account for all these observations. The observations that equiatomic Ti-Al also shows similar grain size dependence (p=4.4) in the power-law creep regime (n=4), and this dependence disappears at slightly higher Al concentrations (51Al) are interesting. These results indicate first that the grain size dependence in the power-law creep is not unique to MoSi$_2$, but may be present in other stoichiometric compounds, and second that the dependence is sensitive to defect density and available diffusion paths for the creep recovery process, in addition to the strain history. High sensitivity to the grain size variations, should be born in mind in evaluating the effect of other variables, particularly if the change in the variables also affects the grain size. The effect of other variables could be overshadowed by the grain size effects when the grain size is $< 20 \, \mu m$.

4. Summary and Conclusions

Detailed evaluation of Creep behavior of MoSi$_2$ is presented. Grain size is shown to influence even in the power-law creep with n = 4. The grain size exponent, p, is in both the Newtonian and power-law regimes is around 4.3, much higher than that expected from pure Coble or Nabarro-Herring creep. The grain size effects in the power-law creep is also history dependent, since prior creep deformation eliminated the effect. The transition stress from the Newtonian to the power-law creep itself varies inversely with grain size by a power of 2. Such large grain size effects are not commonly observed except in equiatomic TiAl. The conventional mechanisms that account for the grain size effects under creep fall short in explaining such large grain size dependence in the power-law regime and the associated strain-history effect. Further investigations are required before the mechanisms governing the phenomenon are fully understood.

5. References

1. Maxwell, W.A. 1952, Some Stress-Rupture and Creep Properties of Molybdenum Disilicide in the Range of 1600-2000°F, NACA RME52D09.
2. Gibbs, W.S., Petrovic J.J. and Honnel, R.E.,1987, Ceram. Engrg. Sci. Proc. Vol.8, p. 645.
3. Sadananda, K., Jones, H., Feng, C.R.., Petrovic, J.J. and Vasudevan, A.K.,1991, Ceram. Eng. Sci. Eng. Proc., Vol.12, pp.1671-1678.
4. Vasudevan, A.K., Petrovic, J.J. and Sadananda, K., 1991, 12th Riso Conf., Hasen, N. et al. (eds.), p. 707.
5 Umakoshi, Y., 1991, Bull. Jpn. Inst. Metals. Vol. 30, p. 72.
6. Aikin, Jr. R.M., 1991, Ceram. Eng. Sci. Proc., Vol.12, pp.1643-55.
7 Vasudevan A.K.and Petrovic, J.J., 1992, Mater. Sci. Engrg, Vol.A155, pp. 1-18.
8. Petrovic, J.J., Bhattacharya, A.K., Honnel, R.E., Mitchell, T.E.and Wade, R.K., 1992, Mater. Sci. Engrg, Vol.A155, pp.259-266.
9. Sadananda, K., Feng, Jones, H. and Petrovic, J.J., 1992, Mater. Sci. Eng., Vol. A155, pp. 227-239.
10. Weiderhorn, S.M. Gettings, R.J. Roberts, D.E., Ostertag C., and Petrovic, J.J., 1992, Mat. Sci. Engrg., Vol. A155, pp. 217-226.
11. Bose, S., 1992, Mat. Sci. Engrg., Vol. A155, pp. 217-226.
12. Sadananda K. and Feng., C.R. in Proc. Processing, Fabrications, and Manufacturing of Composite Materials, edited by Srivatsan T.S. and Lavernia, E.J., 1992, American Society of Mechanical Engineers, New York, MID, Vol.35, pp. 23-245.
13. Deve, H.E., Weber, C.H. and Maloney, N., 1992, Mat. Sci. Engrg., Vol. A153, p. 668
14. Mason D.P. and Van Aken, D.C., 1992, in High Temperature Ordered Intermetallics, edited by Baker, I., Doralia, R., Whittenberger J.D. and Yoo, M.H, 1992, Mater. Res. Soc. Proc. Vol. 286, Pittsburgh, pp.1129-1134.
15. Patrick D.K. and Van Aken, D.C. 1992, ibid., pp 1135-1141.
16.Bieler, T.R.,Whittenberger J.D. and Luton, M.J. 1992, in High Temperature Ordered Intermetallics, ibid, pp 1149-1154;
17. Feng C.R. and Sadananda, K. ibid. pp. 1155-1160.
18 Sadananda K. and Feng, C.R. 1993, Mater. Sci. Engrg., Col. A170, pp. 199-214.
19. Ghosh, A.K. and Basu, A. 1993, in Critical Issues in the Development of High Temperature Structural Materials, ed. by Stoloff N.S., et al., TMS Publication, Warrendale.
20. Suzuki M., Nutt S.R. and Aiken, Jr., R.M., 1993, Mater. Sci. Engrg., Vol.A162, p. 73.
21. French J.D. and Weiderhorn, S.M., 1993, Presented at the 1993 MRS Fall meeting, Boston, MA.
22. Sadananda K. and Feng, C.R. 1994, "A Review of Creep of Silicides and Composites", Mat. Res. Soc. Symp. Vol.322, pp. 157-173.
23. Feng, C.R. and Sadananda, K. 1995, "Grain Size Effect On The Creep Behavior Of Monolithic $MoSi_2$", Mat. Res..Symp. Proc. Vol. 364, pp.1053-1058.
24. Sadananda K. Feng, C.R., Mitra, R., and Deevi S.C., 1999, Mat. Sci. Engrg. Vol.A261, pp.223-238.
25. K. Maruyama, K., Takahashi T. and Oikawa, H. 1992, Mat. Sci. Engrg. Vol.A153, p. 433.
26. Nagai, N., Takahashi T. and Oikawa, H. 1990, J. Mater. Sci., Vol. 25, p. 629.

27. Hokey, B.J., Weiderhorn, S.M., Liu, W., Baldoni J. G. and Buljan, S.T.,1991, J. Mater. Sci. Vol.26, p. 3931.
28. Nieh, T.G., Wadsworth, J., Grensing, F.C. and Yang, J.M., 1992, J. Mater. Sci. Vol.27, pp. 2660-2664.
29. Nieh, T.G., Wadsworth, J., Chow, T.C., Owen D. and Chokshi, A.H.1993, J. Mater. Res. Vol. 8, pp. 757-763.

CREEP OF MAGNESIUM

Creep Behavior and Deformation Substructures of Mg-Y Alloys containing Precipitates

M. Suzuki, T. Kimura, H. Sato*, K. Maruyama and H. Oikawa**

Department of Materials Science, Tohoku University, Sendai, 980-8579, Japan
* Now with Department of Intelligent Machines and System Engineering,
Faculty of Science and Technology, Hirosaki University, Hirosaki, 036-8561, Japan
** Now with College of Industrial Technology, Amagasaki, 661-0047, Japan

Abstract

Compressive creep properties of Mg-Y alloys containing various amount of yttrium, having various initial microstructures were investigated at 530K. The effect of initial microstructures on the minimum creep rate is very slight, but the effect of yttrium content is significantly. On the other hand, the primary stage of creep is affected by the initial microstructure. Low creep rate was observed at the beginning of the creep by introducing fine precipitates within grain interiors.

TEM observations have been revealed that fine precipitates are formed in the supersaturated solid-solution during creep and this precipitation process is rapidly because the diffusivity of yttrium in magnesium is high and the precipitates are nucleated on dislocation lines. The fine precipitates are effective to improve the primary creep behavior, but the initial microstructure does not affect on the minimum creep rate. The reason of the slight effect of the initial microstructure on the minimum creep rate is considered to be the low stability of the pseudo-equilibrium phase β' at high temperature. Dislocations are the most important nucleation sites for the fine β' precipitates, so that the coupling of straining (pre-creep) and aging can enhance the precipitation and decrease significantly the strain-rate in the primary stage of creep.

Keywords: magnesium-yttrium alloys; precipitation hardening; creep; microstructural stability

1. INTRODUCTION

Magnesium-based alloys have been applied to many portable electronics because magnesium is the lightest metal among practical structural metallic materials. Recent trials to improve the strength of magnesium have been reported in many literatures [1,2,3,4,5]. Microstructural refinement is one of the most efficient methods to improve strength and ductility of magnesium-based alloys at room temperature because Hall-Petch slope of magnesium is grater than that of aluminum [6, 7]. Practically, most of the recent new high strength magnesium-based alloys are strengthened by the grain refinement; for example, die casting [1,2], thixomolding [3], extrusion [3], equal channel angular extrusion [3] and so forth. At high temperature, however, fine-grain microstructure is no longer useful to improve strength and accelerates creep rate due to the grain boundary sliding. The improvement of high temperature strength of magnesium-based alloys is an important subject for expansion of the practical application field of magnesium, for example, aircraft and automobile applications.

It has been reported in many literatures that the addition of rare earth elements (RE) improves significantly mechanical properties of magnesium at high temperature; not only tensile properties [8, 9, 10] but also creep properties [8, 11, 12, 13]. The precipitation hardening is generally considered the most important mean for strengthening of Mg-RE alloys. The process of precipitation from the supersaturated solid solution in Mg-Y system and effects of the precipitates on the strength have been reported in [14,15,16]. The strength can be increases by the homogeneous nucleation of pseudo-equilibrium phases within grains during low-temperature aging [15,16,17]. There are some reports that these pseudo-equilibrium phases tend to nucleate along dislocation lines and/or twin boundaries [13,17]. On the other hand, the equilibrium phase has little effect on hardness because they tend to precipitate on grain boundaries and grow rapidly [16].

In general, these fine pseudo-equilibrium phases are important for strengthening in Mg-RE systems, but these precipitates are not stable at high temperature. In long-term high-temperature deformation (creep), the dynamic structural change during deformation cannot be ignored.

In this paper, creep behavior of Mg-Y alloys containing precipitates has been investigated and tried to improve creep properties of these alloys by coupling of heat-treatment and straining (pre-creep).

2. EXPERIMENTAL PROCEDURES

The materials used in this study were three kinds of Mg-Y binary alloys. The ingots were hot-rolled at 723K~823K to plates. Their chemical compositions are listed in Table 1.

Table 1 Alloys used

Alloy designation	Y (mol%)	Mg (mol%)	Heat-treatment	Microstructure/Precipitates
1.6Yg	1.62 (5.68mass%)	Bal.	HA	grainboundary precipitates
2.4Yg	2.39 (8.28mass%)	Bal.	HA	grainboundary precipitates
2.4Yi			LA	Precipitates are introduced along grainboundary and grain interior
3.2Ys	3.15 (10.7mass%)	Bal.	SS	supersaturated solid solution
3.2Yi			LA	Precipitates are introduced along grainboundary and grain interior
3.2Yc			CA	Precipitates are introduced along grainboundary and grain interior
Notes	SS: solution-treatment HA: solution-treatment → aging at 560Kfor 48h LA: solution-treatment → aging below 500K for 48~96h → aging at 540K for 24~48h CA: solution-treatment → pre-creep→ aging at 540K for 10h			

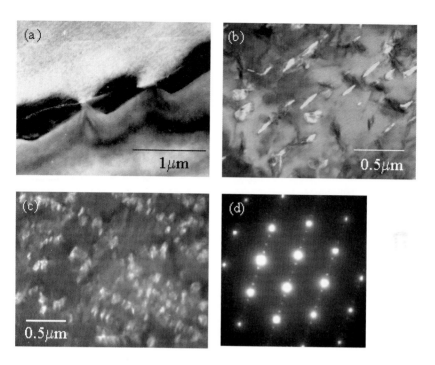

Figure 1. TEM micrographs taken after heat treatments of Mg-Y alloys.
(a) 2.4Yg, (b) 3.2Yi, (c) 3.2Yc, (d) diffraction pattern from (c)

All the alloys were solution-treated in an argon atmosphere at 760~800K followed by water-quenching. Some of them were aged and/or pre-crept to clarify the effect of initial microstructures. The aging conditions are also listed in Table 1. The average grain size after the solution-treatment was about 50μm for all alloys and it does not change during aging and/or pre-creep. Pre-creep was carried out under a constant stress of 120MPa at 550K until a strain of 0.03, it took about 5 hours.

TEM micrographs taken after heat-treatments are shown in Figure 1. Many coarse precipitates are observed along grain-boundaries, but no precipitate is introduced within grains in high-temperature aged (HA) alloys. This coarse precipitates are considered to an equilibrium phase, β ($Mg_{24+x}Y_5$, α-Mn type crystal structure)[18]. On the other hand, in low-temperature two-step aged (LA) alloy, many fine precipitates are introduced uniformly within grains (Fig.1 (b)). After the pre-creep and low-temperature aging (CA), dislocations are introduced uniformly within grains and many fine precipitates are formed along these dislocation lines (Fig. 1(c)). These fine precipitates are a pseudo-equilibrium phase β' having bco structure (base-centered orthorhombic structure) [15] and these precipitates strongly tend to nucleate along dislocations [13]. These fine precipitates within grains have a kind of coherency with the matrix as seen in Fig. 1 (d). Only one kind of heat treatment was

150

performed on Mg-1.6mol%Y because no static and dynamic precipitates within grain interior were introduced during aging or creep in this investigation.

Compressive creep specimens (2mm×2mm×3mm, parallelepipeds) were held at the test temperature for 10.8ks before creep tests to stabilize temperature of the testing system. Creep tests were carried out at 530K under constant stresses in air.   Crept specimens were water-quenched under load for TEM observations.   The foil specimens for TEM were prepared by the twin-jet electropolishing. Deformation substructures in crept specimens were observed by using a transmission electron microscope operated at an acceleration voltage of 200kV.

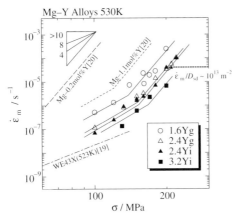

Figure 2. Stress dependence of the minimum creep rate of Mg-Y alloys.

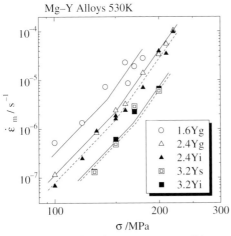

Figure 3.Effects of the heat treatment condition on the minimum creep rate of Mg-Y alloys

3. RESULTS

3-1 Creep behavior

The stress dependence of the minimum creep rate of Mg-Y alloys at 530K is shown in Figure 2.   Also shown in this figure are the data of a Mg-Y-RE heat resistant alloy WE54X (Mg-1.82mass%Zr-0.52mass%Mm-5.14mass%Y-1.71mass%Nd) at 523K [19] and Mg-0.2mol%Y and Mg-1.1mol%Y solid solution alloys at 530K [20].   The Creep strength increases with increasing yttrium content, the stress exponent n of Mg-Y alloys is 10 in the low stress region, which is higher than that of WE54X ($n \sim 4$) [19].   A transition in the stress dependence of the minimum creep rate is observed.   The value of n becomes larger ($n > 10$) in the high stress region.   The transition stresses of Mg-1.6mol%~3.2mol%Y alloys are 140~170MPa.   The Mg-1.1mol%Y solid solution alloy also shows a similar trend with a transition stress of 120MPa.

The effects of heat treatment condition on the minimum creep rate of Mg-2.4mol%Y and Mg-3.2mol%Y are shown in Figure 3.   It is recognized that the aging condition influences very slight on the minimum creep rate though the initial microstructure is quite different among specimens (see Fig.1).   The effect of the yttrium content on the minimum creep rate is much larger

151

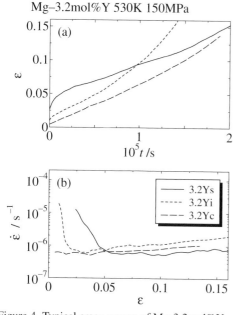

Mg–3.2mol%Y 530K 150MPa

Figure 4. Typical creep curves of Mg-3.2mol%Y alloy. (a) strain-time curves, (b) strain rate-strain curves

than that of the initial microstructures.

Typical creep curves of Mg-3.2mol%Y alloys having different initial microstructures at 530K are shown in Figure 4. The minimum creep rates of these three specimens are almost the same. On the other hand, the features of the primary creep of them are obviously different among the three initial microstructures. Alloy 3.2Ys, solution-treated only, has a large primary stage of creep and the creep rate decrease gradually with increasing strain. On the other hand, in low-temperature aged 3.2Yi alloy, large decrease of the creep rate in the primary stage is observed and the strain at the minimum creep rate is about 0.025. This value is about the half of that of 3.2Ys. The creep rate is very low at the beginning of the creep in the alloy 3.2Yc, pre-crept and aged alloy. The initial plastic strain upon loading is also very small in 3.2Yc as compared with that of the supersaturated solid-solution alloy 3.2Ys. A slight acceleration in creep rate is observed in aged 3.2Yi and 3.2Yc, but a long steady-state stage is observed in the supersaturated solid-solution alloy 3.2Ys.

3-2 Microstructures

TEM micrographs (dark field images) at the minimum creep rate of Mg-3.2mol%Y alloys are shown in Figure 5. The incident beam direction B in these photos is parallel to the [0001] direction. It is clear that the microstructure of the supersaturated solid solution alloy 3.2Ys changes significantly. Many precipitates (bright contrast in Fig. 5(a)) are introduced within a grain, which align on the dislocation lines. On the contrary, morphology and distribution of precipitates in two-step aged alloy 3.2Yi are similar to those of the initial one (Fig 1(b)). The deformation microstructure of pre-crept alloy 3.2Yc is different from the crept other two alloys (3.2Ys and 3.2Yi) and the initial microstructures of 3.2Yc. The diffraction patterns from the crept and the initial alloys are also different each other (Fig. 1(d) and Fig. 5(f)). The weak streaks are observed in the diffraction pattern from the crept specimen (Fig 5(f)) and many fringes are introduced in the precipitates (Fig. 5(d),(e)).

4. DISCUSSION

4-1 Creep behavior

In this investigation, the transition of the stress exponent n is observed in all Mg-Y

152

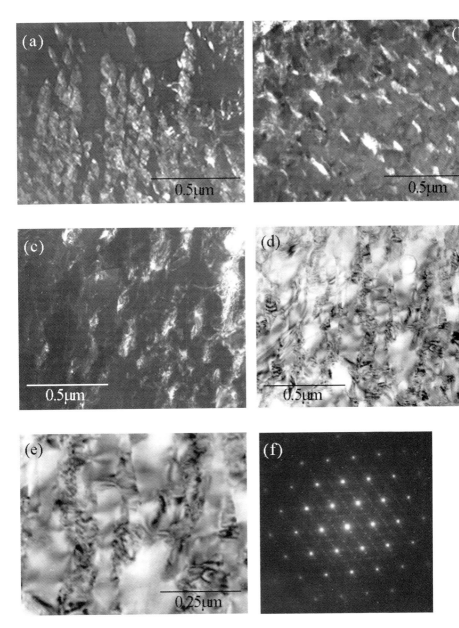

Figure 5. TEM micrographs of crept specimens at the minimum creep rate.
The incident beam direction B is parallel to the [0001].
(a) 3.2Ys, (b) 3.2 Yi, (c) 3.2 Yc (dark field), (d) 3.2 Yc (bright field)
(e) high magnification of a part of (d)
(f) diffraction pattern from (d)

alloys having precipitates, in a single-phase alloy (Mg-1.1mol%Y) as well. The transition stress tends to increase with increasing yttrium content and n under high stresses is more than 10. Such large stress exponent ($n > 10$) in the lower stress region has been reported in many dispersion-hardened alloys and this behavior is explained by the concept of the threshold stress [21]. In the present alloys, however, this transition is not due to the threshold behavior of the precipitation strengthening because the transition is also observed in solid solution Mg-1.1mol%Y [20]. Therefore, this transition of n is considered to show a change in the deformation mechanism. The transition of n has been observed also in Mg-Y binary alloys at 550K [13] and Mg-Y-RE alloys [8]. Henning et al. observed the transition of the stress exponent in Mg-6mass%Y-4mass%Nd at 563~633K[8] and attributed this transition to the break-out of the Cottrell solute atmosphere. In the present results, however, the stress exponent under lower stresses is much higher than that reported by Henning et al. ($n = 2~3$) and the transition stress is much higher than their results (about 40MPa at 563K) [8]. Therefore, the transition of the stress exponent in the present Mg-Y alloys does not correspond to the break-out of the Cottrell solute atmosphere. According to the creep results of the same alloys at 550K [20], the transition of stress exponent can be explained by the power-law breakdown. In this study, however, the normalized strain-rate of $\dot{\varepsilon} / D_{sd}$ (where the self-diffusion coefficient of pure magnesium D_{sd} at 530K is calculated [22, 23] to about 4.08×10^{-18} $m^2 s^{-1}$) at this temperature is much lower than 10^{13} m^{-2} (see Fig.2), which is considered to be the critical value of the power-law breakdown [24]. Therefore, it is difficult to conclude the transition is due to the power-law breakdown. Henning et al. [8] have been reported the 0.2% proof stress of 220MPa in Mg-6mass%Y-4mass%Nd aged alloy. The yield stresses of the present alloys are not known at moment, but a possible explanation of this transition may be yielding. Under lower stresses, the stress exponent is about 8 and is slightly higher than 5 [25], which is considered to be the case where the rate controlling mechanism is the dislocation climb. In basically, nevertheless, the deformation in this lower stress region is considered to be controlled by the dislocation climb mechanisms.

4-2 Effects of initial microstructure on creep

It is hard to observe the effects of the initial microstructure on the minimum creep rate from Fig. 3, but the effects of the yttrium content are significant. The microstructure changes obviously during the primary stage in these three alloys. The dynamic precipitation from the supersaturated solid-solution is fairly fast and the microstructure of 3.2Ys at the minimum creep rate (Fig. 5(a)) is quite similar to the initial microstructure of 3.2Yc (Fig.1(c)). The low-temperature aged microstructure seems to be stable during creep (Fig. 5(b)), but many fringes are introduced during creep within precipitates as seen in 3.2Yc (Fig. 5(e)). The distribution of the precipitates in 3.2Yc becomes uniform at the minimum creep rate. These results indicate clearly that the initial microstructure is not important for the minimum creep rate because the structural change during the primary creep is significant.

The feature of primary stage of creep, on the contrary, is affected significantly by the initial microstructure. The decrease of the creep rate of the low-temperature aged alloy 3.2Yi at the early stage of creep is significant as compared with that of the supersaturated solid-solution alloy 3.2Ys. This fact suggests the fine precipitates act effectively as the pinning points of the dislocation. In pre-crept and low temperature aged alloy 3.2Yc, the creep rate is very low at the beginning of creep, which is close to the steady-state value. The fine precipitates are also introduced in the initial microstructure of 3.2Yi and they are introduced uniformly within grain. On the contrary, in 3.2Yc, most of the precipitates are aligned along dislocation lines though the precipitates of 3.2Yi and 3.2Yc are the same; pseudo-equilibrium phase β'. It is considered that precipitates are introduced easily on dislocation lines to accommodate the misfit strain between precipitates and the matrix. By

154

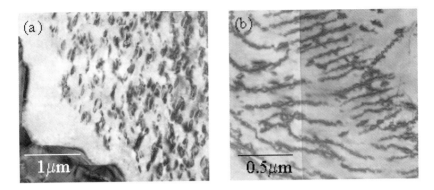

Figure 6. TEM micrographs of low temperature aged Mg-2.4mol%Y (2.4Yi)
The incident beam *B* is parallel to the [0001].
(a) Starting microstructure (b) at the minimum creep rate (100Mpa)

Figure 7. TEM micrographs of Mg-2.4mol% Y alloy.
The incident beam direction B is parallel to the [0001].
(a) 2.4Yg, aged at 550K 48h for 48h
(b) 2.4Yg, at the minimum creep rate (550K, 100Mpa, crept for 10h)

these precipitates, dislocations are also pinned strongly.   Therefore, the distribution of the precipitates is also important to shorten the primary creep stage.   Moreover, some dislocations are introduced during pre-creep.   The dislocation density of pre-crept specimen is roughly 1×10^{13} m^{-2}.   From this result, the effect of work hardening (pre-strain of 4%) is remarkable, too.

4-3 Stability of the precipitates
4-3-1 Strengthening by the unstable phase

It is clear that the stability of the pseudo-equilibrium phase β' is low during creep. The basic reason of this low stability of β' phase is the high diffusivity of yttrium in magnesium [11].   The roles of dislocations are also important in the precipitation and the desolution of β' phase during creep.   Dislocations act as main nucleation sites in the precipitations, but also act as the fast diffusion paths for yttrium solute atoms in the desolution.   Figure 6 shows the initial and the crept (100MPa, 530K) microstructures of 2.4Yi.   It is clear that the dislocations gather up the precipitates along their lines during creep (compare Fig 6(b) with Fig. 6(a)).   On the other hand, the precipitates in 3.2Yc spread out during creep (compare Fig. 5(b) with Fig. 6(a)).   Therefore, the distribution of the precipitates is affected complicatedly by the mobility of dislocation and the diffusivity of yttrium.

4-3-2 Effects of straining on precipitation

Figure 7 shows the effects of the straining on the precipitation of β' phase in Mg-2.4mol%Y alloy.   After the static aging at 550K for 48h (Fig 7(a)), some coarse precipitates appear at grain boundaries, but no precipitation is introduced within grain interior.   In the strained (crept) alloys, however, there are many precipitates (dot-like contrast in Fig.7 (b) [14]) within grains although the thermal exposure time of crept specimen (10h) is shorter than that of static aged alloy.

In static aging, it takes long time to introduce fine precipitates within grains as shown in Table 1.   By the coupling of pre-creep and aging, the total thermal exposure time can be shortened significantly to introduce enough fine precipitates for resisting creep deformation. Moreover, by these treatments, the primary creep stage becomes small.   Therefore, the pre-straining is an attractive method to improve creep properties of Mg-Y precipitation-hardened alloys.

5. CONCLUSIONS

Creep behavior of Mg-Y alloys containing precipitates were investigated at 530K. The results are summarized as follows:
(1) Creep resistance increases with increasing yttrium content.   On the other hand, initial microstructure has little effect on the minimum creep rate.
(2) The transition of the stress exponent n is observed around 150MPa.   This transition is explained neither by the break-out from the Cottrell solute atmosphere, nor by the power-law breakdown behavior.   The mechanism of this transition is to be solved.
(3) The pseudo-equilibrium phase β' precipitates are easily introduced on dislocation lines within grains.   The precipitation can be enhanced efficiently by coupling of straining and aging.
(4) The dynamic precipitation during creep is fast and significant to determine the minimum creep rate.   Therefore, the minimum creep rate hardly depends on the initial microstructure.   On the other hand, the primary stage of creep and the initial plastic strain up on loading are greatly affected by the microstructures of grain interior.   The decrease of the strain rate is significantly by introducing fine precipitates within grain interior.

156

Acknowledgements
This research has been partly supported by Grant-in-Aid for the Priority Group of Platform Science and Technology for Advanced Magnesium Alloys (Grant #11225202), from Ministry of Culture, Science and Education. Materials used in this investigation were kindly provided by TOYOTA Motor Corporation.

References

[1]   S. Koike, K. Washizu, S. Tanaka, T. Baba and K. Kikawa, SAE Technical paper, (2000), #2000-01-1117
[2]   M. O. Pekguleryuz and J. Renaud, *Magnesium Technology 2000*, TMS, (2000), p.279
[3]   T. Tsukeda, A. Maehara, K. Saito, M. Suzuki, J. Koike, K. Maruyama and H. Kubo, *Magnesium Technology 2000*, TMS, (2000), pp.395
[4]   T. Mohri, M. Mabuch, N. Saito and M. Nakamura, Mater. Sci. Eng., **A257**(1998), 287
[5]   S. Kamado, T. Ashie, Y. Oshima and Y.Kojima, Materials Science Forum, **350-351**(2000), 55
[6]   R. Armstrong, I. Codd, R. M. Douthwaite and N. J. Petch, Philos. Mag., **7**(1962), 45
[7]   H. Nagahara, K. Ohtera, K. Higashi, A. Inoue and T. Masumoto, Philos. Mag. Lett., **67**(1993), 225
[8]   W. Henning and B. L. Mordike, *Strength of Metals and Alloys,* (1985), Pergamon Press, p.803
[9]   Y. Kojima and S. Kamado, *Kinzoku,* **63**(1993), 19
[10] H. Karimzadeh, J. M. Worrall, R. Pilkington and G. W. Lorimer, *Magnesium Technology*, (1986), The Inst. of Metals, p.138
[11] B. L. Mordike, W. Henning, *Magnesium Technology*, (1986), p.54
[12] M. Ahmed, R. Pilkington, P.Lyon and G. W. Lorimer, *Magnesium Alloys and Their Applications,* (1992), *Garmisch-Parkenkirchen*, pp.251
[13] M. Suzuki, H. Sato, K. Maruyama and H. Oikawa, Mater. Sci. Eng. **A252** (1998), 248
[14] K. Gradwell: Ph. D. Thesis, University of Manchester, (1985)
[15] T. Sato, *Materia* (J. Japan inst. Matels), **38** (1999), 294
[16] M. Ahmed, G. W. Lorimer, P.Lyon and R. Pilkington, *Magnesium Alloys and Their Applications* (1992), Gramish-Parkenkirchen, p.301
[17] L. Y. Wei, G. L. Dunlop and H. Westengen : J. Mater. Sci., **31**(1996), 387
[18] T. B. Massalski, H. Okamato, P. R. Subranamian, L. Kacprzak (Eds.), *Binary Phase Diagrams,* 2nd ed., ASM, Ohio, USA, (1990), p.2566
[19] H. Karimzadeh, J. M. Worrall, R. Pilkington and G.W. Lorimer, *Magnesium Technology*, (1986), p.138
[20] M. Suzuki, K. Maruyama and H. Oikawa, *Unpublished data*
[21] R. W. Lund and W. D. Nix, Metall. Trans., **6A**(1975), 1329
[22] J. Combronde and G. Brebec: Acta Metall. 19(1971), 1393
[23] S. Fujikawa, *Kei-kinzoku* (J. Japan Inst. of Light Metals), **42**(1992), 822
[24] O. D. Sherby and P. M. Burke: Prog. Materl Sci., **13**(1967), 325
[25] A. K. Mukherjee, J. E Bird and J. E. Dorn: Trans. ASM, **62**(1969), 155

Detailed examinations of creep live improvement by high-temperature pre-creep in binary magnesium-aluminum alloys at around 0.5-0.6Tm

Hiroyuki Sato [1*], Kota Sawada [2], Kouichi Maruyama and Hiroshi Oikawa [3]
Department of Materials Science, Graduate School of Engineering,
Tohoku University, Aramaki-aoba02, Aoba-ku, Sendai, 980-8579, JAPAN

[1*] Now with Department of Machines and System Engineering, Faculty of Science and Technology, Hirosaki University, Bunkyo-3, Hirosaki, Aomori, 036-8561, JAPAN
[2] Graduate Student, Now with NRIM, Tsukuba, JAPAN
[3] Now with College of Industrial Technology, Amagasaki, JAPAN

*Corresponding Author; E-mail: g4sato@mech.hirosaki-u.ac.jp, Tel.&Fax: +81-172-39-3673

Abstract

Effects of high-temperature pre-creep on creep life in magnesium-aluminum solution hardened alloys have been investigated. Creep life, at $0.55T_m$(T_m: the absolute melting temperature) are drastically improved by pre-creep treatment, which is given at higher temperature, $0.7T_m$[1]. Although creep rate in tertiary creep is affected by the pre-creep treatment, the samples show almost the same minimum creep rate. Microstructural observations show that the suppression of linkage of inter-granular cavities at grain boundaries increase creep life and rupture strain. The significance of pre-creep treatment on creep life of the alloys is presented and the high-temperature pre-creep treatment is considered as a promising method to improve the creep life of magnesium alloys. Detailed observations of the shape of grain boundaries and growth of cavities are shown in this paper.

1. INTRODUCTION

Magnesium is important light material to produce lightweight machines and its high temperature characteristics are one of important features to be investigated to expand its applications. Magnesium-aluminum solid solutions are typical solution strengthened magnesium alloys, whose creep behavior has been investigated [1-4]. In general, ductility of hcp materials is considered smaller than that of cubic alloys due to the crystallographic features, and magnesium-aluminum alloys show smaller creep ductility in the same way. Therefore, development of methods to improve creep ductility and creep life is an important subject. Otsuka reported that pre-creep improves creep life in fcc aluminum-magnesium solution hardened alloys [5], but detailed information about effects of pre-creep is still lacking in non-cubic materials.

Creep characteristics of hcp magnesium-aluminum solid solution alloys strongly depend both on stress and on temperature. At the temperatures from about 500K to 650K(0.5-0.6Tm), power law creep behavior of the alloys are essentially similar to that of aluminum-magnesium alloys, *i.e.*, the rate-controlling mechanism of creep is understood as the dragging of solute atmosphere around a dislocation in a lower stress range, and as the climb of dislocations in a higher stress range [2-4,6]. Creep rates are essentially controlled by diffusion in this temperature range. In case of hcp metals, *i.e.*, magnesium-aluminum alloys, however, the deformation behavior changes at the temperatures about 700K and higher. At temperatures, the rate controlling mechanism is understood as the cross-slip of dislocations, and the apparent activation energy of creep is larger than that for diffusion [2,4,7]. Changes in deformation mechanisms in magnesium alloys may give a possibility of modifications of microstructure by heat-mechanical treatments.

It has been reported that pre-creep condition affects creep behavior, and the temperature of pre-creep strongly affects the creep lives [1]. Based on the previous report, pre-creep given at $0.7T_m$ anomalously improves creep life, while pre-creep given at lower temperatures reduces creep lives. At the temperatures above 700K, nonbasal slip becomes easy, so that the von-Mises' criterion is satisfied. Pre-creep given at these temperatures, where uniform deformation is possible, improves ordinary creep lives. In this report, effects of high temperature pre-creep given at the higher temperature on creep lives at 550K of Mg- (3-5)mol%Al are focused and changes of microstructure during creep are investigated in detail.

2. MATERIALS AND EXPERIMENTS

Materials used in this experiments were Mg-2.97mol%Al and Mg-4.93mol%Al solid solution alloys. These materials are designated as Mg-3mol%Al and Mg-5mol%Al. The main impurities and their concentration (mass%) were Zn:0.02, Fe:0.008 and Mn:0.006, respectively. Tensile specimens with a gauge length of 20mm and a diameter of 5mm were machined from hot-extruded rods, and were annealed for 7.2ks at 800K in an argon atmosphere. Flat surfaces were machined in some specimens and scratched after annealing to measure magnitudes of GBS. Equi-axed grains of about 0.2mm were obtained. Prior to ordinary creep tests, high temperature tensile pre-creep was conducted at T_p=700K, then air-cooled without unloading to room temperature. Ordinary tensile creep tests were conducted in the same machine in air. Samples without pre-creep treatment were also examined for comparison. Experimental procedures are schematically illustrated in figure 1. Microstructural observations were conducted in Mg-5mol%Al. Morphologies of grain boundaries were observed by OM, and dislocation microstructures were observed after ordinary creep test by TEM. SEM observations were also conducted to observe distribution of cavities initiated in ordinary creep in same specimen with and without pre-creep treatment. Prior to SEM observations, crept samples ware notched and impact stressed by Charpy test machine at room temperature.

3. RESULTS AND DISCUSSION

3.1 Creep rates, time to failure and rupture strain

Figure 2 shows typical examples of ordinary creep curves and strain rate-time curves in samples with and without pre-creep treatment [1]. When pre-creep are given at T_p=700K, time to failure, t_f, is about two times that of the sample without pre-creep. Pre-creep stress, σ_p, also affects creep life, but the effects are relatively smaller than that of pre-creep temperature. Remarkable improvements of the creep life appear in a specimen pre-crept at T_p=700K, σ_p=10MPa and ε_p=6%. Although, creep rate is not affected by pre-creep in the primary and the secondary stages, the creep rate decreases in the ternary stage. Under the conditions investigated, no weakening is observed in samples pre-crept at T_p=700K in all experimental conditions. The trends were the same and independent on solute concentration. Figure 3 shows magnitude of improvement in creep rupture strain as a function of the applied stresses in ordinary creep. The magnitude of the improvement in the creep life depends on the applied stress σ_c in ordinary creep. During ordinary creep, the creep life of pre-crept samples is two times of that of samples without pre-creep (figure 2), at the stress range between

σ_c=30MPa and 40MPa, while the improvement is about 40% at σ_c=60MPa.

3.2 Microstructural changes during pre-creep treatment

When pre-creep is applied, microstructures of the samples are changed as shown in figure 4. The shape of grain boundaries is affected by pre-creep. Grain boundaries are almost straight after annealing (a), but grain boundaries are curved by pre-creep at T_p=700K(b). With these shape changes in grain boundaries, creep life t_r and ε_r are improved as shown in figure 2 and figure 3.

3.3 Changes in Microstructural Features during Ordinary Creep

3.3.1 Cavity growth during ordinary creep

In Mg-(3-5)mol%Al solid solution alloys, creep lives are controlled by cavity growth on grain boundaries. During ordinary creep, cavitations and its linkage proceed at grain boundaries. The effects of the shape of grain boundaries are shown in figure 4(c,d), taken at the tertiary creep. Cavities at grain boundaries become larger cracks in samples without pre-creep (c), while cavities remain small with pre-creep (d) even at larger creep strain in ordinary creep. Although the cavities are nucleated on grain boundaries in both samples with and without pre-creep, the cavities maintain discrete in pre-crept specimen during ordinary creep. Growth behavior of cavities during ordinary creep is shown in figure 5. Number of cavities, total area fraction of cavities, and average area of each cavity were evaluated as a function of creep strain. While the number of cavities per unit area increases with increasing creep strain in specimen both with and without pre-creep and is not so much different with each other, the total area fraction of cavities drastically restrained in pre-crept specimen (figure 5, left). The average size of each cavity increases from primary stage of ordinary creep in both cases, but the average size in pre-crept samples evidently smaller than that of the specimen without pre-creep. Effects of pre-creep on suppression of cavity growth depends on the applied stress, and are prominent at the stress of 40MPa, where the rupture strain is increased more than 50% by pre-creep (figure 5, right). Creep life to failure and rupture strain, are controlled mainly by growth of each cavity, but not by total number of cavities. Figure 6 shows morphologies of cavities at the end of primary stage in ordinary creep. Specimens are notched after ordinary creep test, and then fractured by Charpy test machine to observe micro cavities initiated on brain boundaries. As shown on figure 5, the total number of cavities per unit area is not affected by pre-creep treatment. The average sizes of micro cavities formed on boundaries, d_{av}, are almost the same between specimen with and without pre-creep. This result also suggests that the nucleation of cavities is not affected by pre-creep treatment.

3.3.2 Effect of GBS on creep life in ordinary creep

As described in previous section, restraint of grain boundary sliding is major mechanism of creep life improvement in fcc aluminum-magnesium solid solutions [5]. In case of hcp magnesium-aluminum solid solution alloys, grain boundary sliding and its restraint play relatively miner role to the improvement of creep life. Magnitude of deviation of scratch lines is small in ordinary creep and is not affected by pre-creep. In all cases of experiments, 70 percent of observed grain boundaries show no grain boundary sliding. Figure 7 shows magnitude of grain boundary sliding after ordinary creep of boundaries which show sliding. Although the average sliding length tends to be somewhat smaller with pre-creep, the magnitude of GBS is not significant to affect cavitations to rupture.

3.3.3 Effect of pre-creep on dislocation structure in ordinary creep

The pre-creep conditions and ordinary creep conditions in Mg-Al alloys used in this experiments, corresponds to the stress range where Alloy-type creep appear [2-4]. In the deformation condition where the Alloy-type behavior appears, the rate controlling mechanism is understood as the viscous glide of a dislocation with solute atmosphere around it. When dislocations move viscously, strain-rates can be represented by the Orowan's equation, $\dot{\varepsilon} = \phi \rho_m b \overline{\upsilon}$. Here, ϕ, ρ_m and $\overline{\upsilon}$ are the geometrical factor, the mobile dislocation density, and the mean dislocation velocity which is proportional to the mean effective stress, $\overline{\tau}^*$, respectively. Under the applied stress σ, the mean internal stress $\overline{\tau}^*$, is connected with the mean internal stress, $\overline{\sigma}_i$, as $\overline{\tau}^* = \phi(\sigma - \overline{\sigma}_i)$, and the mean internal stress $\overline{\sigma}_i$ has a square-root proportionality with a dislocation density ρ, i.e., $\overline{\sigma}_i = \alpha\phi^{-1}Gb\sqrt{\rho}$. Assuming that the dislocation density ρ is the same with the mobile dislocation density ρ_m, strain rates are described by the following equation.

$$\dot{\varepsilon} = \frac{\phi^4}{\alpha^2 b} GB\left(\frac{\sigma}{G}\right)^3 (1-r)r^2 \tag{1}$$

Here, B is the mobility of dislocations, α is a numerical constant and r is the ratio of the mean internal stress to the applied stress, $r = \overline{\sigma}_i/\sigma$. Under the same applied stress, strain rate is controlled by the dislocation density. Thus, the dislocation density and mean internal stress related to the dislocation density control apparent creep rate.

Figure 8 shows dislocations observed in secondary stage of samples both with and without pre-creep treatment. Homogeneous distribution of dislocations is observed in both cases, and it is supported that the rate controlling mechanism is the viscous glide of dislocations. The shape of dislocations are almost the same in samples with and without pre-creep, and suggests that dislocations introduced in pre-creep treatment give minor effect on dislocation structure and creep rate in ordinary creep. Figure 9 also shows minor effect of pre-creep on dislocation density ρ, during secondary and ternary stage of ordinary creep. Dislocation density was measured at the strain, which show the same strain rate in both conditions. It is shown that the dislocation density takes almost the same value at the same creep rate in both secondary and tertiary stage of ordinary creep. Dislocation density takes almost the equal value at the same creep rate in specimen both with and without pre-creep. Thus, based on the glide controlled rate controlling mechanism, pre-creep treatment has no connection with dislocation structures. Differences in dislocation density and creep rate at the same creep strain in both specimens with and without pre-creep in ternary stage, are caused by difference in mean stress level of matrix caused by different growth behavior of cavities.

3.4 Long-term Strengthening by Grain Boundary Shape Control

As described above, the shape of grain boundaries affects long-term strength in magnesium aluminum binary alloys at high temperatures. Strengthening by the shape control of grain boundaries in long-term creep is one method, which is different from methods that focus on the dislocation motion. In the present investigation, the improvement of creep life at temperatures $0.5\text{-}0.6T_m$ is prominent in samples pre-crept at temperatures above $0.7T_m$. In hcp materials, in which both deformation mode and mechanisms change at around $0.7T_m$, heat-mechanical treatment above $0.7T_m$ might be one general way to modify microstructures and long-term strength of materials.

Otsuka reported that pre-creep improves creep life in fcc Al-Mg solid solutions, also [5]. It was concluded that the retardation of grain boundary sliding on serrated boundaries, which are caused by pre-creep, improves the creep life of the fcc alloy. In the present investigation, however, cavities are formed preferentially on grain boundaries perpendicular to the stress axis, and no remarkable grain boundary sliding is recognized during ordinary creep. Cavitations caused by grain boundary sliding are not dominant in magnesium solid solutions. It should be emphasized that dislocation microstructure is not affected by the high temperature pre-creep treatment. Neither creep-rates in the primary nor the steady-state creep are affected by the pre-creep. Therefore, the improvements of creep life in magnesium alloys by pre-creep are caused simply by the difficulty of the linkage of grain boundary cavities.

4. CONCLUSION

1. High temperature pre-creep given at 700K improves both creep life and rupture strain in ordinary creep of Mg-Al solid solution alloys at 550K, without any weakening effect.
2. Curved grain boundaries formed by high temperature pre-creep inhibit the linkage of grain boundary cavities, and thus resist rupture.
3. In contrast to fcc solid solutions, GBS play minor role in growth of grain boundary cavities in ordinary creep in hcp magnesium-aluminum solution hardened alloys.
4. Dislocation microstructures and rate controlling mechanisms are also not affected by pre-creep treatment.

Acknowledgments

This research has been partly supported by Grant-in-Aid for Scientific Research (Grant #07555474 and #10650685) from The Ministry of Education, Science and Culture JAPAN. Materials used were donated by Chuo-Kosan Co. Ltd., Japan and were hot extruded by Hitachi Densen Co. Ltd., Japan.

References

[1] SATO, H., Sawada, K., Maruyama, K. and Oikawa, H., Improvement of creep rupture life by high temperature pre-creep in magnesium-aluminum binary solid solutions, Mater. Sci. Eng., A, 2001, in press.
[2] VAGARALI, S. S. and LANGDON, T. G., Deformation Mechanisms in H.C.P. Metals at Elevated Temperatures. — II. Creep Behavior of Mg-0.8%Al Solid Solution Alloy, Acta Metall., 1982, 30, 1157.
[3] SATO, H. and OIKAWA H., Transition of Creep Characteristics of HCP Mg-Al Solid Solutions at 600-650K, Strength of Metals and Alloys, ICSMA 9, Haifa Israel, 1991, 463 (Freund Publishing House, London).
[4] SATO, H., KOYAMA, M., OIKAWA, H., Deformation Characteristics of HCP Mg-Al Solid Solutions at 600-750K, Science and Engineering of Light Metals, RASELM '91, Sendai Japan, 1991, 109 (Japan Institute of Light Metals).
[5] OTSUKA, M., Effect of Pre-creep on the Creep Life and Ductility of Polycrystalline Al-5.7mol%Mg Alloy, Aspects of High Temperature Deformation and Fracture in Crystalline Materials, JIMIS-7, Nagoya Japan, 1993, 221 (Japan Institute of Metals).
[6] OIKAWA, H., Creep Behavior of Simple Solid Solutions of Aluminum, Hot Deformation of Aluminum Alloys, 1991, (TMS, Warrendale), 153.
[7] YOSHINAGA, H. and HORIUCHI, R., On the Nonbasal Slip in Magnesium Crystals, Trans JIM, 1963, 5, 14.

HIGH-TEMPERATURE CREEP RESISTANCE IN MAGNESIUM ALLOYS AND THEIR COMPOSITES

M. Pahutova[1], V. Sklenicka[1], K. Kucharova[1], M. Svoboda[1] and T.G. Langdon[2]

[1]Institute of Physics of Materials, Academy of Sciences of the Czech Republic, CZ-616 62 Brno, Czech Republic
[2]Departments of Aerospace & Mechanical Engineering and Materials Science, University of Southern California, Los Angeles, CA 90089-1453, U.S.A.

ABSTRACT

A comparison between the creep characteristics of squeeze-cast magnesium alloys AZ 91 and QE 22 reinforced with 20 vol. %Al$_2$O$_3$ short fibres and unreinforced AZ 91 and QE 22 matrix alloys shows that the creep resistance of the reinforced materials is considerably improved compared to the matrix alloys. It is suggested that the creep strengthening in the short fibre composites arizes mainly from the existence of a load transfer effect and a threshold stress. By contrast, investigations of the creep behaviour of a particulate QE 22 - 15 vol. %SiC composite and its unreinforced QE 22 matrix alloy prepared by power metallurgy revealed no substantial increase in the creep strength of the composite. This unexpected result has been explained by the microstructural changes induced by the presence of particle reinforcement and by creep loading in the composite matrix.

1. INTRODUCTION

The major current growth in the use of magnesium alloys is in the high volume commercial automotive sector where the major incentive is weight saving to maximise fuel economy and minimise emissions. To achieve further substantial increases in usage in the automotive industry, magnesium alloys must be utilised in engine and transmission components. These applications require better high temperature strength and creep resistance than is possible with currently available commercial magnesium alloys. In fact, the creep resistance of magnesium alloys is rather limited at temperatures above 400 K. However, a considerable improvement in the creep properties of magnesium alloys can be potentially achieved by non-metallic reinforcement (metal matrix composites - MMCs). The most fundamental issue in the creep behaviour of these composites is a determination of the mechanism(s) by which the creep rate of the composite is reduced by reinforcing the creeping matrix with less-creeping or non-creeping short fibres or particles. When the mechanism is clarified, one can design new MMCs with higher creep resistance by tailoring constituent parameters of the matrix alloy and short fibre and/or particle phases.

To date, comparatively little research has been carried out on the creep behaviour of Mg-based fibre and/or particulate reinforced composites [1-10]. The present study was therefore initiated to perform experiments which were conducted on representative magnesium alloys (AZ 91 and QE 22) and their composites in order to compare their creep resistance. The objective of the present research is a further attempt to clarify the creep strengthening in discontinuously-reinforced magnesium alloys.

2. EXPERIMENTAL MATERIALS AND PROCEDURES

All experimental materials used in the study were fabricated at the Department of Materials Engineering and Technology, Technical University of Clausthal, Germany. Short fibre reinforced and unreinforced blocks of the most common alloy AZ 91 (Mg-9wt%Al-1wt%Zn-0.3wt%Mn) and the high strength silver-containing alloy QE 22 (Mg-2.5wt%Ag-2.0wt%Nd rich rare earths-0.6wt%Zr) were produced by squeeze casting. The fibre preform consisted of planar randomly distributed δ-alumina short fibres (Saffil fibres from ICI, 97% Al_2O_3, 3%SiO_2 ~ 3µm in diameter with varying lengths up to an estimated maximum of ~ 150 µm). The final fibre fraction after squeeze casting in both composites was about 20 volume %. For convenience, the composites are henceforth designated AZ 91-20 vol% Al_2O_3(f) and QE 22-20vol%Al_2O_3(f) where f denotes fibre. An unreinforced AZ 91 matrix alloy and its composite were subjected to a T6 heat treatment (anneal for 24h at 688K, air cool and then age for 24h at 443K). An examples of typical microstructure of the AZ 91-20vol% Al_2O_3(f) composite is shown in Fig. 1a. The QE 22 monolithic alloy and its composite were given the following T6 heat treatment: anneal for 6h at 803 K, air cooling and ageing for 8h at 477 K. For comparison, experiments were also conducted using commercial purity Mg (cp-Mg).

The SiC particle reinforced and unreinforced QE 22 alloys were fabricated by powder metallurgy [8]. The QE 22 + 15vol%SiC composite was prepared from gas-atomized metal alloy powders of various sizes (ASTM sieve sizes 320 and 600 corresponding to mean particle diameters of 30 and 10 µm, respectively) and shapes (bulky particles - BL, rounded particles - HD) of SiC particles: microstructure is shown in Fig. 1b. Both materials were investigated in an as-received state after extrusion and a T6 heat treatment.

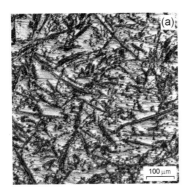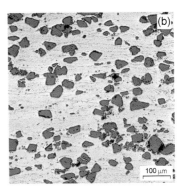

Figure 1. Optical micrographs showing microstructure of: (a) the AZ 91-20vol%Al_2O_3(f) composite, and (b) the QE 22-15vol%SiC(p) composite (320BL).

Flat tensile creep specimen having gauge lengths of 25 mm and cross-sections 3 x 3,2 mm were machined from the blocks so that the longitudinal specimen axes were parallel to the plane in which the long axes of the fibres were preferentially situated for the squeeze-cast composites or parallel to the extrusion direction for the powder metallurgy materials. Constant stress tensile creep tests were carried out at temperatures from 423 to 523 K with the testing temperature continuously monitored and maintained constant to within ±0.5 K of the desired value. The applied stresses ranged from 10 to 200 MPa. Creep tests were performed in purified argon in tensile creep testing machines with the nominal stress maintained constant to within 0.1% up to a true strain of about 0.35. The creep elongations were measured using a linear variable differential transducer and they were continuously recorded digitally and computer processed. Almost all of the specimens were run to final fracture.

Following creep testing, samples were prepared for examination by transmission electron microscopy (TEM). Observations were performed using a Philips CM 12 TEM/STEM transmission electron microscope with an operating voltage of 120 kV, equipped with EDAX Phoenix x-ray microanalyzer. Fractographic details were investigated using light microscopy and scanning electron microscopy (Philips SEM 505 microscope).

3. EXPERIMENTAL RESULTS

The process used to manufacture the composite can potentially affect the matrix microstructure in such a way as to modify the composite strengthening and thus the creep resistance. For this reason, the creep results for squeeze cast and powder metallurgy composites will be presented separately.

3.1. SQUEEZE CAST AND SHORT-FIBRE REINFORCED MATERIALS

3.1.1. Creep Results

The creep data of commercially pure magnesium (cp-Mg), the AZ 91 alloy, and the AZ 91-20vol%Al$_2$O$_3$(f) composite and the QE 22-20vol% Al$_2$O$_3$(f) composite at 473 K are shown in Fig. 2, where the minimum creep rate $\dot{\varepsilon}_m$ is plotted against the applied stress σ on a logarithmic scale.

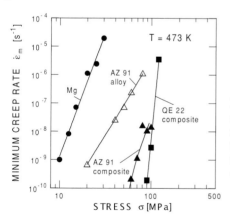

Figure 2. Stress dependences of minimum creep rates for commercially pure magnesium, the AZ 91 monolithic alloy, the AZ 91-20vol%Al$_2$O$_3$(f) composite and the QE 22-20vol%Al$_2$O$_3$(f) composite at 473 K.

Inspection of the creep data in Fig. 2 leads to three observations. First, the AZ 91 alloy exhibits better creep resistance than commercially pure magnesium. Second, the AZ 91-20vol%Al$_2$O$_3$(f) composite exhibits better creep resistance than the AZ 91 monolithic alloy over the entire stress range used in these experiments; the minimum creep rate for the composite is about two to three orders of magnitude lower than for the unreinforced alloy. Third, the creep resistance of the QE 22-20vol%Al$_2$O$_3$(f) composite seems to be essentially equal to the creep resistance of the AZ 91-20vol%Al$_2$O$_3$(f) composite. The stress dependences of the minimum creep rates for both composites and reinforced alloys are different and this is clearly demonstrated in Figs. 3a,b. While the shapes and therefore the apparent stress exponents, $n = (\partial \ln \dot{\varepsilon} / \partial \ln \sigma)_T$, for the monolithic alloys slightly decrease with decreasing applied stress, the curvatures for the composite increase with decreasing applied stress. Such an increase in the apparent stress exponent at low stresses is usually considered to be indicative of the presence of a threshold stress representing a lower limiting stress below which creep cannot occur [4, 10].

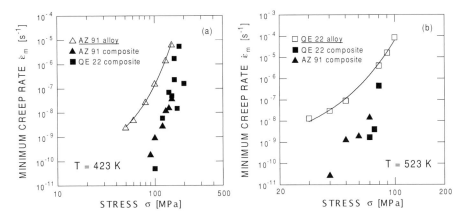

Figure 3. Stress dependences of minimum creep rates for (a) the AZ 91 alloy, the AZ 91-20vol%Al$_2$O$_3$(f) and the QE 22-20vol% Al$_2$O$_3$(f) composites at 423 K, and (b) the QE 22 alloy, the AZ 91-20vol%Al$_2$O$_3$(f) and the QE 22-20vol% Al$_2$O$_3$(f) composites at 523 K.

The double logarithmic plots of the time to fracture t_f as a function of the applied stress σ at a temperature of 473 K are shown in Fig. 4 for the same materials as in Fig. 2. It is clear from these plots that the creep life of the AZ 91-20 vol%Al$_2$O$_3$(f) composite is an order of magnitude longer than the unreinforced AZ 91 alloy. However, this difference consistently decreases with increasing applied stress and there is a tendency for the reinforcement to have no significant effect on the lifetime at the higher stresses. In fact, inspection of Fig. 4 reveals that at stresses higher than 200 MPa the creep life of the composite is essentially equal to that of the monolithic AZ 91 alloy. The same conlusion can be drawn from Fig. 5. Independent of the testing temperature, both composites (AZ 91-20vol%Al$_2$O$_3$(f) and QE 22-20vol% Al$_2$O$_3$(f)) exhibit superior lifetime compared to their unreinforced alloys. It should be noted that no substantial difference in lifetime was found between both composites under the same creep loading conditions.

The presence of the reinforcement leads to a substantial decrease in the creep plasticity [7]. The values of the strains to fracture in both composites are only 1-2%, independent of stress and temperature. By contrast, the values of the strain to fracture in the monolithic alloys are markedly higher, typically 10 - 15% in the AZ 91 alloys and up to 30% in the QE 22 alloy.

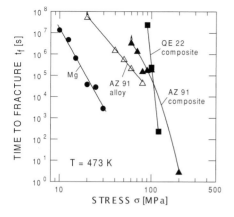

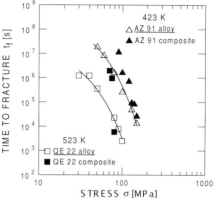

Figure 4. Stress dependences of times to fracture for commercially pure magnesium, the AZ 91 monolithic alloy, the AZ 91-20vol%Al$_2$O$_3$(f) composite and the QE 22-20vol% Al$_2$O$_3$(f) composite at 473 K.

Figure 5. Stress dependences of times to fracture for the AZ 91 alloy and the AZ 91-20vol%Al$_2$O$_3$(f) composite at 423 K, and the QE 22 alloy and QE 22-20vol% Al$_2$O$_3$(f) composite at 523 K.

3.1.2. Microstructural Investigations

The microstructure of the as-cast AZ 91 alloy consists of the β-phase (Mg$_{17}$Al$_{12}$ and/or Mg$_{17}$(Al, Zn)$_{12}$) intermetallic compounds in a matrix of a magnesium solid solution. The standard T6 heat treatment causes partial homogenization of the microstructure and at the same time prevents, at least partially, precipitation of the intergranular β-phase (Mg$_{17}$Al$_{12}$) particles. The particles of the β-phase (Mg$_{17}$Al$_{12}$) can be divided into two types: continuous precipitates of bcc Mg$_{17}$Al$_{12}$ platelets with a basal growth habit and massive discontinuous (cellular) precipitates, as can be seen in Fig. 6a. The particles of the β-phase created by continuous precipitation are responsible for age hardening whereas those grown discontinuously are detrimental to the age hardening response of the alloy. The microstructure of the AZ 91 - 20 vol%Al$_2$O$_3$(f) composite is more complex than that of the unreinforced matrix alloy and the T6 heat treatment was adjusted to minimize matrix-fibre reactions in the composite. The microstructures of the composite and its unreinforced alloy are similar with respect to the two types of β-phase (Mg$_{17}$Al$_{12}$) precipitates. At several sites, the fibres are interconnected by massive discontinuous β-phase "bridges", see Fig. 6b.

However, the most frequent morphology of the β-phase precipitates in the composite is in the form of continuous Mg$_{17}$Al$_{12}$ platelets. A detailed investigation by TEM revealed that the continuous Mg$_{17}$Al$_{12}$ platelets present in the matrix of the composite are substantially larger than in the alloy. This is attributed to the relative lack of Al in solid solution as a consequence of the pronounced precipitation of the massive β-phase between the alumina fibres. Using selected area diffraction, it was established that the precipitate/matrix crystallographic orientation relationship in the composite is identical with that for the monolithic alloy. A thicker reaction zone in contact with the fibres was identified as magnesia (MgO) particles, see Fig. 6c. An increase in the dislocation density in the composite, especially in the vicinity of the fibres, is probably caused by the thermal mismatch between the fibres and the matrix.

The significant microstructural changes observed after creep are a coarsening of the continuous β-phase and a partial dissolution of the discontinuous massive β-phase ("bridges") in the composite. This dissolution effect is probably responsible for an additional fine continuous precipitation of $Mg_{17}Al_{12}$ particles during creep of the composite.

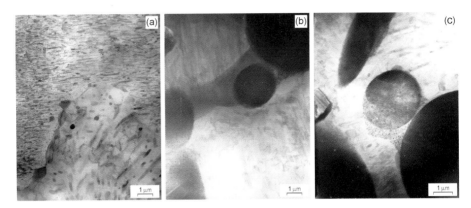

Figure 6. TEM micrographs of (a) AZ 91 alloy showing continuous and discontinuous precipitates, (b) and (c) AZ 91-20vol%Al_2O_3(f) composite showing β-phase bridges between fibres and MgO particles in the vicinity of fibres, respectively.

Detailed microstructural investigations of the squeeze cast QE 22 monolithic alloy and the QE 22 - 22vol% Al_2O_3(f) composite have been reported recently by Kiehn et al. [11]. Thus, the microstructural changes caused by the creep exposures in the monolithic QE 22 alloy and QE 22 - 20vol%Al_2O_3(f) were particularly studied. In the composite, the fibres act as nucleation centres in the precipitation process promoting, for example, precipitation of Al_2Nd or Ag compounds. The Al content in the matrix is enhanced due to the decomposition of the preform binder. The evolution of the particle population inside the grains involves the formation of Al_2Nd-like cubic particles in creep at 423 K. At temperatures above 423 K, these particles are substituted by hexagonal β-phase and/or tetragonal $Mg_{12}Nd$ precipitates. The precipitation process in the grain interiors of the unreinforced alloy is different involving only a change of the morphological features of tetragonal semicoherent $Mg_{12}Nd$ particles existing in the alloy after the T6 heat treatment. It should be noted that the preferential formation of second phase particles along the fibre-matrix interface will reduce the concentration of alloying elements in the matrix so that there is a consequent decrease in the density of obstacles to dislocation motion.

Creep behaviour and creep resistance can be substantially influenced by the development of creep damage and fracture processes. Fractographic investigations of both composites failed to reveal either substantial creep fibre cracking and breakage or any debonding at the interfaces between the fibres and the matrix due to creep. The examples of etched 20vol%Al_2O_3(f) composites after creep are shown in Fig. 7a,b.

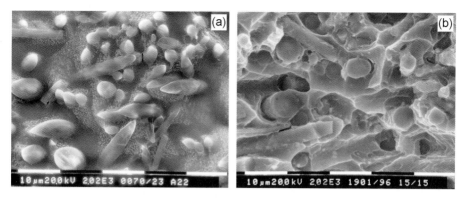

Figure 7. SEM micrographs of (a) AZ 91-20vol%Al₂O₃(f), etched metallographic section, and (b) QE 22-20vol%Al₂O₃(f), creep fracture surface.

3.2. SiC PARTICLE REINFORCED QE 22 COMPOSITE FABRICATED BY POWDER METALLURGY

Magnesium AZ 91 and QE 22 alloys reinforced with alumina short fibres exhibit an excellent creep resistance but their applications may be hindered by their high price. These problems can be avoided by reinforcing the matrix alloy with relatively inexpensive SiC and/or Al₂O₃ particulates. Such particulate-reinforced MMCs present perhaps more modest improvements in creep properties than the fibre-reinforced counterparts, but they can be formed into useful shapes using conventional metal working processes such as powder metallurgy methods and extrusion and they exhibit near-isotropic properties. Very recently, an investigation conducted on the creep behaviour of a 15 vol % silicon carbide particulate reinforced PM magnesium AZ 91 and QE 22 alloys has shown [8] that the reinforcing effect of SiC particles on the creep resistance is not uniform and depends strongly on the matrix alloy. While the creep resistance of the AZ 91 alloy was increased by the particle reinforcement, the creep resistance of the reinforced QE 22 decreased with a decrease in the particle size. The possible explanation of such an unexpected trend in the creep behaviour was a strong motivation for the following more detailed study of the QE 22 - 15 vol% SiC composite.

The creep data of the QE 22 monolithic alloy and the QE 22-15vol%SiC(p) composite are shown in Fig. 8 where the minimum creep rate is plotted against the applied stress on a logarithmic scale. Inspection of the creep data leads to two observations. First, the unreinforced QE 22 alloy exhibits better creep resistance than the composite. Second, as depicted in Fig. 8, a T6 heat treatment has a detrimental effect on the creep resistance.

The QE 22 alloy microstructure after the T6 heat treatment was found to be rather complex. Large and medium size precipitates of Mg₃(Ag, Nd), medium size rounded precipitates of α Nd (hP4-La, a = 0.3658 nm, c = 1.179 nm), rod-like precipitates of complex chemical composition including Zr and Ni, very small GP zones (Mg-Nd) and very small MgO particles were present as the main secondary phases. Two types of Mg₃(Ag, Nd) precipitates were found: large ones, 0.2 - 0.5 μm in diameter, located mostly at triple points and medium size ones, 50 - 100 nm, located both at grain boundaries and in the grain interiors, see Fig. 9a. The chemical composition of the large precipitates was determined to be 76Mg-24Ag (at.%) and the composition of the medium size precipitates was 75Mg-17Ag-8Nd (at.%). Selected area electron diffraction verified that the large precipitates are Mg₃Ag (orthorombic, $a = 1.424$, $b = 1.421$, $c = 1.466$ nm, sometimes denoted as $Mg_{54}Ag_{17}$). Only one type of the very small GP zones, responsible for precipitation hardening of the alloy, was found. This is attributed to solid solution deplection of Ag due to pronounced precipitation of Mg₃Ag.

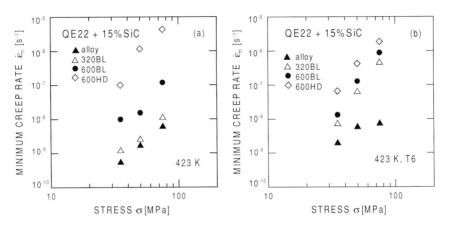

Figure 8. Stress dependences of minimum creep rates for QE 22 +15vol%SiC composite at 423 K in (a) as received state and (b) after T6 heat treatment.

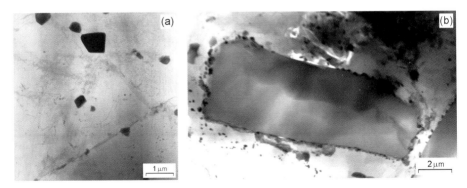

Figure 9. TEM micrographs of (a) QE 22 alloy documented various phases present, and (b) QE 22 + 15vol%SiC composite, showing precipitates at SiC/matrix interface.

The matrix of the QE 22 + 15vol%SiC composite after T6 heat treatment contained all phases present in the QE 22 alloy except that the coherent GP zones were missing. A pronounced precipitation of a Nd-rich phase occurred at SiC/matrix interfaces, as documented in Fig. 9b. It is evident that the SiC/matrix interface acts as a nucleation substrate for this phase. The lack of GP zones in the QE 22 + 15vol%SiC composite after T6 heat treatment can be attributed to matrix depletion of Nd due to precipitation of Nd-rich phases at the interfaces. During creep, the precipitates within the grains grow substantially whereas the interfacial precipitates remain unaffected or they are even partially dissolved. An EDX analysis of the particles at the interfaces has not provided fully reliable results regarding their chemical composition due to the small size of the particles compared to the foil thickness and roughness at the SiC edges.

Nevertheless, it was demonstrated that they are extensively Nd-rich. The average chemical composition of the particles with a negligible Si peak in the X-ray spectra was 91Mg - 9Nd (at.%) which fits well for $Mg_{12}Nd$. The particles with the highest measured content of Nd have a composition of 48Mg-12Nd-40Si(at%) which suggests that another Nd-rich phase forms at the SiC/matrix interface during creep.

4. DISCUSSION

Composite creep strengthening occurs by either direct or indirect mechanisms. Direct composite strengthening is due to load transfer from the matrix to the reinforcements. Load transfer is accompanied by a redistribution of stresses in the matrix and this lowers the effective stress for creep. Indirect composite strengthening occurs when the presence of the ceramic reinforcements or the process used to manufacture the composite affects the matrix microstructure in such a way as to modify the creep resistance. Potential microstructural effects include changes in dislocation arrangements, acelerated ageing, microstructural decomposition, a matrix compositional variation and/or reinforcement transformation.

Both of the squeeze-cast short-fibre reinforced composites exhibit better creep resistance than the monolithic matrix alloys. In the presence of load transfer, the creep data may be succesfuly reconciled by putting the ratio of the creep rates of the composite, $\dot{\varepsilon}_c$, and the matrix alloy, $\dot{\varepsilon}_a$, at the same loading conditions equal to a factor given by $(1-\alpha)^n$, where α is a load transfer coefficient having values lying within the range from 0 (no load transfer) to 1 (full load transfer). The values of α inferred from the data in Fig. 3 using n = 3 [4, 10] are within the range 0.7 to 0.9. Nardone and Prewo [12] suggested a modified shear-lag model by considering the load transfer effect at the end of short fibres and various reinforcement geometries and arrangements (volume fraction, aspect ratio). The values of α predicted from the modified shear-lag model with 20 vol% of short-fibre reinforcement and an experimentally observed fibre aspect ratio (diameter/length) of ~ 50 are roughly about 0.8. The predicted values are in reasonable agreement with the experimental values of α inferred from the present analysis. Thus, these results indicate that direct composite strengthening controls creep behaviour of the short-fibre magnesium composites when the matrix microstructure is constant and stable and the composites indicate good fibre/matrix interface bonding. Indirect composite strengthening in the short-fibre composites may be caused by the dispersion of fine particles in the matrix of the composites inhibiting dislocation motion and resulting in a threshold stress which increases the creep resistance [4, 10].

It should be noted that indirect reinforcement effects generally produce strengthening but they can also produce weakening. This seems to be the case for the SiC reinforced QE 22 alloy-based composites. The reduction in their creep resistance can be explained by differences in the matrix microstructure. The matrix in the composites exhibited a coarser precipitate structure and pronounced precipitation of Nd-rich phases at the SiC/matrix interface.

5. CONCLUSIONS

The creep resistance of squeeze cast AZ 91 and QE 22 magnesium alloys reinforced with 20 vol %Al_2O_3 short fibres is shown to be considerably improved compared to unreinforced matrix alloys. The direct and indirect strengthening effects of short fibre reinforcement arise mainly from (a) effective load transfer and (b) the existence of a threshold stress, respectively.

172

Direct strengthening dominates the creep behaviour of the composites due to good fibre/matrix interfacial bonding. An inferior creep resistance of the particle reinforced powder metallurgy QE 22 - 15 vol%SiC(p) composite by comparison to the monolithic alloy is explained by differences in the composite matrix microstructure.

ACKNOWLEDGEMENTS

Financial support for this work was provided by the Grant Agency of the Academy of Sciences of the Czech Republic under Grant A 2041902 and the Project ASCR K1010104, by the Grant Agency of the Czech Republic under Grant 106/99/0187 and by the National Science Foundation of the United States under Grant No. INT-9602022.

REFERENCES

[1]   Mordike B.L., Lukas P., 1997, Creep Behaviour of Magnesium Alloys Produced by Different Techniques, in Proc. 3rd Int. Magnesium Conference, ed. G.W. Lorimer, IOM, London, pp. 419-429.

[2]   Mordike B.L., Kainer K.U., Sommer B., 1997 Investigation of the Creep Behaviour of Short-Fibre Reinforced Magnesium Alloys, ibid, pp. 638-646.

[3]   Li Y., Sklenicka V., Langdon T.G., 1999, Creep Properties of an AZ 91 Magnesium-Based Composite, in Creep Behaviour of Advanced Materials for the 21st Century, eds. R.S.Mishra et al, TMS, Warrendale, pp.171-178.

[4]   Li Y., Langdon T.G., 1999, Creep Behaviour of an AZ 91 Magnesium Alloy Reinforced with Alumina Fibres, Metall. Mater. Trans. A, Vol. 30A, pp. 2059-2066.

[5]   Pahutova M., Brezina J., Kucharová K., Sklenicka V., Langdon T.G., 1999, Metallographic Investigation of Reinforcement Damage in Creep of an AZ 91 Matrix Composite, Mater. Letters, Vol. 39, pp. 179-183.

[6]   Sklenicka V., Pahutova M., Kucharova M., Svoboda M., Langdon T.G., 2000, Creep of Reinforced and Unreinforced AZ 91 Magnesium Alloy, Key Eng. Materials, Vols. 171-174, pp. 593-600.

[7]   Pahutova M., Sklenicka V., Kucharova K., Brezina J., Svoboda M., Langdon T.G., 2000, Creep Strength and Ductility of an AZ 91 Alloy and its Composite, in Magnesium 2000, ed. E. Aghion and D. Eliezer, Magnesium Res. Inst., Ltd, Beer-Sheva, Israel, pp. 285-292.

[8]   Moll F., Kainer K.U., Mordike B.L., 1998, Thermal Stability and Creep Behaviour of Particle Reinforced Magnesium Matrix Composites, in Magnesium Alloys and their Applications, eds. B.L.Mordike and K.U.Kainer, Werkstoff-Informationsgesellschaft, Frankfurt, pp. 647-652.

[9]   Svoboda M., Pahutova M., Brezina J., Sklenicka V., Moll F., 2000, Microstructure and Creep Behaviour of SiC Particulate Reinforced QE 22 Composite, in Magnesium Alloys and their Applications, ed. K.U.Kainer, Wiley-VCH Verlag GmbH, Weinheim, pp. 234-239.

[10] Sklenicka V., Pahutova M., Kucharova K., Svoboda M., Langdon T.G., 2000, Flow Mechanisms in Creep of an AZ 91 Magnesium-Based Composite, ibid, pp. 246-251.

[11] Kiehn J., Smola B., Vostrý P., Stulikova J., Kainer K.U., 1997, Microstructure Changes in Isochronally Annealed Alumina Fibre Reinforced Mg-Ag-Nd-Zr Alloy, phys. stat. sol. (a), Vol. 164, pp.709-723.

[12] Nardone V.C., Prewo K.M., 1986, On the Strength of Discontinuous Silicon Carbide Reinforced Aluminium Composites, Scripta Metall., Vol. 20, pp. 43-48.

STRENGTHENING PRECIPITATE PHASES AND CREEP BEHAVIOUR OF Mg - RARE EARTH ALLOYS

M.Svoboda, M.Pahutova, K.Kucharova and V.Sklenicka

Institute of Physics of Materials, Academy of Sciences of the Czech Republic, CZ-616 62 Brno, Czech Republic

1. ABSTRACT

It is the purpose of the present paper to report results of a preliminary characterisation of the combine effect of gadolinium and scandium on the creep strength of magnesium alloys. A comparison between the creep characteristics of binary Mg-Sc (up to 19wt%Sc) alloys and a complex Mg5.5wt%Gd0.3wt%Sc1.5wt%Mn alloy prepared by squeeze casting shows that the creep resistance of the MgGdScMn alloy is considerable improved compared to the binary Mg-Sc alloys. It is suggested that superior creep resistance of complex MgGdScMn alloy arises mainly from sequence of precipitation processes during the creep exposure. A reduction of the expensive alloying element Sc by the addition of Gd demonstrates the development potential for creep resistant magnesium alloys.

2. INTRODUCTION

The use of light weight magnesium alloy parts in the transportation industry for applications exposed to elevated temperatures has been limited by the inferior creep resistance of the standard die casting magnesium alloys. Consequently, there has been a rapid growth in interest during the last decade in the development of high-strength light magnesium alloys for elevated temperature [1-3]. It has been known that addition of rare earth (RE) elements to magnesium alloys has a beneficial effect on the creep resistance [2-5]. However, the role of individual alloying elements on the precipitation sequence and resulting creep properties is not fully understood [5-9].

Magnesium forms solid solutions with a number of RE elements and the magnesium-rich regions of the respective binary systems all show simple eutectics. As a consequence, the alloys have good casting characteristics and presence of the relatively low melting point eutectics as networks in grain boundaries tends to supress microporosity. In the as cast condition, the alloys generally have cored α grains surrounded by grain boundary networks. Ageing causes precipitation to occur within the grains and the generally good creep resistance they display is attributed both to the strengthening effect of the precipitate and to the presence of the grain boundary phases which reduce grain boundary sliding.

The most successful magnesium alloys developed to date have been those based on Mg-Y-Nd system [1,5,8], identified as WE 54 (Mg-5.1 wt%Y-3.3wt%RE-0.5wt%Zr) and WE 43 (Mg-4.0wt%Y-3.3wt%RE-0.5wt%Zr). These alloys were result of a development programme, which was initiated 15 years ago to find replacements for the high-strength silver containing alloy QE 22 (Mg-2.5wt%Ag-2.0%RE-0.6%Zr) and the lower-strength but excelent creep resistant alloy HZ 32 which contained radioactive thorium [1]. The strength of WE 54 and WE 43 alloys is achieved essentially via precipitation strengthening and both alloys maintain useful properties at temperatures as high as 573 K.

Although the WE alloys are still regarded as the state-of-the-art for high temperature magnesium alloys, there is considerable interest in other alloy possibilities. Considerable effort has been given by Mordike et al. [10-14] to evaluate the properties of Sc containing magnesium alloys. It has been found that magnesium tertiary MgScMn (up to 15wt%Sc, 1.5wt%Mn), and quaternary MgScCeMn (up to 4 wt%Ce) alloys exhibit high creep resistance at elevated temperatures and their mechanical properties are comparable to those of WE 43 alloy. A T5 heat treatment leads to the steady state creep rates of two orders of magnitude lower than that of WE 43 at 623 K [10]. This was shown to be due to the precipitation of the Mn_2Sc phase in both alloy systems and in addition of $Mg_{12}Ce$ phase in quaternary alloys. However, the high cost of Sc makes alloys with high levels of Sc impractical. Thus, work should be concentrated on alloy systems in which stable precipitates can be produced at lower Sc content. Gd, Y and Zr have been considered for this purpose to achieve a large quantity of suitable precipitations to improve creep properties using a minimum of expensive alloy element additions. These element combinations Mg-Mn-(Sc,Gd,Y,Zr) form a variety of quaternary systems.

The present study was therefore initiated to perform experiments which were conducted on binary Mg-Sc alloys and complex MgSc(Gd)Mn alloy in order to compare their creep resistance. The objective of the present research is a further attempt to clarify the role of strengthening precipitate phases and the combined effect of scandium and gadolinium in creep of mgnesium alloys.

3. EXPERIMENTAL MATERIALS AND PROCEDURES

All experimental alloys used in this study were fabricated at the Department of Mechanical Engineering and Technology, Technical University of Clausthal, Germany. The alloys were produced by squeeze casting under the following conditions: melt temperature ~ 1100 K, stamp temperature ~ 460 K, form temperature ~ 460 K, applied pressure ~ 100 MPa. The starting materials for Mg-Sc binary alloys were HP magnesium (purity of 99.9 wt%) and metallic scandium of 99.99wt%. The procedure adopted was to stir scandium into the magnesium melt. The nominal and the actual Sc contents (as determined by atomic absorption spectrometry) were as follows: MgSc4 (3.23wt.%), MgSc8 (7.42wt%), MgSc12 (11.1wt.%), MgSc14 (14.5wt%) and MgSc19 (19.1wt%). The alloys were not subjected to any heat treatment [15]. The chemical compositions of more complex magnesium alloys were the following: MgGdScMn: 4.64wt%Gd, 0.26wt%Sc, 1.53wt%Mn [6], Mg balance and QE 22: 2.5wt%Ag, 2.0wt%Nd rich rare earth, 0.6wt%Zr, Mg balance [16].

The microstructure of the MgGdScMn alloy in the as-cast state is shown in Fig.1. No macrostructural defects such as pores and lunker were revealed by optical metallography. There are two phases present, Fig. 1b, the light one with higher Gd content than the dark one. The cuboidal precipitates of pure fcc Gd were also scarcely found, see Fig. 1c. Further, the MgGdScMn alloy was subjected to a solution heat treatment by means of isothermal annealing at 773 K for 6 h. The QE 22 alloy was given the following T6 heat treatment: anneal for 6 h at 803 K, air cooling and ageing for 8 h at 477 K [16].

Flat tensile creep specimens having gauge lengths of 25 mm and cross-sections 5 x 3.2 mm were machined from the blocks. The constant stress tensile creep tests were carried out at temperature 523 K and with the testing temperature continuously monitored and maintained constant to within ± 1.0 K. The applied stress ranged from 15 to 80 MPa. The creep tests were

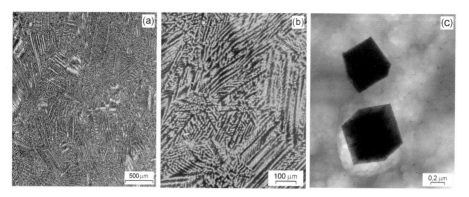

Figure 1. Microstructure in as-cast state: (a) no macroscopic defect, OM, (b) detail of dendrites, SEM, and (c) cuboids of pure fcc Gd, TEM.

performed in purified argon in tensile creep testing machines, making it possible to keep the nominal stress constant to within 0.1% up to a true strain of about 0.35. The creep elongations were measured using a linear variable different transducer and they were recorded digitally continuously and computer processed.

Following creep testing, samples were prepared for examination by transmission electron microscopy. Observations were performed using a Philips CM 12 TEM/STEM transmission electron microscope with an operating voltage of 120 kV, equipped with EDAX Phoenix X-ray microanalyser.

4. RESULTS AND DISCUSSION

Figure 2a shows standard creep curve of the MgGd5Sc0.3Mn1.5 alloy at 70 MPa in the form of strain, ε, versus time, t. The creep test was run to final fracture of the creep specimen. As demonstrated by the figure, the value of the strain to fracture for the alloy is only ~ 0.025. The alloy exhibits markedly longer tertiary stage. In fact, the shape of creep curve shown in Fig. 2a does not clearly indicate the individual stages of creep. However, this standard ε vs t curve can be easily replotted in the form of the creep rate $\dot{\varepsilon}$, versus time, t, as shown in Fig. 2b. It is apparent that curve does not exhibit a well-defined steady state. In fact, this stage is reduced to an inflection point of the $\dot{\varepsilon}$ versus t curve. The primary stage is fairly short and the minimum creep rate $\dot{\varepsilon}_m$ is reached within a period less than 100 h. Very short primary stage is followed by a lengthy tertiary stage. It is important to know at which value of the total strain a minimum creep rate condition is established. As depicted in Fig. 2c, where the data are replotted as the rate $\dot{\varepsilon}$ against strain ε, the minimum creep rate $\dot{\varepsilon}_m$ is established at a strain of about 0.006.

The creep tests were conducted by two different methods. The first method was a monothonic standard creep test at constant stress during the whole creep exposure up to final fracture of the sample (Fig. 2a). The second method was based on the stress-change technique (e.g. progressive loading during the creep of a single sample). Once a minimum creep condition was established at the given stress level, the stress was further increased till the sample

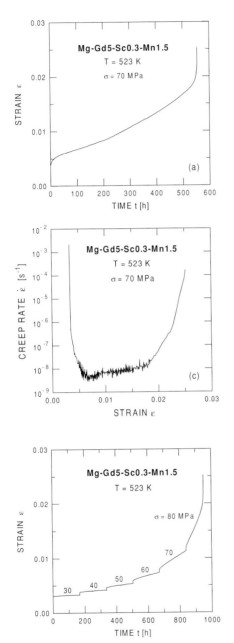

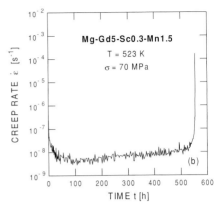

Figure 2a,b,c. Different creep curves at 523 K for the MgGdScMn alloy (constant stress creep test).

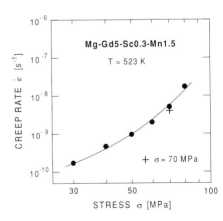

Figure 3. Creep curve at 523 K for the MgGdScMn alloy using the stress-change technique.

Figure 4. Stress dependence of minimum creep rate for the MgGdScMn alloy at 523 K (● the stress-change technigue, + constant stress creep test).

fracture - Fig. 3. It should be stressed that the values of $\dot{\varepsilon}$ inferred from both techniques were in a resonable agreement (Fig.4).

The creep data of binary Mg-Sc alloys, a complex MgGdScMn alloy and the QE 22 alloy at 523 K are shown in Fig. 5, where minimum creep rate $\dot{\varepsilon}_m$ is plotted against the applied stress σ on a logarithmic scale. To complete a set of the data the extrapolated creep results of Buch [17] on the WE 43 alloy (in as-cast state after sand-casting) were inserted in Fig. 5. Inspection of the creep data in Fig. 5 leads to two observations. First, a complex MgGdScMn alloy exhibits better creep resistance than binary MgSc alloys independent of their Sc content. However, the amount of Sc content (4-19wt%) showed a strong effect on the minimum creep rate of the Mg-Sc alloys. Second, a complex MgGdScMn alloy exhibits superior creep resistance to that of convential creep resistant QE 22 and WE 43 alloys. This demonstrates the development potential for MgGdScMn system.

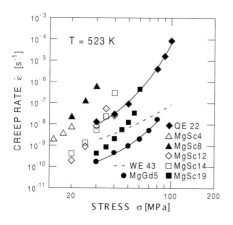

Figure 5. Stress dependences of minimum creep rate for binary Mg-Sc alloys, the QE 22 alloy [16], the WE 43 alloy [17] and complex MgGd5 alloy.

Very recently, an investigation conducted by Buch and Mordike [11] on the binary Mg-Sc system has shown that the microstructure of as-cast (squeeze-casting) alloys (the concentration range from 3 to 19wt%Sc) was inhomogeneous and consisted of β - scandium precipitates in a Mg solid solution and peritectically solidified residual Mg melt. Heat treatment experiments showed that a homogenisation of the alloys was possible at high temperature if conducted for a long time period. Subsequent ageing treatments produced no hardening, although the formation of Mg-Sc precipitates in other study was found [2,18]. The precipitates formed in <11$\bar{2}$0> directions in the basal plane and only in a low volume fraction. We can conclude that the formation of Mg-Sc precipitates in the binary Mg-Sc system at lower Sc concentrations seems to be insufficient to produce effective precipitation hardening and a substantial increase in the creep resistance (Fig.5). Thus, in order to produce Mg-Sc alloys for commercial applications, further development by the addition of one and/or two alloying elements is in progress.

Recent development based on the Mg-Sc-(Ce)Mn system indicated very promising creep results even for the alloys with Sc concentrations ranging from 6 to 9 wt% only and with 1wt%Mn. Microstructural examinations revealed Mn_2Sc precipitates on grain boundaries and in the Mg-Sc solid solution in both as-cast state and after T5 treatment. The addition of Ce led to the formation of $Mg_{12}Ce$ and $Mg_{12}Ce_2$ precipitates situated preferentially on grain boundaries due to the low solubility of Ce in magnesium. The alloys showed a strong annealing response due to the formation of Mn_2Sc, $Mg_{12}Ce$ and/or $Mg_{17}Ce_{12}$ precipitates. The low diffusivity of the alloying elements limits the overagein process.

The important demand for any creep resistant magnesium alloy to be fully competitive is its cost. Thus, further improvement of the creep properties and cost reduction of high-price Sc metal initiated a search for additional alloying elements. Gadolinium has been already shown to promote significant improving of the creep resistance of magnesium alloys [6]. At present the prices of scandium and gadolinium are $ 3500 and $ 180 per kilogramme, respectively. The precipitation of the metastable c-base centred orthorhombic (cboc) phase in Mg-Gd alloys (over 10 wt%Gd) results in considerable age hardening [19].

Figure 6 shows the microstructure of a complex MgGdScMn alloy after annealing at 773 K for 6 h. A fine dispersion of Mn_2Sc and/or MgScMn particles is inhomogeneously distributed in the interior of the grains. Coarser particles of stable Mg_5Gd phase were foud at some grain boundaries. In the very limited extent the square-shaped undissolved coarse particles of pure Gd with fcc structure were observed, Fig. 6b.

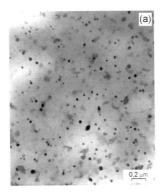 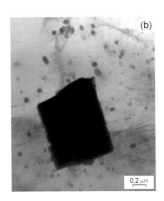

Figure 6. TEM micrographs showing microstructure of MgGdScMn alloy after annealing 6 h at 773 K: (a) roughened precipitates of Mn_2Sc and cboc phase, (b) undissolved pure fcc Gd.

The microstructure of a complex MgGdScMn alloy after creep exposure at 523 K and 70 MPa (time to fracture ~ 550h) is shown in Fig. 7. In comparison to as-anneal state (773 K/6 h) the size, morphology and number of precipitates are significantly different. The grain boundaries are decorated by two types of Mg-Gd phases, Fig. 7a. The coarse particles, appearing as gray objects with the chemical composition of Mg-15Gd (at.%) which fits well for complex structure with a large cubic cell Mg_5Gd (a = 2.23 nm, space group F-43m). This fact was further verified by means of SAED, see inlay in Fig. 7a where the diffraction patterns for B = [013] are given. The fine grain boundary particles, appearing as black ones in Fig. 7a revealed chemical composition near Mg-45Gd. Careful investigation by TEM revealed that the most frequent precipitates in the grain interiors are fine MgScMn particles and plate-like particles of the Mg_5Gd phases originated predominantly in areas of former Gd-rich phase, see Fig. 1b. The detail of tiny precipitates in the grain interior in Fig. 7b suggests that these precipitates nucleate at dislocations. The plate-like precipitates of Mg_5Gd have orientational relationship

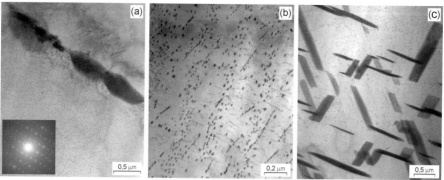

Figure 7. TEM micrographs showing main microstructural features of MgGdScMn alloy after creep: (a) two types of precipitates at grain boundaries, (b) tiny precipitates and (c) plate-like ones in the grain interiors.

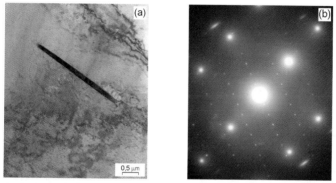

Figure 8. Orientation relationship of the plate-like Mg$_5$Gd precipitate to the α-Mg matrix in the MgGdScMn alloys after creep: (a) TEM micrograph and (b) SAED for B = [$\overline{2}4\overline{2}3$] of Mg and [102] of Mg$_5$Gd.

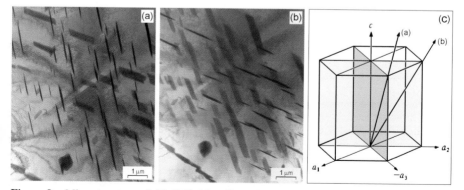

Figure 9. Microstructure of MgGdScMn alloy: (a) and (b) TEM micrographs showing Mg$_5$Gd plate-like precipitates in two different tilts of foil, marked in a sketch (c) of α-Mg hexagonal cell by the appropriate letters.

to the α-Mg matrix as documented in Fig. 8. This orientation relationship can be clearly understood from Fig. 9, where two TEM micrographs of the identical area taken by different foil tilts are shown. The directions of incident electron beam in the hexagonal cell of α-Mg for micrographs in Fig. 9a and b, respectively, are indicated in Fig. 9c. When the beam goes along the direction marked as (a) in Fig. 9c, the plate-like precipitates lying in the dark gray plane appear to be thinner than those lying in the light gray plane. After the foil tilt changes in a such a way that the beam goes along the direction marked (b) in Fig. 9c, the opposite effect in the precipitate thickness appears, as can be seen in Fig. 9b. It is assumed that the plate-like precipitates of such an orientation are able to hinder substantialy the basal slip in the alloy.

It should be noted that the creep exposure of 550 h at 523 K is considerably longer than the time of the T5 heat treatment. It is thus plausible that the precipitation sequence during the creep test involving especially prismatic precipitate phase(s) plays a decisive role in creep strengthening of MgGdScMn alloy and is responsible for superior creep properties of this alloy (Fig.5).

5. CONCLUSIONS

A comparison between the creep characteristics of binary Mg-Sc alloys and a complex MgGdScMn alloy shows that the creep resistance of the MgGdScMn alloy is considerably improved compared to the Mg-Sc alloys. It is suggested that the creep strengthening in the MgGdScMn alloys arise mainly from the precipitation sequence during creep. The results indicate that a plate-like precipitate plays a decisive role in the creep strengthening of a complex MgGdScMn alloy.

6. ACKNOWLEDGEMENTS

Financial support for this work was provided by the Grant Agency of the Czech Republic under Grant 106/99/0187 and by the Grant Agency of the Academy of Sciences of the Czech Republic under Grant A 204 1902 and the Project ASCR K 1010104. We thank Drs. Ivana Stulikova and Bohumil Smola (Faculty of Mathematics and Physics, Charles University, Prague, Czech Republic) for several helpful discussions.

REFERENCES

[1] King J.F., 2000, Development of Practical High-Temperature Magnesium Casting Alloys, in Magnesium Alloys and their Applications, ed. K.U.Kainer, Wiley-CH, DGM, Weinheim, pp. 14-22.
[2] Buch v.F., Schumann S., Aghion E., Bronfin B., Mordike B.L., Bamberger M., Eliezer D., 2000, Development of a Low-Cast, Temperature - and Creep-Resistant Magnesium Die-Casting Alloy, ibid., pp. 23-28.
[3] Pettersen K., Westengen H., Skar J.I., Videm M., Wei L.-Y., 2000, Creep Resistant Mg Alloy Development, ibid., pp. 29-34.
[4] Weiss D., Aghion E., Eliezer D., 2000, High Temperature Properties of Mg-RE-Mn Alloys, in Proc. 2nd Israeli Int. Conference on Magnesium Science and Technology, eds. E. Aghion and D. Eliezer, Magnesium Res. Institute Ltd., Beer-Sheva, Israel, pp. 181-189.

[5]   Apps P.J., Lorimer G.W., 2000, Precipitation Process in Magnesium-Heavy Rare Earth Alloys during Ageing at 300°C, in Magnesium Alloys and Their Applications, ed. K.U.Kainer, Wiley-CH, DGM, Weinheim, pp. 53-58.

[6]   Smola B., Stulikova I., Pelcova J., Buch v.F., Mordike B.L., 2000, Ageing Response of Mg-Rare Earth Alloys with Low Scandium Content, ibid., pp. 92-97.

[7]   Rokhlin L.L., Nikitina N.I., 1994, Magnesium-Gadolinium and Magnesium-Gadolinium-Yttrium Alloys, Z. Metallkd., Vol. 85, pp. 819-823.

[8]   Nie J.F., Mudde B.C., 1999, Precipitation in Magnesium Alloy WE 54 During Isothermal Ageing at 250°C, Scripta Materialia, Vol. 40, pp. 1089-1094.

[9]   Nie J.F., Muddle B.C., 2000, Characterisation and Strengthening Precipitate Phases in Mg-Y-Nd Alloy, Acta Materialia, Vol. 48, pp. 1691-1703.

[10] Mordike B.L., Buch v.F., 2000, Development of High Temperature Creep Resistant Alloys, in Magnesium Alloys and their Applications, ed. K.U.Kainer, Wiley-CH, DGM, Weinheim, pp. 35-40.

[11] Buch v.F., Mordike B.L., 1998, Microstructure, Mechanical Properties and Creep Resistance of Binary and More Complex Magnesium Scandium Alloys, in Magnesium Alloys and this Applications, eds. B.L.Mordike and K.U.Kainer, Werkstoff - Informationsgesellschaft, Frankfurt, Germany, pp. 145-150.

[12] Vostry P., Stulikova I., Smola B., Buch v.F., Mordike B.L., 1998, Resistivity Changes Due to Redistribution of Sc and Gd Atoms in Mg-Base Alloys, ibid., pp. 333-338.

[13] Stulikova I., Smola B., Pelcova J., Buch v.F., Mordike B.L., 2000, Ageing Characteristics of Sc Rich Complex Magnesium Alloys, in Proc 2nd Israeli Int. Conference on Magnesium Science and Technology, eds. E. Aghion and D.Eliezer, Magnesium Res. Institute Ltd., Beer-Sheva, Israel, pp. 218-225.

[14] Buch v.F., Leitzan J., Mordike B.L., Pisch A., Schmidt-Fetzer R., 1999, Development of Mg-Sc-Mn alloys, Mater.Sci.Eng., Vol. A 263, pp. 1-7.

[15] Buch v.F., Mordike B.L., 1998, Microstructure, Mechanical Properties and Creep Resistance of Binary Magnesium-Scandium Alloys, in Proc. 1st Israeli Int. Conference on Magnesium Science and Technology, eds. E. Aghion and D. Elizier, Magnesium Res. Institute Ltd., Beer-Sheva, Israel, pp. 163-168.

[16] Pahutova M., Sklenicka V., Kucharova K., Svoboda M., Langdon T.G., the paper of this conference.

[17] Buch v.F., 1999, Entwicklung hochkriechbestandiger Magnesiumlegierungen des Typs Mg-Sc (-X-Y), Mg-Gd und Mg-Tb, Ph.D. thesis, TU Clausthal, Germany.

[18] Vostry P., Stulikova I., Smola B., Kiehn J., Buch v.F., Mordike B.L., 1998, Application of Electrical Resistivity Measurements in the Development of New Mg-Base Alloys, in Proc 1st Israeli Int. Conference on Magnesium Science and Technology, eds. E.Aghion and D.Elizier, Magnesium Res. Institute Ltd., Beer-Sheva, Israel, pp. 88-93.

[19] Vostry P., Smola B., Stulikova I., Buch v.F., Mordike B.L., 1999, Microstructure Evolution in Isochronally Heat Treated Mg-Gd Alloys, Phys. stat. sol (a), Vol. 175, pp. 491-500.

CREEP OF ALUMINIUM I

Low Stress Creep of Aluminium at Temperatures near to T_m

K. R. McNee, H. Jones and G. W. Greenwood

Department of Engineering Materials, University of Sheffield,
Mappin Street, Sheffield, S1 3JD, UK

Abstract

The rate of deformation of relatively large grained materials under low stresses near their melting temperature is generally identified with Harper-Dorn creep but inconsistencies have been reported. The specific characteristics of this form of creep are a linear stress dependence, an activation energy close to the value for lattice self-diffusion, a low dislocation density and a creep rate independent of grain size. The deformation appears to be dominated by dislocation motion, although the means by which dislocation multiplication takes place remains a subject of debate with no mechanism gaining widespread acceptance.

High temperature creep tests on 99.99%, 99.998% and 99.999% pure aluminium have been performed for up to 800 hours at temperatures of 873 to 923K (0.94 to $0.99T_m$) and stresses of 0.025MPa to 0.21MPa using conventional tensile testpieces. Tests carried out both in air and in vacuum in highly stable and sensitive equipment allowed direct measurement of the extension of the testpieces. Accurate measurement of strain rates down to $5 \times 10^{-10} s^{-1}$ over the test periods gave values significantly less than those expected from the operation of Harper-Dorn creep.

1 INTRODUCTION

Newtonian viscous flow of material tested under creep conditions at low stresses is well documented. Some experimental evidence can only be interpreted as the operation of diffusional creep. However, a thorough examination of the low stress creep behaviour of aluminium near its melting point (T_m) by Harper and Dorn (1957) was not consistent with diffusional creep at large grain sizes, the creep rate being up to three orders of magnitude faster than expected from Nabarro-Herring creep. Good agreement with these results for aluminium was subsequently found by Barrett et al (1972), Mohammed et al (1973), and Ardell and Lee (1986). Experiments on Al+3%Mg (Murty et al, 1972) and Al+5%Mg (Yavari et al, 1982) have also shown close agreement with the Harper-Dorn results for pure aluminium, although transition to power law creep occurred at a higher value with increasing magnesium content. In addition Harper-Dorn creep behaviour has been used to describe the high temperature properties of many other metallic and non-metallic systems.

Mohammed and Ginter (1982), however, examined five distinct aluminiums with different purities and/or preparation routes and found that some demonstrated creep consistent with, albeit slightly slower than, the expected Harper-Dorn creep whereas three of their materials showed no evidence of any transition to Newtonian creep at low stresses. Burton (1972) also examined the behaviour of aluminium at low stresses near T_m and found a grain size dependence of creep rate such that at the large grain sizes normally associated with Harper-Dorn creep behaviour, no creep was detected while at much smaller grain sizes the creep rate was found to be faster than Harper-Dorn creep, after incorporation of a threshold stress related to the strength of the oxide film.

Most recently Blum and Maier (1999) further examined the creep behaviour near T_m and found no Harper-Dorn creep in their one test at a stress below the transition stress to power-law creep. The characteristics of Harper-Dorn creep have been summarised by Yavari et al (1982) and in more detail by Nabarro (1989), but the mechanism together with an explanation of what is completely suppressing Harper-Dorn creep under certain conditions remains in dispute.

The present work set out to re-examine the low stress creep behaviour of aluminium with a range of purities near T_m, with the intention of examining the surface of crept specimens for any indication of the deformation mechanism, together with establishing what, if any, conditions can give rise to Harper-Dorn creep.

2 EXPERIMENTAL PROCEDURE

Experiments were carried out initially on commercially available 99.999% pure aluminium. During the course of the work 99.99% and 99.998% purity material was also obtained. The materials were supplied in cold worked sheet form and testpieces of the dimensions shown in Figure 1 were made. In addition some 99.999% material was gas melted and recast and some testpieces made. A small amount of iron (0.04±0.01%) was subsequently detected in this material attributable to partial dissolution of the thermocouple sheath during melting.

The majority of the tests were performed in apparatus originally designed for testing coils (coil rig) (Eaton et al, 1993) which allowed testing in both air and vacuum ($<10^{-5}$torr), but some additional tests were also carried out in a much larger vertical tube furnace in an arrangement similar to that used by Harper and Dorn (1957). Samples were loaded and unloaded by the addition and removal of weights during testing. Specimen extension was measured by linear variable differential transducers (LVDTs) of an unguided armature type such that they give rise to negligible resistance to travel. The general arrangement for testing is shown in Figure 2. The load application and transducer assemblies subjected the material during heating to a small stress of magnitude 0.025MPa or 0.036MPa in the coil rig and 0.036MPa in the larger rig. Prior to testing, specimen faces were polished to a 1µm finish with diamond paste. The majority of the samples were annealed at the test temperature in the test rig, the others being annealed at 913K prior to insertion in the rig. Temperatures are considered to be accurate to ±5K and the temperature remained stable over each test period to ±1K.

3 RESULTS

3.1 Creep data

The mean grain size of all the plate samples after annealing was ~ 5mm, such that the grains were through the full 2mm thickness with some grains also extending across the 5mm width of the gauge length. The range of test conditions and their associated minimum creep rates are seen in Table 1. Figure 3 shows a creep curve for a test in vacuum at 913K with an applied stress of 0.1MPa. This curve is typical of the majority of creep tests carried out in the present work in both air and vacuum. No steady state was achieved and the creep rate is considerably slower than Harper-Dorn creep after a strain of 0.002 (0.2%). After 0.3% strain the creep rate has fallen below $1\times10^{-9}s^{-1}$ compared with the rate of $\sim4\times10^{-8}s^{-1}$ expected from Harper-Dorn creep. In one test at 873K with an applied stress of 0.1MPa steady state creep did occur and a strain of 1% was achieved at a rate of $6\times10^{-9}s^{-1}$ (Figure 4). This isolated result could not be repeated at the same stress level although a strain of 1.3% was eventually reached by heating a specimen to 913K under an applied stress of 0.12MPa. No data logging facility was available on this test but the average strain rate was calculated as $1\times10^{-8}s^{-1}$. Figure 5 compares the results of Harper and Dorn (1957), Barrett et al (1972) and Blum and Maier (1999) for 901 to 925K together with minimum creep rates measured in the present work at 873 to 913K. Fewer points are plotted below a stress of 0.1MPa as the minimum creep rates were generally too low for the sensitivity of our equipment (i.e. less than $1\times10^{-10}s^{-1}$). The minimum creep rates in the present work were not well defined due to the lack of a steady state with the vast majority of tests tending towards a creep rate close to the limit of sensitivity of the equipment after a small strain (<0.5%). Because of the lack of a steady state creep and the considerable scatter in results no clear stress or temperature dependence could be found. In fact following a step increase of the temperature on some samples, the creep rate was found to fall.

At stresses above 0.1MPa, the relevant reported transition stress from Harper-Dorn to power law creep at 920K (Harper and Dorn, 1957 and Barrett et al, 1972), the creep rates in the present work were also found to be slower than those reported previously, although a steady state was sometimes achievable at stresses between 0.1 and 0.2MPa. Above 0.2MPa, failure of the testpiece was found to occur at low (<1%) strains. Figure 6 shows the creep curve for a test at successive stress levels of 0.036, 0.16 and 0.21MPa. After a short duration (14h) at 0.21MPa the material necked to failure.

Pre-annealing, prior cold working, and prior straining above the transition stress were all ineffective in generating Harper-Dorn creep behaviour; in fact no significant strain was achieved in any test below 0.1MPa. No significant differences in mechanical behaviour were detected between the three different purities of material examined or the remelted material.

3.2   Specimen surface observations after testing

Upon removal from testing in air at temperatures between 873 and 923K, the testpieces (with one exception) were found to have remained shiny. Grain boundaries were clearly visible and some evidence of grain boundary migration could be seen. On the samples strained to 1% at 0.1MPa and 1.3% at 0.12MPa slip bands could be seen with the naked eye in most of the grains over the entire grain surface. At higher stresses slip bands were only found on one or two grains and were contained within a parallel sided band within the grains.

Figures 7a to c show SEM images of slip bands where the material has necked to failure giving rise to a symmetrical razor blade type appearance on both fracture surfaces. This failure occurred within one grain and was not associated with any grain boundaries. Figures 8a to c show slip bands and cross slip within one grain on a testpiece which was unloaded before failure occurred. The remelted impure testpiece did not remain shiny during testing and no grain boundaries were visible after the test.

3.3   Surface grain and defect structure

The following samples were repolished and etched in 45%HCl, 15% HNO$_3$, 10%HF and 30%H$_2$O:
1. As received cold worked 99.999% Al.
2. Testpiece after testing at 923K.
3. Testpiece from recast material.
4. Annealed sample left for several months after polishing.

The etching procedure clearly showed the grain structure for all the samples and in addition gave rise to etch pits in 3 and 4. No etch pits were visible in 1 and 2. No detectable difference in grain size or shape was found between the chemical etch and the thermal etching which occurred during creep testing.

4   DISCUSSION

It is clear from the results of the present work together with the results of Burton (1972), Mohammed and Ginter (1982) and Blum and Maier (1999) that so-called Harper-Dorn creep does not always arise in what appears to be nominally the same material as that upon which the characteristics of Harper-Dorn creep have been based.

A number of workers have measured dislocation densities using etch pits which gave rise to Mohammed and Ginter's proposal that the occurrence of Harper-Dorn creep was dependent on dislocation density such that Harper-Dorn creep does not occur at a dislocation density(ρ) greater than $\sim 2 \times 10^9 \text{m}^{-2}$. As noted by Nabarro (1999), the micrographs given in some papers do not agree with the densities tabulated. In fact three distinct problems can be identified with the dislocation densities reported to be associated with Harper-Dorn creep.

(i)   Reported values from etch pit measurement

The values given by Yavari et al (1982) of $\sim 5 \times 10^7 \text{m}^{-2}$ were subsequently revised to $\sim 5 \times 10^9 \text{m}^{-2}$ by Owen and Langdon (1996). A similar increase by two orders of magnitude to $8 \times 10^9 \text{m}^{-2}$ at a stress of 0.07MPa, would appear to be appropriate for the dislocation densities reported by Barrett et al (1972), based on examination of the one micrograph provided.

(ii)   Reliability of the etch pit technique

Due to the low dislocation densities involved in creep of a material near its melting point, etch pit counting should provide the most viable means of measurement. Meyrick (1989), however, has questioned the reliability of this technique for aluminium since it has been shown to be very sensitive to the time elapsed between polishing and etching, small levels of iron impurity, and to grain orientation. Consequently the dislocation density derived using this technique is likely to be an underestimate. While the etching of samples in the present work was carried out primarily to confirm grain sizes (and therefore not with the etchant composition optimised for etch pitting), it is noteworthy that only the sample with iron impurity and the sample that had undergone prolonged oxidation gave rise to etch pits.

(iii)   Comparison of TEM observations and etch pit optical micrographs.

Mohammed and Ginter (1982) and Yavari and Langdon (1982) give dislocation densities using the etch pit technique together with TEM images to show dislocation structure. The etch pits in Yavari and Langdon indicate a dislocation spacing of ~15μm (ρ~5x10^{9}m^{-2}), whereas the TEM images indicate a spacing of <<1μm (ρ>10^{13}m^{-2}). The TEM images of deformed material in Mohammed and Ginter (1982) appear to show a dislocation spacing of <<1μm (ρ>10^{13}m^{-2}) whereas the etch pits indicate a dislocation spacing of ~70μm (ρ~ 2x10^{8}m^{-2}). The extent to which the dislocations seen in the TEM images were introduced during sample preparation of such a soft material is not clear.

The dislocation density during Harper-Dorn creep is therefore much less clearly defined than suggested by Mohammed and Ginter. Although it remains possible that the diverse behaviour of apparently identical materials could be related to dislocation density or structure, the relevant dislocation structure has yet to be reliably determined and there is as yet no clear indication of a discrete value of dislocation density applicable to Harper-Dorn creep.

In the present work every attempt was made to reproduce Harper-Dorn creep behaviour in aluminium, including testing three grades of aluminium, using two different test arrangements, and a variety of prior heat treatments and mechanical treatments. No detectable steady state strain rate was found at stresses below 0.1MPa. At 0.1MPa and above steady state creep did occur in some tests and a strain of over 1% was achievable, but the creep rates were generally lower than those previously reported. Attempts to determine the stress dependence of creep at stresses above 0.1MPa were further hindered by the tendency of the material to fail by localised slip at low strains (Figures 7a-c). The spacing between slip bands can be seen to be >10μm in both Figures 7 and 8 indicating widely spaced active slip systems. No evidence for this single crystal type failure was given in the earlier reports on the creep behaviour of aluminium near its melting point. It is thought that such failure may be an indication that the operative dislocation density may have been lower in the present work than in these previous studies. Of additional interest is the observation that the creep rate sometimes fell when the test temperature was increased, which may also indicate that the dislocation density was falling to a value too low for Harper-Dorn creep to occur in the present work.

Earlier evidence that Harper-Dorn creep can occur is compelling although the rate controlling mechanism, together with the necessary microstructural conditions, remains to be established and the creep rate it can sustain remains limited. Its rate in aluminium at 0.1MPa and 923K,when it is reported to occur, is ~5x10^{-8}s^{-1}. Since Harper-Dorn creep has been found to be dependent only on stress and temperature this should be regarded as the fastest deformation that Harper-Dorn creep can sustain in aluminium. In an aluminium-magnesium alloy the transition stress to power law creep is closer to 0.5MPa while the maximum temperature would be reduced to 876K and might be expected to lead to a maximum creep rate of ~1x10^{-7}s^{-1}. Such values would be two orders of magnitude faster than Nabarro-

190

Herring creep in the extreme conditions of high temperature and a large grain size. The two creep modes would then give the same creep rate when the average grain size is reduced to 400µm. The grain size dependence of diffusional creep ultimately allows considerably faster creep rates than can be achieved by Harper-Dorn creep in most polycrystalline materials where the grain size is below this value.

5 CONCLUSIONS

A reinvestigation of low stress creep of large grained aluminium near its melting point has shown strain rates that generally decreased with strain and that were substantially lower than expected for Harper-Dorn creep. Similar results were obtained for four different purities of material subject to a variety of prior thermal and mechanical treatments. The evidence for an earlier proposal that Harper-Dorn creep requires a dislocation density below a critical value has been re-examined and it is concluded that the dislocation density or structure associated with Harper-Dorn creep has yet to be reliably determined, and its mechanism remains to be established.

Acknowledgement.

This work is supported by a UK EPSRC Research Grant Award (GR/N00296).

REFERENCES

Ardell, A.J. and Lee, S.S, 1986, "*A Dislocation Network Theory of Harper-Dorn Creep-I. Steady State Creep of Monocrystalline Al,*" Acta Metall., **34**, 2411-2423.

Barrett, C.R., Muehleisen, E.C. and Nix, W.D., 1972, "*High Temperature-Low Stress Creep of Aluminium and Al-0.5Fe,*" Mater. Sci. Eng., **10**, 33-42.

Blum, W. and Maier, W, 1999, "*Harper-Dorn Creep – a Myth?*" Phys.Stat.Sol.(a), **171**, 467-474.

Burton, B, 1972, "*The Low Stress Creep of Aluminium near to the Melting point: the influence of Oxidation and Substructural Changes,*" Phil. Mag., **25**, 645-659.

Eaton, T.W., Greenwood, G.W. and Knight, D.T., 1993, "*Equipment to Determine Some Mechanical and Physical Properties of Materials at Elevated Temperatures,*" Eur. J. Phys., **14**, 234-242.

Harper, J. and Dorn, J.E., 1957, "*Viscous Creep of Aluminum near its Melting Temperature,*" Acta Metall., **5**, 654-665.

Mohamed, F.A., Murty, K.L. and Morris, J.W., 1973, "*Harper-Dorn Creep in Al, Pb, and Sn,*" Metall. Trans., **4**, 935-940

Mohamed, F.A. and Ginter, T.J., 1982, "*On the Identification of Creep Processes at Low Stresses,*" J.Mater. Sci., **16**, 2890-2896.

Murty, K.L., Mohamed, F.A. and Dorn, J.E., 1972, *"Viscous Glide, Dislocation Climb and Newtonian Viscous Deformation Mechanisms of High Temperature Creep in Al-3Mg,"* Acta Metall., **20**, 1009-1018.

Nabarro, F.R.N., 1989, *"The Mechanism of Harper-Dorn Creep,"* Acta Metall., **37**, 2217-2222.

Owen, D.M. and Langdon, T.G., 1996, *"Low Stress Creep Behaviour: An Examination of Nabarro-Herring and Harper-Dorn Creep,"* T.G., Mater. Sci. Eng. A., **216**, 20-29.

Yavari, P, Miller, D.A., and Langdon, T.G., 1982, *"An Investigation of Harper-Dorn Creep, 1-Mechanical and Microstructural Characteristics,"* Acta Metall, **30**, 871-879

Table 1 – Full extent of testing and minimum creep rates for all tests on 99.999% (testpieces 1-12), 99.998% (4n8_) and 99.99% (4n) aluminium.

Testpiece 1	Temp °C	Stress MPa	Strain rate s^{-1}	Strain
1A VAC	640	0.025	nil	nil
VAC	640	0.036	nil	nil
1B VAC	640	0.025	1×10^{-10}	1×10^{-5}
VAC	640	0.1	1.2×10^{-9}	1.8×10^{-3}
1C VAC	640	0.025	2×10^{-10}	1×10^{-5}
VAC	640	0.12	6×10^{-10}	1.7×10^{-3}
1D AIR	610	0.025	Nil	nil
AIR	610	0.12	2.8×10^{-9}	4×10^{-3}
1E Cold worked - Heated under load				
AIR	635	0.1	3×10^{-10}	1.3×10^{-3}

Testpiece 4	Temp °C	Stress MPa	Strain rate s^{-1}	Strain
4 AIR	600	0.025	nil	nil
AIR	600	0.1	6×10^{-9}	8×10^{-3}
4B AIR	630	0.025	nil	nil
AIR	630	0.1	1.3×10^{-9}	1.7×10^{-3}
VAC	630	0.1	5×10^{-10}	9×10^{-4}
AIR	630	0.1	3×10^{-10}	3×10^{-4}
4C AIR	640	0.036	-	0.0003
AIR	640	0.16	4×10^{-10}	2×10^{-4}
AIR	640	0.21	1.3×10^{-8}	1×10^{-3}
AIR	640	0.16	nil	nil
AIR	640	0.21	1.7×10^{-8}	4×10^{-4}

Testpiece 5	Temp °C	Stress MPa	Strain rate s^{-1}	Strain
5 VAC	640	0.025	-	nil
VAC	640	0.1	7×10^{-10}	3.3×10^{-3}
5a AIR	610	0.025	5×10^{-10}	2×10^{-4}
AIR	610	0.1	1×10^{-9}	6×10^{-4}
Taken back to room temperature				
VAC	640	0.025	6×10^{-10}	2×10^{-4}
VAC	640	0.1	1×10^{-9}	6×10^{-4}
AIR	610	0.1	9×10^{-10}	6×10^{-4}

Testpiece 6	Temp °C	Stress MPa	Strain rate s^{-1}	Strain
No data-logging – Test in large furnace rig				
6 AIR	640	0.11	9×10^{-9}	1.3×10^{-1}

Testpiece 8	Temp °C	Stress MPa	Strain rate s^{-1}	Strain
AIR	610	0.025	6×10^{-10}	1.5×10^{-4}
AIR	610	0.1	7×10^{-9}	1×10^{-3}
AIR	630	0.1	9.7×10^{-9}	1.3×10^{-3}
VAC	630	0.1	1.9×10^{-9}	1.5×10^{-3}
AIR	630	0.1	1×10^{-9}	6×10^{-4}
Taken back to room temperature				
AIR	630	0.1	1.5×10^{-9}	1.3×10^{-3}
8b AIR repolished	630	0.025	nil	nil
AIR	630	0.1	3×10^{-9}	2.8×10^{-3}

Testpiece 9	Temp °C	Stress MPa	Strain rate s^{-1}	Strain
Heated prior to testing at 6°C/min to 640°C in air with 30 min hold and then furnace cooled to room temp. NB noisy data				
AIR	620	0.025	-	nil
AIR	620	0.12	8.8×10^{-9}	2×10^{-3}
AIR	610	0.12	5.2×10^{-9}	2×10^{-3}
AIR	630	0.12	2.5×10^{-9}	1×10^{-3}
AIR	610	0.12	2.5×10^{-9}	1×10^{-3}
AIR	610	0.12	1.6×10^{-9}	5×10^{-4}

Testpiece 10	Temp °C	Stress MPa	Strain rate s^{-1}	Strain
Heat treated prior to testing. Placed in furnace at 640°C in air with 30 minute hold and then furnace cooled.				
AIR	610	0.025	nil	0.0001
AIR	610	0.12	9.4×10^{-9}	1.4×10^{-3}

Testpiece 11	Temp °C	Stress MPa	Strain rate s^{-1}	Strain
AIR	610	0.025	-	-
AIR	610	0.12	2.7×10^{-9}	1.5×10^{-3}
AIR	630	0.12	1×10^{-9}	5×10^{-3}
AIR	630	0.025	Nil	nil
AIR	630	0.1	3.7×10^{-10}	3×10^{-4}

Testpiece 12	Temp °C	Stress MPa	Strain rate s^{-1}	Strain
Heat treated for 30 hours at 600°C in Ar-5%H_2 prior to testing.				
AIR	630	0.025	nil	nil
AIR	630	0.1	2.5×10^{-10}	1.6×10^{-3}

Testpiece 4n8_1	Temp °C	Stress MPa	Strain rate s^{-1}	Strain
AIR	630	0.025	1.1×10^{-9}	1.5×10^{-3}
AIR	630	0.1	1×10^{-10}	1.3×10^{-3}
VAC	630	0.1	4×10^{-10}	7×10^{-4}
VAC	630	0.025	Nil	nil
VAC	645	0.1	1.4×10^{-9}	2.5×10^{-4}
VAC	650	0.1	2.9×10^{-10}	1.3×10^{-4}
VAC	655	0.1	5×10^{-10}	1.5×10^{-4}
AIR	650	0.1	9×10^{-10}	3×10^{-4}

Testpiece 4n1	Temp °C	Stress MPa	Strain rate s^{-1}	Strain
AIR	630	0.037	5.6×10^{-10}	1.5×10^{-4}
AIR	630	0.1	3×10^{-9}	9×10^{-4}
AIR	630	0.15	2.3×10^{-9}	2.5×10^{-4}
AIR	630	0.037	9.4×10^{-10}	5×10^{-05}
AIR	630	0.16	9.4×10^{-9}	9×10^{-4}
AIR	630	0.21	2.8×10^{-8}	failed

Testpiece 4n3	Temp °C	Stress MPa	Strain rate s^{-1}	Strain
Test in large furnace rig				
AIR	600	0.037	Nil	Nil
AIR	600	0.08	Nil	Nil
AIR	600	0.13	5×10^{-10}	1×10^{-3}
AIR	610	0.13	1.7×10^{-9}	4×10^{-4}
AIR	630	0.13	2.2×10^{-9}	1×10^{-3}
AIR	610	0.13	4×10^{-10}	6×10^{-4}

Testpiece 4n4	Temp °C	Stress MPa	Strain rate s^{-1}	Strain
AIR	610	0.038	Nil	0.0003
AIR	610	0.083	Nil	0.0013
AIR	610	0.126	3.8×10^{-8}	1.6×10^{-3}
AIR	610	0.182	1.58×10^{-8}	1.5×10^{-3}
AIR	610	0.219	6.9×10^{-8}	4.8×10^{-3}
AIR	610	0.126	nil	nil

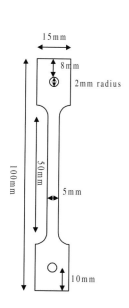

Figure 1 - Testpiece dimensions.

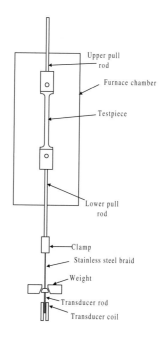

Figure 2 – General arrangement of test rigs.

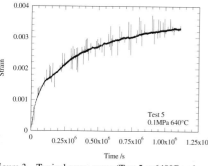

Figure 3 – Typical creep curve (Test 5 at 640°C and
0.1MPa) for 99.999% aluminium in vacuum
showing primary nature of creep with a final creep
rate of $7x10^{-10}s^{-1}$. No steady state was achieved.

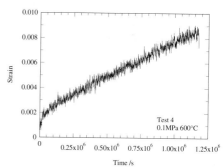

Figure 4 – Creep curve for testpiece 4 at 600°C and
0.1MPa in air showing a steady state creep at a rate
of $6x10^{-9}s^{-1}$ up to a strain of 0.8%.

194

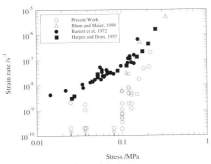

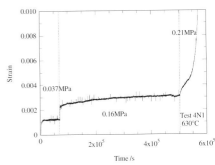

Figure 5 – Comparison of the low stress creep behaviour of aluminium in the present work at 600 to 640°C with results previously reported for 577 to 652°C. Results have been normalised to 920K (647°C) using an activation energy for creep of 149kJmol[-1] (Harper and Dorn, 1957).

Figure 6 – Creep curve for 99.99% aluminium at 630°C showing rapid acceleration of creep rate to failure after a short duration at 0.21MPa.

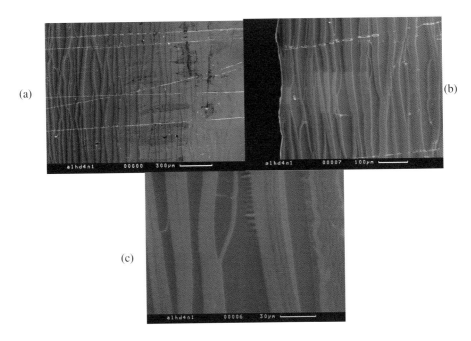

(a)

(b)

(c)

Figure 7 - SEM micrographs of test 4n1 on 99.99% aluminium showing the surface of the necked region after the testpiece failed at 0.21MPa and 630°C. The width of the slip bands increases as the failure tip is approached (a) and are much wider near the tip (b). Slip steps can be seen within some of these bands(c).

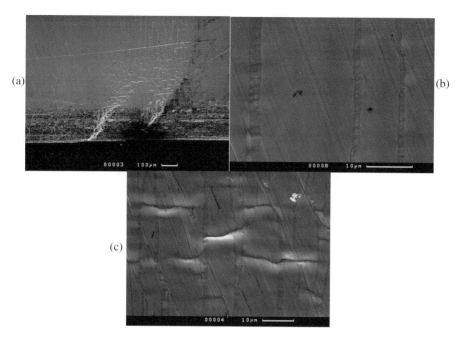

Figure 8 - SEM micrographs of test 4 at 0.21MPa and 630°C on 99.999% aluminium showing the surface of the necked region of the testpiece where localised deformation has taken place. The deformation has been restricted on the bottom edge shown in (a) such that on one side the deformation has given rise to parallel slip steps (b) whereas nearer to the bottom edge (c) evidence of cross slip is also apparent.

CORRELATION BETWEEN CREEP PROPERTIES AND TENSILE TEST RESULTS ON AGED SPECIMENS OF A 6061 ALUMINIUM ALLOY

H.D. Chandler and J-N. Goertzen

School of Mechanical Engineering

University of the Witwatersrand, Johannesburg

PO 2050 WITS

South Africa

Abstract

For many applications, it may be useful to estimate creep properties of an alloy from simpler tests such as tensile tests. Most alloys used for creep service conditions are in a hardened condition and thus creep failure tends to be due to degradation of the microstructure. This paper presents some preliminary results on correlating creep curves and the time to rupture of an aluminium alloy with calculated results based on the softening characteristics of the material.

1. Introduction

Creep testing is usually a time consuming and costly exercise and can generally only be undertaken by specialist laboratories. For many applications, such testing may not be feasible or cost effective and it may be desirable to obtain a guide to creep behaviour using cheaper and more readily available methods. An obvious candidate is, of course, tensile testing. However, there is a danger that the dominant deformation mechanism being measured in the tensile test is different from that occurring under creep conditions [1]. To be confident of its validity, therefore, tensile testing has to be carried out at rates commensurate with those expected under creep conditions, which defeats the object of the exercise [2].

Most alloys used for creep resistant applications are in a hardened form. Creep behaviour is dominated by the tertiary deformation phase in which the accelerating deformation is due to micro-structural degradation and, particularly in the later stages, by void growth and coalescence. Whether the failure life is governed by micro-structural degradation or by void growth and coalescence is perhaps an open question and is dependent on the particular material under consideration. In a metal such as pure copper, the fracture ductility is stress dependent and considerable macroscopic cracking is visible in regions adjacent to the fracture surface and therefore failure seems to be governed by void growth and coalescence. In other materials (including the aluminium alloy, which is the subject of the present investigation), the fracture ductility is fairly constant for a given temperature and the fracture surface relatively smooth with little evidence of macro cracking associated with the vicinity of the fracture surface. In such cases, micro-structural degradation probably governs fracture behaviour and fracture occurs due to exhaustion of the intrinsic ductility of the material.

For the latter type of material, it should be possible to estimate creep behaviour using a suitable deformation equation which contains terms representing the structural state of the material, together with information on the way the structural state changes with time. To determine the latter, long term ageing tests would need to be performed. However, much simpler equipment would be required since a number of specimens could be accommodated in a single furnace, removed at intervals, and subjected to tensile testing. In this paper, some preliminary results are presented on correlating creep test results with tensile data obtained from aged specimens of a 6061 aluminium alloy.

2. Materials and method

The material used was an aluminium alloy 6061 with chemical analysis shown in Table I.

Table I
Chemical analysis of 6061 alloy

Element	Mg	Si	Cu	Fe	Mn	Zn	Al
Amount (wt.%)	0.74	0.61	0.42	0.16	0.29	0.03	Bal.

The material was heat treated to the T6 condition and machined into threaded end specimens having a gauge length of 50 mm and a gauge length diameter of 5 mm.. Tensile testing was carried out at several temperatures and a number of strain rates to obtain values of activation energy and initial structure parameter. These temperatures had to be relatively low to ensure that ageing effects were negligible during the warm-up and test periods. A number of specimens were then aged for different of times at several temperatures. Tensile tests were carried out on these samples to obtain the yield stress and this was used to evaluate the time and temperature dependence of the structure parameter. In parallel to these tests, a number of creep tests were performed under constant force conditions.

3. Analysis and results

The kinetic equation used was proposed to describe deformation where the dominant mechanism for deformation is obstacle-controlled glide [1,3]. The equivalent shear strain rate, $\dot{\gamma}$, is expressed in terms of the equivalent shear stress normalised with respect to the shear modulus, τ, an activation energy, ΔF, and a normalised stress term representing the structural state of the material, $\hat{\tau}$, (often referred to as a 0 K flow stress) as:

$$\dot{\gamma} = v_0 \tau^2 \exp\left[-\frac{\Delta F}{kT}\left(1 - \frac{\tau}{\hat{\tau}}\right)\right] \tag{1}$$

where v_0 is a frequency factor which is often taken to be $10^{11} \sec^{-1}$ [1] with k and T having their usual meanings. To find ΔF and the initial value of the structure parameter $\hat{\tau}_0$ from tensile yield stresses obtained at different temperatures and strain rates, τ_y, Eq. (1) was re-arranged to:

$$\tau_y = \hat{\tau}_0\left[1 - \frac{kT}{\Delta F}\ln\left(\frac{v_0\tau_y^2}{\dot{\gamma}}\right)\right] \tag{2}$$

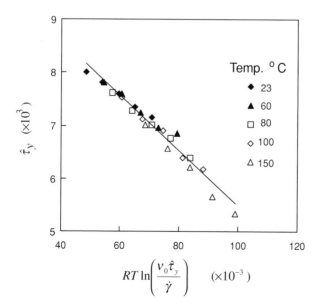

Fig. 1 Tensile yield values of the T6 material at different strain rates and temperatures plotted in terms of Eq. (2).

from which a plot of τ_y vs. $kT \ln(v_0 \tau_y^2 / \dot{\gamma})$ should yield a slope of $-\hat{\tau}_0 / \Delta F$ and intercept $\hat{\tau}_0$ if linear. Results of yield stresses from tensile tests on the T6 material are shown in Fig.1 for several temperatures and axial strain rates ranging from $10^{-1} \sec^{-1}$ to $10^{-5} \sec^{-1}$. Points lie reasonably close to a straight line and indicate values of ΔF corresponding to $0.57 \mu b^3$ (μ is the shear modulus and b the magnitude of the Burgers vector) and $\hat{\tau}_0 = 0.0107$.

The kinetics of the softening process was modelled using standard expressions for precipitate coarsening by the Ostwald ripening process. The precipitate growth rate can be expressed as [4]:

$$\frac{dr}{dt} = \frac{A'}{RT} \exp\left(-\frac{Q}{RT} \right) \tag{3}$$

where Q is the activation energy for diffusion of the precipitate species and the factor A' depends on quantities such as surface energy, solubility of precipitate in the matrix etc.. The structure parameter $\hat{\tau}$ can be introduced by writing it in terms of the precipitate spacing $l(=1/\sqrt{v_f})$ where v_f is the precipitate volume fraction) as:

$$\hat{\tau} = \frac{\alpha b \sqrt{v_f}}{r} \tag{4}$$

where α is a constant. Substituting Eq. (4) into Eq. (3) then yields:

$$\frac{d\hat{\tau}}{dt} = -\frac{A''}{RT} \hat{\tau}^4 \tag{5}$$

where

$$A'' = A \exp(-Q/RT) \tag{6}$$

Eq. (5) can be integrated to give:

$$\hat{\tau} = \left(\frac{A''}{RT} + \frac{1}{\hat{\tau}_0^3} \right)^{-1/3} \tag{7}$$

The factor A in Eq. (6) contains terms that are temperature dependent such as v_f and the solute concentration of the matrix, both of which can be found from the thermal equilibrium diagram of the Al-Mg$_2$Si system. However, for the purposes of the present investigation, it was found that taking A to be constant yielded reasonably satisfactory results. Figure 2 shows results for the drop in values of the structure parameter determined from:

$$\hat{\tau} = \tau \left[1 - \frac{\Delta F}{kT} \ln \left(\frac{v_0 \tau^2}{\dot{\gamma}} \right) \right]^{-1} \tag{8}$$

The solid lines are calculated from Eq. (7). A value of 141 kJ/mol. was estimated for Q [5] and a value for A of 4.5×10^{20} sec$^{1/3}$/mol. gave the fit to experimental results shown in Fig. 2.

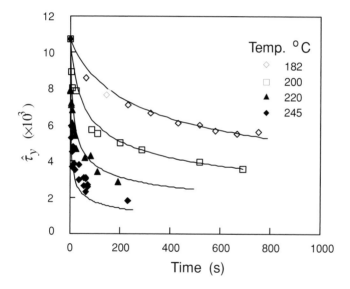

Fig. 2 Tensile yield against time curves for specimens aged at different temperatures.

The kinetic and evolution equations (Eqs. (1) and (5) respectively) together with the parameters from Figs.1 and 2 can be used to construct strain-time curves having tertiary creep characteristics. Creep curves so constructed are found to give higher strain rates and thus shorter rupture times (as defined by setting a limit to the strain achieved) than those observed.

A possible reason for this is that some degree of work hardening is occurring which is not accounted for in the structure evolution equation. Such work hardening is apparent, particularly at higher stresses and lower temperatures where some primary creep deformation is observed. This could be accounted for by introducing a limiting value of the structure parameter, $\hat{\tau}_w$, towards which the material softens. Eq. (5) would then become:

$$\frac{d\hat{\tau}}{dt} = -\frac{A''}{RT}(\hat{\tau}^4 - \hat{\tau}_w^4) \tag{9}$$

As a function of the degree of work hardening, it might be expected that $\hat{\tau}_w$ would be a function of the applied stress and would also evolve with time. For the present purpose, however, it has been assumed that $\hat{\tau}_w$ does not change with time and depends only on the applied stress. Empirically, it was found that $\hat{\tau}_w$ is of approximately the same magnitude as the applied stress. This would make sense in the context of Eq. (1) since a $\hat{\tau}$ value equal to τ would represent a limit to its applicability. Better results are obtainable if $\hat{\tau}_w$ is assigned a slightly higher value than the applied stress and the value used in Eq. (9) was taken to be:

$$\hat{\tau}_w = \hat{\tau} + 2.5 \times 10^{-4} \tag{10}$$

Introducing a non-zero value of $\hat{\tau}_w$ at zero applied stress does, of course, affect the calculated results in Fig. 2 slightly and this has, in fact, been accounted for.

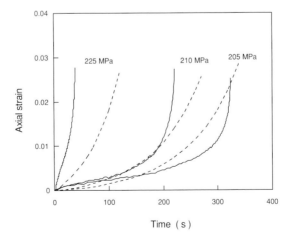

Fig. 3 Constant force creep curves for a temperature of 175 °C. Solid curves are experimental results and broken curves are calculated

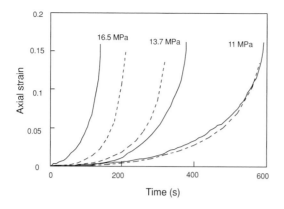

Fig. 5 Constant force creep curves for a temperature of 350 °C. Solid curves are
experimental results and broken curves are calculated

Examples of creep curves calculated from Eqs. (1), (9) & (10) are illustrated in Figs.
3 and 4 together with corresponding experimental results. Since the calculated results do not
account for any primary creep deformation, they tend to represent the experimental curves
better at higher temperatures and lower stresses where tertiary creep is predominant.

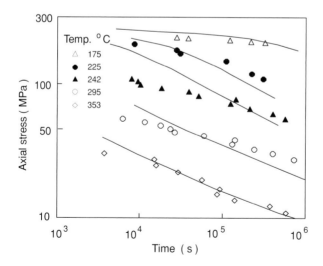

Fig. 4 Axial stress against rupture time for creep tests at several temperatures.
Solid lines are calculated from the model.

Although not reflecting accurately the shape of the curves, calculated results can give an indication of the rupture life if truncated at a suitable value of the strain. Although the strain to failure increases with temperature, because the calculated creep rate becomes quite high, a representative value of strain can be used. Figure 5 shows calculated and experimental axial stress-rupture life curves for tests at several temperatures, the calculated curves being terminated at an axial strain of 6%. There is a tendency (with the exception of the 242 $^\circ$C results) for the estimates to be better at higher temperatures and lower stresses, i.e. where tertiary creep is dominant, and to over-estimate lives at lower temperatures and higher stresses. This is probably due partly due to the fact that evolution of work hardening and thus of $\hat{\tau}_w$ with time is not accounted for but mostly because primary creep has exhausted some of the inherent ductility of the material.

4. Discussion

Modelling creep using a kinetic equation containing a structure parameter term which softens according to established precipitate growth equations is able to give a reasonable estimate of rupture life as well as representing the shape of the creep curve when tertiary creep predominates. However, it does require the introduction of an extra term, $\hat{\tau}_w$, which probably represents a degree of work hardening. At first sight it may seem reasonable to introduce a work hardening term into the structure evolution equation, Eq. (9). However, this would lead to steady state behaviour. Depending on the relative magnitudes of the hardening and softening terms, either primary creep followed by steady state behaviour or tertiary creep followed by steady behaviour would result.

This suggests, perhaps, that it would be more appropriate to model such alloys as composite materials as has been done for work hardened pure metals e.g. [6]. This may take the form of a material in which the precipitates act as the major barrier to deformation but some limited deformation resulting in primary creep and work hardening is possible in the matrix phase, particularly at higher stresses. This would require the use of two kinetic equations plus two evolution equations to describe flow in the matrix and flow governed by precipitates, together with terms describing the relative contribution of each to overall deformation. A further complication would be that this latter quantity would be expected to change during deformation and the relative contribution of matrix deformation would tend to increase with time.

5. Conclusion

Preliminary results are presented from modelling the tertiary creep behaviour of a precipitation hardened aluminium alloy using tensile test results on aged specimens. Reasonable results for both creep curves and rupture times are obtainable provided the evolution equation, based on Ostwald ripening kinetics, is modified by the introduction of an empirical term which accounts for deformation induced hardening during the creep test. Better results are given, as expected, for tests at low stresses and higher temperatures, i.e. where tertiary creep is predominant and primary creep effects small. It is suggested that improvements could be made by adopting a composite model approach.

6. References

[1] Frost, H.J. and Ashby, M.F., 1982, Deformation-Mechanism Maps, Pergamon Press, Oxford, U.K.

[2] Osgerby, S. and Dyson, B.F., 1993, Constant Strain Rate Testing: Prediction of Stress-Strain Curves from Constant-Load Creep Data, Proceedings of the 5th. International Conference on Creep and Fracture of Engineering Structures, Institute of Materials, London, UK, pp53 - 61.

[3] Kocks, U.F., Argon, A.S. and Ashby, M.F., 1975, Thermodynamics and Kinetics of Slip, Progress in Materials Science, Vol.19, pp 1 – 288.

[4] Martin, J.W. and Dougherty, R.D., 1976, Stability of Microstructure in Metallic Systems, Cambridge University Press, Cambridge, U.K.

[5] Smithells, C.J., 1976, Metals Reference Book, 5th Edition, Butterworth, London, U.K.

[6] Nix, W.D. and Ilschner, B., 1979, Mechanisms Controlling Creep of Single Phase Metals and Alloys, Proceedings of the 5th International Conference on the Strength of Metals and Alloys, Aachen, pp 1503 – 1530.

On the Influence of Fibre Texture on Shear Creep Behaviour of Short Fibre Reinforced Aluminium Alloys

G. KAUSTRÄTER [1], A. YAWNY [1]*, J. SCHÜRHOFF [1], B. SKROTZKI [1], G. EGGELER [1] and B. STÖCKHERT [2]

[1] Ruhr-University Bochum, Institute for Materials, 44780 Bochum, Germany
[2] Ruhr-University Bochum, Institute for Geology, 44780 Bochum, Germany
*) on leave from: Centro Atomico Bariloche, Argentina

1 Abstract

The present work investigates creep of a metal matrix composite (MMC) which was produced by squeeze casting. An Al alloy with 11 wt.-% Zn was reinforced using 8 volume per cent of commercially available Saffil fibres (Al_2O_3-fibres). The MMC investigated in the present study has a random planar fibre texture. A double shear creep specimen was used to study creep under biaxial stress conditions. It was found that the addition of fibres results in a strong decrease of creep rate. Fibre reinforcement thus increases creep strength. The addition of fibres resulted in a decrease of creep ductility. Fibre reinforcement not only affected creep rate and creep ductility but moreover changed the characteristics of the stress and temperature dependence of the overall creep process. A striking result of the present investigation was that shear creep rates of specimens with fibre orientations in the shear plane and perpendicular to the shear plane did not differ significantly.

2 Introduction

Reinforcing metals and alloys by ceramic fibres considerably improves creep strength: when fibres are added creep rates decrease. There has been considerable effort to characterise and to understand the high temperature deformation mechanisms of metal matrix composites in general and specifically of short fibre reinforced aluminium alloys which were produced by squeeze casting and which are characterised by a random planar fibre texture [1-12]. It was also shown for uniaxial creep testing (with the uniaxial stress direction parallel to the planar fibres) that the presence of fibres not only results in a decrease of creep rate and a decrease of rupture strain but moreover alters the stress and temperature dependence of the overall creep process [4, 7, 11]. Stress exponents and apparent activation energies of creep of the MMCs generally differ from the values of the unreinforced matrix materials. This could be rationalised on the basis of a microstructural scenario suggested by Dlouhy and co-workers [7], in which three coupled elementary processes control creep: the formation of a work hardened zone around the fibres (loading of fibres by dislocations) [13], fibre breakage (once a critical fibre stress is reached) and a recovery process, during which dislocations move to the fibre ends where they shrink and annihilate.

Most of the previous work, however, was performed in uniaxial tension and from a material mechanics point of view there is a need to consider simple cases of multiaxial loading. Therefore the present investigation sets out to study the creep behaviour of a short fibre reinforced aluminium alloy using double shear creep testing. One objective of the present study is to investigate shear creep behaviour of short fibre reinforced aluminium alloys and compare the

208

results to findings from uniaxial creep testing. Another objective of the present study is to investigate the influence of fibre orientation on shear creep behaviour.

3 Material

The material used in the present study consisted of a matrix alloy with a composition of 88.8 wt.-% Aluminium, 11 wt.-% Zn and 0.2 wt.-% Mg reinforced by 8 volume-% of aluminium oxide ("Saffil") fibres. The average fibre length and fibre thickness were 200 μm and 3 μm, respectively. The aluminium alloy was obtained from VAW, Bonn (Germany). The fibre preform was purchased from Vernaware Ltd. in Bolton (England). The short fibre reinforced aluminium alloy investigated in the present study was produced by squeeze casting using the experimental facilities from EMPA, Thun (Switzerland). The microstructure of the as-processed material of the present study is shown in Figure 1.

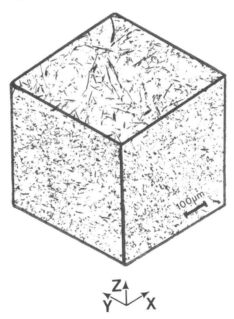

Fig. 1. : As processed material (Al-11%Zn-0.2%Mg + 8 vol.-% fibres). The axis X, Y and Z are parallel to the edges of the fibre preform. More long fibres are visible in the top area of the cube than on the side faces.

All processing details are described in the literature [14, 15]. It has also been explained why an aluminium alloy with 11 wt.-% Zn and 0.2 % Mg - a classic type II creep alloy [16] (and this type of creep behaviour is not affected by the small amounts of Mg) - was used to explore the creep mechanism of the MMC [11]. Squeeze casting was performed using more alloy than required to only fill the fibre preform. After squeeze casting, excess alloy therefore was available for shear creep testing of the matrix material.

4 Shear creep testing

In the present study we use a double shear creep specimen which is well characterised in terms of the homogeneity of the shear stress state in its two shear zones [17]. So far, this technique was successfully used to study creep of single crystal superalloys at temperatures above 1000 °C [18]; here we use the technique to investigate creep of short fibre reinforced MMCs. Figure 2 a schematically illustrates the double shear creep specimen. Figure 2 b shows how the specimen macroscopically deforms under creep conditions. Since the MMC investigated in the present study exhibits a random planar fibre texture the orientations of fibres with respect to the macroscopic shear plane and shear direction are important. Figure 2 c schematically illustrates one of the limiting cases which was considered in the present study. We refer to the shear direction and shear plane in Figure 2 c in terms of the axis u, v and w. Figure 2 c shows a shear specimen where the macroscopic shear plane normal is perpendicular to the axis u and the shear direction is parallel to the axis w. Figure 2 c schematically illustrates a loading scenario where the fibres are oriented perpendicular to the direction of shear. The other extreme is the case where the longitudinal fibres are parallel to the shear plane (not illustrated in Figure 2). These two extremes of shear creep texture are considered in the present study and are referred to as "MMC-perpendicular" (case shown in Figure 2 c) and "MMC-parallel" (fibres parallel to the shear plane) throughout this paper.

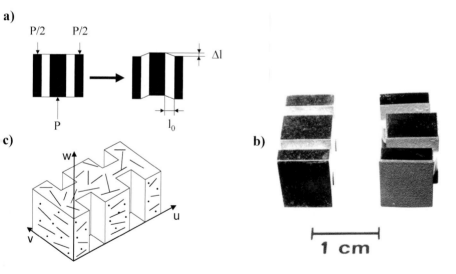

Fig. 2. Double shear creep specimen. (a) Definition of l_0 and Δl. (b) Shear specimen before and after deformation ($T = 320$ °C, $\tau = 10$ MPa, $\gamma = 0.46$). (c) One type of fibre texture investigated in the present study.

Shear creep specimens were produced using spark erosion machining. All details of shear creep testing have been described in the literature [18]. Shear creep tests were performed under constant load at shear stresses ranging from 10 to 20 MPa (MMCs) and 5 to 15 MPa (matrix material). Creep testing was performed at temperatures between 280 and 320 °C. Most importantly, the shear γ is obtained as the ratio of the shear displacement Δl to the width of the shear zone l_0:

$$\gamma = \Delta l / l_0 \tag{1}.$$

The shear creep test conditions used in the present study are summarised in Table 1. In addition to shear creep testing uniaxial tensile creep testing in the same matrix and MMC materials was performed at normal stresses ranging from 5 to 30 MPa and in the temperature range from 240 to 350 °C. The details of tensile creep testing have been described elsewhere [11]. The MMC tensile creep specimens were produced such that the direction of the applied stress was parallel to the plane of the random planar fibres. The creep parameters which were used for uniaxial tensile testing are summarised in Table 2.

Table 1: Shear creep conditions used in the present study.

Material	fibre orientation	shear stress [MPa]	temperature [°C]	total shear strain accumulated
Matrix: Al Zn11 Mg0.2	--	5	300	0.310
		10	300	0.392
		15	300	0.392
		10	280	0.456
		10	320	0.458
MMC: Al Zn11 Mg0.2 + 8 vol-% Saffil fibres	perpendicular to w	10	280	0.236
		10	300	0.397
		10	320	0.235
		15	300	0.489
		20	300	0.361
	perpendicular to u	10	280	0.417
		10	300	0.382
		10	320	0.420
		15	300	0.390
		20	300	0.389

4 Results and Discussion

Figures 3 a and 3 b show creep results presented as shear strain γ vs. time t which were obtained for the lowest (280 °C) and the highest (320 °C) temperature for the three materials considered in the present study (matrix material, MMC - perpendicular and MMC - parallel). Figure 3 c and 3 d show the same results plotted as logarithm of shear rate log $\dot{\gamma}$ vs. shear γ. From a mechanistic point of view plots of the shear creep rate vs. shear can be more directly interpreted in terms of overall hardening or softening behaviour.

Table 2: Tensile creep conditions used in the present study. The fibre orientations were in a plane which also contained the direction of the applied tensile stress.

Material	tensile stress [MPa]	temperature [°C]	strain rate at minimum	strain at minimum
Matrix: Al Zn11 Mg0.2	15	240	$3.47 \cdot 10^{-9}$	0.001
	15	260	$2.93 \cdot 10^{-8}$	0.001
	15	280	$2.33 \cdot 10^{-6}$	0.006
	15	300	$2.42 \cdot 10^{-4}$	0.011
	5	320	$5.74 \cdot 10^{-6}$	0.011
	10	320	$1.03 \cdot 10^{-4}$	0.013
	15	320	$3.88 \cdot 10^{-4}$	0.016
MMC: Al Zn11 Mg0.2 + 8 vol.-% Saffil fibres	10	300	$1.32 \cdot 10^{-9}$	0.015
	20	300	$5.60 \cdot 10^{-7}$	0.017
	30	300	$2.14 \cdot 10^{-5}$	0.014
	20	320	$2.03 \cdot 10^{-6}$	0.014
	20	350	$1.04 \cdot 10^{-5}$	0.015

Figure 3 reveals two important results. Firstly, it clearly shows that fibre reinforcement results in an increase of shear creep resistance in the whole temperature range. The fibre reinforced materials deform at a rate which is more than three orders of magnitude smaller than the creep rate of the unreinforced matrix material. Secondly, there is no significant effect of fibre orientation on the shear creep behaviour of the MMCs up to well within the secondary creep regime. The creep curves of the fibre reinforced materials show a pronounced primary creep behaviour and reach creep rate minima after shear strains of the order of 10 %. The matrix material deforms very fast and there is a small decrease of creep rate throughout the creep process up to strains of the order of 40 %; but a pronounced primary or tertiary creep behaviour cannot be detected for the matrix material. From MMC creep curves like those shown in Figure 3 it is easy to determine the minimum creep rate data which are traditionally taken to characterise the stress and temperature dependence of the creep process as a whole:

$$\dot{\gamma}_{min} = C_1 \cdot \exp\left(-\frac{Q_{app}}{RT}\right) \cdot \tau^n \qquad (2),$$

$\dot{\gamma}_{min}$ is the shear creep rate, C_1 is a constant, Q_{app} is the apparent activation energy of creep, R is the gas constant, T is the absolute temperature, τ is the shear stress and n is the stress exponent. In case of the matrix material, where no clear creep rate minimum was reached, shear creep rates were determined after a shear value of 10 % (which corresponds to the shear strain values where the MMCs reached their shear rate minima).

212

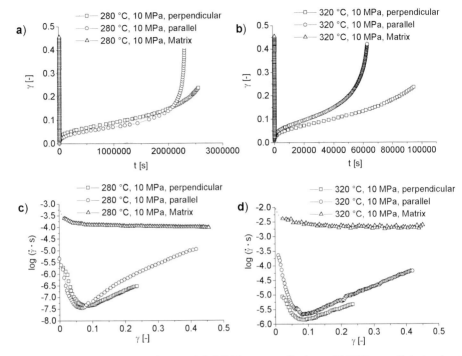

Fig. 3. Shear creep data for matrix material, MMC – perpendicular and MMC - parallel at a shear stress of 10 MPa at the lowest and highest temperature of the test programme (280 °C and 320 °C). (a) $\tau = 10$ MPa, $T = 280$ °C, $\gamma = f(t)$. (b) $\tau = 10$ MPa, $T = 320$ °C, $\gamma = f(t)$. (c) $\tau = 10$ MPa, $T = 280$ °C, $\dot{\gamma} = f(\gamma)$. (d) $\tau = 10$ MPa, $T = 320$ °C, $\dot{\gamma} = f(\gamma)$.

The stress dependence of the shear rate is obtained by plotting the minimum shear rate vs. the shear stress in a log-log-plot, Figure 4. The resulting data points can be represented by straight lines in Figure 4 and the slopes yield the stress exponents n. The matrix material shows a stress exponent near 4 while the two MMCs show stress exponents near 10.

The apparent activation energy of creep Q_{app} is obtained by representing the shear creep data in an Arrhenius plot, where the minimum shear rate is plotted over the inverse of the absolute temperature, Figure 5. The slope of the fitted straight line in an Arrhenius plot yields a negative value which must be multiplied by $- R$ to yield the apparent activation energy of creep.

As is indicated in Figure 5 the matrix material shows an apparent activation energy of 214 kJ/mole while the two composites show higher values of 235 kJ/mole (MMC perpendicular) and 268 kJ/mole (MMC - parallel). We do not present the tensile data on their own because the results look very similar to what was reported previously for the same matrix material with 15 volume percent of fibres. However, it is interesting to compare these uniaxial data with the shear creep data which were presented in this section. We compare our uniaxial and biaxial data using simple von Mises type of reference expressions for stress and strain rate [19]:

$$\sigma_{eff} = \frac{1}{\sqrt{2}} \cdot [(\sigma_1 - \sigma_2)^2 + (\sigma_1 - \sigma_3)^2 + (\sigma_2 - \sigma_3)^2]^{\frac{1}{2}} \tag{3}$$

$$\dot{\varepsilon}_{eff} = \frac{\sqrt{2}}{3} \cdot [(\dot{\varepsilon}_1 - \dot{\varepsilon}_2)^2 + (\dot{\varepsilon}_1 - \dot{\varepsilon}_3)^2 + (\dot{\varepsilon}_2 - \dot{\varepsilon}_3)^2]^{\frac{1}{2}} \tag{4}$$

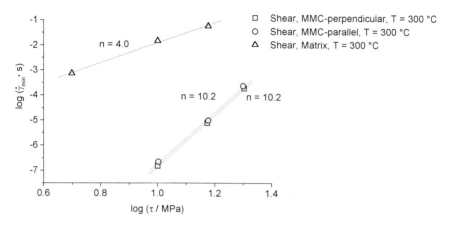

Fig. 4. Stress dependence of the shear rate in a log-log-plot for 300 °C. MMC materials: minimum creep rate data were used. Matrix material: shear creep rates were determined for = 0.10.

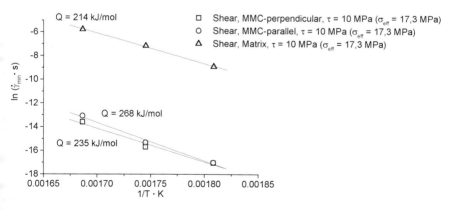

Fig. 5. Temperature dependence of the minimum shear creep rates (MMCs) and 0.10 shear creep rates (matrix material), respectively, in an Arrhenius plot.

With these definitions the effective stress and strain rate data from uniaxial tensile testing correspond to the reported values. In case of shear creep testing, where overall stress state is characterised by three maximum principal stresses. The effective stress is obtained as $\sigma_{eff} = \sqrt{3}\,\tau$. The effective strain rates can be calculated assuming isotropic material behaviour and for the effective strain rate in shear we obtain: $\dot{\varepsilon}_{eff} = \sqrt{3}/3\,\dot{\gamma}$. It is now easy to plot the effective deformation rates vs. the effective stresses in a log-log-plot, Figure 6. The corresponding Arrhenius plot is shown in Figure 7. It can be seen that the simple effective stress and strain rate relations (Equations 3 and 4) bring all data close to common lines.

It is found that neither the stress exponent nor the apparent activation energies for creep depend on whether the matrix or the two MMCs are loaded in tension or shear.

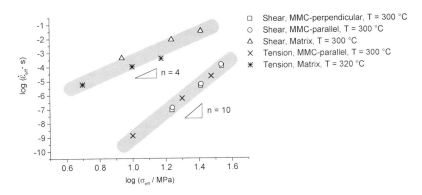

Fig. 6. Effective stress dependence of effective strain in a log-log-plot for the different material and geometries studied in the present work.

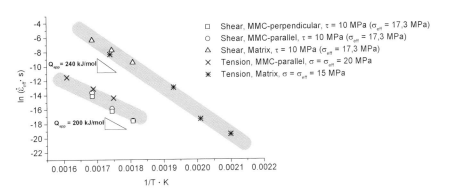

Fig. 7. Temperature dependence of the minimum effective strain rate for the different material and geometries studied in the present work in an Arrhenius plot.

This indicates that multiaxiality does not introduce a new mechanistic element as compared to uniaxial testing. Opposite to what one might intuitively expect, the two MMCs with fibres parallel and perpendicular to the shear plane deform at the same rate. Figure 8 schematically illustrates that those two shear loading conditions are not fundamentally different when normal stresses in the environment of individual fibres are considered.

5 Summary and Conclusions

In the present paper we investigate tensile and shear creep behaviour of a short fibre reinforced Aluminium alloy with 8 volume-% Al_2O_3 fibres and the corresponding matrix alloy. The following results were obtained: (1) Using simple von Mises type reference terms for stress and strain rate all shear and tensile creep data can be presented by common lines in Norton and Arrhenius plots. (2) Stress exponents and apparent activation energies for creep of the MMCs were found to be in agreement with the microscopic view of creep which was put forward by Dlouhy and co-workers [7]. (3) As a striking result the shear creep behaviour of our MMC does not depend on whether the randomly planar oriented fibres are parallel or perpendicular to the shear plane. A preliminary micromechanical explanation is given in the paper. First microstructural results which explain some of the present mechanical findings are reported elsewhere [20].

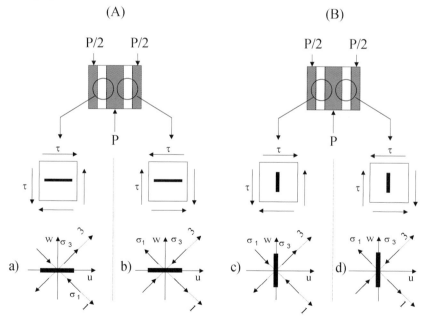

Fig. 8. Schematic drawing of a MMC unit cell together with the directions of the macroscopic principal stresses for the two fibre orientations considered in this work. (A) fibres perpendicular to the shear direction. (B) fibres parallel to the shear plane.

216

6 Acknowledgements
The authors would like to acknowledge funding from the Deutsche Forschungsgemeinschaft (EG 101/5-2).

References

[1] Dragone, T. L., Nix, W. D., 1992, Steady State and Transient Creep Properties of an Aluminum Alloy Reinforced with Alumina Fibers, *Acta metall. mater.,* Vol. **40**, No. 10, pp. 2781- 2791.

[2] Goto, S., McLean, M., 1991, Role of Interfaces in Creep of Fibre-Reinforced Metal Matrix Composites – I. Continuous Fibres, *Acta metall. mater.,* Vol. **39**, No. 2, pp. 153- 164.

[3] Goto, S., McLean, M., 1991, Role of Interfaces in Creep of Fibre-Reinforced Metal Matrix Composites – II. Short Fibres , *Acta metall. mater.*, Vol. **39**, No. 2, pp. 165- 177.

[4] Dlouhy, A., Merk, N., Eggeler, G., 1993, Microstructural Study of Creep in Short-Fibre-Reinforced Aluminium Alloys, *Acta metall. mater.*, Vol. **41**, No. 11, pp. 3245- 3256.

[5] Dlouhy, A., Eggeler, G., 1993, A Quantitative Metallographic Study of a Short Fibre Reinforced Aluminium Alloy, *Prakt. Metallogr.*, Vol. **30**, No. 4, pp. 172- 185.

[6] Eggeler, G., 1994, On the Mechanisms of Creep in Short-Fibres-Reinforced Aluminium-Alloys, *Z. Metallkd.*, Vol. **85**, No.1, pp. 39- 46.

[7] Dlouhy, A., Eggeler, G., Merk, N., 1995, A Micromechanical Model for Creep in Short Fibre Reinforced Aluminium Alloys, *Acta metall. mater.*, Vol. **43**, No. 2, pp. 535- 550.

[8] Cadek, J., Oikawa, H., Sustek, V., 1995, Threshold Creep Behaviour of Discontinuous Aluminium and Aluminium Alloy Matrix Composites: An Overview, *Mat. Sci. Eng. A*, Vol. **190,** pp. 9- 23.

[9] Kucharova, K., Horkel, T., A. Dlouhy, 1999, Creep Anisotropy of Aluminium Alloy-Base Short Fibre Reinforced MMC, in *Creep Behavior of Advanced Materials for the 21st Century* (eds.: Mishra, R. S., Mukherjee, A. K., Murty, K. L.), The Minerals, Metals and Materials Society, Warrendale, pp. 127- 136.

[10] Li, Y., Langdon, T., 1999, Creep Properties of Metal Matrix Composites, ibid., pp. 73- 82.

[11] Kausträter, G., Eggeler, G., 2000, On the Importance of Matrix Creep in High Temperature Deformation of Short Fibre Reinforced Metal Matrix Composites, in *Proceedings EUROMAT 2000* (eds.: Miannay, D., Costa, P., Francois, D., Pineau, A.), Elsevier, Amsterdam, pp. 1285- 1290.

[12] Rösler, J., Bäker, M., 2000, A Theoretical Concept for the Design of High-Temperature Materials by Dual-Scale Particle Strengthening, *Acta mater.*, Vol. **48**, No. 13, pp. 3553- 3567.

[13] Kelly, A., 1971, Particle and Fibre Reinforcement, in *Strengthening Methods in Crystals,* (eds.: Kelly A., Nicholson, R.B.), Elsevier, Amsterdam, p. 433- 484.

[14] Mortensen, A., 1991, Metal Matrix Composites – Processing, Microstructure and Properties, in *Proceedings of the 12th Risø Int. Symp. on Materials Science* (eds.: Hansen, N., Jensen, D.J., Leffers, T., Lilholt, H., Lorentzen, T., Pedersen, A. S., Pedersen, O. B., Ralph, B.), Risø National Laboratory, Roskilde, Denmark, pp. 31- 49.

[15] Roulin, M., 1991, Research Report: FEW 91/008, Alusuisse Lonza Services AG, Neuhausen am Rheinfall, Switzerland.

[16] Nix, W.D., Ilschner, B., 1980, Mechanisms controlling creep of single phase metals and alloys, in *Proceedings of the 5th International Conference on Strength of Metals and Alloys* (edited by P. Haasen, V. Gerold and G. Kostorz,), Pergamon Pres, Toronto, pp. 1503- 1530.

[17] Peter, G., Probst-Hein, M., Kolbe, M., Neuking, K., Eggeler, G., 1997, Finite Element Stress and Strain Analysis of a Double Shear Creep Specimen, *Mat.-wiss u. Werstofftech.*, Vol. **28**, pp. 457 - 464.

[18] Mayr, C., Eggeler, G., Webster, G.A., Peter, G., 1995, Double Shear Creep Testing of Superalloy single crystals at temperatures above 1000 °C, *Mat. Sci. Eng. A*, Vol. **199,** pp. 121 - 130.

[19] Dieter, G. E., 1986, *Mechanical Metallurgy* (3rd edition), McGraw-Hill, Singapore, pp. 87.

[20] Yawny A., Kausträter G., Valek R., Schürhoff, J., Skrotzki, B., Eggeler, G., Stöckhert, B., 2001, Microstructural Assessment of Fibre Damage after Tensile and Shear Creep Deformation of a Short Fibre Reinforced Aluminium Alloy with a Fibre Volume Fraction of 8 %, *Proceedings of this Conference.*

Microstructural Assessment of Fibre Damage after Tensile and Shear Creep Deformation of a Short Fibre Reinforced Aluminium Alloy with a Fibre Volume Fraction of 8 %

A. YAWNY [1]*, G. KAUSTRÄTER [1], R. VALEK **, J. SCHÜRHOFF [1], B. SKROTZKI [1], G. EGGELER [1] and B. STÖCKHERT [2]

[1] Ruhr-University Bochum, Institute for Materials, 44780 Bochum, Germany
[2] Ruhr-University Bochum, Institute for Geology, 44780 Bochum, Germany
*) on leave from: Centro Atomico Bariloche, Argentina
**) on leave from: Institute of Physics of Materials, Brno, Czech Republic

1 Abstract

The present paper reports microstructural results on shear creep behaviour of short fibre reinforced metal matrix composites with an Al – 11 wt.-% Zn matrix and 8 volume percent of Saffil fibres. The Saffil fibres had a typical length of 200 µm and a typical thickness of 3 µm and exhibit a random planar fibre texture. Creep tests were performed under tensile (direction of loading parallel to the random planar fibre plane) and shear (shear plane parallel and perpendicular to the random planar fibre plane) creep conditions. Multiple fibre breakage is a common feature of tensile and shear creep testing and in shear creep testing it does not depend on whether the preferential plane of the fibres is perpendicular or parallel to the shear plane. Even when shear stresses act almost perpendicular to the longitudinal fibre axis, subfibres after fibre breakage move apart in the direction of the longitudinal fibre axis. Under shear creep conditions creep rate minima are reached after larger effective strains than in tensile creep testing. A preliminary explanation of this effect is based on the orientation of the fibres to the direction of the maximum shear stress.

2 Introduction

Reinforcing metals and alloys with ceramic fibres increases their creep strength and therefore there is an interest in metal matrix composites for high temperature applications. It is well known that the creep strength of metallic alloys can be significantly improved by adding ceramic fibres [1, 2] and this was demonstrated for the case of short fibre reinforced aluminium alloys [3]; creep rupture data for unreinforced aluminium alloys and short fibre reinforced materials were reported and the composite materials showed significantly higher rupture lives than the unreinforced matrix materials. Mechanical results from uniaxial creep testing were published in the literature [3 - 6] and the mechanical behaviour was discussed on the basis of microstructural results obtained by transmission electron microscopy, scanning electron microscopy and optical microscopy. In addition, a combined chemical-metallographic-procedure was developed [7] and refined [8] that allowed to assess the kinetics of fibre breakage. The squeeze cast materials were creep deformed to given levels of strain. The creep tests were interrupted. Then the matrix alloy was dissolved using an appropriate chemical solution. The remaining fibre sponge did not exhibit any inherent strength. In fact many isolated fibres were found after the dissolution process. It was then possible to measure fibre lengths distributions and it was shown on several occasions that fibre breakage occurs during processing (infiltration of the fibre preform by the melt under pressure) and during creep [7, 9]. In summary it could be shown that three coupled elementary processes control creep of short fibre reinforced aluminium alloys: loading of fibres (by dislocations which form a work hardened zone around the fibres), unloading of fibres (by a recovery process in the course of which dislocations move to the fibre ends where they shrink and annihilate) and fibre breakage once fibres reach a critical stress. It was shown that it is possible to model creep of short fibre reinforced aluminium alloys on the basis of this microstructural scenario and the model of Dlouhy and co-workers [4] not only rationalised the shape of individual creep curves but moreover correctly predicted the stress and the temperature dependence of the minimum creep rate. All of our previous work was based on microstructural analysis of interrupted uniaxial creep tests and focused on short fibre reinforced materials with a fibre volume fraction of 15 per cent; this materials behaved creep brittle and rupture occurred at total strains of the order of 2 per cent.

Recently we started to work on short fibre reinforced aluminium alloys with a lower fibre volume fraction of 8 per cent. The creep behaviour of these alloys was investigated in tension and under shear loading [10]. We investigated the tensile and shear creep behaviour and placed special emphasis on the orientation of fibres in both types of loading. The MMC had a random planar fibre texture, where all fibres (mean fibre length: 200 µm, mean fibre diameter: 3 µm) were oriented randomly in one plane. Tensile tests were performed with the direction of the tensile stress parallel to the fibre plane; moreover, shear creep tests were performed with fibres parallel to the macroscopic shear plane and with fibres perpendicular to the macroscopic shear plane. The previous mechanical results can be summarised as follows: Using simple von Mises type reference terms for stress and strain rate all shear and tensile creep data could be presented by common lines in Norton and Arrhenius plots. As a striking result the shear creep behaviour of the MMC did not depend on whether the randomly planar oriented fibres were perpendicular or parallel to the shear plane. Stress exponents and apparent activation energies for creep of the MMCs were found to be in agreement with the microscopic view of creep which was put forward by Dlouhy and co-workers [4]. So far no microstructural results on the creep deformed low fibre volume fraction MMC after tensile and shear creep testing [10] were reported. This paper presents the microstructural results for the creep deformed specimens from [10]. Special emphasis is placed on how shear creep loading affects fibre damage accumulation.

3 Material and Creep Behaviour

The microstructure and fibre texture of the MMC material we discuss and investigate in the present study have been described previously [10]. It was obtained by the squeeze casting technique and it is composed of matrix 88.8 wt.-% Aluminium, 11 wt.-% Zn and 0.2 wt.-% Mg reinforced by 8 volume-% of aluminium oxide ("Saffil") fibres. The material is characterised by a random planar fibre texture with random fibre orientations in one common plane. In the present work microstructures after three types of interrupted creep tests were analysed, (i) after uniaxial creep testing (direction of the applied stress parallel to the fibre plane), (ii) after shear creep testing with fibres parallel to the macroscopic shear plane and (iii) after shear creep testing with fibres perpendicular to the macroscopic shear plane. Details of uniaxial and shear creep testing have also been described in [10]. Some of the creep curves from uniaxial and from shear creep testing from [10] are shown in Figure 1, where creep curves are shown as creep rate vs. deformation in log-linear-plots. All uniaxial and shear creep rates from the MMC with 8 volume-% fibres in Figure 1 were normalised by the corresponding minimum creep rates. In Figure 2 one tensile and one shear creep curve with similar von Mises type of effective stresses are plotted. In Figure 2a, the von Mises type of effective strain rate vs. a von Mises type of effective strain is shown for both cases. In Figure

2b, the curves are redrawn representing the normalised effective strain rate vs. the von Mises type of effective strain normalised by the reference strain at the minimum (Figure 2b). We used here the following types of reference terms [11]:

$$\sigma_{eff} = \frac{1}{\sqrt{2}} \cdot [(\sigma_1 - \sigma_2)^2 + (\sigma_1 - \sigma_3)^2 + (\sigma_2 - \sigma_3)^2]^{\frac{1}{2}} \tag{1}$$

$$\dot{\varepsilon}_{eff} = \frac{\sqrt{2}}{3} \cdot [(\dot{\varepsilon}_1 - \dot{\varepsilon}_2)^2 + (\dot{\varepsilon}_1 - \dot{\varepsilon}_3)^2 + (\dot{\varepsilon}_2 - \dot{\varepsilon}_3)^2]^{\frac{1}{2}} \tag{2}$$

$$\varepsilon_{eff} = \left[\frac{2}{3} \cdot \left(\varepsilon_1^{\,2} + \varepsilon_2^{\,2} + \varepsilon_3^{\,2} \right) \right]^{\frac{1}{2}} \tag{3}$$

Note that Equation 1 yields $\sigma_{eff} = \sqrt{3}\,\tau$ and $\sigma_{eff} = \sigma$ for the shear and the uniaxial case, respectively; therefore the shear stress τ which would precisely match an uniaxial and effective stress of 20 MPa is 11.5 MPa. For easy comparison of the different creep curves shown in Figure 2 we use the simple Norton law reported in our previous work with a stress exponent of 10.2 to adjust the shear creep data for $\tau = 10$ MPa to $\tau = 11.54$ MPa.

Figure 1 indicates that a common type of creep behaviour characterises both tensile (Figure 1a) and shear creep behaviour (Figure 1b) independent of the level of the applied stress. Figure 2 shows that more effective strain is needed to reach the shear rate minimum than to reach the tensile creep minimum (Figure 2a). Other than that, there is no basic difference between tensile and shear creep deformation as far as the shape of individual creep curves is concerned (Figure 2b).

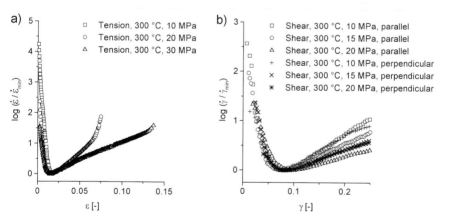

Fig. 1: Creep curves of 8 volume-% short fibre reinforced aluminium alloys as reported in [10] at different stress levels in tension and shear. Here we normalise the creep rates by the minimum creep rate in each test. (a) Uniaxial tests. (b) Shear creep tests.

222

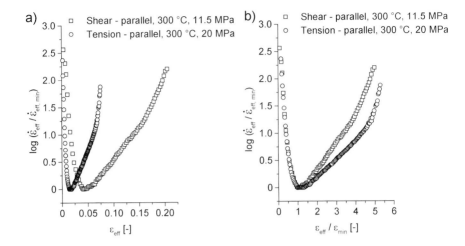

Fig. 2: Creep curves of one tensile and one shear creep test of 8 volume - % fibre reinforced aluminium alloys as reported in [10]. (a) Normalised effective strain rate vs. effective strain. (b) Normalised effective strain rate vs. ratio of effective strain to effective strain at minimum.

4. Microstructural Investigations

Microstructures were investigated using optical and scanning electron microscopy (OM and SEM). OM in combination with polarised light was used to study the grain structure of the unreinforced aluminium alloy and the MMC. After the mechanical polishing procedure anodic etching was used to reveal the grain structure. Anodic etching was performed at room temperature using Barker's reagent (14 % HBF_4 in 86 % H_2O) between 80 and 180 seconds at 15 V in a Struers Electropol for both OM and SEM specimens. SEM was used to study details of fibre damage during creep. This work was performed using a JEOL 840 LaB_6 SEM. Pictures were taken using secondary electrons and back-scattered electrons. In one case, a field emission SEM of type LEO-Gemini 1530 equipped with an orientation imaging hardware and software which allowed to analyse electron back scatter patterns was used.

Fibre breakage kinetics was studied following the procedures outlined in [7, 8]: between 20 and 200 mm^3 of short fibres reinforced composite material was exposed to 100 ml of concentrated HCl for 24 hrs at room temperature. The exposure to the concentrated acid dissolved all of the metallic matrix material and leaves the fibres unattacked. In case of the as processed material this first matrix extraction procedure left the fibre preform intact. A second exposure to concentrated HF attacks the SiO_2 binder without interacting with the fibres. Then the fibre preform disintegrates. In the crept material states a second treatment with HF was not required: the fibre preform had already disintegrated after the HCl attack. The fibres and subfibres that were thus isolated from the MMC were then removed from the solutions by careful paper filtration or by a centrifugal separation procedure and cleaned with distilled water. The fibres were then detached from the paper filter by immersing the paper with the fibres in ethanol or by replacing ethanol for water after the centrifugal separation procedure. Thus a suspension of fibres in ethanol was obtained. Drops of this suspension were brought onto a microscope slide where they dried. The fibre lengths were then analysed in an optical microscope. The length of between 500 and 600 fibres was evaluated for each material state using the software DIAna [12].

5. Microstructural Results

5.1 Grain Morphology: We first present a number of optical micrographs which represent parts of montages and reveal the grain structure of our materials in the double shear specimen (Figure 3).

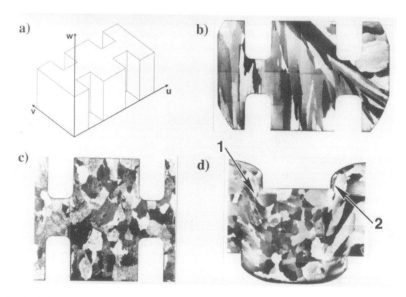

Fig. 3: Definition of directions and grain structures in matrix and MMC material. (a) Directions u, v and w in double shear specimen. (b) Grain structure in the u, v-plane of the double shear specimen for matrix material before creep. (c) Grain structure in the u, v-plane of the double shear specimen for composite material before creep. (d) Grain structure in the u, w-plane of the deformed double shear specimen for matrix material ($\tau = 5$ MPa, $T = 300$ ^{0}C, $\gamma = 120$ %).

The as received material clearly shows the largest grain size, Figure 3b. It can be seen in Figure 3c that the grain sizes in the short fibre reinforced aluminium alloys are smaller. Figure 3d shows that the grain microstructures evolves during creep; it is not certain whether the small grains visible in the centre of the shear zone (arrow 1) in Figure 3d are associated with subgrain formation (Barker's reagent also reveals low angle subgrains boundaries) or to dynamic recrystallisation. EBSP measurements suggest that the small grains in Figure 3d represent creep subgrains rather than recrystallised grains. This is supported by the fact that dynamic recovery (instead of dynamic recrystallization) is favoured in metals of high stacking fault energy, such aluminium and its alloys, due to easy dislocation climb and cross-slip [13]. No further effort was made to elaborate this aspect of matrix creep. Moreover Figure 3d shows very small grains near the loading section of the shear specimen (arrow 2) where plastic deformation and shape change of the specimen clearly suggest that the macroscopic

224

stress state at this extreme deformation is no longer pure shear. No emphasis was placed on the explanation of this effect in the present paper. As was shown earlier [14], reasonable homogeneous shear stress distributions are obtained for shear values smaller than 0.10.

5.2 Fibre Damage: Figure 4 shows a fibre which underwent several rupture events in a shear creep test. The SEM micrograph suggests that the three clearly visible individual fibre breakage events occurred at different stages of the creep process. The big gap between the subfibres in the upper left of the micrograph corresponds to an early rupture event followed by a separation of the subfibres during further creep (arrow 1). We interpret the fine fibre crack at the lower right of the micrograph (arrow 3) as the most recent rupture event. Like in uniaxial creep [4, 7], high temperature deformation of our MMC is associated with multiple fibre breakage. Most importantly, fibres move away from each other in the direction of the longitudinal fibre axis while the shear stress acts almost perpendicular to the fibre.

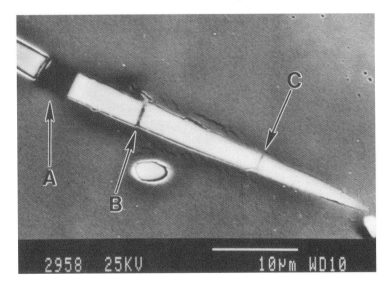

Fig. 4: SEM-micrograph of Saffil fibre after shear creep deformation ($\tau = 20$ MPa, $T = 300\ ^0$C, $\gamma = 0.36$). Three fibre breakage events can clearly be identified (arrow A: early, arrow B: intermediate, arrow C: late).

We would like to emphasise the fact that the small crevices parallel to the fibre need not be associated with debonding; instead they could be due to the preferential chemical attack associated with etching in areas with elevated dislocation densities (work hardened zones around fibres in the Dlouhy model [4]). Figure 5 shows that broken subfibres can move apart more than 20 μm. A trace without matrix material connects the two subfibres. Here it must be mentioned that the anodic attack used for etching of the cross section was quite severe and therefore the width of the crevices where matrix material is missing may be related to the duration of etching. However, it also seems reasonable to assume that the material flow at 300 °C was not capable of closing the gaps between the subfibres opposite to what was observed by Dlouhy and co-workers [3] for an aluminium alloy at 350 °C.

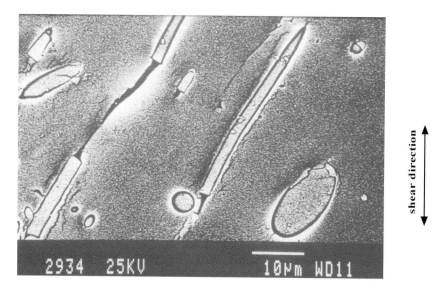

shear direction

Fig. 5: SEM-micrograph providing microstructural evidence for subfibres moving apart for more than 20 µm during shear creep deformation ($\tau = 15$ MPa, $T = 300\ ^0$C, $\gamma = 0.39$).

Finally we present the results which were obtained for fibre length distributions in the as received material and after tensile (Figure 6a) and shear creep deformation (Figure 6b). All crept material states were creep deformed to large plastic deformations. It is clear from Figure 6 that in all cases creep is associated with a significant decrease of fibre length. In case of the shear creep tests, the final fibre length does not depend on whether shear occurred parallel or perpendicular to the randomly planar oriented fibres. Moreover, the fibre length distributions after large strain tensile and shear creep deformation look very similar.

6 Discussion

An important finding of the present work is that even under shear creep conditions fibres break into subfibres and move apart in longitudinal direction as is shown in Figure 4, even when the shear stress acts almost perpendicular to the fibre axis. Fibres break as is schematically illustrated in Figure 7a and not (how one might intuitively expect) like in Figure 7b. This is a strong confirmation of the Dlouhy model [4], where stresses are induced into the fibres by dislocations in the work hardened zone around the fibres and not by macroscopic material flow in the sense of Figure 7b. All features of fibre breakage are very much the same as observed in uniaxial testing [3, 7]. Fibre breakage comes to an end as soon as fibres are smaller than a critical aspect ratio [4]. This is why after large creep strains all size distributions look similar, Figure 6.

226

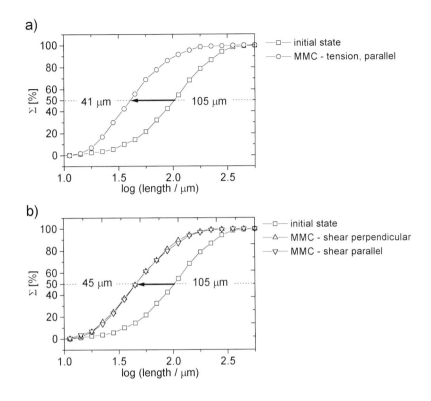

Fig. 6: Cumulative frequency curves of logarithmic fibre lengths. (a) As received material state and tensile creep deformed MMC. (b) As received material state (same as before) and shear creep deformed MMC-perpendicular and MMC-parallel.

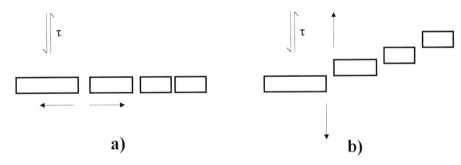

Fig. 7: Schematic illustration of the relative displacement of the subfibres after fracture under shear creep conditions. (a) Observed in this work. (b) Expected, based in the macroscopic material flow.

However, there is an important difference between uniaxial and shear testing: tensile creep tests reach the creep rate minima after effective strains of 1 % while it takes 5 % of effective strain to establish the minimum creep rate under shear conditions. In order to rationalise this difference we consider the three simplified scenarios of dislocation–fibre interactions which are schematically illustrated in Figure 8. Figure 8a shows the Dlouhy-scenario which was explained in detail in the literature [4]. Dislocations which are driven by a resolved shear stress (resulting from an applied tensile stress) form loops around the fibre. These loops induce stresses into the fibre. We now use this simplified single crystal matrix scenario to discuss dislocation fibre interactions for shear creep loading for our two materials, MMC-perpendicular (Figure 8b) and MMC-parallel (Figure 8c). Our model dislocations are now directly driven by the applied shear stress. For MMC-perpendicular (Figure 8b) we note that the dislocation which moves from the left to the right can temporarily induce stress into the fibre. However, it can pass the fibre without forming a loop around the fibre. The dislocation that moves upwards will form a loop around the fibre but it does not induce tensile stress into the fibre. The two dislocations of the same microscopic slip system change their role in the scenario associated with MMC-parallel, Figure 8c, resulting in the same overall effect. The scenarios in Figure 8b and c clearly are not realistic in the sense that fibres are not precisely oriented in single crystal matrices; and so the dislocations will not only move parallel or perpendicular to the longitudinal fibre axis as was assumed in Figure 8b and c. But the figures helps to understand why more strain is needed to form a work hardened zone under the shear creep conditions considered in the present study as compared to tensile testing (Figure 8a).

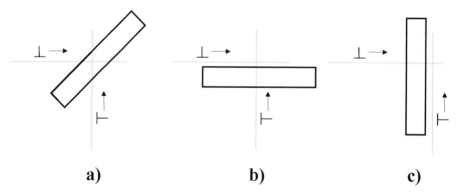

a)   **b)**   **c)**

Fig. 8: Schematic illustration of dislocation fibre interactions. (a) Fibre loading in uniaxial tension as described by Dlouhy and co-workers [3]. The dislocations are driven by a resolved shear stress and form a work hardened zone near the fibre from where tensile stresses are induced into the fibre. (b) Dislocation-fibre interaction in shear testing of MMC-perpendicular. (c) Dislocation-fibre interaction in shear testing of MMC-parallel.

7 Summary and Conclusions

The present study considers tensile and shear creep deformation of short fibre reinforced aluminium composites. The following conclusions can be drawn: (1) Creep processes in short fibre reinforced MMCs with a random planar fibre texture subjected to tensile loading (with the stress direction parallel to the fibre plane) and to shear loading (with the shear plane parallel and perpendicular to the fibre plane) are governed by the same basic processes: loading of fibres with dislocations, recovery and fibre breakage. This represents a

228

confirmation of the microstructural scenario proposed by the Dlouhy-model of creep in short fibre reinforced MMCs [4]. (2) However, more plastic strain is needed to reach the creep rate minimum under both shear conditions as compared to tensile loading. This can be explained by a geometrical argument: under tensile creep conditions fibres form obstacles to dislocation motion and dislocation fibre interactions result in an early stress transfer from the matrix to the fibre. Under shear creep conditions the combination of both types of fibre textures and dislocation crystallography results in a more sluggish build up of the work hardened zone and therefore creep rate minima are observed after larger effective strains.

8 Acknowledgements

The authors would like to acknowledge funding from the Deutsche Forschungsgemeinschaft (SFB 526-D4).

References

[1] Chawla, K. K., 1998, *Composite Materials, Science and Engineering* (2nd Edition), Springer-Verlag, New York.

[2] Clyne, T.W., Withers, P.J., 1993, *An Introduction to Metal Matrix Composites*. Cambridge University Press, Cambridge.

[3] Dlouhy, A., Merk, N., Eggeler, G., 1993, Microstructural Study of Creep in Short-Fibre-Reinforced Aluminium Alloys, *Acta metall. mater.*, Vol. **41**, No. 11, pp. 3245- 3256.

[4] Dlouhy, A., Eggeler, G., Merk, N., 1995, A Micromechanical Model for Creep in Short Fibre Reinforced Aluminium Alloys, *Acta metall. mater.*, Vol. **43**, No. 2, pp. 535- 550.

[5] Phillips, J., Staubach, M., Skrotzki, B., Eggeler, G., 1997, Effect of Geometry and Volume Fraction of Fibers on Creep in Short Fiber Reinforced Aluminum Alloys, *Mat. Sci. Eng. A*, Vol. **234- 236**, pp. 401- 405.

[6] Kausträter, G., Eggeler, G., 2000, On the Importance of Matrix Creep in High Temperature Deformation of Short Fibre Reinforced Metal Matrix Composites, in *Proceedings EUROMAT 2000* (eds.: Miannay, D., Costa, P., Francois, D., Pineau, A.), Elsevier, Amsterdam, pp. 1285- 1290.

[7] Eggeler, G., 1994, On the Mechanisms of Creep in Short-Fibres-Reinforced Aluminium-Alloys, *Z. Metallkd.*, Vol. **85**, No.1, pp. 39- 46.

[8] Valek, R., Dlouhy, A., Assesment of Ceramic Reinforced Damage during Creep in Al-Alloy Base MMCs, to be published in *Scripta mater.*.

[9] Kausträter, G., Skrotzki, B., Eggeler, G., 2000, A Quantitative Metallographic Study of Fibre Fracture during Processing and Creep of a Short Fibre Reinforced Aluminium Alloy, *Prakt. Metallogr.*, Vol. **37**, No.11, pp. 619- 630.

[10] Kausträter G., Yawny, A., Schürhoff, J., Skrotzki, B., Eggeler, G., Stöckhert, 2001, On the Influence of Fibre Texture on Shear Creep Behaviour of Short Fibre Reinforced Aluminium Alloys, *Proceedings of this Conference*.

[11] Dieter, G. E., 1986, *Mechanical Metallurgy* (3rd edition), McGraw-Hill, Singapore, pp. 87.

[12] Duyster, J., DIAna 1.2, 1991- 1997, Ruhr-Universität Bochum.

[13] Humphreys, F. J., Hatherly, M., 1996, Recrystallization and Related Annealing Phenomena, Pergamon, U.K., pp. 363.

[14] Peter, G., Probst-Hein, M., Kolbe, M., Neuking, K., Eggeler, G., 1997, Finite Element Stress and Strain Analysis of a Double Shear Creep Specimen, *Mat.-wiss u. Werstofftech.*, Vol. **28**, pp. 457- 464.

CREEP OF ALUMINIUM II

DEFORMATION BEHAVIOR OF A 2219 ALUMINUM ALLOY

R.Kaibyshev[1], O.Sitdikov[1], I.Mazurina[1], D.R. Lesuer[2]

[1]Institute for Metals Superplasticity Problems, Khalturina 39, Ufa 450001, Russia,
e-mail: sitdikov@imsp.da.ru, fax: +7 3472 253856
[2]Lawrence Livermore National Laboratory, L-342 P.O. Box 808, Livermore, CA 94551, USA,
e-mail: lesuer1@llnl.gov, fax: 510 422 96 33.

Abstract

The deformation behavior of a 2219 aluminum alloy was studied in the temperature range from 250 to 500°C. The results, which cover four orders of magnitude in strain rate, give an increasing trend for the apparent stress exponent and the apparent activation energy with a temperature decrease. It is shown that the 2219 aluminum alloy exhibited a threshold behavior, like aluminum alloys produced via powder metallurgy technique. The introduction of a threshold stress into the analysis leads to stress exponent of ~7 and value of true activation energy of about 80 kJ/mol in the temperature range 300–450°C. At T=500°C the true stress exponent, n, is equal 5 and true activation energy is equal 144 kJ/mol. Strong temperature dependence of normalized threshold stress takes place. Value of an energy term, Q_o, was found to be equal about 37 kJ/mol in the temperature range 250–450°C and tended to reduce at higher temperatures. Operating deformation mechanism is discussed in terms of transition from low temperature climb to high temperature climb with increasing temperature.

1. Introduction

The deformation behavior of Al-Cu alloys has been extensively studied in two different temperature ranges [1-5]. The first temperature interval is the low temperature range from 100 to 250°C where copper precipitates provide strengthening due to the interaction between precipitates and dislocations [1-3]. This work was motivated by the importance of creep properties of Al-Cu alloys under service conditions. The second temperature interval is the range of premelting temperatures (500–580°C) where copper exists in solid solution [4,5]. Mohamed et al. [4,5] had reported the examination of viscous dislocation glide phenomenon. No attempts have been made to examine the creep behavior of an Al-Cu alloy in the intermediate temperature range from 250 to 500°C.

The present study was carried out to provide detailed information on the deformation behavior of a 2219 aluminum alloy at these intermediate temperatures. This temperature interval is widely used for working the 2219 aluminum alloy and examination of deformation mechanisms operating in this alloy is important to determine the optimum conditions of warm and hot working and to establish new thermomechanical processing as well as to discuss its physical basis for grain refinement during equal channel angular extrusion [6]. The 2219 alloy contains Al_3Cr and Al_3Zr dispersoids. It is well known that aluminum alloy produced by powder metallurgy techniques exhibits threshold behavior due to the presence of nanoscale alumina particles [7-10]. At the same time, other authors [11] have shown that an Al-5%Cr-2%Zr alloy produced by rapid solidification exhibited threshold behavior. It is therefore expected that the 2219 alloy may exhibit threshold behavior in the temperature range where coarse particles of θ-phase (Al_2Cu) are present. At present, an experimental evidence in support of this expectation is not available.

2. Materials and Experimental Technique

The 2219 aluminum alloy with the chemical composition Al-6.4%Cu-0.3%Mn-0.18%Cr-

0.19%Zr-0.06%Fe was manufactured at the Kaiser Aluminum - Center for Technology by direct chill casting. The alloy was then homogenized at 530°C for 6 hours and cooled, slowly, in a furnace to provide an equilibrium two phase structure. The alloy had an initial grain size of about 120 μm. Compression specimens 9 mm in diameter and 11 mm high were machined from the alloy ingot. The samples were deformed in compression at constant initial crosshead speed using a Scheck RMS-100 testing machine. Tests were carried out over the temperature range 250-500°C with step 50°C at the initial strain rate in the range from $1.5 \cdot 10^{-6}$ s^{-1} up to $3 \cdot 10^{-2}$ s^{-1}. At the initial strain rate of $3 \cdot 10^{-4}$ s^{-1} tests were performed with step 25°C. For surface examination, specimens were mechanically polished and treated with boric acid to prevent oxidation. The specimens were deformed to 15% at strain rates of $3 \cdot 10^{-4}$ s^{-1} and $3 \cdot 10^{-2}$ s^{-1} at 250°C and 500°C. A SEM JSM-840 was used to observe the surface features. Samples were also examined using TEM. For these examinations, samples were cut from the gauge section of deformed specimens and thinned to about 0.25 mm. Discs with 3 mm diameter were cut and electropolished to perforation with a Tenupol-3 twinjet polishing unit using a 20% nitric acid solution in methanol at -38°C and 20V. The thin foils were examined using a JEOL-2000EX TEM with a double-tilt stage at an accelerating potential of 160 kV.

3. Results
3.1 The shape of σ-ε curves
True stress- true strain curves in the temperature range 250-500°C at a fixed strain rate of $3 \cdot 10^{-4}$ s^{-1} (Fig.1a) and at fixed temperatures of 250, 300 and 500°C and different initial strain rates ($1.5 \cdot 10^{-6}$ to $3 \cdot 10^{-2}$ s^{-1}) (Fig.1b) are presented in Fig.1. All curves exhibit well-defined steady state. A minor strain hardening in the initial stage of plastic flow takes place at 250°C.

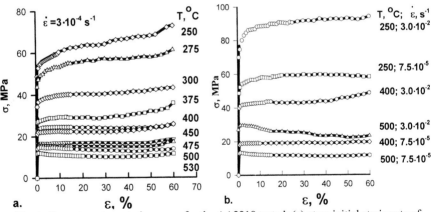

Fig. 1. True stress-true strain curves for the AA2219 tested: (a) at an initial strain rate of $3 \cdot 10^{-4}$s^{-1} in the temperature range 250-530°C; (b) at different initial strain rates.

3.2 The variation of steady-state stress with strain rate
Fig. 2a shows the variation of strain rate with flow stress plotted on logarithmic axes. It is seen that the experimental data in the plot can be represented accurately by straight lines. The results demonstrate that the power-law equation [12] appears valid from 250 to 500°C:

$$\dot{\varepsilon} = A \cdot \left(\frac{\sigma}{G} \right)^{n} \cdot \exp\left(\frac{-Q}{R \cdot T} \right), \tag{1}$$

where $\dot{\varepsilon}$ is the strain rate, A is a constant, n is the stress exponent, σ is the steady state flow

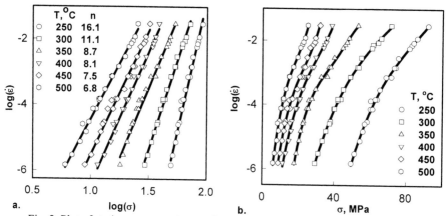

Fig. 2. Plot of strain rate *vs* steady state flow stress in (a) logarithmic scale; (b) semi-logarithmic scale.

stress, G is the shear modulus, Q is the activation energy for plastic deformation, R is the gas constant and T is the absolute temperature. The slope of the lines provides the stress exponent n. The values of n are indicated on the figure. As it is seen, the stress exponent decreases from n~16.1 at the lowest temperature of 250°C to n~6.8 at the highest temperature of 500°C. However, in the temperature range 250-350°C, the values of the stress exponent lying in range from 9 to 16 are essentially high and it can be assumed that deformation behavior of the 2219 aluminum alloy can be described in terms of the exponential relation [12]:

$$\dot{\varepsilon} = B \cdot \exp(\beta\sigma) \cdot \exp\left(\frac{-Q}{R \cdot T}\right), \qquad (2)$$

where B is a constant and β is a coefficient. In order to check this presumption, the semi-logarithmic plot is presented in Fig.2b. No satisfactory linear fit is observed even in the temperature range 250-350°C, where the regression coefficient r was found to be lower (r=0.95) in comparison with the logarithmic plot where value of r is equal to 0.99. This result is unusual because it is well known [13] that the data can be equally well described by equations (1) and (2) in the case when n≥8 and $\beta \leq 0.065 MPa^{-1}$. Thus, a power-law relationship is appropriate at all temperatures and strain rates examined.

3.3 The activation energy for plastic deformation

The apparent activation energy for plastic deformation, Q_a, was determined by the use of the procedure described by Pickens et al. [14]. Algebraically, Eqn. (1) can be converted to:

$$\ln\left(\frac{\sigma}{G}\right) = \ln\left(\frac{\dot{\varepsilon}}{A}\right)^{\frac{1}{n}} + \frac{Q_a}{R \cdot n} \cdot \frac{1}{T}. \qquad (3)$$

The values of Q_a were obtained graphically by logarithmic plotting of the normalized stress, σ/G, against the inverse of the absolute temperature and taking the slope of the data to be $Q_a/(R \cdot n)$ (Fig.3a) for different values of the initial strain rate. In the calculations, the shear modulus for pure Al was used which can be defined as a function of temperature as [10]:

$$G=(3.022 \cdot 10^4)-16 \cdot T , MPa. \qquad (4)$$

The dependence of $\ln(\sigma/G)$ on 1/T is linear at all temperatures examined. The slope of these lines tends to decrease with increasing strain rate. As a result, there is a strain rate dependence of the apparent activation energy values (Fig.3b). This observed temperature

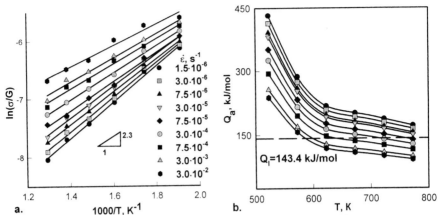

Fig. 3. (a) Shear modulus compensated applied stress *vs* inverse absolute temperature;
(b) temperature dependencies of the apparent activation energy, Q_a.

dependence of the apparent activation energy is unusual and in contrast with data for pure Al [15] and other metallic materials. For $\dot{\varepsilon}=3\cdot10^{-4}$ s^{-1}, the values of Q_a sharply decrease from $\approx$326 kJ/mol at T=250°C to 151 kJ/mol at T=400°C. With further increases in temperature to 500°C, the values of Q_a tend to decrease to the value of activation energy for lattice diffusion in pure aluminum (Q_l=143.4 kJ/mol [10]). Thus, there is a well-defined decrease in Q_a with increasing temperatures.

3.4 Surface observations

Surface observations showed that temperature and strain rate did not have a significant effect on dislocation glide (Figs.4 and 5). At T=250°C and $\dot{\varepsilon}=10^{-4}$ s^{-1}, long slip features belonging to two systems of dislocation are observed (Fig.4a). Steps on slip lines and the wavy shape of slip features are evidence of cross-slip. Dislocation glide is essentially uniform. An increase in strain rate up to 10^{-2} s^{-1} results in intense localization of dislocation glide on the microscale level (Fig.4b). The slip lines form bands of localized deformation.

Uniform deformation takes place at T=500°C (Figs.5a and 5b). Short and wavy slip lines being evidence for operation of cross-slip on a single system are observed both near the initial boundaries and inside the grains. Thus, a decrease in temperature and an increase in strain rate result in localization of dislocation glide, only. At all temperatures and strain rates examined cross-slip on multiple slip systems occurs. Particles of secondary phases provide barriers for

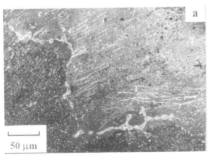

Fig. 4. Deformation relief of AA 2219: ε= 15%, T=250°C, (a) $\dot{\varepsilon}=10^{-4}$ s^{-1}; (b) $\dot{\varepsilon}=10^{-2}$ s^{-1}.

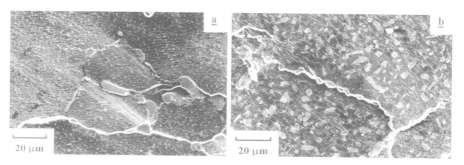

Fig. 5. Deformation relief of AA 2219: ε=15%, T=500°C, (a) $\dot{\varepsilon}=10^{-4}s^{-1}$; (b) $\dot{\varepsilon}=10^{-2}\,s^{-1}$.

the propagation of slip. Slip lines bypass these particles. These observations suggest that secondary phases are responsible for initiating extensive cross slip.

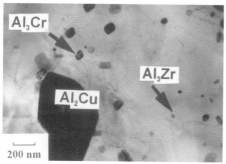

Fig. 6. TEM structure of AA2219 deformed at 475°C.

3.5 Microstructure observation

Second-phase particles (Fig.6) were identified by TEM analysis as Al_3Cr and Al_3Zr with average sizes 0.1 μm and 20 nm, respectively. It is known that the Al_3Zr particle boundaries can be both coherent as well incoherent [17]. Coherent boundaries exhibit coherent stress fields which are distinguished by TEM as non-uniform diffraction contrast near a dispersoid. It was observed that at least 85pct of the Al_3Cr and Al_3Zr dispersoids exhibited incoherent boundaries.

4. Discussion

An inspection of the deformation behavior of the 2219 aluminum alloy suggests that this material could exhibit threshold behavior. This possibility follows from the observations by Li et al. [18] that threshold behavior typically results in an anomalous dependence of flow stress on strain rate. It is seen from Fig.2a that a progressive decrease in strain rate with decreasing applied stress is observed. It results in an increasing stress exponent, n, with decreasing strain rate and unrealistically high values of the apparent activation energy in the low temperature range where high values of the stress exponent are observed. Thus, the deformation behavior of the 2219 Al alloy cannot be described by a single expression of the form given by Eqn (1) and is described by an equation accounting for the threshold stress [19]:

$$\dot{\varepsilon}=A\left(\frac{\sigma-\sigma_{th}}{G}\right)^n\exp\left(\frac{-Q_c}{R\cdot T}\right), \qquad (5)$$

where σ_{th} is threshold stress, $\sigma-\sigma_{th}$ is an effective stress driving the deformation observed and Q_c is a "true" activation energy for plastic deformation. The possibility of threshold behavior in this alloy is examined in the next section.

4.1 Estimation of threshold stresses

A standard procedure was used to determine the threshold stress [7]. The experimental data at a single temperature were plotted as $\dot{\varepsilon}^{1/n}$ against the steady state flow stress σ on a double-linear scale for all temperatures examined. Data in the strain rate interval were best fitted to a straight line by varying the n values. The n value exhibiting the highest regression coefficient in a linear fit was taken as the true stress exponent value. The intercept on the stress axis yields the σ_{th} value, which, in principle, is independent of the applied stress.

Values of threshold stress obtained and the true stress exponent values, n, are summarized in Table 1 as a function of testing temperature. Inspection of Table 1 shows that stress exponents of 7, 6 and 5 yield the best linear fit between $\dot{\varepsilon}^{1/n}$ and σ at T=250°C, in the temperature range T=300 - 450°C, and at T=500°C, respectively. The values of threshold stress can be as

Table 1. Threshold stresses and n values at different temperatures.

T, °C	250	300	350	400	450	500
n	7	6	6	6	6	5
σ_{th}, MPa	38.0	19.8	9.7	5.2	3.3	2.7

much as 75% of the value of the applied stress at T=250°C and the lowest strain rate examined and tend to decrease with increasing temperature to 10% at T=500°C. Notably, the values of threshold stress obtained are, at least, two times lower than values reported at a temperature of 250°C for Al-6%Cu -0.6%Mg alloy additionally alloyed by 1.8%Mn, 0.1%Cr, 0.4%Ag, 0.4%Zr, 0.2%V and 0.25%Ti [1] and lower in comparison with data for Al-5%Cr-2%Zr alloy by factors of 1.5-4 at similar temperatures [11]. The threshold stresses obtained are essentially similar to those reported for the 2124, 6061 and 2024 aluminum alloys produced by powder metallurgy techniques [8,9,18] in the temperature interval from 350 to 400°C.

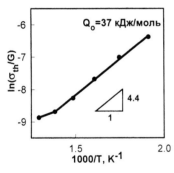

Fig. 7. Plot of logarithmic of the normalized threshold stresses *vs* reciprocal absolute temperature.

Fig.7 shows the temperature dependence of the modulus-normalized threshold stress. It is seen that the data follows the equation [7-10,18]:

$$\frac{\sigma_{th}}{G} = B_o \exp\left(\frac{Q_o}{R \cdot T}\right), \qquad (6)$$

where B_o is a constant; Q_o is an energy term representing the activation energy for a dislocation to overcome an obstacle. The value of Q_o inferred from the slope of the plot in the temperature range 250-450°C in Fig.7 is 37 kJ/mol and tends to decrease sharply with increasing temperature. This energy term value is slightly higher in comparison with values of the energy term reported for Al2124 (27.5 kJ/mol) [18], Al6061 (19.3 kJ/mol) [8] and Al2024 (23 kJ/mol) [9] produced via powder metallurgy technique.

4.2 Activation energy

To determine the activation energy for deformation, the normalized effective stresses ((σ-σ_{th})/G) were plotted as a function of 1/T on semi-logarithmic scales at fixed strain rates (Fig.8a) [14]. Three temperature intervals (250-300°C, 300-450°C and 450-500°C), distinguished by slope value, k, can be found. In these temperature intervals the values k are ~0.8, ~1.6 and ~3.5, respectively. Notably in the temperature ranges revealed, the data in Fig. 8a shows very little strain rate dependence, which is indicative of the validity of the threshold stress values determined.

Values of true activation energy (Q_c) of plastic deformation were computed assuming that Eqn (5) adequately describes the deformation behavior of the 2219 Al alloy. Results are shown in Fig. 8b. It is seen that the Q_c values ranging from 73 to 92 kJ/mol depending on

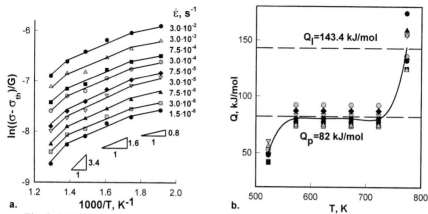

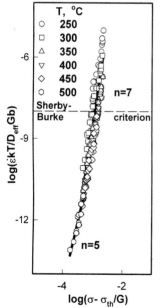

Fig. 8. (a) Shear modulus compensated effective stress *vs* inverse absolute temperature;
(b) temperature dependence of the true activation energy, Q_c.

strain rate were measured in the temperature range T=300-450°C. The average value of Q_c in this temperature range is about 80 kJ/mol. This value is similar to that for pipe diffusion in Al (Q_p=82 kJ/mol [10]). At T=500°C the true activation energy for plastic deformation, Q_c, is close to the value of 144 kJ/mol. This value is similar to the activation energy for self-diffusion in Al (143.4 kJ/mol) [10]. At T=250°C, the average value of Q_c was calculated to be about 50 kJ/mol. Thus, the incorporation of a threshold stress in the activation energy analysis provides realistic values of the true activation energy. A classic temperature dependence of the true activation energy [15] was found in the 2219 aluminum alloy as shown in Fig. 8b. The activation energy tends to increase up to the value of activation energy for pipe-diffusion Q_p at T≤300°C [20]. No significant variation in Q_c value is observed in the temperature range 300-450°C. Further increases in temperature result in an increase of the true activation energy up to a value of activation energy for lattice self-diffusion.

4.3 An examination of normalized deformation data

Figure 9 shows a double logarithmic plot of the normalized strain rate, $\dot{\varepsilon}kT/D_{eff}Gb$, vs the normalized effective stress, $(\sigma-\sigma_{th})/G$ for the 2219 aluminum alloy. The Burger's vector was taken as b=2.86·10^{-10} m and an effective diffusion coefficient [20] was used:

$$D_{eff}=f_l D_l + f_p D_p , \qquad (7)$$

where D_l and D_p are the lattice self-diffusion and the pipe-diffusion coefficients, and f_l and f_p are the fractions of atoms participating in lattice and pipe- diffusion, respectively. The lattice self-diffusion coefficient was calculated as [10]:

$$D_l =1.86·10^{-4}\exp\left(\frac{143.4}{R·T}\right) \text{ m}^2/\text{s} . \qquad (8)$$

Fig. 9. Normalized strain rate vs normalized stress for the AA2219.

238

The value of f_l was assumed to equal 1. The term associated with the contribution of pipe-diffusion to the total diffusion process was determinated for pure Al according to data presented in [20] as:

$$f_p D_p = 10 \left(\frac{a_c}{b^2}\right) \cdot \left(\frac{\sigma}{G}\right)^2 \cdot D_p, \tag{9}$$

$$a_c D_p = 7.0 \cdot 10^{-25} \exp\left(\frac{-82}{R \cdot T}\right), \text{ m}^4/\text{s}, \tag{10}$$

where a_c is the area of cross-section of the dislocation core.

It is seen from Fig. 9, that there are two different intervals for plastic deformation of the 2219 Al alloy. At T=500°C and at $\dot{\varepsilon}kT/(D_{eff}Gb)<10^{-11}$ there is a well-defined power-low relationship with the stress exponent, n, of 5. The analysis presented in [12,20], allows to conclude that for n, which is in the range from 3 to 5, and for Q_c, which is equal to Q_l, the climb of edge dislocations is the most probable rate-controlling mechanism for creep and creep deformation is controlled by lattice diffusion. Therefore, on the basis of relative agreement of the measured and anticipated values of n and Q_c, a high-temperature dislocation climb mechanism of some type can control the creep behavior of 2219 Al alloy at T ≥500°C. I.e. this is a region for true hot deformation of the 2219 aluminum alloy where an extensive dislocation rearrangements by climb take place. At lower temperatures, 250-450°C, the slope of straight line dependence $\dot{\varepsilon}kT/(D_{eff}Gb)$ vs $(\sigma-\sigma_{th})/G$ is equal 7. An inspection of the normalized deformation data shows that a classic transition from high temperature climb controlled by vacancy diffusion through the lattice to low temperature climb, which is controlled by vacancy diffusion along dislocation cores, takes place with temperature decrease

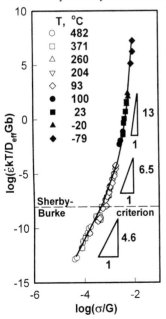

Fig. 10. Normalized strain rate vs normalized stress for Al from datum points from [15].

from 500 to 450°C. This transition results in an increase in the stress exponent from n=5 to n=7 in accordance with the rule n=n+2 [15]. Concurrently, a decrease in true activation energy for plastic deformation from $Q_c = Q_l$ to $Q_c = Q_p$ takes place. It is indicative that transition from high temperature climb, controlled by lattice diffusion to low temperature climb, controlled by pipe-diffusion takes place [12,20,21]. This transition results in significant increase in rate of dislocation climb and restricts an ability of lattice dislocation to rearrange. As a result, this transition can be considered as an incipient point for progressive transition from hot deformation to warm deformation. As it is seen from surface observations the cross-slip plays an important role in dislocation rearrangements in the temperature range examined. Notably, the region of low temperature climb is sheared by the Sherby –Burke criterion ($\dot{\varepsilon}kT/(DGb)\approx10^{-8}$), which according to [21] is a criterion for the beginning of power law breakdown. However, no breakdown of power-law behavior is observed at this value of normalized strain rate.

The transition from hot deformation to warm deformation in the 2219Al initiates at a normalized effective stress of $(\sigma-\sigma_{th})/G\approx6\cdot10^{-4}$. Notably, this transition takes place at a similar value of normalized stress in pure aluminum (Fig.10 is replotted Fig.7a from [15]). Thus, a range of hot deformation in the 2219 aluminum alloy is observed at temperatures more than $200°C$ higher than that for pure aluminum [15] due to enhanced flow stress. A process controlling the deformation behavior in the 2219 aluminum alloy is low temperature climb in the temperature range $300-450°C$ and high temperature climb at $T>450°C$. The temperature increase enhances dislocation rearrangements by climb or cross-slip. As a result, dislocation glide tends to become more uniform with increasing temperature.

4.4 Origin of threshold stress

It seems that in the 2219 aluminum alloys the incoherent Al_3Cr and Al_3Zr particles are the origin of the threshold stress, i.e., the 2219 aluminum alloy exhibits threshold behavior as observed in dispersion strengthened alloys [7,8] because of the nanoscale particles of transition alloying elements.

These incoherent particles act as barriers to dislocation glide. Fig.11 schematically represents a modification of well-known second Weertman's model [12]. A dislocation glides on a crystallographic plane, encounters an incoherent particle, overcomes the dispersoid by climb and detaches from this particle. The following dislocation glide results in the formation of a dislocation dipole consisting of two dislocations with opposite signs. Climb of these dislocations toward one another leads to their annihilation. Therefore, the deformation behavior of the 2219 alloy is controlled by two sequential processes of an interaction between a dislocation and an obstacle and dislocation climb. The first process determines threshold behavior, and the second one controls general deformation behavior.

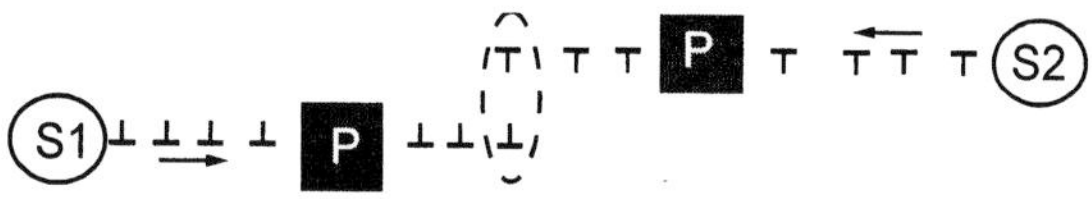

Fig. 11. A scheme of deformation mechanism consisting of two sequential processes: interaction of a dislocation with a dispersoid and dislocation annihilation by classic Weertman's model: S1 and S2 are the dislocation sources; P are the particles of secondary phases.

5. Conclusion

1. The deformation behavior of the 2219 aluminum alloy was investigated at temperatures from 250 to $500°C$ in strain rate range from $1.5 \cdot 10^{-6}$ up to $3 \cdot 10^{-2}$ s^{-1}. Analysis of experimental data of the 2219Al revealed the presence of a threshold stress at temperatures examined. A temperature dependence of threshold stress with the energy term, Q_o, of 37 kJ/mol was found in the temperature range $250-450°C$.

2. An incorporation of threshold stresses into analysis of deformation behavior allows two intervals of plastic deformation to be revealed. A transition from warm deformation range with value of true stress exponent, n, of 7 to range of hot deformation with value of true stress exponent, n, of 5 takes place at temperature about $450°C$ and normalized effective stress $(\sigma-\sigma_{th})/G \cong 6 \cdot 10^{-4}$. Notably, this transition takes place at a temperature approximately $200°C$ higher in the 2219 Al alloy than in pure aluminum.

3. The true activation energy for plastic deformation, Q_c, of about 49 ± 5 kJ/mol, 80 ± 5 kJ/mol and 144 ± 13 kJ/mol were found in the temperature intervals $250-300°C$, $300-450°C$ and at $500°C$, respectively.

4. Dislocation cross-slip on multiple systems occurs. Temperature decrease and strain rate increase result in localization of dislocation glide.

5. Incoherent nanoscale dispersoids in the 2219 Al alloy originate the threshold behavior of this alloy.

240

Acknowledgments

This work was performed in collaboration with the Lawrence Livermore National Laboratory under the auspices of the U.S. Department of Energy under contract No. W-7405-ENG-48.

6. References

[1].Moris, M.A., 1992, Creep Deformation of an Al Alloy with Intermetallic Particles, Phil.Mag., Vol.65, pp. 943-960.

[2].Kazanjian, S., Wang, N., Starke, E., 1997, Creep behavior and microstructural stability of Al-Cu-Mg-Ag and Al-Cu-Li-Mg-Ag alloys, Mater.Sci.Eng., Vol.A234-236, pp. 571- 574.

[3].Wang, J., Wu, X., Xia, K., 1997, Creep behaviour at elevated temperatures of an Al-Cu-Mg-Ag alloy, Mater.Sci.Eng., Vol.A234-236, pp. 287-290.

[4].Chaudhury, P.K., Mohamed, F.A., 1987, Creep and Ductility in an Al-Cu Solid-Solution Alloy, Metall.Trans., Vol.18A, pp. 2105-2115.

[5].Chaudhury, P.K., Mohamed, F.A., 1988, Creep Characteristics of an Al-2wt.% Cu Alloy in the Solid Solution Range, Mater.Sci.Eng., Vol.A101, pp. 25-30.

[6].Kaibyshev, R., Sitdikov, O., Mazurina, I., Lesuer, D.R., 2000, Continuous Dynamic Recrystallization in Warm Equal Channel Angular Extrusion of AA2219, Scr.Mater. (submitted for publication).

[7].Mohamed, F.A., Park, K.T., Lavernia, E.J., 1992, Creep Behavior of Discontinuous SiC-Al Composites, Mat. Sci. Eng., Vol.A150, pp. 21-35.

[8].Park, K.T., Lavernia, E., Mohamed, F., 1994, High-Temperature Deformation of 6061`Al, Acta Metall. Mater., Vol.42, pp. 667-678.

[9].Kloc, L., Spigarelli, S., Cerri, E., Evangelista E., Langdon, T.G., 1997, Creep Behavior of an Al 2024 Alloy Produced by Powder Metallurgy, Acta Mater., Vol.45, No:2, pp.529-540.

[10].Li, Y., Langdon, T.G., 1999, A Unified Interpretation of Threshold Stresses in the Creep and High Strain Rate Superplasticity of Metal Matrix Composites, Acta Mater., Vol.47, pp. 3395-3403.

[11].Brahma, A., Gerique, T., Torralba, M., Leiblich, M., The threshold stress in a rapidly solidified Al-5Cr-2Zr alloy, Mater.Sci.Eng., Vol.A246, No.1, pp. 55- 60.

[12].Chadek, J., 1994, Creep in Metalic Materials, Academia, Praque, p. 302.

[13].Drury, M.R., Humphreys, F.J., 1986, The Development of Microstructure in Al-5%Mg during High Temperature Deformation, Acta Metall., Vol.34, No.11, pp. 2259-2271.

[14].Pickens, R., Langan, T.J., England, R.O., Liebson, M., 1987, A Study of the Hot Working Behaviour of SiC-Al Alloy Composites and their Matrix Alloys by Hot Torsion Testing, Metall. Trans., Vol.18A, No.2, pp. 303-312.

[15].Luthy, H., Miller, A., Sherby, O., 1980, The Stress and Temperature Dependence of Steady State Flow at Intermediate Temperature for Pure Policrystallyne Al, Acta Metall. Vol.28, No.2, pp.169-182.

[16].Zong, B.Y, Derby, B., 1997, Creep Behavior of a SiC Particulate Remforced Al-2618 Metal Matrix Composite, Acta Mater., Vol.45, No.1, pp. 41-49.

[17].Commercial aluminum alloys, Moscow, Metallurgy, 1984, p. 396

[18].Li, Y., Nutt, S.R., Mohamed, F.A., 1997, An Investigation of Creep and Substructure Formation in 2124 Al, Acta Mater., Vol.45, pp. 2607-2620.

[19].Mohamed, F., 1988, On the Threshold Stress for Superplastic Flow, J. Mater.Sci.Lett., Vol.7, pp. 215-217.

[20].H.Frost, M.Ashby, 1982, Deformation-Mechanisms Maps, Pergamon Press, p. 328

[21].Raj, S., Langdon, T., 1989, Creep Behavior of Copper at Intermediate Temperatures – I. Mechanical Characteristics, Acta Metall., Vol.37, No.2, pp. 843-852.

Creep Behaviour of the Particle Strenghtened Aluminium Alloy AA2024

A. Kranz[1], E. El-Magd[1]

[1] Department of Materials Science (LFW), RWTH Aachen University of Technology
Augustinerbach 4, D-52062 Aachen, Germany

Keywords: Constitutive Equations, Internal Back Stress, Precipitation Strengthening, Damage Evolution, Heat Treatment, Modelling.

Abstract

The microstructure evolution and creep behaviour are investigated on the aluminium alloy AA2024 in two different heat treatment conditions: solution annealed and overaged. The creep rate is considered as a function of an effective stress intensity which is the difference between the applied stress σ and the internal deformation resistance including an internal back stress σ_i and a precipitation particle resistance σ_p. Damage is taken into consideration by a damage function D. The evolution equations of the internal variables σ_i, σ_p and D are determined by simple structure-mechanical models verified by experimental procedures. The internal back stress σ_i is determined using Strain-Transient-Dip-Test technique. The evolution of particle resistance σ_p is estimated from the kinetics of precipitation which depends on the dislocation density and the Ostwald-ripening. The damage variable D is determined as a function of the creep strain in metallographic investigations on the tested material.

1 Introduction

Precipitation hardened aluminium alloys like AA2024 (Duralumin, AlCuMg2, 3.1355) are commonly used for lightweight constructions. In different applications, e.g. in the automotive industry, aluminium alloy parts are exposed to temperatures under service conditions. For this purpose an exact analysis of mechanical properties is necessary, especially at temperatures higher than 0.4 of the absolute melting temperature. Constitutive equations based on a combination of overstress and threshold stress concept can allow an adequate description of the materials behaviour if successive damage is taken into consideration.
The current value of the strain rate depends on the current values of the applied stress σ, the internal back stress σ_i, the particle deformation resistance σ_p, the material creep resistance σ_F and the degree of damage D. The creep rate can be represented by a modified power law [1, 2, 3, 4]:

$$\dot{\varepsilon} = \dot{\varepsilon}_0 \left[\frac{\sigma - (\sigma_i + \sigma_P)}{\sigma_F (1-D)} \right]^n \tag{1}$$

$$\sigma_i = f\left(\varepsilon,\dot{\varepsilon},\sigma,S_i\right) \qquad \sigma_p = f\left(\sigma,V_p,d,T\right) \qquad D = f\left(\varepsilon\right)$$

Internal Back Stress   Particle Resistance   Damage

dislocations   precipitation and ripening   cavitations

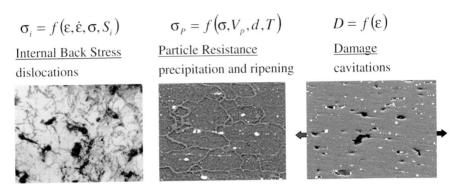

Fig.1 Modified power creep law which consists of internal back stress, particle stress and damage

The internal back stress σ_i is an anisotropic deformation resistance which developes with increasing deformation. It represents a kinematic hardening and is determined by the mobile dislocation density. The particle stress σ_p accounts for the interaction between mobile dislocations and precipitates and depends on the distance between precipitates which may change in the course of creep. The parameter σ_r is considered as a reference stress and set to 1 MPa in this study.

2 Experimental Procedure

The chemical composition is given in
Table 1:

Table 1. Chemical composition of AA2024.

2024 T351								
Element:	Si	Fe	Cu	Mn	Mg	Cr	Zn	Ti	Al
Weight-%:	0,11	0,24	4,25	0,63	1,38	0,12	0,081	0,037	Rest

The creep-test specimens measure 5 mm in diameter and 50 mm in gauge length. They are machined from hot extruded 10 mm solid circular bars delivered in the T351 heat treatment condition (solution annealing, water quenching, tensile strengthened of 3% and ageing). Two further heat conditions are investigated:
- Solution annealed (W): 1 hour at 495°C, water quenched.
- Highly Overaged (O): 24 hours at 400°C, air cooled at room temperature.

The microstructure of AA2024 in different heat treatment conditions is given in Fig.2. The grain size remains constant (mean grain size = 19 μm).

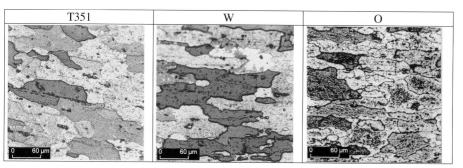

T351	W	O

Fig.2 Micrographs of AA2024 for different heat treatments: T351, solution annealed (W) and overaged (O).

For each heat treatment condition several creep tests are carried out. For determination of damage the first creep test is kept under load until rupture at a creep strain ε_f. Other creep tests are stopped after reaching predefined strain values $0 < \varepsilon/\varepsilon_f < 1$ in order to obtain crept specimens with different degrees of damage. The internal back stress is determined in Strain-Transient-Dip-Tests as described in [5].

3 Creep Model

3.1 Internal Back Stress

Previous investigations [6,7,8] showed that the internal back stress is a function of the dislocation density ρ:

$$\sigma_i = \alpha Gb\sqrt{\rho} \qquad (2)$$

where G is the modulus of rigidity, b the Burgersvector and α is a constant. The evolution equation of the dislocation density for fcc-materials introduced by KOCKS and MECKING [8]

$$\frac{d\rho}{d\varepsilon} = k_1\sqrt{\rho} - k_2\rho \qquad (3)$$

can be used to evaluate the variation of the internal back stress according to

$$\frac{d\sigma_i}{d\varepsilon} = \frac{C_1}{\varepsilon_s}(\sigma_{is} - \sigma_i) \qquad (4)$$

which is validated experimentally in [9].

In eq. (4), $\sigma_{is} = \alpha Gbk_1/k_2$ the quasi-stationary value of the internal back stress, ε_s is the creep strain at the end of the primary creep stage and C_1 is a constant. Under constant stress and temperature, the internal back stress increases in the primary stage according to

$$\frac{\sigma_i}{\sigma_{is}} = 1 - \exp\left[-C_1 \cdot \frac{\varepsilon}{\varepsilon_s}\right] \qquad (5)$$

This relationship is shown in Fig.3a for AA2024 "Cast". For different stresses the quasi-stationary value σ_{is} reaches a temperature dependent saturation value σ_{iss}, which can be described with the following exponential function:

$$\sigma_{iss}(T) = k_0 \cdot \exp(\frac{\beta}{T}) \tag{6}$$

where k_0 and β are material constants.

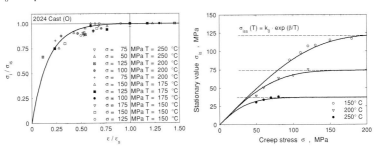

<u>Fig.3</u> a) Evolution of the internal back stress σ_i in the primary creep stage (Cast „O")

b) Quasi-stationary value of the internal back stress σ_{is} as a function of σ [5]

3.2 Particle Strengthening

The variation of the particle deformation resistance σ_p during creep exposure was studied in [1] and validated on austenitic steel. It is considered, that σ_p depends on the current mean value of the interparticle spacing L yielding

$$\sigma_p = \frac{S}{L} \tag{7}$$

where S is a material constant.

This relationship is based on the theory of BROWN and HAM, reported by MARTIN [10], which assumes that the high temperature yield stress of particle-hardened material is controlled by local climb of dislocations over the particles.

From geometry considerations it can be shown that L varies with V_p and d according to the relationship

$$L = \sqrt{\frac{\pi}{6V_p}} \cdot d \tag{8}$$

The particle diameter increases due to OSTWALD ripening [11] an can be represented by the function

$$d = \left[d_0^{\mu} + Ct \right]^{1/\mu} \approx (Ct)^{1/\mu} \tag{9}$$

where C is a function of temperature according to the relation $C = C_0 \exp\left[-Q_1 / RT\right]$. The volume fraction is dependent on the number N_p of particles per unit volume and the particle diameter. At a given temperature, the value of N_p can be written as

$$N_p = N_{p0} + f(t) \cdot g(t) \tag{10}$$

where N_{p0} is the initial value. The function $f(t)$ accounts for the initiation of new nuclei whereas the reduction of supersaturation in the coarse of the precipitation process is considered by introducing the function $g(t)$, which is represented by a hyperbolic function

$$g(t) = \left[1 + (Bt)^{\nu}\right]^{-m} \tag{11}$$

where B is a function of temperature and creep strain.

The initiation function $f(t)$ depends on the temperature and the dislocation density. The latter is a function of creep strain $\varepsilon(\sigma, T, t)$ and $f(t)$ can be written as

$$f(t) = \left\{\left[a_1 f_1(T) + a_2 (\sigma^{n_2} e^{-Q_2/RT})^{M_2}\right] \cdot t\right\}^M \tag{12}$$

where Q_2 is the creep activation energy. The volume fraction of the precipitates,

$$V_p = N_p \frac{\pi}{6} d^3 \tag{13}$$

follows from eqs. (9)-(12) in (13) reading

$$V_p = \frac{\pi}{6}\left[a_1 f_1(T) + a_2 (\sigma^{n_2} e^{-Q_2/RT})^{M_2}\right]^M C^{3/\mu} t^{M+3/\mu} \cdot \left[1 + (Bt)^{\nu}\right]^{-m} \tag{14}$$

Assuming that with increasing time V_p approaches asymptotically a final value $V_{p\infty}$ which depends only on the initial supersaturation but is not a function of time or stress, eq. (14) is to be rewritten as

$$V_p = V_{p\infty}\left[\frac{(Bt)^{\nu}}{1 + (Bt)^{\nu}}\right]^m \tag{15}$$

where following relations exist between the parameters of the functions $f(t)$ and $g(t)$:

$$m = \frac{1}{\nu}\left(M + \frac{3}{\mu}\right) \tag{16}$$

$$B = \left\{\frac{\pi}{6 V_{p\infty}}\left[a_1 f_1(T) + a_2 (\sigma^{n_2} e^{-Q_2/RT})^{M_2}\right]^M C^{3/\mu}\right\}^{1/(m\nu)} \tag{17}$$

As M must be positive if $f(t)$ is to increase with time according to eq. (12), $m\nu > 3/\mu$. With $\mu = 3$, the product $m\nu$ is greater than 1. From eqs. (8), (9), (15) in eq. (7) the particle resistance can be written as

$$\sigma_p = \frac{S}{C^{1/\mu}}\sqrt{\frac{6}{\pi}V_{p\infty}}\left[\frac{(Bt)^{\nu}}{1 + (Bt)^{\nu}}\right]^{m/2} t^{-1/\mu} \tag{18}$$

σ_p reaches a maximum value $\sigma_{p\max}$ after time t^*, given by

$$t^* = \frac{1}{a^{\nu}B} \tag{19}$$

where $a = 2/\nu m\mu - 2$ and the maximum particle resistance follows the relation

$$\sigma_{p\max} = \frac{S}{C^{1/\mu}}\sqrt{\frac{6}{\pi}V_{p1}} \cdot a^{-m/2} \cdot t^{*-1/\mu} \tag{20}$$

The variation of σ_p with creep time is given by the relationship

$$\frac{\sigma_p}{\sigma_{p\max}} = \left[\frac{a+1}{1 + a\left(\dfrac{t^*}{t}\right)^{\nu}}\right]^{m/2}\left(\frac{t^*}{t}\right)^{1/\mu} \tag{21}$$

The time t^* defined by the time point at the minimum creep rate in the case of particle strengthened material is shown in Fig.4.

246

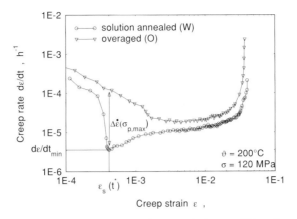

Fig.4 Comparison of creep behaviour for solution annealed $\sigma_p = f(t)$ and overaged ($\sigma_p \approx 0$) material

3.3 Determination of Damage Evolution

An example of quantitative analysis of the damage evolution for the aluminium alloy AA2024 is presented including statistical results taken from interrupted creep tests.

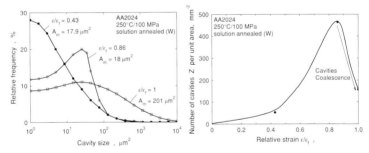

Fig.5 Statistical distribution of cavity size (a) and total number of cavities Z per unit area (b) at different relative strains $\varepsilon/\varepsilon_f$.

The statistical distribution of the void size is shown in Fig.5a. At a relative small strain of $0.43\,\varepsilon_f$, the fraction of small voids is much higher than the fraction of the large ones. With increasing creep strain, the fraction of large voids increases which indicates that the rate of void growth is relative high compared with the rate of initiation of new voids. The number of voids per unit area (Fig.5b) increases first with increasing relative strain $\varepsilon/\varepsilon_f$ and is dramatically reduced immediately before fracture due to void coalescence.

Load Direction

a)   b)

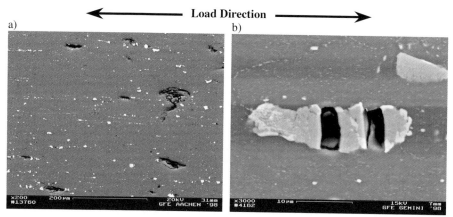

<u>Fig.6</u> (a,b) Scanning Electron Microscope: Polished Specimen, T351, 250°C/100 MPa
a) Overview ($\varepsilon = 0.0337$) ; b) Broken Precipitate ($\varepsilon = 0.0278$)

The SEM-micrograph in Fig.6a shows voids (dark areas) and precipitates (white spots) in different sizes. Fig.6b represents a broken precipitate. This points to a damage mechanism which develops voids by fracture of precipitates that are fixed strongly to the matrix [12]. The results obtained from metallographic investigations were used to describe damage functions (D). Different formulations can be applied for the evolution function $dD/d\varepsilon$:

A modified KACHANOV-ROBOTNOV [13,14] relation reads:

$$\frac{dD}{d\varepsilon} = \frac{a}{(1-D)^m} \Rightarrow D = 1 - (1 - \frac{\varepsilon}{\varepsilon_f})^{\frac{1}{m+1}} \tag{22}$$

Further functions are a power law $dD/d\varepsilon = aD^n$ leading to $D = (\varepsilon/\varepsilon_f)^{1/(1-n)}$ [15] or a linear function $dD/d\varepsilon = a + b\varepsilon$ indicating $D = [\exp(b \cdot \varepsilon/\varepsilon_f)]/[\exp(b)-1]$.

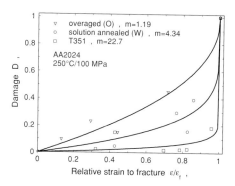

<u>Fig.7</u> Damage parameter $D = A_D/A_{Df}$ determined metallographically as a function of the relative creep strain $\varepsilon/\varepsilon_f$ and described with eq. (22)

Fig.7 presents the damage function for different heat treatment conditions of alloy AA2024. The damage parameter is determined metallographically as the ratio of the void area A_D at an arbitrary creep strain ε to the void area A_{Df} at fracture.

4 Application to creep tests

Fig.8 shows the evolution of the internal variables for the solution annealed as well as the overaged material during creep exposure at 200 °C with 140 MPa. The implementation of eqs. (5),(21),(22) into the power-creep-law of eq. (1) leads to a good description of the experimental data. Especially the influence of particle resistance on the minimum creep rate and - in combination with different damage behaviour - the differences in fracture time between the heat treatments is obvious. In Fig.9 the model is applied to experimental data in a temperature range of 150-200°C for different stresses and heat treatments.

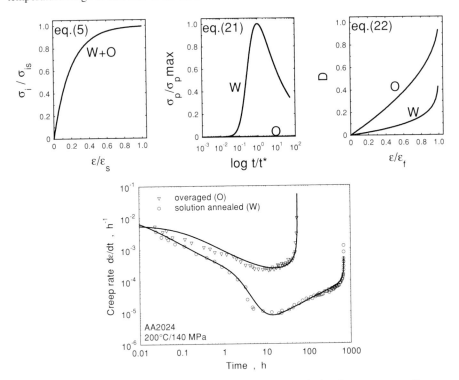

Fig.8 Evaluation of eqs. (5),(21),(22) and comparison of the calculated creep curve with creep results for solution annealed (W) and overaged (O) specimens (200°C/140 MPa)

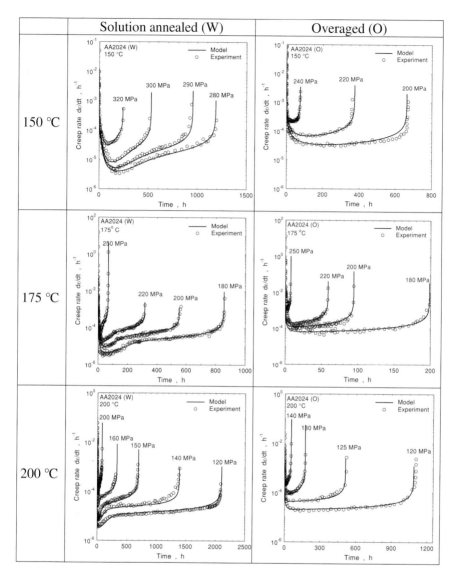

Fig.9 Application of the model to experimental data in a temperature range of 150-200°C for different stresses and heat treatments

Acknowledgements

The authors gratefully acknowledges the financial support of the Deutsche Forschungsgemeinschaft (DFG) within the Collaborative Research Center 370 „Integrated Modelling of Materials".

250

References

[1] E. El-Magd, G. Nicolini and M.M. Farag, (1996), Effect of carbide precipitation on the creep behaviour of alloy 800 HT in the temperature range 700-900°C, Met. & Mat. Transactions, USA, 27A 3, pp. 747-756.

[2] J. Dünnwald, E. El-Magd, (1996), Description of the creep behaviour of the precipitation-hardened material Al-Cu-Mg Alloy 2024 using finite element computations based on microstructure mechanical models, Computational Materials Science 7 pp. 200-207.

[3] J. Dünnwald, E. El-Magd, M. Deuper, L. Löchte, G. Gottstein, (1997), Betrachtung der Ausscheidungskinetik in Al-Cu-Mg (AA2024) mit äußerer Spannung sowie deren Auswirkung auf das Kriechverhalten, Werkstoffwoche 1996 in Stuttgart, Tagungsband zu Symposium 7, "Materialwissenschaftliche Grundlagen", DGM-Verlag, pp. 23-28.

[4] E. El-Magd and C. Shaker, (1991), Ein strukturmechanisches Modell für das Kriechverhalten metallischer Werkstoffe bei zeitveränderlicher Spannung, Mat.-wiss. u. Werkstofftech. 22, pp. 56-62.

[5] E. El-Magd, A. Kranz: Determination of internal back stress of Aluminium AA2024 under creep load; Mat.-wiss. u. Werkstofftech. 31 (2000) 1, S. 96-101

[6] B. Ilschner: Mechanisches Verhalten bei hoher Temperatur, Festigkeit und Verformung bei hoher Temperatur, DGM Informationsgesellschaft Verlag, (1989)

[7] C.N. Ahlquist, W.D. Nix: The Measurement of Internal Stresses During Creep of Al and Al-Mg Alloys, Acta metall., 19(1971)4, S. 373-385

[8] H. Mecking, U.F. Kocks: Kinetics of Flow and Strain-Hardening, Acta metall., 29(1981)11, S. 1865-1875

[9] E. El-Magd, C. Shaker: Determination of Internal Back Stress under Time-Dependent Creep-Conditions, in: „Creep and Fracture of Engineering Materials and Structures", Ed.: B. Wilshire und R.W. Evans, Inst. Metals, London, (1990), S. 119-129

[10] J.W. Martin: Micromechanisms in Particle-Hardened Alloys, Cambridge University Press, Cambridge, 1980, pp 162-187.

[11] W. Ostwald: Über die vermeindliche Isotropie des roten und des gelben Quecksilberoxyds und die Oberflächenspannung fester Körper, Zeitschrift für physikalische Chemie, 34(1900)10, S. 495-503

[12] J. Gurland, J. Plateau, (1963), Trans. ASM, Nr. 56, pp.442.

[13] L.M. Kachanov, (1960), The theory of creep, Engl. Transl., Ed.: A. J. Kennedy, Boston Spa, Wetherby.

[14] Y.N. Rabotnov, (1969), Creep problems in structural members, Engl. Trans., Ed.: F. A. Leckie, North Holland Publishing Company, Amsterdam.

[15] E. El-Magd and S. Pantelakis, (1984), Anwendung der Schadensakkumulationstheorie zur Kriechbeanspruchung reiner Metalle, Metall 38, pp. 314 – 318.

Corresponding author
Ansgar Kranz
email:
web: www.rwth-aachen.de/lfw

Application of the θ projection concept to Hiduminium RR58 data.

J.S. Robinson[1] J.T. Evans[2] J. Lammas[3]

[1] Department of Materials Science and Technology, University of Limerick, Ireland.
[2] Department of Mechanical, Materials and Manufacturing Engineering, University of Newcastle upon Tyne, Newcastle upon Tyne, NE1 7RU, UK
[3] HDA Forgings Ltd. , Redditch, Worcestershire, B97 6EF, UK

1   Abstract

The Thematic Network CREEPAL (BRRT-CT98-5101) was established in 1998 to collect and disseminate information on the long-term creep and thermal mechanical cycling behaviour of aluminium alloys. One early observation the network made was that very limited good quality creep data for aluminium alloys was available in the public domain. Through CREEPAL, the opportunity has arisen to analyse a large experimental data set for the alloy Hiduminium RR58 (2618A) provided by HDA Forgings Ltd. (UK). This company was intensively involved in the development of this main structural material for the Anglo-French Concorde supersonic aircraft in the late 1950s and 1960s. 66 creep curves were made available for analysis. While now dated, the creep curves are extensive with systematic variation of variables such as ageing treatment, stress and test temperature. The θ projection concept has been applied to data for forged Hiduminium RR58 bar. The creep data has been analysed to obtain the four θ parameters. It is found that the equation $\varepsilon = \theta_1\left(1 - e^{-\theta_2 t}\right) + \theta_3\left(e^{\theta_4 t} - 1\right)$ gives an accurate fit to all of the creep strain versus time data for the alloy. Correlations between the θ parameters and stress and temperature are considered.

2   Introduction

2.1   The CEC Thematic Network CREEPAL (BRRT-CT98-5101)

The objectives of the CREEPAL thematic network initiated in 1998 are;

    i. to increase the amount of creep data in the public domain,
    ii. to define rules for best practice testing and data acquisition,
    iii. to assess the microstructural basis of the resistance to creep and stress relaxation,
    iv. to quantify characterization and microstructural modelling of creep.

A significant programme of creep testing was carried out on the Anglo-French Concorde main structural material, Hiduminium RR58 (subsequently registered as 2618A) in the 1960's by HDA Forgings Ltd. UK, but relatively little of this data has been published. In subsequent years various different ways of interpreting creep data have emerged. Thus it is worth re-examining this creep data in the light of some of these developments.

2.2   Next generation SST

A target design requirement for the fuselage of the next generation of supersonic transport is for creep deformation to be less than 0.1% after 10^4 hours at 110 °C under a stress of 100 MPa. This creep strain is small - only about 70% of the elastic strain at this load and the question arises of aluminium alloy performance in this application. For this aircraft, which was more of commercial possibility in the late 1980 and early 90s than it is now, new materials were considered essential for the fuselage. Advanced aluminium alloys and/or organic composites were both candidates

[1,2]. The CEC funded BRITE-EURAM "NEWAL150" programme initiated in 1992 aimed to develop aluminium base alloys that could demonstrate superior creep and damage tolerance performance compared to 2618A. It has been reported that a modified 2650 alloy exhibited much improved performance in creep [3,4].

2.3   Historical development of Hiduminium RR58

Prior to Concorde, aluminium alloys had been used successfully in aero-engines in cast and wrought forms at temperatures rarely below 200°C. Short design lives up to 1,000 hours had allowed this exploitation even up to temperatures approaching 350°C. The initial development of RR58 arose from the requirement in the late 1930s and early 1940s for an aluminium base alloy capable of operating in the 200-250°C range. The target application at this time was cold section components of the Whittle gas turbine and the first forged impellers made in the alloy were fitted to the Gloster Meteor Mk1, which first flew in 1943. The actual composition of RR58 was developed as a low silicon (improved castability) variant of RR59, an aero-engine piston alloy used extensively during the Second World War. RR58 was selected as the skinning material for the Concorde in the early 1960s over the other candidate alloys 2024, 2219 and 2020 due to its superior combination of mechanical properties.

2.4   Physical metallurgy of Hiduminium RR58

The physical metallurgy of RR58 is well understood [5] with strengthening occurring though the precipitation sequence $\alpha_{ss}\rightarrow GPB\rightarrow S'\rightarrow S(Al_2CuMg)$. Al_2CuMg is orthorhombic and retains coherency in <100> directions over a large range of particle sizes. In Al-Cu-Mg alloys containing Fe but no Ni, the Al_3CuFe phase forms, and for alloys with Ni but no Cu, the AlCuNi phase forms. Both deplete the matrix of Cu and the ageing response is diminished. The Fe and Ni addition in RR58 forms the intermetallic phase $Al_9(Fe,Ni)$ which contains only small amounts of Cu. The $Al_9(Fe,Ni)$ intermetallic is usually randomly dispersed as particles 1-3μm in diameter, Figure 1.

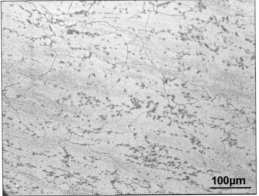

Figure 1 Typical microstructure of forged RR58.

The exact role of the intermetallic particles is still unclear but there is some evidence that the particles nucleate screw dislocations during the quenching stage due to differing thermal contraction with the matrix. These dislocations interact with vacancies forming sessile helical dislocations, which in turn act as potent nucleation sites for the S-phase during age hardening.

3   Analysis

Until relatively recently theories of creep have focussed almost exclusively on the steady-state creep rate in determining the effects of temperature and applied stress. Whilst this approach has some merit for pure materials, there is has been a growing realisation that it is not ideal for practical applications. Indeed some authorities question the validity of the concept of steady state creep in commercial alloys [6,7]. These doubts arise because of the intrinsic metastability of strong alloys. They are formulated and then heat-treated to produce a spectacular increase in yield strength over the pure form of the base metal. But the sub-structural strengthening mechanisms are unstable over long periods at elevated temperatures, which of course are the conditions under which creep occurs. In the θ factor approach [8] the concept of steady-state creep is abandoned and is replaced by the idea that the minimum creep rate is simply observed as a transition between Stage I (decelerating) and Stage III (accelerating) creep behaviour. In this paper we re-examine the RR58 data using the θ parameter method

The starting point for the θ parameter method is that the general form of a three-stage creep curve can be represented by the an equation of the form

$$\varepsilon = \theta_1\left(1 - e^{-\theta_2 t}\right) + \theta_3\left(e^{\theta_4 t} - 1\right) \tag{1}$$

where ε is the creep strain, t is the time and θ_α ($\alpha = 1, 2 .. 4$) are positive parameters. Equation 1 can be fitted to experimental data but its virtue (over, say, a spline fit) is that it provides a strain that increases monotonically with time, as required.

If it has no other virtue, this template for curve fitting can provide a highly condensed representation of creep data that is easily stored and manipulated. For instance, if required the minimum creep rate can be written in terms of the θ parameters in the following form,

$$\dot{\varepsilon}_{min} = \theta_1\,\theta_2\left[\frac{\theta_1\,\theta_2^2}{\theta_3\,\theta_4^2}\right]^{\frac{-\theta_2}{\theta_2+\theta_4}} + \theta_3\,\theta_4\left[\frac{\theta_1\,\theta_2^2}{\theta_3\,\theta_4^2}\right]^{\frac{\theta_4}{\theta_2+\theta_4}} \tag{2}$$

4   Experimental Procedures

Data was considered for RR58 specimens with the following heat treatment. Sections from forgings were solution treated at 530°C for 20 hours and then quenched into boiling water. The specimens were aged for 20 hours at 200°C. Creep tests were then carried out at temperatures between 100 and 350°C. The creep data were presented as hand drawn creep curves. To facilitate analysis these curves were digitised using software (Grab It, V1.51 from Datatrend Software, Inc www.datatrendsoftware.com). The RR58 creep curves were found to be consistent with those recently reported for 2618T61 [9].

5   Results

The strengthening sub-structures of quenched and aged aluminium alloys are metastable. The change in short-term yield strength with extended soaking at elevated temperature is shown in Figure 2. The level of applied stress used in the creep tests at any particular temperature was below the short-term yield strength shown in Figure 2.

254

Equation (1) was fitted to the creep data. Examples are shown in Figure 3 where the points represent the experimental digitised data and the continuous curves are the fitted θ parameter curves.

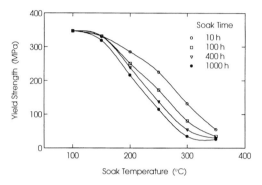

Figure 2 Short-term yield strength as a function of soaking temperature showing the effect of soaking time.

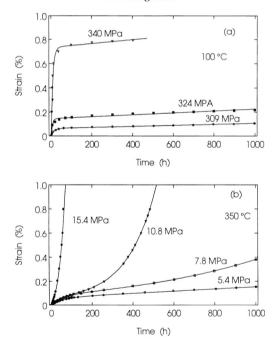

Figure 3 Creep curves at (a) 100°C and (b) 350°C showing experimental digitised data (points) and fitted curves.

In cases where steady-state creep is a dominating process in creep the Monkman-Grant relation is observed to apply. This states that the product of minimum or steady-state creep rate and the time to failure t_c is constant, i.e.

$$\dot{\varepsilon}_{min} t_c = \text{const.} \tag{4}$$

Figure 4 shows a Monkman-Grant plot for the present material, where t_c is taken as the time to achieve 0.1% strain. The best-fit relation from (4) is also shown by the solid curve (const. = 5.83×10^{-4}). The data for temperatures between 250 and 350°C show a reasonable fit to (4). The data for 100 and 150°C are clearly out of line with the Monkman-Grant relation.

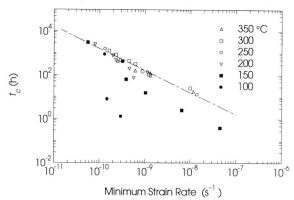

Figure 4 Monkman-Grant plot showing the correlation between the time to achieve a fixed strain (0.1%) and the minimum strain rate.

Power law creep is conventionally represented by the following general relation between steady-state creep rate $\dot{\varepsilon}_{ss}$ and applied stress [910]

$$\dot{\varepsilon}_{ss} = A \frac{D\mu b}{kT} \left(\frac{\sigma}{\mu} \right)^n , \tag{3}$$

where D is the volume or pipe diffusion coefficient, μ is the shear modulus, b is the magnitude of the Burgers vector, k is Boltzmann's constant and T is the temperature. The stress σ is the applied stress, A and n are dimensionless constants. Analysis of the present results (not shown here), identifying $\dot{\varepsilon}_{ss}$ with $\dot{\varepsilon}_{min}$ and using eqn. (3), failed to produce a consistent basis for further analysis since apparent values of the exponent n were found to vary widely with temperature. This is taken to support Wilshire's view that $\dot{\varepsilon}_{min}$ has no particular theoretical significance in characterising the creep of metastable alloys. Further analysis of the results was thus focussed on the experimentally determined θ parameters. However, it is noted that a rudimentary Arhennius analysis of the results suggested an activation energy of between 145 and 188 kJ mol^{-1} for the creep process. These values can be compared with an activation energy of 142 kJ mol^{-1} for solid state diffusion in aluminium [11]. Given the inaccuracies inherent in attempting an Arhennius analysis in cases where the exponent n varies with temperature, identifying the activation energy for diffusion $Q = 142$ kJ mol^{-1} with the activation energy for creep can be justified. This value is used in the temperature-compensating factor used below.

From eqn. (1) the strain rate can be written in terms of the θ factors

$$\dot{\varepsilon} = \theta_1 \theta_2 \, e^{-\theta_2 t} + \theta_3 \theta_4 \, e^{\theta_4 t} \; . \tag{4}$$

Thus the products $\theta_1\theta_2$ and $\theta_3\theta_4$ as well as the θ parameters θ_2 and θ_4 can be regarded as having a significant effect in controlling the strain rate. Thus these parameters may be anticipated to vary systematically with stress and temperature. However, it is difficult not to accept that the strain rate in creep is diffusion controlled. In that case, a major component of the effect of temperature can be factored out by multiplying the strain rate by the compensating factor $\exp(Q/RT)$, where R is the gas constant. In the present case the parameters θ_2 and θ_4 and the products $\theta_1\theta_2$ and $\theta_3\theta_4$ are temperature compensated in the plots in figures 5 and 6.

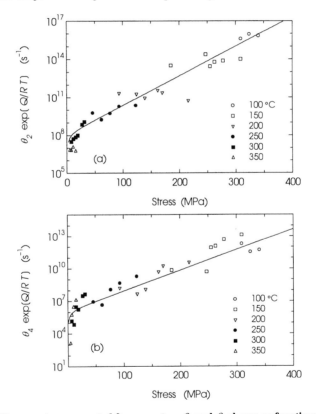

Figure 5 Temperature corrected θ parameters θ_2 and θ_4 shown as functions of stress.

Plots of the temperature corrected parameters θ_2 and θ_4 as a function of stress are shown in figure 5. Although the temperature compensated parameters generally show a monotonic increase with stress, there is considerable scatter about the trend lines. More consistent results were obtained with the temperature compensated products.

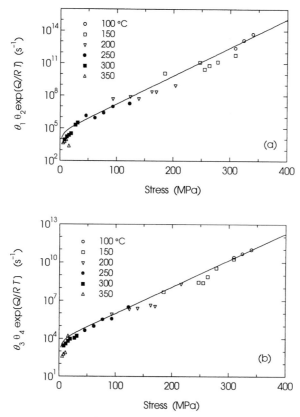

Figure 6 Temperature corrected θ parameter products $\theta_1\theta_2$ and $\theta_3\theta_4$ shown as functions of stress.

The plots in figure 6 show the temperature corrected products $\theta_1\theta_2$ and $\theta_3\theta_4$ as a function of stress. In this case the trend lines were drawn in the following way. If P represents a temperature compensated product, the equation

$$P = A\sinh\left(\sigma / B\right), \tag{5}$$

where A and B are constants, was fitted to the points in figure 6. For the $\theta_1\theta_2$ product, $A = 7.84\times10^4$ s^{-1} and $B = 16.6$ MPa. For the $\theta_3\theta_4$ product $A = 1.56\times10^4$ s^{-1} and $B = 20.8$ MPa.

258

6   Discussion

A consideration of the creep data for alloy 2818A shows the inadequacy of conventional analysis based on the concept of steady state creep rate. For much of the time during creep at temperatures above 100 °C, deformation occurs in Stage I or Stage III and the concept of steady-state creep may not be applicable. The resulting complexity of the creep phenomenon requires a different approach, such as provided by the θ parameter method of analysis.

Equation 1 has been shown to be capable of fitting a wide range of creep curves. Thus, at the very least, it is a useful means of storing and smoothing creep data. Further, since the equation is a monotonically increasing function of time it can be used for generating strain rates from strain-time data. This contrasts, say, with polynomial fitting functions that cannot be safely used to generate differentials. Beyond these practical advantages there is the question of extent to which the θ parameters reflect fundamental creep processes. An important feature of the θ projection concept is that, in addition to describing the shapes of individual creep curves through equation (1), for many metallic and ceramic materials each θ parameter varies systematically with stress and temperature [12,13,14,15]. In the present investigation it was found that the most clear-cut correlations with stress and temperature were obtained with the products $\theta_1\theta_2$ and $\theta_3\theta_4$ (Fig. 6).

In addition, fitting a trial function to the temperature compensated products versus stress data suggests the following relations

$$\theta_1\theta_2 = A_1 \sinh(\sigma/B_1)\exp[-Q/RT],$$
$$\theta_3\theta_4 = A_3 \sinh(\sigma/B_3)\exp[-Q/RT], \tag{6}$$

where A_1, A_3, B_1, and B_3 are constants and Q is the activation energy for bulk diffusion.

7   Conclusions

(i) Creep data for precipitation hardened Hiduminium RR58 has been examined using the θ parameter method.

(ii) The conventional power law description of steady-state creep does not provide an adequate basis for analysis of creep in the alloy.

(iii) The equation $\varepsilon = \theta_1\left(1 - e^{-\theta_2 t}\right) + \theta_3\left(e^{\theta_4 t} - 1\right)$ can be fitted to the experimental creep curves. Thus, the potentially large amount of experimental data in a creep curve can be represented in a compact form as four θ values.

(iv) It was found that the temperature compensated products $\theta_1\theta_2 \exp[Q/RT]$ and $\theta_3\theta_4 \exp[Q/RT]$ increased monotonically with applied stress.

8   Acknowledgements

The authors wish to acknowledge the support of the European Commission for the support of this work through the CREEPAL Thematic Network (BRRT-CT98-5101), and HDA Forgings Ltd. for supplying the original creep data.

9   References

[1] Barbaux, Y., Guedra-Degeorges, D., Cinquin, J., Fournier, P. and Lapaset, G., 1996, The long term elevated temperature behaviour of materials: a key issue for the next SST, Proceedings of ICAS Conference, Sorrento, Italy, 8-13 September, pp.937-942.
[2] Barbaux, Y., Pons, G. and Lapasset, G., 1995, *New creep resistant aluminium alloys for the future supersonic civil transport aircraft*, EUROMAT 95, September 25-28, Padova/Venice, Italy.
[3] Barbaux, Y., 1997, *Future trends for aircraft structural materials*, Proceedings of the 5th European Conference on Advanced Materials and Processes and Applications. 21-23 April 1997, EUROMAT, Zwinjndrect, Maastrict, The Netherlands, Vol.4, pp.559-566.
[4] Pantelakis, S., Kyrsanidi, A., El-Magd, E., Dunnwald, J., Barbaux, Y. and Pons, G., 1999, *Creep resistance of aluminium alloys for the next generation supersonic civil transport aircrafts*, Theoretical and applied fracture mechanics, Vol.31, No.1, pp.31-39.
[5] Polmear, I.J., 1989, Light Alloys, Edward Arnold.
[6] Evans, R.W., Parker J.D. and Wilshire, B., 1982 in *Recent Advances in Creep and Fracture of Engineering Materials and Structures*. Ed. by B Wilshire and D R J Owen, Pineridge Press, Swansea, pp.135-184.
[7] Dyson, B.F., 1999, in *Creep Behaviour of Advanced Materials*. Ed by R. S. Mishra et al. TMS Warrendale pp 3-12.
[8] Evans, R.W. and Wilshire, B., 1996 in *Engineering Application Through Scientific Insight*, Ed. by E. D. Hondros and M. McLean, The Institute of Materials, London, 155-172.
[9] Kaufman, J.F, 1999, *Properties of aluminium alloys: tensile, creep and fatigue data at high and low temperatures*. ASM International, Materials Park, OH 44073-0002, USA.
[10] Frost, H.J. and Ashby M.F, 1982, Deformation Mechanism Maps, The Plasticity and Creep of Metals and Ceramics, Pergamon Press.
[11] Lundy, T.S. and Murdoch J.F., 1962, J. Appl. Phys. 33, pp.1671-1676.
[12] Evans, R.W., Scharning, P.J. and Wilshire, B.1985 in Creep Behaviour of Crystalline Solids (Eds. B. Wilshire and R.W. Evans) Pineridge Press, Swansea.
[13] Brown, S.G.R., Evans, R.W. and Wilshire, B.1986, *Creep Strain and Creep Life Prediction for the Cast Nickel-Based Superalloy IN-100*, Mater. Sci. Engng., 84, pp. 147
[14] Brown, S.G.R., Evans, R.W. and Wilshire, B.1987, *New Approach To Creep Of Pure Metals With Special Reference To Polycrystalline Copper*, Mater. Sci. Tech., 3, pp. 23
[15] Evans, R.W., Fadalalla, A.A. and Wilshire, B., 1990, *Prediction of Long-Term Creep Rupture Properties for an Aluminium Alloy for Airframe Applications*, Proc. 4th International Conference on Creep and Fracture of Materials and Structures, Institute of Metals.

CREEP AND CREEP FRACTURE OF THE ALUMINIUM ALLOY, 2124-T851

B Wilshire and H Burt

Department of Materials Engineering,
University of Wales, Swansea SA2 8PP, UK

ABSTRACT

To provide information relevant to the selection of airframe materials for future high-speed civil transport, the tensile creep and creep fracture properties recorded at 373, 423 and 463K are analysed for the aluminium alloy 2124-T851. Comparisons with long-term property values available for this alloy then characterize the stress rupture behaviour within the temperature ranges relevant to airframe design, showing that 2124-T851 exhibits creep and creep rupture strengths superior to Hiduminium RR58 and 2618.

1. INTRODUCTION

For over a decade, major research programmes have focussed on aluminium alloys for high-temperature airframe applications, aiming specifically to identify materials which may prove superior to the current Concorde alloy, Hiduminium RR58. In this context, at Swansea, several investigations have been concerned with characterization of the creep and creep fracture properties of a range of candidate aluminium alloys, initially in partnership with the Royal Aerospace Establishment, Farnborough, (now DERA) and, currently, through the EC Creepal Consortium.

As part of the RAE programme, one of the candidate materials included was alloy 2124-T851 (1). In the present study, the creep behaviour patterns observed for this product are analysed, initially adopting traditional power-law approaches and then considering the applicability of the θ methodology developed to quantify creep curve shapes. In addition, to provide information relevant to airframe applications, the data derived from short-term tests completed for alloy 2124-T851 are compared
(a) with long-term creep property values determined independently for this material (1, 2) and
(b) with results available for Hiduminium RR58 (3) and the near-equivalent alloy 2618 (2).

Table I   Compositions of Aluminium Alloys (wt %)

	Cu	Mg	Mn	Si	Fe	Cr	Ti	Zn	Ni
2124 (actual)	3.76	1.33	0.49	0.02	0.08	0.02	-	0.01	
2124 (min)	3.80	1.20	0.30	-	-	-	-	-	
2124 (max)	4.90	1.80	0.90	0.20	0.30	0.10	0.15	0.25	
2618 (min)	1.90	1.30	-	0.10	0.90	-	0.04	-	0.90
2618 (max)	2.70	1.80	-	0.25	1.30	-	0.10	0.10	1.20
RR58 (min)	2.25	1.35	-	0.18	0.90	-	-	-	1.00
RR58 (max)	2.70	1.65	-	0.25	1.20	-	0.20	-	1.30

2.   EXPERIMENTAL PROCEDURES

Alloy 2124-T851 has a nominal composition of Al-4.4%Cu–1.5%Mg–0.64%Mn, with a low iron and silicon content, ie a high purity version of 2024. Table I gives the chemical analysis of the material tested, together with the composition limits specified, plus the composition ranges defined for Hiduminium RR58 and 2618.

The alloy was supplied in the form of 63mm plate, with the T851 designation indicating plate which was solution treated at 768 ± 3K, water quenched and cold worked by stretching to 1.5% strain, then aged at 463K for over 12 hours. The as-received alloy had a mean grain diameter of 0.1 to 0.2mm, with some large particles of MnAl and a finer dispersion of $MnAl_6$ particles. The heat-treatment then resulted in a fine distribution of lath-like S precipitates ($Al_2Cu.Mn$) lying on (210) planes in <001> directions throughout the grains, with narrow precipitate-free zones (~130nm wide) at the grain boundaries. Moreover, X-ray analysis showed no evidence of a strong crystallographic texture.

Testpieces, with gauge lengths of 25.4mm and diameters of 4mm, were machined such that the gauge lengths were parallel to the plate rolling direction. These specimens were tested in tension using high-precision constant-stress machines (4).

3.   EXPERIMENTAL RESULTS

At all stresses studied at temperatures of 373, 423 and 463K, normal creep curves were displayed, ie after the initial strain on loading at the creep temperature, the creep rate decayed continuously during the primary stage, reaching a minimum value before accelerating during the tertiary stage which led to fracture. Yet, while the θ methodology seeks to quantify the shapes of the full creep strain/time curves, with conventional power-law approaches, the only parameters normally derived from each creep curve are the minimum creep rate ($\dot{\varepsilon}_m$), the time to fracture (t_f) and the creep ductility (ε_f).

Power Law Representation of Creep Data

In the vast majority of studies, the creep properties of metals and alloys are discussed by reference to the dependence of the minimum creep rate on stress (σ) and temperature (T) using power-law equations of the form

$$\dot{\varepsilon}_m = A\sigma^n \exp{-Q_c / RT} \tag{1}$$

where the values of A, the stress exponent (n) and the activation energy for creep (Q_c) are considered to be constant within defined stress/temperature regimes.

With pure metals, at stresses giving easily-measurable creep rates at temperatures above about half of the melting point, n is usually around 4 to 6, with Q_c less than or equal to the activation energy for lattice self diffusion. In contrast, with most particle-hardened alloys, much larger values of n and Q_c are often recorded. Thus, over the stress ranges considered for alloy 2124-T851, n values of 40, 20 and 8 are found at 373, 423 and 463K respectively (Figure 1). Then, because n decreases with increasing temperature, Q_c is stress dependent, seemingly increasing from values approaching the activation energy expected for lattice diffusion of aluminium in the alloy matrix (~140 $kJmol^{-1}$) as the test durations increase.

As with the stress/creep rate data (Figure 1), the stress and temperature dependences of the rupture life are also high for alloy 2124-T851 (Figure 2). This result would be expected because, for many metals and alloys, the rupture life depends inversely on the minimum creep rate as

$$\dot{\varepsilon}_m .t_f = \text{constant (M)} \tag{2}$$

Yet, compared with the n values determined from the $\log\sigma/\log\dot{\varepsilon}_m$ plots (Figure 1), the equivalent gradients for the $\log\sigma/\log t_f$ relationships are -40, -14 and −7 at 373, 423 and 463K respectively (Figure 2). Thus, while $\varepsilon_m .t_f$ appears to be constant at 373K (with M ≅ 0.03 in equation 2), $\varepsilon_m .t_f$ decreases with decreasing stress at 423 and 463K (Figure 3). However, as found for a number of materials (4,5), at all temperatures studied for alloy 2124-T851,

$$\dot{\varepsilon}_m .t_f = X\varepsilon_f \tag{3}$$

with X ≅ 0.5, as illustrated in Figures 4 and 5.

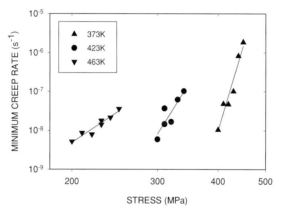

Figure 1. The stress dependences of the minimum creep rate for alloy 2124-T851 at 373, 423 and 463K.

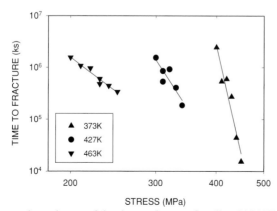

Figure 2. The stress dependences of the time to fracture for alloy 2124-T851 at 373, 423 and 463K.

264

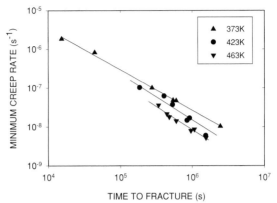

Figure 3. The dependences of the times to fracture on the minimum creep rate for alloy 2124-T851 at 373, 423 and 463K.

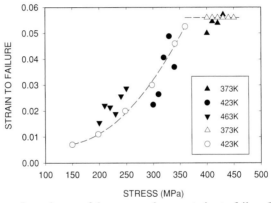

Figure 4. The stress dependences of the measured creep strains to failure for alloy 2124-T851 (closed symbols), together with ε_f values predicted from Figure 3 (open symbols).

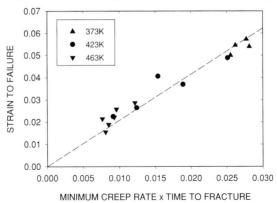

Figure 5. The dependences of the product of the minimum creep rate and rupture life ($\varepsilon_m.t_f$) on the creep ductility (ε_f) for alloy 2124-T851 at 373, 473 and 463K.

Creep Data Comparisons

To evaluate alloy 2124-T851 as a candidate airframe material, the present data can be compared with information available for RR58 (3) and 2618 (2). However, the temperatures now considered for 2124-T851 do not coincide with those chosen in the creep studies completed for those other alloys, except at 423K. Even so, the logσ/log$\dot{\varepsilon}_m$ plots at 423K indicate that 2124-T851 has a creep resistance better than 2168 and RR58. (Figure 6).

Unfortunately, the creep rate values presented in Figure 6 for 2124-T851 were derived from creep tests with a maximum duration of less than 1000 hours but, in the selection of materials for long-term service under creep conditions, stress rupture rather than creep tests are generally performed. For this reason, the logσ/log t_f plots in Figure 7 include results obtained at 423K for tests lasting up to 100,000 for 2124-T851 (2), but only up to 1000 hours for 2618 (2) and even shorter times for RR58 (3). Yet clearly, the stress rupture relationships in Figure 7 are consistent with the creep rate data in Figure 6, suggesting that 2124-T851 has creep and creep rupture strengths superior to 2618 and RR58.

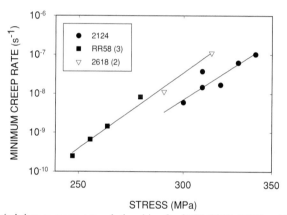

Figure 6. Stress/minimum creep rate relationships for 2124-T851, 2618 and RR58 at 423K.

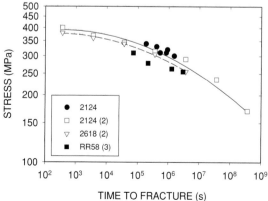

Figure 7. Stress rupture properties of 2124-T851, 2618 and RR58 at 423K.

Theta Representation of Creep Data

When normal creep curves are analysed using the θ Projection Concept (4), the most widely used form of the θ equations describes the variation in creep strain (ε) with time (t) as

$$\varepsilon = \theta_1\left(1 - \exp\left(-\theta_2 t\right)\right) + \theta_3\left(\exp\left(\theta_4 t\right) - 1\right) \tag{4}$$

where θ_1 and θ_3 define the extents of the primary and tertiary stages with respect to strain, while the rate parameters, θ_2 and θ_4, relate to the curvatures of the primary and tertiary components respectively.

For alloy 2124-T851, it has been shown (1) that the shapes of individual creep curves can be described using equation 4, so it would seem reasonable to assume that determination of the stress dependences of the θ parameters at each creep temperature should allow quantitative prediction of long-term data. However, extrapolation was not straight-forward for two reasons.

(a) Alloy 2124-T851 showed significant specimen-to-specimen scatter (1). Moreover, creep curves were recorded only over narrow stress ranges, giving rupture lives differing by little more than an order of magnitude at each creep temperature (Figure 2). Clearly, when using θ data for extended extrapolation, tests spanning several decades of rupture times should be completed, with the maximum test duration being well over a thousand hours. Especially with materials displaying specimen-to-specimen scatter, analysing θ data covering only narrow stress ranges decreases confidence in the accuracy of the long-term estimates.

(b) While equation 4 adequately describes normal creep curve shapes, to calculate rupture times, it is necessary to define the point of failure i.e. under uniaxial test conditions, the time to fracture (t_f) can be estimated as the time taken for the accumulated creep strain to reach the limiting creep ductility (ε_f). With many materials, such as most steels and superalloys, the creep ductilities are reasonably large (usually > 5%) and do not vary sharply with stress. Furthermore, in low stress tests, the primary strains are low and the curve shapes are tertiary dominated. So, in low stress tests when equation 4 describes the continuous acceleration in creep rate during extended tertiary stages, little error in life prediction results if the estimates of ε_f are not precise. However, with alloy 2124-T851, the creep ductility appears to decrease with decreasing stress, even over the narrow stress ranges covered experimentally (Figure 4). Clearly, ε_f cannot extrapolate to zero, but it is difficult to predict low-stress ε_f values in a theoretically justifiable manner. Thus, a predictive knowledge of the factors determining ε_f remains a key requirement for low-ductility materials.

Although long-term stress-rupture predictions have been made using the θ methodology (1), extrapolation techniques are less necessary when actual data has been gathered at low stresses. In the case of aluminium alloys for supersonic airframe applications, the relevant service temperatures are within the range 373 to 423K and, for alloy 2124-T851, long-term property values are available from several sources at these temperatures (1,2). Thus, Figure 8 includes information generated at RAE, Farnborough, for three separately – produced batches of 2124-T851 plate at 373 and 423K. Given the specimen-to-specimen and batch-to-batch scatter expected for this material, plus the inevitable variations in testing procedures in different laboratories, the result sets incorporated in Figure 8 reveal consistent patterns of behaviour.

4.   DISCUSSION

Power law methods and the θ concept offer fundamentally different approaches to the identification of the processes governing strain accumulation and damage development during creep. With power-law relationships, differences in n and Q_c values are interpreted on the basis that different creep mechanisms are dominant in different stress/temperature regimes. Traditionally, with pure metals, a change from n ≅ 1 at low stresses to n ≅ 4 to 6 at intermediate stresses is interpreted as a transition from diffusional creep to dislocation creep mechanisms, with the high-stress regime in which n increases rapidly to values well above 4 being defined as 'power-law breakdown.'   With particle-hardened alloys, the situation becomes more complex, because n values substantially greater than four can extend over several decades of strain rate before reaching the so-called 'breakdown' region. (Figure 6).

When power-law approaches are used to describe the stress dependence of the creep rate (equation 1), because t_f increases as $\dot{\varepsilon}_m$ decreases (equations 2 and 3), the gradients of the logσ/log$\dot{\varepsilon}_m$ and logσ/log t_f plots are obviously related.   However, Figure 8 shows the logσ/log t_f plots curve continuously over extended stress ranges.   Clearly, over any limited range of stress, the curves can appear as straight lines (Figures 1 and 2), but the measured gradients then depend on the precise stress ranges selected.   It should also be noted that Figure 8 includes the UTS values for 2124-T851 at 373 and 423K.   Obviously, the UTS represents the upper stress limit of logσ/log t_f plots.   Thus, the logσ/log t_f plots curve sensibly towards the UTS at both temperatures (Figure 8), making a decision on the 'power-law breakdown' conditions somewhat arbitrary.

In contrast to power-law approaches, the θ equations see logσ/log$\dot{\varepsilon}_m$ and logσ/log t_f plots as continuous curves, with the gradients (n) normally increasing gradually and systematically with increasing stress, both above and below the 'breakdown' condition.   Because the curvature of these plots are viewed as an inevitable consequence of the complex dependence of the 'minimum' creep rate on the varying shapes of the creep curves, no transitions in creep mechanism need then to be invoked, i.e. at temperatures of practical interest, dislocation processes are considered to be dominant at all stress levels.

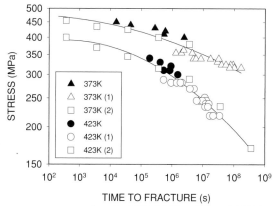

Figure 8. Long-term stress rupture data for 2124-T851 at 373 and 423K.

Unfortunately, for the reasons outlined earlier, the scatter in the θ data derived from the creep curves recorded over limited stress ranges for 2124-T851 did not allow accurate extrapolation to low stress test conditions (1). Moreover, the problem of using the θ methodology to predict long-term creep rupture lives was exacerbated by uncertainties over the creep ductilities expected at low stresses. However, the present study suggests a possible way of estimating long-term ε_f values. Thus, the $\log \dot{\varepsilon}_m$/log t_f plots can be extrapolated to determine $\varepsilon_m t_f$ for various rupture lives (Figure 3) and therefore for various stresses (Figure 8), using equation 3 to calculate the corresponding ε_f values (Figure 5). The resulting estimates suggest that, while ε_f is reasonably constant at ~6% at 373K (comparable with the elongation values (2) obtained from hot tensile tests), the σ/ε_f plot curves towards a limiting strain to failure of ~0.5% under long-term test conditions at 423K (Figure 4).

5.   CONCLUSIONS

Alloy 2124-T851 has been shown to display creep and creep fracture strengths superior to those found for Hiduminium RR58 and the near-equivalent alloy, 2618. Moreover, results obtained from a variety of sources provide a consistent set of stress rupture data, which span the full stress/temperature ranges relevant to high-temperature airframe applications.

6.   ACKNOWLEDGEMENTS

The authors wish to record their gratitude for the support received under the EC Creepal Programme, Contract Number BRRT – CT98 – 5101.

7.   REFERENCES

1.   Evans, R.W., Fadlalla, A.A., Wilshire, B., Butt, R.I. and Wilson, R.N., 1990, 'Prediction Of Long-Term Creep Rupture Properties For An Aluminium Alloy For Airframe Applications', in Proc. Fourth Inter. Conf. on 'Creep and Fracture of Engineering Materials and Structures', (Ed. B. Wilshire and R.W. Evans), The Institute of Metals, London, pp 1009-1016.
2.   Metals Handbook, 1979, **2**.
3.   Robinson, J.S., Private Communication.
4.   Evans, R.W. and Wilshire, B., 1985, 'Creep of Metals and Alloys', The Institute of Metals, London.
5.   Wilshire, B. and Carreño F., J. Eur. Ceram. Soc., 2000, 20, 463.

CREEP OF ALUMINIUM III

Creep behaviour of an Al-2024 PM –15% SiC Composite

Stefano Spigarelli*, Marcello Cabibbo*, Enrico Evangelista* and Terence G.Langdon**

* INFM/ Department of Mechanics, University of Ancona, I-60131 Ancona, Italy
** Departments of Aerospace & Mechanical Engineering and Materials Science,
University of Southern California, Los Angeles, CA 90089-1453, U.S.A.

Abstract

The creep response of a 2024 + 15% SiC composite was investigated at 548, 573 and 603 K. An analysis of the dependence of the minimum creep-rate on the applied stress clearly revealed the existence of a threshold stress. A comparison between creep data obtained for the unreinforced alloy and the composite showed the two materials exhibit essentially similar behaviour. In particular, there is a conventional power law behaviour when the applied stress is replaced by the effective stress. A best correlation was obtained between experimental data and a constitutive equation by assuming that the threshold stress increases with the time of exposure during creep, and therefore with decreasing applied stress, because of the effect of dynamic precipitation during the creep process.

1. Introduction

In the last two decades, several investigators have analysed the high-temperature creep response of aluminium-based discontinuously reinforced composites [1-20]. In these materials, when the minimum creep rate $\dot{\varepsilon}_m$ is represented as a function of the applied stress σ, the apparent stress exponent $n_a = \partial \log \dot{\varepsilon}_m / \partial \log \sigma$ increases when the applied stress decreases. This behaviour has been associated with the presence of a threshold stress, σ_0, which decreases with increasing temperature [21]. Once the applied stress is replaced by the difference between the applied stress and the threshold stress, the value of the true stress exponent n is close to the theoretical values of n = 4 - 5 for climb-controlled creep [18,19,22,23] or n = 3 for creep controlled by viscous glide [15-18]. The true activation energy for creep is close to the activation energy for self diffusion in Al for climb control or to the activation energy for interdiffusion of the solute atoms in the case of dilute solid solution alloys where creep is controlled by viscous glide [24]. A problem that has not yet been fully addressed is the role of the reinforcement since the threshold stresses are higher in the composites than in the unreinforced alloys. It is established that no plastic flow occurs in the ceramic reinforcements and plastic deformation is controlled by flow in the matrix. In this context, the threshold stress has frequently been associated with the presence of fine oxide particles in the matrix of the composites [3] or in the unreinforced alloys [25] produced by powder metallurgy (PM). The interaction between fine oxides and dislocations in PM composites with a pure-Al matrix [11,23] is a reasonable explanation for the existence of a threshold stress since no other precipitates are present. Nevertheless, the presence of a threshold stress has been clearly identified also in alloys produced by ingot metallurgy [18]. It can be concluded that, in the absence of oxides, other particles, for example fine precipitates formed during ageing or creep, generate a threshold stress of lower but not negligible magnitude.

The aim of the present study was to investigate the creep response of a 2024PM composite reinforced with 15% SiC particulates and to make a comparison with the creep behaviour of the corresponding unreinforced alloy tested under the same conditions.

2. Experimental procedures

Experiments were conducted on an Al-2024 composite reinforced with 15 pct of irregularly shaped SiC particulates. The composite was produced by powder metallurgy; the powders were compacted, sintered at 773 K and then extruded at 723 K. The chemical composition of the powders is given in Table I; the chemical composition of a 2024PM alloy, previously tested after a similar heat-treatment and in the same experimental conditions [26,27], is also reported for comparison. Prior to creep testing, cylindrical specimens machined from the bars were heat treated at 673 K for 5h and then slowly cooled at 60 K/h to produce a stable microstructure. As in the case of the unreinforced alloy, after heat treatment the grain size of the composite remained fine (typically ~3 μm). Tensile creep tests were conducted in air under constant load at absolute testing temperatures, T, in the range from 548 to 603 K.

Table I: Chemical composition of the composite matrix (wt.%). The chemical composition of the unreinforced 2024PM alloy [26,27] is included for comparison.

Material	Si	Fe	Cu	Mg	Mn	Al
Unreinforced alloy	0.24	0.26	4.42	1.47	0.56	Bal
Composite matrix	0.01	0.03	3.99	1.60	0.61	Bal

3. Results

Figure 1 shows representative strain-rate vs strain curves obtained by testing the composite at 573 K. The steady state in this material, as in the unreinforced alloy and in many other composites, is not well defined but consists essentially of a short region of minimum creep rate. The curves obtained at the other temperatures were substantially similar in shape with a short primary stage, a minimum-creep-rate region and then a prolonged tertiary stage.

Figures 2 and 3 plot the minimum creep-rate as a function of the applied stress for the composite and the unreinforced alloy. The general shape and similarity of these two sets of curves indicate the presence of threshold stresses; extrapolation to 10^{-10} s^{-1} suggests the magnitudes of the threshold stresses are larger in the composite than in the unreinforced alloy.

Figure 4 shows the microstructure of the unreinforced alloy (Figures 4a and 4b) and of the composite (Figure 4c) after creep testing. A comparison of Figures 4a and 4b clearly demonstrates that long exposure at high temperature leads to a precipitation of fine particles (~50-60 nm in diameter). An analysis of the microstructure of the composite revealed an increase in the number per unit volume of precipitated particles although, possibly due to different kinetics of precipitation, they appear coarser than in the unreinforced alloy (Figure 4c).

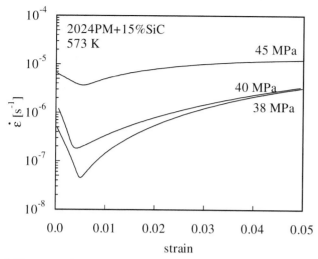

Figure 1. Representative creep curves obtained by testing the composite at 573 K.

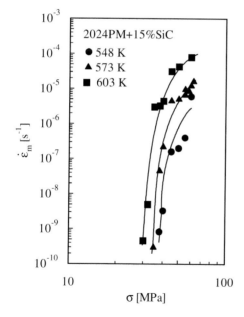

Figure 2. Minimum strain rate dependence on applied stress for the 2024PM+15%SiC composite.

274

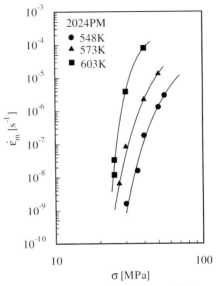

Figure 3. Minimum strain rate dependence on applied stress for 2024PM.

A calculation of the number of particles per unit volume (N_V) and of their average size (d) in the sample tested at 603 K and 32 MPa gave $N_V = 7.3 \times 10^{18}$ m^{-3} and d = 210 nm. The same calculation [26,27] for the unreinforced alloy tested at 603 K and 25 MPa gave $N_V = 1.8 \times 10^{19}$ m^{-3} and d = 60 nm. Figure 5 plots the variation in hardness, measured on the heads of the crept samples, as a function of time of exposure; as expected, the hardness initially increases as a consequence of the precipitation and then subsequently decreases due to overageing.

The nature of the precipitates was not investigated in detail since the ageing response of the 2024 is well known. Two different ageing sequences occur separately or simultaneously in this alloy: these processes involve the precipitation of stable Al_2Cu (θ) and Al_2CuMg (S) phases, preceded by the formation of Guinier-Preston (GP) zones, metastable θ'' and semicoherent θ' and S'. Ageing at elevated temperatures (for example at 700 K) results in the presence of stable precipitates (θ and S) in equilibrium with α solid-solution; on the other hand, calorimetric studies have indicated that precipitation of $\theta'' + \theta' + S'$ occurs during ageing of a solution-treated 2024 alloy in the temperature range between 518 and 558 K [28].

4. Discussion

Figures 2-3 show that both the composite and the unreinforced alloy exhibit the typical behaviour of a material in which there is a threshold stress. A detailed analysis of the microstructural evolution in the unreinforced alloy during creep [26] can therefore be used to obtain useful information on the creep mechanisms occurring in the composite.

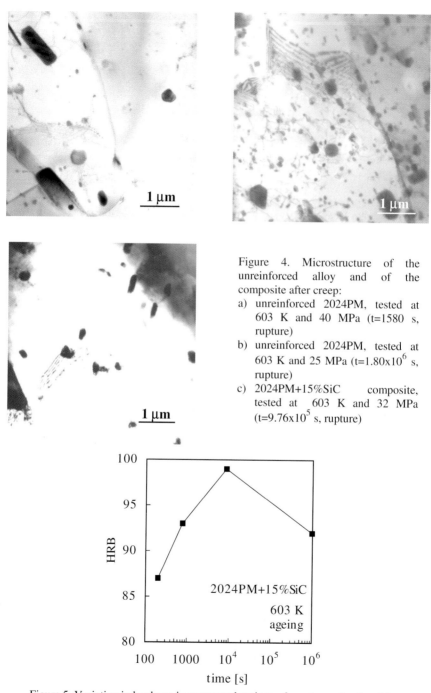

Figure 4. Microstructure of the unreinforced alloy and of the composite after creep:
a) unreinforced 2024PM, tested at 603 K and 40 MPa (t=1580 s, rupture)
b) unreinforced 2024PM, tested at 603 K and 25 MPa (t=1.80×10^6 s, rupture)
c) 2024PM+15%SiC composite, tested at 603 K and 32 MPa (t=9.76×10^5 s, rupture)

2024PM+15%SiC

603 K ageing

Figure 5. Variation in hardness in unstressed regions of crept samples (heads).

4.1 Threshold stress in the unreinforced alloy

The minimum creep-rate dependence on applied stress and temperature for materials that exhibit a threshold-like behaviour is expressed by the relationship:

$$\dot{\varepsilon}_m = A \frac{DGb}{kT} \left(\frac{\sigma - \sigma_0}{G} \right)^n \tag{1}$$

where D is the appropriate diffusion coefficient $(= D_0 \exp(Q/RT))$, D_0 is a frequency factor, Q is the activation energy for creep, G is the shear modulus, b is the Burgers vector, k is Boltzmann's constant, T is the absolute temperature, n is the true stress exponent (equal to 4 - 5 for climb controlled creep and 3 for creep controlled by viscous glide of dislocations dragging solute atom atmospheres) and A is a dimensionless constant. In the case of the unreinforced alloy, the use of Eqn.1 gave a creep exponent close to 5, while the diffusion coefficient was equivalent to the coefficient for self diffusion in Al (D = 1.84 x 10^{-4} exp(Q/RT) m^2 s^{-1}, with Q=143.4 kJ mol^{-1}). Dynamic precipitation during creep increases N_V, and this leads to a corresponding decrease in the interparticle spacing and therefore an increase in the Orowan stress [26,27]. Since the threshold stress was assumed to be a fraction of the Orowan stress, σ_{Or}, the direct dependence of the Orowan stress on time of exposure gives an indirect dependence of the threshold stress on the applied stress. This dependence can be described by a phenomenological equation in the form:

$$\sigma_0 = \sigma_{0i} + a \exp(-s\sigma) = \alpha' \sigma_{Or} \tag{2}$$

where σ_{0i} is the threshold stress corresponding to the initial distribution of precipitates and a, s and α' are temperature-dependent parameters; the last term of Eqn.2 indicates that the threshold stress was found to be proportional to the Orowan stress calculated by taking into account the interparticle spacing of fine intragranular precipitates [27]. Least-squares analysis of the minimum strain-rate vs applied-stress data in Fig.3 was used to calculate σ_{0i}, a and s, assuming n = 3 or n = 4.4 in Eqn.1. With n = 3, the threshold stress was approximately constant but the calculation of the activation energy gave unrealistically high values of Q. By contrast, with n = 4.4 the threshold stress varied as shown in Fig.6 (solid curves) and the activation energy for creep was very close to the anticipated value of 143 kJ mol^{-1}.

4.2 Threshold stress calculation for the composite

In the composite material the occurrence of creep also results in the precipitation of additional particles. Although the precipitation kinetics may be different from that of the unreinforced alloy [28], the same general form of Eqn.2 can be used to describe the indirect dependence of threshold stress on applied stress. After substituting Eqn.2 into Eqn.1 with n = 4.4, least-square analysis was performed to obtain the best-fit values of σ_{0i}, a and s for each set of data plotted in Fig.2. The results of this calculation are presented in Fig.6 (broken lines); as expected, the threshold stress at any given temperature is larger in the composite than in the unreinforced alloy but in both cases σ_0 increases with decreasing applied stress (i.e. with increasing time of exposure). Figure 7 shows the calculation of the activation energy for creep for an effective stress of 10 MPa; the slope of the curve (Q/R) gives an activation energy for creep of 145 kJ mol⁻¹, almost equivalent to the activation energy for self diffusion in Al. It can thus be concluded that creep in the composite is also controlled by high-temperature climb, a result that can be easily predicted by considering the very low amounts of Cu and Mg in solid solution after heat treatment and dynamic precipitation.

Figure 7 plots the temperature normalised creep rate for the composite and the unreinforced alloy; the two sets of data overlap on a single line of slope close to 4.4. Figure 8 shows a comparison between the line of Fig. 7 and the temperature-normalised creep-rate for the 2124+20%SiC composite [22] where the threshold stress was recalculated by taking n = 4.4.

4.3 Nature of threshold stress and role of reinforcement

Recent studies [3,25] have convincingly demonstrated that in materials produced by powder metallurgy the threshold stress is generated by the presence of fine oxide particles resulting from the production route. This conclusion is reasonable, since it explains why the threshold stress is higher in PM materials than in ingot alloys of similar composition; on the other hand, it has been shown that a

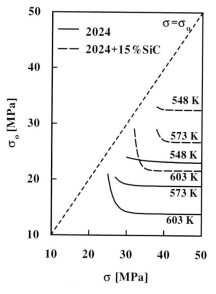

Figure 6. Variation of threshold stress as a function of applied stress (Eqn.2).

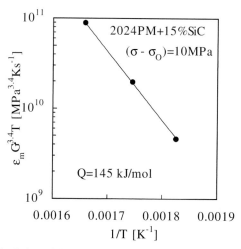

Figure 6. Calculation of the activation energy for creep for the composite.

threshold stress exists also in composites produced by conventional casting techniques [18]. Similarly, the study of the 2024PM alloy indicates that precipitation during creep leads to an increase in threshold stress; thus, in this case also the fine oxide particles introduced by the manufacturing processes may be part, but not all, of the source of the threshold stress. In fact, Eqn.2 reveals that the threshold stress can be related to the Orowan stress that may be calculated directly from the number per unit volume of fine particles (either oxide particles introduced during processing or precipitates). In particular, for the unreinforced alloy this calculation gives α' = 1.0, 0.84 and 0.64 at 548, 573 and 603 K, respectively. These results were obtained by calculating the number of particles at the time corresponding to the minimum-creep rate (see Table II for an example).

For the composite tested at 603 K and 32 MPa, the calculation gives an Orowan stress at rupture of 79 MPa (Table II). Since the time corresponding to the minimum-creep rate was less than 10% of the time to rupture, it is reasonable to assume the Orowan stress corresponding to the minimum strain rate is different. Clearly, a reliable estimation of the Orowan stress at the time corresponding to this minimum creep rate cannot be obtained without an extensive microstructural analysis. Nevertheless, an examination of data reported in Table II gives qualitative but significant information: under those conditions, relatively similar values of the threshold stress (19.2-23 MPa and close to 25 MPa for the unreinforced alloy and the composite, respectively) correspond to an Orowan stress at rupture of similar, but not identical, magnitude (71-89 and 79 MPa, respectively).

Another source of debate is the role of the reinforcements: indeed, load transfer from the matrix to the ceramic particles is frequently considered a major effect in the presence of discontinuous reinforcements [20]. Load transfer will result in an expression for the effective stress, σ_e, of the form:

$$\sigma_e = (1-\alpha)\sigma - \sigma_0 \qquad (3)$$

where σ_0 is the threshold stress acting on dislocations in the matrix, and α is the load-transfer coefficient. Equation 2 can thus be transformed into:

$$\dot{\varepsilon}_m = A\frac{DGb(1-\alpha)^n}{kT}\left(\frac{\sigma-\sigma^*_0}{G}\right)^n \qquad (4)$$

where $\sigma^*_0 = \sigma/(1-\alpha)$ is the apparent threshold stress. When the threshold stress is calculated by conventional procedures or by the method used in the present study (substitution of Eqn.2 into Eqn.1), the values obtained are more correctly defined as the relevant values of the apparent threshold stress. Since it is expected that A is

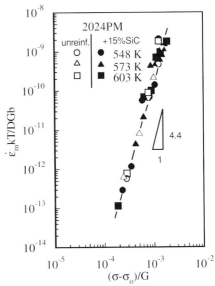

Figure 7. Temperature-normalised minimum creep rate vs modulus-compensated effective stress.

the same for both the unreinforced alloy and the composite, the magnitude of the load-transfer coefficient can then be calculated from the expression:

$$\frac{\dot{\epsilon}_{composite}}{\dot{\epsilon}_{alloy}} = (1 - \alpha)^n \qquad (5)$$

Comparison of Eqn.5 with Fig.7 shows that for the investigated composite $\alpha = 0$, i.e. the reinforcement has no additional strengthening effect, at least in terms of a load-transfer mechanism. The same behaviour was observed when the 2124PM unreinforced alloy was compared with the 2124PM+10%SiC composite [29,30]. Also in that case the threshold stress calculated in the composite was larger than in the unreinforced matrix.

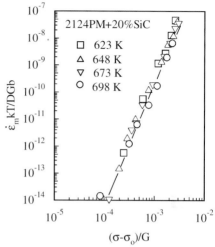

Figure 8. Temperature-normalised minimum creep rate as a function of modulus-compensated effective stress for 2124+20%SiC [22].

Table II. Calculated number per unit volume of particles and related Orowan stress for the unreinforced alloy and the composite at 603 K; the calculated values at times corresponding to minimum creep rate, for the unreinforced alloy, are based on the analysis of particle distribution reported earlier [26,27].

material	condition	t at $\dot{\epsilon}_{min}$ [s]	N_v at $\dot{\epsilon}_{min}$ [m^{-3}]	σ_{Or} at $\dot{\epsilon}_{min}$ [MPa]	$t_{rupture}$ [s]	N_v at rupture [m^{-3}]	σ_{Or} at rupture [MPa]
2024PM	25 MPa	2.1×10^5	1.9×10^{19}	36*	1.09×10^6	5.8×10^{19}	71*
		1.1×10^5	1.4×10^{19}	30*	1.8×10^6	8.1×10^{19}	89*
2024PM+ 15SiC	32 MPa	9.0×10^4	n.c.	n.c.	9.76×10^5	7.3×10^{18}	79**

* calculated with d=60 nm.
** calculated with d=210 nm

5. Conclusions

The creep response of a 2024PM composite reinforced with 15%SiC particulates was investigated in the temperature range between 548 and 603 K; the material exhibits a creep behaviour that indicates the presence of a threshold stress. Incorporation of a threshold stress into the minimum strain rate dependence on applied stress gives a true stress exponent close to 4.4 and an activation energy for creep close to 143.4 kJ mol^{-1}. As in the case of the unreinforced alloy, it was necessary to assume that the magnitude of the threshold stress increases in the composite with decreasing stress (i.e. with increasing time of exposure in creep). This increase is related to the precipitation of additional fine particles during creep. It is thus concluded that the threshold stress is generated by the interaction between particles (both pre-existing oxides and precipitated phases) and mobile dislocations.

6. References

1. Nieh, T.G., *Metall. Trans.* 1984, **15A**, 139.
2. Mishra, R.S. and Pandey, A.B., *Metall.Trans.* 1990, **21A**, 2089.
3. Park, K.-T., Lavernia, E.J. and Mohamed, F.A., *Acta metall.mater.* 1990, **38**, 2149.
4. Mohamed, F.A., Park, K.-T. and Lavernia, E.J., *Mater.Sci.Engng.* 1992, **A150**, 21.
5. Pandey, A.B., Mishra, R.S. and Mahajan, Y.R., *Acta metall.mater.* 1992, **40**, 2045.
6. Gonzales-Doncel, G. and Sherby, O.D., *Acta metall.mater.* 1993, **41**, 2797.
7. Cadek, J. and Sustek, V., *Scripta metall.mater.* 1994, **30**, 277.
8. Park, K.-T. and .Mohamed, F.A., *Scripta metall.mater.* 1994, **30**, 957.
9. Cadek, J., Sustek, V. and Pahutova, M., *Mater.Sci.Engng.* 1994, **A174**, 141.
10. Park, K.-T. and Mohamed, F.A., *Metall. Mater.Trans.* 1995, **26A**, 3119.
11. Cadek, J., Oikawa, H. and Sustek, V., *Mater.Sci.Engng.* 1995, **A190**, 9.
12. Pandey, A.B., Mishra, R.G., Paradkar, A.G. and Mahajan, Y.R., *Acta mater.* 1997, **45**, 1297.
13. Li, Y. and Langdon, T.G., *Metall. Mat.Trans.* 1997, **28A**, 1271.
14. Li, Y. and Langdon, T.G., *Scripta mater.*1997, **36**, 1457.
15. Ma, Y. and Langdon, T.G., *Mater.Sci.Engng.* 1997, **A230**, 183.
16. Li, Y. and Langdon, T.G., *Acta mater.* 1997, **45**, 4797.
17. Li, Y. and Langdon, T.G., *Metall. Mater.Trans.* 1998, **29A**, 2523.
18. Li, Y. and Langdon, T.G., *Mater.Sci.Engng.* 1998, **A245**, 1.
19. Li, Y. and Langdon, T.G. *Acta mater.* 1998, **46**, 1143.
20. Li, Y. and Langdon, T.G., *Acta mater.* 1999, **47**, 3395.
21. Gibeling J.C. and Nix W.D., *Mater.Sci.Engng.* 1980, **45**, 123.
22. Cadek, J., Pahutova M. and Sustek V., *Mater.Sci.Engng.* 1998, **A246**, 252.
23. Cadek J, *Acta techn. CSAV*, 1993, **38**, 651.
24. Yavari P., Mohamed F.A. and Langdon T.G., *Acta metall.* 1981, **29**, 1495.
25. Park, K.-T., Lavernia, E.J. and Mohamed, F.A., *Acta metall.mater.* 1994, **42**, 667.
26. Kloc L., Cerri E., Spigarelli S., Evangelista E. and T.G.Langdon, *Mater.Sci.Engng.*1996, **A216**, 161.
27. Kloc L., Cerri E., Spigarelli S., Evangelista E. and T.G.Langdon, *Acta mater.*, 1997, **45**, 529.
28. Badini C., Marino F. and Verné E., *Mater.Sci.Engng.*, 1995, **A191**, 185.
29. Li, Y. and Mohamed F.A., *Acta mater.* 1997, **45**, 4775.
30. Li.Y, Nutt R.S. and Mohamed F.A., *Acta mater.*, 1997, **45**, 2607.

Fibre Breakage during Creep of Short Fibre Reinforced Al-7Si-3Cu-1Mg Alloy

R. Válek [1)], K. Kuchařová [2)], A. Dlouhý [2)]

[1)] Chemical Faculty, University of Technology, Brno, Czech Republic
[2)] Institute of Physics of Materials, Academy of Sciences of the Czech Republic, Žižkova 22, 616 62 BRNO, Czech Republic
e-mail: , fax: ++420 5 41212301

Abstract

A metal matrix dissolution technique has been improved to assess a breakage of short fibres during creep of Al-7Si-3Cu-1Mg alloy based metal matrix composite (MMC). Short reinforcing Saffil fibres (15vol%), originally embedded in the metal matrix (85vol%), were separated from the chemical solution using centrifugal forces and the evaluation of their length was assisted by a PC-based image analysis procedure.

Results show unambiguously that short ceramic fibres do break during creep at 623 K. The distribution of the fibre length after 2% of creep strain peaks at the length value that is approximately half of the one observed before creep. This result has not only been found for the population of fibres originally situated close to the fracture surface but also for the fibre population extracted from the crept specimen gauge length far from fracture. Consequently, it is concluded that the fibre breakage process is rather uniform in the specimen gauge length during creep of the investigated MMC. It is also suggested that a close link exists between observed anisotropy of creep in the fibre textured MMCs and the fibre breakage process. These results are in agreement with the microstructural scenario and the model of creep in MMCs proposed earlier [1,2].

1. Introduction

Since more than twenty years, basic mechanisms of high temperature deformation and damage accumulation in short fibre reinforced metal matrix composites (MMCs) have been investigated, for a literature survey see [1,2]. This experimental study concentrates on still open issues associated with the quantitative assessment of damage accumulation. Particularly, the breakage of short fibres during creep is investigated in detail.

There has been a long-term discussion about artefacts related to the preparation of MMC metallographic cross-sections that could mask a ceramic reinforcement damage of real interest associated mainly with various types of thermo-mechanical loading [3-7]. The specimen preparation-related damage (SPRD) is mainly due to the grinding and polishing of metallographic cross-sections. It is therefore important to develop a method which suppresses the SPRD and which, in the same time, offers a fair chance to quantify the real damage of the reinforcement introduced during material processing and service. The development and first preliminary results of the method are presented in this contribution. The method could be particularly useful in cases when high aspect ratio fibres (length/diameter) are used as the matrix reinforcement since they are especially prone to the SPRD.

The reliability and efficiency of the new method is tested performing dissolution experiments before and after creep of Al-7Si-3Cu-1Mg alloy reinforced with 15vol% of short Al_2O_3 fibres. The relation between observed anisotropy of creep and the extent of fibre breakage is investigated.

2. Experimental details

2.1. Experimental alloy and creep specimens

The alloy Al-7Si-3Cu-1Mg (in wt%) reinforced with 15vol% of Al_2O_3 short fibres (Saffil) was produced by Alusuisse-Lonza using the squeeze casting processing route [8]. Dimensions of the blocks were 100×140×40 mm. Inside the block, the reinforced MMC part occupied a volume with dimensions 80×120×15 mm. The mean length and diameter of the individual cylindrical fibres were 200 μm and 4 μm, respectively. Due to a preparation of the fibre preforms [8], the orientation of fibres in space within the reinforced part of the cast block was not completely random [4]. The preferential orientation of fibres was characterized as a random planar fibre texture (RPFT) [4] where the longitudinal axes of fibres tend to be oriented in the horizontal plane of the cast block. However, no further angular correlation of long fibre axes in the horizontal plane of the block was detected [4].

Tensile (gauge length 25 mm, cross-section 5×3.2 mm^2) and cylindrical compression (diameter 6 mm and height 12 mm) creep specimens were cut out of both, the matrix alloy and the reinforced part using the spark erosion technique. Two different orientations of the specimen axis were selected with respect to the horizontal plane of RPFT. Specimens the axis of which was normal to the horizontal plane are denoted as **TN** (tension-normal) and **CN** (compression-normal) throughout this study. Similarly, specimens with the axis parallel to the horizontal plane are further referred to as **TP** (tension-parallel) and **CP** (compression-parallel). In what follows, the abbreviations TN, CN, TP and CP are also used to refer to the corresponding type of creep test. In passing we note that the limited height of the reinforced volume in the cast block (15 mm) required special design of the TN specimens. Therefore, tensile creep technique has been developed in which miniature specimens with the gauge length 8 mm and the cross-section 4×2.5 mm^2 were tested.

2.2. Creep testing

The four types of specimens described above which corresponded to all four possible combinations of the two axis orientations and the two loading modes were creep tested at 623 K in a purified argon atmosphere. The miniature TN specimen was the only exception when creep was performed under the constant load, in all other experiments the constant stress was applied during creep. True strain-time readings were continuously recorded by the PC-based data acquisition system. The testing temperature was maintained constant within ±1 K along the specimen gauge length and invariable during the tests. These measures yielded a good reproducibility of the creep results.

2.3. Metallographic methods

The metal matrix dissolution method is based on a procedure proposed earlier [5] in which the metal matrix is solved out in a suitable chemical solvent while the solvent does not react with ceramic phases of MMC. The specimen cut either out of the initial squeeze cast block or out of the crept sample was cleaned for 15 minutes in an acetone bath using the ultrasonic cleaner. Specimens then were exposed to a concentrated solution of NaOH (10 mol/l) at room temperature for 72 hours. The solution attacked and dissolved aluminium alloy but did not react with Al_2O_3 (fibres) and SiO_2 (binder).

The fibres were isolated from the solution in a centrifugal separator. In the first step, the suspension resulting from the matrix dissolution was poured into tubes, introduced into the separator and centrifuged with an angular velocity either 1000 or 3000 rpm for 20 min. In the next step, the solution situated above fibres was replaced by distilled water and the centrifuging procedure was repeated until the solution above fibres was pH neutral. Finally, the segregated fibres were immersed into ethanol. A drop of the suspension was deposited on a microscope slide (NEOPHOT 22, ZEISS) and micrographs were taken in reflected light. A typical micrograph is presented in Fig.1.

These micrographs were re-drawn onto transparent foils. The layering option of the CorelDRAW 9.0 program enabled the separation in cases where an overlap of fibres was observed in the original micrograph. The re-drawn microstructures were subjected to the PC-assisted image analysis (DIPS 5.0) in which the fibre length was measured for a sufficient sample size of fibres (typically around 500) in each investigated microstructural state.

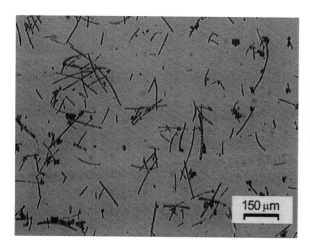

Figure 1: Saffil fibres centrifugal separated from the matrix solution of the initial cast block, immersed in ethanol and deposited on the slide of a light microscope.

Foils for transmission electron microscopy (TEM) were thinned and perforated in an ion mill EDWARDS IBT 200. The foils were investigated in Philips CM12 STEM operating under the accelerating voltage of 120 kV. Standard imaging and diffraction techniques were used to assess the specimen microstructure. The phase composition was studied in STEM mode with EDAX.

3. Results

3.1. Creep behaviour

Compression creep curves of Al-7Si-3Cu-1Mg alloy reinforced with 15vol% of Al_2O_3 short fibres are presented in Fig. 2 in co-ordinates creep rate versus creep strain. The creep curves were obtained at 623 K and in the applied stress range 30-60 MPa. Figures 2a and b illustrate respectively the results of CP and CN creep. A brief inspection of Fig. 2 shows that the compression creep curves exhibit a sharp creep rate minimum early in the creep life at approximately 2% of creep strain. This minimum separates the primary creep stage during which creep rate decreases more than 4 orders of magnitude and the tertiary stage in which creep rate steadily increases and approaches a value measured in the reference tests for the unreinforced matrix alloy. The upper strain limit of 40% represents the strain level up to which it is possible to maintain the constant applied stress with our compression creep machines.

A summary of Al-7Si-3Cu-1Mg matrix and MMC creep data obtained in this study for all four combinations of the specimen axis orientation and the loading modes is presented as a plot of minimum creep rate versus applied stress in Fig. 3. The first important result shown in

284

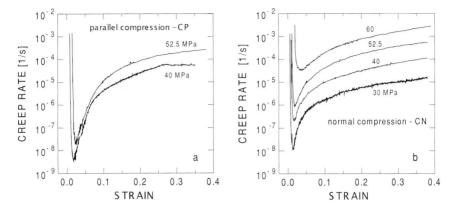

Figure 2: Creep curves obtained at 623 K for the Al-7Si-3Cu-1Mg alloy reinforced with 15vol% of short Al_2O_3 fibres (Saffil). The results correspond respectively to two combinations of the compression loading and the orientation of the specimen axis with respect to the fibre texture plane, parallel CP in part (a) and normal CN in part (b).

this plot is the data set of the unreinforced matrix Al-7Si-3Cu-1Mg alloy. All the matrix data fall within a narrow band delimited by the shaded region in Fig. 3 regardless whether, at the given applied stress, the specimen was tested in tension or in compression or whether the specimen axis was normal to or parallel with the horizontal plane of the cast block. This result suggests that the matrix exhibits a complete isotropy of creep properties in the range of external conditions investigated in the present study.

In contrast, the minimum creep rates for the corresponding MMC material are clearly dependent on both, the loading mode and the orientation of the specimen axis with respect to the horizontal plane of the cast block (RPFT plane [4]). Each of the four different combinations of the MMC specimen orientation and the loading mode is represented by a distinct dependence in Fig.3. At the applied stress of 40 MPa, where data for all four types of MMC tests are available, the CP specimen exhibits the highest creep strength followed by

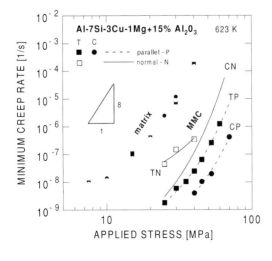

Figure 3: Summary of minimum creep rate vers. applied stress data obtained in this study at 623 K. The anisotropy of the MMC creep is clearly demonstrated. The matrix alloy does not exhibit the anisotropy of creep rate, the data for different specimen orientations fall in the narrow shaded band.

TP, CN and TN specimens. The minimum creep rate obtained in the fastest TN test is approximately two orders of magnitude higher comparing to the slowest CP test. Nevertheless, the TN specimen clearly creeps much slower as compared to the unreinforced matrix at 40 MPa. As it is further documented in Fig. 3, a similar trend can be observed also for other applied stresses in the investigated stress range.

3.2. Fibre breakage

Results of the matrix dissolution experiments are presented as cumulative distributions of fibre length in Figs. 4 and 5. The influences of the separator angular speed, creep conditions and the sampling along the creep specimen were studied.

Distribution curves of fibre length in the material state after MMC squeeze casting are shown in Fig.4 a. The distributions were obtained in two independent experiments where the separation of fibres in the first experiment was performed at the angular speed of 1000 rpm while in the second experiment the angular speed of 3000 rpm was used. In both cases the average length of fibres (100 µm) as well as the shape of the distribution are identical. This result shows that the separation procedure does not introduce any preparation-related artefacts like the fibre breakage.

The similar experiment with the angular velocity of 1000 rpm was performed for the tensile specimen crept 2% at 623 K and 45 MPa. Results are presented in Fig.4 b where the fibre length distributions corresponding to the fracture part of the testing rod and the gauge length part situated 24 mm off the fracture are compared with the distribution obtained for the undeformed material of the specimen head. Apparently, both distributions measured in the gauge length are shifted left towards smaller fibre length classes. Fibre length close to the fracture surface is slightly lower as compared to the distribution obtained for the gauge length part. The characteristic average value of the fibre length in the fracture part is 50 µm while in the gauge length part the average length of 70 µm was found. However, both distributions obtained in the gauge length are distinctly different as compared to the fibre length distribution obtained for the specimen head. This result suggests that extensive fibre breakage takes place during creep.

Distribution curves of fibre length in the material state after MMC squeeze casting and in the

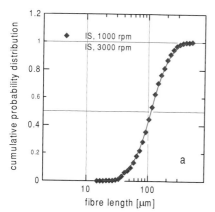 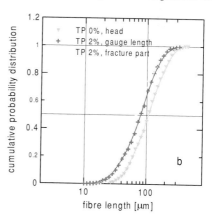

Figure 4: Cumulative fibre length distributions obtained for the initial as-cast state (IS) of the MMC alloy **(a)** and after 2% of tensile creep (TP) at 623 K and 45 MPa **(b)**. The influence of the separator rotation speed is shown in **(a)**, the influence of the sample position in the crept specimen is presented in **(b)**.

286

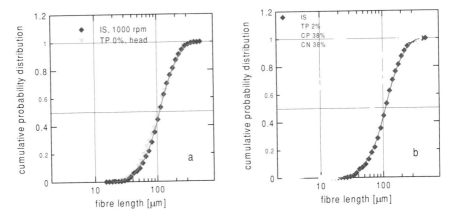

<u>Figure 5:</u> Cumulative fibre length distributions obtained for the initial as-cast state of the MMC block and after tensile and compression creep at 623 K. **(a)** the influence of the thermal exposure in the specimen head, **(b)** the influence of creep strain and the combination of creep mode and the specimen axis orientation (experiments TP, CP and CN described in section 2).

specimen "head" of the specimen crept at 623 K and 45 MPa are compared in Fig.5 a. The material taken from the specimen head was exposed only to temperature changes associated with the heating and cooling to the creep testing temperature and back. The head part did not experience any relevant stress applied externally. As compared to the as cast MMC state, the distribution after the thermal exposure is negligibly shifted left towards shorter fibre lengths. Fibre damage during the thermal exposure could result from different thermal expansion coefficients of the metal matrix and ceramic fibres.

Finally, Fig.5 b illustrates how the different orientation of creep specimens with respect to the RPFT plane, different loading modes (tension or compression) and different strains influence the intensity of fibre breakage. Four curves shown in Fig.5 b were obtained for initial as-cast MMC and after creep at 623 K and (i) 45 MPa, 2% in tension, (ii) 40 MPa, 38% in compression – experiment CP and (iii) 40 MPa, 38% in compression – experiment CN. The Fig. 5 clearly demonstrates that the distribution curves of the fibre length shift towards small length classes not only with increasing creep strain (curves obtained after 2 and 38% of strain) but the length of fibres also depends on the orientation of the specimen axis with respect to the RPFT plane (curves which correspond to CP and CN experiment). Thus the fibre breakage during creep depends on the preferential orientation of fibres with respect to the loading axis as well as on the amount of creep strain.

3.3. TEM
Transmission electron microscopy provided an independent evidence of the fibre breakage during creep. Two TEM micrographs shown in Fig.6 document the microstructure of the specimen crept in tension at 623 K and 52.5 MPa in an early creep stage (0.5% creep strain). Numerous narrow cracks in fibres like the one shown in Fig.6 a were observed in this material state. The width of cracks in fibres was generally smaller than 100 nm what makes them almost undetectable during observation by conventional light microscope. In places with high local stress concentrations multiple breakage events occurred which apparently initiated and progressed in a co-operative manner; such situation is shown in Fig.6 b. We note that when looking at TEM foils of the initial MMC state after squeeze casting no single breakage event similar to those shown in Fig. 6 was found in transparent parts of the foils.

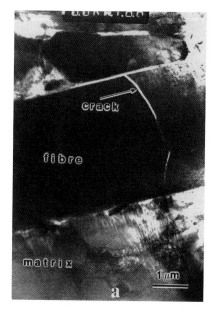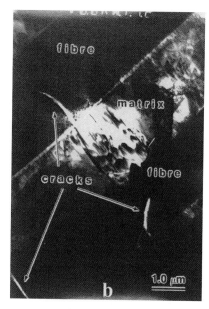

Figure 6: TEM evidence of fibre breakage processes (a) in early stage of TP creep (0.5% creep strain) at 623 K and 52.5 MPa. Multiple cracks can be detected in the bottom fibre in part (b).

High dislocation densities were present after creep in both, MMC matrix and close to the fibre matrix interface. However, the accumulation of dislocations was still discernable in the matrix close to the fibres. One such situation is documented in Fig.7 a after 38% creep strain in compression at 623 K under the applied stress of 40 MPa. A work hardened zone (see also [1,2]) in Fig.7 a extends several microns from the fibre-matrix interface into the matrix interior. The Al₂Cu particles observed in the micrograph of Fig.7 a coarsen during creep. However, in contrast to the creep behaviour of the unreinforced matrix, this process was shown unimportant as far as the creep strength of the MMC is concerned [1]. The other important feature is presented in Fig.7 b, which displays the microstructure of the same material state as that in Fig.7 a. The highly dislocated area denoted by an arrow in Fig.7 b is situated in the matrix next to the fibre crack. Apparently, the unloading of the fibre due to the breakage event transferred stresses back to the matrix and promoted the local matrix plasticity in the vicinity of the fibre rupture.

4. Discussion

Experimental results presented in this study support a microstructural scenario proposed earlier [1,2] in which three basic interrelated processes determine the macroscopic response of MMC to the TP creep loading. These processes are: (i) the load transfer to fibres associated with the formation of work hardened zones around fibres, (ii) the recovery mechanism which reduces the dislocation density in the work hardened zones and (iii) fibre breakage triggered when the tensile stress induced into the fibres from their work hardened zones exceeds a critical value. The results presented in this contribution suggest that the three basic processes play an important role also during creep in other two geometric situations,

288

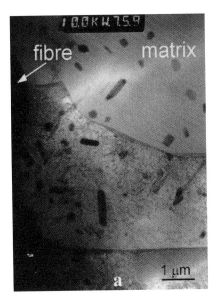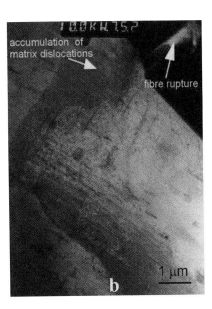

Figure 7: TEM evidence showing the accumulation of dislocations at the fibre-matrix interface (a) and the highly dislocated areas in the matrix next to the crack in fibre (b) after 38% of CN creep strain at 623 K and 40 MPa.

when compression is applied normal and parallel to the plane of RPFT (experiments CN and CP). The experimental evidence is mainly based on the shape of creep curves which all exhibit the steep decrease of creep rate in primary creep, see Fig.2 of this study and also [9]. The pronounced deceleration of creep rate associated with strains as small as 1 or 2% must be microstructurally related to severe constrains which cause a fast evolution of internal stresses opposing the matrix creep. The creation of work hardened zones (WHZs), experimentally confirmed for TP creep earlier [1] and for CN creep in this study (Fig.7 a), represents such a hardening mechanism that efficiently redistributes stresses in the MMC microstructure.

A combination of fibre breakage and dislocation climb controlled recovery had been shown to contribute to the increase of creep rate after the creep rate minimum [2]. In contrast to the hardening mechanism, the fibre breakage and recovery process redistribute stresses back to the matrix (see Fig.7 b of the present study) and this steadily increasing driving force causes the steady acceleration of creep. The metallographic results obtained in this investigation indicate the relevant intensity of fibre breakage during CN and CP creep (Fig.5 b). This also supports the argument that the microstructural scenario proposed for TP creep is at least partially transferable to other studied combinations of the specimen axis orientation and the loading mode.

Admitting that the minimum creep rate of MMCs results from the interplay among the three mentioned microstructural processes, the observed anisotropy of minimum creep rate can be accounted for qualitatively in the following way. Driving forces for each of the three microstructural processes depend on the overall stress state which evolves in the MMC during creep as a result of the loading mode and the particular distribution of fibres. Figure 8 schematically illustrates how different stress states originate in the case of compression creep. The left fibre in Fig. 8 represents a relevant number of fibres which are loaded by

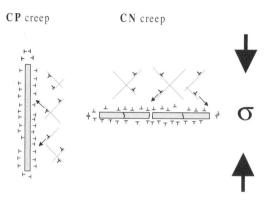

Figure 8: Schematic representation of additional stresses induced into the fibres from work hardened zones during compression creep. The parallel fibre on the left is loaded by the additional compression while the perpendicular fibre on the right experiences the additional tensile stress which causes the fibre breakage.

additional compression stresses induced from WHZ during CP creep due to the fact that these fibres are oriented close to the compression axis.

On the other hand, large majority of fibres is loaded by additional tensile stresses during CN creep since they approximately lie in the RPFT plane and are thus normal to the compression axis (the right part of Fig.8). Since the ceramic fibres loaded by the additional tension stress are much more prone to rupture as compared to the additional compression loading, more intensive fibre breakage could be expected in the CN creep. However, the more intensive fibre breakage means that the balance of hardening, recovery and damage processes required at the point of minimum creep rate is shifted towards higher values of creep rate. This is the reason why higher values of minimum creep rate are observed during CN creep in comparison to CP creep. In passing we note, that the proposed explanation is also in line with the more intensive fibre breakage observed after CN creep in Fig.5 b.

5. Summary and conclusions

A metal matrix dissolution technique has been improved and applied to assess a breakage of short fibres resulting from different modes of creep in Al-7Si-3Cu-1Mg alloy based metal matrix composite (MMC). Based on the obtained experimental results, following conclusions can be drawn.

1) The dissolution technique does not cause any additional damage to the ceramic MMC reinforcement and is therefore suitable for an artefact-free assessment of ceramic reinforcement rupture due to various types of thermo-mechanical loading.

2) The dissolution technique provided unambiguous evidence that ceramic fibres do break during creep of Al-7Si-3Cu-1Mg commercial alloy reinforced with 15vol% of short Saffil fibres.

3) The investigated reinforced Al-base alloy exhibited rather uniform distribution of fibre breakage events throughout the specimen gauge volume. The intensity of the fibre breakage was only slightly enhanced close to the rupture surface.

4) The fibre breakage starts early in creep life and its intensity depends on the accumulated creep strain as well as on the particular combination of the loading mode and the orientation of the specimen axis with respect to the fibre texture plane. With increasing creep strain individual fibres undergo several rupture events.

5) The microstructural scenario of creep in short fibre reinforced MMCs proposed for TP creep [1,2] is at least partially transferable to other studied combinations of the specimen axis orientation and the loading mode.

Acknowledgement

The financial support from the Ministry of Education, Youth and Sports (contract no. OC P3.50) is highly acknowledged. The authors also thank to Mr. J. Březina, Mrs. L. Adamcová and Mr. M. Daniel for the assistance with the light microscopy investigation, the preparation of final TEM micrographs and TEM foils.

References

[1] Dlouhý, A., Merk, N., Eggeler, G., 1993, A microstructural Study of Creep in Short Fibre Reinforced Aluminium Alloys, Acta Metall. Mater., Vol. 41, pp. 3245-3256.
[2] Dlouhý, A., Eggeler, G., Merk, N., 1995, A micromechanical Model for Creep in Short Fibre Reinforced Aluminium Alloys, Acta Metall. Mater., Vol. 43, pp. 535-550.
[3] Komenda, J., Henderson, P.J., 1991, Microstructure and Creep Properties of an Al Alloy/Al_2O_3 Fibre Composite, in: Proceedings of the 12th Risø International Symposium on Materials Science, editors: N. Hansen et al., Risø National Laboratory, Roskilde, Denmark, pp. 449-454.
[4] Dlouhý, A., Eggeler, G., 1993, A Quantitative Metallographic Study of a Short Fibre Reinforced Aluminium Alloy, Pract. Metallography, Vol. 30, pp. 172-185.
[5] Eggeler, G., 1994, On the Mechanism of Creep in Short Fibre Reinforced Aluminium Alloys, Z. Metallkd., Vol. 85, pp. 39-46.
[6] Nicholas, T., Castelli, M.G., Gambone, M.L., 1997, Fibre Breakage in Metal Matrix Composites – Reality or Artifact, Scr. Mater., Vol. 36, pp. 585-592.
[7] Horkel, T., Březina, J., Dlouhý, A., 1997, Metallographic Cross-Sections of Aluminium Alloy-Base Metal Matrix Composites,, Pract. Metallography, Vol. 34, pp. 162-174.
[8] Roulin, M., 1991, Untersuchungen an den mit 15 Vol.% verstärkten Legierungen AlSi7Cu3 und AlSi7Cu3Mg, Report FWE 91/008, Neuhausen am Rheinfall.
[9] Kuchařová, K., Horkel, T., Dlouhý, A., 1999, Creep Anisotropy of Aluminium Alloy-Base Short Fibre Reinforced MMC, in: Creep Behavior of Advanced Materials for the 21st Century, editors: R. S. Mishra, A. K. Mukherjee and K. L. Murty, TMS, Warrendale, USA, pp. 127-136.

HIGH TEMPERATURE DEFORMATION OF AN
Al 6061- Al$_2$O$_3$ PARTICULATE METAL MATRIX COMPOSITE

Stefano Spigarelli *, Elisabetta Gariboldi**and Terence G.Langdon***

* INFM/Department of Mechanics, University of Ancona, Ancona I-60131, Italy
** Department of Mechanics, Politecnico di Milano, Milano I-20133, Italy
*** Departments of Aerospace & Mechanical Engineering and Materials Science
University of Southern California, Los Angeles, CA 90089-1453, U.S.A.

Abstract

The high temperature deformation of an Al 6061-20% Al$_2$O$_3$ particulate-reinforced composite was analysed by comparing three different sets of data obtained by testing the same material under conditions of either creep (using a constant load in tension or a constant stress in shear) or torsion (using a constant strain rate). By incorporating a back stress into the analysis, it is possible to determine the value of the true stress exponent. The results show the exponent is close to 5 at high stresses and close to 3 at low stresses, suggesting a transition from climb-control at the higher stresses to a viscous glide process at low stresses where dislocations drag solute atom atmospheres. It is shown that the stress corresponding to the transition from the viscous glide to the climb-controlled regions is consistent with the expectations for the transition from class A to class M behaviour. The analysis indicates the back stress, σ_0, is equivalent to a true threshold stress when the applied stress exceeds a limiting value but it is proportional to the applied stress at stresses below this limit. The creep behaviour of this composite is essentially in agreement with the expectations from earlier creep analyses obtained from the testing of dilute Al-Mg alloys and this similarity supports the proposal that creep in the composite is controlled by deformation within the matrix material.

Keywords: Aluminium, creep mechanisms, metal matrix composites, threshold stresses

1. Introduction

Numerous studies have been conducted to date in order to determine the characteristics of high-temperature deformation in discontinuously-reinforced aluminium-based metal matrix composites [1-20]. In these materials, a logarithmic representation of the minimum creep rate, $\dot{\varepsilon}$, as a function of the applied stress, σ, often leads to a plot that exhibits a marked curvature [3,7-20]. This behaviour has been observed not only in composites but also in some Al-based unreinforced alloys [21,22] and it is substantially different from the creep behaviour of pure metals and dilute solid solution alloys.

In general, the steady-state creep rate of pure metals and simple single-phase alloys depends upon the applied stress and temperature through a relationship of the form:

$$\dot{\varepsilon} = \frac{ADGb}{kT} \left(\frac{\sigma}{G} \right)^n \qquad (1)$$

where $D = D_0 \exp(-Q_0/RT)$ is the diffusion coefficient, D_0 is a frequency factor, Q_0 is the true activation energy for creep, R is the gas constant, T is the absolute temperature, G is the shear modulus, b is the Burgers vector, k is Boltzmann's constant, n is the stress exponent and A is a dimensionless constant. For pure metals and class M alloys [23], creep is controlled by high temperature climb and under these conditions the stress exponent is close to 5 and D is the coefficient for self-diffusion in the metal lattice (D_L). For class A alloys, creep is controlled by a viscous glide process where dislocations drag solute atom atmospheres and n = 3 and D = $\tilde{D}$ which is the coefficient for diffusion of the solute element in the matrix alloy. In practice, when the applied stress is sufficiently high, the dislocations are able to break away from their solute atom atmosphere so that there is a breakdown in class A behaviour and climb again dominates with n = 5 and D = D_L.

The curvature in the logarithmic plots of the strain rate vs applied stress, which is typical of composites or materials where obstacles such as particles or dispersoids restrict the movement of dislocations, may be rationalised by introducing the concept of a threshold stress. According to this model, the deformation is then driven by an effective stress equal to $\sigma - \sigma^*_0$, where σ^*_0 is a temperature-dependent threshold stress that delineates a lower limiting strain rate. In this analysis, Eq.1 is then modified to the form:

$$\dot{\varepsilon} = \frac{ADGb}{kT} \left(\frac{\sigma - \sigma^*_0}{G} \right)^n \qquad (2)$$

The introduction of a temperature-dependent threshold stress justifies the high values of n observed experimentally at the lowest stresses and also the high and stress-dependent values of the apparent activation energy for creep, defined as

$$Q = \left[\frac{\partial \ln \dot{\varepsilon}}{\partial (-1/kT)} \right]_\sigma \qquad (3)$$

The values of the threshold stresses may be calculated by plotting $\dot{\varepsilon}^{1/n}$ against σ on linear axis and extrapolating linearly to zero creep rates [24]. However, this method has the disadvantage that it requires a knowledge of the value of the true stress exponent n. Alternatively, the threshold stress may be estimated by directly extrapolating the creep data to very low strain rates (for example, 10^{-10} s^{-1} [14]). Following these procedures, the creep behaviour of several composites have been extensively analysed including the Al-6061 matrix alloy reinforced with alumina [15,16,18] or SiC particles [3,8,10].

In parallel with these investigations on creep, other authors [25-27] have investigated the hot formability of the same materials. In these studies, the material is generally tested by torsion at equivalent strain rates ranging from 10^{-3} to 10 s^{-1}. Although the mechanisms controlling deformation are probably the same in creep and torsion, nevertheless the use of very high strain rates leads to a breakdown in the conventional power-law behaviour as given by Eq.1. For this reason, results obtained in these hot-working studies are generally described by expressing the dependence of the flow stress on strain rate and temperature by means of the relationship:

$$\dot{\varepsilon} = A' \left[\sinh(\alpha\sigma) \right]^{n'} \exp(-Q/RT) \tag{4}$$

where A', α and n' are stress and temperature-independent constants and Q is the apparent experimental activation energy. The experimental value of Q is frequently close to, but almost invariably higher than, the activation energy for self-diffusion in Al (143.4 kJ mol^{-1}).

The objective of the present study was to develop a unified description to incorporate three different sets of data obtained by testing of the Al 6061-20% Al_2O_3 composite under conditions of shear (constant stress creep), tension (constant load creep) and torsion (constant strain rate).

2. Experimental

The metal matrix composite used in the present investigation was fabricated via ingot-metallurgy by Duralcan USA (San Diego, California). The material was reinforced with 20% Al_2O_3 particulates having average dimensions close to 20 µm.

Double-shear [28] creep tests were carried out at temperatures of 623, 673 and 773 K [16] and additional creep tests were conducted in tension at 523 and 423 K [29]. In addition, torsion testing was performed at temperatures between 648 and 773 K. The equivalent stress and strain were calculated by means of the von Mises yield criterion by using the relationships $\sigma = \tau\sqrt{3}$ and $\varepsilon = \gamma/\sqrt{3}$. These same equations were also used to convert shear creep data into the form of equivalent stress and strain rate. All the experimental flow curves in torsion exhibited a peak in the strain range between 0.2 and 0.3 and the analysis of the dependence of flow stress on strain rate and temperature was performed by measuring σ at the peak.

It should be noted that the creep samples were tested in a T6 conditions (with a solution treatment at 803 K and aging to peak hardness at 448 and 443 K for the shear and tension creep specimens, respectively) whereas the torsion tests were performed on the material in an as-extruded or substantially overaged state. This difference may be especially relevant for tests at low temperatures but it is of only minor significance for the tests conducted at the highest temperatures of 773-673 K in the present study. Thus, the relatively long times necessary for heating the samples and stabilising the furnace before creep testing at temperatures between 623 and 773 K (i.e. at temperatures well above the ageing temperature) would lead to extensive overageing or even partial dissolution of the precipitates. For this reason, the effect of any differences in the initial microstructure is assumed to be negligible.

3. Experimental results and Discussion

The typical creep curves exhibited a short primary stage, followed by a short secondary stage or rather a minimum-creep rate, and finally an extended tertiary stage. In the case of the equivalent stress-equivalent strain curves obtained by testing in torsion at constant strain rates, the flow stresses decreased with increasing temperature and at a given strain rate the flow stress increased up to a maximum value and then gradually decreased with increasing strain. Figure 1 plots the minimum equivalent strain rate or the testing strain rate for the torsion tests (both indicated as $\dot{\varepsilon}$) as a function of the equivalent applied stress (or the peak flow stress in torsion). The various datum points align on lines of slopes varying from 5.7 to 12 for temperatures between 773 and 423 K. These results seem to suggest that the conventional power law, as given by Eq.1, remains valid over the entire range of experimental conditions presented in Fig.1 and therefore over approximately 10 order of magnitude of strain rate. However, this observation is unrealistic because it is well known that, in face-centred cubic metals, power-law breakdown occurs at normalised creep rates of the order of $\dot{\varepsilon}/D = 10^{13}\,m^{-2}$ [30] where $D = D_L$ for pure metals or class M alloys. The coefficient for self-diffusion D_L can be expressed in the form [31]

$$D_L = 1.86 \times 10^{-4} \exp(-143.4/RT)\quad m^2\,s^{-1} \tag{5}$$

and it is therefore easy to estimate the limit for power-law breakdown. Figure 2 shows the experimental data included in the power-law regime.

There was an earlier extensive analysis of the creep data obtained in shear at 773, 673 and 623 K [16]. It was found that the highest stress levels, corresponding to class M behaviour and identified as region I in Fig. 2, gave high creep rates due to the break-away of dislocations from their solute atom atmospheres. Based on this observation, the preliminary analysis was restricted to data falling within the low stress region, corresponding to class A behaviour and identified as region II in Fig. 2. Thus, at intermediate stresses the curvature of the strain rate vs stress plots clearly implies the presence of a threshold stress so that the dependence of the minimum creep rate on the applied stress can be described by means of Eq.2, where σ^*_0 may be calculated at 773, 673 and 623 K by plotting $\dot{\varepsilon}^{1/n}$ vs σ and taking $n = 3$ for the data in region II. Figure 3 illustrates the dependence of strain rate on the effective stress, $\sigma - \sigma^*_0$. Thus, the two regions identified earlier [16] can be easily distinguished. In particular, the slope of the lines in region I is close to 5 while in region II it is close to 3 at temperatures of 623 to 773 K.

The calculation of the threshold stress at 423 K is more complicated. At these low temperatures, the equilibrium concentration of solute elements is low so that the dislocations are able to break away from their solute atom atmosphere. Thus, it may be assumed initially that, in the range of applied stresses investigated experimentally, creep is climb-controlled. On the other hand, most of the tests conducted at 423 K lie above the condition of power-law breakdown. As a result, the threshold stress has not been calculated from Eq.2 but using a modification of Eq.4 of the form:

$$\dot{\varepsilon} = A'' \frac{DGb}{kT} \sinh\left[\alpha'' \frac{\sigma - \sigma^*_0}{G}\right]^n \tag{6}$$

with $n = 5$ and $D = D_L$. Since Eq.6 reduces to Eq.2 when $(\sigma - \sigma^*_0)$ is low, this form of constitutive equation should be equivalent, in principle, to Eq.4 for materials that have a threshold stress. The data in region I at 773 and 673 K were interpolated by means of Eq.6 and the value of the threshold stress calculated to give $\alpha'' \approx 370$. Having calculated α'', the values of the threshold stresses at 723 and 423 K were obtained as the value giving the best fit to the data by means of Eq.6.

The values of the threshold stresses so estimated were very close or slightly higher than the lowest values of the applied stress. Indeed the values of $(\sigma - \sigma*_0)$ are very low or even negative at the lowest values of the applied stress at 673 and 423 K. This behaviour may be rationalised by considering that, when the applied stress approaches the value of the threshold stress, there is a transition in the particle-by-passing mechanism. Since it is unrealistic to abruptly set the threshold stress equal to zero, a reasonable assumption is that for these data the threshold stress is proportional to the applied stress. This interpretation is consistent with the model proposed by Lagneborg [32] where the new length of dislocation created in climbing over the obstacle depends upon the curvature of the gliding dislocations and therefore upon the magnitude of the applied stress. Thus, for applied stresses lower than the Orowan stress, σ_{Or}, the model of Lagneborg [32] leads to a back stress given by $\sigma*_b = K_0 \sigma$ $(K_0 = 0.7)$; whereas when the applied stress exceeds the Orowan stress, the back stress becomes constant and therefore assumes the significance of a real threshold stress. This model implies that the dependence of the minimum creep rate on the applied stress assumes a typical S-shape; this behaviour has been observed in other composites [33] and in aluminium alloys reinforced with incoherent particles [34]. By assuming that, for the low-stress data in region II, $\sigma*_0 = \sigma*_b = K_0 \sigma$ and these points align on the same lines as for the higher stress data in region II, the value of $\sigma*_0$, now defined as a back stress instead of a true threshold stress, can be calculated to give K_0 close to 0.8. Figure 4 plots the variation of threshold stress with temperature and applied stress calculated using this procedure.

The activation energy for creep in region I was obtained by plotting A" Db/k in Eq. 6 as a function of 1/T as shown in Fig.5(a) and, as expected, the value estimated for the activation energy (~148 kJ mol^{-1}) is very close to the activation energy for self-diffusion in aluminium.

The activation energy for creep in region II was calculated by plotting $\dot{\epsilon} G^{n-1} T$ vs $1/T$ for an effective stress of 5 MPa as shown in Fig.5(b), and this plot gave an activation energy for creep of ~130 kJ mol^{-1} which is in agreement with creep controlled by viscous glide and an activation energy equal to the value for diffusion of Mg in an aluminium matrix (~130.5 kJ mol^{-1} [35]).

4. Conclusions

The high temperature deformation of an Al 6061-20% Al$_2$O$_3$ composite was analysed by comparing three different sets of data obtained by testing the same material using different procedures and at different temperatures. By introducing a back stress into the analysis, it is demonstrated that the true stress exponent is close to 5 in the high stress region and close to 3 at low stresses. These results, and the estimated values for the activation energies for creep, are consistent with climb-controlled creep at the higher stresses and control by viscous glide at the lower stresses. It is shown that the back stress exhibits the nature of a well-defined threshold stress above a certain level of the applied stress but it is proportional to the applied stress below this level. The results of this analysis support the proposal that creep in the particulate-reinforced composite is controlled by deformation in the matrix .

296

Acknowledgements

This work was supported by CNR (Consiglio Nazionale delle Ricerche) and MURST (Ministero dell'Università e della Ricerca Scientifica) in Italy and by the U.S. Army Research Office under Grant DAAD19-00-1-0488.

References

1. Nieh, T.G., *Metall. Trans.* 1984, **15A**, 139.
2. Mishra, R.S. and Pandey, A.B., *Metall.Trans.* 1990, **21A**, 2089.
3. Park, K.-T., Lavernia, E.J. and Mohamed, F.A., *Acta metall.mater.* 1990, **38**, 2149.
4. Mohamed, F.A., Park, K.-T. and Lavernia, E.J., *Mater.Sci.Engng.* 1992, **A150**, 21.
5. Pandey, A.B., Mishra, R.S. and Mahajan, Y.R., *Acta metall.mater.* 1992, **40**, 2045.
6. Gonzales-Doncel, G. and Sherby, O.D., *Acta metall.mater.* 1993, **41**, 2797.
7. Cadek, J. and Sustek, V., *Scripta metall.mater.* 1994, **30**, 277.
8. Park, K.-T. and .Mohamed, F.A., *Scripta metall.mater.* 1994, **30**, 957.
9. Cadek, J., Sustek, V. and Pahutova, M., *Mater.Sci.Engng.* 1994, **A174**, 141.
10. Park, K.-T. and Mohamed, F.A., *Metall. Mater.Trans.* 1995, **26A**, 3119.
11. Cadek, J., Oikawa, H. and Sustek, V., *Mater.Sci.Engng.* 1995, **A190**, 9.
12. Pandey, A.B., Mishra, R.G., Paradkar, A.G. and Mahajan, Y.R., *Acta mater.* 1997, **45**, 1297.
13. Li, Y. and Langdon, T.G., *Metall. Mat.Trans.* 1997, **28A**, 1271.
14. Li, Y. and Langdon, T.G., *Scripta mater.*1997, **36**, 1457.
15. Ma, Y. and Langdon, T.G., *Mater.Sci.Engng.* 1997, **A230**, 183.
16. Li, Y. and Langdon, T.G., *Acta mater.* 1997, **45**, 4797.
17. Li, Y. and Langdon, T.G., *Metall. Mater.Trans.* 1998, **29A**, 2523.
18. Li, Y. and Langdon, T.G., *Mater.Sci.Engng.* 1998, **A245**, 1.
19. Li, Y. and Langdon, T.G. *Acta mater.* 1998, **46**, 1143.
20. Li, Y. and Langdon, T.G., *Acta mater.* 1999, **47**, 3395.
21. Park, K.-T., Lavernia, E.J. and Mohamed, F.A., *Acta metall.mater.* 1994, **42** , 667.
22. Kloc, L., Spigarelli S., Cerri, E., Evangelista E. and Langdon, T.G., *Acta mater.* 1997, **45**, 529.
23. Yavari, P. and Langdon, T.G., *Acta metall.* 1982, **30**, 2181.
24. Lagneborg, R. and Bergman, B., *Metal Sci.* 1976, **10**, 20.
25. Yu, D. and Chandra, T., in *Advanced Composites Materials*, T.Chandra and K.Dhingra eds., The Minerals, Metals and Materials Society, Warrendale, PA, 1993, p.1073.
26. Xia, X., Sakaris, P. and McQueen, H.J., *Mater.Sci.Techn.* 1994, **10**, 487.
27. Evangelista, E., Forcellese, A., Gabrielli, F. and Mengucci, P., in *Hot Deformation of Aluminum alloys*, T.G.Langdon, H.D.Merchant, J.G.Morris and M.A.Zaidi eds., The Minerals Metals and Material Society, Warrendale, PA, 1991, 121.
28. Chirouze, B.Y., Schwartz, D.M. and Dorn, J.E., *Trans. Quart. ASM* 1967, **60**, 51.
29. Gariboldi, E. and Vedani, M., *Advanced Engineering Materials*, 2000, **2**, No 11, 737.
30. Sherby, O.D. and Burke, P.M., *Prog.Mater.Sci.* 1967, **13**, 325
31. Mohamed, F.A. and Langdon, T.G., *Metall. Trans.* 1974, **5**, 2339.
32. .Lagneborg, R., *Scripta metall.* 1973, **7**, 605.
33. Cadek, J., Kucharova, K. and Sustek, V., *Scripta mater.* 1999, **40**, 1269
34. Goto, S., in *Creep and Fracture of Engineering materials and Structures*, B.Wilshire and R.W.Evans eds., The Institute of Metals, London, U.K., 1984, p.131.

35. Rothman, S.J., Pearson, N.L., Nowicki, L.J., and Robinson, L.C., *Phys. Stat. Solidi (b)* 1974, **63**, K29.

Figures

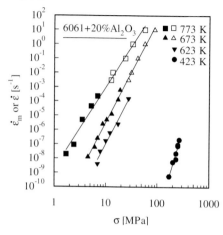

Figure 1. Strain rate as a function of applied stress: open points are torsion data and the solid points show creep in shear at the three highest temperatures [16] and tensile creep at 423 K.

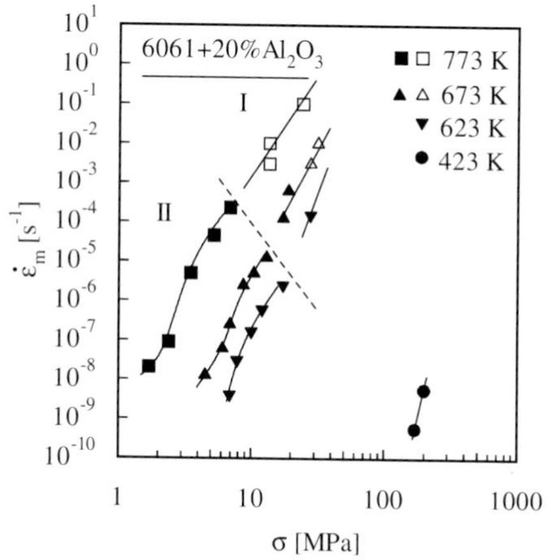

Figure 2. Strain rate vs. applied stress showing the division into two regions.

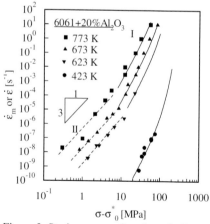

Figure 3. Strain rate as a function of effective stress

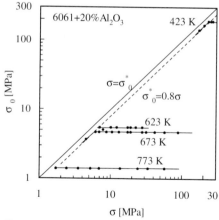

Figure 4. Variation of σ^*_0 with applied stress.

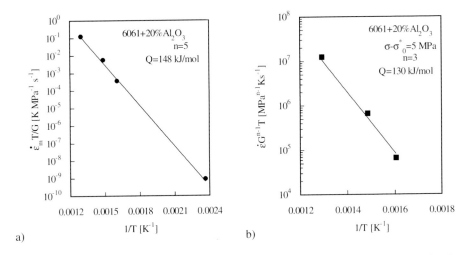

Figure 5.Calculation of the activation energy for high-temperature deformation in (a) region I and (b) region II.

CREEP OF INTERMETALLICS

Factors Influencing the Accelerating Creep Strain Rate of Near γ-TiAl Intermetallics

J. Beddoes, W.R. Chen, D.Y. Seo

Department of Mechanical & Aerospace Engineering
Carleton University
Ottawa, Canada

Abstract

The creep behaviour of fully lamellar γ-titanium aluminide intermetallics is characterized by a primary stage of decreasing strain rate, leading to a strain rate minimum, followed by an extensive period of accelerating strain rate until rupture. The tensile creep response of two fully lamellar γ-TiAl compositions is presented. The accelerating strain rate is associated with microstructural instabilities and intergranular void formation. The effect of β precipitation along lamellar interfaces and grain boundaries on the tertiary creep response is highlighted. The microstructural instabilities during creep of the fully lamellar structure include dissolution and decomposition of the α_2 lamellae, shear band formation, and lamellae spheroidization. β precipitation/growth can also occur during creep loading. The results indicate that α_2 decomposition is a major factor causing the initial increase in strain rate following the strain rate minimum. However, sufficient β precipitation can compensate for the loss of α_2/γ lamellae caused by α_2 decomposition. The major form of creep damage is the formation and coalescence of intergranular voids, promoted by the presence of intergranular β particles.

1. Introduction

The low density, high melting temperature, good elevated temperature strength, and oxidation and creep resistance make γ-TiAl intermetallics potential candidate materials for some gas turbine applications [1-6]. Heat treatment of γ-TiAl is effective in producing a variety of microstructures, with the $\alpha_2+\gamma$ fully lamellar microstructural state of Figure 1 exhibiting the best creep resistance [7-13], primarily because lamellar interfaces hinder dislocation motion. The creep response of the fully lamellar condition, Figure 2a, is similar to that exhibited by many engineering alloys. However, as evident in Figure 2b, a secondary region in which the creep strain rate is constant does not occur. Rather, creep of fully lamellar γ-TiAl is typified by a period of decreasing strain rate, leading to a minimum strain rate, followed by an increasing strain rate leading to failure. In this paper, the period of increasing strain rate following the minimum strain rate is referred to as tertiary creep. Usually, microstructural instabilities or the formation and coalescence of voids or cracks cause the increasing strain rate during tertiary creep. For the binary γ-TiAl fully lamellar structure of Figure 1, tertiary creep is associated with microstructural instabilities including lamellar spheroidization and shear band formation, as well as intergranular void nucleation and growth [8,14].

The creep resistance of fully lamellar γ-TiAl intermetallics can be further improved by the addition of ternary or quaternary alloying elements. Potentially, improved properties may be due to solute effects or the formation of third phase particles. Titanium silicides, nitrides,

Figure 1. Fully lamellar $\alpha_2+\gamma$ structure of binary Ti-48%Al.

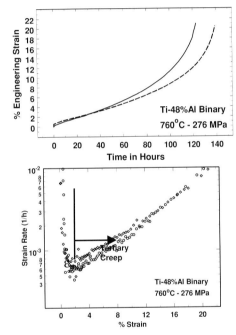

Figure 2. Creep properties of fully lamellar Ti-48%Al.

and carbides all can improve creep resistance [15-17], and the formation of β particles reduces primary creep strain [18]. Furthermore, appropriate control of heat treatment parameters to prevent the formation of massive γ or Widmanstätten γ structures improves the creep resistance of the fully lamellar microstructure. Hence, appropriate alloying and microstructural control of the fully lamellar structure can lead to creep properties vastly improved over those of the binary alloy shown in Figure 1. However, in these more complex γ-TiAl alloys the factors causing the onset of tertiary creep and the consequent increasing strain rate have not been well characterized. Therefore, this paper presents the creep behaviour of two γ-TiAl alloys heat treated to the fully lamellar condition, with particular attention to the microstructural and damage mechanisms occurring during tertiary creep.

2. Materials & Experimental Procedure

The creep of the two alloys listed in Table 1 is presented in this paper. The ternary TiAl+W alloy was produced by hot isostatically pressing gas atomized powder. TiAl+WMoSi was investment cast and hot isostatically pressed. The processing details are presented in [19,20].

The heat treatments of Table 1 were applied to produce fully lamellar microstructures. The design rationale for these heat treatments is described in [18,20]. Subsequently, some samples were aged at 850°C or 950°C. All samples for heat treatment were wrapped in Ta foil and encapsulated in quartz tubes backfilled with Ar. Microstructures were examined using standard metallographic techniques.

Table 1
Nominal Chemical Composition (at%) of γ-TiAl Alloys (bal. Ti)

Alloy	Al	Nb	Mn	W	Mo	Si	Fully Lamellar Heat Treatment
TiAl+W	48.4			1.9			1 h/1400°C fc to 1280°C ac
TiAl+WMoSi	47.0	1.98	0.94	0.5	0.5	0.21	20 h/1320°C + 20 min/1380°C fc to 1220°C ac

ac - air cool, fc – furnace cool

Tensile creep samples were prepared by low stress grinding to a straight gauge length of 22 mm and a gauge diameter of 4 mm. Constant load creep tests were conducted in air at 760°C, with an initial stress of either 207 MPa or 276 MPa. Creep strain was measured with an LVDT equipped extensometer attached to grooves in the specimen shoulders, providing a strain resolution of at least $\pm 5 \times 10^{-4}$. Strain measurement was initiated prior to specimen loading. Selected creep tests were interrupted prior to rupture for examination of creep deformed microstructures, by rapidly cooling the creep sample while still fully loaded.

3. Results

3.1 Heat Treated Microstructures

After the heat treatments of Table 1, the resulting fully lamellar structures of both alloys optically appear similar to that of Figure 1. Virtually no β particles exist along lamellar interfaces of either alloy, Figure 3a, although fine intergranular β particles do exist, Figure 3b. Ageing at 850°C or 950°C precipitates fine β, and titanium silicides in TiAl+WMoSi, along lamellar interfaces, Figure 3c, and causes coarsening of the pre-existing intergranular β, Figure 3d. An important difference between the two alloys is that after the same ageing treatment, the β interlamellar precipitate particles in TiAl+W are finer, but more dense, than those in TiAl+WMoSi, Figure 4. Presumably, since of the elements listed in Table 1, W is the most effective β stabilizer, the precipitation difference evident in Figure 4 is due to the greater W content of TiAl+W. Concurrently, with the precipitation reactions, partial dissolution of α2 lamellae occurs during ageing, as evident by the discontinuous lamellae in Figure 3c. α2 dissolution is believed to be due to the diffusion of W, and in the case of TiAl+WMoSi diffusion of W and Mo, from the α2 lamellae into the precipitating β. Further details of the heat treated microstructures are presented in [21,22].

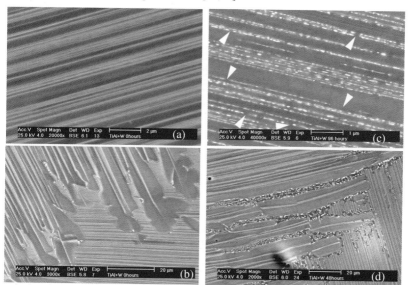

Figure 3. Scanning electron microscope backscattered electron images of TiAl+W (a) & (b) after heat treatment of Table 1, (c) after aging 96 h at 950°C, (d) after aging 48 h at 950°C. (a) & (c) are lamellar regions, (b) & (d) are intergranular regions. Arrow in (c) highlight discontinuous lamellae.

304

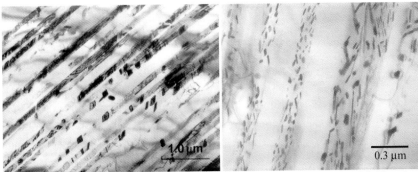

Figure 4. Interlamellar precipitation after ageing 24 h at 950°C (a) TiAl+WMoSi and (b) TiAl+W.

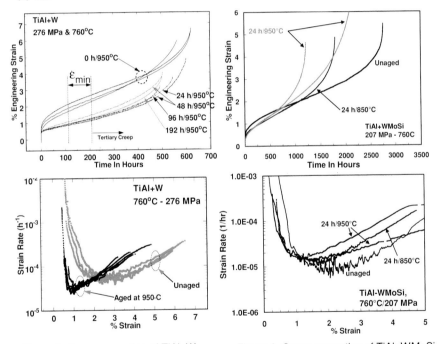

Figure 5. Creep properties of TiAl+W.　　　Figure 6. Creep properties of TiAl+WMoSi.

3.2 Creep Properties

The creep properties of the two alloys are shown in Figures 5 and 6. As a result of a combination of lower stress, and possibly a more creep resistant composition, the creep life of TiAl+WMoSi is longer than TiAl+W. For both alloys ageing at 850°C or 950°C reduces the creep life. However, the creep life of TiAl+W is ostensibly independent of the ageing time. For both alloys an important influence of ageing is an earlier onset of tertiary creep.

An important result is decreased primary strain for the aged conditions. For TiAl+W all aged conditions have approximately half the primary strain of the unaged state leading to a lower creep strain throughout the creep life, Figure 5. The difference is less marked for TiAl+WMoSi, but nevertheless ageing reduces the primary strain. The influence of ageing on primary creep is discussed elsewhere [21-23].

3.3 Creep Deformed Microstructures

Two major microstructural changes occurring during creep are the formation of discontinuous α_2 lamellae and concurrently, limited β precipitation/growth. In samples previously aged, in which α_2 decomposition has already begun, Figure 3c, significant dissolution of α_2 lamellae is apparent even at strains close to the minimum strain rate, Figure 7, and dislocation interactions with β particles, located on prior α_2 lamellae occur. After rupture numerous discontinuities in the α_2 lamellae of TiAl+W are evident, Figure 8. In unaged samples, in which prior α_2 decomposition during ageing has not occurred, discontinuous α_2 are not readily apparent until tertiary creep. However, in unaged samples the low β particle

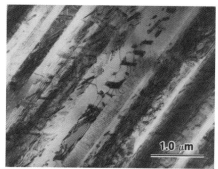

Figure 7. TiAl+WMoSi aged 24 h/850°C after 1.8% creep strain, showing dislocation interaction with β particles along prior α_2 lamellae.

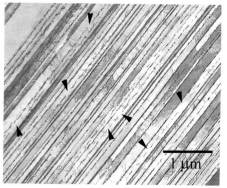

Figure 8. TiAl+W aged for 96 h/950°C after creep fracture. Arrows identify discontinuous α_2.

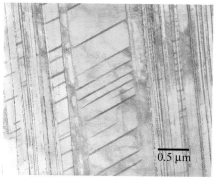

Figure 9. Unaged TiAl+W after creep fracture.

306

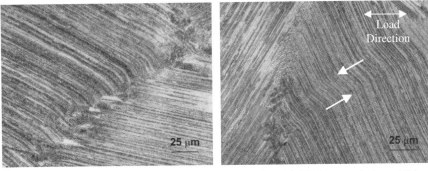

Figure 10. TiAl+WMoSi aged 24 h/850°C interrupted after (a) 1.8% creep strain and (b) during tertiary creep.

Figure 11. Unaged TiAl+WMoSi intergranular creep fracture.

density means that once α_2 become discontinuous, the resulting coarser γ lamellae are sites for dislocation and twinning activity, Figure 9.

In aged samples of TiAl+WMoSi shear across lamellar interfaces occurs at strains close to the minimum strain rate, Figure 10a, and with increasing creep strain shear bands form across lamellae, Figure 10b. In contrast, in unaged samples similar shearing is not observed at strains close to the minimum strain rate and only becomes evident at tertiary creep strains. Similar shear bands have been observed as a result of creep in other fully lamellar γ-TiAl alloys [8,24].

For all conditions rupture is predominantly intergranular, Figure 11. However, unaged conditions exhibit less intergranular void formation, Figure 12, and interlamellar delamin-ation, evident by comparing Figure 11 with 13.

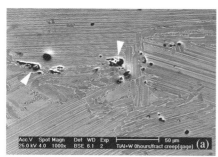 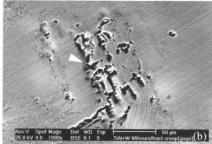

Figure 12. SEM backscattered images of TiAl+W (a) unaged, and (b) aged 96 h/950°C after creep fracture. Arrows identify intergranular voids.

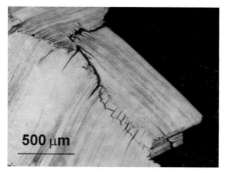

Figure 13. Grain boundary and lamellar delamination in TiAl+WMoSi aged 24 h/850°C after creep.

4. Discussion

After the heat treatments of Table 1, the $\alpha_2+\gamma$ fully lamellar structures evolve into a three phase $\alpha_2+\gamma+\beta$ structure during subsequent ageing or creep exposure. With increasing ageing time and/or creep exposure β precipitation along lamellar interfaces and α_2 decomposition becomes more extensive. Intergranular β is coarser than interlamellar β, since the lamellar grain boundaries are preferred locations for β precipitation. Figures 5 and 6 illustrate that these microstructural changes influence creep resistance. In particular, the three phase $\alpha_2+\gamma+\beta$ state has shorter rupture life, but decreased primary creep strain. The latter affect is discussed elsewhere [21,23].

During the $\alpha_2+\gamma \rightarrow \alpha_2+\gamma+\beta$ evolution, the α_2 lamellae decompose into a discontinuous morphology. Decomposition of α_2 lamellae during creep of other fully lamellar γ-TiAl alloys has been reported [25,26]. In samples aged prior to creep, the $\alpha_2+\gamma \rightarrow \alpha_2+\gamma+\beta$ evolution will be further advanced. The effect of the α_2 decomposition is somewhat different in the two compositions investigated.

In TiAl+WMoSi, despite the concurrent increase in interlamellar β particle density, α_2 decomposition reduces the creep resistance of the lamellar structure, since the γ lamellae width increases, thereby increasing the dislocation free slip distance and making twinning easier, as evident by the increased dislocation and twinning activity illustrated in Figure 9. A ramification of the increased γ lamellar spacing due to α_2 decomposition is easier shear across lamellar grains of the type shown in Figure 10, which occurs earlier during creep of the aged conditions than the unaged state. Thus, in TiAl+WMoSi α_2 decomposition causes the creep strain rate to increase at a lower strain, causing the earlier onset of tertiary creep evident in Figure 6.

In contrast, in TiAl+W the effect of α_2 decomposition is less pronounced than in TiAl+WMoSi. Despite the development of discontinuous α_2 lamellae during ageing, as evident in Figure 2, the tertiary creep is ostensibly the same for the aged and unaged conditions, Figure 5. Therefore, it appears that for the TiAl+W composition the formation of a finer, but denser, interlamellar distribution of β precipitates, Figure 4, compensates for concurrent removal of α_2/γ lamellar interfaces. Consequently, the free dislocation slip distance is remains ostensibly unchanged.

In light of the preceding discussion, it appears that the α_2+γ→α_2+γ+β evolution during ageing, and subsequent creep loading, has a greater effect on TiAl+WMoSi than TiAl+W. Tungsten is a strong β stabilizer and therefore, in both alloys α_2 decomposition is believed to be related to W diffusion from the α_2 lamellae into the precipitating β [21,22], consistent with previous results indicating stabilisation of α_2 lamellae by W solute [27]. Therefore, the large difference in total creep life as a function of ageing for TiAl+WMoSi compared to TiAl+W is attributed to the lower W content of the former, Table 1, and the consequent less stable α_2 lamellae and/or less effective β particle strengthening.

After creep fracture of the aged conditions, more voids are evident along grain boundaries, Figure 12, and greater interlamellar delamination occurs, Figure 13, suggesting that β particles make it easier for cracks to nucleate and propagate along interfaces. This may be associated with incompatible deformation between the γ lamellae and β particles, causing stress concentrations at γ/β interfaces. As such, it appears that final fracture is caused by growth and coalescence of intergranular voids, the nucleation of which is promoted by β particles.

From the foregoing it is suggested that in the aged conditions of TiAl+WMoSi it is the earlier decomposition of α_2 lamellae that is the major factor causing the earlier onset of tertiary creep and the shorter creep life. The effect of α_2 decomposition is not as great in TiAl+W due to more effective β particle strengthening as a result of the higher W content. Lamellar delamination and the formation of voids along lamellar grain boundaries, leading to intergranular crack propagation is the main cause of creep fracture.

5. Summary & Conclusions

The influence of third phase β precipitation via ageing at 850°C or 950°C on the creep response of two near γ-TiAl intermetallic compositions has been investigated. The major conclusions are:
1. Creep of near γ-TiAl intermetallics is characterized by a period of decreasing strain rate leading to a minimum strain rate, followed by increasing strain rate to rupture.
2. The formation of interlamellar and intergranular β precipitates during ageing reduces the primary creep strain, but shortens the rupture life.
3. An important microstructural change concurrent with β precipitation during ageing is dissolution of the α_2 lamellae, which continues during creep loading.

4. In TiAl+WMoSi, α_2 dissolution during creep reduces the obstacles to dislocation slip and is thought to be responsible for the increase in strain rate following the strain rate minimum.

5. Addition of sufficient W, and possibly other β stabilizing alloying elements, can largely offset the effect of α_2 decomposition, through the formation of β particles that impede dislocation motion.

6. β precipitation along lamellae and grain boundaries act to weaken these interfaces by promoting lamellar delamination and the formation of intergranular voids.

From an application viewpoint the current results indicate that γ-TiAl microstructures must be specified with knowledge of the creep design criteria. It appears that if maximum time to 0.5% or 1% creep is paramount, then β precipitation via ageing is beneficial, however if maximum rupture life is the design objective, then β precipitation should be avoided.

Acknowledgements

This work was made possible through funding from the Natural Sciences and Engineering Research Council of Canada and support via project JHR-01 of the Structures, Materials and Propulsion Laboratory, Institute for Aerospace Research of the National Research Council of Canada (SMPL-IAR-NRC). The authors are grateful for access to the microstructural preparation and analysis facilities of SMPL-IAR-NRC.

References

[1] Lipsitt, H.A., 1985, Titanium Aluminides: An Overview, High-Temperature Ordered Intermetallic Alloys, eds. C.C. Koch, et al., Mat. Res. Soc. **39**, pp351-364.

[2] Kim, Y-W., 1989, Intermetallic Alloys Based on Gamma Titanium Aluminide, JOM, **41**, pp24-30.

[3] Kim, Y-W., Froes, F.H., 1990, Physical Metallurgy of Titanium Aluminides, High Temperature Aluminides & Intermetallics, eds. S.H. Whang, et al., TMS, pp465-492.

[4] Kim, Y-W., Dimiduk, D.M., 1991, Progress in the Understanding of Gamma Titanium Aluminides, JOM, **43**, pp40-47.

[5] Kim, Y-W., 1994, Ordered Intermetallic Alloys, Part III: Gamma Titanium Aluminides, JOM, **46**, pp30-39.

[6] Chesnutt, J.C., Hall, J.A., Lipsitt, H.A., 1995, Titanium Intermetallics - Present and Future, Titanium'95: Science and Technology, eds. P.A. Blenkinsop, et al., IoM, pp168-175.

[7] Tönnes, C., Rösler, J., Baumann, R., Thumann, M., 1993, Influence of Microstructure on the Tensile and Creep Properties of Titanium Aluminides Processed by Powder Metallurgy, Structural Intermetallics, eds. R. Darolia, et al., TMS, pp241-245.

[8] Triantafillou, J., Beddoes, J., Wallace, W., 1996, Creep Properties of Lamellar Near γ-Titanium Aluminides, Can. Aero. and Space J., **42**, pp108-115.

[9] Es-Souni, M., Bartels, A., Wagner, R., 1995, Creep behaviour of near γ-TiAl base alloys: effects of microstructure and alloy composition, Mat. Sci. Eng., **A192/193**, pp698-706.

[10] Worth, B.D., Jones, J.W., Allison, J.E., 1995, Creep Deformation in Near γ-TiAl: Part I. The Influence of Microstructure on Creep Deformation in Ti-49Al-1V, Met. Mat. Trans., **26A**, pp2947-2959.

[11] Viswanathan, G.B., Vasudevan, V.K., 1995, Microstructural Effects on Creep Properties of a Ti-48Al Alloy, Gamma Titanium Aluminides, eds. Y-W. Kim, et al., TMS, pp967-974.

310

[12] Schwenker, S.W., Kim, Y-W., 1995, The Creep Behavior of Gamma Alloy Ti-46.5Al-3Nb-2Cr-0.2W (Alloy K5) in Two Microstructural Conditions, Gamma Titanium Aluminides, eds. Y-W. Kim, et al., TMS, pp985-992.

[13] Bhowal, P.R., Konkel, W.A., Merrick, H.F., 1995, The Effect of Processing and Microstructure on Properties of Ti-47Al-3Nb-1W and 46Al-2.5Nb-2Cr-0.2B γ-Titanium Aluminides", Gamma Titanium Aluminides, eds. Y-W. Kim, et al., TMS, pp787-794.

[14] Beddoes, J., Zhao, L., Triantafillou, J., Au, P., Wallace, W., 1995, Effect of Composition and Lamellar Microstructure on Creep Properties of P/M Near γ-TiAl Alloys, Gamma Titanium Aluminides, eds. Y-W. Kim, et al., TMS, pp959-966.

[15] Yun, J.H., Oh, M.H., Nam, S.W., Wee, D.M., Inui, H., Yamaguchi, M., 1997, Microalloying Effects in TiAl+Mo Alloys, Mat. Sci. Eng., **A239-240**, pp702-708.

[16] Viswanathan, G.B., Kim, Y-W., Mills, M.J., 1999, On the Role of Precipitation Strengthening in Lamellar Ti-46Al Alloy with C and Si Additions, Gamma Titanium Aluminides, eds. Y-W. Kim, et al., TMS, pp653-660.

[17] Herrouin, F., Hu, D., Bowen, P., Jones, I.P., 1998, Microstructural Changes during Creep of a Fully Lamellar TiAl Alloy, Acta Mat., **46**, pp4963-4972.

[18] Beddoes, J.,Zhao, L., Chen, W.R., Du, X., 1999, Creep of Fully Lamellar Near γ-TiAl Intermetallics, High Temperature Ordered Intermetallics VIII, eds. E.P. George, et al., Mat. Res. Soc. **552**, ppkk1.1.1-kk1.1.12.

[19] Wallace, W., Zhao, L., Beddoes, J., Morphy, D., 1993, Densification and Microstructural Control of Near γ-TiAl Intermetallic Powders By HIP'ing, Hot Isostatic Pressing93, eds. L. Delaey et al., pp99-108.

[20] Chen, W.R., Zhao, L., Beddoes, J., 1999, Microstructural modification of investment cast Ti-47Al-2Nb-1Mn-0.5Mo-0.5W-0.2Si alloy, Gamma Titanium Aluminides, eds. Y-W. Kim, et al., TMS, pp323-330.

[21] Seo, D.Y., Beddoes, J., Zhao, L., Botton, G., 2000, The Influence on the Microstructure and Creep Behaviour of a γ-TiAl+W Alloy, submitted to Mat. Sci. Eng.

[22] Chen, W.R., Beddoes, J., Zhao, L., 2000, Effect of Ageing on the Tensile and Creep Behaviour of a Fully Lamellar Near γ-TiAl Alloy, submitted to Mat. Sci. Eng.

[23] Beddoes, J., Seo, D.Y., Zhao, L., 2001, Relationship Between Tensile and Primary Creep Properties of Near γ-TiAl Intermetallics, Int. Symp. on Deformation & Microstructure in Intermetallics, TMS, New Orleans.

[24] Hayes, R.W., McQuay, P.A., 1994, "A First Report on the Creep Deformation and Damage Behavior of a Fine Grained Fully Transformed Lamellar Gamma TiAl Alloy", Scripta Met. Mat., **30**, pp259-264.

[25] Nam, S.W., Cho, H.S., Hwang, S-K., Kim, N.J., 2000, Investigation of Primary and Secondary Creep Deformation Mechanism of TiAl, Metals & Materials, **6**, pp287-292.

[26] Du, X-W., Zhu, J., Zhang, X., Cheng, Z.Y., Kim, Y-W., 2000, Creep Induced $\alpha_2 \rightarrow \beta_2$ Phase Transformation in a Fully Lamellar TiAl Alloy, Scripta Mat., 43, pp597-602.

[27] Larson, D.J., Liu, C.T., Miller, M.K., 1997, Microstructural Characterization of Segregation and Precipitation of $\alpha_2+\gamma$ Titanium Aluminides, Mat. Sci. Eng., A239-240, pp220-228.

9th International Conference on Creep and Fracture of Engineering Materials

Tensile and creep behaviour of newly developed multiphase titanium aluminides

J.M. Franchet[1], L. Germann[3], A.K. Gogia[2], D. Banerjee[2] and J.L. Strudel[3]

[1] *Snecma Moteurs, YKOG2, 291, avenue d'Argenteuil, B.P. 48, 92234 Gennevilliers Cedex (France)*
[2] *Defence Metallurgy Research Laboratory, P.O. Kanchanbagh, Hyderabad, 500 058 (India)*
[3] *Centre des Matériaux, UMR CNRS 7633, Ecole Nationale Supérieure des Mines de Paris, B.P. 87, 91003 Evry Cedex (France)*

Abstract

The tensile and creep behavior of three titanium alloys hardened by the ordered orthorhombic phase was examined between 20°C and 650°C. The first two alloys, with 22 Al-25 Nb also contained 1% Mo, and 0.5 %Si was added to one of them. The last one with 20 Al-24 Nb-1.5 Mo had a larger fraction of the B2 matrix, in an attempt to improve ductility. Various microstructures were developed by appropriate thermomechanical treatments.

Several forging sequences and subsequent heat treatments were carried out. The resulting microstructures were examined by OM and SEM as well as by EBSD techniques in order to relate not only the morphology, size and distribution of the phases but also the scale of the crystalline coherency, to the mechanical performance of the various materials. Three phase $B2+\alpha_2+O$ microstructures were obtained with various sizes, volume fraction and morphologies of α_2 precipitates and O-laths. Tensile tests were carried out between RT and 650°C. The extent of primary creep and the values of secondary creep rates were investigated and compared at 550°C and 650°C. Creep rupture appeared to be generally transgranular at 550°C and turned predominantly intergranular at 650°C. Si additions, if improving the creep resistance, tend to reduce the tensile ductility of these alloys, whereas Mo has a beneficial effect on both the tensile strength and the creep resistance without affecting adversely the ductility. The best combinations of composition and optimised microstructures for these newly developed alloys, seem to exhibit monotonic mechanical properties comparable to those of IN718 when corrected by the density ratio.

1. INTRODUCTION

Titanium aluminides have received considerable attention recently as potential materials for high temperature engineering parts and structures [1, 2]. Several compositions in the range Ti-22 Al-25 Nb with 1 %Mo and 0.3 %Si additions have been developed and patented [3, 4]. Several authors [4, 5] have pointed out that a final prolonged heat treatment in the temperature range 550°C to 650°C reduces primary creep strain and lowers the secondary creep rate. Carisey has shown that such heat treatment decreases room temperature ductility but increases the strength and flow stress of the material at all temperatures, and has attributed these effects to the growth of several variants of very fine O-laths in the B2 matrix corridors [4].

The ternary phase diagram for these alloys has been studied by several authors [2, 5-7], and the kinetics of phase evolution during thermomechanical treatments have been taken into account in order to improve the creep response of the material [5]. Based on these results, isopleth sections of the ternary diagram such as that made for 22% Al can be drawn (Figure 1). Yet, the introduction of a quaternary β-prone element such as Mo which partitions more strongly to the B2 phase than to the O phase [8], tends to shift the phase boundaries towards lower temperatures (as indicated by the dotted lines, Figure 1), thus opening the field for appropriate heat treatments after forging. Silicon was also introduced in these alloys, as a minor element capable of improving creep resistance [9]. Carefully controlled isothermal forging sequences in the presence of a dispersed equiaxed α_2 phase were explored in this work, in order to generate in the material a scale of microstructure in the 20 to 200 µm range, meaningful for the plastic behavior as well as the generation of damage in the material.

2. EXPERIMENTAL PROCEDURES

The three alloy compositions studied appear in Table 1. The effect of Si content can be estimated by comparison of S3 and S4, the effect of Mo content by comparing S4 and S7.

material	Al	Nb	Mo	Si	Ti	T_β (°C)	density
S3	22	25	1	-	bal.	1055	5.34
S4	22	25	1	0.5	bal.	1065	5.34
S7	19.5	24	1.4	0.2	bal.	1060	5.39

Table 1 : Nominal composition of alloys (atomic %).

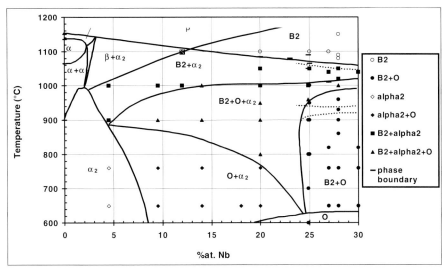

Figure 1: Isopleth section of Ti-Al-Nb system at 22 at.% Al. Data from experimental results [4, 5, 10-14] and domain boundaries based on Sadi's thermodynamic calculations [6].

The double VAR cast and the upset forging in hot dies above the β transus temperature of the alloys were done at DMRL. Subsequent 1-step and 2-step forging was implemented at Snecma under isothermal conditions and under controlled argon atmosphere. Final heat treatments were carried out under argon atmosphere (Table 2).

Microstructural observations by MO and SEM was done on polished and etched samples, and EBSD mapping on electropolished samples on a ZEISS DSM 982 GEMINI.

Tensile testing and creep testing were carried out in air at various temperatures on machines equipped with extensometers capable of micron resolution and stability over 24 h.

TMP	forging	heat treatment
C	1-step	T_β- 25°C/1h + 300°C/h - 900°C/24h + AC - 550°C/48h + AC
L	2-step	T_β-100°C/2h + AC - 850°C/24h + FC - 650°C/24h + AC
K	2-step	T_β-100°C/2h + 300°C/h - 850°C/24h + FC - 650°C/24h + AC

Table 2: Thermomechanical processing schedules (AC: air-cooling, FC: furnace cooling).

3. RESULTS AND DISCUSSION

In consideration of the various requirements regarding the mechanical performances as well as the chemical resistance set by the potential industrial users of this class of material, a wide combination of compositions, forging conditions, and thermal treatments were explored and led to a variety of microstructures [8]. In this study, three different alloy compositions are considered, together with two different forging schedules. The 1-step forging scheduled was performed at T_β-100°C, in an isothermal press and under controlled atmosphere at the strain rate of $10^{-3}s^{-1}$ and a total average strain of one. The 2-step forging schedule was carried out at slightly below T_β, at high strain rate, followed by a second forging sequence at a lower strain rate and a slightly lower temperature. The various microstructures obtained will be described and examined in relation with their corresponding tensile and creep properties.

3.1 Microstructures resulting from processing and heat treatments

Large flattened grains of the B2 phase appear unrecrystallized in most of the forged pancakes, especially after 1-step forging (Figure 2 a). On a finer scale (Figure 2 b), a dense dispersion of needles of O phase is filling most of the matrix grains with α_2 equiaxed particles 0.2 to 1 µm in size, dispersed in the background (5 to 20 µm apart). Similar microstructures are observed in alloy S4 (Figure 2 d), containing an addition of 0.5% Si. However, after 2-step forging followed by a heat treatment involving higher cooling rates (treatment L), a tendency for finer basket weave microstructures develops (Figure 2 c) note also that equiaxed α_2 particles are larger in this microstructure. In an alloy with more Mo and less Al, such as S7, a similar thermomechanical treatment leads again to and even finer microstructure (Figure 2 e). Finally, if a controlled cooling rate of 300°C per second is applied (treatment K), a coarser microstructure (Figure 2 f), more likely to resist to high temperature creep is formed [2]. Notice also the larger proportion of B2 phase in alloy S7 than in alloy S3 and S4.

Although 2-step forging seems to lead to the development of microstructures similar to 1-step forging (Figure 3 a), careful examination of cross sections in SEM by EBSD technique reveals the presence of recrystallized zones (Figure 3 b and d) in this type of material, which was not detected after 1-step forging. The cube texture, which appears in red on Figure 3 d, is present in the population of recrystallized grains, but not dominant. Mapping grain misorientations (Figure 3 c) reveals an average grain size of 20 to 15 µm, which is indicative of the ability of the B2 phase to undergo fine grain recrystallization when the process is controlled by the presence of a hard phase.

Microstructures shown on Figure 2 c, d and f were obtained from back scattered electrons. Since the contrast is related to the average atomic number of the phases, we can see that despite the prolonged heat treatment at 650°C, the α_2 nodules did not transform into O phase, their refined needle basketweave arrangement appears as a result of the high molybdenum content in these alloys; however the ultra fine precipitation of multiple variants of O-laths in the B2 channels is not visible at this magnification, has been in evidenced in previous work [4].

314

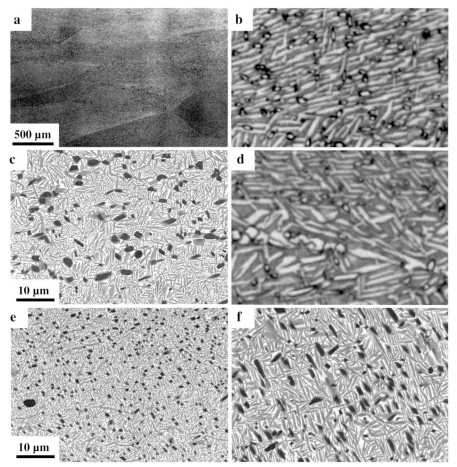

Figure 2: Typical microstructures resulting from thermomechanical processing.
a, b) alloy S3 C [4], c) alloy S4 L, d) alloy S4 C, e) alloy S7 L and f) alloy S7 K
a, b, d) Optical micrographs, c, e, f) SEM micrographs using BE.

3. 2 Tensile properties between 25°C and 650°C

The tensile strengths of alloy S3 and S4 after 1-step forging, as a function of temperature are rather similar as shown on Figure 4. Hence, the silicon addition does not alter significantly the tensile properties of this class of alloys. Both the yield stress and UTS data obtained on 2-step forged material alloy S4 (Figure 3), are higher than those resulting from 1-step forging especially above 300°C. Alloy S7 with the lower volume fraction of O phase, compares favorably with alloys S3 and S4 in the 2-step forged condition (Figure 4) if strengthened by a fine microstructure (treatment L). On the other hand, when a coarser microstructure is present (treatment K), the strength is dropping by about 200 MPa.

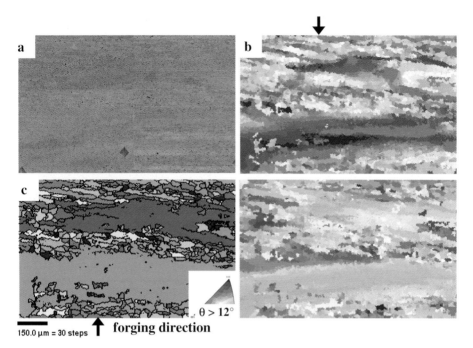

*Figure 3: **a**) Backscattered electron micrograph of 2-step forged alloy S4,*
***b, d**) Lattice directions of β grains: **b**) along radial direction of the pancake*
*and **d**) along forging direction, are given in the standard triangle,*
***c**) Orientation imaging: dark lines correspond to misorientation larger than 12°.*

As is well known for these alloys, their room temperature ductility is very limited. This statement is certainly true for alloys S3 and S4 in the 1-step forged condition (Figure 5 a); yet, after 2-step forging, alloy S4 can exhibit a significant improvement of its room temperature ductility and that of S7 even reaches 5% (Figure 5 b). Hence, by optimizing both composition and processing schedule of quaternary alloys, their room temperature ductility can be raised to a manageable level. Above room temperature, the ductility of all microstructures rises above 5% in most cases, although in the 2-step forged condition a significant improvement over the 1-step forged material, is observed with a peak around 400°C or 500°C, resulting mostly from extended necking of the test piece. Notice that a coarser lath structure (S7 K) tends to shift the ductility peak toward higher temperatures than the fine lath structure (S7 L); similarly, for an alloy with a higher proportion of O phase and a lower molybdenum content (S4), the ductility peak is shifted toward lower temperatures and significantly enhanced. These observations can be interpreted as suggesting that the B2 phase plays a major role at room temperature, whereas the plastic behavior of the O phase controls the high temperature plasticity.

The mechanical properties, at room temperature of this class of alloys are summarized on Figure 6 by plotting the UTS as a function of the total plastic strain. Ternary alloys such as 22-25 are deprived of the solid solution hardening provided by Mo (alloys S3 and S4), and alloy 24-15 is too low in Nb. On the other hand, the ductility of quaternary alloys is improved by 2-step forging, but the high volume fraction of O phase strengthens the material at the expense of its elongation to rupture.

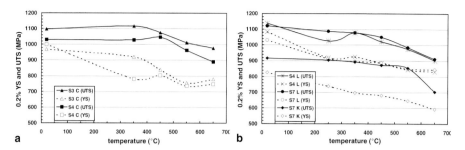

Figure 4: Yield Stress and Ultimate Tensile Strength vs temperature curves.

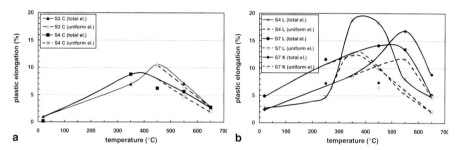

Figure 5: Ductility vs temperature curves.

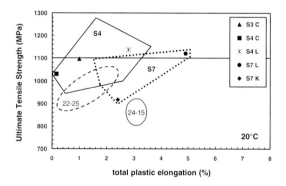

Figure 6: Summary of tensile properties at Room Temperature.

3. 3 Creep properties

As shown in Table 2, all alloys were heat treated for a fairly long time below 700°C. Previous studies [4, 15] have clearly indicated that a very dense population of fine O-laths belonging to several variants, is developing in the B2 corridors, thus hardening the B2 matrix. Keeping in mind potential industrial applications in this temperature range, we considered that stabilizing these materials was necessary before characterizing their mechanical behavior.

Creep curves at 550°C and 650°C are presented on Figure 7 as strain rate versus plastic strain, with the time to rupture mentioned. Primary creep (plotted on Figure 9 as a function of applied stress) extends to almost 1% strain and occupies most of the actual lifetime of the material at 550°. Steady state creep rate become well established at 650°C, the temperature of the last stabilizing heat treatment applied to the material. Note also, that despite a decent lifetime, the total elongation to rupture is of the order of about 1% at 550°C and reaches at most 4% at 650°C. Tertiary creep is not observed in these alloys; instead, steady state creep is interrupted by a cracking process probably related to environmental interactions.

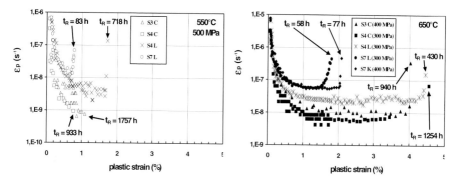

Figure 7: Creep rate vs plastic strain curves at 550°C and 650°C.

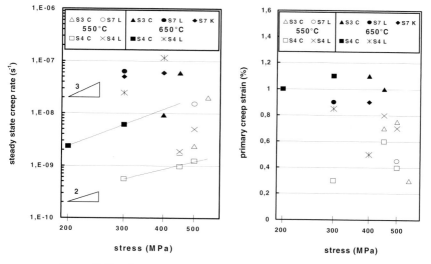

Figure 8: Steady state creep rate
vs stress at 550°C and 650°C

Figure 9: Primary creep strain
vs stress at 550°C and 650°C

In the 1-step forged and heat treated state S3 C and S4 C have fairly similar properties and are presenting the highest creep resistance. Their steady state creep rates (Figure 8) are among the lowest and the extent of their primary creep is minimal (Figure 9). Their strain rate sensitivity ranging between 2 and 3.3 , is remarkably low for highly strengthened alloys as observed previously [2]. If the beneficial effect of the Si addition is not obvious, that of Mo has been established earlier by comparison with a 22-25 alloy [2, 4].

In the 2-step forged and heat treated state S4 L is somewhat less creep resistant than in the 1-step forged state, yet its primary creep strain appears slightly reduced at 650°C. Combining a lower proportion of O phase with a 2-step forging treatment (alloy S7), still reduces further the creep resistance of the material.

318

In order to compare the creep behavior of titanium aluminides with that of nickel base alloys, density corrected Larson-Miller plots of the stress for rupture and for 1% strain are presented on Figure 10. Creep rupture properties of this new class of alloys compare favorably with that of high temperature titanium base alloys, and are only slightly lower than those of nickel base disk alloys of the new generation like N18. The more meaningful parameter for disk material is the value of the LM parameter to reach 1% strain. Again titanium aluminides, judged with this criterion, are only slightly lower than their nickel base equivalents, especially in the range 500°C to 600°C.

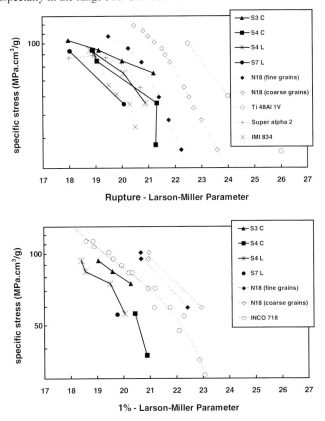

Figure 10: Comparative Larson-Miller diagrams of Ni superalloys and Ti alloys [15-17].

Looking at the rupture surface after creep at 650°C, we see that even in the absence of specimen necking, the large β grains themselves are necking, one more evidence of importance as crystallographic entities playing a major role in the plasticity, as well as in the rupture of these materials (Figure 11 a). In the 1-step forged material (Figure 11), the other sign of ductility appears on the ultra-fine level of the few microns, where cusps of B2 matrix are formed around fractured O-laths (Figure 11 b). On the other hand, on 2-step forged materials (Figure 12 a and c), the necking of β grains is less marked, and another meaningful scale of plasticity is revealed in the range 30 to 300 μm (Figure 12 c), with multiple crack branching taking place on that scale. The improved ductility of this material observed in tensile tests is partially preserved here and is attributed to an homogenization of plastic flow on that intermediate scale [18]. However, if the Al content of the material is decreased at the same time the Nb content is raised as in alloy S7, the necking of individual B2 grains may disappear to the benefit of a general reduction of area of the specimen (Figure 12 b), but the smoothness of the rupture surface (Figure 12 d) indicates that the rapid crack propagation is probably linked now, with a strong interaction with environment. In all cases, crack initiation always seems to take place on the outside hardened layer of the test piece, where oxygen stabilizes a 50 to 200 μm thick α_2 layer.

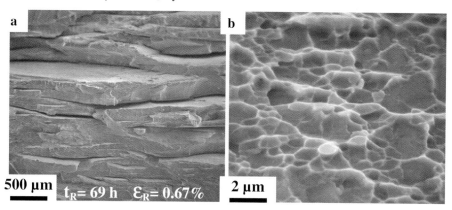

Figure 11: Fractographs of creep rupture surface of alloy S3 C tested at 550°C / 500 MPa.

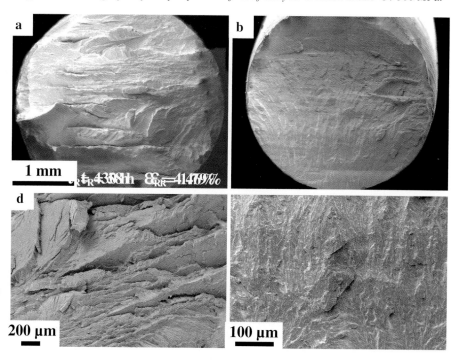

*Figure 12: Fractographs of creep rupture surface of **a, c**) alloy S4 L and **b, d**) alloy S7 L tested at 650°C / 300 MPa.*

4. CONCLUSIONS

The aluminum content of niobium rich (22-23 at%) titanium aluminides can be lowered from 22 to 20 at% in order to increase the volume fraction of the B2 matrix over that of the α_2 + O phases. Furthermore, the B2 phase can effectively be solid solution hardened by about 1% molybdenum without any loss of ductility.

Controlled multi-step isothermal forging just below the β transus temperature, can produce microstructures on the meso-scale (20 to 200 μm) which will interact positively with the plastic behaviour as well as the development of damage in the material. The size and distribution of the equiaxed α_2 nodules seemed to be responsible for this effect.

It was shown, that a careful choice of composition and appropriate thermomechanical treatments can lead to the development of titanium aluminides having a flow stress in the range of 1000 MPa up to 500°C, and associated with a room temperature ductility of 3-5%.

However, a primary creep amplitude of 0.2 to 1.2%, at 550°C as well as at 650°C has frequently been observed in these alloys and is acting as a strong deterrent for potential users.

The density corrected steady state creep rate compares favorably with that of the best forged nickel base alloys in the 500°C-600°C range. However, the fine scale microstructures designed for best tensile resistance and ductility are less appropriate for creep resistance and conversely. A necessary compromise must be reached for each specific application.

Ternary creep is not observed in these alloys, instead the steady state creep stage appears to be interrupted by the sudden propagation of cracks leading to rupture. Several features of the rupture surface tend to reveal the strong interaction with the environment.

Acknowledgements

We are deeply indebted to the Defence Metallurgical Research Laboratory for manufacturing the alloys and to Snecma for performing the isothermal forging schedules. We acknowledge the financial support of Snecma and Turbomeca in France and of DRDO in India.

References

[1] Y.W. Kim et al., *High Temperature Aluminides and Intermetallics*, 1990, pp. 465-492.
[2] A.K. Gogia et al., *Intermetallics*, vol. 6, 1998, pp. 741-748.
[3] T. Carisey, D. Banerjee et al., *french patent No 97.16057, 1997.*
[4] T. Carisey, *PhD thesis*, Ecole des Mines de Paris, Centre des Matériaux, 1998.
[5] C.J. Boehlert et al., *Structurals Intermetallics*, 1997, pp. 795-804.
[6] F. Sadi, to be published in *Mat. Sc. & Eng.*, 2001.
[7] K. Muraleedharan et al., *Intermetallics*, vol. 3, 1995, pp. 187-199.
[8] L. Germann et al., *SNECMA annual report*, 2000, Centre des Matériaux, Evry.
[9] P.K. Sagar et al., *Mat. Sc. & Eng.*, vol. 192/193, 1995, pp. 799-804.
[10] S.G. Kumar et al., *J. of Phase Equilibria*, vol. 15, 1994, pp. 279-284.
[11] J. Kumpfert et al., *Structurals Intermetallics*, 1997, pp. 895-904.
[12] R. Kainuma et al., *Intermetallics*, vol. 2, 1994, pp. 321.
[13] C.G. Rhodes et al., *Structurals Intermetallics*, 1993, pp. 45-52.
[14] R.J.T. Penton et al., *Mat. Sc. & Eng.*, vol. A153, 1992, pp. 508-513.
[15] M.Y. Nazmy, *Phys Metall & Process of Interm Comp*, Stoloff et al., 1996, pp. 95-125.
[16] G.F. Harrison et al., *Adv. Perform. Mat.*, 1996, pp. 263-278.
[17] F.H. Froes, *Phys Metall & Process of Interm Comp*, Stoloff et al., 1996, pp. 297-345.
[18] S. Li et al., *5th Int. Conf. on Structural & Functional Intermet.*, Vancoover, July 2000.

Microstructurally Based Understanding of the Ti-52at%Al Creep Strength

A. Dlouhý, K. Kuchařová

Institute of Physics of Materials, Academy of Sciences of the Czech Republic, Žižkova 22, 616 62 BRNO, Czech Republic
e-mail: , fax: ++420 5 41212301

Abstract

Creep data obtained for an equiaxed-γ Ti-52at%Al intermetallic compound at temperatures 1050, 1000 and 1150 K and applied stresses within 50-300 MPa are summarized. It is shown that after approximately 10% of transition strain, the creep behaviour of the pure-γ phase can be rationalized in terms of substructure controlled creep. The transition strain is analysed in view of new microstructural observations. It has been verified again that the pre-heating treatment has a pronounced effect on the creep strain accumulation kinetics in the transition strain period. Microstructural changes associated with the pre-heating and subsequent creep are investigated by transmission electron microscopy (TEM). It is suggested that oxide particles, the size of which exceeds 1 μm, facilitate the nucleation of twins. The pinning of dislocations by fine H-phase particles (~10 nm in size) is discussed.

1. Introduction

The pure-γ TiAl phase represents a base of many near-γ TiAl alloys which are considered for near term high temperature applications. Consequently, creep in pure-γ TiAl has received significant attention in last two decades [1-7]. Loiseau and Lasalmonie [1] first reported on enhanced creep strength of the pure-γ TiAl associated with sharp creep rate minima (SCRMa) in early stages of creep. Oikawa and co-workers observed similar minima and investigated the temperature and stress dependence of creep rate as well as the corresponding microstructural evolution [2,3]. The Sendai group also first described the influence of grain size on the creep strain accumulation kinetics [4,5]. Lu and Hemker [6] performed temperature change tests to assess the apparent activation energy associated with creep in single phase-γ TiAlMn alloy. Two quantitative studies on strain induced microstructural changes during creep of pure-γ TiAl [7] and near-γ two-phase alloy [8] indicated that nucleation of deformation twins (DTs) in equiaxed microstructures contributed to the SCRMa.

Loiseau and Lasalmonie [1] attributed the SCRMa in pure-γ TiAl to the heterogeneous precipitation of secondary phases during creep. Similar mechanism was also documented in two-phase γ/α_2 alloy after long-term creep [9]. It was only recently recognized that oxygen segregation to dislocations and related heterogeneous precipitation of oxide phases could account for the yield stress anomaly observed in the same temperature range where the sharp creep rate minima were reported [10]. Consequently, in spite of the considerable effort spent and progress made so far, the mechanisms that govern the strain accumulation kinetics during early stages of creep in the pure-γ TiAl remain unclear. The present study assesses recent experimental creep and microstructural data and relates them to the fundamental deformation mechanisms. Deformation twinning and interaction between oxide particles and dislocations are considered as potential sources of transition strain and SCRMa in early stages of creep in pure-γ TiAl.

2. Experimental details

The pure-γ Ti-52at%Al phase was plasma-melted in IRC Birmingham. Final HIP treatment at 1523 K/150 MPa/4 h resulted in an equiaxed grain microstructure (mean grain size 200 µm). Tensile (gauge length 25 mm and cross-section 4×3.2 mm^2) and cylindrical compression (diameter 5 mm and height 12 mm) creep specimens were cut out of the HIPped material using the spark erosion technique. The oxygen content in the initial state of the alloy did not exceed the level of 700 wt ppm which represents the atomic concentration below $1.6 \cdot 10^{-3}$.

Creep tests were performed in tension and compression under the constant applied stress in purified argon. The specimens were introduced into the cold furnace and the argon atmosphere gradually replaced the inside air. Specimen heating to the testing temperature in the range 1050-1150 K required the time period of 5 hours. Then the specimen was held at the testing temperature for 3 hours before loading to ensure the thermal equilibrium in the specimen-machine system. This period was adequately prolonged in pre-heating experiments reported later. The testing temperature was maintained constant within ±1 K along the specimen gauge length and invariable during the tests. True strain-time readings were continuously recorded by the PC-based data acquisition system.

Table 1: Creep data of Ti-52at%Al reported in this study.

specimen no.	mode	T [K]	stress [MPa]	rate min [s-1]	time min [s]	time min [h]	strain min [%]
TA 20	c	1050	100	1,3E-08	1,54E+05	42,8	0,56
TA 7	c	1050	150	5,0E-08	5,91E+04	16,4	0,71
TA 19	c	1050	250	1,4E-05	*	*	*
TA 8	c	1100	50	6,5E-09	3,95E+05	109,7	0,83
TA 4	c	1100	70	2,2E-08	6,11E+04	17,0	0,35
TA 3	c	1100	100	1,1E-07	3,94E+04	10,9	1,15
TA 12	c	1100	100	1,0E-07	3,10E+04	8,6	0,72
TA 1	c	1100	150	4,5E-07	9,57E+03	2,7	0,99
TA 2	c	1100	200	2,6E-06	3,18E+03	0,9	1,30
TA 14	c	1100	225	1,3E-06	2,35E+03	0,7	0,64
TA 10	c	1100	225	2,7E-05	*	*	*
TA 9	c	1100	250	5,5E-05	*	*	*
TA 5	c	1100	300	7,0E-05	*	*	*
TD 10	t	1100	60	1,9E-08	6,10E+04	16,9	0,53
TD 2	t	1100	70	7,5E-09	3,50E+05	97,2	0,63
TD 9	t	1100	80	4,0E-08	4,00E+04	11,1	0,45
TD 1	t	1100	100	9,5E-08	3,50E+04	9,7	0,66
TD 7	t	1100	100	1,0E-07	2,80E+04	7,8	0,64
TD 12, a0	t	1100	100	7,2E-08	3,34E+04	9,3	0,64
TD 15, a0	t	1100	100	8,5E-08	3,01E+04	8,4	0,52
TD 11, a10	t	1100	100	3,8E-07	1,04E+04	2,9	0,98
TD 14, a10	t	1100	100	2,8E-07	1,44E+04	4,0	0,72
TD 13, a100	t	1100	100	9,5E-07	8,92E+03	2,5	1,31
TD 17, a100	t	1100	100	9,5E-07	3,92E+03	1,1	0,62
TD 6	t	1100	120	7,0E-07	*	*	*
TD 3	t	1100	150	4,7E-07	*	*	*
TA 23	c	1150	50	6,0E-08	6,69E+04	18,6	0,91
TA 22	c	1150	70	2,0E-07	4,30E+04	11,9	1,58
TA 21	c	1150	100	1,3E-06	3,50E+03	1,0	0,90
TA 6	c	1150	150	9,0E-06	*	*	*
TA 24	c	1150	250	1,7E-04	*	*	*

* without a well defined creep rate minimum; c-compression, t-tension; a0, a10 and a100 additional annealing 0, 10 and 100 h at 1100 K before the creep test.

A special care was given to the preparation of TEM foils to avoid any damage due to thinning operations. Slices cut out of the creep specimen gauge length were ground on emery papers to the thickness of 0.3 mm. From this point on only chemical methods were applied. The double jet TENUPOL equipment was used for final perforation of the foil. The electrolyte $HClO_4$ (5%) and methanol (95%) was kept at -50°C. The foils were investigated in Philips CM12 STEM operating at 120 kV.

3. Results

3.1. Creep data

A set of tensile and compression creep data obtained at three temperatures 1050, 1100 and 1150 K and in the applied stress range 50 – 300 MPa is presented in Table 1. The Tab. 1 summarizes data on the minimum creep rate, the time and the strain accumulated to the point of the minimum creep rate. Compression creep curves measured at 1100 K are plotted in Fig.1 in co-ordinates creep rate vers. creep strain. Depending on the applied stress level, two distinct creep curve types are observed. At stresses below 220 MPa, the curves exhibit the SCRM. The average creep strain accumulated to the SCRM (all the tests of Table 1 considered) is 0.7%. On the other hand, creep curves with a normal metal-type primary transition are observed at high stresses above 225 MPa. These curves show a steady state in which creep rate stays constant for more than 20% of strain.

How abrupt the transition is from the low stress (curves with SCRMa – below 220 MPa) to the high stress region (curves without SCRMa – above 225 MPa) is demonstrated in Fig.2. At the same applied stress 225 MPa, one specimen exhibited the high stress behaviour while the other specimen deformed in the low stress mode. Thus the transition between the two stress regions clearly has a threshold character which is sensitive to the specimen microstructure. The mode of creep loading (tension or compression) does not influence the creep curve shape in the strain range of SCRMs. As can be seen in Fig.3, the tensile creep only deviates from the compression one close to the rupture in tension where the final rupture event takes over the

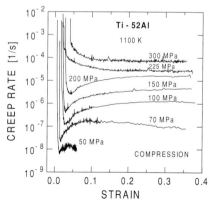

Figure 1: Ti-52Al compression creep curves obtained at 1100 K and in the applied stress range 50-300 MPa. The curves exhibit clear minimum of creep rate for applied stresses below 220 MPa.

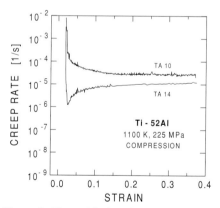

Figure 2: The ambiguity of the creep curve shape observed at 1100 K and 225 MPa. The presence of two distinct behaviour types at the same stress indicates a sharp threshold-like transition between high and low stress regions.

324

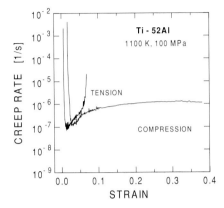

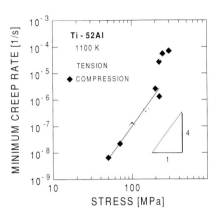

Figure 3: Comparison of creep curves obtained in tension and compression. There is almost no difference in creep strain accumulation kinetics up to the advanced stage of tertiary creep in tension.

Figure 4: Applied stress - minimum creep rate plot for 1100 K. The break in the dependence is associated with the extinction of SCRMa for applied stresses above 225 MPa. The dashed line represents creep rates at 30% strain.

control of deformation kinetics. Therefore, the acceleration of the creep rate after SCRMa is not associated with the damage accumulation in the form of voids or cavities which would grow in tensile experiment and which would be absent during compression.

The applied stress dependence of minimum creep rate is presented in Fig.4. Three features of the plot should be highlighted. (i) There is a creep rate gap at the stress threshold (225 MPa) which separates the high and low stress regions. This gap reflects the fact that the SCRMa suddenly vanish (see also Figs.1 and 2) once the stress exceeds the threshold value. (ii) The tensile data that, due to the intrinsic brittleness of the alloy are available only in the low stress region, agree well with the compression results. (iii) The stress exponent n is close to 4 in both stress regions. In passing we note that the creep strain accumulated to the SCRM in the low stress region is generally below 1% (Table 1) while in the high stress region the "minimum" creep rate is in fact the steady state creep rate that is usually attained at strains higher than 10%. The dashed line in Fig.4 represents these steady state creep rates measured at 30% strain.

3.2. Apparent activation energy and characteristic microstructures

Along with the dependence of creep rate on the applied stress, the temperature dependence of creep rate plays a central role in the classical analysis of steady state creep. It has been shown on many occasions that the temperature dependence of steady state creep rate in pure metals and solid solution alloys results from the lattice self diffusion controlled creep [11]. There also are many examples demonstrating the failure of the steady state creep rate – lattice self diffusion formalism mainly in cases of more complicated high temperature resistant materials [6]. In the present study, the traditional steady state creep formalism is slightly extended with the aim to identify the domains of applied stress and creep strain in which the traditional analysis would have some justification. Consequently, in the present study, the apparent activation energy Q_{app} is defined (see also Takahashi et al. [2]) as

$$Q_{app}(\varepsilon) = R \cdot T^2 \cdot \frac{\partial(\ln[\dot{\varepsilon}(\varepsilon)])}{\partial T}\bigg|_{\sigma,\varepsilon} . \tag{1}$$

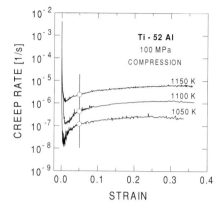

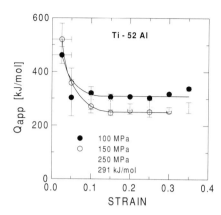

<u>Figure 5:</u> Ti-52Al compression creep curves obtained at 100 MPa and three different temperatures. The vertical line at 5% strain indicates one of the sections made to construct the $Q_{app}(\varepsilon)$ dependence of Fig.6.

<u>Figure 6:</u> Dependence of the apparent activation energy on creep strain. At strains higher than 10%, the Q_{app} is fairly constant for all the three investigated stresses and close to the value of 291 kJ/mol reported for Ti diffusion [12].

Figure 5 illustrates how the dependence $Q_{app}(\varepsilon)$ is constructed from the set of creep curves obtained at temperatures 1050, 1100 and 1150 K and at constant applied stress 100 MPa. Particularly, a vertical line $\varepsilon=5\%$ in Fig.5 intersects the creep curves in three highlighted points, each representing a pair of values in terms of temperature and strain rate. These pairs are subjected to a standard analysis using Eq.1 that yields the apparent activation energy at the indicated strain $\varepsilon=5\%$. Similarly, the apparent activation energy is assessed at other strain levels resulting in the 100 MPa curve of Fig.6. The same method also provides the other two curves shown in Fig.6 for applied stresses 150 and 250 MPa.

The $Q_{app}(\varepsilon)$ dependence in Fig.6 exhibits two key features. First, at creep strains higher than approximately 10%, the apparent activation energy is fairly constant for all the three applied stresses. In this strain range, the experimental values of Q_{app} are not far from the activation energy of 291 kJ/mol for Ti diffusion in γ-TiAl phase reported by Kroll and coworkers [12]. The changes of the apparent activation energy with the applied stress are not systematic and should be probably ascribed to the experimental scatter. Second, below 10% strain, the apparent activation energy increases considerably with the decreasing strain and attains values higher than 500 kJ/mol in the strain range between 0.5 and 1% where SCRMa are situated (Table1). These high Q_{app} values are rather independent of the applied stress for applied stresses lower than 225 MPa.

Therefore, the fairly constant creep rates, the apparent activation energy close to the one for lattice diffusion and the stress exponent close to 4 all suggest that the pure γ-TiAl phase exhibits a metal type creep behaviour [11] for strains exceeding 10%, or in other words, that creep is substructure controlled. TEM micrographs in Fig.7 a and b show the characteristic dislocation structure in the Ti-52at%Al after 35% of compression strain at 1100 K and 70 MPa. Indeed, the microstructure is composed of well developed subgrains with many knitting in and out reactions. Even in this low applied stress region DTs, indicated by arrows in Fig.7 b, are observed in the microstructure. The offsets in places where the DTs intersect the subgrain boundaries prove that DTs nucleated and spread through the grain volume later in creep life after the subgrain boundaries had already been formed.

3.3. Influence of pre-annealing on the sharp creep rate minimum

It has long time been known that the pre-annealing of the pure-γ TiAl specimen at the testing temperature before loading substantially influences the shape of the resulting creep curve [1]. In this study three tests were performed in which specimens were held at the testing temperature (1100 K) for three different times 0, 10 and 100 h before tensile creep loading. Results of this experiment are shown in Fig.8. It is evident that with increasing pre-annealing time the SCRM gradually ceases and the creep rate and ductility increases. In order to study microstructural consequences of pre-annealing, the same set of experiments was repeated and creep tests were interrupted at strains corresponding to the minimum creep rate. The respective microstructures were subjected to the TEM investigation. In parallel, the oxygen content in the specimens was analyzed after the creep test interruption. Results of this analysis showed unambiguously the increasing amount of oxygen in the specimens (1170 ± 210, 1500 ± 320 and 3280 ± 1280 wt ppm) with increasing time of the pre-annealing (0, 10 and 100 h, respectively). The oxygen pick-up occurred in spite of the fact that the furnace was continuously filled with purified argon throughout the test duration (see also the section 2).

TEM performed after the interruption of creep tests revealed two key microstructural features presented in Figs. 9 and 10. Figure 9 illustrates a typical phenomenon found in the samples pre-annealed 10 and 100 h. The oxide particle, the size of which is in the 1 – 10 µm range, serves as an effective stress concentrator such that, at the strain where the creep test was interrupted for the TEM investigation, DTs nucleate at the particle and spread into the grain interior.

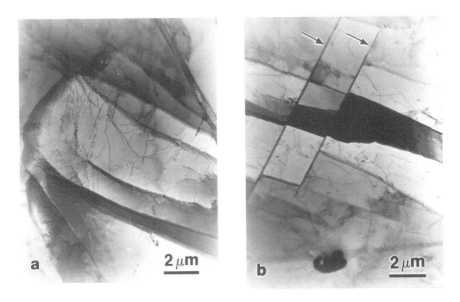

Figure 7: Subgrains (a) and deformation twins (b) in the Ti-52at%Al microstructure after 35% of compression creep strain at 1100 K and 70 MPa.

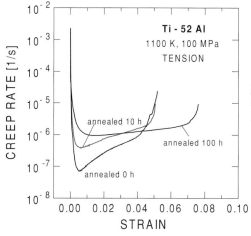

Figure 8: Ti-52Al tensile creep curves obtained at 1100 K and 100 MPa after pre-annealing for 0, 10 and 100 hours.

The other key microstructural feature observed in the specimen without the pre-annealing is shown in the weak beam TEM micrograph of Fig.10. Here the small oxide particles (perhaps H-phase type [10]) form a continuous chain denoted as A that decorated a dislocation during primary creep. Branches B of the decorated dislocation are just unzipping from the particle chain as can be seen in Fig. 10, where the arrow marks the detachment point. The interactions between dislocations and oxide particles were also observed in specimens which underwent the pre-annealing (10 and 100 h). In all the three investigated samples only a certain part of the overall dislocation density was covered by particles.

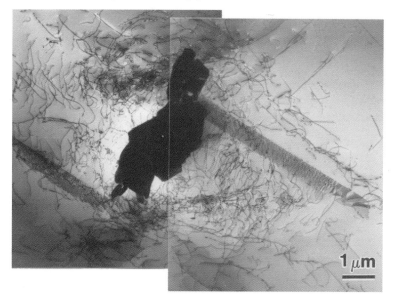

Figure 9: Nucleation of deformation twins at the oxide particle. Interrupted tensile creep test (1100 K, 100 MPa, strain 0.7%, pre-annealing 10 hours).

328

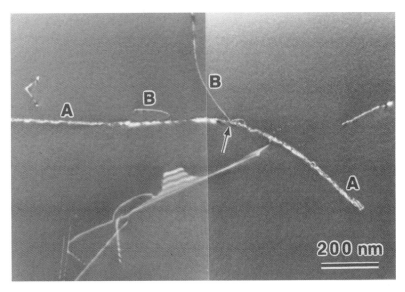

<u>Figure 10:</u> Unzipping of dislocations (B) from the chain of small oxide particles (A). Interrupted tensile creep test (1100 K, 100 MPa, strain 0.5%, pre-annealing 0 hours).

4. Discussion

At present there seems to be a little doubt that creep of pure-γ TiAl is substructure controlled for creep strains exceeding 10% at temperatures above 1000 K. This conclusion is validated not only by the results presented in this study, but also by the earlier investigations in which stress exponents in the range 4-5 and a well developed subgrain structure were reported [2,3,13]. Attempts to relate the apparent creep activation energies to diffusion controlled processes have been much less conclusive. In the present study the value $Q_{app} \sim 275$ kJ/mol found in the high strain range is in reasonable agreement with the activation energy for Ti diffusion in the γ-TiAl phase [12]. However, a review of the literature by Lu and Hemker [6] shows that the measurements of Q_{app} usually yield values in the range 300-360 kJ/mol. These values are higher, but not much higher, then the value obtained in the present study and the value reported for Ti diffusion [12]. Therefore, with certain cautiousness we conclude that, after the transition strain, creep in pure-γ TiAl could be accounted for on the basis of diffusion controlled mechanisms.

While the relatively straightforward interpretation of creep behaviour can be suggested for strains exceeding the transition strain, much more effort would be needed to clarify satisfactorily the situation in the low stress-low strain regime where SCRMa occur. Results of the present and earlier investigations [2-5,7,8] strongly support the view that the damage accumulation and the dynamic recrystallization, which had formerly been proposed to rationalize the transition strain, can be ruled out. In contrast, two quantitative microstructural studies [7,8] showed that deformation twinning only provides a significant contribution to the overall creep strain beyond the creep rate minimum. In primary creep, where the creep rate steeply decreases (at strains lower than the one which corresponds to the SCRM – Table 1) the twinning activity was found limited also by other authors [6,14]. Since DTs represent an important source of plastic displacement parallel to the c-axis of the tetragonal L1$_0$ lattice, it has been suggested that the insufficient twinning activity in primary creep causes an internal stress redistribution which, on one hand, hinders the easy glide of ordinary dislocations and, on the other hand, supports the nucleation of DTs [8]. In this view, the SCRMa result from the superposition of decelerating glide of dislocations and increasing kinetic contribution due to the nucleation and spreading of DTs.

The interpretation of SCRM proposed in the original paper by Loiseau and Lasalmonie [1] relied on a completely different mechanism. These authors ascribed the transition strain to the heterogeneous precipitation of Ti_2Al particles on dislocations during primary creep. In their view, the precipitating particles reduce significantly the density of mobile dislocations what results in the rapid hardening. Only when equilibrium volume fraction of particles is formed the freshly created dislocations are not affected by precipitation and can effectively contribute to strain. The acceleration of creep was also attributed to the particle coarsening that allows the breakaway of trapped dislocation lines from the particle chains.

The interpretation proposed by Loiseau and Lasalmonie [1] slightly contradicts the results of the present study since the growth of creep rate with increasing oxygen content experimentally observed in this study would not be expected based on the mechanism suggested in [1]. However, there is an increasing body of evidence [10] that, in the studied temperature range, dislocation cores become enriched with interstitial impurity elements like oxygen and, later in creep life, the oxide phases precipitate heterogeneously and pin dislocations where available [9]. A similar type of oxide precipitation has been documented in Fig. 10 of the present study. Therefore, a systematic quantitative analysis of the oxygen segregation to dislocations under creep conditions is needed to assess the significance of this mechanism for the enhanced creep strength of pure-γ TiAl in the transition strain range.

5. Summary and conclusions

Tensile and compression creep data obtained for Ti-52at%Al intermetallic compound at temperatures 1050, 1100 and 1150 K and in the applied stress range 50-300 MPa have been presented and analysed in view of the fundamental deformation processes. Based on the obtained results, following conclusions can be drawn.

1) An applied stress threshold of 225 MPa exists at 1100 K which delimits two stress regions with distinctly different creep behaviour. At stresses below the threshold creep curves exhibit the sharp creep rate minimum (SCRM) situated below 1% of creep strain while above the stress threshold a normal metal type – steady state creep behaviour is observed.
2) In the low stress region, a transition strain of the order of 10% is needed to change the creep response of the Ti-52at%Al compound from the SCRM regime back to the standard metal type behaviour. This standard behaviour is characterized by the stress exponent n=4 and the apparent activation energy Q_{app}~275 kJ/mol which is close to the activation energy of Ti diffusion in γ phase.
3) Transmission electron microscopy confirmed that well developed deformation substructure in the form of subgrains is present in the samples crept behind the range of transition strain.
4) At present, two microstructural mechanisms could account for the observed stress threshold, the regime of transition creep strain and SCRMa. The deformation twinning, which is activated only after some stress redistribution associated with primary creep, contributes to the plastic strain parallel with the c-axis of the tetragonal $L1_0$ lattice and thus relaxes the elastic incompatibility strains set in early stages of creep life. On the other hand, interactions of all dislocation types with oxygen in solid solution and the heterogeneous precipitation of

330

oxide phases during creep could explain the SCRMa in a close analogy to yield points observed due to the strain ageing in constant strain rate tests.

Acknowledgement

The financial supports from the Ministry of Education, Youth and Sports (contract no. OC 522.100) and from Grant Agency AS CR (contract no. S2041001) are highly acknowledged. The authors also thank to Mrs. L. Adamcová and Mr. M. Daniel for the assistance with the preparation of final TEM micrographs and TEM foils.

References

[1] Loiseau, A., Lasalmonie, A., 1984, Influence of the Thermal Stability of TiAl on Its Creep Behaviour at High Temperatures, Mat. Sci. Eng., Vol. 67, pp. 163-168.
[2] Takahashi, T., Nagai, H., Oikawa, H., 1989, Creep Behaviour of Polycrystalline Intermetallic Ti-53.4mol.%Al, Mat. Sci. Eng. A, Vol. 114, pp. 13-20.
[3] Ishikawa, Y., Oikawa, H., 1994, Structure Change of TiAl during Creep in the Intermediate Stress Range, Mat. Trans., JIM., Vol. 35, pp. 336-345.
[4] Takahashi, T., Nagai, H., Oikawa, H., 1990, Effects of Grain Size on Creep Behaviour of Ti-50mol.%Al Intermetallic Compound at 1100 K, Mat. Sci. Eng. A, Vol. 128, pp. 195-200.
[5] Hamada, M., Oikawa, H., 1996, Effect of Grain Size on High-Temperature Creep Behaviour of Ti-53mol.%Al Intermetallics in the Low Stress Region, Mat. Trans., JIM., Vol. 37, pp. 1607-1610.
[6] Lu, M., Hemker, K., J., 1997, Intermediate Temperature Creep Properties of Gamma TiAl, Acta mater., Vol. 45, pp. 3573-3585.
[7] Dlouhý, A., Kuchařová, K., Horkel, T., 2000, Microstructural Evolution Associated with Creep Rate Minima in Pure- and Near-γ TiAl Intermetallics, in: Creep and Fracture of Engineering Materials and Structures, Proceedings of the CFEMS 8, eds. T Sakuma and K. Yagi, Key Eng. Mat., Vols. 171-174, Trans Tech Publications, Uetikon-Zuerich, pp. 693-700.
[8] Dlouhý, A., Kuchařová, K., Březina, J., 2001, Dislocation Slip and Deformation Twinning Interplay during High Temperature Deformation in γ-TiAl Base Intermetallics, Mat. Sci. Eng., in press.
[9] Oehring, M., Ennis, P.J., Appel, F., Wagner, R., 1997, Microstructural Changes during Long-Term Tension Creep of Two-Phase γ-Titanium Aluminide Alloys, MRS Symp. Proc., Vol. 460, pp. 257-262.
[10] Gregori, F., Penhoud, P., Veyssiere, P., 2001, Extrinsic Factors Influencing the Yield Stress Anomaly Behaviour of Al-Rich γ-TiAl, Phil. Mag. A, in press.
[11] Sherby, O.D., Burke, P.M., 1968, Mechanical Behaviour of Crystalline Solids at Elevated Temperature, Prog. Mat. Sci., Vol. 13, pp. 323-390.
[12] Kroll, S., Mehrer, H., Stolwijk, N., Herzig, C., Rosenkranz, R., Frommeyer, G., 1992, Titanium Self-diffusion in the Intermetallic Compound γ-TiAl, Z. Metallkd., Vol. 83, pp. 591-595.
[13] Lipsitt, H.A., Shechtman, D., Schafrik, R.E., 1975, The Deformation and Fracture of TiAl at Elevated Temperatures, Metall. Trans. A, Vol. 6, pp. 1991-1996.
[14] Morris, M.A., Leboeuf, M., 1997, II. Deformed Microstructures during Creep of TiAl Alloys: Role of Mechanical Twinning, Intermetallics, Vol. 5, pp. 339-354.

Creep Behavior of in-situ TiB Fiber Reinforced Ti Matrix Composite

Kenshi KAWABATA, Eiichi SATO and Kazuhiko KURIBAYASHI

The Institute of Space and Astronautical Science

3-1-1 Yoshinodai, Sagamihara, 229-8510, Japan

tel: (81) 42-759-8266   fax: (81) 42-759-8461

E-mail: kawabata@materials.isas.ac.jp

Abstract

The creep behavior of inclusion bearing materials was analyzed through considering two accommodation processes, diffusion and plastic, which are inevitable for the materials to continue creep deformation. The creep experiments were performed using a model material, in-situ TiB fiber reinforced Ti matrix composite, which have a good interfacial bonding, the moderate diffusional accommodation rate and no fine oxide dispersions. A sigmoidal curve of $\dot{\varepsilon}$ and σ relation in a double logarithmic plot was observed, which indicated the presence of three deformation region: plastic accommodation, diffusional accommodation control, and complete diffusional accommodation regions. The activation energies in the plastic accommodation and diffusional accommodation control regions were close to those of volume and interface diffusion, respectively.

1. Introduction

Dispersing second phases or reinforcements into metallic matrices is one kind of the methods which strengthen metallic materials. The composite materials fabricated by this method offer unique combinations of high temperature mechanical properties often unattainable from individual constituents of the materials. Creep resistance is one such important property, since the structural materials are required to increase the maximum temperature limit to which they can be used for extended periods.

The understanding of the contribution of the inclusions to strengthen composite materials, in particular at steady state creep condition, has been established in the following two ways: the threshold stress concept based on dislocation theory and the load transfer concept based on continuum mechanics.

First, for dispersion strengthening alloys, such as for TD-Ni [1, 2], Al-Be [3] and P/M Al matrix composites [4-6], which contain sub-micron size particles, the high values of stress exponent have been observed and attributed to the threshold stress. The threshold stress has been explained by the Orowan mechanism or the void strengthening mechanism, and the latter is, now, believed to be operating at high temperatures [3].

On the other hand, composite materials, which contain coarser inclusions such as whiskers, have been analyzed to show the stress exponent identical to the monolithic matrix material by the shear-lag model [7], the self-consistent potential method [8], the unit cell model [9], micromechanics [10], and the finite element method [11]. However the experimental result of these predictions are not available except the early study on Ag-W composite [12]. The reason of the luck of experiment is that the usual composites such as Al-SiC [13] show degradation of interfacial debonding during creep and also contain large amounts of fine oxide particles, resulting in the absence of the steady state creep and the existence of a high threshold stress, respectively.

The above two concepts predict quite different creep behavior, especially in the value of the stress exponent, and thus they have been discussed separately. However we believe that considering the accommodation processes of the misfit strain between the inclusions and the matrix, which is inevitable for the composites to continue creep deformation, we can unify the two strengthening mechanisms. At high temperatures, two kinds of accommodation processes can be considered: diffusional accommodation [14, 15] by interfacial sliding and diffusion, and plastic accommodation [7-11] by heterogeneous flow of the matrix. While only the diffusional accommodation has been considered in the dispersion strengthening mechanism, only the plastic accommodation has been considered in the load transfer mechanism.

The aim of the present study is to discuss creep mechanism of composite materials through the accommodation processes using a model composite which shows the two types of the deformation behavior based on the two accommodation mechanisms. In order to that, we fabricate the composite having the moderate diffusional accommodation rate which lies in the experimentally measurable range.

2. Experimental procedures

We selected Ti-TiB in-situ composite [16, 17] as a model material which have a moderate diffusional accommodation rate in the measurable range. Ti-15vol.%TiB rods with a diameter 10 mm were supplied by Toyota Central R&D Lab., Inc. The rods were fabricated by P/M method followed by extrusion so that TiB whiskers were well oriented. The size of the whiskers were determined by microstructural observation on a deep-etched surface by a hydrofluoric acid and nitric acid solution for one hour.

The creep tests were performed under compressive condition in order to avoid debonding of the interface during creep deformation. Cylindrical specimens with $\phi 8$ mm x 10 mm were cut from the extruded rods with the compression axis parallel to the extrusion direction. The creep test was carried out at 923, 1023 and 1123 K under the constant crosshead speed, and the strain rate was measured by eddy-current displacement censors attached to the top of lower and upper push rods.

3. Results

Figure 1 shows the microstructure of Ti-15vol.%TiB composite of longitude and transverse sections. The photograph indicates good orientation of the TiB whiskers toward the extruded direction. The size of the whiskers is identified that the diameter and length are 10 μm and 50 μm, respectively.

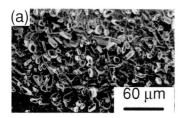 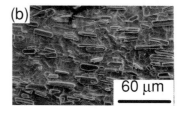

Figure 1. SEM micrograph of the Ti-15vol.%TiB composite:
(a) transverse and (b) longitude sections.

The diffusional accommodation rate for the inclusion of the present size at each testing temperature lies in the measurable range, 5×10^{-8} s^{-1} $< \dot{\varepsilon} < 1 \times 10^{-3}$ s^{-1} and 2 MPa $< \sigma < 100$ MPa, where $\dot{\varepsilon}$ and σ are strain rate and applied stress, respectively. The details are discussed in § Discussion. Here we confirm that the composites have the moderate diffusional accommodation rates for the present study.

Figure 2 shows $\dot{\varepsilon}$ and σ of the composites at each temperature under the steady state condition on a double-logarithmic scale. For a comparison, the predicted diffusional accommodation rate at each temperature is also plotted by a broken line. Figure 2 indicates a sigmoidal curve of $\dot{\varepsilon}$ and σ relation at the highest temperature and the upper parts of sigmoidal curves at the lower two temperatures. In the high stress region, the power low creep behavior with the stress exponent of 4.5 is observed at all temperatures. This value is in good agreement with the value, 4.3, of the monolithic Ti for power law creep [19]. As the stress decreases below the calculated diffusional accommodation rate, the data points at all temperatures deviate from the power law lines and show lower values of stress exponent. It can be considered as the transition from the plastic accommodation control region to the diffusional accommodation control region. At the highest temperature, the data show another

334

transition from low to high values of the stress exponent. It can be considered as the transition from the diffusional accommodation control region to the complete diffusional accommodation region.

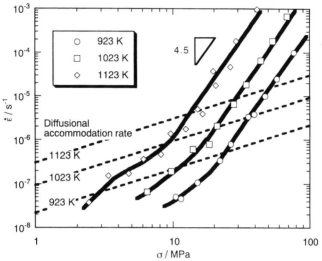

Figure 2. Deformation behavior of Ti-15vol.%TiB composites at 923, 1023 and 1123 K.

4. Discussion

Considering the deformation behavior of composite materials through the two accommodation processes, we obtain a schematic drawing of $\dot{\varepsilon}$ and σ relation in a double logarithmic scale (Fig. 3). Figure 3 indicates the deformation behavior of the monolithic matrix and the composite materials. The latter is drawn based on the straight lines of complete diffusional accommodation (1), diffusional accommodation control (2), plastic accommodation control (4) and threshold stress.

When the strain rate of the composite is lower than the diffusional accommodation rate (region I), the strain mismatch is accommodated completely by diffusion, and the interaction between the dislocations and the inclusions becomes attractive. Thus the threshold stress by the void strengthening mechanism [3] appears.

In the region (region II) where the applied stress is sufficiently higher than the threshold stress and the strain rate is lower than the diffusional accommodation rate, the strain mismatch

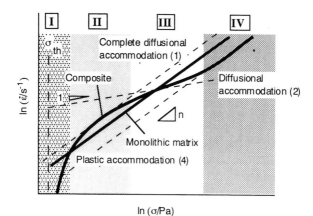

In (σ/Pa)

Figure 3. Schematic drawing of the deformation behavior of the composite

is accommodated completely and the inclusions support no external load. In this region, dispersion weakening given by the following equation appears [15]:

$$\dot{\varepsilon}_{comp} = (1 - f)^{-n} \dot{\varepsilon}_M, \qquad (1)$$

where $\dot{\varepsilon}_M$ and n are the strain rate and the stress exponent of the monolithic matrix materials, respectively.

When the strain rate of the composites reach the diffusional accommodation rate, the deformation becomes controlled by the accommodation processes. In the region (region III) where the diffusional accommodation rate is higher than the plastic accommodation rate, the deformation is mainly controlled by the diffusional accommodation, and in the opposite situation (region IV) the deformation is mainly controlled by the plastic accommodation.

The diffusional accommodation rate is given by [18]

$$\dot{\varepsilon}_{diff} = \frac{5(1 - v)}{2\tau_i \mu(7 - 5v)} \sigma, \qquad (2)$$

where μ and v is the shear modulus and poison ratio of the matrix respectively, and τ_i is accommodation time, given by [18]

$$\tau_i = (\frac{V}{512})\alpha \left\{ \frac{3\mu + \mu* + 3K*}{\mu(\mu* + 3K*)} \right\} \frac{kT}{D_i \delta \Omega}, \qquad (3)$$

where V, α, $\mu*$ and $K*$ are the volume, the aspect ratio, the shear modulus and the bulk modulus of the inclusions respectively, and k, T, D_i, δ and Ω are Boltzmann's constant, the absolute temperature, the interfacial diffusion coefficient, the thickness of the interface and the atomic volume of the matrix, respectively. The parameters using in the present study is shown in Table 1. The plastic accommodation rate given by [8]

$$\dot{\varepsilon}_{plast} = (1 - f)^g \dot{\varepsilon}_M, \qquad (4)$$

Table 1. Data used in the prediction of the diffusional accommodation for α-Ti and TiB

Parameters of the matrix [19]	Parameters of the inclusions [20]
$\mu = (4.95 \times 10^4 - 25\,(T\,/\,K)\,)$ MPa	$\mu* = 140$ GPa
$v = 0.34$	$v* = 0.16$
$D_i = 7.5 \times 10^{-5} \exp(\,(-121\ kJ/mol)\,/\,RT)$ m^2/s	
$\delta = 2b = 2 \times 2.89 \times 10^{-10}$ m	
$\Omega = 1.26 \times 10^{-29}$ m^3	

where g is the constant depend on α and n and is given in Table 1 of ref. [8]. The value of g increases significantly with increasing the aspect ratio of the inclusions.

The theoretical reconstruction of the creep behavior has been performed as follows, whose results are shown in Fig. 4 for 1123 K and Table 2 for all temperatures.

First, the creep behavior of the imaginary matrix, $\dot{\varepsilon}_M$, is estimated so that it satisfy eqs. (1) and (4). It is because the deformation behavior of the imaginary matrix predicted in the present study is different from that of the monolithic matrix, since the inclusions change the microstructure of matrix from the monolithic matrix [21]. The creep behavior in the plastic accommodation region and the complete diffusional accommodation region is then theoretically determined using values of f, α and n.

Second, since the theoretical diffusional accommodation seems to deviate from results, the modified diffusional accommodation rates are fitted from the results.

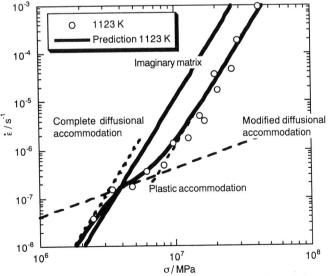

Figure 4. Comparison of the result and theoritical prediction

Table 2 Estimation of the deformation behavior of imaginary matrix, plastic accommodation, and complete diffusional accommodation used in drawing Fig.4.

Deformation behavior		923 K	1023 K	1123 K
Diffusional accommodation	calculated	$\dot{\varepsilon} = 2.31 \times 10^{-8} \sigma$	$\dot{\varepsilon} = 9.83 \times 10^{-8} \sigma$	$\dot{\varepsilon} = 3.26 \times 10^{-7} \sigma$
	modified	$\dot{\varepsilon} = 3.40 \times 10^{-9} \sigma$	$\dot{\varepsilon} = 1.00 \times 10^{-8} \sigma$	$\dot{\varepsilon} = 4.00 \times 10^{-8} \sigma$
Imaginary Matrix		$\dot{\varepsilon} = 3.17 \times 10^{-12} \sigma^{4.5}$	$\dot{\varepsilon} = 2.45 \times 10^{-11} \sigma^{4.5}$	$\dot{\varepsilon} = 3.70 \times 10^{-10} \sigma^{4.5}$
Plastic accommodation		$\dot{\varepsilon} = 3.26 \times 10^{-13} \sigma^{4.5}$	$\dot{\varepsilon} = 2.52 \times 10^{-12} \sigma^{4.5}$	$\dot{\varepsilon} = 3.80 \times 10^{-11} \sigma^{4.5}$
Complete diffusional accommodaiton		$\dot{\varepsilon} = 6.59 \times 10^{-12} \sigma^{4.5}$	$\dot{\varepsilon} = 5.09 \times 10^{-11} \sigma^{4.5}$	$\dot{\varepsilon} = 7.69 \times 10^{-10} \sigma^{4.5}$

($\dot{\varepsilon}$ in s^{-1} and σ in MPa)

The resulting deformation behavior of the composite is drawn in Fig. 4, which shows only the data at 1123 K. The experimental data agrees quite well with the predicted lines.

It is expected that the activation energies of the plastic accommodation region and the transition regions are given by those of volume and interface diffusion, respectively. Figure 5 shows the Arrhenius plots for the plastic accommodation region (40 MPa) and the transitional region (9 MPa). The activation energies of the plastic accommodation and the transition regions are estimated as 209 and 135 kJ/mol, and are close to those for volume diffusion of Ti (241 kJ/mol) and the interfacial diffusion of Ti (121 kJ/mol) [19], respectively. It also confirms that the deformation behavior shown in the present study reveal the transition between the dispersion strengthening mechanism and the load transfer mechanism.

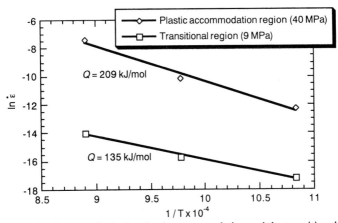

Figure 5. Activation energies in the plastic accommodation and the transitional region.

338

In the present study, we do not observe the threshold stress clearly. Using the spacing between the inclusions, the threshold stress is expected to be around 1 MPa [3], which is much lower than the measured range.

5. Conclusions

In the present study, we discussed the deformation behavior of the dispersion strengthening mechanism and the load transfer mechanism through the accommodation processes of the misfit strain between the matrix and the inclusions used by the model composites, Ti-15vol.%TiB, which have the moderate diffusional accommodation rate.

A sigmoidal curve of $\dot{\varepsilon}$ and σ relation in a double logarithmic plot is observed, which indicates three deformation region: plastic accommodation, diffusional accommodation control, and complete diffusional accommodation regions. The activation energies in the plastic accommodation and diffusional accommodation control were close to those of volume and interface diffusion, respectively.

Acknowledgments

We wish to acknowledge Dr. Takashi Saito, Toyota Central R&D Labs., Inc. for fabricating the specimens. This research was supported by Research Fellowships of the Japan Society for the Promotion of Science for Young Scientists.

References

[1] R. W. Lund and W. D. Nix, 1976, High Temperature Creep of Ni-20Cr-2ThO$_2$ Single Crystals, *Acta Met.*, vol. 24, pp. 469-481.

[2] J. H. Hausselt and W. D. Nix, 1977, Dislocation Structure of Ni-20-2ThO$_2$ after High Temperature Deformation, *Acta Met.*, vol. 25, pp. 595-607.

[3] Y.-H. Yeh, H. Nakashima, H. Kurishita, S. Goto and H. Yoshinaga, 1990, Threshold Stress for High-Temperature Creep in Particle Strengthened Al-1.5 vol.%Be Alloys, *Mater. Trans., JIM*, vol. 31, pp. 284-292.

[4] K-T. Park, E. J. Lavernia and F. A. Mohamed, 1990, High Temperature Creep of Silicon Carbide Particulate Reinforced Aluminum, *Acta Metall. Mater.*, vol. 38, pp. 2149-2159.

[5] F. A. Mohamed, K-T. Park and E. J. Lavernia, 1992, Creep behavior of discontinuous SiC-Al Composites, *Mater. Sci. Eng.*, vol. A150, pp. 21-35.

[6] J. Cadek, H. Oikawa, V. Sustek, Threshold Creep Behavior of Discontinuous Aluminium and Aluminium Alloy Matrix Composites: an overview, *Mater. Sci. Eng.*, vol. A190, pp. 9-23.

[7] A. Kelly and K. N. Street, 1972, Creep of discontinuous fiber composites II. Theory for

the steady-state, *Proc. Roy. Soc. Lond.*, vol. A328, p. 283-293.

[8] B. J. Lee and M. E. Mear, 1991, Effect of Inclusion Shape on The Stiffness of Nonlinear Two-Phase Composite, *J. Mech. Phys. Solids.* vol. 39, pp. 627-649.

[9] K. S. Ravichandran and V. Seethraman, 1993, Prediction of Steady State Creep Behavior of Two Phase Composite, *Acta Metall. Mater.*, vol. 41, pp. 3351-3361.

[10] E. Sato, T. Ookawara, K. Kuribayashi, and S. Kodama, 1998, Steady-State Creep in an Inclusion Bearing Material by Plastic Accommodation, *Acta Mater.*, vol. 46, pp. 4153-4159.

[11] G. Bao, J. W. Hutchinson and R. M. McMeeking, 1991, Particle Reinforcement of Ductile Matrices against Plastic Flow and Creep, *Acta Metall. Mater.*, vol. 39, pp. 1871-1882.

[12] A. Kelly and W. R. Tyson, 1996, Tensile Properties of Fiber Reinforced Merals–II. Creep of Silver-Tungsten, *J. Mech. Solids*, vol. 14, pp. 177-186.

[13] T. Morimoto, T. Yamaoka, H. Liholt, M. Taya, 1988, Second Stage Creep of SiC Whisker / 6061 Aluminum Composite at 573 K, *Trans. Am. Soc. Mech. Eng.*, vol. 110, pp. 70-76.

[14] R. C. Koeller and R. Raj, 1978, Diffusional relaxation of stress concentration at second phase particles, *Acta Met.*, vol. 26, pp.1551-1558.

[15] T. Mori, Y. Nakashima, M. Taya and K. Wakashima, 1997, Steady-state creep rate of a composite : two-dimensional analysis, *Phil. Mag. Lett.*, vol. 75, pp. 359-365.

[16] Z. Y. Ma, S. C. Tjong and L. Gen, 2000, in-situ Ti-TiB Metal-Matrix Composite Prepared by a Reactive Pressing Process, *Scripta Mater.*, vol. 42, pp. 367-373.

[17] T. Saito, 1995, A Cost-Effective P/M Titanium Matrix Composite for Automobile Use, *Adv. Perform. Mater.*, vol. 2, pp. 121-144.

[18] S. Onaka, T. Okada, and M. Kato, 1991, Relaxation Kinetics and Relaxed Stresses Caused by Interface Diffusion around Spheroidal Inclusions, *Acta Metall. Mater.*, vol. 39, pp. 971-978.

[19] G. Malakondaiah and P. R. Rao, 1981, Creep of Alpha-Titanium at Low Stresses, *Acta Met.* vol. 29, pp. 1263-1275.

[20] R. R. Atri, K. S. Ravivhandran and S. K. Jha, 1999, Elastic Properties of in-situ Processed Ti-TiB Composites Measured by Impulse Excitation of Vibration, *Mater. Sci. Eng.*, vol. A271, pp. 150-159.

[21] S. Ranganath and R. S. Mishra, 1996, Steady State Creep Behaviour of Particulate-Reinforced Titanium Matrix Composites, *Acta Mater.*, vol. 44, pp. 927-935.

CREEP OF COMPOSITES

STATE OF THE ART OF DAMAGE-CREEP IN CERAMIC MATRIX COMPOSITES

J.L. CHERMANT, G. BOITIER, S. DARZENS, G. FARIZY, J. VICENS

LERMAT, URA CNRS 1317, ISMRA, 6 Bd Maréchal Juin, 14050 Caen Cedex, France

ABSTRACT

Creep tests were performed with ceramic matrix composites in the 1173-1673K temperature and 25-400 MPa stress domains. As no dislocations and no diffusion phenomena are activated, and as brittle damages are observed in this experimental domain, the use of the damage mechanics and the support of multiscale observations by SEM and TEM gave access to the micromechanism(s) involved under creep solicitation. All this information gives strong argument for the existence of a damage-creep mechanism.

1. INTRODUCTION

Ceramic matrix composites reinforced by continuous ceramic fibers, CMC, offer many advantages in a field where un-protected metals or metallic alloys cannot be used anymore: low density, high strength, high chemical resistance to severe environment, un-brittle behavior, low emission and noise, ... So this class of materials permits yielded performance gains including higher thrust-to-weight ratio, longer engine lives, fully reusible parts for space transportation systems, ... Thus applications for terrestrial and aeronautical gas turbines and combustor liners, space transportation, fuel rocket engines, brakes for aircrafts and terrestrial vehicles, ... are presently proposed [1-9].

To develop new parts in CMCs, the bureau design needs toughness and statistical datæ, and also life time parameters in order to accede to correct predictions under stress, creep and fatigue solicitations. The scope of this paper is to give the state of the art of the creep of CMCs and to highlight that classical creep mechanisms do not work as temperature and stress conditions are not high enough to develop dislocation motion or diffusion phenomena.

2. MATERIALS AND TECHNIQUES

Many ceramic matrix composites with a monolithic ceramic or a glass-ceramic matrix were investigated in our laboratory : SiC_f-MLAS [10, 11], SiC_f-YMAS [12] with A = Al, L = Li, M = Mg, Y = Y, S = SiO_2, and C_f-SiC [13], SiC_f-SiC [14,15] and SiC_f-SiBC [16].

CMCs with a glass-ceramic matrix were fabricated from a slurry of the adequate glass powders which embedded Nicalon NLM 202 SiC_f plies or clothes and then hot-pressed, by

Aérospatiale (Saint-Médard-en-Jalles, France) for SiC$_f$-MLAS composites and by ONERA (Establishment of Palaiseau, France) for SiC$_f$-YMAS ones.

All the CMCs with a monolithic ceramic matrix were fabricated by SEP-SNECMA (now SNECMA, Division Moteurs et Fusées, Saint-Médard-en-Jalles, France) by a more or less complex chemical vapor infiltration (CVI) process [17,18]. A thin layer of pyrolytic carbon was deposited on all the fiber preforms (NLM 202 and Hi-Nicalon SiC$_f$ fibers or ex PAN C$_f$ fibers). SiC$_f$-SiBC composites are made of self sealing multilayered matrix.

A Schenck Hydropuls PSB 100 servohydraulic machine (Darmstadt, Germany) was most often used to perform creep tests under a partial pressure of argon, between 1273K and 1673K, at a stress up to 250 MPa. It was equipped with an airtight fence and an induction furnace (AET and Célès, respectively Meylan and Lautenbach, France). Strain was measured with two resistive extensometers (Schenck), inside the furnace. Dogbone tensile specimens were used, with a length of 200 mm and a thickness between 2 and 5 mm, and a width in the gauge length between 8 and 16 mm. Temperature was measured both with two W-Rh 5/26 % thermocouples and an optical pyrometer IRCON Mirage (Niles, USA). Tests were performed taking particularly care especially of the load frame alignment, the thermal gradient and its stability, the temperature and strain measurements and the pressure variation [19,20].

SEM observations were made with a Jeol 6400 (Jeol, Tokyo, Japan) and TEM with a Jeol 2010 and a Topcon EM 002B, both equipped with EDS analysis, (Figure 1).

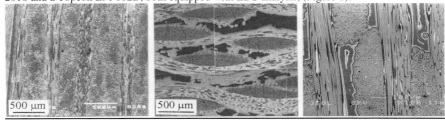

a)   b)   c)

Figure 1: SEM micrographs of a (0-90°)$_6$ SiC$_f$-MLAS (a), 2.5D C$_f$-SiC (b) and SiC$_f$-SiBC (c).

3. RESULTS

3.1. Macroscopical and microscopical approaches

Some creep curves (strain-time, ε-t) are presented in Figure 2 for different types of CMCs. To be sure of the presence of a steady state creep regime, one plots the creep rate as a function of strain or time, $\dot{\varepsilon}$-ε or t. The existence of a plateau is a way to confirm a true stationary stage. For each CMCs, two different states have been evidenced: a primary and a stationary or pseudo-stationary stage.

At that stage and in this investigated domain one can only say that: there is no tertiary stage, that class of materials presents a good creep resistance (between 10^{-8}-10^{-9}.s^{-1}) in a domain where un-protected superalloys cannot be used, and the architecture plays an important role.

To access to more information and to the creep mechanism(s) involved requires microstructural observations and analysis.

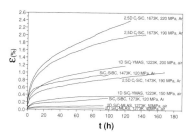

Figure 2: Some creep curves (strain-time, ε-t) for different CMCs tested at different temperatures and stresses, in argon or in air.

Due to the brittle character of the ceramic matrices, SEM observations reveal the presence of many classical brittle damages in such materials: fiber/matrix debonding, matrix microcracking, fiber and yarn pull-out, fiber and yarn bridging, fiber and yarn fracture, microcracks along the longitudinal fibers or yarns in the fiber/matrix interphase, microcracks at the interface of different matrix layers, ... Figure 3 illustrates some of these features. Some of them have been investigated from a semi-empirical/theoretical approach by many authors in order to accede to some specific parameters which can highlight the micromechanism(s) at work [21-25].

All these SEM observations have been also confirmed by TEM and HREM analysis, as one can see on Figure 4a,b and c. That evidences the role of both the fiber/matrix interfaces (which are most often interphases) and of yarn/yarn interfaces. When the pyrolytic carbon presents its turbostratic shape, it allows the development of lenticular pores parallel to the carbon planes (pores // to the applied stress, σ, in longitudinal yarns and ⊥ to σ in transverse yarns). These carbon planes can bridge the microcracks by the presence of carbon ribbons (Figure 4c), at the nanoscopic scale [26].

3.2. Damage mechanics approach

As in our field of investigation (1173-1473K and 25-500 MPa for CMCs with a glass-ceramic matrix, and 1173-1773K and 50-400 MPa for those with a "monolithic" ceramic matrix) there is no dislocation motion and no diffusion phenomena, so diffusion-creep and/or dislocation-creep cannot be the operating mechanism(s).

Thus, as many damages were evidenced, one has used the damage mechanics proposed by Kachanov [27] and Rabotnov [28], and extended to anisotropic materials by Ladevèze [29,30], leading to the damage parameter, D, (D = 1 - E/E_0, with E and E_0 respectively the elastic modulus of the damaged material at a given time and of the un-damaged material).

So during creep tests, un-loading and reloading loops were regularly performed in order to follow one of the elastic modulus and calculate the damage parameter. Figure 5 presents some damage curves as a function of time, D-t, in the case of 2.5D C_f-SiC and SiC$_f$-SiBC composites.

This macroscale approach involves the definition of damage kinematics so that a description of the damage state can be made with a minimum of state variables. The tensile rupture at room temperature of a 2D SiC$_f$-SiC has correctly been described by Gasser [31,32] and its creep by Rospars [15,33]. Rospars has shown that the damage parameter follows the same change with the inelastic strain (i.e. the damage caused by creep, not considering the static damage due to the loading), whatever the stress level is (Figure 6). So this evolution appears stress-dependent, and the creep damage is directly related to the viscoplastic strain, ε_{vp}.

346

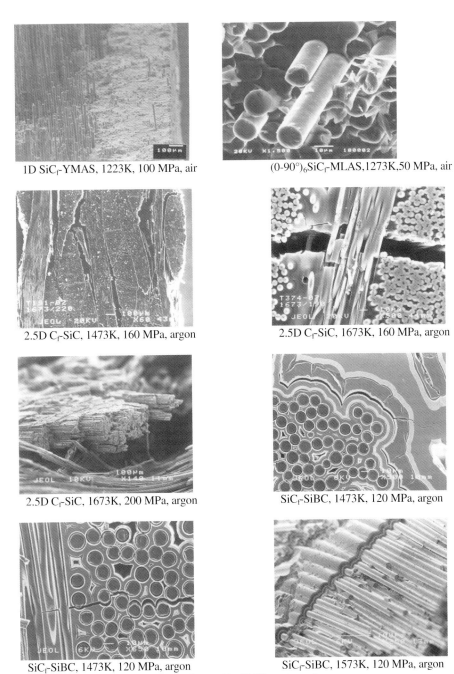

1D SiC$_f$-YMAS, 1223K, 100 MPa, air

(0-90°)$_6$SiC$_f$-MLAS,1273K,50 MPa, air

2.5D C$_f$-SiC, 1473K, 160 MPa, argon

2.5D C$_f$-SiC, 1673K, 160 MPa, argon

2.5D C$_f$-SiC, 1673K, 200 MPa, argon

SiC$_f$-SiBC, 1473K, 120 MPa, argon

SiC$_f$-SiBC, 1473K, 120 MPa, argon

SiC$_f$-SiBC, 1573K, 120 MPa, argon

Figure 3: Different damage features observed in CMC crept specimens.

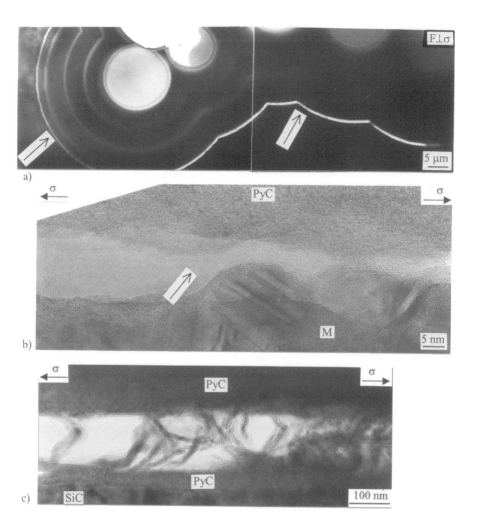

Figure 4: TEM micrograph of the deviation of a microcrack at matrix/matrix interface (a) and HREM micrograph of a microcrack in pyrolytic carbon (b), in SiC$_f$-SiBC after creep tests at 1473K, under 200 MPa, in tension, under argon ; TEM micrograph showing the bridging of a microcrack at the fiber/matrix interphase (c), in a 2.5D C$_f$-SiC after creep tests at 1673K, under 220 MPa, in tension under argon.

4. DISCUSSION

Whatever the type of plots is, D-ε_{in} or D-t, the damage evolution presents always two regimes: ① a very important increase at the beginning, i.e. during the loading and probably the first hours of the primary stage ; then ② an important change in the slope which becomes very low. That evidences the presence of two different mechanisms or regimes in the damage process.

348

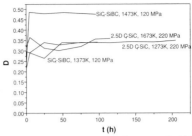

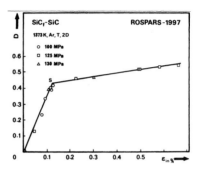

Figure 5: Damage evolution as function of time, D-t, for 2.5D C$_f$-SiC and SiC$_f$-SiBC specimens, creep tested in tension, under a partial pressure of argon, at different temperatures and stresses.

Figure 6: Creep damage as a function of the inelastic strain, D-ε_{in}, for 2D SiC$_f$-SiC specimens creep tested in tension under argon, between 100 and 130 MPa.

If step by step observations are made during creep tests (dynamic creep observations) on the same zones, one has shown [11,13,15,16] that for such materials there are two morphological steps during the creep deformation: ① formation and development of a matrix microcrack array, or development of a matrix microcrack array for materials with some microcracks in the as-received state (that is due to the difference in the thermal expansion coefficient of the fibers and the matrix, as for C$_f$-SiC) until saturation, which corresponds to point S of Figure 6; then one only observes their openings, and one of the main cracks becomes the major crack leading to the rupture of the materials (Figure 7). Using automatic image analysis methods, the surface area of the microcracks and of the yarns, and the opening of the microcracks were quantitatively measured [13,16,34], which confirms the proposed microcracking mechanism.

This opening damage process is time dependent and, consequently, can be considered as a mechanism of slow crack growth type [35]. In the case of SiC$_f$-SiBC composites, one has shown [16] that this mechanism is governed by the creep of the longitudinal yarns and precisely by the SiC fibers. The complex plots of the damage parameters as a function of time (Figure 5) can be explained from accurate observations: after the first step of damage accumulation until the saturation, the observed decrease of the damage parameter can be attributed to architectural distortions, which lock the microcracks from closing upon un-loading, leading to a reduction of the elastic strain and the subsequent apparent stiffening of the composite [36]. Then due to larger architectural distortions, no locks can arise and it leads to a new increase of the elastic strain and damage parameter.

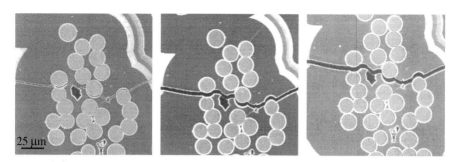

Figure 7: Evidence of the matrix microcrack opening during creep tests, at 1473K under 120 MPa and argon, of a SiC$_f$-SiBC composite, after 1, 10 and 50 h of creep (SEM micrographs).

Based on these mesoscopical, microscopical, and nanoscopical observations, and on the evolutions of the damage parameter, we can propose and draw the deformation and rupture mechanism for CMCs with a 2D and 2.5D architecture, presented in Figure 8, [16], where the development of the matrix microcrack arrays takes into consideration the fiber/matrix, matrix/matrix and matrix/yarn interfaces :

① development or creation and development of matrix microcracks in transverse yarn, until saturation, (Figure 8b),

② under stress and temperature there is the opening of the microcracks leading to a partial debonding between longitudinal yarns and matrix of the transverse yarns (Figure 8c),

③ these important inter-yarn cracks will appear along the longitudinal fibers, allowing in the case of CMCs with SiC fibers the fiber creep (if the applied temperature is higher than 1373K), leading to the rupture of the material (Figure 8d).

That illustrates that we have called at LERMAT the damage-creep mechanism [24,37].

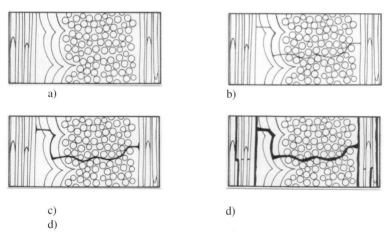

a)   b)

c)   d)
d)

Figure 8: Different steps of the rupture of a 2D or 2.5D CMCs.

Some semi-empirical models based on the fiber and matrix damages, interfacial shear strains, fiber and/or yarn bridging, viscous fiber/matrix interface and on fiber creep have recently been proposed by several authors [38-44]. From our experimental results and observations one has to try to use them, or to use the damage elasto-viscoplastic model [29,30,33] in order to model, from the mechanical point of view, the creep behaviour of these CMCs.

5. CONCLUSION

Different tensile creep results were presented in the case of ceramic matrix composites. As in the field of temperature and stress investigation there is no dislocation motion and no diffusion phenomena, one has used the damage mechanics. The results obtained where confirmed by multiscale observations – meso-, micro-, nano-scopic– and a damage-creep mechanism has been proposed based first on the development of a matrix microcracking array, and secondly on the opening of the microcracks mainly in the transverse yarn by a process of slow crack growth type.

ACKNOWLEDGEMENTS

This work has been supported by SEP-SNECMA (now SNECMA, Division Moteurs et Fusées, Saint-Médard-en-Jalles, France), and the Ministère de l'Education Nationale, de la Recherche et de la Technologie (MENRT) and Région de Basse-Normandie for two fellowships (GB and SD). The authors wish to warmly thank Drs E. Pestourie, J.P. Richard and J.M. Rougès from SNECMA, Mr H. Cubero for his helpful assistance to perform creep tests in the "best" conditions", and Mrs L. Chermant for the morphological analysis of the microcrack evolution.

REFERENCES

1. Shirouzu, M., Yamamoto, M., 1996, "Overview of the HYFLEX project", Amer. Inst. Aeron. Astronautics, **4524**, pp1-8.
2. Spriet, P., Habarou, G., 1997, "Applications of CMCs to turbojet engines : overview of the SEP experience", in: CMMC 96, San Sebastian, Spain, Sept. 9-12, ed. by Fuentes, M., Martinez-Esnaola, J.M., Daniel, A.M., Key Eng. Mat., **vol 127-131,** pp 1267-1276.
3. Labbé, P., 1998, "Les freins d'Airbus sur nos voitures ? Bientôt !", C.E.A. Technologies, **40**, p 8.
4. Brockmeyer, J.W., 1999, "Ceramic matrix composite applications in advanced liquid fuel rocket engine turbomachinery", J. Eng. Gas Turbines & Power, Trans. ASME, **115**, pp 58-63.
5. Nishio, K., Igashira, K.I., Take, K., Suemitsu, T., 1999, "Development of a combustor liner composed of ceramic matrix composite, CMC", J. Eng. Gas Turbines & Power, Trans. ASME, **121**, pp 12-17.
6. Ohnabe, H., Masaki, S., Onozuka, M., Miyahara, K., Sasa, T., 1999, "Potential application of ceramic matrix composites to aero-engine components", Composites Part A, **30A**, pp 489-496.
7. Trabandt, U., Wulz, H.G., Schmid, T., 1999, "CMC for hot structures and control surfaces of future launchers", in: HT-CMC II, Osaka, Japan, Sep. 6-9, ed. by Niihara, K., Nakano, K., Sekino, T., Yasuda, E., CSJ Series. Ceram. Soc. Jap., **3**, pp 445-450.
8. Renz, R., Krenkel, W., 2000, "C/C-SiC composites for high performance emergency brake systems", in: "Composites : from Fundamentals to Exploitation", ECCM 9, 4-7 June, Brighton, UK, ECCM 9 CD ROM C 2000, IOM Communications Ltd.
9. Staehler, J.M., Zawada, L.P., 2000, "Performance of four ceramic-matrix composite divergent flap inserts following ground testing on an F110 turbofan engine", J. Amer. Ceram. Soc., **83**, pp 1727-1738.

10. Kervadec, D., 1992, "Comportement en fluage sous flexion et microstructure d'un SiC-MLAS 1D", Thèse de Doctorat of the University of Caen.

11. Maupas, H., 1996, "Fluage d'un composite SiC_f-MLAS 2D en flexion et en traction", Thèse de Doctorat of the University of Caen.

12. Doreau, F., 1995, "Microstructure, morphologie et comportement en fluage de composites SiC_f-YMAS unidirectionnels", Thèse de Doctorat of the University of Caen.

13. Boitier ,G., 1997, "Comportement en fluage et microstructure de composites C_f-SiC 2,5D", Thèse de Doctorat of the University of Caen.

14. Abbé, F., 1990, "Fluage en flexion d'un composite SiC-SiC 2D", Thèse de Doctorat of the University of Caen.

15. Rospars, C., 1997, "Modélisation du comportement thermomécanique de composites à matrice céramique: étude du SiC-SiC 2D en fluage", Thèse de Doctorat of the University of Caen.

16. Darzens, S., 2000, "Fluage en traction sous argon et microstructure de composites SiC_f-SiBC", Thèse de Doctorat of the University of Caen.

17. Christin, F., Naslain, R., Bernard, C., 1979, "A thermodynamic and experimental approach of silicon carbide CVD. Application to the CVD-infiltration of porous carbon-carbon composites", Proceedings of the 7th International Conference on CVD, edited by Sedwick, T.O., Lydtin, H., The Electrochemical Society, Princeton, pp 499-514.

18. Goujard, S., Vandenbulke, L., 1994, "Deposition of Si-B-C materials from the vapor phase for applications in ceramic matrix composites", Ceram. Trans., **46**, pp 925-935.

19. Boitier, G., Maupas, H., Cubero, H., Chermant, J.L., 1997, "Sur les essais de traction à longs termes à haute température", Rev. Comp. Mat. Avancés, **7**, pp 143-172.

20. Boitier, G., Cubero, H., Chermant, J.L., 1999, " Some recommendations for long term high temperature tests", in: High Temperature Ceramic Matrix Composites, HT-CMC3, CSI Series, Publications of the Ceramic Society of Japan, **vol. 3**, pp 309-312.

21. Thouless, M.D., Sbaizero, O., Sigh, L.S., Evans, A.G., 1989, "Effects of interface mechanical properties on pull-out in a SiC-fibers-reinforced lithium aluminium silicate glass-ceramic", J. Amer. Ceram. Soc., **72**, pp 525-532.

22. Spearing, S.M., Evans, A.G., 1992, "The role of fiber bridging in the delamination resistance of fiber-reinforced composites", Acta Met. Mat., **40**, pp 2192-2199.

23. Curtin, W.A., 1995, "Toughnening of crack bridging in heterogeneous ceramic", J. Amer. Ceram. Soc., **18**, pp 1313-1323.

24. Chermant, J.L., 1995, "Creep behavior of ceramic matrix composites", Sil. Ind., **60**, pp 261-273.

25. He, M.Y., Wissuchek, D.J., Evans, A.G., 1997, "Toughening and strengthening by inclined ligament bridging", Acta Mat., **45**, pp 2813-2820.

26. Boitier, G., Chermant, J.L.,Vicens, J., 1999, "Bridging at the nanoscopic scale in 2.5D C_f-SiC composites", Appl. Comp. Mater., **6**, pp 279-287.

27. Kachanov, L., 1958, "Rupture time under creep conditions", Izv. Akad. Nauk. SSR, **8**, pp 26-31.

28. Rabotnov, M., 1989, "Creep Problem in Structural Members", North-Holland, Amsterdam.

29. Ladevèze, P., 1983, "Sur une théorie de l'endommagement anisotrope", Report n°34, Laboratoire de Mécanique et Technologie of Cachan, France.

30. Ladevèze P., 1993, "On an anisotropic damage theory", in: Failure Criteria of Structured Media, ed. by Boehler, J.P., Balkema, Rotterdam, pp 355-363.

31. Gasser, A., 1994, "Sur la modélisation et l'identification du comportement mécanique des composites céramique-céramique à température ambiante",Thèse de Doctorat of ENS of Cachan.

Modelling of Non-Linear Creep in Bi-polymer Composites

D.W.A. Rees,

Dept of Systems Engineering, Brunel University, Uxbridge,
Middlesex UB83PH (e-mail D.W.Rees@brunel.ac.uk)

Abstract
The Nutting law, commonly used to model creep in metals but originally proposed for non-metals, was employed here for non-linear creep of polymer composites. The creep observed within series and parallel dissimilar polymer arrangements is characterised within the time and stress exponents of this law. While the law remains valid over a wide stress range the time interval is restricted to where the visco-elastic strain rate diminishes with expired time as in primary metal creep. A marked change in the stress exponent appears under stress levels high enough to promote an unstable geometry as with necking and very rapid viscous flow. Here the accompanying strain rates increase in a manner reminiscent of teriary metal creep. The phenomenological approach is potentially useful for modelling creep of composites. The total strain appears with an instantaneous strain added to a creep strain that is characterised by one coefficient and two exponents: one of stress and the other of time. The instantaneous strain is exponential in stress. For creep, constant time and stress exponents appear within limited ranges. This description employs far fewer material constants than a classical mathematical approach to non-linear visco-elasticity [1]. As a consequence of this, superposition principles are advanced for when creep curves may or may not be geometrically similar under incremental loading.

1. Introduction
The theory of linear visco-elastic creep has been well represented mathematically though it is accepted that few polymers conform to linearity over a wide range of stress [2]. In fact, the stress index within an isochronous stress-strain relation for common polymers will usually exceed unity. It will be shown here that the same is true for series and parallel combinations of different polymers. This observation invalidates the application of the linear superposition principle. The non-linear mathematical theory [1] requires a very large number of coefficients and it is not surprising that a simpler empirical approach may be preferred to describe the obsereved deformation particularly for design purposes. Nutting [3] proposed a power law function for the dependence of creep strain upon stress and time within pitch and asphalt. A similar equation was later employed for polystyrene and pvc [4]. The basic form is similar to that proposed by Andrade and Norton for high temperature creep in metals and is generally accepted to apply over a restricted time interval, i.e. the primary creep regime. The two exponents are readily found from a repeated loading-unloading sequence.

The present paper shows that creep in four polymers: nylon (N66), polypropylene (PP), low and high density polyethylene (LDPE and HDPE) conforms to the Nutting relation where the stress exponent can reach values exceeding 2. The time exponents lie in the range 0.15 - 0.25. A similar relation applies to bi-polymer composites in both series or parallel arrangements. For example, in a parallel bonded combination of PP and N66 the stress and time exponents are 1.5 and 0.25 respectively. Their series combination yields these exponents as: 1.37 and 0.23. The instantaneous strain conforms consistently with an exponential relation in which a stress asymptote is approached at large strain.

Finally, it is shown how the derived relation is adapted to provide a prediction of the cumulative creep strain under stepwise loadings. The stress and time functions re-appear within a non-linear superposition principle. For this to apply it is essential that a geometric similarity pertains to individual creep curves for the range and time interval of interest.

2. Experimental

A load-unload series of experiments were conducted at room-temperature in which an 80 hour period of creep was followed by a similar period of recovery. This enabled the exponents in time and in stress to be found from a single test on a single polymer or a polymer composite. Both parallel and series arrangements of different polymers composites were loaded in this manner at room temperature until they lost load bearing capacity. Individual polymer testpiece dimensions were nominally 10×4 mm in cross-section and 65 mm in gauge length. The displacement in this length was recorded continously during periods of creep and recovery with a 100 mm stroke displacement transducer. The signal was amplified and demodulated for display on a chart recorder. Loads were applied to the end of a lever in a Denison creep machine and transmitted, with a 10:1 magnification, to the enlarged ends of a vertically aligned testpiece through vice grips and universal joint connections. To allow for the large displacements encountered in these materials the lever was raised initially so that it could pivot on its knife edge for the full rotation allowed. When this rotation became exhausted during creep by the extending length of the testpiece, the bottom grip was lowered by adjusting its keyed screw support with a hand wheel. The effect was to lift the lever from its stop back to its initial position. During re-setting the loads were supported and thereafter the load was re-applied to continue the test.

The deformation behaviour of polymer constituents was established from creep tests on individual polymers to the stated dimensions. Their load versus extension behaviour was further determined from tensile testing in a Dartec machine at a rate of 1 mm/min. Parallel bi-polymer composites were tested with their interfaces in a bonded and unbonded condition to give a section 10×8 mm. Cyanoacrylate and epoxy resins were employed as bonding agents. Series assemblies with 10×4 mm section were produced by cutting the standard testpieces in half and connecting the two halves of dissimilar materials together with a clamp.

3. Analysis of Creep Curves

The general form of the expression describing total strain ε under a given stress σ in the chosen composite arrangements has the form:

$$\varepsilon = \varepsilon_0 + A\sigma^n t^m \tag{1}$$

where $n > 1$, $m < 1$ and ε_o is the instantaneous strain that may take one of two forms. If Hooke's law was obeyed then $\varepsilon_o = \sigma/E$. However, some of the composites displayed Hookean elasticity only for a very limited stress range. With increased stress the non-linear dependence of ε_0 upon σ conformed closely to an exponential relationship:

$$\sigma = \frac{E}{q}\left(1 - e^{-q\varepsilon_o}\right) \tag{2}$$

in which $E = d\sigma/d\varepsilon_o$ is the initial gradient for $\varepsilon_o = 0$ and $q = E/\sigma_\infty$ for $\varepsilon_o = \infty$. To apply eq (2) a stress asymptote σ_∞ was assumed. To determine m, eq (1) is re-written as follows:

$$\varepsilon_c = a\,t^m \tag{3a}$$

where $\varepsilon_c = \varepsilon - \varepsilon_o$ and $a = A\sigma^m$. Thus, by plotting the creep strain component of the total strain versus time elapsed from each load step on logarithmic scales, m becomes the gradient of the linear plot. This assumes that m is independent of σ, resulting in a series of parallel lines. This was sensibly true for the loads employed. The appendix shows that random variations in gradient m were due more to the inaccuracy in the estimate of ε_o. Using a single average m - value enabled n to be determined from isochronous logarithmic plots between the creep strain for times $t = 1.67, 5, 20$ and 50 h and the applied stress level. Equation (3a) is then applied as:

$$\varepsilon_c = b\sigma^n \tag{3b}$$

where $b = A\, t^m$. Thus, the gradient n and the intercept b allows a calculation of the coefficient $A = b/\, t^m$ for each of the chosen times. Taking the lines to be parallel when identifying a single n - value required that an average A - value be employed in the Nutting law. A graphical procedure was applied to the results for each composite. A sample of the analysis leading to the A, E, m, n and q - values within eqs(1) and (2) will follow in section 5.

4. Non-Linear Superposition Principles
In an incremental load test without periods of recovery, the Nutting law may be applied within either of two superposition schemes. The time and stress functions are generalised to give the creep strain, i.e. an equation describing a monotonic creep curve of a *non-linear solid*, as:

$$\varepsilon_c = \phi(t)\psi(\sigma) \tag{4}$$

Now let increasing total stress levels σ_1, σ_2, σ_3, σ_4 etc, be applied at respective times $t = 0, t_1, t_2, t_3$ etc, as shown in Fig. 1.

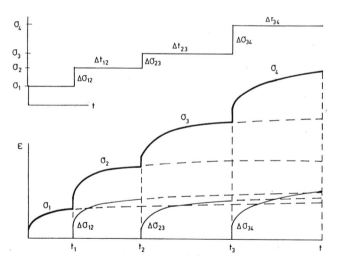

Figure 1 Superposition of creep responses to incremental loading

By the *linear superposition principle*, the net creep strain at time t is given as:

$$\varepsilon(t) = \phi(t)\psi(\sigma_1) + \phi(t - t_1)\psi(\sigma_2 - \sigma_1) + \phi(t - t_2)\,\psi(\sigma_3 - \sigma_2) + \phi(t - t_3)\psi(\sigma_4 - \sigma_3) + ... \tag{5a}$$

$$\varepsilon(t) = \phi(t)\psi(\sigma_1) + \phi(t - t_1)\psi(\Delta\sigma_{12}) + \phi(t - t_2)\,\psi(\Delta\sigma_{23}) + \phi(t - t_3)\psi(\Delta\sigma_{34}) + ... \tag{5b}$$

Equation (5b) is simplified for stepped loading with equal increments in stress: $\Delta\sigma = \sigma_1 = \Delta\sigma_{12}$ $= \Delta\sigma_{23} = \Delta\sigma_{34}$ and in time: $\Delta t = t_1 = \Delta t_{12} = \Delta t_{23} = \Delta t_{34}$, as:

$$\varepsilon(t) = \sum_{i=1}^{N} \phi[t - (i - 1)\Delta t]\, \psi(\Delta\sigma) \tag{5c}$$

where i refers to the step number. For the Nutting functions $\phi(t) = A' t^m$ and $\psi(\sigma) = A''\sigma^n$, eqs(5a,b) are written respectively as:

$$\varepsilon(t) = A\, t^m \sigma_1^n + A(t - t_1)^m (\sigma_2 - \sigma_1)^n + A(t - t_2)^m (\sigma_3 - \sigma_2)^n + A(t - t_3)^m (\sigma_4 - \sigma_3)^n + \ldots \tag{6a}$$

$$\varepsilon(t) = At^m (\Delta\sigma)^n + A(t - \Delta t)^m (\Delta\sigma)^n + A(t - 2\Delta t)^m (\Delta\sigma)^n + A(t - 3\Delta t)^m (\Delta\sigma)^n + \ldots \tag{6b}$$

where $A = ab$. An instantaneous elastic strain accompanies each stress increment so that $\Delta\varepsilon_e = \Delta\sigma/E$ must be added to each term to give the total strain at time t. Alternatively, we may add a single elastic term $\varepsilon_e = N\Delta\sigma/E$ for N equal stress increments $\Delta\sigma$.

In the Persoz [5] non-linear superposition principle the absolute stress levels are used so that eq(6a) would become:

$$\varepsilon(t) = A\, t^m \sigma_1^n + A(t - t_1)^m (\sigma_2^n - \sigma_1^n) + A(t - t_2)^m (\sigma_3^n - \sigma_2^n) + A(t - t_3)^m (\sigma_4^n - \sigma_3^n) + \ldots \tag{7}$$

As the n-value increases within eqs(6a) and (7), so the latter provides the greater strain. Within eq(6b), the implication is that creep curves for each $\Delta\sigma$ are identical making it an easier scheme to apply geometrically. That is, the ordinates of the creep curve under $\Delta\sigma$ may be added to the creep strain at the current stress level in the manner of Fig. 1. For linear material $n = 1$ and both schemes reduce to Boltzmann's linear superposition principle [6].

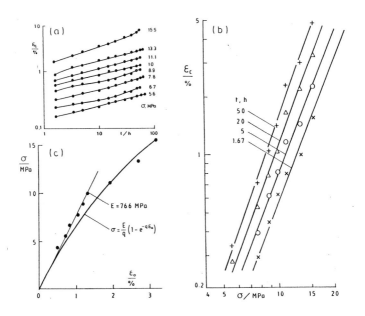

Figure 2 Application of Nutting's equation to creep in polypropylene

5. Results and Discussion
5.1 Single Polymers

A sample set of experimental data for polypropylene is shown in Figs 2a-c. Stress increments of 1.1 MPa were applied from a level of 5.6 to 15.5 MPa with intervening periods of recovery. Allowing recovery to occur results in loading that approximates to conducting a separate creep test at each stress level but with the economy and consistency of material response assured from within a single testpiece. In Fig. 2a, eq(3a) has been applied to the creep strain data. This shows that the m-values, i.e. the gradients, remain fairly constant at 0.24. Other single polymers of nylon and polyethylene were suitably represented by a power time law in which m lay betwen 0.15 and 0.25 (see Table 1). In Fig. 2b, eq(3b) has been applied to the creep strain data for the times shown lying within the creep interval. The isochronous data reveals that the stress exponent, i.e. the inverse gradient, remains constant over the stated stress range at $n = 2.55$. This is clearly in marked contrast to taking a unity value for n when linear visco-elasticity is assumed to apply at low stress levels . Other polymers revealed similar departures with their n-values lying between 1.26 and 2.97 (see Table 1). Figure 2c shows that the instantaneous strain for PP is Hookean initially with $E = 766$ MPa but thereafter conforms more closely to eq(2) with $q \approx 30$.

5.2 Parallel Polymer Composites
5.2.1 Parallel Unbonded

When two dissimilar polymer samples are clamped at their ends to carry the load they share the load and suffer the same strain during creep. However, during the recovery phase the strains are allowed to occur at different rates within the unbonded gauge lengths. A bowing occurs between the two materials at high stress levels which is subsequently eliminated in creep. We again assume that each creep curve so found is representative of that from a single application of each total stress level and persue a similar analysis as before. Thus, for an N66-LDPE parallel combination (see Figures 3a,b) we find from the gradients average time and stress exponent of $m = 0.18$ and $n = 2.52$ respectively.

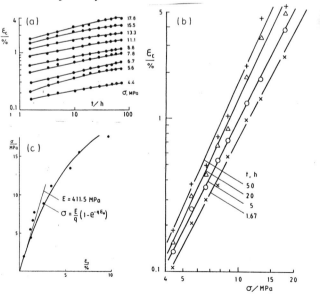

Figure 3 Application of Nutting's equation to creep in a parallel connection

Figure 3c shows that the instantaneous strain can be approximated with eq(2) in which $E = 411.5$ MPa and $q = 18.5$. Thus, for the individual polymers N66 and LDPE we have a creep strain dependence upon time and stress of the respective forms:

$$\varepsilon_c = 4.04 \times 10^{-6} \, \sigma^{2.97} \, t^{0.245}$$

$$\varepsilon_c = 1660 \times 10^{-6} \, \sigma^{1.263} \, t^{0.162}$$

while for their parallel combination:

$$\varepsilon_c = 22.34 \times 10^{-6} \, \sigma^{2.52} \, t^{0.18}$$

These show a trend in which A decreases as n and m increase. However, such trends were not apparent with the other parallel polymer arrangements. Table 1 shows that despite obtaining the exponents for each individual polymer, it was not obvious how to combine them theoretically to provide Nutting's exponents for the remaining parallel composites. At best the experiments reveal a range of n and m-values within which the exponents for each test will fall. It appears that the empirical representation of creep strain within the composite is best determined experimentally for the given combination of its constituents.

Table 1 Moduli and Exponents in the Nutting Law

Arrangement	Material	E/MPa	A/10^{-6}	n	m
SINGLE POLYMER	N66	780	4.05	2.97	0.245
	HDPE	238	174	2.26	0.16
	PP	766	15.3	2.55	0.236
	LDPE	51	1660	1.263	0.162
UNBONDED PARALLEL TYPE	N66+LDPE	412	22.34	2.52	0.180
	LDPE+HDPE	48	1315	1.52	0.153
	PP+HDPE	434	21.4	2.714	0.254
	N66+ HDPE	518	74.67	1.78	0.204
	LDPE+PP	164	747	1.21	0.21
	PP+N66	646	57.8	1.84	0.188
SERIES TYPE	N66+LDPE	200	87.73	2.633	0.253
	N66+HDPE	733	304	1.513	0.216
	LDPE+PP	156.3	365	1.952	0.205
	PP+N66	385	286	1.367	0.234

5.2.2 Parallel Bonded

The influence of a third layer of adhesive between two parallel polymers will alter both the creep and recovery behaviour of the composite. For example, compared to the unbonded case considered paragraph 5.2.1, the stiffening influence of a layer of cyanoacrylate adhesive between the N66 and LDPE altered the average time and stress exponents to $m = 0.256$ and $n = 1.36$. The instantaneous loading strain remained purely elastic with $E = 362$ MPa for a comparable stress range. A similar influence was found for PP and N66. For the remaining parallel combinations referred to in Table 1, the exponents n amd m were either both reduced (PP + HDPE) or increased (HDPE + N66 and HDPE + PP) by the bonding agent.

5.2.3 Series Connected

Four series combinations were tested for which a sample of data for N66 and polypropylene is given in Fig. 4a - c. The creep strain, as measured in both materials across the clamp, conforms to the Nutting relation in which m and n remained constant for the creep time interval under the loading that was applied. Figure 4c shows that the loading strain is purely elastic with $E = 385$ MPa. The stress level did not exceed 11 MPa and this composite remained approximately linearly visco-elastic with $n = 1.37$. This was not, however, a general feature of the lower load bearing capacity of a series arrangement since other materials show n-values approaching 3. Table 1 again shows that the behaviour of the constituent materials do not reflect within the composite in an obvious manner other than having n amd m values that lay within comparable ranges to a parallel arrangement but for a lesser stress range.

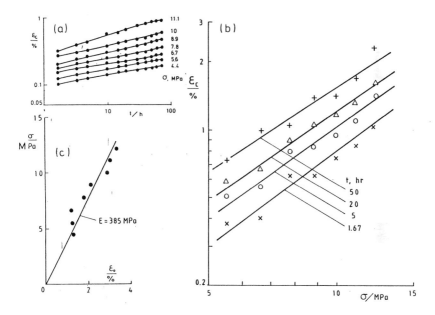

Figure 4 Application of Nutting's equation to creep in a series connection

6. Superposition Principle

An incremental loading was applied to single samples of N66 and HDPE and in a parallel combination for extended creep periods without the alternating periods of recovery. The results are shown in Fig. 5 for which the predictions from eq(6b) are overlaid. A creep time interval approaching 180 h was allowed so that the creep rate approached zero. The assumption is then made that eq (1) will represent the family of creep curves for the total applied stress levels enabling n and m to be found in the usual manner. This lends itself to a geometrical interpretation for finding exponents in the Nutting equation when loading incrementally in the absence of recovery. Referring to the inset figure 5 the creep curve under a total stress $\sigma_1 + \sigma_2$ is derived from stress increments σ_1 and σ_2 by summing the strain within similar time intervals. Thus for $\Delta t_1 = \Delta t_2$ the creep strain becomes $\varepsilon_{c1} + \varepsilon_{c2}$ at a similar time under $\sigma_1 + \sigma_2$. Added to this is the sum of the instantaneous strains: $\varepsilon_{o1} + \varepsilon_{o2}$ as shown, which may bear either a Hookean dependence upon σ or that described by eq(2).

360

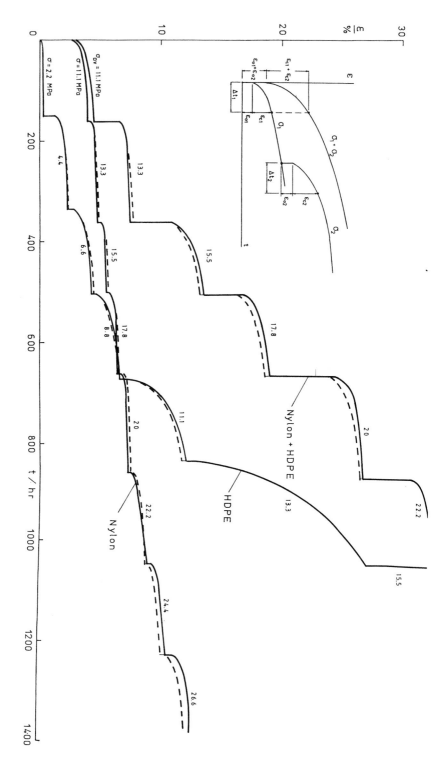

Figure 5 Application of superposition to incremental creep in N66 and HDPE

It follows that the n and m so found will describe identical incremental creep curves when the stress increments are the same. These may be translated to corresponding times at which the stress change occurs to produce the predictions shown as broken lines. The instantaneous strains, as derived from eq(2), have been added to the creep strains. The accuracy shown reflects the geometrical similarity implied by this linear superposition method. Clearly, for higher stress levels in HDPE this similarity appears to break down. However, the accompanying loss in cross-sectional area promotes an inequality in the true stress increments even though the loads are incremented by similar amounts. Equation (7) would be a more appropriate non-linear superposition principle to employ where there is not a geometrical similarity between creep curves when the true stress increments are maintained constant.

7. Conclusion

The Nutting relation has proven to be a useful means of representing the creep observed in simple polymer composites. In principle this phenomenological approach can be employed to describe creep in any composite without the need to account for contributions from its constituents whether they be linear or non-linear in nature. The stress exponents found show that most polymers and composites display non-linear visco-elasticity. Moreover it appears, from multi-step loading programmes on a single testpiece of a given polymer, that the time and stress exponents remain sensibly constant over its useful stress range. This concurs with the implicit assumption of a geometric similarity in the individual creep curves which makes possible the extension of a Boltzmann's linear superposition principle to a non-linear visco-elastic solid.

References

1. Lockett, F.J. Non-linear visco-elastic solids, Academic Press, 1972
2. Ogorkiewicz, R.M. Engineering Propreties of Thermoplastics, Wiley Interscience, 1976.
3. Nutting, P.G. Proc ASTM 21, 1921, 1162-1171.
4. Buchdahl, R. and Nielsen, L.E. Jl Applied Physics, 22, 1951, 1344-1349.
5. Persoz, B. Cashier Groupe Franc. Etudes Rheol, 1959, 4(1), 41-44.
6. Boltzmann, L. Progress Ann Physik, 7, 1876.

Appendix - Influence of ε_0 upon m

The initial estimate of the instantaneous strain ε_0 is crucial to the determination of the time exponent m. Normally one seeks to establish ε_0 for time $t = 0$ but obviously some time will elapse before the onset of creep strain. Ideally one would need to capture the initial displacement with a rapid chart speed so that an instantaneous value can be approximated with greater accuracy. Applying a pre-determined displacement rate to such a recording would ensure consistency when finding ε_0 for many creep curves provided the loading rate is constant. However, the loading rate will not be a constant when lowering a number of deadweights incrementally on to a hanger by hand in order to attain the applied stress level. Here the best that can then be done is to use one's judgement as to when time-dependent creep begins giving some assessment of the influence of error in making this judgement. The following method is suggested. Let us assume two descriptions from eq(3a) of the same creep curve resulting from different estimates of ε_0. On the double logarithmic plots in Figures 6a,b this has resulted in different gradients m amd m' as shown. Note that different intercept values: $\log a$ and $\log a'$, are made with the ordinate corresponding to $t = 1s$, where $\log t = 0$.

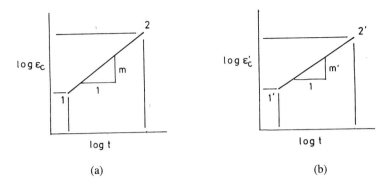

(a) (b)

Figure 6 Geometrical interpretations of the creep law $\varepsilon_c = \varepsilon - \varepsilon_0 = at^m$

Taking two points on the lines for similar times t_1 and t_2 the gradients become:

$$m = \frac{\log \varepsilon_{c2} - \log \varepsilon_{c1}}{\log t_2 - \log t_1} \tag{8a}$$

$$m' = \frac{\log \varepsilon'_{c2} - \log \varepsilon'_{c1}}{\log t_2 - \log t_1} \tag{8b}$$

Dividing eq(8b) by (8a) gives:

$$\log\left(\frac{\varepsilon'_{c2}}{\varepsilon'_{c1}}\right) = \frac{m'}{m} \log\left(\frac{\varepsilon_{c2}}{\varepsilon_{c1}}\right) = \frac{m'}{m} \log\left(\frac{\varepsilon_2 - \varepsilon_o}{\varepsilon_1 - \varepsilon_o}\right) \tag{9}$$

Noting that the total stains ε_1 and ε_2 at the respective times t_1 and t_2 in each of Figs 6a and 6b must be the same, the creep strains within the first term of eq(9) become $\varepsilon'_{c1} = \varepsilon_1 - \varepsilon'_o$ and $\varepsilon'_{c2} = \varepsilon_2 - \varepsilon'_o$. With these substitutions the first and third terms within eq(9) leads to a dependence relation between the instantaeous strain and the time exponent for the two levels of total strain:

$$\varepsilon'_o = \frac{\varepsilon_2 - \left(\frac{\varepsilon_2 - \varepsilon_o}{\varepsilon_1 - \varepsilon_o}\right)^{\frac{m'}{m}} \varepsilon_1}{1 - \left(\frac{\varepsilon_2 - \varepsilon_o}{\varepsilon_1 - \varepsilon_o}\right)^{\frac{m'}{m}}} \tag{10a}$$

where from eq(9):

$$\frac{m'}{m} = \frac{\log\left(\frac{\varepsilon_2 - \varepsilon'_o}{\varepsilon_1 - \varepsilon'_o}\right)}{\log\left(\frac{\varepsilon_2 - \varepsilon_o}{\varepsilon_1 - \varepsilon_o}\right)} \tag{10b}$$

Equations(10a,b) allow the correct estimate of ε_o for when $m'/m = 1$.

Deformation Mechanisms in Metal Matrix Composites at elevated temperatures

H. Iwasaki[1], T. Mori[1], M. Mabuchi[2] and K. Higashi[3]

[1]College of Engineering, Department of Materials Science and Engineering, Himeji Institute of Technology, Shosha, Himeji, Hyogo 671-2201, Japan

[2]National Industrial Research Institute of Nagoya, Hirate-cho, Kita-ku, Nagoya 462-8510, Japan

[3]College of Engineering, Department of Metallurgy and Materials Science, Osaka Prefecture University, Gakuen-cho, Sakai, Osaka 599-8531, Japan

Abstract

Tensile tests were carried out over the wide temperature range of 713 ~ 843 K and the deformation characteristics and the fracture mechanisms were investigated for the high strain rate superplastic Si_3N_{4p}/Al-Cu-Mg (2124) composite. Analysis through the threshold stress concept revealed that the deformation behavior can be divided into three regions from the viewpoint of the activation energy: Region I of 713 ~ 758 K where the activation energy is 97 kJ/mol, Region II of 773 ~ 803 K where the activation energy is 943 kJ/mol, and Region III of 818 ~ 843 K where the activation energy is 146 kJ/mol. The high values in Regions II and III are attributed to the presence of a liquid phase. The analysis of the activation energy suggests that the mechanisms of high strain rate superplasticity in a solid state including no liquid for the composites is the same as those for metals. The deformation characteristics, however, are drastically changed by the presence of a liquid phase.

1 Introduction

It has been demonstrated [1-11] that many aluminum matrix composites with discontinuous reinforcement materials exhibit superplastic behavior. In particular, some composites showed superplasticity at strain rates greater than 10^{-2} s^{-1}. High strain rate superplasticity is very attractive for commercial applications because one of the major problems in the current superplastic forming technique is the very slow forming rates of typically $10^{-5} \sim 10^{-3}$ s^{-1}.

The deformation mechanisms of high strain rate superplasticity in the composites are currently the subject of some debate. Mishra et al.[12-14] noted that the activation energy values for high strain rate superplasticity in the Al-Cu-Mg (2124) matrix composites are higher than that for lattice diffusion of the matrix, and an inverse grain size dependence and an inverse particle size dependence provide a good correlation. This indicates that the deformation mechanisms of high strain rate superplasticity for the composites are different from those of superplasticity for metals. Recently, Koike et al.[15] revealed by in-situ TEM observations that partial melting occurs at the matrix/reinforcement interfaces at elevated temperatures in some aluminum matrix composites exhibiting high strain rate superplasticity. Analyses based on extensive tensile testing data [16] revealed that the activation energy is equal to that for lattice diffusion of the matrix and the grain size exponent is close to 2 in the temperature range below the partial melting temperature, however, the activation energy is increased in the temperature range above the partial melting temperature for many high strain rate superplastic aluminum matrix composites except for the Al-Cu-Mg (2124) matrix composites.   Therefore, it is suggested that the deformation mechanisms of high strain rate superplasticity in a solid state including no liquid for the composites are the same as those for metals, however, the deformation characteristics are drastically changed by the presence of a liquid phase.   In addition, it was shown [8,17,18] that a maximum elongation is attained at the temperature close to or slightly above the partial melting temperature.   This suggests that the liquid phase plays an important role in the high strain rate superplasticity for the composites.   It was reported [19-21] that the development of cavities is limited by the presence of a liquid phase for the composites.   Therefore, a liquid phase may assist in the relaxation of the stress concentrations caused around reinforcements and thereby limit development of internal cavitation.   In order to understand the role of the liquid phase during high strain rate superplasticity, it is important to investigate the relationship between the deformation characteristics and the fracture mechanisms in the states including liquid and no liquid.

As already mentioned, there is disagreement in the activation energy value for the high strain rate superplasticity between the Al-Cu-Mg matrix composites and other aluminum matrix composites [16].   For the Al-Cu-Mg alloys, pseudo-ternary eutectics consisting of Al, Mg and Cu are often formed.   Complicated phases consisting of Al, Mg and Cu may give rise to a very small volume of liquid in the Al-Cu-Mg matrix composites which is too small to be measured by DSC.   Therefore, it is important to reassess the activation energy in the Al-Cu-Mg matrix composites based on the testing data over a wider temperature range.   In the present paper, tensile tests are carried out over the wide temperature range of 713 ~ 843 K and the deformation characteristics and the fracture mechanisms are investigated in the states including liquid and no liquid for a high strain rate superplastic Si_3N_4pAl-Cu-Mg (2124) composite.

2   Experimental procedure

A fine-grained Al-Cu-Mg (2124) matrix composite reinforced with 20 vol.% Si_3N_4 particulates was processed by hot extrusion.   The diameter of the Si_3N_4 particulates is less than 1 μm.   A schematic illustration of the fabrication processes is shown in Fig. 1.

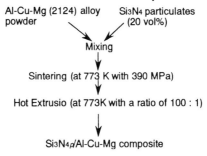

Fig. 1 Schematic illustration of the fabrication processes for the fine-grained Al-Cu-Mg (2124) matrix composite reinforced with 20 vol.% Si_3N_4 particulates

The extrusion temperature was 773 K.   In a previous study [9], the fine-grained Al-Cu-Mg (2124) matrix composites reinforced with 20 vol.% Si_3N_4 were processed through a similar fabrication procedure.   However, the composite used in the present investigation is not the same as the one in the previous study [9].   The $Si3N4p$ particulates were distributed with reasonably uniformly.   No cracks and no cavities were observed prior to straining.   An apparent mean matrix grain size was investigated from a measurement of more than 150 grains in the samples annealed for 1.8 ks at a given temperature, and the mean matrix grain size was given by multiplication of the apparent mean matrix grain size by a constant (=1.74).   The grain size is listed in Table 1.   The DSC experiment [8] was carried out for the as-extruded composite.   The heating rate was 10 K/min.   The weight of the sample for the DSC investigation was 30.1 mg.

Table 1 The variation in grain size with temperature

Temperature (K)	Grain Size (μm)
713	1.50
723	1.50
733	1.50
743	1.50
758	1.50
773	1.55
783	1.82
788	1.95
793	2.22
803	2.35
818	2.75
848	3.54

Tensile specimens with a gauge length of 5 mm and a gauge diameter of 2.5 mm were machined from the as-extruded bars.   Two types of tensile tests were conducted.   The first type of tensile test was a change-in-strain-rate test, which was carried out over the wide temperature range of 713 ~ 843 K to determine the stress exponent and the activation energy for plastic flow.   This test was carried out over small strain increments.   The second type of tensile test was a constant stress test, which was carried out at 723, 773, 783, 793 and 843 K to investigate the elongation and fracture mechanisms.   The tensile axis was positioned parallel to the extrusion direction and the samples required about 1.8 ks to equilibrate at the test temperature prior to straining for all tests.

3   Results

3.1 DSC investigation

The DSC experimental data from 650 to 873 K are shown in Fig. 2. An almost flat area was found, and then a sharp endothermic peak and finally a continuous endothermic curve appeared.   The sharp endothermic peak is attributed to partial melting at the matrix/reinforcement interfaces [15].   The partial melting temperature estimated from
The sharp endothermic peak is 782 K, which is determined from the intercept of the two dottedlines shown in Fig. 2.   However, inspection of the DSC data in Fig. 2 revealed that a

small endothermic reaction starts to occur at 752 K. This indicates that very small partial melting starts to occur at about 752 K, which is too small to recognize its reaction as a sharp endothermic peak in the DSC investigation. This will be discussed later. The final continuous endothermic reaction probably results from melting in the interior of the grains.

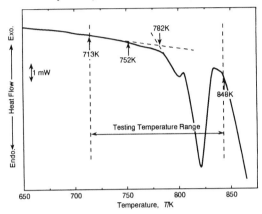

Fig. 2 The DSC experimental data from 650 to 873 K.

3.2 Deformation characteristics

The variation in strain rate as a function of stress from 713 ~ 843 K is shown in Fig. 3.

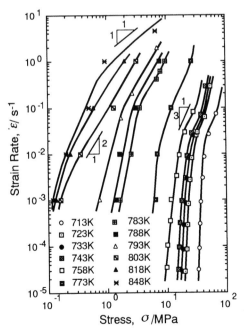

Fig. 3 The variation in strain rate as a function of stress at 713 ~ 843 K.

The strain rate increases with increasing stress at all temperatures. In the equation $\dot{\varepsilon} = A\sigma^n$, n is the slope of the curve $(= d(\log\dot{\varepsilon})/d(\log\sigma))$, where $\dot{\varepsilon}$ is the strain rate, A is a constant incorporating structure and temperature dependencies, σ is the stress and n is the stress exponent. The stress exponent values are very high in a low strain rate range below 10^{-2} s^{-1}, in particular, the stress exponent values are $10 \sim 20$ at the low temperatures of $713 \sim 773$ K. However, the stress exponent decreases with increasing stress and the minimum value is about 3 at $713 \sim 773$ K, about $2 \sim 2.5$ at $783 \sim 818$ K and about 1 at 843 K. It is noted that the minimum value of the stress exponent at each temperature tends to decrease with increasing temperature and a very low stress exponent of about 1 is attained at 843 K. The fact that $n = 1$ suggests that Newtonian flow or lubricated flow occurs at 843 K.

The stress exponent values tended to increase with decreasing stress. This is attributed either to the presence of a threshold stress or to deformation mechanism changes. Many experimental results [13,14,22-24] showed that a change in the strain rate sensitivity exponent results from a threshold stress, not the deformation mechanism changes for the high strain rate superplastic materials. A threshold stress may be estimated by extrapolation to zero strain rate for a line which the data give as σ versus $\dot{\varepsilon}^{1/n}$ on a double-linear scale [25].

Mabuchi and Higashi [24] showed the plots of σ versus $\dot{\varepsilon}^{1/2}$, $\dot{\varepsilon}^{1/3}$, $\dot{\varepsilon}^{1/5}$ and $\dot{\varepsilon}^{1/8}$ and they noted that the value of $n = 2$ is correct for the stress exponent for superplastic deformation, though the apparent value of the stress exponent is 3. In the present investigation, the minimum values of the stress exponent are about 3 at $713 \sim 773$ K, about $2 \sim 2.5$ at $783 \sim 818$ K and about 1 at 843 K. Hence, the threshold stresses were calculated from the plots of σ versus $\dot{\varepsilon}^{1/2}$ at $713 \sim 818$ K and from the plot of σ versus $\dot{\varepsilon}$ at 843 K. The plots of σ versus $\dot{\varepsilon}^{1/2}$ at $713 \sim 818$ K are shown in Fig. 4.

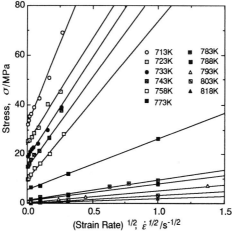

Fig. 4 The plots of σ against $\dot{\varepsilon}^{1/2}$ at $713 \sim 818$ K.

The linearity of the data in this plot indicates that the value of $n = 2$ is correct for the stress exponent from $713 \sim 818$ K. It can be seen from Fig. 4 that the threshold stress decreases with an increase in temperature. A similar trend has been reported in previous papers [13,14,22-24].

The activation energy for superplasticity in the composites can be analyzed based on the data shown in Figs. 3 & 4. When $(\sigma - \sigma_{th})/G$ is constant, the activation energy for superplasticity is given by [16]

$$Q = \frac{d\ln[\dot{\varepsilon}(d/b)^p]}{d\ln(1/T)} \tag{1}$$

where σ_{th} is the threshold stress, G is the shear modulus, Q is the activation energy for superplasticity, b is Burgers vector, d is the grain size, p the grain size exponent and T is the absolute temperature. Many theoretical models on superplasticity [26-29] predict $p = 2$. Recently, it was shown that the grain size exponent is experimentally about 2 for high strain rate superplasticity in the Si3N4w/Al-Mg-Si composite. Hence, p was taken to be 2 in the present investigation. The variation in $\dot{\varepsilon}(d/b)^2$ as a function of $1/T$ is shown in Fig. 5.

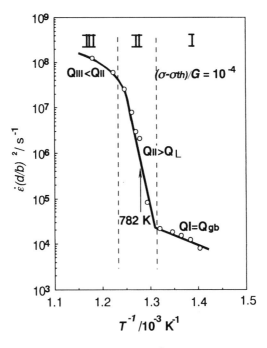

Fig. 5 The variation in $\dot{\varepsilon}(d/b)^2$ as a function of $1/T$.

It should be noted that the superplastic behavior can be divided into three regions from the viewpoint of the activation energy: Region I of $713 \sim 758$ K where the activation energy is 97 kJ/mol, Region II of 773 ~ 803 K where the activation energy is 943 kJ/mol, and Region III of 818 ~ 843 K where the activation energy is 146 kJ/mol. The value in Region I is close to that for the grain boundary diffusion of aluminum (= 84 kJ/mol [30]. On the other hand, the activation energy in Region II is much higher than that for lattice diffusion of aluminum (= 142 kJ/mol [30]), and the activation energy value decreases in Region III. It should be noted that a drastic change in the activation energy occurs from 758 ~ 773 K which is lower than the partial melting temperature of 782 K as determined by DSC.

A constant stress tensile test was conducted to failure in the temperature range of 723 ~ 843 K. The applied stress is determined such that the effective stress, $\sigma - \sigma_{th}$, may be almost the same and the stress exponent may be low during the testing conditions based on the data in Figs. 3 & 4. The results of the test are listed in Table 2. It is experimentally and theoretically [31] recognized that an elongation increases with increasing m. In the present investigation, however, the elongation is very low (= 18 %) at 843 K where a high m of about 1 is attained. In addition, a large elongation of 180 % is attained at 773 K. On the other hand, an elongation at 723 K is very low (= 24 %), though a relatively high m of about 0.3 is attained at both temperatures of 723 and 773 K. It should be noted that a high m does not always give rise to a large elongation for the composite. A large elongation of 550 % is attained at 783 K which is very close to the temperature (= 782 K) where the sharp endothermic peak appears in the DSC investigation. On the other hand, the elongation decreases at the higher temperature of 793 K. Premature fracture at 793 and 843 K is probably associated with too much liquid.

Table 2 The mechanical properties by constant stress tensile tests in the Si3N4p/Al-Cu-Mg composite

Temperature (K)	Applied Stress (MPa)	Threshold Stress (MPa)	$\sigma - \sigma_{th}$	Strain Rate (s^{-1})	m value	Elongation (%)
723	32.7	26.1	6.6	3×10^{-2}	0.3	24
773	12.0	5.4	6.6	1×10^{-1}	0.3	180
783	8.0	1.4	6.6	6×10^{-1}	0.4	550
793	7.2	0.6	6.6	1.9	0.4	278
843	6.6	0.07	6.53	4.4	1	18

4   Discussion

In general, the activation energy for superplasticity in metals is equal to that for the lattice diffusion or grain boundary diffusion [32]. However, Mishra et al. [12-14] showed that the activation energy for high strain rate superplasticity in the Al-Cu-Mg matrix composites is higher than that for lattice diffusion and they noted that the high value is attributed to interface diffusion. However, it was reported [16] that the activation energy for high strain rate superplasticity is equal to that for lattice diffusion in a temperature range below the partial melting temperature, and the apparent value of the activation energy is drastically increased by the presence of a liquid phase for many aluminum matrix composites except for the Al-Cu-Mg matrix composites. In the present investigation, the activation energy in Region I, which is the relatively low temperature range of 713 ~ 758 K, is close to that for the grain boundary diffusion of aluminum. The effective diffusion concept [13,24] indicates that the dominant diffusion process during superplasticity over a low temperature range tends to be grain boundary diffusion. Therefore, the fact that the activation energy in Region I is close to that for the grain boundary diffusion of aluminum is probably because of the low testing temperature range of 713 ~ 758 K. Analyses of the activation energy in the present and previous investigations [16] revealed that the activation energy for superplasticity in a solid state including no liquid for all the aluminum matrix composites is equal to that for the lattice diffusion or grain boundary diffusion of the matrix. This indicates that the deformation mechanisms of high strain rate superplasticity in the solid state including no liquid for the composites are the same as those for metals.

It should be noted that the drastic change in the activation energy value occurs at 758 ~ 773 K which is lower than the partial melting temperature of 782 K as determined by DSC. It was shown that the apparent value of the activation energy for superplastic flow is drastically increased by the presence of a liquid phase [16]. It is therefore suggested that for the Al-Cu-Mg matrix composites, partial melting occurs at temperatures lower than the temperature where a sharp endothermic peak appears in the DSC investigation. It is note that a small endothermic reaction starts to occur at 752 K as shown in Fig. 2. This suggests that very small partial melting occurs at about 752 K, but it is too small to recognize the reaction as a sharp endothermic peak in the DSC investigation. In the present investigation, the composite was processed by hot extrusion at 773 K. If partial melting occurs at about 752 K, partial melting should occur during hot extrusion at 773 K. Recently, Jeong et al. [33] investigated the nature of the matrix/reinforcement interfaces of the high-strain-rate superplastic Si3N4w/Al-Mg-Si composite using high-resolution electron microscopy and they found the formation of new phases, which were epitaxially grown, with a FCC structure at the interfaces. These reaction phases are attributed to partial melting. Hence, it is worthwhile to investigate the interfaces in the as-extruded sample for evidence of partial melting occurring during hot extrusion at 773 K.

5 Summary

Tensile tests were carried out over the wide temperature range of 713 ~ 843 K and the deformation characteristics and the fracture mechanisms were investigated in the high str... rate superplastic Si3N4p/Al-Cu-Mg (2124) composite. Analysis through the threshold stress concept revealed that the deformation behavior can be divided into three regions from the viewpoint of the activation energy: Region I of 713 ~ 758 K where the activation energy is 97 kJ/mol, Region II of 773 ~ 803 K where the activation energy is 943 kJ/mol, and Region III of 818 ~ 843 K where the activation energy is 146 kJ/mol. The high values in Regions II & III are attributed to the presence of a liquid phase. The analysis of the activation energy suggests that the mechanisms of high strain rate superplasticity in a solid state including no liquid for the composites is the same as those for metals. By the presence of a liquid phase, however, the deformation characteristics are drastically changed.

References

1. Nieh, T.G., Henshall, C.A., and Wadsworth, J., 1984, Superplasticity at high strain rates in a SiC whisker reinforced Al alloy, Scripta Metall., Vol. 18, pp. 1405- 1408.

2. Mahoney, M.W., and Ghosh, A.K., 1987, Superplasticity in a high strength powder aluminum alloy with and without SiC reinforcement, Metall. Trans. A, Vol. 18A, pp. 653- 661.

3. Pilling, J., 1989, Superplasticity in aluminium base metal matrix composites, Scripta Metall., Vol. 23, pp. 1375- 1380.

4. Imai, T., Mabuchi, M., Tozawa, Y., and Yamada, M., 1990, Superplasticity in β-silicon nitride whisker-reinforced 2124 aluminium composite, J. Mater. Sci. Lett., Vol. 9, pp. 255- 257.

5. Xiaoxu, H., Qing, L., Yao, C.K., and Mei, Y., 1991, Superplasticity in a SiCw-6061 Al composite, J. Mater. Sci. Lett., Vol. 10, pp. 964- 966.

6. Mabuchi, M., Higashi, K., Okada, Y., Tanimura, S., Imai, T., and Kubo, K., 1991, Very high strain-rate superplasticity in a particulate Si3N4/6061 aluminum composite, Scripta Metall. Mater., Vol. 25, pp. 2517- 2522.

7. Higashi, K , Okada, T., Mukai, T., Tanimura, S., Nieh, T.G., and Wadsworth, J., 1992, Superplastic behavior in a mechanically alloyed aluminum composite reinforced with SiC particulates, Scripta Metall. Mater., Vol. 26, pp.185- 190.

8. Mabuchi, M., and Higashi, K., 1994, Thermal stability and superplastic characteristics in Si3N4/Al-Mg-Si composties, Mater. Trans. JIM, Vol. 35, pp.399-405.

9. Mabuchi, M., Higashi, K., and Langdon, T.G., 1994, An investigation of the role of a liquid phase in Al-Cu-Mg metal matrix composites exhibiting high strain rate superplasticity, Acta Metall. Mater., Vol. 42, pp. 1739- 1745.

10. Chan, K.C., Han, B.Q., and Yue, T.M., 1996, Constitutive equations for superplastic deformation of SiC particulate reinforced aluminum alloys, Acta Mater., Vol. 44, pp. 2515- 2522.

11. Han, B.Q., and Chan, K.C., 1997, High-strain-rate superplasticity of an Al6061-SiCw composite, Scripta Mater., Vol. 36, pp. 593- 598.

12. Mishra, R.S., and Mukherjee, A.K., 1991, On superplasticity in silicon carbide reinforced aluminum composites, Scripta Metall. Mater., Vol. 25, pp. 271-275.

13. Mishra, R.S., Bieler, T.R., and Mukherjee, A.K., 1995, Superplasticity in powder metallurgy aluminum alloys and composites, Acta Metall. Mater., Vol. 43, pp. 877-891.

14. Mishra, R.S., Bieler, T.R., and Mukherjee, A.K., 1997, Mechanism of high strain rate superplasticity in aluminium alloy composites, Acta Mater., Vol. 45, pp. 561- 568.

15. Koike, J., Mabuchi, M., and Higashi, K., 1995, In situ observation of partial melting in superplastic aluminum alloy composites at high temperatures, Acta Metall. Mater., Vol. 43, pp. 199- 206.

16. Mabuchi, M., and Higashi, K., 1996, Activation energy for superplastic flow in aluminum matrix composites exhibiting high-strain –rate superplasticity, Scripta Mater., Vol. 34, pp. 1893-1897.

17. Mabuchi, M., and Higashi, K., 1994, Thermal stability in a superplastic Si3N4(p)/Al-Mg composite, Mater. Sci. Eng., Vol. A179/180, pp. 625- 627.

18. Higashi, K., Nieh, T.G., Mabuchi, M., and Wadsworth, J., Effect of liquid phases on the tensile elongation of superplastic aluminum alloys and composties, Scripta Metall. Mater., Vol. 32, pp. 1079- 1084.

19. Iwasaki, H., Mabuchi, M., Higashi, K., and Langdon, T.G., 1996, The development of cavitation in superplastic aluminum composites reinforced with Si3N4, Mater. Sci. Eng., Vol. A208, pp. 116- 121.

20. Iwasaki, H., Mabuchi, M., and Higashi, K., 1996, Cavitation and fracture in high strain rate superplastic Al alloy/Si3N4(p) composites, Mater. Sci. Tech., Vol. 12, pp. 505-512.

21. Iwasaki, H., Mabuchi, M., and Higashi, K., 1997, The role of liquid phase in cavitation in a Si3N4p/Al-Mg-Si composite exhibiting high-strain-rate superplasticity, Acta Mater., Vol. 45, pp. 2759- 2764.

22. Higashi, K., Nieh, T.G., and Wadsworth, J., 1995, Effect of temperature on the mechanical properties of mechanically-alloyed materials at high strain rates, Acta Metall. Mater., Vol. 43, pp. 3275- 3282.

23. Mabuchi, M., and Higashi, K., 1995, Constitutive equation of a superplastic Al-Zn-Mg composite reinforced with Si3N4 whisker, Mater. Trans. JIM, Vol. 36, pp. 420-425.

24. Mabuchi, M., and Higashi, K., 1996, High-strain-rate superplasticity in magnesium matrix composites containing Mg2Si particles, Philos. Mag. A, Vol. 74, pp. 887-905.

25. Mohamed, F.A., 1983, Interpretation of superplastic flow in terms of a threshold stress, J. Mater. Sci., Vol. 18, pp. 583 -592.

26. Mukherjee, A.K., 1971, The role controlling mechanism in superplasticity, Mater. Sci. Eng., Vol. 8, pp. 83-89.

27. Gifkins, R.C., 1976, Grain-boundary sliding and its accommodation during creep and superplasticity, Metall. Trans. A, Vol. 7A, pp. 1225-1232.

28. Burton, B., 1983, The characteristic equation for superplastic flow, Philos. Mag. A, Vol. 48, pp. L9- L13.

29. Hayden, H.W., Floreen, S., and Goodell, P.D., 1972, The deformation mechanisms of superplasticity, Metall. Trans., Vol. 3, pp. 833- 842.

30. Frost, H.J., and Ashby, M.F., 1982, Deformation-Mechanism Maps, Pergamon Press, Oxford, p. 21.

31. Hart, E.W., 1967, Theory of the tensile test, Acta Metall., Vol. 15, pp. 351-355.

32 Sherby, O.D., and Wadsworth, J., 1989, Superplasticity-recent advances and future directions, Prog. Mater. Sci., Vol. 33, pp. 166-221.

33. Jeong, H.-G., Hiraga, K., Mabuchi, M., and Higashi, K., Interface structure of Si3N4-whisker-reinforced Al-Mg-Si alloy (Al alloy 6061) composites studied by high-resolution electron microscopy, Philos. Mag. Lett., Vol. 74, pp. 73 -80.

SUPERALLOYS I

Modelling and Validation of Anisotropic Creep Deformation in Single Crystal Superalloys: Variable and Multiaxial Loading

H.C.Basoalto, M.G.Ardakani, B.A.Shollock and M.McLean
Department of Materials
Imperial College of Science, Technology and Medicine
Prince Consort Road, London SW7 2BP

Abstract

Anisotropic creep in single crystal superalloys under steady isothermal and uniaxial loading has previously been described by a number of crystallographic slip-based models(1,2,3); the Imperial College approach, based on the concepts of *Continuum Damage Mechanics* (CDM), is particularly effective in accounting for the progressive increase in creep rate that occurs over most of the creep life of the material. This model has been extended to account for deformation under (a) changing uniaxial tensile stresses and variable temperatures, (b) multiaxial loading and (c) through-zero cyclic axial loading. Theoretical simulations are compared with experimental measurements on the Second Generation SX alloy CMSX4 for step-changes in load and for Bridgman notch creep specimens. Characterisation of the spatial heterogeneity of lattice rotation resulting from creep, using Electron Back Scatter Diffraction, allows the active slip systems in multiaxial stresses to be identified. These are consistent with the assumptions of the model.

Introduction

The development of single crystal superalloys for use as blading for gas turbines has resulted in a significant increase in the temperature capability of the materials, that has led to improved engine performance. However, the strong crystallographic anisotropy of these alloys has created challenges in effectively representing their creep performance and, of more importance, in extrapolating from a realistic experimental database to account for likely service performance. Among the practical requirements are:

Design The development of computer aided design methods requires an effective constitutive description of stress/strain evolution, rather than conventional stress rupture and minimum creep rate parametric representations of creep performance. This is difficult for isotropic materials; the introduction of crystallographic anisotropy complicates the issue significantly.

Life Prediction Gas turbines experience complex service cycles, which depend on their particular application – fighter aircraft, commercial air-transport, electricity generation. It is important to be able to estimate lives of components in the variable and multiaxial loading experienced in service.

Quality Control The developments in alloy chemistry that have given the improved creep performance have also made control of the single crystal processing more difficult. It is important to be able to establish the deviations from the specified crystal orientation that will give acceptable performance.

It is impractical to establish an experimental database that will describe stress/strain/time behaviour for all possible eventualities of crystal orientation, constant and variable stress/temperature, load/strain control, and multiaxiality.

Intensive research effort has been devoted to developing creep models that will allow predictions of these features from simple databases.

The present paper describes the extension of a model of anisotropic creep of single crystal superalloys that has been developed at Imperial College (and previously at NPL) to complex loading. It compares model predictions with experimental measurements for various types of variable and multiaxial loading. It compares the predicted changes in crystal orientation due to plastic strain with experimental measurements of microcrystallinty in order to validate the model.

The Creep Model

The details of the basic creep model have been presented previously (1,4) and will only be summarised briefly here. The creep rate is expressed in terms of two state variables. The first is a 'hardening' state variable, S, that describes primary creep as the partitioning of stress between the γ matrix and the γ' particles. The second state variable is a damage term, ω, that results in a progressive loss in creep strength (i.e. tertiary creep). For nickel-based superalloys the dominant damage has been shown to be accumulation of mobile dislocations (5).

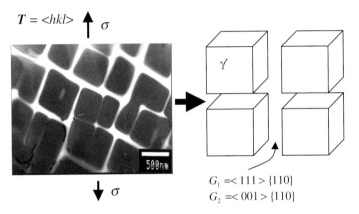

$$G_1 = <111>\{110\}$$
$$G_2 = <001>\{110\}$$

Figure 1 Basic geometry of a Nickel-based superalloy

If a single crystal superalloy is subject to a uniaxial stress σ in a direction $T = <hkl>$ (Figure 1), there will be a virtual glide force $f = \tau b$ on a dislocation, where τ is the shear stress acting on the allowed slip system and b is the Burger's vector. When the activate slip systems are denoted by the unit vectors $\left(m^{(k)}, n^{(k)}\right)$, with $m^{(k)}$ being the slip direction and $n^{(k)}$ is the slip plane normal, the shear stress on the kth slip system is given by

$$\tau^{(k)} = n_i^{(k)}\sigma_{ij}m_j^{(k)} \tag{1}$$

where σ_{ij} is the stress tensor. For single crystal superalloys we assume that two families of slip systems are activated: the octahedral system $\{111\}<1\bar{1}0>$ and cube system $\{001\}<110>$. There is evidence of the activation of a second slip system, $\{111\}<\bar{2}11>$ particularly at low creep temperatures (6,7); it would be straightforward to accommodate this in the model. The constitutive equations for anisotropic deformation developed by Ghosh et. al (1) and Pan et. al (4) for each slip system are:

$$\dot{\gamma}^{(k)} = \dot{\gamma}_0^{(k)}(1 - S^{(k)})(1 + \omega^{(k)})$$
$$\dot{S}^{(k)} = \dot{\gamma}_0^{(k)} H^{(k)}(1 - S^{(k)}/S_{ss}^{(k)})$$
$$\dot{\omega}^{(k)} = \beta^{(k)}\dot{\gamma}^{(k)} \tag{2}$$

for all $\{111\} < 1\,\overline{1}\,0 >$ and $\{001\} < 110 >$ slip systems. $\dot{\gamma}_0^{(k)}, H^{(k)}, S_{ss}^{(k)}$, and $\beta^{(k)}$ are the model parameters which differ for each family of slip systems.

Following previous studies, including those of Ghosh *et. al* (1) and Pan *et. al* (4), the model parameters have been determined for individual creep curves by least squares fitting procedures; the variation of each parameter with stress and temperature has then been determined. The model parameters are expressed as functions of stress and temperature and individual creep curves are generated from this optimised set of materials parameters.

Extension of the model to conditions of either partial or complete strain controlled deformation requires the partitioning between elastic and creep strain to be specified. This can be expressed for axial loading:

$$\dot{\sigma} = E_{<hkl>}(\dot{\varepsilon}_T - \dot{\varepsilon}_c) \tag{3}$$

$$\frac{1}{E_{<hkl>}} = S_{11} + [S_{44} - 2(S_{11} - S_{12})](h^2k^2 + k^2l^2 + l^2h^2)$$

where $E_{<khl>}$ is Young's modulus in a specified crystal direction $< hkl >$ (see Nye(8)), S_{ij} the compliance tensor, $\dot{\varepsilon}_T$ is the total strain rate, and $\dot{\varepsilon}_c$ is given by

$$\Delta\varepsilon = \frac{1}{2}\left[\Delta J + \Delta J^T\right] \tag{4}$$

$$\Delta J_{ij} = \sum_{s \in G} \Delta\gamma^{(k)} n_i^{(k)} m_j^{(k)} \tag{5}$$

where ΔJ_{ij} is the Jacobian tensor. If the appropriate boundary conditions are specified, the coupling of the constitutive equations (2) with equation (3) allows the simulation of any complex uniaxial mechanical deformation.

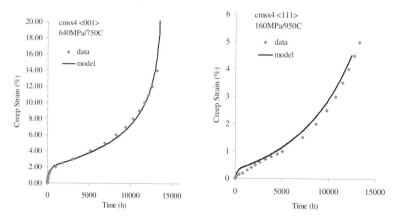

Figure 2 Comparison of calculated and measured creep curves

Calibration of the Model

The parameters of the model have been determined by analysis of a database of constant stress tensile creep curves in the temperature range 750 to 1050°C for specimens of CMSX4 with orientations within 6° of <001> and <111>. The minimum requirement of the model is to adequately represent these creep data. Figure 1 shows typical creep curve comparisons for two different test conditions. The model represents the experimental creep data well over the entire stress and temperature range for which data are available. Since the calculated curves are generated from an averaged parameter database and are not fits of the individual creep curve data to the model, the agreement between calculated and measured creep curves cannot be closer than the intrinsic scatter of the creep data, which is typically ±20% in life.

Model Predictions and Comparisons with Measurements.

In this section we calculate the creep deformation for crystal orientations and testing conditions that were not used in calibrating the model. These are true predictions, rather than fits of the model to curves, and indicate the scope for extrapolation beyond the range of the input data. In this context, we consider extrapolation to be extension to complex loading conditions, rather than to longer times.

(i). Complex Orientations

Since the strain is calculated by summing the shear strains on all allowed slip systems, the model can predict creep curves for axial stresses in arbitrary crystal directions. Figure 2 compares the calculated and measured curves for a near <110> specimen. The model predicts the creep behaviour well to 0.5% strain and adequately to 2% strain. However, it fails to account for the rapid acceleration in creep rate leading to failure at a relatively low strain (4% compared with >25% for <001> and <111>.)

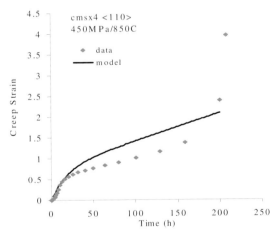

Figure 3 Simulation of creep curve for <110> specimen of CMSX4 tested at 450MPa and 850°C.

In the case of <001> and <111> axial loading, multiple slip systems are symmetrically activated with the result that the crystal orientation and specimen cross-sectional shapes are essentially invariant.   For low- and non-symmetric orientations there can be large crystal rotations and shape changes which are predicted by the model. Henderson et al.(9) and Pan et al.(10) have used these phenomena as a much more direct validation of the model than comparison of strain/time curves. It is clear that deformation on $\{111\}<1\bar{1}0>$ dominates for axial stresses within about $40°$ of <001>. The situation is less clear for other orientations; activation of multiple slip systems makes discrimination between $\{001\}<110>$ and $\{111\}<\bar{2}11>$ glide difficult. There is little evidence of dislocation activity on $\{001\}<110>$ and $\{111\}<\bar{2}11>$ is only significant at low temperature and high stresses where γ'cutting is prevalent. However, the dual glide system formalism appears to be required over the entire temperature range. The apparent cube slip is likely to result from constrained deformation in the narrow γ channels between the $\{001\}$ faces of the relatively rigid γ' cubes.

(ii).   **Step changes in stress and/or temperature.**

With models that express creep strain as a function of time, calculation for a constant stress/temperature condition is straightforward. However, if the loading conditions are altered it is necessary to introduce an additional rule to account for the change in strain rate at the point where the load changes. Translation between the creep curves for the two loading conditions at a constant strain (strain hardening) is normally assumed. In the CDM approach, the creep rate is an explicit function of the state variables $S^{(k)}$ and $\omega^{(k)}$; changes in loading condition can be taken into account without introducing additional rules. This becomes particularly important when changes in microstructure due to thermal processes make an important contribution to the creep behaviour.(11)

Figure 3 shows the calculated and measured strain/time plots for two <111> creep tests in which the test conditions were changed between low stress/high temperature and high stress/low temperature, but with the test conditions in the reverse order. The conditions have been selected to have quite differently shaped creep curves for constant stress/temperature testing. For the particular example shown, 780MPa is outside the range of stresses at 750°C in the database from which the model was calibrated, and consequently constitutes a true prediction. The comparison for the case where a high temperature creep phase was applied first is well within the range of experimental error. For the reverse order of loading, the agreement is less good. The main reason for this discrepancy is that the model in its present form is informed by relatively long-term creep data and is over-estimating the creep rate in the high stress/low temperature condition. It is of interest to compare the present predictions with those of the strain hardening and life fraction (Robinson Equation) estimates (Table 1). We have particularly chosen one condition that gives a distinct primary creep regime and one that is dominated by tertiary creep. The strain hardening construction and CDM calculation predict diametrically opposite effects. Strain hardening considers the effects leading to primary creep to be equally damaging to those in the low strain rate regime leading to tertiary creep. This ignores the athermal contribution to primary creep that is recoverable. The CDM approach is consistent with experimental evidence that creep damage in the secondary and tertiary stages is more deleterious. The present model represents the experimental trend to within $\pm20\%$ in life. Although not precisely accounting for the experimental observations, it is significantly better than the strain hardening and life fraction estimates.

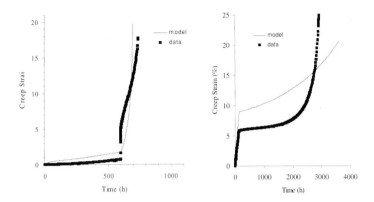

Figure 4 Step changes in stress and temperature for <111> specimen of CMSX 4: (a) 250MPa/950C to 780MPa/750C, and (b) 780MPa/750C to 300 MPa/900C.

Table 1: Cumulative life normalised to the Life-Fraction (Robinson Equation) expectation for step loaded creep tests: experimental results and model predictions.

Test Conditions	Experimental Result	Theoretical Expectation		
		Robinson Equation	Strain Hardening	CDM model
250MPa/950°C/600h + 780MPa/750°C to failure	0.612	1	1.213	0.515
780MPa/750°C/137h + 300MPa/900°C to failure	1.112	1	0.638	1.315

(iii).   **Multiaxial stresses**

Double notch Bridgman creep specimens with <001> and <111> axial orientations were tested at 850°C with net-section stresses in the range 600-850MPa. Figure 3 compares the times to rupture for uniaxial and the multiaxial creep tests of <001> and <111> single crystal of CMSX4 at 850°C. This clearly shows the effect of the multiaxial stress state is to extend the creep life by about an order of magnitude.

The crystallographic slip model, with parameters established by analysis of a uniaxial creep database, has been used to simulate the deformation of Bridgman notch specimens. The FE analysis was carried using the commercial package ABAQUS, and the model was interfaced using a modified creep user-subroutine. An example of the output in Figure 6 shows a transverse section through the minimum notch diameter for a <001> specimen tested at 400MPa/850°C at an early stage of creep. This shows that high strains have accumulated at the notch surface relaxing the stress there and transferring the load to the centre of the specimen. Moreover, there is considerable anisotropy of deformation around the perimeter of the notch.

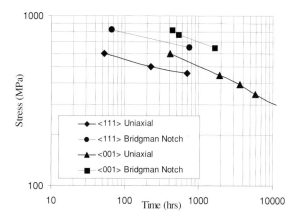

Figure 5 Lifetime data for CMSX4 <111> and <001> orientations under uniaxial and multiaxial loading conditions at 850°C.

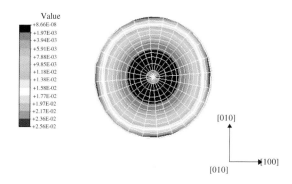

Figure 6 Bridgman notch simulation of accumulated creep strain for <001> orientation tested at 400MPa/850°C.

Longitudinal and transverse section of the uniaxial and multiaxial creep tested samples were examined using the EBSD (electron back-scattered diffraction) technique in conjunction with a JEOL 840 scanning electron microscope (SEM) equipped with SINTEF hardware to map the spatial distribution of crystal orientation over these sections. CHANNEL software was used to analyse the three Euler angles ϕ_1, ϕ, and ϕ_2, measured by indexing Kikuchi bands. These Euler angles are used by the software for displaying orientation data in the form of inverse pole figure (IPF), pole figure (PF) or orientation map (OM). Details are given in References (6,12).

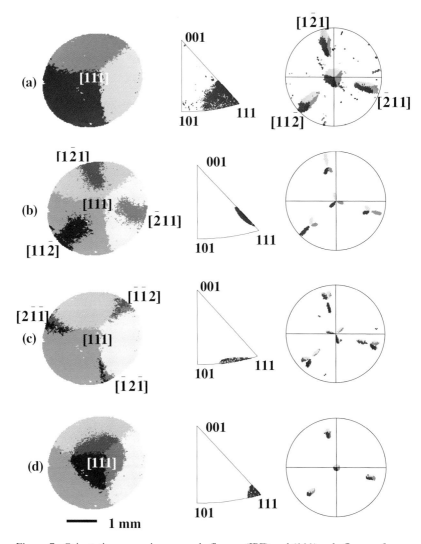

Figure 7. Orientation maps, inverse pole figures (IPF) and (111) pole figures for different areas of a transverse section, near the fracture surface, of a Bridgman-notch [111] single crystal tested at 850°C/820 MPa.. (For dark contrast areas.) (a) overall distribution, (b, c) near-notch regions and (d) central region.

Transverse sections through the minimum diameter of the notch were prepared by polishing one fracture surface until a flat surface was obtained which was electro-polished. Figure 7 shows the results of EBSD analysis of the transverse section of a <111> specimen tested at 820MPa/850°C. An orientation map, inverse pole figure (IPF) and (111) pole figure for the entire section are shown. The starting orientation was within 2° of <111> over the entire specimen. Uniaxial creep testing of <001> and <111> specimens shows no significant changes in average orientation, although a spread of $\pm 5^0$ can develop. By contrast, the Bridgman notch specimens show a deviation from the original <111> of $\pm 20^0$. The crystal rotations vary systematically around the circumference of the section. In transverse radii pointing towards <211> the rotation on the IPF is from <111> to <001> (Figure 7(b)) and for radii pointing to <011> the rotation is from <111> towards <101> (Figure 7(c)). At the centre of the section there is a small spread of orientations, but they are all within $\pm 5^0$ of <111> (see Figure 10(d)). The rotations observed are consistent with the assumed shear on (001) planes in <110> directions and would be equivalent to >25% local strain at the notch root. Similar observations on <001> specimens show large rotations with four-fold symmetry consistent with shear parallel to octahedral planes.

(iv). Fatigue and Creep –Fatigue interactions

Although the present model has been developed and calibrated for optimum description of relatively long-term creep behaviour, the same formalism can be used to calculate time dependent deformation during low cycle or thermo-mechanical fatigue. Other models, such as that of the Ecoles de Mines group(13), have a more sophisticated hardening evolution law specifically aimed at effective fatigue modelling. Moreover, these fatigue-oriented models are calibrated with high strain rate LCF data. The present form of the Imperial College model has a relatively simple hardening law, and work is in hand to extend the model to address this aspect. However, in its present form the model incorporates an interaction between hardening and damage development that gives an indication of how the fatigue and creep damages are likely to interact. The damage is taken to be non-reversible, so that it progressively increases during both tensile and compressive deformation; we set the strain softening damage variable to vary linearly with the absolute value of the creep rate:

$$\dot{\omega} = C|\dot{\varepsilon}| \qquad (7)$$

Figure 8(a) shows the calculated effect of various levels of creep strain accumulated at 400MPa/850ºC on the plastic strain range generated in a relatively mild LCF test. The virgin material essentially develops elastically; the plastic strain per cycle is predicted to increase by over an order of magnitude as a result of 10% creep strain. It

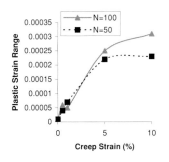
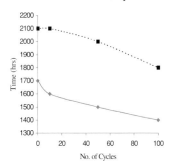

Figure 8(a) Effect of prior creep exposure at 400MPa/850ºC on the plastic strain range for $\dot{\varepsilon} = 3.6\ h^{-1}$ and $2\Delta\varepsilon_T = 0.008$ assuming non-reversible damage, $\dot{\omega} = C|\dot{\varepsilon}|$.

(b) Effect of strain-controlled cyclic loading on the time to reach 5% and 10% strain (400MPa/850ºC). $\dot{\varepsilon} = 3.6\ h^{-1}$ and $2\Delta\varepsilon_T = 0.008$.

is important to understand how the LCF response will change during the service life a material. Figure 8(b) shows the predicted effect of LCF cycles on the creep life of the alloy. In the case shown, 100 LCF cycles is predicted to reduce the times to achieve 5% or 10% creep strain by about 20%, which is about the same as the scatter in the creep data.

Further development and validation are in progress to establish a unified creep/LCF/thermo-mechanical deformation model.

Conclusions

1. A model of anisotropic creep of single crystal superalloys has been developed and calibrated with a tensile creep database for <001> and <111> specimens of CMSX4.
2. The model has been validated by comparing calculated and experimental creep curves for other orientations and for changing loads.
3. The model has been implemented in finite element code to simulate deformation in notch creep specimens. Heterogeneous crystal rotations have been measured in the notched specimens using EBSD and compared with the model expectation.
4. The potential of the model for estimating creep-fatigue interactions has been discussed.

Acknowledgements
The work was made possible by support from the Engineering and Physical Science Research Council (Grant Numbers GR/J02667, GR/K19358; Visiting Fellowship GR/L67042) and BRITE EURAM III Project BE 96-3911. The authors thank Drs Maldini and Marchionni of Tempe for permission to use their creep data in this analysis.

References

1 R.N.Ghosh, R.V.Curtis and M.McLean, Acta Met. Mater. **40,**(1990), p. 1977

2 L.Meric,P.Poubanne, and G.Cailletaud, Trans. ASME, **113, (**1991) p.162.

3 N. Ohno, T. Mizuno, H.Kawaji, and I.Okada, Acta metall.mat.,**40, (**1992),p. 567.

4 L-M.Pan, B.A.Shollock and M.McLean,Proc.R.Soc.Lond.A,**453**,(1997),1689-1715.

5 B.F.Dyson and M.McLean, ISIJ International **30** (1990), p.802

6 M.G.Ardakani, M.McLean and B.A.Shollock, Acta Mater.**47**, (1999), 2593-2602.

7 N.Matan, D.C.Cox, P.Carter, M.A.Rist, C.M.F.Rae and R.C.Reed, Acta Mater. **47**,(1999), 1549-1563..

8 J.F.Nye in *Physical properties of Crystals*, Oxford University Press (1957).

9 M.B.Henderson, L.M.Pan, B.A.Shollock and M.Mclean, Proc. Euromat, 1995, Symp. D, 25/9-28/9, Padua/Venice,1-10.

10 L-M.Pan, I.Scheibli, M.B.Henderson and M.McLean, Acta.Mater. **43**, (1995), 1375-1384.

11 B.F.Dyson and M.McLean, in *Microstructural Stability of Creep Resistant Alloys for High Temperature Plant Applications,* ed. A.Strang *et al.*, Inst. Of Mater. (1998).

12 B.A.Shollock, J.Y.Buffiere, R.V.Curtis, M.B.Henderson and M.McLean, Scripta Mater. 36, (1997) 1471-1478.

13 L.Remy, Brite-Euram Project BE 96-3911, Final report, Jan 2001.

Kinetics of γ-channel widening in super alloy single crystals under conditions of high temperature and low stress creep

K. Serin, G. Eggeler
Ruhr-University Bochum, Institute for Materials,
D - 44780 Bochum, Germany

1 Abstract

In the present study we investigate the kinetics of γ-channel widening using transmission and scanning electron microscopy in combination with quantitative image analysis. It is shown that the increase of channel width can be attributed to a multi atom diffusion process through the γ-channels and that the γ-channel widening can be rationalized by a parabolic rate law. It is found that a higher level of external stress accelerates the kinetics of rafting; the effect is significant but small. From an Arrhenius plot of the parabolic rate constants the apparent activation energies for γ-channel widening are determined for uniaxial (<001>-tensile) and biaxial ({011}<01-1>-double shear) testing. They were found to be 481 and 447 kJ/mole, respectively.

2 Introduction

Rafting is a well known microstructural instability which occurs under creep conditions in single crystal super alloys [1-5]. For appropriate loading conditions in high temperature and low stress uniaxial (<001>-tensile) and biaxial ({011}(<01-1>-double shear) testing [6] and appropriate microstructural parameters (crystallography, lattice misfit, morphology), the γ'-cubes form rafts and the γ-channels widen. In uniaxial <001>-tension, negative misfit alloys develop γ'-rafts perpendicular to the axis of the applied stress σ_1. Recently [6], rafting was studied under shear loading conditions ($\sigma_1 = -\sigma_3$, $\sigma_2 = 0$) at high temperature (above 1000 °C) and low stress (below 100 MPa) and an example of raft formation under these creep conditions is given in Figure 1.

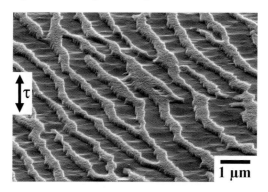

Figure 1: Example for a rafted γ/γ'-morphology which was observed in CMSX 4 after 315 hours of shear creep loading of the macroscopic crystallographic shear system {011}<01-1> at a shear stress of 50 MPa and a temperature of 1080 °C.

For the macroscopic crystallographic shear system {011}<01-1> it was found that γ'-rafts form in an angle of 45° to the macroscopic shear direction. This is not unexpected because (like in the case of uniaxial loading) rafts form perpendicular to the maximum principal stress σ_1 [6]. Recently it was shown that there is no difference between γ-channel widening in uniaxial <001>-tension and biaxial {011}<01-1>-double shear loading as long as the maximum principal stresses in tension and shear are the same [7]. The results reported in [7] were based on a quantitative metallographic method which is described in the literature [8] and which uses a line intersection method to quantify rafting kinetics by evaluating a series of interrupted creep tests.

The method described in [8] is based on the evaluation of around 100 individual γ-channel widths per crept material state. Individual γ-channel widths, w_n in Figure 2, follow a log-normal-distribution.

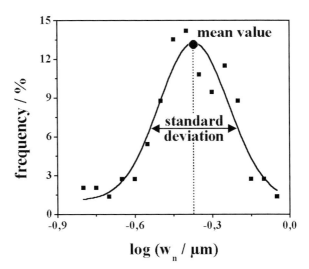

Figure 2: Log-normal-distribution of γ-channel widths after 315 hours shear creep deformation of the macroscopic crystallographic shear system {011}<01-1> at a shear stress of 50 MPa and a temperature of 1080 °C.

As is illustrated in Figure 2 the quantitative metallographic data can be well represented by a numerically fitted Gauss-function which yields the mean value and the standard deviation of the logarithmic distribution. By simply linearizing the logarithmic mean value and its upper and lower bound (determined by the logarithmic standard deviation) we obtain representative values for the γ-channel widths and its microstructural scatter [8]. The data which were obtained by this procedure [8] show that the γ-channel widening can be rationalized in terms of a parabolic rate law, Figure 3.

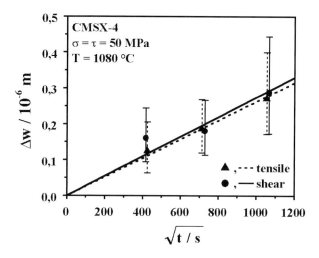

Figure 3: Evolution of the widths of γ-channels during high temperature (T=1080 °C) and low stress (σ=τ=50 MPa). It can be seen that there is no significant difference in γ-channel widening between uniaxial <001>-tension (full triangles) and biaxial {011}<01-1>-double shear (full circles) for a constant maximum principal stress.

It can be noted that the error bars in Figure 3 are non symmetric with respect to the mean value and this results from the linearization of the logarithmic data shown in Figure 2. The data points in Figure 3 can be well represented by a straight line through zero. This yields a simple relation of type:

$$\Delta w = C_2 \cdot \sqrt{t} \tag{1},$$

where Δw is the increase of channel width due to cross-channel-diffusion, t is the time and C_2 is the parabolic rate constant.

For both data sets in Figure 3 values for C_2 can be obtained from the mean values by using a least square fit method. In order to obtain a feel for the scatter of these C_2-values straight lines through zero were also fitted through the upper and lower limits of the error bars of Figures 3 yielding values for C_2-max and C_2-min. The results of this evaluation are summarized in Table 1.

Table 1: Values for parabolic rate constants and their upper and lower bounds obtained from the experimental data shown in Figure 3.

experiment	C_2 (from Equation 1) in $10^{-10} \cdot [m/\sqrt{t}]$	C_2 –max in $10^{-10} \cdot [m/\sqrt{t}]$	C_2 –min in $10^{-10} \cdot [m/\sqrt{t}]$
<001>-tensile	2.62	3.90	1.64
{011}<01-1>-shear	2.75	4.19	1.67

So far our results and our procedures were reported in the literature [7,8]. The present work has two objectives: Its first objective is to find out whether the level of the external stress affects the parabolic rate constant C_2. The second objective of the present paper is to assess the temperature dependence of γ-channel widening.

3 Experiments

Material: The single crystal superalloy CMSX-4 used in the present investigation had a standard heat treatment and was received in the form of cylindrical <001>-cast rods. Details of heat treatment and chemical composition of the material are given elsewhere [9].

Creep testing: Tensile creep tests were performed in <001>-direction using miniature creep specimens [10]. Shear creep testing was performed using the shear specimen geometry developed by Mayr et al. [11] and considering the macroscopic crystallographic shear system {011}<01-1>. All details on mini-tensile and shear creep testing were reported in the literature [10, 11]. The present paper reports microstructural results which were obtained from 24 interrupted creep tests. The creep conditions of the test programme are summarized in Table 2. For each creep condition three interrupted creep tests were performed.

Table 2: Tensile and shear creep conditions used in the present study.

experiment	temperature [°C]	stress [MPa]	time of creep exposure [h]
<001>-tensile	1020	50	50; 143; 302
<001>-tensile	1050	50	47; 149; 304
<001>-tensile	1080	50	50; 140; 308
<001>-tensile	1050	75	51; 142; 293
{011}<01-1>-shear	1020	50	55; 160; 301
{011}<01-1>-shear	1050	50	50; 150, 302
{011}<01-1>-shear	1080	50	48; 147, 315
{011}<01-1>-shear	1050	75	48; 167; 288

Metallography: After creep testing metallographic cross sections were prepared and micrographs like the one shown in Figure 1 were obtained using a field emission scanning electron microscope (LEO-Gemini 1530). The SEM-based method used in the present study has a higher resolution than the method which we used previously [7]. A detailed description of our metallographic procedure which requires input from both TEM and SEM measurements is given in the literature [8].

4 Results and Discussion

We first consider the influence of the external stress on γ-channel widening. The corresponding results are presented in Figure 4a and b.

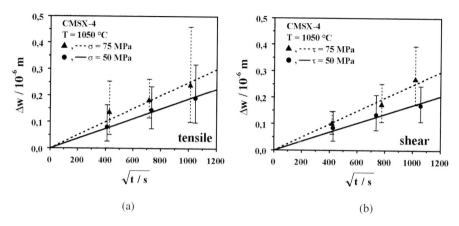

(a)   (b)

Figure 4: Parabolic rate laws for γ-channel widening in CMSX-4 at 1050 °C at two levels of external stress. (a) <001>-tensile creep testing (σ_1= 75 and 50 MPa), (b) {011}<01-1>-shear creep testing (τ=75 MPa and 50 MPa).

The results in Figure 4 prove again that all data sets can be represented by simple parabolic rate laws. At higher external stresses (both in <001>-tension and {011}<01-1>-shear) γ-channels widen faster than at lower external stresses. Figure 4 shows that this effect is significant but small. It is not unexpected that the external stress has an influence on the kinetics of rafting because it does affect the chemical potential of diffusing atoms (thermodynamic effect) as well as the concentration of thermal vacancies which are needed for diffusion (kinetic effect) [12, 13]. The accelerating effect of an external stress on the rate of γ-channel widening is observed for both <001>-tension (Figure 4a) and {011}<01-1>-shear creep experiments (Figure 4b). In agreement with Peng and co-workers [16] the effect of stress is small and it seams reasonable to conclude that from an engineering point of view it can be neglected. Our two types of tests (tensile and shear) were performed for the same maximum principal stress. Comparing the mechanical data from both types of experiments on the basis of a simple effective strain type of argument (e.g. Dieter [14], p. 87) shows that the amount of effective strain accumulated in the shear creep experiments was significantly higher than in tension. This will be reported in more detail elsewhere [15]. But since there is only very little difference in the channel widening behaviour observed in tension (Figure 4a) and shear (Figure 4b) we conclude that stress is more important than strain in affecting channel widening behaviour and the underlying diffusion processes.

We now consider the temperature dependence of channel widening. The experimental results obtained for three temperatures are presented in Figure 5a (<001>-tensile creep) and Figure 5b ({011}<01-1>- shear creep). The data points in Figure 5 are presented without error bars but scatter in all cases is of the same order of magnitude as for the results reported in Figure 4. Figure 5 clearly demonstrates that channels widen faster as temperature increases for both types of testing. We present the rate constants (which were obtained as the slopes of the straight lines in Figure 5) in an Arrhenius plot, Figure 6a (tensile tests) and Figure 6b (shear tests). Each point in Figure 6 represents three interrupted creep tests. In Figure 6 we do

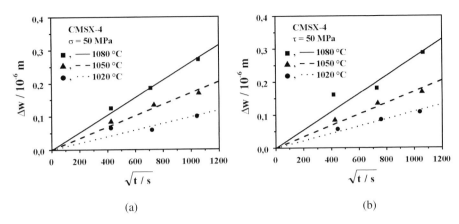

(a)   (b)

Figure 5: Parabolic rate laws for γ-channel widening in CMSX-4 at σ_1=50 MPa at three temperatures. (a) <001>-tensile creep testing, (b) {011}<01-1>-shear creep testing.

not simply use the parabolic rate constants C_2 from Equation 1. Instead we use the constant C_1 which appears in the well known derivation of the parabolic rate law:

$$\frac{d\Delta w}{dt} = \frac{C_1}{\Delta w} \qquad (2),$$

The reason for taking C_1 rather than C_2 is that C_1 is directly related to diffusion [13]; and C_1-values can be obtained from C_2-results:

$$C_1 = \frac{C_2^{\,2}}{2} \qquad (3),$$

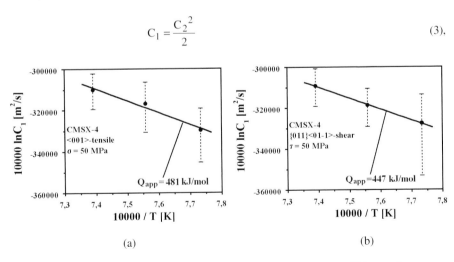

(a)   (b)

Figure 6: Arrhenius plots of the rate constants C_1 from Equation 3. (a) <001>-tensile creep data, (b) {011}<01-1>- shear creep tests.

The slopes of the straight lines in Figure 6 represent the negative ratios of the apparent activation energies of γ-channel widening, Q_{app}, and the gas constant R. Fitting straight lines through the data points in Figure 6 yields apparent activation energies of 481 kJ/mol and 447 kJ/mol for the temperature dependence of the rate constant C_1 of channel widening under <001>-tensile and {011}<01-1>-shear creep loading, respectively.

The data presented in Figure 6 require careful interpretation. Each point represents a quantitative metallographic evaluation of three interrupted creep tests and the results are therefore reliable. The error bars shown in Figure 6 indicate that the values of the apparent activation energies reported above should be considered as exhibiting uncertainties of at least 30%. Therefore considering the experimental scatter (which is merely due to the scatter in measured channel width distribution and not to uncertainties in stress, temperature or time!), the temperature dependencies for γ-channel widening in <001>-tensile and {011}<01-1>-shear creep are not significantly different.

We conclude that the kinetics of γ-channel widening expressed in terms of C_1 (Equation 3) shows a very strong temperature dependence which is characterized by 464 ± 139 kJ/mol. This value is higher than values reported for the apparent activation energy of self-diffusion of Ni in Ni (284 kJ/mol) [17]. Further work is required to interpret our channel widening results in terms of the underlying diffusion processes and to understand the role of channel widening in the temperature dependence of the overall creep process.

Summary and Conclusions

The present work investigates the kinetics of γ-channel widening in the single crystal superalloy CMSX-4 under <001>-tensile and {011}<01-1>-shear creep conditions at temperatures above 1000 °C. The following results were obtained:

(1) γ-channel widening can be fully accounted for in terms of a parabolic rate law where the increase of channel width Δw depends on the square root of time $\Delta w = C_2 \cdot \sqrt{t}$.

(2) An increase in external stress results in faster γ-channel widening. The effect is small but significant in both <001>-tensile and {011}<01-1>-shear.

(3) The present work reports the temperature dependence of γ-channel widening which is represented by a constant C_1 from $d\Delta w / dt = C_1 / \Delta w$. C_1 can be obtained from the measurement of C_2-values as $C_1 = \sqrt{C_2} / 2$. Arrhenius plots of C_1 vs. 1/T yield apparent activation energies of 481 kJ/mol for <001>-tensile and 447 kJ/mol for {011}<01-1>-shear creep. Further work is required to rationalize C_1 on the basis of the underlying diffusion processes and to understand the role of channel widening in the temperature dependence of the overall creep process.

Acknowledgement

The authors acknowledge funding from the Deutsche Forschungsgemeinschaft under contract Ko 1508/2-3 and from the European Community under contract Brite-Euram BRPR-CT96-0224.

References

[1] J. K. Tien, S. M. Copley, Metall.Trans., 2, 543 (1971)

[2] J. K. Tien, R. P. Gamble, Metall.Trans., 3, 2157 (1972)

[3] A. Pineau, Acta Metall., 24, 559 (1976)

[4] T. M. Pollock and A. S. Argon, Acta metall.mater., 42, 1859 (1994)

[5] H. Mughrabi, W. Schneider, V. Sass and C. Lang, Proc.10th. Int. Conf. Strength of Materials, eds. Oikawa, Jap. Inst. Met., Sendai (Japan), 705 (1994)

[6] M. Kamaraj, C. Mayr, M. Kolbe and G. Eggeler, Scripta Mat., 38, 589 (1998)

[7] M. Kamaraj et al., accepted for publication in Mat. Sci. Eng. A (to appear shortly)

[8] K. Serin et al., accepted for publication in Practical Metallography (to appear shortly)

[9] K. Serin, Dipl.-Ing.-Thesis, Ruhr-Universitaet Bochum, (October 1998)

[10] M. Kolbe, J.Murken, D.Pistelok, G.Eggeler, H.-J.Klam, Mat.-wiss. u. Werkstofftech., 30, 465 (1999)

[11] C. Mayr, G.Eggeler, G.A.Webster, G.Peter, Double shear creep testing of superalloy single crystals at temperatures above 1000°C, Mat. Sci. Eng., A199 (1995) pp. 121-130

[12] R.A. Swalin, Thermodynamics of Solids, Second Edition, John Wiley & Sons, New York 1972

[13] P. Shewmon, Diffusion in Solids, Second Edition, TMS, Warendale 1989

[14] G.E. Dieter, Mechanical Metallurgy, McGraw Hill, London 1988

[15] K. Serin, G. Eggeler , to be published

[16] Z.F. Peng et al., Scripta Materialia, 42, 1059-1064 (200)

[17] J. Askill, Tracer Diffusion Data for Metals and Simple Oxides, plenum Press, New York, NY, 1970

A COMPARATIVE STUDY OF STRAIN RELAXATION AND STRESS RELAXATION IN NICKEL-BASE SUPERALLOY IN-738LC

N. K. Sinha

Institute for Aerospace Research, National Research Council Canada
Building M-13, Montreal Rd. Complex, Ottawa, Ontario, Canada K1A 0R6
E-mail: nirmal.sinha@nrc.ca

ABSTRACT

Short-term closed-loop controlled "constant-strain" stress relaxation tests (SRT) and "constant-stress" strain relaxation and recovery tests (SRRT) were carried out successively on the same specimen of a nickel-base superalloy, IN-738LC, at 1000°C. Both techniques have been found to be suitable for evaluating the initial microstructure dependent flow properties of a material. However, the SRT tends to give a higher strain rate at a given stress level than that given by the SRRT. This is attributed to the contribution of the delayed elastic strain rate (due to the grain-boundary shearing mechanisms and hence grain size dependent) to the total strain rate during the SRT. The delayed elastic strain is not negligible and its role in the process of relaxation cannot be ignored. SRRT, in contrast, is capable of separating delayed elastic strain and viscous strain from the total strain. SRRT is therefore more powerful than SRT and is easier to perform.

INTRODUCTION

The precipitation strengthened alloys based on Ni-Cr with additions of Co, Al, Ti, W etc. obtain their high strength from a fine dispersion of the ordered fcc phase N$_3$ (Ti, Al), or γ' (gamma prime) which precipitates in the Ni-rich fcc, γ (gamma) matrix [1, 2]. Cobalt reduces the solubility of Al and Ti in the Ni-Cr matrix and improves strength at high temperatures as well as the workability of these alloys. Other elements are added to improve microstructural stability, strength and deformation characteristics. The creep-rupture life of these alloys can be increased significantly by the careful control of composition and microstructure.

The same features that improve high temperature performance of nickel-base superalloys also make them a) difficult to repair by welding and b) vulnerable to structural degradation, such as coarsening of the γ' precipitates and the oxidation of nickel and carbon producing sub-surface grain-boundary defects during service life. Liquidation zone cracks (LZC), and heat-affected zone (HAZ) cracks develop due to thermal stresses generated by volumetric changes associated with the dissolution and re-precipitation of γ' phase during welding.

A project was undertaken to study the weldability of 'difficult-to-weld' commercially cast nickel-base superalloys, such as IN-738LC. Pre-weld heat treatment was used in order to modify the characteristics of the γ' particles that would be responsive to the relaxation of any residual stresses introduced during welding. Stress relaxation test (SRT) at a constant tensile strain of 0.0025, imposed in about one second, was used to evaluate the flow properties of the microstructurally modified materials. Strain relaxation and recovery tests (SRRT), developed by the author [3], were also conducted on the same specimens used for stress relaxation tests, under constant tensile stress to the level of stresses encountered in stress relaxation tests. Both of these types of tests were performed at 1000°C. The objectives of this paper are to describe the test method developed for conducting both SRT and SRRT on the same specimen and to make a comparison between the two sets of results obtained for the stress dependence of strain rate.

EXPERIMENTAL PROCEDURE

Uniaxial, tensile constant-strain stress relaxation tests and constant-stress strain relaxation (creep) and recovery tests were performed at $1000°C \pm 1°C$ on commercially prepared vacuum cast, hot isostatically pressed (HIPed) and heat treated nickel-base superalloy, IN-738LC. Tensile specimens with a diameter of 6.4 mm, a uniform gauge section of 44 mm and a total length of 89 mm were prepared from heat-treated (in-house) bars (19 mm diameter by 146 mm length). Results obtained on specimens subjected to 'standard' heat treatment [2, 4] will be primarily presented here. This heat treatment consists of solution anneal ($1125°C$ /2h/gascool) followed by aging ($850°C$/24h/gascool). It produces a duplex γ' structure corresponding to the optimum properties for industrial applications of this superalloy. An example of a bimodal distribution of large cubical (normal or primary) γ' particles with mean diameter of 0.6 µm and small spherical (cooling or secondary) γ' particles with mean diameter of 0.08 µm is shown in Fig. 1. Grain-boundary morphology is shown in Fig. 2. It also shows that the grains were large in the material used in this series of test. Consequently, there were only a few grains within the cross-sectional area of the specimens – an undesirable, but unavoidable condition. In-house heat treatments did not change the size of the grains, but modified the morphologies of γ' in a significant manner.

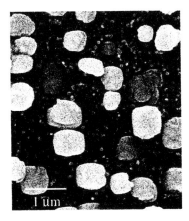

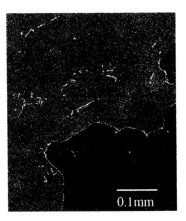

Figure1. Scanning electron micrograph showing bimodal distribution of large cubical (primary) γ' particles and small spherical (secondary) γ' particles in IN-738LC.

Figure 2. Optical micrograph showing carbide precipitates along grain boundaries between two large grains.

A computer controlled universal servo-hydraulic closed-loop test machine (MTS) was used. Hydraulically loaded water-cooled stainless steel grips were used to hold the specimens, but specially made extenders were added to the load train to reduce the cooling effect of the water-cooled grips on the temperature distribution in the specimen. A three-zone furnace was used to provide the constant temperature environment. Three thermocouples were used to monitor and control the temperature of the three zones of the furnace. The furnace was calibrated using one IN-738LC specimen with three thermocouples spot-welded on the uniform section. A MTS Model 632.42C-10 air-cooled extensometer with a pair of ceramic rods and a gauge length of 25 mm was used to monitor the axial strain in the gauge section of the specimen. Extensometer contact loads of about 150 grams were used.

Both SRT and SRRT were carried out on the same specimen. Test durations were purposely kept very short in order to maintain a constant microstructure (or keep it as close as possible to the initial conditions). To evaluate the effects of a given heat treatment on the mechanical properties and the repeatability of the tests, experiments were performed on pairs of specimens.

Stress Relaxation Test (SRT)

The SRT was carried out first. It was performed under a constant gauge-section strain of 0.0025. Using a strain rate of 2.5×10^{-3} s^{-1}, the full load corresponding to the total strain of 0.0025 was applied in one second. The strain was then held constant. The procedure is based on optimizing the initial loading path and subsequent control of strain during the hold time [5]. To capture the strain history during the loading and initial rapid stress relaxation period, strain- and stress-time readings were taken at intervals of 0.1s or less. The strain was then held constant for 900 s to induce a significant (about 50%) stress relaxation. The load was then removed suddenly to a negligible level (to keep the computer-controlled system running) and the strain was monitored for a period long enough to observe no more strain recovery. This test procedure induced a small permanent strain (typically 0.001 or less) at the end of the tests and thereby prevented large-scale changes in the microstructure.

Strain Relaxation & Recovery Test (SRRT)

A strain relaxation and recovery (SRRT) test was performed by using force control to apply the full load, equivalent to the level of stresses encountered during stress relaxation tests, in about one second or less and holding the load constant for 900 s. This step, as mentioned earlier, was taken to prevent large-scale changes in the microstructure. As in the stress relaxation tests, the load was then removed suddenly to a negligible level and the strain was monitored for a period until there was no more strain recovery. To capture the strain history during the rapid loading and unloading period, strain-time as well as stress-time readings were taken at intervals of 0.1 s or less. The test was then repeated on the same specimen for a different stress level.

Since the SRT covered a large range of stresses from about 250 MPa to 130 MPa, the SRRT were carried out at around 160 MPa, 150 MPa and 140 MPa for loading times of 900 s. Because of the number of tests involved, high stresses were avoided for SRRT in order to keep the accumulated permanent strain in the specimen as small as practically possible. The SRRT was then repeated at around 160 MPa for two or three other loading times in the range of 300 s to 1200 s.

RESULTS

Figure 3 shows the histories of strain-time and stress-time for the SRT carried out on a specimen designated as Stan1 at 1000°C ± 1°C. In this case, the strain was held constant for 900 seconds, a period long enough to induce a noticeable amount of stress relaxation, from 290 MPa to 125 MPa and a permanent strain of 0.0012. The Young's modulus, E, was determined from the linear stress-strain relationship obtained for the rapidly increasing stress during the loading period, as well from the rapidly decreasing stress during the unloading period. The two values agreed well with each other. The average E was found to be 116.5 GPa for this specimen.

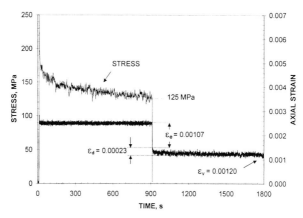

Figure 3. Stress-time and strain-time histories for stress relaxation test (SRT) at constant strain of 0.0025 on specimen No.1 (Stan1) of cast, HIPed and 'standard' heat-treated IN-738LC at 1000C.

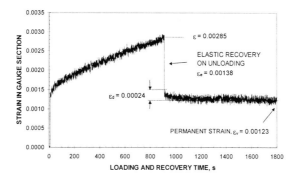

Figure 4. Strain-time records for strain relaxation and recovery test (SRRT) on specimen No.1 (Stan1) of cast, HIPed and 'standard' heat-treated IN-738LC at 1000C for 161 MPa.

Figure 3 shows that the imposed strain fluctuated slightly ($\pm$ 0.0001) around the mean value of 0.0025. These fluctuations, caused primarily due to the noise associated with the system, were sufficient to produce noticeable changes in the stress. For E = 116.5 GPa, a scatter of $\pm$ 0.0001 in strain is expected to develop a stress change of $\pm$ 11.6 MPa. Moreover, as the furnace control tried to maintain the temperature, it produced visible variation in the amount of radiation. This also added thermal strains and stresses to the specimen and the extensometer rods.

The strain relaxation (creep) during the constant load application period and strain recovery on unloading for the SRRT test at 161 MPa on specimen No.1 (Stan1) is shown in Fig. 4. This test was performed immediately after completing the SRT presented above in Fig. 3. The elastic strain for 161 MPa and E of 116.5 GPa was expected to be 0.00138. As can be seen in Fig. 4, the strain recorded immediately after the load application and the elastic rebound immediately on unloading seem to agree extremely well with this expectation. Note the significant amount (0.0012) of accumulated viscous or permanent strain, ε_v.

The stress relaxation process is usually described by assuming a simple relationship between the imposed strain ε_o, the elastic strain, ε_e, and the inelastic strain ε_p [6-8]:

$$\varepsilon_o = \varepsilon_e + \varepsilon_p \qquad (1)$$

Since ε_o is constant ($d\varepsilon_o/dt = 0$) during a constant-strain SRT, the rate of change of ε_e and ε_p with time can be described by rewriting Equation (1) as:

$$d\varepsilon_p/dt = - d\varepsilon_e/dt = - (d\sigma/dt)/E \qquad (2)$$

where σ is the stress at a given time, t, and E is Young's modulus.

The stress dependence of strain rate, $d\varepsilon_p/dt$, obtained from the experimental results shown in Fig. 3 and Equation (2) is plotted in Fig. 5 along with the results obtained for the average viscous strain rate, $d\varepsilon_v/dt$, from the SRRT tests for three levels of stress including the results shown in Fig. 4.

The average viscous strain rate for SRRT is given by:

$$d\varepsilon_v/dt = \varepsilon_v/t \qquad (3)$$

where t is the time during which the stress is maintained constant and ε_v is the permanent strain measured after unloading and recovery period.

Figure 6 shows the results obtained for the second specimen (Stan2) that was subjected to the same standard heat treatment as used for Stan1. Actually, both specimens yielded practically the same results. Figures 5 and 6 indicate clearly that both SRT and SRRT yield results that are comparable to each other. Both set of results may be described by the usual power law:

$$d\varepsilon_v/dt = A (\sigma/\sigma_1)^n \qquad (4)$$

where σ_1 is the unit stress (= 1 MPa), n is the stress exponent and A may be interpreted as the strain rate for unit stress.

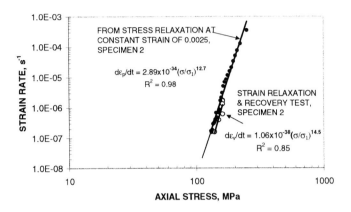

Figure 5. Stress dependence of strain rate for specimen 1 (Stan1) of IN-738LC at 1000°C from stress relaxation test (SRT) and strain relaxation and recovery tests (SRRT).

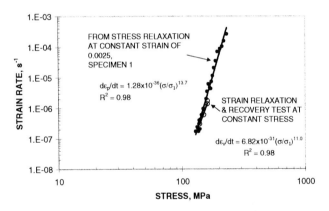

Figure 6. Stress dependence of strain rate for specimen 2 (Stan2) of IN-738LC at 1000°C from stress relaxation test (SRT) and strain relaxation and recovery tests (SRRT).

The stress exponents shown in Fig. 5 and Fig. 6 are high but agree well with results reported in the literature for this type of nickel-base superalloy [9-12]. SRT on the two specimens yielded n as 13.7 and 12.7 respectively. SRRT provided the corresponding values of 11 and 14.5 respectively. Of course, a larger scatter is noticed in the results of SRRT because of the small range of stresses and a limited number of data points.

DISCUSSION

One common feature that may be seen in both Fig. 5 and 6 is fact that the data points obtained from SRRT tend to be on the right side of the results obtained from SRT. This means that the SRRT tend to give a lower strain rate, at a given stress, than that given by the SRT. This feature was noticed, without any exception, in all the results obtained so far from specimens that were heat treated in many different ways to make them softer as part of the weldability improvement project.

Although Equation (1) is assumed commonly in analyzing the results obtained from stress relaxation tests, it is well known that it does not really apply to polycrystalline behaviour at temperatures greater than about 0.4 T_m, where T_m is the melting point in Kelvin. Grain-boundary sliding (shearing), gbs, mechanism (s) play a dominant and complicated role at high temperatures. After reviewing available experimental results on the dependence of gbs on stress, strain, temperature, grain size and time, this author concluded that a measure of the contribution of the gbs strain to the total strain could be determined from the delayed elastic strain, des [13]. He then proposed a simple constitutive relationship in the usual familiar form:

$$\varepsilon_o = \varepsilon_e + \varepsilon_d + \varepsilon_v \tag{5}$$

but the delayed elastic strain, ε_d, was expressed as a function of grain size, stress, Young's modulus, time, temperature, stress exponent, state of the grain boundaries, etc. It was shown that ε_d could have the same activation energy as that of viscous flow due to the mobility of lattice dislocations.

For conditions of constant-strain stress relaxation processes, Equation (5) can be rewritten as:

$$d\varepsilon_v/dt = - d\varepsilon_e/dt - d\varepsilon_d/dt = - d\sigma/dt)/E - d\varepsilon_d/dt \tag{6}$$

If the quantity - $d\sigma/dt)/E$, is denoted as the "Global Strain Rate", then the Equation (6) can be written as:

$$\text{Global Strain Rate} = d\varepsilon_v/dt + d\varepsilon_d/dt \tag{7}$$

Thus the Global Strain Rate (GSR) given by a SRT is always greater than the viscous (permanent or plastic) strain rate. Viscous strain rate can be determined from the GSR in a SRT only when the delayed elastic strain rate is zero or negligible. Figure 3 shows that the accumulated ε_d during the relaxation process is small, but not negligible. It was obtained by subtracting the final viscous strain and the computed elastic strain (from E and the remaining stress at the time of unloading) from the imposed strain. How to evaluate $d\varepsilon_d/dt$ at a given time (hence stress level) during the stress relaxation after completing a SRT is still the biggest challenge.

Interpreting the results of SRRT is, however, relatively simpler than those of SRT. For a given constant stress, the accumulated ε_v in a SRRT can be evaluated, as illustrated in Fig. 4, by measuring the permanent strain and determining the average rate by using Equation (3). If required, ε_d can also be evaluated by subtracting the viscous strain and the calculated elastic strain from the total strain before unloading. The delayed elastic strain of 0.00024 in Fig. 4 is estimated by subtracting ε_e (0.00138) and ε_v (0.00123) from the total strain of 0.00285. This is 8.4% of the total strain and 20% of the viscous strain.

The stress dependence of delayed elastic strain and viscous strain, obtained from SRRT for 161 MPa, for a loading time of 900 s is shown in Fig. 7 and Fig.8 respectively. It is not clear at this moment why the specimen 2 exhibited lower values of delayed elastic strain than those obtained for the specimen 1, but Fig. 7 shows that ε_d increases with stress (almost linearly) for both the specimen, as expected [13]. The accumulated viscous strain, ε_v, however, increases nonlinearly with the increase in stress in accordance with the power law.

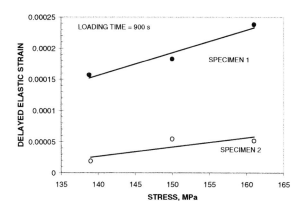

Figure 7. Stress dependence of delayed elastic strain at 900 s at 1000°C for cast, HIPed and standard heat-treated IN-738LC.

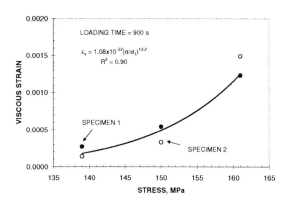

Figure 8. Stress dependence of viscous or permanent strain at 900 s at 1000°C for cast, HIPed and standard heat-treated IN-738LC.

It should be mentioned here that ε_d in Equation (5) was shown in reference [13} to be inversely related to grain size. Consequently, the contribution of delayed elastic strain to the total strain is expected to increase with the decrease in grain size. Since the grains were very large in the cast IN-738LC materials used here and since only a few grains were involved in a specimen, delayed elastic strain is expected to be small. Figure 3 for SRT and Fig. 4 for SRRT corroborate well with this hypothesis. It is well known that the intergranular microstructural features, such as the size and volume fraction of Υ'' precipitates and grain-boundary morphology influence the flow properties in general. Very few studies, however, have been conducted to examine the delayed elastic effect in case of nickel-base superalloys [14]. Figure 9 shows an example of large differences in the results obtained from SRT and SRRT on a microstructurally modified (for improved weldability) IN-738LC. Note the large variation of delayed elastic strain with stress for this material (Fig. 10).

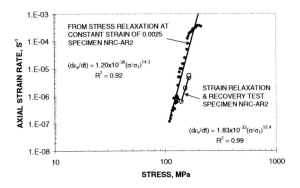

Figure 9. Stress dependence of strain rate for a microstructurally modified specimen of IN-738LC at 1000°C from stress relaxation test (SRT) and strain relaxation and recovery test (SRRT).

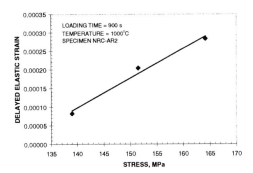

Figure 10. Stress dependence of delayed elastic strain at 900 s at 1000°C for cast, HIPed and microstructurally modified IN-738LC corresponding to Fig. 9.

SUMMARY

An experimental technique has been developed for conducting isothermal short-term tensile "constant-strain" stress relaxation tests (SRT) and "constant-stress" strain relaxation and recovery tests (SRRT), in succession, on the same specimen at the same elevated temperature.

The technique was applied to a nickel-base superealloy, IN-738LC at a temperature of 1000°C. The stress relaxation tests were conducted at a strain of 0.0025 imposed in about one second. The duration of the tests was 900 s, during which the stress decreased from an initial stress of about 250 MPa to 125 MPa for specimens with standard heat treatment. The strain relaxation tests were conducted at stresses of about 160 MPa, 150 MPa and 140 MPa for 900 s followed by strain recovery period of 900 s or longer.

Both techniques have been found to be suitable for evaluating the initial microstructure dependent flow properties of a material.

The "Global Strain Rate" for a given stress level, that can be estimated from a SRT, tends to be higher than that given by a SRRT. This difference has been attributed to the contribution of delayed elastic strain (associated to the grain-boundary shearing mechanisms) to the total strain and the effects of intragranular features, such as γ'' precipitates and grain-boundary morphology, on the flow properties in general and delayed elastic response in particular.

Stress relaxation tests are ideal for evaluating the ability of a material either to accommodate local stress concentrations, encountered during welding for example, or to absorb temporarily excessive stress regimes generated during certain operational conditions of engineering components. Under these situations, both delayed elastic and viscous flow operate at the same time as part of the relaxation process. However, one should be careful in applying the results from SRT to determine the matrix flow properties of materials in general, particularly for materials with fine grain sizes. SRT are often performed to determine the stress dependence of plastic or viscous strain rate of materials, and relate them to minimum creep rate, microstructural features and micromechanics at the scale of dislocations. In these situations, the role played by the mechanisms controlling delayed elasticity could add complications to the interpretation of results for conditions (such as fine-grained materials) in which delayed elastic effect is not negligible.

ACKNOWLEDGEMENTS

The author acknowledges the technical assistance provided by R. Kearsey, D. Morphy and T. Terada during the tests.

REFERENCES

[1] Betteridge, W. and Heslop, J., 1974, The Nimonic Alloys and Other Nickel-Base High-temperature Alloys, Edward Arnold Publishers Limited, London, Chapters 1 to 7.

[2]E. Ross, E.W. and Sims, C.T., 1987, Nickel-Base Alloys, in "Superalloys II" (Ed. C. T. Sims, N. S. Stoloff and W. C. Hagel), A Wiley-Interscience Publication, John Wiley & Sons, New York, pp. 97-133.

[3] Sinha, N.K., 2001, Short Strain Relaxation/Recovery Tests for Evaluating Creep Response of Nickel-base Superalloys like IN-738LC, J. Mater. Sci. L. (in press)

[4] Stevens, R. A. and Flewitt, P.E.J., 1979, The Effect of γ' Precipitate Coarsening During Isothermal Aging and Creep of the Nickel-base Superalloy IN-738, Mat. Sci. Eng. Vol. 37, pp. 237-247.

[5] Sinha, N.K. and Kearsey, R., 2001, Optimizing Deformation Path for Stress Relaxation Tests on Superalloys at High Temperatures, Proc. Turbo Expo-2001, The 46th ASME International Gas Turbine & Aeroengine Technical Congress, Exposition and Users Symposium, June 4-7, 2001, New Orleans.

[6] Murty, G.S., 1973, Stress Relaxation in Superplastic Materials, J. Mat. Sci. L., Vol. 8, pp.611-614

[7] Saint-Antonin, F. and Strudel, J. L., 1990, Stress Relaxation in a Ni Base Superalloy After Low Initial Straining, in "Creep and Fracture of Engineering Materials and Structures", (Ed. B. Wilshire and R. W. Evans), The Institute of Metals, London, pp. 303-312.

[8] Woodford, D.A., Van Steele, R.D. and Stiles., 1993, Design for performance – A new Conceptual Approach to High Temperature Deformation and Fracture, In "Creep and Fracture of Engineering Materials and Structures", (Ed. B. Wilshire and R. W. Evans), The Institute of Materials, London, pp.603-612.

[9] Evans, R.W. and Wilshire, B.,1985, Creep of Metals and Alloys, The Institute of Metals, London, Chapters 3, 4, pp. 69-153.

[10] An, S.U., Wolf, H., Vogler, S. and Blum,W., 1990, Verification of the aeffective aastress amodel for Creep of NiCr22Co12Mo at 800°C, in "Creep and Fracture of Engineering Materials and Structures", (Ed. B.Wilshire and R.W.Evans), The Institute of Metals, London, pp.81-95.

[11] Davies, C.K.L., Older, A.G., Stevens, R.N., 1990, Internal Stresses in the Creep of Nimonic 91, in "Creep and Fracture of Engineering Materials and Structures", (Ed. B.Wilshire and R. W. Evans), The Institute of Metals, London, pp.97-107.

[12] Nabarro, F.R.N. and de Villiers, H L., 1995, in "The Physics of Creep", Taylor & Francis, London, Chapter 7, pp.255-293.

[13] Sinha, N.K., 1979, Grain Boundary Sliding in Polycrystalline Materials, Phil. Mag. A., Vol.40, No.6, pp. 825-842.

[14} Luping, V, and Gabrielli, F., 1979, Effect of Grain Size, Particle Size and γ' Volume Fraction on Strain Relaxation in Ni-Cr Base Alloys, Mat. Sci. Eng., Vol. 37, pp. 143-149.

Deformation Behaviour and Development of Microstructure during Creep of a nickel-base Superalloy

Y.H. Zhang and D.M. Knowles

Rolls Royce University Technology Centre
Department of Materials Science and Metallurgy
University of Cambridge, England

ABSTRACT

The creep behaviour and deformation microstructures of nickel-base superalloy C263 at 800°C have been investigated. The steady-state creep rate of the material follows the power law relation at stresses ranging from 120 to 250 MPa, with a power law exponent value of 5.5. Transmission electron microscope observation indicates that dislocation structures were not affected by the applied stresses used. Dislocation move mainly on parallel slip bands in a dissociated manner during steady state creep. When a unit matrix dislocation approaches to γ/γ' interfaces, it dissociated into a leading partial 1/3<112> and a trailing partial 1/6<112>. The leading partial dislocation shears both γ matrix and γ' precipitates, creating extended stacking faults while the trailing partial was pinned at the interfaces by high energy anti-phase boundary (APB). As the γ' particles coarsened during prolonged high temperature creep, dislocation bowing, looping around γ' particles and climbing became predominant mechanism. Furthermore, prolonged high temperature exposure resulted in microstructure instability at grain boundaries. η phase began to precipitate from carbides at grain boundaries and grew in a lamella form with γ phase. There was a γ' phase depletion zone surrounding the η phase. This has promoted grain boundary migration towards η phase and consequently the frequent presence of η phase along grain boundaries. Grain boundary migration away from η phase has also been observed due to large plastic deformation near and the reduced carbide content locally at grain boundaries. Fracture surface observation revealed a predominantly intergranular fracture surface with a considerable amount of creep cavities.

1. INTRODUCTION

Nimonic C263 is a polycrystalline Ni-base superalloy that is strengthened by $Ni_3(Al,Ti)$ γ' precipitates and solid solution elements chromium, cobalt and molybdenum. It is widely used in non-rotating components such as combustion chambers, casing, liners, exhaust ducting and bearing housing in aeroengines. The choice of this material is based primarily on a combination of excellent fabricability, weldability and good resistance to oxidation [1]. When it is used in combustion chambers, it can be subjected to severe regimes of thermomechanical fatigue and creep deformation.

Considerable efforts have been made in the past and are continuing to develop physically based models to describe creep behaviour and life prediction [2,3]. It is well known that the creep strength is a sensitive function of microstructure. For a given alloy, the microstructural parameters such as grain size, strengthening precipitate size and volume fraction, stacking fault energy, grain boundary carbides and microstructure stability can all significantly influence creep strength and ductility. To establish a model for creep life prediction or creep damage evaluation, it is essential to fully understand the deformation and damage mechanisms, such as dislocation interaction with strengthening particles, particle coarsening, grain boundary cavity formation, microstructural instability etc, before a realistic and physically based model can be assembled.

As part of an ongoing programme addressing combustor lifing in turbine engines, creep deformation and damge behaviour of superalloy C263 at 800°C has been studied in this work. Creep tests at several stress levels were undertaken and the microstructutres of the crept specimens at different creep strain were analysed using transmission electron microscopy (TEM) and scanning electron microscopy (SEM). Although this material is widely used, reports concerning its creep behaviour and its microstrural development during high temperature straining are limited in the open literature.

2. EXPERIMENTS

The C263 alloy was provided by Haynes as 14.8mm diameter bars in a standard heat treatment. This involves solutioning at 1150°C for 2 h, quenching and ageing at 800°C for 8h followed by air cooling which produced an average HV hardness of 275. The chemical composition of the alloy is given in Table 1. Microstructural observation revealed that the material has a mean-linear-intercept grain size of about 104 μm, an average γ' precipitate size of about 22 nm, many annealing twins and almost continuous precipitation of $M_{23}C_6$ carbides at grain boundaries. A typical microstructure of the alloy is shown in Fig. 1 where a γ' depletion zone near grain boundary carbides and fine γ' precipitates within grains can be seen. The γ' volume fraction was determined as ~20% using a Semper image processing program (Synoptics Ltd). $M_{23}C_6$ carbides are present along grain and twin boundaries with a finer size at the latter and are coherent with one of the neighbouring grains exhibiting the typical cube-cube relationship, i.e. $(100)_\gamma //(100)_{carbide}$, $[001]_\gamma // [001]_{carbide}$.

Table 1. Nominal chemical composition (wt. %) of C263 alloy.

Ni	Co	Cr	Fe	Mo	Mn	Si	Ti	Al	C
Bal.	20	20	0.7	5.8	0.6	0.4	2.15	0.45	0.06

Creep tests were carried out under constant load control at 800°C at an initial stress range from 120 to 250 MPa using cylindrical creep specimens with a gauge length of 28.0mm and a diameter of 5.64mm. The specimens were heated via a three-zone furnace and the temperature was controlled to ± 1.0°C over the specimen gauge length using three thermocouples and a three-term temperature controller. Strain was measured using linear variable displacement transducers capable of detecting displacement of 5×10^{-4}mm.

For detailed microstructural analysis of deformation and damage mechanisms, thin foils for TEM were prepared from specimens either crept to the steady-state strain rate and then cooled down to room temperature under load, or crept to failure. Slices of 0.4 mm thickness were cut from the gauge length, ≥ 3.0 mm away from the fracture surface. After grinding and polishing to a thickness of about 100 μm, the foils were prepared by the twin jet polishing technique in a solution consisting of 10% perchloric acid in ethanol at about −10°C. TEM foils were examined in a JEOL 2000 FX electron microscope with an

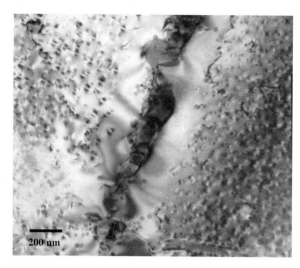

Fig. 1, TEM micrograph of the as-received material showing grain boundary carbide distribution, fine γ′ precipitates and the γ′ depletion zone near carbides.

operating voltage of 200 kV. Two-beam condition was always used in dislocation structure analysis. JEOL 820 SEM was used to characterise the creep crack initiation sites and fracture surfaces.

3. RESULTS AND DISCUSSION

3.1. Creep curves

The creep curves at four stress levels are shown in Fig. 2. The creep test at stress σ=120 MPa did not run to failure. The material exhibited limited creep rupture strain at 800°C (≤ 4%). This is in contrast to its good tensile ductility of about 20% elongation at this temperature [1]. A plot of log steady-state creep strain rate, including those from interrupted creep tests, versus log applied stress is shown in Fig. 3. It can be seen that they can be fitted to a power law relation and the stress exponent was determined to be 5.5. This plot suggests that the same creep deformation mechanism operated during steady state creep at all stress levels used in the present study. The value of the power law exponent was similar to those obtained in commercial superalloys with similar microstructure as C263 alloy [3-5].

3.2. Microstructural studies

3.2.1. Within grains

408

The dislocation structure of the crept specimens either interrupted during steady-state creep or run till final failure were investigated with TEM. The exposure time and creep strain of the interrupted creep specimens are summarised in Table 2. The observation of crept specimens revealed that at all stress levels used, dislocation deformation was inhomogeneous and its activity was mainly confined to planar slip bands during steady-state creep. At comparable creep strain, it was found that dislocation density increased and inhomogeneous deformation was more prominent with increasing applied stress, but the main feature of dislocation structures in both interrupted and failed creep specimens were similar under all the stresses used in this study.

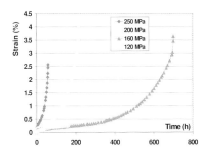

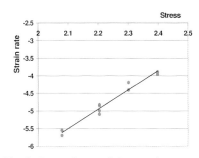

Fig. 2, Creep curves of C263 alloy at 800°C at four stress levels. Note that the creep test at σ=120 MPa was interrupted during steady-state creep.

Fig. 3, Dependence of the steady-state creep rates on stresses at 800°C (log scale).

In each grain dislocation movements were mainly confined to one set of {111} planes, which was determined by tilting to edge-on condition. Through specimen tilting operations, it was found that the slip bands were mainly composed of planar faults as can be seen from the typical micrographs shown in Figs. 4 and 5 taken from interrupted creep tests at σ =120 and 250 MPa respectively. This suggests that partial dislocation movements play an important role in the early stage of creep. Fringe contrast can be seen in both γ matrix and γ' particles and they extended widely (up to 4 μm). Since no fringe contrast was observed in γ' particles when its superlattice diffraction spot was used [6,7], this suggests that the planar faults in γ' phase were stacking faults, not APB. The presence of extended stacking faults covering both γ matrix and γ' precipitates is indicative of the alloy having a low stacking fault energy in both phases.

Table 2. Details of the interrupted creep specimens sectioned for TEM analysis.

Stress level (MPa)	Test time (h)	Creep strain (%)	Life fraction (%)
σ=120	164	0.28	N/A
σ=160	144	0.27	20.6
σ=250	12	0.12	20.7

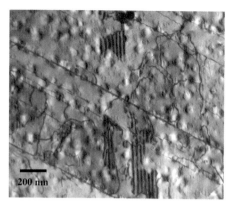

Fig. 4, TEM micrograph showing the inhomogeneous deformation of dislocations localised in slip bands in an interrupted creep specimen, σ =120 MPa.

Fig. 5, TEM micrograph showing parallel slip bands (over a twin boundary) in an interrupted creep specimen, σ =250 MPa.

The Burgers vectors of the partial dislocations bounding a stacking fault were analysed. This was performed by examining the dislocation contrast under various ±**g** vectors. The criterion is that: a partial is invisible when **g.b** = 0, ±1/3, -2/3, +3/4, and visible when **g.b** = integer, +2/3, -4/3 [6,8]. The stacking fault, shown in Fig. 6, was identified to be of extrinsic nature by examining the outmost fringe contrast in bright and dark field images. The Burgers vectors of the top and bottom partial dislocations were determined to be a/3 [11$\bar{2}$] and a/6 [$\bar{2}$11], respectively. Dislocation loops were often found surrounding γ′ particles containing fault fringes near a stacking fault, see Fig. 7 These dislocation loops were of

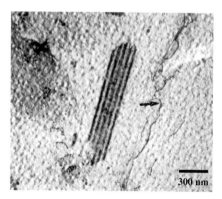

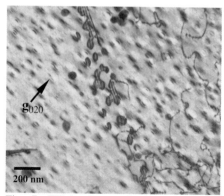

Fig. 6, SESF and partial dislocations in the interrupted creep specimen with σ =250 Mpa, **g** = 0$\bar{2}$0, BD~[001].

Fig. 7, TEM micrograph showing fringed particles being surrounded by partial dislocation loops in an interrupted creep specimen at σ =160 Mpa.

partial dislocation character since the loops disappeared when it was viewed with an opposite **g** vector.

Based on these results as well as the observations of dislocation structures in other crept Ni-base superalloys [9-12], it appears that a unit matrix dislocation dissociated according to the following reaction:

$$a/2\,[01\,\overline{1}] = a/3\,[11\,\overline{2}] + a/6\,[\,\overline{2}\,11] \qquad (1)$$

It appears that when a unit matrix dislocation approaches γ' particles, it is blocked by the creation of high energy APB in the particles. It is then dissociated according to the above mode. The leading partial of a/3 <112> sweeps through both γ matrix and γ' particles, producing stacking faults in both phases. The trailing partial a/6 <112> was, however, pinned at the interfaces of matrix and γ' precipitates since its shearing of γ' involves creating APB energy. It can only move by looping the γ' particles while in the meantime its glide through the matrix eliminates the stacking fault in the γ phase.

TEM observation indicates that the γ' volume fraction does not change during the creep test time, but the average size of γ' particles can increase considerably. For example, at a stress of 160 MPa for 700 hours, the average γ' particle size has increased from an original diameter of ~22 nm to ~110 nm whilst the γ' volume fraction remained ~20%. The coarsening of γ' phase resulted in a change in the interaction of dislocations with these particles. Although dislocation shearing of γ' particles can still be seen occasionally by dislocation pairs in these specimens, evidence of dislocation bypass through bowing and climbing has been extensively observed as can be seen in Fig. 8. With increasing coarsening of γ' size, dislocation deformation was also found to become more homogeneous, in agreement with Gibbons and Hopkins's work [4].

3.2.2. At grain boundaries

During the process of creep testings, η phase with a hexagonal close-packed structure, began to precipitate from grain boundaries. It exhibits a plate-like structure. These precipitates have a specific orientation relation with the γ matrix as determined from selective electron diffraction pattern: $[011]_\gamma$ // $[2\,\overline{1}\,\overline{1}\,0]_\eta$, and $(11\,\overline{1})_\gamma$ // $(0001)_\eta$. This relationship is consistent with a previous report [13]. It was found that there is a γ' depletion zone surrounding these η plates.

η phase plates were often found to be present along grain boundaries. Since η phase plates always precipitate on {111} planes of the γ matrix, it is implied that the grain boundary migrates towards nucleated η phase plates, resulting in a grain boundary segment locally along the η phase plate. This can be seen in Fig. 9 where the grain boundary migration towards two closely precipitated η phase has produced a stepped grain boundary between the two plates. The migration of the grain boundary is driven by the energy change associated with difference in the dislocation density and chemical compositions. Although grain boundary migration occurred locally, it reduced the pinning effects by grain boundary carbides and makes grain boundary slide more possible. It has been suggested that this kind of grain boundary can provide an easy path for crack propagation or voids formation [14]. This has indeed been observed as can be seen in Fig. 10, which shows microcrackings at the interface of η and γ phases in the laminar structure. It is noticed that the local high strain resulted in a strong contrast in the γ phase near the interface.

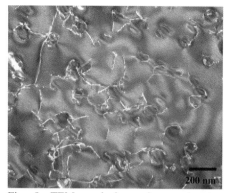 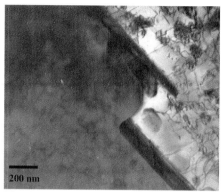

Fig. 8, TEM weak beam image showing dislocation looping and climbing characteristics in creep when γ′ coarsened.

Fig. 9, η phase plates present along grain boundary as a result of grain boundary migration.

Reports on the effect of η precipitation at grain boundaries on creep property are not consistent. On one hand, it can improve creep strength if it promotes a serrated grain boundary formation [15]. On the other hand, it decreases creep life by planar distribution along grain boundaries [16]. The high Ti/Al ratio makes the alloy unstable with respect to η precipitation. The present work suggests that the presence of η phase at grain boundaries can deteriorate creep strength in three ways. First, since it promotes local grain boundary migration, some grain boundary carbides will be left behind, reducing the beneficial pinning effect by carbides. Secondly, it creates microcracking (Fig. 10) at the interfaces which provides an easier crack initiation site and propagation path. Thirdly, the γ′ depletion zone surrounding η phase is parallel to the grain boundary and is therefore prone to localised deformation.

With increasing creep time, another microstructure change at grain boundaries occurred involving carbides. Although they were still coherent with one of neighbouring grains, their size increased and number density decreased resulting in a discontinuous carbide distribution along grain boundaries as can be seen in Fig. 11. The presence of carbides at grain boundaries serves as barriers preventing grain boundary migration and gliding at high temperatures. With increasing separation between carbides, grain boundary migration became possible. The micrograph in Fig. 11 suggests that the grain boundary has migrated away from the carbides driven by the reduction in dislocation density either side of the boundary. The boundary curvature is a result of carbides still pinning some points of the boundary. With grain boundary migration, some carbides were left in the intragranular locations, resulting in a higher density of carbides present in the region of the grain boundaries. This event was not expected at a temperature as low as 800ºC for this alloy. This effectively weakened the grain boundary strength through the ease of grain boundary migration and sliding and promoted

412

diffusion along the grain boundaries. This is consistent with the observation on specimen cross section that the carbide density was much higher on transverse grain boundaries (perpendicular to the loading direction) as can be seen from a SEM micrograph in Fig. 12. It indicates that carbides moved through carbon diffusion from grain boundaries parallel to loading direction to those vertical to loading direction, the same observation as reported before [15,17]. At these boundaries, creep cavities and cavity induced creep cracks were also observed.

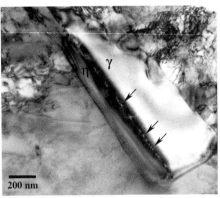

Fig. 10, Laminar structure of η and γ precipitates at the grain boundary. Microcracking occurred at their interface, from a ruptured specimen at σ =160 MPa.

Fig. 11, TEM micrograph showing grain boundary migration and the retardation of its movement by carbides, from a ruptured specimen at σ =160 MPa.

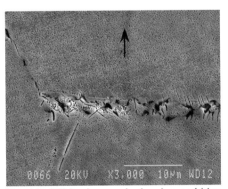

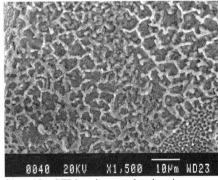

Fig. 12, SEM micrograph showing carbides, platelet η phase and creep cavities at a grain boundary vertical to the loading direction (arrow direction).

Fig. 13, SEM micrograph showing creep cavities on fracture surface, σ =160 MPa.

SEM fracture surface observation revealed that all creep tested specimens fractured predominantly in an intergranular manner. Cavities with high density were often observed on grain boundaries nearly vertical to the loading direction, see Fig. 13. This is consistent with the above cross section observation. The low creep ductility of the material at 800°C is believed to be mainly caused by grain boundary cavity formation and linkage. This view is further supported by the recent work of room temperature prestraining in tension followed by high temperature creep [18]. It was observed that with increasing prestraining, creep rupture strain was consistently reduced. Concomitantly, grain boundary cavity density increased.

CONCLUSIONS

1, Shearing of γ' particles by partial dislocations in slip bands was the predominant mechanism during early creep when γ' precipitate size was small. A unit matrix dislocation dissociates into a leading partial a/3 <112> and a trailing partial a/6 <112>.

2, With increasing γ' coarsening during creep test dislocation bowing and climbing became predominant deformation mechanism and the deformation became more homogeneous.

3, With increasing high temperature exposure, platelet η phase began to precipitate at grain boundaries. Grain boundary migration occurred either because of less pinning effect of grain boundary carbides or η phase precipitation.

4, The low creep ductility of the material was caused by grain boundary cavity formation and linkage during creep. This was enhanced by the η phase distribution along grain boundaries.

ACKNOWLEDGEMENTS

The authors wish to thank EPSRC for financial support for Y.H.Z. and Rolls-Royce Plc for the provision of materials. Helpful discussions with Mr. S. Williams (Rolls-Royce Plc) and Dr Q. Chen are appreciated. The provision of laboratory facilities in the Department of Materials Science and Metallurgy at the University of Cambridge by Professor A.H. Windle is acknowledged.

REFERENCES

1, W. Betteridge and J. Heslop ED., The Nimonic Alloys, Edward Arnold Publishers Limited, London, 1974.

2, B. Dyson, Journal of Pressure Vessel Tech., in press.

3, A. Miller, J. Eng. Mater. Tech., **98** (1976) 97.

4, T.B. Gibbons and B.E. Hopkins, Metal Science, **18** (1984) 273.

5, C.K.L. Davies, A.G. Older and R.N. Stevens, J. Mater. Sci., **27** (1992) 5365.

6, J.W. Edington, Electron Diffraction in the Electron Microscope, Vol.2. MacMillan Publishers, London, 1972.

7, K. Trinckauf and E. Nembach, Phil. Mag. A, **65** (1992) 1383.

8, B.H. Kear, G.R. Leverant and J.M. Oblak, Trans. of ASM, **62** (1969) 639.

414

9, A.J. Huis in't Veld, G. Boom, P.M. Bronsveld and J. Th. M. De Hosson, Scripta Metallurgia, **19** (1985) 1123.

10, M. Condat and B. Decamps, Scripta Metallurgica, **21** (1987) 607.

11, W.W. Milligan and S.D. Antolovich, Metall. Trans., **22A** (1991) 2309.

12, U. Glatzel and M. Feller-Kniepmeier, Scripta Metall. Mater., **25** (1991) 1845.

13, L. Remy, J. Laniesse and H. Aubert, Mater. Sci. Eng., **38** (1979) 227.

14, G.K. Bouse, in Superalloys 1996, Proc. 8th Int. Symp. On Superalloys, R.D. Kissinger, D.J. Deye, D.L. Anton, A.D. Cetel, M.V. Nathal, T.M. Pollock and D.A. Woodford, Eds., Warrendale, PA, The Metallurgical Society AIME, 1996, pp.163.

15, Y. Zhang and F.D.S. Marquis, in Superalloys 1996, Proc. 8th Int. Symp. On Superalloys, R.D. Kissinger, D.J. Deye, D.L. Anton, A.D. Cetel, M.V. Nathal, T.M. Pollock and D.A. Woodford, Eds., Warrendale, PA, The Metallurgical Society AIME, 1996, pp.391.

16, T. Shibata, Y. Shudo, T. Takahashi, Y. Yoshino and T. Ishiguro, in Superalloys 1996, Proc. 8th Int. Symp. On Superalloys, R.D. Kissinger, D.J. Deye, D.L. Anton, A.D. Cetel, M.V. Nathal, T.M. Pollock and D.A. Woodford, Eds., Warrendale, PA, The Metallurgical Society AIME, 1996, pp. 627.

17, S. Kihara, J.B. Newkirt, A. Ohtomo and Y. Saiga, Metall. Trans., **11A** (1980) 1019.

18, Y.H. Zhang and D.M. Knowles, to be presented to the 10th International Conference on Fracture (ICF-10), Honolulu, Hawaii, Dec. 2001.

SUPERALLOYS II

DEFORMATION BEHAVIOR OF A COMMERCIAL Ni-20%Cr ALLOY

R.Kaibyshev, N.Gajnutdinova, V.Valitov

[1]Institute for Metals Superplasticity Problems, Khalturina 39, Ufa 450001, Russia,
e-mail: rustam@anrb.ru, phone/fax: +7 (3472) 253856

Abstract

Deformation behavior of a commercial Ni-20%Cr alloy containing dispersoids was studied in the temperature range from $600^\circ C$ to $950^\circ C$ over five orders of magnitude of strain rate. It is shown that the Ni-20%Cr alloy exhibits a threshold behavior, like that in superalloys produced via mechanical alloying technique. It was demonstrated, by incorporating a threshold stress into the analysis, that the alloy exhibited three characteristic modes of deformation behavior. At the normalized strain rate, $\varepsilon kT/(D_l Gb)$ less than ranging 10^{-8} the value of the true stress exponent, n, is ~4. The true activation energy for plastic deformation, Q_c, in this range is 285 ± 30 kJ/mol. In the normalized strain rate range $10^{-8} - 10^{-4}$, the value of the true stress exponent, n, is ~6, and Q_c value tends to decrease to a value of 175 ± 30 kJ/mol. This change in deformation behavior with decreasing temperature was interpreted in terms of transition from high temperature climb, controlled by lattice diffusion, to low temperature climb, controlled by pipe-diffusion. At $\varepsilon kT/(D_l Gb)=10^{-4}$ the power-law breakdown is observed and at higher normalized strain rates there is a range of exponential creep. A temperature dependence of normalized threshold stress was found in the temperature range 750-$950^\circ C$. Value of the energy term, Q_o, was found to be about 18.5 kJ/mol and tends to increase at lower temperature range.

1. Introduction

Ni-20%Cr solid solution is used as a base alloy in the development of superalloys. There is a great interest in assessing the potential of superalloys for use as structural materials for high-temperature application. The most important targets in development of new superalloys is increment in service temperatures and an increase in duration of its service period with high reliability. Such assessment requires systematic investigation of the creep properties of superalloys but also a close comparison between deformation behavior of a multiphase superalloy and pseudo-single phase Ni-20%Cr solid solution. In addition, a detailed information on deformation behavior of Ni-20%Cr is important to develop the optimum conditions of new thermomechanical processing to produce ultrafine grain structure in wrought superalloys. It will provide a strong enhancement in workability of these materials. Accordingly, systematic investigations on deformation behavior of Ni-20%Cr alloy are essential to the characterization of the creep strength of superalloys and to develop physical basis for grain refinement during forging operations in these materials.

Thus, the aim of the present study is to report the deformation behavior of commercial grade Ni-20%Cr alloy with initial coarse grained structure in a wide temperature range.

2. Materials and Experimental Technique

The Ni-20%Cr alloy with the chemical composition Ni–21%Cr–1.1%Si–0.3%Mn–0.75%Fe – 0.31%Al – 0.08%Ti – 0.35%Cu – 0.05%C, was manufactured at the Beloryatzk Metallurgical Work and purchased as a hot rolled square rod with a face size of 80mm. The alloy had an initial grain size of about 80 μm. Compression specimens 10 mm in diameter and 12 mm in heigh were machined from the alloy ingot. The samples were deformed in compression with lubricant at constant initial crosshead velocity using a Schenck RMS-100 testing machine. Tests were carried out over the temperature range 600-950°C with step 50°C at an initial strain rate in the range from $1.5 \cdot 10^{-6}$ s^{-1} up to $5 \cdot 10^{-2}$ s^{-1}. Strain-rate-jump tests were carried out to provide an additional information of the strain rate effect on mechanical properties. Samples were also examined by using TEM. For these examinations, samples were cut from the gauge section of deformed specimens and thinned to about 0.25 mm. Discs with 3 mm diameter were cut and electropolished to perforation on a Tenupol-3 twinjet polishing unit by using an electrolyte of 10pct perchloric acid in butanol at ambient temperature and 60V. The thin foils were examined using a JEOL-2000EX TEM with a double-tilt stage at an accelerating potential of 160 kV.

3. Results
3.1 The shape of σ-ε curves

Typical true stress- true strain curves in the temperature range 600-950°C at a fixed strain rate of $7 \cdot 10^{-4}$ s^{-1} are presented in Fig.1. It is seen that at low temperatures a significant strain hardening takes place at the initial stage and is followed by a steady-state. An increase in testing temperature causes a reduction in duration of the initial stage and a reduction in the strain-hardening coefficient. The initial stage becomes neglegible at temperatures higher than 800°C. At T=900°C and $\varepsilon = 7 \cdot 10^{-4}$ s^{-1} the steady-state is attained after a true strain of 0.04. The steady-state stresses were used for analysis of deformation behavior of the Ni-20%Cr alloy.

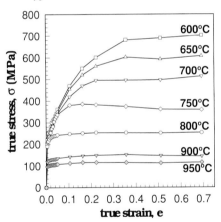

Fig.1. Typical true stress-true strain curves for the Ni-20%Cr alloy at an initial strain rate of $7 \cdot 10^{-4}$s^{-1} at different temperatures.

3.2 The variation of steady-state stress with strain rate

Fig. 2a shows the variation of strain rate with flow stress plotted on a double logarithmic scale for each of temperatures examined. It is seen, that the data obtained from strain rate jump test show excellent agreement with those obtained at constant initial strain rate. It is seen that the experimental data lying to the left from dashed line in the plot can be represented accurately by straight lines at strain rates higher than 10^{-5} s^{-1} and can be described in terms of the power-law [1]:

$$\dot{\varepsilon} = A \cdot \left(\frac{\sigma}{G}\right)^{n} \cdot \exp\left(\frac{-Q}{R \cdot T}\right), \qquad (1)$$

where ε is the strain rate, A is a constant, n is the stress exponent, σ is the steady state flow stress, G is the shear modulus, Q is the activation energy for plastic deformation, R is the gas constant and T is the absolute temperature, at temperatures ranging from 700 to 950°C in a wide temperature range. The stress exponent, n, which represent the slope of the plots tends to minor increase from 6 to 7 with decreasing temperature from 950 to 700°C. In the temperature range 600-650°C, the values of the stress exponent lying in range from 9 to 12 are essentially high and it can be assumed that deformation behavior of the Ni-20%Cr alloy can be described in terms of the exponential relation [1]:

$$\varepsilon = B \cdot \exp(\beta\sigma) \cdot \exp\left(\frac{-Q}{R \cdot T}\right), \qquad (2)$$

where B is a constant and β is a coefficient. It is apparent from Fig.2 that the datum points lying to the right from the dashed line exhibit the best linear fit in case of the semi-logarithmic plot (Fig.2b). Therefore, the exponential relationship (2) adequately describes the deformation behavior of Ni-20%Cr in the range of low temperatures.

The Fig.2a also shows that in the temperature range 700-950°C at strain rates less than 10^{-5} s^{-1} the stress exponent, n, tends to increase with decreasing strain rate.

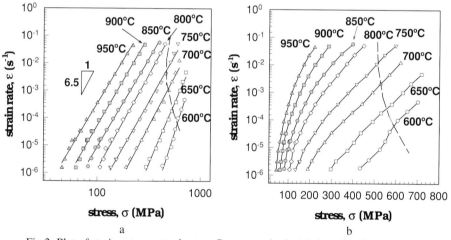

Fig.2. Plot of strain rate *vs* steady state flow stress in double logarithmic scale. Broken line deviates the region of exponential creep from region of power-law. (a) Double logarithmic scale. (b) Semi-logarithmic scale.

3.3 The activation energy for plastic deformation

Algebraically, Eq. 1 can be converted [4] to

$$\ln\sigma = \ln(\varepsilon/A)^{1/n} + \frac{Q}{Rn} \cdot \frac{1}{T}, \qquad (3)$$

420

The apparent activation energy for plastic deformation, Q_a, was obtained graphically by logarithmic plotting of the normalized stress compensated by non-linear temperature dependence of the shear modulus [5], σ/G, against the inverse of the absolute temperatures and taking the slope of tangent to be $Q/(Rn)$ (Fig. 3) for different strain rates.

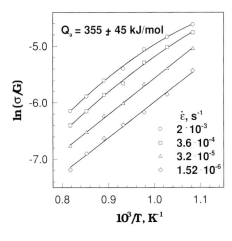

It is seen, that calculated values of Q_a, lying in the range from 300 to 400 kJ/mol, are higher than the activation energy for lattice diffusion of Ni atoms into Ni-20%Cr solid solution (Q_l=285kJ/mol) [5].

Fig. 3. Shear modulus-compensated flow stress *vs* inverse of absolute temperature.

3.4 Microstructure observation

TEM observation of deformed specimens shows that microstructure features depend on temperature. At T≥750°C a uniform distribution of lattice dislocation is observed. Observation of helicoidal dislocations is indicative on an extensive dislocation climb in the Ni-20%Cr alloy at these temperatures (Fig.4a).

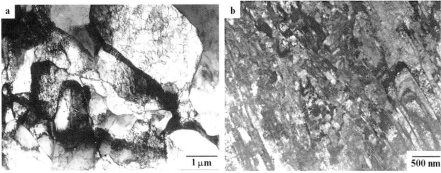

Fig. 4. TEM micrographs of the Ni-20%Cr alloy after deformation: (a) T=900°C, $\dot{\varepsilon}$ =7·10^{-4}s^{-1}, ε=50%, (b) T=600°C, $\dot{\varepsilon}$ =7·10^{-4}s^{-1}, ε=70%.

Temperature decrease leads to localization of dislocation glide. It is seen (Fig.4b), that at low temperatures (≤700°C) formation of deformation bands is observed.

4. Discussion

An inspection of the deformation behavior of the Ni-20%Cr alloy suggests that this material could exhibit threshold behavior. The deformation behavior resembles that of dispersion strengthened alloys (DS) [2] with regard to (i) the progressive decrease in strain rate with decreasing applied stress in the range of lowest strain rates; (ii) the increase in the stress exponent with decreasing applied stress; (iii) the enhanced values of the activation energy for plastic deformation. This type of deformation behavior can be interpreted in terms of a threshold stress, σ_{th}. In this case, the steady-state strain rate, $\dot{\varepsilon}$, of materials is represented by relationship of the form

$$\dot{\varepsilon} = A \cdot \left(\frac{\sigma - \sigma_{th}}{G} \right)^n \cdot \exp\left(\frac{-Q_c}{R \cdot T} \right), \qquad (4)$$

where σ is the applied stress, G is the shear modulus, n is the true stress exponent, Q_c is the true activation energy for plastic deformation, R is the gas constant, T is the absolute temperature and A is a dimensionless constant. Plastic deformation is driven by the effective stress σ_{eff}, equal to $\sigma - \sigma_{th}$.

4.1 Examination of the presence of a threshold stress

A standard procedure described in [3] was used to determine the threshold stress. The experimental data at a single temperature were plotted as $\dot{\varepsilon}^{1/n}$ against the steady state flow stress σ on a double linear scale for all temperatures examined.

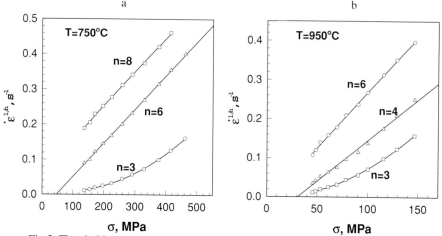

Fig.5. Threshold stress in Ni-20%Cr. (a) The linear extrapolation method for estimating the threshold stress at T=750°C using values for n of 3, the datum points exhibit curvature (concavity);6, datum points provide the best linear fit datum points; 8, datum point exhibit curvature (convexity). (b) The linear extrapolation method for estimating the threshold stress at T=950°C using values for n of 3, the datum points exhibit curvature (concavity); 4, datum points provides the best linear fit datum points; 6, datum point exhibit curvature (convexity).

If the datum points are obeyed by equation (4) and the threshold stress is constant for each testing temperature, the data can be fitted to a straight line by varying the n values.

Extrapolation of this line to zero strain rates yields the σ_{th} value. The n value exhibiting the highest regression coefficient in a linear fit is taken as the true stress exponent value. Fig.5 shows the typical plots $\varepsilon^{1/n}$ vs σ. At temperatures of 950°C and 750°C the stress exponent value of 4 and 6 yields the best linear fit between $\varepsilon^{1/n}$ and σ, respectively. At higher and lower values of stress exponent, n, the datum points exhibit curvature. Notably the experimental points from region of exponential creep at T≤750°C were discarded.

Table 1 summarizes the threshold stresses and true stress exponent values calculated. It is seen that the true stress exponent tends to increase from 4 to 7 with decreasing temperatures from 950°C to 650°C. It is apparent that the estimated threshold stress decreases with increasing temperature. In the DS materials the threshold stress is obeyed by the relationship of the form [3]:

$$\frac{\sigma_{th}}{G} = B_o \exp\left(\frac{Q_o}{R \cdot T}\right), \tag{5}$$

where B_o is a constant; Q_o is the energy term representing the activation energy for a dislocation to overcome an obstacle.

Table 1. Threshold stresses and n values at different temperatures.

T, °C	650	700	750	800	850	900	950
n	7	6	6	5.5	5	4.5	4
σ_{th}, MPa	114.8	57.6	49.5	45.8	40.5	38.5	30.8

The variation of normalized threshold stress, σ_{th}/G, with temperature in semi-logarithmic scale is shown in Fig.6. Two different temperature dependencies of threshold stress obeyed by relationship (2) can be distinguished. It is seen, that in the temperature range 950-700°C the threshold stresses fit to a line for which the value of Q_o may be estimated as ~18.5 kJ/mol. In the temperature range 650-700°C the threshold stresses exhibit the linear temperature dependence with the Q_o value of about 100 kJ/mol. A transition in temperature dependence of threshold stress is observed with decreasing temperature with inflection point at T=700°C.

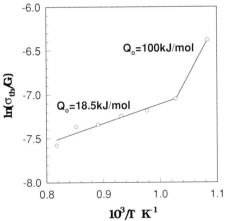

Fig.6. Semi-logarithmic plot of the normalized threshold stress vs the resiprocal of absolute temperature.

4.2 Activation energy

To determine the activation energy for deformation, the normalized effective stresses $((\sigma-\sigma_{th})/G)$ were plotted as a function of $1/T$ on semi-logarithmic scales at a fixed strain rate (Fig.7a) [4]. Two temperature intervals (650-700°C, 700-950°C), distinguished by slope value, k, can be found. In these temperature intervals the values k are ~7.35, and ~3.1, respectively. Notably in the temperature ranges revealed, the data in Fig. 7a shows very little strain rate dependence, which is indicative of the validity of the threshold stress values determined.

Values of true activation energy (Q_c) of plastic deformation were calculated assuming that Eqn. (4) adequately describes the deformation behavior of the Ni-20%Cr alloy and presented in Fig. 7b. It is seen that the Q_c value is 175±30 kJ/mol in the temperature range 650-700°C. This value is slightly higher than that for pipe-diffusion of Ni atoms into Ni-Cr solid solution (Q_p=170 kJ/mol) [5]. In the temperature range 700-950°C the value of true activation energy, Q_c, is 285±30 kJ/mol. This value is essentially close to activation energy for lattice diffusion of Ni atoms into Ni-Cr solid solution (Q_l=285 kJ/mol) [5]. Therefore, an increase in Q_c value with increasing temperature can be interpreted in terms of transition from dislocation climb controlled by pipe-diffusion to dislocation climb controlled by lattice diffusion [1,6].

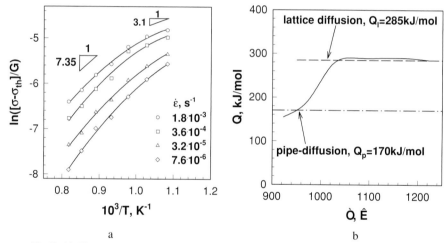

a   b

Fig.7. (a) Shear modulus-compensated effective stress, $(\sigma-\sigma_{th})/G$ *vs* inverse of absolute temperature in semi-logarithmic scale. (b)The true activation energy for plastic deformation, Q_c, as a function of inverse of absolute temperature.

4.3 An examination of normalized deformation data

Figure 8 shows a double logarithmic plot of the normalized strain rate, $\dot\varepsilon\, kT/D_lGb$, vs the normalized effective stress, $(\sigma-\sigma_{th})/G$ for the Ni-20%Cr alloy. The Burger's vector was taken as $b=2.5\cdot10^{-10}$ m [5]. The $D_l=1.6\cdot10^{-4}$ exp (-285/RT) is a lattice diffusion coefficient of Ni atoms into Ni-Cr solid solution [5]. It is seen, that the Ni-20%Cr alloy exhibits three characteristic modes of deformation behavior. First, at the low normalized strain rate, $\dot\varepsilon\, kT/(D_lGb)<10^{-8}$, there is a well-defined power-low relationship with the stress exponent ~4. Taking into account the aforementioned values of the true activation energy this deformation behavior can be in consistent with high temperature dislocation climb controlled by lattice self-diffusion. Second, at moderate normalized strain rate, $\dot\varepsilon\, kT/(D_lGb)$, lying in the range from 10^{-8}, to 10^{-4}, the slope of linear dependence $\dot{\varepsilon}\,kT/(D_1Gb)$ *vs* $(\sigma-\sigma_{th})/G$ is ~6. The classic relationship $n_{lt}=n_{ht}+2$ is observed at this transition. Inspection of figs.7b and 8 shows that datum points fitted to the line with a slope of about ~6 lie in the stress range where values of the true activation energy are close to activation energy for pipe-diffusion. Such transition can be interpreted in terms of transition from high temperature climb controlled by lattice diffusion to low temperature climb controlled by pipe-diffusion along the dislocation cores. A drop in a value of the true activation energy at this transition is indicative about strong decrease in rate of dislocation rearrangments. It is known [7] that the ability of lattice dislocations to rearrange via cross-slip is restricted due to a low value of stacking fault energy. The dislocation climb plays the most important role in dislocation rearrangments in this material. As a result, a decrease in the rate of dislocation climb effects strongly deformation behavior of this material. The inflection point of $\dot{\varepsilon}\,kT/(D_1Gb)=10^{-8}$ is a well-defined boundary between hot deformation and warm deformation in Ni-20%Cr. Probably, this is the reason for exact match between this point and the Sherby-Burke criterion.

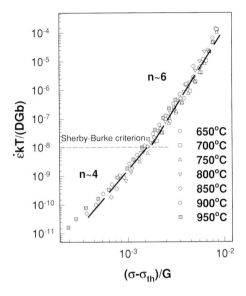

Fig.8. Normalized strain rate $\dot{\varepsilon}\,kT/(D_1Gb)$, *vs* normalized effective stress, $(\sigma-\sigma_{th})/G$.

A transition from the power-law (4) to exponential creep (5) takes place at the normalized strain rate of about 10^{-4}. Therefore, the breakdown of the power-law takes place at the strain rate which is higher by four orders than the strain rate for the Sherby-Burke criterion ($\dot{\varepsilon}\,kT/(D_1Gb)=10^{-8}$).

5. Conclusions

1. The deformation behavior of the Ni-20%Cr commercial alloy was studied in the temperature range 600-950°C over five orders of magnitude of strain rate. The Ni-20%Cr alloy exhibits the threshold behavior, like that in dispersion strengthened (DS) alloys.
2. Analysis of the deformation data of the Ni-20%Cr alloy revealed the presence of the threshold stress, σ_{th}, that depends strongly on temperature according to the equation:

$$\frac{\sigma_{th}}{G} = B_o\exp\left(\frac{Q_o}{RT}\right),$$

where B_o is a constant, and Q_o is the energy term, G is the shear modulus. Two types of the temperature dependence of threshold stress distinguishing by values of energy term were found. In the temperature ranges 650-700°C and 700-950°C the Q_o value is equal to 100 and to 18.5 kJ/mol, respectively.

3. By incorporating the threshold stress into the analysis, it was shown that the true exponent, n, decreases from ~6 at the intermediate temperatures to ~4 at high temperatures. This transition takes place at the normalized strain rate $\varepsilon\, kT/(D_1Gb)=10^{-8}$ which matches exactly the Sherby-Burke criterion Transition to exponential creep was found at the normalized strain rate of about 10^{-4}.

4. The true activation energy for plastic deformation, Q_c, was found to decrease, from a value of about $285\pm30kJ/mol$ at temperatures ranging from 750 to 950°C to the value of about $175\pm30kJ/mol$ at T=650-700°C. Deformation behavior of the Ni-20%Cr alloy in range of the power-law was interpreted in terms of transition from low temperature climb to high temperature climb.

6. References

[1]. Chadek, J., 1994, Creep in Metallic Materials, Academia, Prague, p. 302.

[2]. Lund, R.W., Nix, W.D., 1976, High temperature creep of Ni-20Cr-2ThO$_2$ single crystals, Acta Metall., vol.**24**, p.469.

[3]. Mohamed, F.A., Park, K.T., Lavernia, E.J., 1992, Creep Behavior of Discontinuous SiC-Al Composites, Mat. Sci. Eng., Vol.A150, pp. 21-35.

[4]. Pickens, R., Langan, T.J., England, R.O., Liebson, M., 1987, A Study of the Hot Working Behaviour of SiC-Al Alloy Composites and their Matrix Alloys by Hot Torsion Testing, Metall. Trans., Vol.18A, No.2, pp. 303-312.

[5]. Frost, H., Ashby, M., 1982, Deformation-Mechanisms Maps, Pergamon Press, p. 328.

[6]. Raj, S., Langdon, T., 1989, Creep Behavior of Copper at Intermediate Temperatures – I. Mechanical Characteristics, Acta Metall., Vol.37, No.2, pp. 843-852.

[7]. Fridel, J., 1964, Dislocations, Pergamon Press.

Creep deformation of Ni-20mass%Cr single crystals with [001], [011] and [-111] orientations

Yoshihiro Terada[1], Daiji Kawaguchi[2] and Takashi Matsuo[1]

[1] Department of Metallurgy and Ceramics Science, Tokyo Institute of Technology,
Meguro-ku, Tokyo 152-8552, Japan
[2] Graduate student, currently at HONDA R&D Co., LTD, Asaka-shi,
Saitama 351-8555, Japan

Abstract

To characterize the effect of stress on creep of γ single phase Ni-20mass%Cr single crystals with [001], [011] and [-111] orientations, the creep tests were carried out at 1173 K in the stress range of 14.7 - 98.0 MPa for the single crystals. In the higher stresses above 49.0 MPa, the decreasing ratio of creep rate during transient stage is negligible and the accelerating creep is prematurely initiated for the [001] and [-111] single crystals, whereas in the lower stresses below 49.0 MPa the decrease in creep rate during transient stage is pronounced and the onset of accelerating creep is retarded for every orientation. In the lower stresses, the transient creep stage is the most pronounced in the [001] single crystal and the least significant in the [-111] single crystal. The microstructures of the single crystals creep-ruptured at the higher stresses above 49.0 MPa are characterized by the heterogeneity in every orientation. In the higher stresses above 49.0 MPa dynamically recrystallized grains are observed heterogeneously for the [011] and [-111] single crystals and twins are running oblique to the stress axis for the [001] single crystal, while in the lower stresses below 49.0 MPa the dynamic recrystallization occurs homogeneously in the whole gage portion for every orientation.

1. Introduction

Creep of the Ni-20mass%Cr single crystals with γ single phase is affected by the crystallographic orientation corresponding to the stress axis. In our previous study, we investigated the creep curve for the Ni-20mass%Cr single crystals with the [001], [011] and [-111] orientations at 1173 K - 29.4 MPa [1], where the stress is low enough to appear the creep characteristics of the single crystal [2]. The characteristics of creep curves for the single crystal are the extended transient creep stage and the retarded onset of accelerating creep stage. The [001] single crystal shows the most extended transient creep stage, resulting in the longest rupture life and the largest rupture strain among the three. The [-111] single crystal has the smallest creep rate just after loading, however the accelerating creep is initiated at the quite small strain. The creep rate for the [011] single crystal is the largest, and the time to initiate the accelerating creep is the shortest among the three. The microstructures of the creep-ruptured single crystals are characterized by dynamically recrystallized grains. The dynamic recrystallization occurs homogeneously in the whole gage portion in the [001] and [011] single crystals, while fine grains are heterogeneously distributed in the [-111] single crystal. It is noteworthy that many twins are introduced in the [001] single crystal.

Creep curve and microstructure of the Ni-20mass%Cr single crystals depend also on the applied stress. We conducted the creep tests for the Ni-20mass%Cr single crystals at 1173 K in the wide stress range of 19.6 - 98.0 MPa, where the crystallographic orientations of the single crystals corresponding to the stress axis were located within the standard stereographic triangle [3]. It was indicated that the transient creep stage becomes pronounced and the onset of accelerating creep stage is retarded with decreasing the stress in the lower stresses below 49.0 MPa, while at the higher stresses above 49.0 MPa the slight decrease in creep rate during transient stage is detected. The dynamic recrystallization occurs homogeneously in the whole gage portion at the lower stresses, while the microstructures turn to heterogeneous at the higher stresses; that is, the appearance of dynamically recrystallized grain boundaries is limited to the center of the specimen.

In this study, the creep for the Ni-20mass%Cr single crystals with the [001], [011] and [-111] orientations has been investigated at 1173 K in a wide stress range of 14.7 - 98.0 MPa. The microstructures of the [001], [011] and [-111] single crystals subjected to creep deformation have been also characterized using the creep-ruptured specimens.

2. Experimental

A Ni-20mass%Cr alloy, designated as Ni-20Cr in this paper, was prepared for this study. The chemical composition of the alloy is shown in **Table 1**. The alloy was melted in a high frequency induction furnace in a vacuum and cast into a 4 kg ingot in an argon atmosphere. The ingot was hot-forged into rod with a diameter of 13 mm. The single crystals with [001], [011] and [-111] orientations were successfully grown by a modified Bridgman furnace in a flow of purified argon and then homogenized at 1523 K for 36 ks. The crystal orientation corresponding to the stress axis was determined by the Laue back-reflection X-ray technique. Only the crystals within three degrees of the desired orientations were employed for creep tests. Full size creep specimens with the gage portion of 30 mm in length and 6 mm in diameter were machined from the crystals.

Table 1. Chemical composition (mass%) of crystals prepared in this investigation.

Alloy	C	Si	Mn	Cr	Ni
Ni-20Cr	0.001	0.49	0.30	19.3	Bal.

Tensile creep tests were carried out at 1173 K under the constant stresses in the range of 14.7 - 98.0 MPa using the single-lever type creep machines. During the creep tests, stress was kept constant by adjusting the auxiliary weights in an accuracy within one per cent against initial stress. Creep strain was automatically recorded using linear variable-differential transducers via extensometer fitting onto annular ridges machined at both ends of the specimen gage portion. Microstructures subjected to creep deformation were observed by the optical and the transmission electron microscopes.

3. Results

3.1. Creep characteristics

Creep rate - time curves of the [001] single crystals are shown in **Fig. 1**, where the creep rate is a true strain rate [4]. At 29.4 MPa, the creep rate initially remains constant and then decreases to a minimum creep rate, followed by accelerating. The minimum creep rate is 1/40 smaller than that of the plateau region at the beginning of the creep test. The marked decrease in creep rate during the transient stage disappears and the plateau region is followed by the abrupt creep acceleration at the higher stresses above 49.0 MPa. Note that the rupture strains of the [001] single crystals exceed 1.5 in every stress.

Creep rate - time curves of the [011] single crystals are shown in **Fig. 2**. In each stress, the curve basically shows a similar shape; the creep rate initially remains constant and then decreases to a minimum creep rate, followed by accelerating. The minimum creep rate is about 1/5 smaller than that of the plateau region at the beginning of the creep test in every stress. The rupture strain is around unity at the higher stresses, while it abruptly decreases to 0.1 at the lower stresses below 19.6 MPa. Note that the creep curve at 98.0 MPa was not obtained, because the instantaneous plastic strain introduced at the stress application exceeded 0.8.

Creep rate - time curves of the [-111] single crystals are shown in **Fig. 3**. As in the case of [001] single crystal, the decrease in creep rate during the transient stage observed at the lower stress disappears at 68.6 MPa also for the [-111] single crystal. The rupture strain of the [-111] single crystals is around 0.4, irrespective of stress.

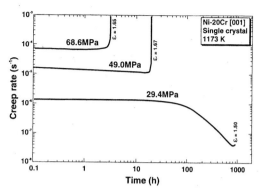

Fig. 1. Creep rate - time curves of the [001] single crystal of Ni-20Cr tested at 1173 K under the constant stresses of 29.4, 49.0 and 68.6 MPa.

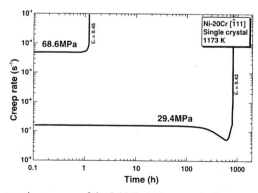

Fig. 2. Creep rate - time curves of the [011] single crystal of Ni-20Cr tested at 1173 K under the constant stresses between 14.7 and 68.6 MPa. An anomalous creep occurs at the beginning of the accelerating stage at the lower stresses below 29.4 MPa.

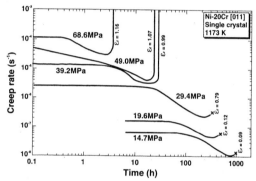

Fig. 3. Creep rate - time curves of the [-111] single crystal of Ni-20Cr tested at 1173 K under the constant stresses of 29.4 and 68.6 MPa.

430

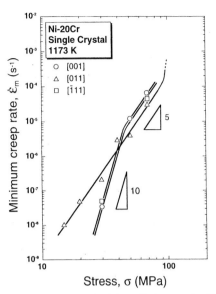

Fig. 4. Variation of minimum creep rate with applied stress for Ni-20Cr single crystals with [001], [011] and [-111] orientations at 1173 K.

The stress - minimum creep rate curves of the Ni-20Cr single crystals with the [001], [011] and [-111] orientations are summarized in **Fig. 4**. For the [011] single crystal, the stress exponent of the minimum creep rate, n, is five in the whole stress region investigated in this study. The n value for the [001] single crystal is about ten at the lower stresses below 49.0 MPa, whereas it is five at the higher stresses above 49.0 MPa. The increase in n value at the lower stress region for the [001] single crystal would be caused by the expansion of transient creep stage in the lower stress as indicated in Fig. 1. One certainly finds that the plots for the [-111] single crystal fall well on the line of [001] single crystal.

3.2. Microstructures

Optical micrographs of the [001] single crystals creep-ruptured at 1173 K - 49.0 and 68.6 MPa are shown in **Fig. 5**. Note that the observation was conducted on the sections of the creep specimen parallel to the stress axis. Dynamically recrystallized grains are observed homogeneously in the whole gage portion in the ruptured specimen at 29.4 MPa, as reported in our previous article [1]. However, in the ruptured specimen at 49.0 MPa, twins are running in the zigzag way oblique to the stress axis **(Fig. 5(a))** and dynamically recrystallized grains were quite restricted in the vicinity of the ruptured portion. At 68.6 MPa, twins observed in the gage portion looks straight in shape, and they cross each other **(Fig. 5(b))**.

Optical micrographs of the [011] single crystals creep-ruptured at 1173 K - 49.0 and 68.6 MPa are shown in **Fig. 6**. At 49.0 MPa, dynamically recrystallized grains with equi-axed in shape generate in the whole gage portion. The grain diameter is approximately 70 μm in the vicinity of the ruptured portion **(Fig. 6(a))**. The diameter of dynamically recrystallized grains monotonically increases with a distance from the ruptured portion. In the gage portion far distant from the ruptured portion, subboundaries are usually detected in the large size of dynamically recrystallized grains **(Fig. 6(b))**. In the creep-ruptured specimen at 68.6 MPa, subboundaries aligned perpendicular to the stress axis cover in the whole gage portion, and dynamically recrystallized grains are not observed at all **(Fig. 6(c))**.

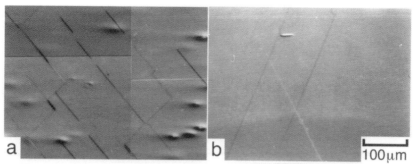

Fig. 5. Optical micrographs of the [001] single crystals of Ni-20Cr creep-ruptured at 1173 K under the constant stresses of 49.0 (a) and 68.6 MPa (b). The stress axis corresponds to the horizontal direction.

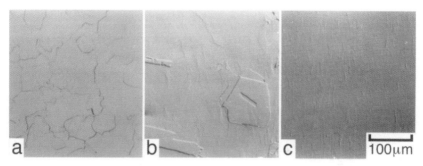

Fig. 6. Optical micrographs of the [011] single crystals of Ni-20Cr creep-ruptured at 1173 K under the constant stresses of 49.0 (a,b) and 68.6 MPa (c). The stress axis corresponds to the horizontal direction

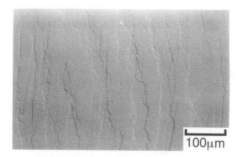

Fig. 7. Optical micrograph of the [-111] single crystal of Ni-20Cr creep-ruptured at 1173 K - 68.6 MPa. The stress axis corresponds to the horizontal direction.

432

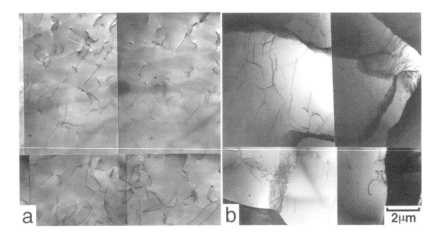

Fig. 8. Transmission electron micrographs of Ni-20Cr single crystals with [001] (a) and [011] orientations (b) creep-ruptured at 1173 K - 49.0 MPa.

For the [-111] single crystal creep-ruptured at 1173 K - 68.6 MPa, subboundatries cover in the whole gage portion as shown in **Fig. 7**. It is noted that a few dynamically recrystallized grain boundaries were introduced parallel to the stress axis typically in the vicinity of the ruptured portion. From the above results of the optical micrographs of the creep-ruptured specimen, one can find that the microstructures at higher stresses are characterized by the heterogeneity in every orientation. Dynamically recrystallized grains and subboundaries are generated heterogeneously for the [011] and [-111] single crystals, while twins are introduced for the [001] single crystal.

Figure 8 shows the transmission electron micrographs of the [001] and [011] single crystals creep-ruptured at 49.0 MPa, where twins are predominantly observed in the [001] single crystal by the optical microscope and subboundaries cover the gage portion in the [011] single crystal. It is noted that the thin films are extracted from the gage portion far distant from the ruptured portion, where cracks are scarcely included in both specimens. Well-established subgrains with the diameter of 7 μm are observed for the [011] single crystal. On the contrary, dislocations are homogeneously distributed and no subboundaries are observed for the [001] single crystal.

4. Discussion

By summarizing the creep curves of the Ni-20Cr single crystals with the [001], [011] and [-111] orientations, one can find the following characteristics. Firstly, the decreasing ratio of creep rate during transient stage is negligible and the accelerating creep is prematurely initiated for the [001] and [-111] single crystals in the higher stresses above 49.0 MPa, whereas the decrease in creep rate during transient stage is pronounced and the onset of accelerating creep is retarded in the lower stresses below 49.0 MPa. Secondly, the reduction of transient creep at the higher stresses is not detected for the [011] single crystal. And thirdly, the decreasing ratio of creep rate during transient stage in the lower stresses is the most pronounced in the [001] single crystal, while it is the least significant in the [-111] single crystal. To characterize the creep curves of the [001], [011] and [-111] single crystals, we pay attention to the ratio of transient creep duration to rupture life, t_m/t_r, and the decreasing ratio of creep rate during transient stage, $\dot{\varepsilon}_m/\dot{\varepsilon}_I$.

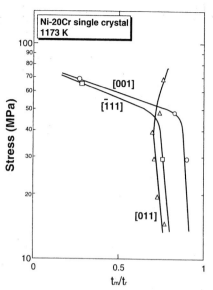

Fig. 9. Correlation between applied stress and t_m/t_r for Ni-20Cr single crystals with [001], [011] and [-111] orientations, where t_m is the time showing the minimum creep rate and t_r is the rupture life.

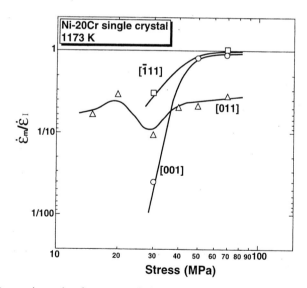

Fig. 10. Decreasing ratio of creep rate during transient creep stage, $\dot{\varepsilon}_m/\dot{\varepsilon}_I$, as a function of applied stress for Ni-20Cr single crystals with [001], [011] and [-111] orientations, where $\dot{\varepsilon}_m$ is the minimum creep rate and $\dot{\varepsilon}_I$ is the creep rate of plateau region at the beginning of the creep test.

The ratio of transient creep duration to rupture life, t_m/t_r, is summarized for the [001], [011] and [-111] single crystals in **Fig. 9**, where t_m is the time showing the minimum creep rate and t_r is the rupture life. For the [011] single crystal, the value of t_m/t_r is around 0.7 in every stress, indicating that the creep is predominantly occupied by transient creep stage rather than accelerating creep stage. The value of t_m/t_r is about 0.9 in the lower stresses below 49.0 MPa for the [001] single crystal, while it drastically decreases with increasing the stress in the higher stresses above 49.0 MPa. The value is 0.27 at 68.6 MPa for the [001] single crystal, indicating that the accelerating creep stage predominantly occupies the creep. The drastic decrease in t_m/t_r is observed in the higher stresses above 49.0 MPa also for the [-111] single crystal.

The decreasing ratio of creep rate during transient stage, $\dot{\varepsilon}_m/\dot{\varepsilon}_I$, is summarized as a function of stress for the [001], [011] and [-111] single crystals in **Fig. 10**, where $\dot{\varepsilon}_m$ is the minimum creep rate and $\dot{\varepsilon}_I$ is the creep rate of plateau region at the beginning of creep test. For the [001] single crystal, the decreasing ratio of creep rate is quite small in the stresses above 49.0 MPa, while the decreasing ratio is drastically pronounced with decreasing the stress below 49.0 MPa. The value of $\dot{\varepsilon}_m/\dot{\varepsilon}_I$ reaches 1/40 at 29.4 MPa for the [001] single crystal. The decreasing ratio is pronounced below 49.0 MPa also for the [-111] single crystal, while the magnitude is less significant than that for the [001] single crystal. The value of $\dot{\varepsilon}_m/\dot{\varepsilon}_I$ is 1/3 at 29.4 MPa for the [-111] single crystal. The changing manner for the [011] single crystal is rather complicated, however it is mentioned that the value of $\dot{\varepsilon}_m/\dot{\varepsilon}_I$ ranges between 1/3 and 1/10 in every stress.

In conclusion, in the higher stresses above 49.0 MPa the decreasing ratio of creep rate during transient stage is negligible and the creep duration is predominantly occupied by accelerating stage for the [001] and [-111] single crystals, whereas in the lower stresses below 49.0 MPa the decrease in creep rate during transient stage is pronounced and the creep duration is predominantly occupied by transient stage in every orientation. In the lower stresses, the transient creep stage is the most pronounced in the [001] single crystal and the least significant in the [-111] single crystal. The shortening of transient creep stage at higher stresses may be caused by the loss of dislocation tangle through the induction of twins in the course of creep for the [001] single crystal and by the heterogeneous formation of dynamically recrystallized grain boundaries for the [-111] single crystal. To clarify the formation mechanism of twins during creep typically identified for the [001] single crystal would be a future subject.

5. Conclusions

To characterize the effect of stress on creep of γ single phase Ni-20mass%Cr single crystals with [001], [011] and [-111] orientations, the creep tests were carried out at 1173 K in the stress range of 14.7 - 98.0 MPa for the single crystals. The results are summarized as follows:

1. In the higher stresses above 49.0 MPa, the decreasing ratio of creep rate during transient stage is negligible and the accelerating creep is prematurely initiated for the [001] and [-111] single crystals, whereas in the lower stresses below 49.0 MPa the decrease in creep rate during transient stage is pronounced and the onset of accelerating creep is retarded for every orientation. In the lower stresses, the transient creep stage is the most pronounced in the [001] single crystal and the least significant in the [-111] single crystal.

2. The microstructures of the single crystals creep-ruptured at the higher stresses above 49.0 MPa are characterized by the heterogeneity in every orientation. In the higher stresses above 49.0 MPa dynamically recrystallized grains are observed heterogeneously for the [011] and [-111] single crystals and twins are running oblique to the stress axis for the [001] single crystal, while in the lower stresses below 49.0 MPa the dynamic recrystallization occurs homogeneously in the whole gage portion for every orientation.

References
[1] Miyazawa, H., Takaku, R., Kawaguchi, D., Terada, Y., Matsuo, T., 2000, Creep and Evolution of Dynamic Recrystallization in γ Single Phase Single Crystals Located at Poles in Standard Streo-Triangle, Key Eng. Mater., Vol.171-174, pp.577-584.
[2] Matsuo, T., Takahashi, S., Ishiwari, Terada, Y., 2000, Creep in Single Crystals of γ Single phase Ni-20Cr Alloy and Evolution of Dynamic Recrystallization, Key Eng. Mater., Vol.171-174, pp.553-560.
[3] Terada, Y., Ishiwari, Y., Matsuo, T., 2000, Stress Dependence of Evolution of Dynamic Recrystallization in γ Single Phase Single Crystal, Key Eng. Mater., Vol.171-174, pp.585-592.
[4] Hertzberg, R.W., 1996, *Deformation and Fracture Mechanisms of Engineering Materials, 4th edn.*, Wiley, New York.

Yoshihiro Terada, e-mail: terada@mtl.titech.ac.jp, fax: +81-3-5734-2874
Daiji Kawaguchi, e-mail: daiji.kawaguchi@mail.a.rd.honda.co.jp, fax: +81-48-462-2957
Takashi Matsuo, e-mail: tmatsuo@mtl.titech.ac.jp, fax: +81-3-5734-2874

Stress Dependence of Strain Attained to Rafting of γ' phase in Single Crystal Nickel-based Superalloy, CMSX-4

Nobuhiro MIURA, Yoshihiro KONDO and Takashi MATSUO*
The National Defense Academy, 1-10-20 Hashirimizu Yokosuka Kanagawa 239-8686, Japan
*Tokyo Institute of Technology, 2-12-1 Oookayama Meguro Tokyo 152-8552, Japan

Abstract

The stress dependence of the strain attained to the rafted structure of the γ' phase was investigated using a single crystal nickel-based superalloy, CMSX-4. The strain attained to the rafted γ' is designated as the rafted strain. The rafted strain is confirmed as the strain at which the ratio of the length of the γ' plate to the thickness of the plate becomes the maximum value. Creep tests were done at 1273K, 100-400MPa. At the stresses higher than 250MPa, the cuboidal γ' turned to the rafted structure at the strain of 0.1 and more. At the stresses less than 200MPa, the rafted strain becomes the smaller value of 0.01. Corresponding to the drastic change in the rafted strain between the two specimens crept at 200 and 250MPa, the marked difference in the rafted structure is obtained as follows. The γ / γ' interface is waved in the specimen crept at 250MPa, while in the specimen crept at 200MPa the shape of the interface turns to straight in shape. The aspect ratio of the rafted γ' decreases monotonously with decreasing the stress. The difference in the features of the rafted γ' between the two specimens is discussed, correlated with the difference in the rafted strain.

1. Introduction

By subjecting to creep deformation, the cuboidal γ' in the [001] orientated single crystal nickel-based superalloy turns its shape to the rafted γ' whose plates are oriented normal to the stress axis. There are many reports presented the formation of the rafted structure of the γ' phase. Nagai et al. examined the morphological change in the γ' phase subjected to creep deformation in a wide stress range at 1273K in the nickel-based superalloy, Inconel 713C, and indicated that the rafting of the cuboidal γ' did not occur in the specimens at the stresses less than 100MPa[1]. In contrast to this, the occurrence of rafting of the γ' phase at the stresses less than 100MPa was confirmed in the CMSX-4 crept at 1323K[2] in the works done in the research group on strengthening of heat resisting steel and alloy of ISIJ.

A few researchers showed the stress dependence of the rafted strain. Henderson measured the aspect ratio of the γ' phase in CMSX-4 crept at 1223K-155MPa, and found the rafted strain to be about 0.01[3]. Moreover, the rafted strains in the two single crystals of nickel-based superalloys, SC-16 and a Ni-Al-Mo-Ta alloy, were compared with that of CMSX-4 and indicated that the rafted strains in these two single crystals were larger than that of CMSX-4[4, 5]. From these reports, it is supposed that the rafted strain would be smaller in the specimen crept at the lower stress and in the specimen with higher creep resistance. In this study, to confirm the feature of the rafting of the γ' phase, the change in the rafted strain with the decreasing the stress was investigated by conducting the creep tests in the wide stress range of 100 to 400MPa for CMSX-4. And the rafted strains in IN-100 were also measured in the wide stress range of 60 to 137MPa and compared with the rafted strains in CMSX-4.

2. Experimental Procedure

A single crystal alloy of CMSX-4 (having the chemical composition in weight per cent ; 6.4Cr, 9.3Co, 5.5Al, 0.9Ti, 6.3Mo, 6.2Ta, 6.2W, 2.8Re, 0.1Hf, balance Ni) was prepared in the form of bars 13mm in diameter by a directional precision casting. The exact orientations were determined by the Laue back-reflection technique. Tensile axes of the single crystals for this study were within 5deg of the [001] orientation. After employing an eight step solution treatment (1550Kx7.2ks → 1561Kx7.2ks → 1569Kx10.8ks → 1577x10.8ks → 1586Kx7.2ks → 1589Kx7.2ks → 1591Kx7.2ks → 1594Kx7.2ks → GFC) and a three step aging heat treatment (1413Kx21.6ks → 1353Kx21.6ks→1144Kx72ks→AC), the specimens for the creep tests with a gauge diameter of 6mm and a gauge length of 30mm were machined. The creep tests were carried out at 1273K in the stress range of 100 to 400MPa under the constant load. Creep strain was measured automatically through the linear variable differential transformers (LVDT's) attached to an extensometer. The creep tests were interrupted at the certain times ranging from the transient creep stage to the accelerating creep stage. All the creep interruption were conducted by cooling using compressed air under load. Microstructural examinations by a scanning electron microscopy (SEM; Hitachi S-4200) were carried out on the specimens sectioned parallel to (100) planes. Specimens for the SEM observation were prepared metallographically and electroetched with a supersaturated oxialic acid aqueous solution. The ratio of the average width to the length in the γ' phase was defined as the aspect ratio which was measured by the image processor-analyzer.

3. Results

3.1 Microstructure of the as-heat treated specimen

The scanning electron micrograph of the as-heat treated CMSX-4 is shown in Fig. 1. The cuboidal γ' is regularly arrayed in the γ matrix. The average edge length of the cuboidal γ' is about 0.5μm. The mean thickness of the γ channel is approximately 0.1μm. The volume fraction of the γ' phase is estimated as 78 per cent. There were no eutectic γ' phase.

3.2 Creep rate-time curves of CMSX-4 at 1273K

The creep rate-time curves of CMSX-4 at 1273K in the stress range of 100 to 400MPa are shown in Fig. 2. All creep curves consist of the transient creep stage and the accelerating creep stage. The open circles on each curve show the times where the creep tests were interrupted. The creep interrupted specimens were prepared to examine the rafting of the cuboidal γ'. For this object, the creep interrupted tests were done over a wide range of creep life from the latter half of the transient creep stage to the end of the accelerating creep stage.

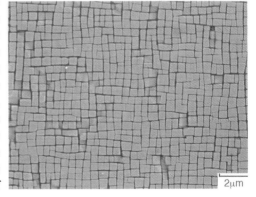

Fig.1. Schanning electron micrograph of a single crystal nickel-based superalloy, CMSX-4.

3.3 Change in the rafting process of the cuboidal γ′ with increasing the stress

The scanning electron micrographs of the specimens creep-interrupted at 1.08×10^6, 7.20×10^6 and 1.66×10^7s under the stress of 100MPa are shown in Fig.3. The stress axis is the horizontal direction of the photos. By subjecting to the creep for 1.08×10^6s (Fig.3-(a)), at the latter half of the transient creep stage, the cuboidal γ′ is still remained, whereas some of the cuboidal γ′ contact each other in the direction normal to the stress axis. In the specimen crept for

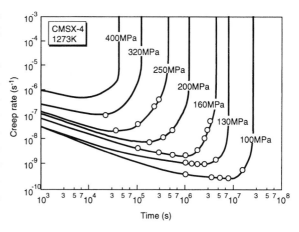

Fig.2. Creep rate-time curves of CMSX-4 at 1273K in the stress range of 100 to 400MPa. The open circles on each curve show the times where the creep tests were interrupted.

7.20×10^6s, at the time of the minimum creep rate, the rafted γ′ structure forms perpendicular to the stress axis as shown in Fig.3-(b). In the specimen crept for 1.66×10^7s (Fig.3-(c)), in the middle of the accelerating creep stage, the rafted γ′ turns to the waved one by connecting with the rafted γ′ in the direction parallel to the stress axis. As the result, the matrix turns from the γ to the rafted γ′.

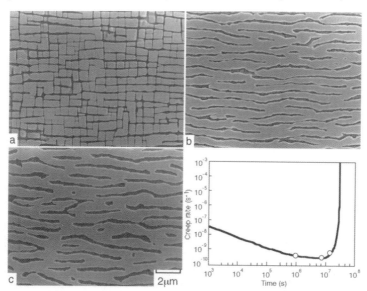

Fig.3. Scanning electron micrographs of the specimens crept at the stress of 100MPa for (a) 1.08×10^6, (b) 7.20×10^6 and (c) 1.66×10^7s. The stress axis is the horizontal direction of the photos.

440

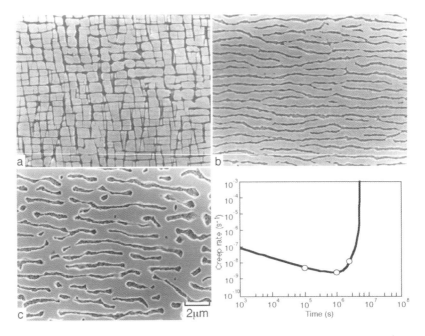

Fig 4 Scanning electron micrographs of the specimens crept at the stress of 160Mpa for (a) 1.08 x 10^5, (b) 1.08 x 10^6 and 2.52 x 10^{6}s.

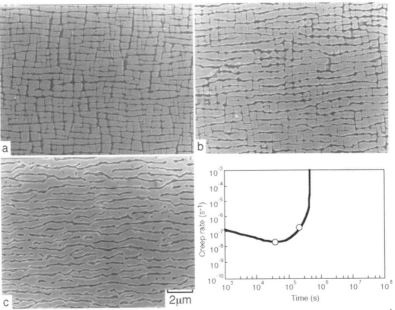

Fig. 5 . Scanning electron micrographs of the specimens crept at the stress of 250MPa for (a) 3.60 x 10^4, (b) 2.52 x 10^5 and (c) creep ruptured.

The scanning electron micrographs of the specimens crept at the higher stress of 160MPa for 1.08x10⁵, 1.08x10⁶ and 2.52x10⁶s are shown in Fig.4. By subjecting to the creep for 1.08x10⁵s (Fig.4-(a)), at the latter half of the transient creep stage, the γ′ phase remains cuboidal in shape. In the specimen crept for 1.08x10⁶s, at the time of the minimum creep rate, the rafted γ′ structure forms in the direction perpendicular to the stress axis as shown in Fig.4-(b). In the specimen crept for 2.52x10⁶s (Fig. 4(c)), at the end of the accelerating creep stage, the rafted γ′ turns to its shape into the more complex one by connecting with each other in the direction parallel to the stress axis. Obviously, the turning of the matrix from the γ to the rafted γ′ occurs.

The scanning electron micrographs of the specimens crept at the highest stress of 250MPa for 3.60x10⁴, 2.52x10⁵s and the creep-ruptured specimen are shown in Fig.5. By subjecting to the creep for 3.60 x10⁴s (Fig.5-(a)), at the time of the minimum creep rate, the γ′ phase remains cuboidal in shape. For the specimen tested for 2.52x10⁵s, in the middle of the accelerating creep stage, Fig5-(b), the γ′ phase remains cuboidal in shape, whereas some of the cuboidal γ′ contact each other in the direction normal to the stress axis. In the creep-ruptured specimen, Fig5-(c), the cuboidal γ′ turns to the rafted structure. At 250MPa, the γ - γ′ interface is waved, and the thickness of the γ channel is narrow compared with those of the specimens tested at 100 and 160MPa. In this way, the marked change in the rafting of the γ′ phase occurs with increasing the stress.

3.4 Change in the aspect ratio of the rafted γ′ with increasing the testing time

The aspect ratio of the γ′ phase is measured to distinguish the difference in the rafting of the γ′ with increasing the stress. The aspect ratio is defined as the ratio of the average width to the length in the rafted γ′.

The aspect ratio of the γ′ phase of the creep-interrupted and ruptured specimens is plotted as a function of the creep testing time as shown in Fig.6. The aspect ratio of the γ′ phase of the as-heat treated specimen is approximately 1. In every stress, the aspect ratios increase with the creep testing time, and attained the maximum value. However, the maximum values of the aspect ratio do not remain, and soon the aspect ratios of the γ′ phase decrease with increasing the creep testing time. The maximum value of the aspect ratio is 4 at 100MPa, and it decreases with increasing the stress and attains 2.2 at 320MPa. The time at which the aspect ratio reaches to the maximum value, termed as the time to raft, the time to rupture and the time to the minimum creep rate are plotted as a function of the stress, together with the time to raft of the polycrystalline nickel-based superalloy, IN-100, reported

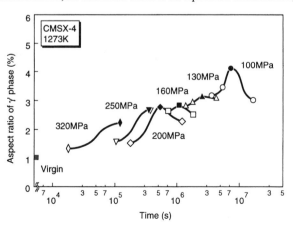

Fig.6. Relation between the aspect ratio of the γ′ phase and the creep testing time.

442

previously [6] as shown in Fig.7. At 320MPa, the rafted γ' is formed at the end of the accelerating creep stage. However, the difference between the time to raft and the time to the minimum creep rate decreases with decreasing the stress. The rafted γ' structure forms during the accelerating creep stage at the stresses of 320 to 200MPa and it forms at the time of the minimum creep rate at 160MPa. And at the stresses less than 160MPa, the rafted γ' forms at the transient creep stage. The correlation between the time to raft and the stress of CMSX-4 shows linear behavior on a log-log scale, and the slope of the curve is minus third. The value agrees well with that for IN-100. However, the time to raft of CMSX-4 is one order of magnitude larger than that of IN-100. Therefore, it is obvious that the formation of the rafted structure of CMSX-4 is inhibited in comparison with that of IN-100.

4. Discussion
4.1 Stress dependence of the rafted strain

The rafted strain is characterized by the strain at which the aspect ratio of the γ' phase reaches the maximum value. The rafted strain is plotted as a function of the stress as shown in Fig.8. At the stresses less than 200MPa, the cuboidal γ' turns to the rafted structure at the strain of 0.01. However, at the stresses higher than 250MPa, the rafted structure attains at the strain of 0.1 or more and the rafted strain increases with increasing the stress. At 320MPa, the cuboidal γ' turns to the rafted structure at the end of the accelerating creep stage, at the strain of 0.35. With decreasing the stress from 250 to 200MPa, the drastic change in the rafted strain is confirmed. Consequently, at the stresses higher than 250MPa, the rafted strain strongly depends on the stress. However, at the stresses less than 200MPa, the rafted strain is independent of stress and the rafting is attained at the strain of 0.01.

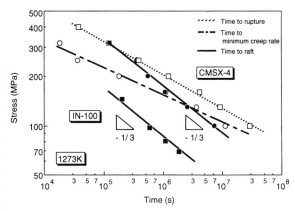

Fig.7. Relation between the stress and the time to raft, the time to rupture and the time to the minimum creep rate, compared with the time to raft of the polycrystalline nickel-based superalloy, IN-100, reported previously[6].

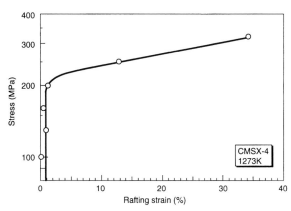

Fig.8. Change in the rafted strain with the stress.

4.2 Stress dependence of the aspect ratio

The maximum aspect ratio is plotted as a function of the stress in the stress range of 100 to 320MPa, together with the typical SEM micrographs, as shown in Fig.9. In the specimen crept at 100MPa, the maximum aspect ratio of the rafted γ' is 4 which is highest value in all of the specimens, and the shape of the γ / γ' interface is straight. However, the maximum aspect ratio of the rafted γ' decreases with increasing the stress and the γ / γ' interface turns to wave. In the specimen creep-ruptured at 320MPa, the maximum aspect ratio attains to 2.2 and the shape of the γ / γ' interface is waved.

Then, the reason why the maximum aspect ratio of the rafted γ' and the high completion of the rafted γ' are obtained at a lower stress side is discussed. It is well known that the crystal lattice of the single crystal nickel-based superalloy is rotated during the creep deformation[7],[8]. The difference in the crystal orientation at the vicinity of the fracture surface and that of the screw portion of CMSX-4, crept at 1273K in the stress range of 100 to 400MPa has been reported previously. And it was indicated that the crystal lattice was rotated during the creep deformation at the stresses higher than 130MPa[8]. The difference in the crystal orientation of the adjacent $\gamma - \gamma'$ phase and the same area of the γ and γ' phase of CMSX-4, creep-ruptured at 1273K in the stress range of 200 to 320MPa was reported previously. And it was indicated that the difference in the crystal orientation of the adjacent $\gamma - \gamma'$ phase and the same area of the γ and γ' phase decreased with decreasing the stress[9]. These reports indicated that the rotation of the crystal lattice during the creep deformation did not occur at the lower stress side.

As mentioned earlier, at the stresses less than 200MPa, the cuboidal γ' turns to the rafted structure at the strain of 0.01. However, at the stresses higher than 250MPa, the rafted structure attains at the strain of 0.1 or more and the rafted strain increases with increasing the stress. At 320MPa, the cuboidal γ' turns to the rafted structure at the end of the accelerating creep stage, at the strain of 0.35.

From these results, at the lower stress side, the rafted γ' is formed after a little creep deformation and it is supposed that the rotation of the crystal lattice is also little and the cuboidal γ' has regularly lined up. Therefore, the large aspect ratio of the rafted γ' and the high completion of the rafted γ' is obtained, and the γ / γ' interface is straight (Fig.10-(a)). On the other hand, the higher stress side, the rafted γ' is formed after large creep deformation, and the rotation of the crystal lattice is also large, and the cuboidal γ' seems to turn to the random direction. Therefore, it is suggested that the small aspect ratio of the rafted γ' and the low completion of the rafted γ' are obtained, and the γ / γ' interface is waved (Fig.10-(b)).

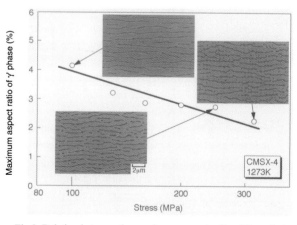

Fig.9. Relation between the maximum aspect ratio of the rafted γ' and the stress, together with the typical scanning electron micrographs.

444

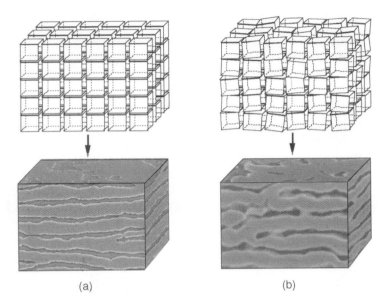

Fig. 10. Schematic illustrations of the cuboidal γ′ arrangements just before the rafting.
(a) Lower stress side and (b) higher stress side

5. Conclusions

By using the single crystal nickel-based superalloy, CMSX-4, the creep interrupting test were conducting the stress range of 100 to 400MPa at 1273K. The aspect ratio of rafted plates of the γ′ phase was evaluated as a function of the testing time and the stress. The relation between the rafted strain and the stress was investigated. The results can be summarized in the following.

1) The rafted γ′ is formed perpendicular to the stress axis at the stresses less than 320MPa. The rafted structure is formed during the accelerating creep stage at the stresses of 320 to 200MPa and it is formed at the minimum creep rate stage at 160MPa. At the stresses less than 160MPa, the cuboidal γ′ turns to the rafted structure at the transient creep stage.

2) The aspect ratio of the γ′ phase increases with increasing the creep testing time and attains the maximum value, and then it decreases with increasing the creep testing time.

3) At the stresses less than 200MPa, the rafted γ′ attains at the strain of 0.01 and the rafted strain is independent of stress. However, at the stresses higher than 250MPa, the rafted γ′ attains at the strain of 0.1 or more, and the rafted strain is dependent on the stress.

4) The maximum aspect ratio of the rafted γ′ is 4 at 100MPa and decreases with increasing the stress, and attains 2.2 at 320MPa. The shape of the γ / γ′ interface is straight at 100MPa, while the shape of the interface is waved at 320MPa.

5) From these results, at the lower stress side, the rafted γ′ is formed after a little creep deformation and it is supposed that the rotation of the crystal lattice is also little and the cuboidal γ′ has regularly lined up. Therefore, the large aspect ratio of the rafted γ′ and the high completion of the rafted γ′ is obtained, and the γ / γ′ interface is straight. On the other hand, the higher stress side, the rafted

structure is formed after large creep deformation, and the rotation of the crystal lattice is also large, and the cuboidal γ' seems to turn to the random direction. Therefore, it is suggested that the small aspect ratio of the rafted γ' and the low completion of the rafted γ' is obtained, and the γ / γ' interface is waved.

References

[1] H.Nagai, 1995, Effect of Stress on Morphology of γ' phase and Creep Strength Properties of Inconel 713C, Tetsu-to-Hagane, Vol.81, No.6, pp.667-672.

[2] Research group on strengthening of heat resisting steel and alloy of ISIJ, 2000, Final Report of Research Group on Strengthening of Heat Resisting Steel and Alloy, pp.121-228.

[3] P.Henderson, 1994, A Metallographic Technique for High Temperature Creep Damage Assessment in Single Crystal Alloys, Journal of Eng. For Gas Turbines and Power, Vol.121, No.4, pp.687-686.

[4] Mackay, R.A., 1985, The Development of γ-γ' Lamellar Structures in a Nickel-Base Superalloy during Elevated Temperature Mechanical Testing, Metall. Trans. A, Vol.16A, pp.1969-1982.

[5] Mukherji, D., 1997, Mechanical Behavior and Microstructural Evolution in The Single Crystal Superalloy SC16, Acta Metall., Vol.45, pp.3143-3154.

[6] K. Ishibashi, 1994, Raft Structure of the γ' Phase Formed by the Creep Deformation in Nickel-based Superalloy, IN-100, 123rd Committee on Heat Resisting Metals and Alloys Rep., Vol.35, No.3, pp.343-351.

[7] M. McLean, 1992, Anisotropy of High Temperature Deformation of Single Crystal Superalloys – Constitutive Laws, Modelling and Validation, Proc. of the 7th Int'l Symp., Superalloys 1992, pp.609-618

[8] K. Ishibashi, 1996, Long Time Creep Rupture Properties of Single Crystal Nickel-based Superalloy, CMSX-4, 123rd Committee on Heat Resisting Metals and Alloys Rep., Vol.37, No.1, pp.1-10.

[9] T. Minobe, 1997, Dislocation Substructure in Single Crystal Nickel-based Superalloy, CMSX-4, Crept at 1273K, 123rd Committee on Heat Resisting Metals and Alloys Rep., Vol.38, No.2, pp.129-137.

Nobuhiro MIURA :
E-mail : nmiura@nda.ac.jp, FAX : +81-468-44-5900
Yoshihiro KONDO :
E-mail : kondo@nda.ac.jp, FAX : +81-468-44-5900
Takashi MATSUO
E-mail : tmatsuo@mtl.titech.ac.jp, FAX : +81-3-5734-2874

Influence of primary and secondary crystal orientations on strengths of Ni-based superalloy single crystals

K. Kakehi
Dept. of Mechanical Engineering,
Tokyo Metropolitan University,
1-1 Minami-Osawa, Hachioji, Tokyo 192-0397, JAPAN.
Tel : +81-426-77-2723
Fax : +81-426-77-2717
e-mail: kakehi-koji@c.metro-u.ac.jp

ABSTRACT

In this study, by using the single crystals of the experimental superalloy which shows distinct active slip systems, the influence of primary and secondary crystal orientation on creep and fatigue strengths of Ni-based superalloy single crystals was investigated. It was revealed that the strengths of the notched specimens were affected by the crystallographic orientation not only in the tensile direction but also in the thickness direction. The influence of crystallographic orientations and plastic anisotropy on the creep and fatigue strengths of single crystals of the Ni-base superalloy was discussed on the assumption that $\{111\}<101>$ and $\{111\}<112>$ slip systems were activated. In the case of creep strength, the results were in agreement with the assumption of the operation of the $\{111\}<112>$ slip in the primary creep and $\{111\}<101>$ slip in the secondary creep. The notched creep behavior was found to be influenced by the additional aging at 850°C for 20 h, which prohibited activity of $\{111\}<112>$ slip systems. The fatigue lifetime and crack growth behavior depended on both plastic anisotropy caused by arrangement of $\{111\}<101>$ slip systems and the stress state.

1. INTRODUCTION

Single crystal are used for turbine blade of air craft jet engines because of their excellent creep, stress rupture, thermo mechanical fatigue capabilities. The microstructure of a Ni-based superalloy single crystal consists of a γ phase and cuboidal γ' precipitates, whose interfaces are parallel to the $\{001\}$ planes. The dendrite structure grows in the <001> direction in the face-centered-cubic Ni-based superalloy. This orientation provides better creep strength and a low modulus of elasticity that enhances thermal fatigue resistance. Thus, the turbine blades are designed so that their primary orientation is within 10 to 15° of the <001> axis to insure better creep strength and thermal fatigue resistance. The secondary dendrite direction (<010> direction) is usually randomly oriented with respect to the longitudinal direction of the turbine blade. The air-cooled turbine blades, which have a complicated hollow structure, are composed of sections of various thicknesses. Therefore, the mechanical properties of each blade section will depend on plastic anisotropy and the stress state as well as stress in the longitudinal direction. Fatigue crack growth rate was affected by the plastic anisotropy in the crack tip due to the geometric arrangement of the $\{111\}$ slip planes [1,2]. Since a single crystal possesses the intrinsic plastic anisotropy, the strengths of single crystals are influenced by the crystallographic orientations not only in the tensile direction but also in the normal direction of the specimen [3-5]. The important point to note is that control of secondary crystallographic orientation has the potential to significantly increase a component's resistance to fatigue crack growth without additional weight or cost [6]. Although a large number of studies nave been made on effect of primary orientation, little is known about that of secondary orientation [6]. In this study, by using the single crystals of the experimental superalloy which shows distinct active slip systems [7], the influence of secondary orientation on the creep and fatigue strengths was investigated.

2. EXPERIMENTAL PROCEDURES

The chemical composition of the Ni-based superalloy is listed in Table 1. Single crystals of this alloy were grown from the melt by the modified Bridgman method. After the analysis of the crystallographic orientation by the back Laue reflection method, the notched creep [4] and fatigue (Fig. 1) specimens were cut out from the as-grown crystals using a spark cutter. The thicknesses of the notched specimen were determined to obtain the plane stress condition [3]. Crystallographic orientations of the four kinds of specimens with different arrangements of slip systems are shown in Fig. 2. The specimens were subjected to solution treatment at 1255°C for 10 h and aging treatment at 1100°C for 10 h and subsequently to an additional aging at 850°C for 20 h. After each heat treatment, the specimens were cooled in an air-blast. A single-step aging treatment at 1100°C for 10 h resulted in the clear {111}<112> slip

because of numerous hyperfine secondary γ' precipitates in the matrix channel [7]; therefore, the single-step aging treatment was also employed for creep specimens. The average edge lengths were 0.40 μm for the single-aged specimen and 0.39 μm for the double-aged specimen, respectively. Creep and fatigue rupture tests were conducted in laboratory environmental and at 700°C. Creep rupture test was carried out under a nominal stress of 820 MPa for the rectangular-cross-section specimens. In the case of the notched specimens, the stress was equally applied for the initial net cross-sectional area. The tension-zero fatigue test (the stress range of 600 MPa was loaded for the net cross-sectional area) was done at a frequency of 5Hz.

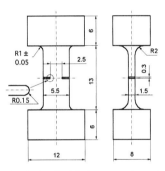

Fig. 1 Notched fatigue specimen (mm).

Table 1 Chemical composition (mass %)

Cr	Mo	Co	W	Ti	Al	V	Ni
10.1	2.50	9.97	0.04	4.75	5.73	0.93	Bal.

3. EXPERIMENTAL RESULTS

3.1. Creep rupture test

Creep curves of the notched specimens are shown in Fig. 3. The creep properties of rectangular-cross-section specimens are shown in Table 2. The rectangular-cross-section specimen with the [001] orientation showed large primary creep strain of 6.0% and steady-state creep. The [011] orientation exhibited appreciably shorter lives and large rupture

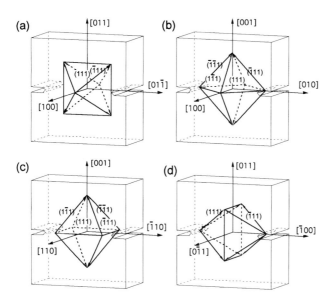

Fig. 2 Four kinds of specimens and arrangement of octahedral slip systems;
(a) orientation A, (b) orientation B, (c) orientation C and (d) orientation D.

elongation because of the absence of strain hardening. In the case of the notched creep,
single-aged specimens showed distinct plastic anisotropy [4]. When the tensile direction is
[001], the rupture lifetime of orientation B was seven times longer than that of orientation C.
In the case of specimens whose tensile orientation is [011], in spite of the poor creep strength
of the rectangular-cross-section specimen (Table 2), orientation D exhibited extremely small
steady-state creep rate and exceptionally long rupture time. Conversely, orientation A showed
larger creep elongation rate and shorter rupture lifetime than orientation D. The rupture
lifetime of orientation D was 162 times longer than that of orientation A. In the single-aged
specimens, it is obvious that the notched-tensile creep strength is influenced by the
crystallographic orientations not only in the tensile direction but also in the thickness
direction.

Table 2 Creep properties of rectangular cross-section specimens (700°C, 820MPa)

Specimen	Primary creep strain (%)	Rupture lifetime (h)	Rupture Elongation (%)
[001] single aged	6.0	520.6	10.1
[001] double aged	1.5	385.1	3.85
[011] single aged	—	5.6	23.5
[011] double aged	—	4.8	10.0

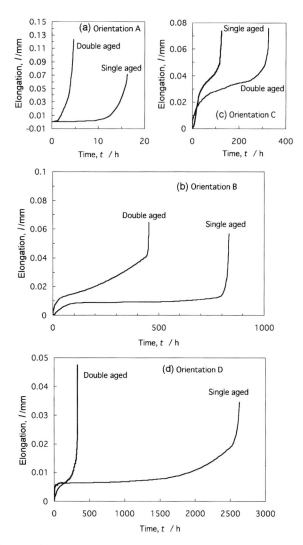

Fig. 3 Influence of aging heat treatment on creep curves in four kinds of notched specimens.

The creep behavior was found to be influenced by the additional aging at 850°C for 20 h. As shown in Fig. 3(a), in orientation A, the single-aged specimen showed an incubation period, but the incubation period disappeared in the specimen aged at 850°C. Both specimens exhibited very short the rupture lifetimes. The creep rupture lifetime was increased by the second aging at 850°C in orientation C. However, both in orientations B and D, the creep rupture lives were decreased by the second aging at 850°C.

3.2. Fatigue test

Crystallographic orientation dependence of fatigue rupture lifetime is presented in Fig. 4. Orientation D showed the longest fatigue lifetime. Orientations B and C have same tensile [001] direction; however, orientation C showed longer fatigue lifetime than orientation B. Figure 5 showed fatigue fracture surfaces. Stage I crack and mode I-dominated stage II crack growth regions were observed. As shown in Fig. 6, as the area fraction of stage II region increased, the fatigue lifetime decreased. Stage II crack propagation took up much of the fatigue lifetime of the thin notched specimens.

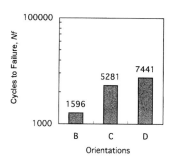

Fig. 4 Orientation dependence of fatigue rupture lifetime.

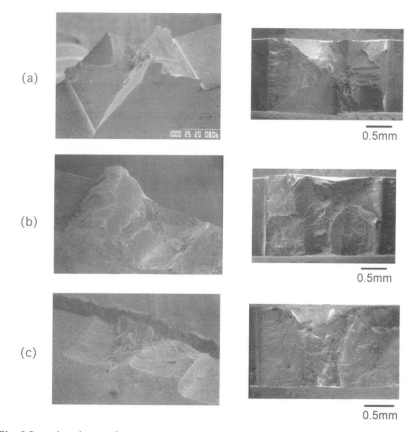

Fig. 5 Scanning electron fractographs showing the cleavage crystallographic fracture by slip-band decohesion and stage II crack propagation; (a) orientation B, (b) orientation C, (c) orientation D.

4. DISCUSSION

Single-aged specimens showed distinct active {111}<112> slip systems and plastic anisotropy [7]. The creep rupture test results of single-aged specimens were in agreement with the assumption of the operation of {111}<112> slip systems during primary creep region and {111}<101> slip systems during secondary creep region [4]. The {111}<112> slip systems in four kinds of specimens are illustrated in Fig. 7. The group of slip systems that results in a contraction of the thickness of the specimen is named the T group, while the group of slip systems that results in a contraction of

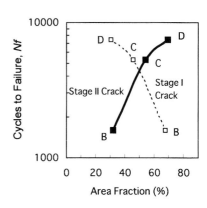

Fig. 6 Relationship between area fraction of stable crack trace on fracture surface and cycles to rupture.

the width of the specimen is named the W group, which necessarily accompanies the contraction in the width direction and would be constrained. If a group of slip systems results in a contraction of both the thickness and width of the specimen, it is named the TW group. In orientation B, even under a complex multiaxial stress state, a lateral distortion in the thickness direction is simultaneously produced with that in the width direction. Even if only one slip system operates, the contraction in the thickness direction necessarily accompanies that in the width direction, therefore, the slip systems belonging to TW group would be constrained. Singe-aged specimen of orientation B has no unconstrained {111}<112> systems but {111}<101> slip systems belonging to T group operating unconstrained (Fig. 2(b)); therefore, it would rapidly enter secondary creep as shown in Fig. 3(b). Whereas, in orientation C, as shown in Fig. 7(c), there are two kinds of slip systems. The slip systems belonging to W group would be constrained; however, those belonging to T group would not be constrained. The poor creep resistance of orientation C (Fig. 3(c)) resulted from a combination of the free operation of two {111}<112> slip systems combined with a lack of activity on the {111}<101> slip systems which would terminate primary creep. In orientation A, none of the slip systems belonging to T group (Figs. 2(a) and 7(a)) could be constrained and the creep strengths were extremely low

The creep behavior was found to be greatly influenced by the additional aging at 850°C for 20 h [7]. In orientation A, as shown in Fig. 3(a), the single-aged specimen showed an incubation period, but the incubation period disappeared in the specimen aged at 850°C. The hyperfine secondary precipitates would retard the dislocations glide in the matrix channel, and resulted in an incubation period in the single-aged specimen. However, the incubation period disappeared in the specimens aged at 850°C because the matrix dislocations will gradually glide in the matrix channel bypassing the large particles, which reprecipitates during the second aging at 850°C. The creep rupture lifetime was increased by the second aging at 850°C in orientation C. The poor creep resistance of single-aged specimen of orientation C resulted from the free operation of two {111}<112> slip systems. As a result of the additional aging at 850°C, the hyperfine γ' precipitates were dissolved into the matrix and resultant large mean surface-to-surface between the cuboidal precipitates inhibited extensive shearing of the γ-γ' structure by the $(\bar{1}11)[1\bar{1}2]$ slip system [7]. Inhibition of operation of {111}<112> slip systems would bring about increase of creep lifetime in the double-aged specimen. Both in orientations B and D, all {111}<112> slip systems are constrained. However in these orientations, the creep rupture lives were decreased by the second aging at 850°C because the other slip systems would be activated; (1) {111}<101> slip systems which shear both γ and γ' phases, and (2) macroscopic cube slip by the motion zigzag dislocation motion in the matrix channel [8]. In orientation D, as shown in Fig. 8(b), cube slip systems belong to T group which is not be constrained. In double-aged specimens, {111}<112> slip systems which show distinct plastic anisotropy were inhibited, instead of the slip systems, isotropic deformation mechanism, consisted of the dislocation motion in matrix channel and {111}<101> shear slip, would be operative during creep. The point is that the second aging at

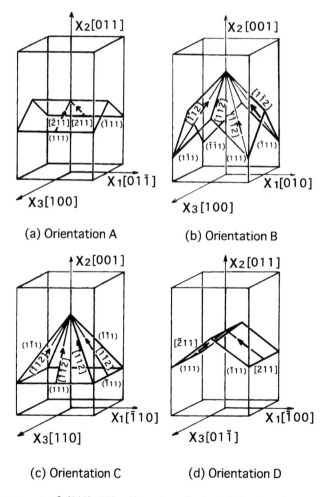

(a) Orientation A   (b) Orientation B

(c) Orientation C   (d) Orientation D

Fig. 7 Arrangement of {111}<112> slip systems for four kinds of specimens.

454

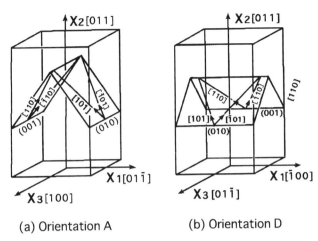

(a) Orientation A   (b) Orientation D

Fig. 8 Arrangement of {001}<101> slip systems for two kinds of specimens.

850°C decreased the extent of plastic anisotropy in creep test.

In the fatigue test, as the area fraction of stage II region increased, fatigue lifetime increased (Fig. 6). This result indicates that stage II crack propagation would take up much of the fatigue lifetime of the thin notched specimens. The transition from stage II crack propagation on (001) plane to stage I crack propagation on {111} planes brings about increase of crack growth rate [9,10]. Telesman and Ghosn showed the relationship between crack growth and the stress intensity [9]. As the stress intensity was increased, area containing facets on {111} planes also became appearance. With a further increase in the stress intensity, the (001) fatigue failure completely disappeared, and was replaced by the {111} fatigue failure. The increase in the ΔK resulted in an increase in the size of the {111} failure facets.

In orientation B, in this study, the cleavage crystallographic-fracture surface by slip-band decohesion was observed and rupture lifetime was the shortest. The reason for the result is that it is difficult to relax stress concentration of crack tip because of arrangement of {111}<101>slip systems in orientation B [5]. The secondary slip with a shear displacement component perpendicular to the crack plane can relax the normal stress of the primary slip, therefore, the difficulty of activating secondary slip in return results in large hydrostatic and normal stresses near the crack tip [11]. The combination of intense coplanar shear and large normal stresses results in easy crack propagation and low toughness [11]. However, in the orientation, under plane strain, all four slip planes were active and crack path zigzagged on two distinct pairs of {111} planes [9]. The difference in fatigue failure appearance between this study and others [2,9] would be due to the stress condition. The resolved shear stress intensity parameter (K_{rss}) was able to predict the microscopic crack path under difference stress states, and that the magnitude of ΔK_{rss} under the plane stress is always approximately twice that of plane strain [9]. Compared with the thick specimens, the cleavage crystallographic-fracture surface by slip-band decohesion would easily occur in the thin specimens used in this study because of the plane stress condition. In orientation C, shear symmetry between ($1\bar{1}1$) and ($\bar{1}11$) planes with respect to the cracking (001) plane would be able to relax stress concentration of the crack front. As a result of stable stage-II cracking, the fatigue lifetime was longer than that of orientation B. When the resolved shear stress is below the critical value needed for a dislocation to cut through the precipitates, the slip becomes confine to the matrix [9]. The localization of the damage to the {111} matrix channel results in a preferential failure in this area, exposing the cuboidal facets of the precipitates and creating a (001) failure appearance. Furthermore, in orientation D, in addition to shear symmetry between (111) and ($\bar{1}$11) slip planes with respect to the cracking (001) plane, there is no slip system belonging to T group, which results in cleavage crystallographic-fracture surface by slip-band decohesion and short fatigue lifetime. Therefore, orientation D exhibited the largest cycles to failure among three kinds of crystallographic specimens.

5. CONCLUSIONS

1. Creep and fatigue strengths of the notched-tensile specimens were influenced by the crystallographic orientations not only in the tensile direction but also in the thickness direction.
2. Orientation D exhibited superior creep and fatigue strengths because both {111}<101> and {111}<112> slip systems are constrained. However operation of macroscopic cube slip decreased creep strength in this orientation.
3. Second-step aging treatment at 850°C decreased the influence of plastic anisotropy on notched creep strength.
4. The fatigue lifetime and crack growth behavior depended on both plastic anisotropy caused by arrangement of {111}<101> slip systems and the stress state.

REFERENCES

[1] M.B. Henderson and J.W. Martin, Acta Metall. Mater., 1996, 44 , pp. 111-126
[2] P.A.S. Reed, X.D. Wu and I. Sinclair, Metall. and Mater. Trans. A, 31A, pp. 109-123 (2000)
[3] K. Sugimoto, T. Sakaki, T. Horie, K. Kuramoto and O. Miyagawa, Metall. Trans. A, 1985, 16A, pp. 1457-1465
[4] K. Kakehi, Metall. and Mater. Trans. A, 31A (2000), pp. 421-430
[5] K. Kakehi, Scripta Materialia, 42 (2000), pp. 197-202.
[6] N.K. Arakere, G. Swanson, Proc. of ASME TURBOEXPO 2000, May 8-11, Munich Germany, 2000, pp.1-16.
[7] K. Kakehi, Metall. and Mater. Trans. A, 30A (1999), pp.1249-1259
[8] D. Bettge and W. Österle, Scripta Materialia, 40 (1999), pp. 389-395.
[9] J. Telesman and L.J. Ghosn, Superalloys 1988, ed. by D.N. Duhl, G. Maurer, S. Antolovich, C. Lund and S. Reichman, TMS, Warrendale, PA, 1988, pp. 615-624.
[10] M. Okazaki, T. Tanabe and S. Nohmi: Metall. Trans. A, 21A (1990), pp.2201-2208.
[11] D.A. Koss and K. S. Chan: Acta Metall. 28(1980), pp. 1245-1252.

SUPERALLOYS III

EFFECTS OF PHOSPHORUS AND BORON ON THE CREEP BEHAVIOR OF WASPALOY

C. G. McKamey, E. P. George, C. A. Carmichael, W. D. Cao,[†] and R. L. Kennedy[†]
Metals and Ceramics Division, Oak Ridge National Laboratory, Oak Ridge, TN 37831-6093
[†]Allvac, An Allegheny Technologies Company, Monroe, NC 28110-0531

ABSTRACT

Five heats of Waspaloy were investigated in this study, including two commercial-purity heats (40-60 ppm P and 50-60 ppm B), one high-purity heat (10 ppm P and <10 ppm B), one heat containing high P and high B (220 and 110 ppm, respectively), and one heat containing low P and high B (10 and 140 ppm, respectively). The grain sizes ranged from 7 to 12 μm. Creep tests were conducted in air at stresses in the range 200-520 MPa and temperatures in the range 730-815°C. Creep strength was highest in the high-B low-P alloy and lowest in the high-B high-P and high-purity alloys with two commercial-purity alloys falling in between. Fracture in the high-purity alloy occurred by separation along the grain boundaries in a direction perpendicular to the stress direction and was accompanied by ductilities of less than about 10%. In contrast, fracture in all of the other alloys was by ductile dimple rupture. Since poor ductility and low strength can both be life-limiting factors in high-temperature applications, the high-B low-P modification of Waspaloy appears to be optimal. Possible mechanisms for the strengthening due to boron are discussed.

INTRODUCTION

Boron is added to many nickel-based superalloys at levels of 50 to 500 wppm. Because of its small size and low solubility in the γ and γ' phases, boron segregates strongly to the grain boundaries where it slows down diffusion and prevents segregation of harmful elements [1-8]. In creep specimens it therefore prolongs rupture life and improves rupture ductility [1-4,9,10]. Boron has been observed to co-segregate with molybdenum, both in nickel-based superalloys [6] and in austenitic stainless steels [11]. An increase of boron over carbon in superalloys results in the precipitation of brittle, blocky, Mo-rich M_3B_2 precipitates both in the matrix and at grain boundaries [10,12,13] and in the displacement of carbon in grain boundary $M_{23}C_6$ precipitates to form $M_{23}(C,B)_6$ precipitates.

Phosphorus, on the other hand, has generally been treated as a harmful trace element and most alloy specifications limit it to low levels (typically <0.015 wt%). It is known to segregate strongly to grain boundaries [14-17] where it causes grain boundary embrittlement at high temperatures [1,18], reduces hot workability [19], and decreases high temperature corrosion resistance [20]. Its effects on specific properties are not understood in any great detail and very little information on the effect of minor phosphorus additions on properties is available.

In recent years, however, a beneficial effect of phosphorus addition has been found in some corrosion-resistant nickel alloys [21-23] and nickel-based superalloys [12,24-28]. In alloy 718, which is primarily strengthened by the presence of coherent ordered γ' and γ'' precipitates, Cao et al. [12,24-26] have found that controlled additions of boron and phosphorus, added together, increase the average rupture life by one order of magnitude [12,24-27]. Their research indicates that there is a strong synergistic interaction between phosphorus and boron, the mechanism of which is still to be determined. These results have been substantiated by others

[27,29]. The data indicate that the effect is alloy specific, e.g., the same levels of phosphorus and boron which improve creep strength in alloy 718 (a Ni-Fe-Cr-based alloy) do not improve it in Waspaloy (a Ni-Co-Cr-based alloy) [12]. Furthermore, stress-rupture life in Waspaloy is longest, not when phosphorus and boron are added together, but when phosphorus is kept as low as possible and boron is at the optimum level (~0.016%).

Results of Auger and atom probe analyses have shown that phosphorus and boron segregate to γ-γ grain boundaries in alloy 718 [24,29-31] where they join with molybdenum and niobium to form M_3P-type phosphides [3,6,30,31]. There is also some evidence that in alloy 718 the γ' phase is slightly stabilized by the addition of boron and phosphorus [3] and phosphorus additions affect the size and distribution of δ-phase particles at the grain boundaries [32]. They proposed that the beneficial effect of phosphorus additions on creep properties arose from the inhibition of diffusion along grain boundaries.

The mechanism(s) by which phosphorus and boron work synergistically to improve creep rupture properties in certain superalloys, but not in others, is still not understood. The purpose of the current research is to more thoroughly characterize the effect of phosphorus and boron additions on the creep-rupture properties of Waspaloy and to discuss the possible mechanisms involved.

EXPERIMENTAL PRODCEDURES

Five microalloyed Waspaloys with a base composition of Ni-19.76Cr-13.44Co-4.26Mo-2.98Ti-1.30Al-0.08Fe-0.01Si-0.037C-0.0006S (wt%) and varying amounts of boron and phosphorus (as shown in Table I) were vacuum induction melted as 23-kg heats, cast as 70 mm diameter electrodes, and further processed into 15 mm diameter bars. The wrought bars were then given the following heat treatment: a solution treatment at 1010°C for 4 h followed by a water quench, aging at 843°C for 4 h followed by an air cool, aging at 760°C for 16 h followed by an air cool to room temperature. The alloys included two commercial alloys (designated G752-2 and WB74), one high purity alloy containing essentially no P or B (designated G757-1), one alloy with high amounts of phosphorus and boron (designated G766-2), and one alloy with high boron but very low phosphorus (designated G949-1).

Creep-rupture specimens with a gage section 3.18 mm in diameter and 17.8 mm in length were machined from the as-heat-treated bars and tested without further heat treatment. Creep rupture tests were conducted in air using lever-arm creep machines with either a 5:1 or 20:1 load ratio. Tests were conducted at various stresses at temperatures of 732 (1350°F), 760 (1400°F), and 815°C (1499°F). Some of the specimens were tested to rupture at a single load, while on other specimens the load was incrementally increased in order to obtain minimum creep rates at several different stresses with the same specimen. The secondary stage minimum creep rates were measured from the straight-line portions of the creep curves

Flat tensile specimens with gage dimensions of approximately 12.7 x 3.2 x 0.75 mm were machined from the wrought bars of alloys G757-1, G766-2, and G949-1. Slow strain rate tensile tests, at temperatures between 600 and 900°C, were performed on a screw-driven Instron machine at a strain rate of 3 x 10^{-5} s^{-1} in a vacuum of 1.3 mPa (10^{-5} torr). The lengths of the specimens were measured before and after testing to determine the percent elongation.

Selected test specimens were metallographically polished through 0.5 μm diamond and then were etched in a solution of 4 parts HCl and 1 part HNO_3. Microstructures of the specimens were studied using optical and scanning electron microscopy (SEM) techniques. Scanning electron microscopy was also used to study the fracture surfaces of creep-tested and tensile-tested specimens.

RESULTS

Figures 1 and 2 show typical microstructures of the alloys tested. The high purity G757 alloy (Fig. 1) exhibited a duplex microstructure with large grains 20-100 μm in diameter in some regions and very fine grains about 10 μm or less in diameter in other regions. Where these bands occurred, they appeared to run for long distances in a direction parallel to the direction of the bars. Optical metallography was performed on four different specimens of G757-1. In these four specimens, large grains occupied 50-80% (estimated visually) of the cross sectional area. The G766-2 alloy, with high B and high P additions, contained a few bands of large grains that made up about 10-15% of the cross sectional area. The bands were normally no larger than about 100 μm in width. Alloy G949-1, with high B and low P (Fig. 2), had a fairly uniform microstructure with a grain size of about 10 μm. Very few bands of large grains were observed in three different specimens analyzed. The two commercial alloys, WB74 and G752-2, each had several bands of large grains across the 3-mm-diameter cross section. These large-grain bands made up about 10% of the cross section.

Scanning electron microscopy was also used to study the microstructures. Figure 3 shows a SEM micrograph of the high purity alloy. The small γ' precipitates visible in the matrix were present in all the alloys with the proper etching. The precipitates along the grain boundaries were rich in Ni and Mo, while large precipitates in the matrix were rich in Ti and Mo. At this level of magnification, there was no obvious difference in the microstructures of the five alloys.

Creep and creep-rupture tests were performed on all five alloys at temperatures of 732 to 815°C. In general, the G766-2 alloy, containing high levels of B and P, was the weakest of the five alloys, while the high B-low P G949-1 alloy was the strongest. Figure 4 compares the stress dependence of the minimum creep rates for the high B-high P, high B-low P, and high-purity alloys. The degree to which the high B-low P alloy was better than the other four alloys became more evident as the test temperature was increased. Figure 5 shows creep curves for the five alloys at 815°C. The high B-high P and commercial purity alloys had short rupture lives because they were relatively week compared to the high B-low P alloy. In contrast, the high-purity alloy fractured prematurely (at short rupture times) because of low ductility rather than poor strength.

Inspection of the fractured specimens of high-purity G757-1 showed that there was no uniform elongation and no localized necking. In order to determine the fracture mode, fractured specimens were mounted horizontally in epoxy and polished and etched using standard metallographic techniques. Figure 6 shows that in the high-purity G757-1 alloy cracks first

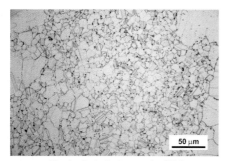

Fig. 1. Optical micrograph of high purity Waspaloy, heat G757-1.

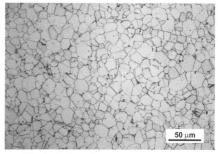

Fig. 2. Optical micrograph of Waspaloy with high B and low P concentrations, heat G949-1.

462

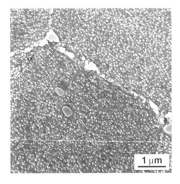

Fig. 3. Scanning electron micrograph of
high-purity Waspaloy, heat G757-1.

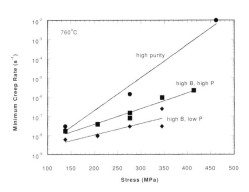

Fig. 4. Minimum creep rate versus stress for high
purity, high B-high P, and high B-low P alloys.

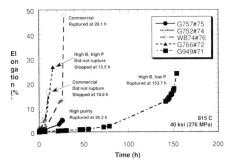

Fig. 5. Creep curves for five Waspaloys
tested at 815°C with an initial stress of 276
MPa (40 ksi).

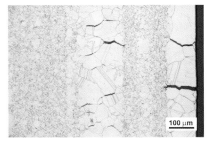

Fig. 6. Optical micrograph showing the fracture
morphology of high-purity Waspaloy creep tested to
rupture at 732°C with an initial stress of 517 MPa.

appeared perpendicular to the loading direction along the grain boundaries of the larger-grained
bands. The small-grained bands served to delay crack propagation to some degree. However,
since the small-grained bands comprised less than 50% of the cross-sectional area, it could not
sustain the specimen for very long. The low elongations that were measured in the G757-1
specimens were reflective of this premature separation along grain boundaries perpendicular to
the stress axis. Several creep tested specimens, tested at various temperatures and stresses, were
metallographically prepared and inspected in this manner and all showed this same fracture
mode.

It was not possible to observe the actual fracture surface of creep-tested specimens
because the tests were conducted in air and the fracture surfaces oxidized after failure. However,
some of the specimens were creep-tested to approximately 5% chart elongation, then cooled to
room temperature under load and fractured in air. The fracture surfaces showed that failure
occurred predominantly along grain boundaries, in support of the results from Fig. 6.

Fig. 7. Optical micrographs showing the fracture morphology of high-purity Waspaloy G766-2 containing high B and high P concentrations creep tested to rupture at 732E with an initial stress of 517 MPa.

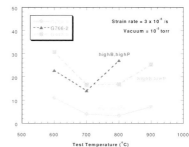

Fig. 8. Tensile elongation versus test temperature for Waspaloys tested in vacuum at a strain rate of 3×10^{-5} s^{-1}.

Figure 7 shows the fractured ends of the high B-high P G766-2 (also typical of fracture in the high B-low P and commercial-grade alloys), creep tested at 732°C and 517 MPa. This alloy had very little large-grained banding and the bands that were present were much smaller in thickness that those in the high-purity G757-1 alloy. A few wedge-shaped cracks were observed on grain boundaries in the large-grained bands, but because these bands were not very wide the cracks did not grow to become the major failure mode as was observed in alloy G757-1.

The results of the slow strain rate tensile tests are shown in Fig. 8. The high-purity G757-1 alloy had low tensile elongations at all temperatures tested, while the G949-1 and G766-2 alloys, both of which contain high levels of B, had much higher ductilities. Note that these curves suggest a strong correlation between the presence of B in the alloy and the tensile elongation. The correlation between the presence of P and elongation is less clear.

Analysis of the fracture modes showed that the G757-1 specimens tested at 700 to 900°C fractured predominantly in a brittle intergranular mode [see Figs. 9(a,b), while the fracture surface of the G757-1 specimen tested at 600°C had a more transgranular appearance. As in the creep-ruptured specimens, fracture always occurred initially along grain boundaries in the large-grained bands, in a direction perpendicular to the load axis. There was no necking visible in any of the G757-1 specimens. The G949-1 and G766-2 specimens, on the other hand, failed by dimpled rupture, as shown in Figs. 9(c,d). Failure was predominantly by transgranular cleavage at 600°C but ductile dimples at all other temperatures.

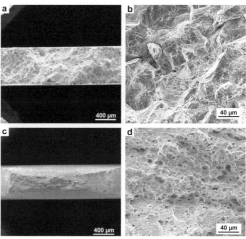

Fig. 9. Scanning electron fractographs showing the fracture modes in (a,b) high-purity Waspaloy G757-1 and (c,d) G949-1 with high B and low P concentrations. Both specimens were tested in tension at 700EC.

DISCUSSION

The results of the current study show that Waspaloy behaves similarly to many other nickel-based superalloys in that, for best creep-rupture properties, P additions should be avoided, while optimal B additions increase creep strength without significantly sacrificing ductility. Over the range of temperatures and stresses used in this study, the high-purity alloy (G757-1) and the alloy containing high levels of both B and P (G766-2) tended to have the shortest creep lives, while the alloy with high B and low P concentrations (G949-1) usually had the longest creep life. These results support those of Cao et al. [12] who showed that the longest creep lives in Waspaloy were produced with minimum P concentrations and an optimum B concentration of about 0.015%. In the current study, the high-purity alloy always exhibited the lowest ductility upon fracture at any particular temperature and stress and always failed by brittle cracking of grain boundaries perpendicular to the stress axis (see Fig. 6). None of the other alloys exhibited this type of brittle fracture. Instead, all the other alloys failed with large elongations and by dimpled rupture. This improvement in fracture behavior was observed even for the two commercial alloys containing B and P additions as low as 0.004-0.006%. In light of the results of Cao et al. [12], as well as the results of the current study, the strengthening in Waspaloy is believed to be due to the B addition, rather than P. The exact mechanism by which the addition of B affects creep strength in Waspaloy is not understood at present. However, it is possible to discuss potential mechanisms in light of the results of this study, the Cao et al. results, and what is generally known about B in nickel-based superalloys.

First, let us discuss the difference in microstructures between the five alloys of the current study. Although the average matrix grain size was similar for all five alloys (7-12 μm, see Table I), the high-purity G757-1 alloy contained many bands of much larger grains (see Figs. 1 and 6), amounting to as much as 50% of the cross section of the specimens that were tested. In general, when grain boundary sliding is the dominant deformation mechanism in creep, larger grains are considered to improve creep life and lower tensile strength [33,34]. However, in the present case, the creep life of G757-1 was lower than that of the other alloys because of premature failure caused by the grain-boundary cracks in the large grained bands (Fig. 6).

As indicated in the introduction, there have been several studies to determine the effect of B and/or P additions on the size, morphology, and composition of the many precipitates present in nickel-based superalloys [3,7,9,13,28,30,31,35]. Many of these have been conducted recently on alloy 718 because of the finding of improved creep life with the addition of an optimum combination of P and B [12,24-27]. Among other things, it has been suggested that additions of these elements can change the size or stability of the γ' phase, prevent agglomeration of $M_{23}C_6$-type precipitates at grain boundaries, or combine with other elements at the grain boundaries to form borides or phosphides which take up space and prevent the formation of other harmful precipitates at grain boundaries. However, none of these studies have shown a direct correlation between any change in the precipitates (including γ', γ', and carbides) and the creep properties. In the current study, too, no obvious difference was noted in the size, morphology, or distribution of any of the γ' or carbide precipitates. If all these studies are correct, and since B additions also improve creep strength in many γ'-free alloys, cobalt-based alloys, and stainless steels [36], then the B strengthening mechanism in Waspaloy probably does not involve changes in these precipitates.

It is well known that both B and P have low solubilities and therefore segregate strongly to grain boundaries in nickel-based alloys [4,6,24,37-39], where, if both elements are present, they will compete for grain boundary sites. There are conflicting opinions about whether B

[3,30,31] or P [40,41] is the stronger segregant. However, it is generally accepted that B, once it has segregated to the grain boundaries, is much more effective in increasing grain boundary cohesion [42]. In the current study, the change from brittle grain boundary separation and low ductilities in the high-purity G757 alloy to ductile failure with large ductilities in the high-B-containing alloys (with or without P additions) suggests that additions of B may enhance the grain boundary cohesion.

Another possible mechanism for the improved creep life of Waspaloy with B addition involves the slowing down of grain-boundary diffusional processes which contribute to creep deformation. Segregation of B (misfitting atoms) to grain boundaries may reduce grain boundary diffusion rates [35] (consistent with findings by Tien and Gamble [43] on the formation of denuded zones in Ni-16Cr-5Al-4Ta) and may therefore affect the Nabarro-Herring mass transport mechanism for creep in superalloys [43-45]. Through a slowing down of diffusional processes, boron could also be suppressing the formation of creep cavities on grain boundaries and at triple points [28], which may be the reason it improves ductility (in addition to strength).

CONCLUSIONS

Creep properties were determined for five Waspaloys microalloyed with various amounts of phosphorus and boron. Creep strength was highest in the high-B low-P alloy and lowest in the high-B high-P and high-purity alloys with two commercial-purity alloys falling in between. Fracture in the high-purity alloy occurred by separation along the grain boundaries in a direction perpendicular to the stress direction and was accompanied by ductilities of less than about 10%. In contrast, fracture in all of the other alloys was by ductile dimple rupture. Since poor ductility and low strength can both be life-limiting factors in high-temperature applications, the high-B low-P modification of Waspaloy appears to be optimal. Possible mechanisms for the strengthening due to boron include increased grain boundary cohesion and slowing down of diffusional processes which contribute to creep.

ACKNOWLEDGEMENTS

This research was sponsored by the Laboratory Technology Research Program, Office of Science, U.S. Department of Energy through a CRADA with Allvac, an Allegheny Technologies Company and by the Division of Materials Science and Engineering, U.S. Department of Energy through contract DE-AC05-00OR22725 with the Oak Ridge National Laboratory managed by UT-Battelle, LLC.

REFERENCES

1. R. T. Holt and W. Wallace, "Impurities and Trace Elements in Nickel-Base Superalloys," Intern. Metals Rev. 21, 1-24 (1976).

2. J. Kameda and A. J. Bevolo, "High Temperature Brittle Intergranular Cracking in High Strength Nickel Alloys Undoped and Doped with S, Zr, and/or B--II. Solute Segregation Analysis," Acta Metall. Mater. 41(2), 527-37 (1993).

3. M. K. Miller, J. A. Horton, W. D. Cao and R. L. Kennedy, "Characterization of the Effects of Boron and Phosphorus Additions to the Nickel-Based Superalloy 718," J. De Physique IV, C5, 6, 241-46 (1996).

4. J. M. Walsh and B. H. Kear, "Direct Evidence for Boron Segregation to Grain Boundaries in a Nickel-Base Alloy by Secondary Ion Mass Spectrometry," Metall. Trans. 6A, 226-29 (1975).

5. R. F. Decker, "Strengthening Mechanisms in Nickel-Base Superalloys," pp. 275-98 in Section VIII: Nickel-Base Superalloys, of Source Book on Materials for Elevated-Temperature Applications, ASM, 1979, compiled by Elihu F. Bradley.

6. L. Letellier, A. Bostel, and D. Blavette, "Direct Observation of Boron Segregation at Grain Boundaries in Astroloy by 3D Atomic Tomography," Scripta Metall. Mater. 30(12), 1503-08 (1994).

7. M. A. Burke, J. Greggi, Jr. and G. A. Whitlow, "The Effect of Boron and Carbon on the Microstructural Chemistries of Two Wrought Nickel Base Superalloys," Scripta Metall. 18, 91-94 (1984).

8. D. Blavette, P. Duval, L. Letellier, and M. Guttmann, "Atomic-Scale APFIM and TEM Investigation of Grain Boundary Microchemistry in Astroloy Nickel Base Superalloys," Acta Mater. 44, 4995-5005 (1996).

9. N. S. Stoloff, "Alloying of Nickel," pp. 371-417 in Alloying, eds. J. L. Walter, M. R. Jackson, and C. T. Sims, ASM International, Materials Park, OH 1988.

10. C. P. Sullivan and M. J. Donachie, Jr., in Section VIII: Nickel-Base Superalloys, of Source Book on Materials for Elevated-Temperature Applications, compiled by Elihu F. Bradley, ASM, 1979, pp. 250-59.

11. Karlsson and H. Norden, J. Phys. C47-11(7), 257 (1986).

12. W. D. Cao and R. L. Kennedy, "Phosphorus-Boron Interaction in Nickel-Base Superalloys," in Superalloys 1996, TMS Warrendale, PA, 589-97 (1996).

13. R. F. Decker and J. W. Freeman, Trans. AIME 218, 277 (1960).

14. C. L. Briant, "Grain Boundary Segregation in the Ni-Base Alloy 182," Met. Trans. 19A, 137-43 (1988).

15. M. Guttman, Ph. Dumoulin, Nguyen Tan-Tai, and P. Fontaine, "An Auger Electron Spectroscopic Study of Phosphorus Segregation in the Grain Boundaries of Nickel Base Alloy 600," Corrosion 37, 416-25 (1981).

16. B. J. Berkowitz and R. D. Kane, "The Effect of Impurity Segregation on the Hydrogen Embrittlement of a High Strength Nickel Base Alloy in H_2S Environments," Corrosion 36(1), 24-29 (1980).

17. D. Kane and B. J. Berkowitz, "Effect of Heat Treatment and Impurities on the Hydrogen Embrittlement of a Nickel Cobalt Base Alloy," Corrosion 36(1), 29-36 (1980).

18. W. Yeniscavich and C. W. Fox, p. 24 in Effect of Minor Elements on the Weldability of High Nickel Alloys, Welding Research Council, 1969.

19. M. Tamura p. 215 in Superalloys, Supercomposites and Superceramics, Akademic Press, Inc., 1989.

20. D. A. Vermilyea, C. S. Tedmon, Jr., and D. E. Broecker, Corrosion 31, 222 (1975).

21. C. G. Bieber and R. F. Decker, Trans. AIME 221, 629 (1961).

22. J. K. Sung and G. S. Was, Corrosion 47, 824 (1991).

23. G. S. Was, J. K. Sung, and T. M. Angeliu, "Effects of Grain Boundary Chemistry on the Intergranular Cracking Behavior of Ni-16Cr-9Fe in High-Temperature Water," Met. Trans. 23A, 3343-59 (1992).

24. D. Cao and R. L. Kennedy, "The Effect of Phosphorous on Mechanical Properties of Alloy 718," in Proc. Int. Symp. Superalloys 718, 625, 706 and Various Derivatives, ed. E. A. Loria, TMS, Warrendale, PA, pp. 463-77 (1994).

25. W. D. Cao and R. L. Kennedy, "Effect and Mechanism of Phosphorus and Boron on Creep Deformation of Alloy 718," pp. 511-20 in Superalloys 718, 625, 706 and Various Derivatives, ed. E. A. Loria, TMS, 1997.

26. R. L. Kennedy, W. D. Cao, and W. M. Thomas, "Stress-rupture Strength of Alloy 718," Adv. Mater. Proc. **149**, 33-35, March 1996.

27. C. G. McKamey, C. A. Carmichael, W. D. Cao, and R. L. Kennedy, , "Creep Properties of Phosphorus+Boron-Modified Alloy 718," Scripta Mater. **38**, 485-91 (1998).

28. E. P. George and R. L. Kennedy, "Trace Element Effects on High Temperature Fracture," in Impurities in Engineering Materials, ed. C. L. Briant, Marcel Dekker, Inc., NY, pp. 225-58 (1999).

29. X. Xie, X. Liu, J. Dong, Y. Hu, Z. Xu, Y. Zhu, W. Luo, Z. Zhang, and R. G. Thompson, "Segregation Behavior of Phosphorus and Its Effect on Microstructure and Mechanical Properties in Alloy System Ni-Cr-Fe-Mo-Nb-Ti-Al," pp. 531-42 in Superalloys 718, 625, 706 and Various Derivatives, ed. E. A. Loria, TMS, 1997.

30. J. A. Horton, C. G. McKamey, M. K. Miller, W. D. Cao, and R. L. Kennedy, "Microstructural Characterization of Superalloy 718 with Boron and Phosphorus Additions," in Superalloys 718, 625, 706 and Various Derivatives, ed. E. A. Loria, TMS, Warrendale, PA, pp. 401-08 (1997).

31. S. J. Sijbrandij, M. K. Miller, J. A. Horton, and W. D. Cao, "Atom Probe Analysis of Nickel-Based Superalloy IN-718 with Boron and Phosphorus Additions," Mat. Sci. & Eng. A250, 115-19 (1998).

32. H. Song, S. Guo, and Z. Hu, "Beneficial Effect of Phosphorus on the Creep Behavior of Inconel 718," Scripta Mater. 41(2), 215-19 (1999).

33. E. G. Richards, J. Inst. Met. 96, 365 (1968).

34. E. W. Ross and C. T. Sims, "Nickel-Base Alloys," Chapter 4, pp. 97-133 in Superalloys II, eds. C. T. Sims, N. S. Stoloff, and W. C. Hagel, John Wiley & Sons, New York, 1987.

35. T. J. Garosshen, T. D. Tillman, and G. P. McCarthy, , "Effects of B, C, and Zr on the Structure and Properties of a P/M Nickel Base Superalloy," Met. Trans. 18A, 69-77 (1987).

36. E. W. Ross and C. T. Sims, "Nickel-Base Alloys," Chapter 4, pp. 97-133 in Superalloys II, eds. C. T. Sims, N. S. Stoloff, and W. C. Hagel, John Wiley & Sons, New York, 1987.

37. A. Choudhury, C. L. White, and C. R. Brooks, Acta Metall. Mater. 40, 57 (1992).

38. M. K. Miller, D. J. Larson, and K. F. Russell, "Characterization of Segregation in Nickel and Titanium Aluminides," pp. 53-62 in Structural Intermetallics, eds. M. V. Nathal, R. Darolia, C. T. Liu, P. L. Martin, D. B. Miracle, R. Wagner, and M. Yamaguchi, TMS, Warrendale, PA, 1977; K. Aoki and O. Izumi, Nippon Kinzoku Gakkaishi 43, 1190 (1979).

39. T. Liu, C. L. White, and J. A. Horton, Acta Metall. 33, 213-29 (1985).

40. E. L. Hall and C. L. Briant, "The Microstructural Response of Mill-Annealed and Solution-Annealed INCONEL 600 to Heat Treatment," Metall. Trans. 16A, 1225-36 (1985).

41. R. M. Kruger and G. S. Was, "The Influence of Boron on the Grain Boundary Chemistry and Microstructure of Ni-16Cr-9Fe-0.03C," Metall. Trans. 19A, 2555-66 (1988).

42. J. Takesugi and O. Izumi, Acta Metall. 33, 1247-58 (1985).

43. J. K. Tien and R. P. Gamble, "The Influence of Applied Stress and Stress Sense on Grain Boundary Precipitate Morphology in a Nickel-Base Superalloy During Creep," Met. Trans. 2, 1663-67 (1971).

44. F.R.N. Nabarro, p. 75 in Report of a Conference on the Strength of Solids, Physical Society, London, 1948.

45. C. Herring, J. Appl. Phys. 21, 437 (1950).

Table I. Chemical Compositions and Grain Sizes of Modified Waspaloy

Type of Alloy	Heat #	Chemistry (wt%)		Grain Size[a]	
		P	B	ASTM#	D (µm)
High Purity	G757-1	0.001	<0.001	10 W/B 6-7	7.9
Commercial	G752-2	0.004	0.006	10	7.2
Commercial	WB74	0.006	0.0046	9	11.9
High B, High P	G766-2	0.022	0.011	9 W/B 6	10.3
High B, Low P	G949-1	0.001	0.014	9	10.3

[a]D is grain size measured by intersection line method. W/B means alloy had a banded structure with the bands being the second grain size shown.

Creep in Oxide Dispersion Strengthened Ni-based Superalloy, MA754, with Different Orientation of Columnar Grain

Nobuhiro MIURA, Yoshihiro KONDO, Michio OKABE* and Takashi MATSUO**
The National Defense Academy, 1-10-20 Hashirimizu Yokosuka Kanagawa 239-8686, Japan
*Daido Steel Co., Ltd., 2-30 Daido-cho Minami Nagoya Aichi 457-8545, Japan
**Tokyo Institute of Technology, 2-12-1 Oookayama Meguro Tokyo 152-8552, Japan

1. Abstract

The features of creep in an oxide dispersion strengthened Ni-based superalloy, MA754, was investigated using seven specimens with the different orientations between the stress axis and the extruded direction, θ. Creep tests were conducted at 1273K, 100-160MPa. The creep rupture life was shortened with an increase in θ. The rupture elongation becomes smaller with a decrease in θ. The creep rate during the transient creep stage was irrespective to θ. Strong θ dependence of the creep life and rupture elongation are only decided by the difference in the time of the onset of accelerating creep. The early onset of accelerating creep in the specimen with larger value of θ correlated with the angle of the grain boundary which is the bottom side in the columnar grain and the stress axis. There was no difference in the dislocation substructures at the constant strain among the specimens with different θ.

2. Introduction

Excellent creep resistance of the oxide dispersion strengthened nickel-based superalloys was due to the existence of the uniform dispersion of oxides such as the Y_2O_3 in the matrix[1]-[3]. Howson et al. compared the creep properties of the oxide dispersion strengthened nickel-based superalloy, MA754, selected parallel and normal to the extruded axis at 1033-1366K in the wide stress range and reported that the creep life, creep rate and ductility of the specimens with the columnar grain parallel to the extruded axis were superior to that of the specimens normal to extrusion axis[4]. Tsukuta et al. carried out the compressive creep tests of the oxide dispersion strengthened nickel-based superalloys, MA754 and MA758, at 1523 to 1623K using the specimen with parallel and normal directions of the extruded axis. And it was elucidated that the creep life of the normal direction specimen was inferior to that of the parallel one[5]. However, the origin of the strong stress direction dependence of creep has not been elucidated[6]-[10]. The effect of the angle between the direction of the columnar grain boundary and the stress direction on the creep rate must be investigated. Creep of ODS alloys prepared by hot extrusion is affected by the microstructural change occurred along the grain boundary. To clarify θ dependence of the creep rate, the rupture life and the rupture elongation will give the evidence of origin. Therefore, the creep rate-time curves are investigated.

In this study, the creep tests of an oxide dispersion strengthened nickel-based superalloy, MA754, were conducted at 1273K, 100-160MPa, using the seven specimens with different orientations between the stress axis and the extruded direction in each 15 degree, and the effect of the orientation between the tensile direction and the extruded one on the creep properties was discussed.

3. Experimental procedure

An oxide dispersion strengthened (ODS) nickel-based superalloy, MA754, (having the nominal

470

chemical composition in weight percent ; 19.67Cr, 0.21Fe, 0.25Al, 0.42Ti, 0.07C, 0.6Y$_2$O$_3$, balance Ni) produced by mechanical alloying, was obtained in the form of 33.4 x 75.2 x 1200mm bar. The seven specimens with different orientations between the longitudinal direction, namely, the stress axis and the

extruded direction in each 15 degree were cut by the electric discharge machine (EDM), as shown in Fig.1. The alloy was tempared as follow : 1588K for 3.6ks / AC. The specimens for the creep tests with a gauge diameter of 6mm and a gauge length of 30mm were prepared.

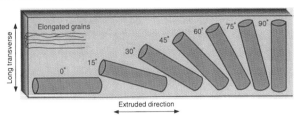

Fig.1. Schematic illustrate of the specimens cut from the bar.

Creep tests were conducted at 1273K in the stress range from 100 to 160MPa under the constant load. Creep strain was measured automatically through linear variable differential transformers (LVDT's) attached to extensometers. The creep tests of the specimen were interrupted at the time of the minimum creep rate. And the creep tests of the specimen with θ of 0 degree were interrupted at the time corresponding to the minimum creep rate of the specimens with θ from 15 to 90 degrees.

Microstructural observations were done by an optical microscopy (OM) and by a transmission electron microscopy for the specimens sectioned parallel to the extruded plane. Specimens of TEM were prepared by electropolishing using the twin jet polisher with a 10% perchloric acid-alcohol solution.

4. Results
4.1 Microstructure of the tempared specimen

The optical micrograph of the tempared MA754 are shown in Fig.2. The mechanical alloying, extrusion and subsequent tempering treatment resulted in the substructure with the elongated grain along the extruded direction.

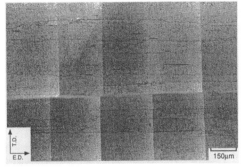

Fig.2. Optical micrograph of the tempared MA754.

The transmission electron micrograph of the tempared MA754 is shown in Fig.3. Finely dispersed Y$_2$O$_3$ oxide particles, relatively large inclusions and a high density of twins are observed in the grains.

4.2 Correlation between θ and creep property

The creep curves of the specimens with θ of 0, 15, 30, 45, 60, 75 and 90 degree at 1273K-130MPa are shown in Fig.4. The rupture life is

Fig.3. Transmission electron micrograph of the tempared MA754.

shortened with an increase in θ. The rupture elongation of the specimen with θ of 0 degree is about 2%. However, the rupture elongation decreases steeply with a decrease in θ. Consequently, the rupture elongation of the specimen with θ of 90 degree becomes 0.2%.

The stress-creep rupture life curves of the specimens with θ of 0, 15, 30, 45, 60, 75 and 90 degree at 1273K are shown in Fig.5. The linear relation is obtained for each curve using log-log scale and all curves are parallel. The creep rupture life is shortened with an increase in θ.

The rupture elongation of the specimens with θ of 0, 15, 30, 45, 60, 75 and 90 degree is plotted as a function of the creep rupture life as shown in Fig.6. The rupture elongation of the specimen with θ smaller than 30 degree is more than 1%, but that of the specimens with θ larger than 45 degree is less than 0.1%. The rupture elongation of the specimens with θ smaller than 30 degree decreases with an increase in the rupture life.

The creep rate-time curves of the specimens with θ of 0, 15, 30, 45, 60, 75 and 90 degree at 1273K-130MPa are shown in Fig.7. At the beginning of the transient creep stage, the creep rate and the decreasing ratio of the creep rate is irrespective of θ. However, with an increase in the time, the drastic change in the creep rate-time curves is appeared in a shorter time in the specimens with larger value of θ. The onset of accelerating creep is fully suppressed in the specimen with θ of 0 degree. Therefore, the specimen with smaller value of θ has the longer transient stage.

The minimum creep rates as a function of the stress in all specimens at 1273K are

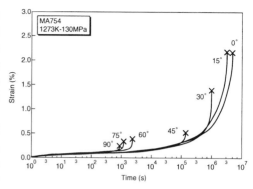

Fig.4. Creep curves of the specimens with θ of 0, 15, 30, 45, 60, 75 and 90 degree at 1273K

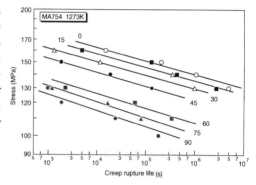

Fig.5. Stress-creep rupture life curves of the specimens with θ of 0, 15, 30, 45, 60, 75 and 90 degree at 1273K.

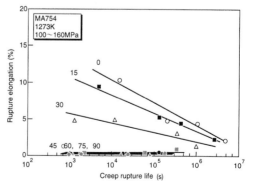

Fig.6. Relation between the rupture elongation and the creep rupture life.

shown in Fig.8. The minimum creep rate of the specimens with larger value of θ is larger. The correlation between the minimum creep rate and the stress is linear on a log-log scale ; the stress exponent, n, of creep rate is larger value of 26. The increasing ratio of the minimum creep rate with an increase in θ is about twice, but there is a gap between the specimens with θ of 45 and 60 degree. This gap is about 10 times.

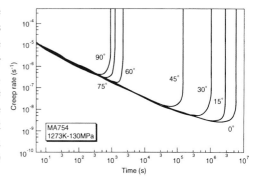

Fig.7. Creep rate-time curves of the specimens with θ of 0, 15, 30, 45, 60, 75 and 90 degree at 1273K-130MPa.

5. Discussion
5.1 Rupture elongation behavior

The difference in θ does not affected substantially on creep rate. The difference in the minimum creep rate is decided by the marked difference in the time of the onset of accelerating creep. The difference in the time of accelerating creep is considerably correlated with the rupture elongation. So, firstly, the reason why the rupture elongation drastically decreases with an increase in θ will be discussed.

The creep rate of the specimens with θ of 0, 45 and 90 degree crept at 130MPa is plotted as a function of the ratio of the time to the creep rupture life as shown in Fig.9. The ratio of the accelerating creep stage to the creep rupture life of the specimen with θ of 0 degree is occupied 40%, but those of the specimens with θ of 45 and 90 degree made up 20 and 10%, respectively. The specimen with smaller value of θ has the larger accelerating creep stage.

The optical micrographs of the vicinity of the fracture surface of the specimens with θ of 0, 45 and 90 degree creep ruptured at 1273K-130MPa are shown in Fig.10. In these photos, the stress axis is vertical. Voids and cavities are observed at the grain boundary of the specimens with

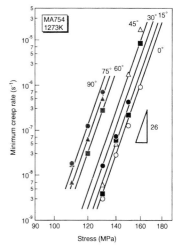

Fig.8. Stress-minimum creep rate curves at 1273K.

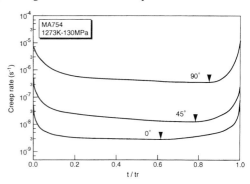

Fig.9. Relation between the creep rate and the ratio of the time to the creep rupture life.

θ of 0 degree (Fig.10 (a)). However, few voids and cavities are formed at the grain boundary of the specimens with θ of 45 and 90 degree (Fig.10 (b and c)). For the specimen with θ of larger than 45 degree, once a void or a cavity is generated at the grain boundary, it immediately propagated along the grain boundary and advanced to crack, and then led to creep rupture. From these supposition, the magnitude of grain boundary area will be directory correlated with cracks and voids and increase with increasing θ. Therefore, the rupture elongation of the specimen with larger value of θ seemed to be smaller than that of the specimen with smaller value of θ.

5.2 θ dependence of the minimum creep rate

As described before, the minimum creep rate is directly related with the time of the onset of accelerating creep. Here, such correlation will be elucidated. What kind of the microstructural change does lead the onset of accelerating creep.

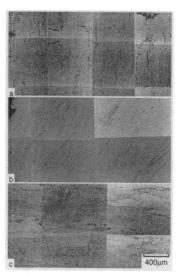

Fig.10. Optical micrographs of the vicinity of the fracture surface of the specimens with θ of (a) 0, (b) 45 and (c) 90 degree.

The transmission electron micrograph of the specimen with θ of 0 degree crept at 1273K-130MPa for 2.52 x 10⁶s, at the time of the minimum creep rate, is shown in Fig.11, where the electron beam direction, B, is closed to [110]. A number of dislocations are restricted their movement by Y_2O_3 oxide particles.

The transmission electron micrograph of the specimen with θ of 45 degree crept for 7.20 x10⁴s, at the time of the minimum creep rate, is shown in Fig.12, where B = [110]. A few dislocations are correlated with Y_2O_3 oxide particles.

The transmission electron micrograph of the specimen with θ of 90 degree crept for 6.00 x 10²s, at the time of the minimum creep rate, is shown in Fig.13, where B = [110]. Few dislocations are observed in the γ matrix. The dislocation density in the γ matrix is lower than those of the other specimens.

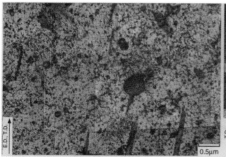

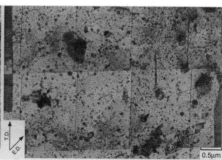

Fig.11. Transmission electron micrograph of the specimen with θ of 0 degree crept at 1273K-130MPa for 2.52x10⁶s, where B=[110].

Fig.12. Transmission electron micrograph of the specimen with θ of 45 degree crept at 1273K-130MPa for 7.20x10⁴s, where B=[110].

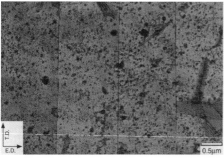

Fig.13. Transmission electron micrograph of the specimen with θ of 90 degree crept at 1273K-130MPa for 6.00x10²s, where B=[110].

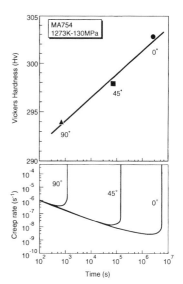

Fig.14. Relation between Vickers hardness of the specimens with θ of 0, 45 and 90 degree and the time.

Vickers hardness of the specimens with θ of 0, 45 and 90 degree is plotted as a function of the time as shown in Fig.14. Hardness increased with an increase in the time. Consequently, it is supposed that an increase in the dislocation density would be corresponded to the increase in strain hardening.

To confirm the difference in dislocation substructures, creep interrupted tests of the specimens with θ of 0 degree were done at the same testing time for 7.20 x10⁴ and 6.00 x 10²s. These two transmission electron micrographs are shown in Fig.15. The dislocation density increases with an increase in the time. The dislocation substructures of the specimens with θ of 0 degree crept for 7.20 x10⁴ and 6.00 x 10²s are similar to those of the specimens with θ of 45 and 90 degree crept for the same testing time (as shown in Figs. 12 and 13), respectively.

Vickers hardness of the specimen with θ of 0 degree is plotted as a function of the time compared with that of the specimens with θ of 45 and 90 degree at the time of the minimum creep rate, as shown in Fig.16. Hardness of the specimen with θ of 0 degree increases with an increase in the testing time. There is no difference among three specimens with θ of 0, 45 and 90 degree at the same testing time.

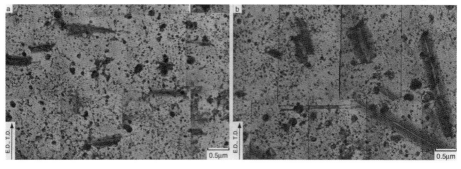

Fig.15. Transmission electron micrographs of the specimens with θ of 0 degree creep interrupted tests at 1273K-130MPa for (a) 7.20x10⁴s and (b) 6.00x10²s, where B=[110].

From these results, the early onset of accelerating creep in the specimen with larger value of θ is not correlated with the dislocation density.

The creep tests were carried out on the specimens oriented parallel and perpendicular to the longitudinal grain direction using a directionally solidified nickel-based superalloy, MAR-M247LC[11]. The creep features of MAR-M247LC are compared in order to confirm the remarkable θ dependence of the creep rate of MA754.

The creep rate-time curves at 1273K-130MPa of the specimens with θ of 0, 45 and 90 degree of MA754 are shown in Fig. 17, compared with those at 1273K-160MPa of MAR-M247LC oriented parallel and perpendicular to the longitudinal grain direction. As described above, the creep rupture life and the creep rate is remarkably dependent on θ in MA754. On the other hand, there are slight differences in the creep rupture life and the creep rate of MAR-M247LC oriented parallel and perpendicular to the longitudinal grain direction. Our examination by the scanning electron microscope provided evidences that intergranular fracture occurs in MA754 and MAR-M247LC. In MAR-M247LC, the presence of the γ' phase and the eutectic γ' phase at the grain boundary seemed to act as suppression of cracking. On the other hand, in MA754, there is no precipitation at the grain boundary.

Consequently, the remarkable θ dependence of the creep rate of MA754 is attributed to easy crack formation correlated with the direction of the grain boundary to the stress axis.

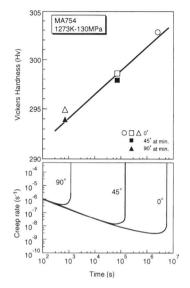

Fig.16. Relation between Vickers hardness of the specimen with θ of 0 degree and the time, compared with that of the specimens with θ of 45 and 90 degree at the time of the minimum creep rate.

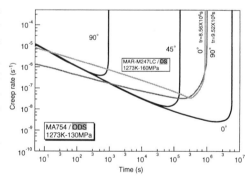

Fig.17. Creep rate-time curves of the specimens with θ of 0, 45 and 90degree of MA754 at 1273K-130MPa, compared with those of MAR-M247LC oriented parallel and perpendicular to the longitudinal grain direction at 1273K-160MPa.

6. Conclusions

Creep behavior in an oxide dispersion strengthened nickel-based superalloy, MA754, in the stress range from 100-160MPa at 1273K is investigated using the seven specimens with the different

476

orientations between the stress axis and the extruded direction, θ.

The following conclusions are obtained.

1) The creep rupture life is shortened with an increase in θ. The rupture elongation of the specimen with θ smaller than 30 degree is more than 1%, but that of the specimens with θ larger than 45 degree is less than 0.1%.

2) At the beginning of the transient creep stage, the creep rate and the decreasing ratio of the creep rate is irrespective of θ. However, the onset of accelerating creep is appeared earlier in the specimen with larger value of θ and the minimum creep rate of the specimens with larger value of θ is larger.

3) The specimen with smaller value of θ has the larger accelerating creep stage. Voids and cavities are observed at the grain boundary of the specimen with θ of 0 degree. However, few voids and cracks are formed at the grain boundary of the specimens with θ of 45 and 90 degree.

4) The dislocation density at the time of the minimum creep rate decreases with an increase in θ. There was no difference in the dislocation substructures.

5) There is no difference in hardness among three specimens with θ of 0, 45 and 90 degree at the same testing time, and hardness of the specimen with θ of 0 degree increases with an increase in the testing time. Therefore, the early onset of accelerating creep in the specimen with larger value of θ is not correlated with dislocation density.

6) Consequently, the remarkable θ dependence of the creep rate of MA754 is attributed to easy crack formation correlated with the direction of the grain boundary to the stress axis.

Reference

[1] Inco Alloys International, Inconel alloy MA754, IncoMAP

[2] W.D. Klopp, 1992, Nickel Base Alloys, MA754, Aerospace Structural Metals Handbook, Code4106, pp1-29.

[3] A.H. Cooper, 1984, Attractive Dislocation and Particle Interactions in ODS Superalloys and Implications, Proc. Of the 5th Inter. Conf. Superalloys1984, pp.357-366.

[4] T.E. Howson, 1980, Creep and Stress Rupture of Oxide Dispersion Strengthened Mechanically Alloyed Inconel Alloy MA754, Metall. Trans. A, Vol.11A, pp.1599-1607.

[5] K. Tsukuta, 1990, High Temperature Strength of Mechanically Alloyed Oxide Dispersion Strengthened Ni-Base Superalloys, Electric Furnace Steel, Vol61, No.2, pp93-101.

[6] J.D. Whittenberger, 1977, Creep and Tensile Properties of Several Oxide Dispersion Strengthened Nickel Base Alloys, Metall. Trans. A, Vol.8A, pp.1155-1163.

[7] H. Kawai, 1982, High Temperature Property of a Oxide Dispersion Strengthened Nickel-based Heat Resisted Alloy MA754, 123rd Committee on Heat Resisting Metals and Alloys Rep., Vol.23, No.2, pp.21-29.

[8] J.J. Stephens, 1985, The Effect of Grain Morphology on Longitudinal Creep Properties of INCONEL MA754 at Elevated Temperatures, Metall. Trans. A, Vol.16A, pp.1307-1324.

[9] J.J. Stephens, 1984, Creep and Fracture of Inconel MA754 at Elavated Temperatures, Proc. of the 5th Inter. Conf. Superalloys1984, pp.327-334.

[10] B. DeMestral, 1996, On the Influence of Grain Morphology on Creep Deformation and Damage Mechanisms in Directionally Solidified and Oxide Dispersion Strengthened Superalloys, Metall. Trans. A, Vol.27A, pp.879-890.

[11] N.Miura, 1998, Morphology of γ′ phase of Directionally Solidified Ni-baed Superalloy, MAR-M247LC, Crept for Longitudinal and Transverse Section, CAMP ISIJ, Vol.11, p.1105.

Nobuhiro MIURA :
E-mail : nmiura@nda.ac.jp, FAX : +81-468-44-5900
Yoshihiro KONDO :
E-mail : kondo@nda.ac.jp, FAX : +81-468-44-5900
Michio OKABE :
E-mail : D3200@so.daido.co.jp, FAX : +81-52-611-9412
Takashi MATSUO
E-mail : tmatsuo@mtl.titech.ac.jp, FAX : +81-3-5734-2874

Creep-fatigue behavior in a Ni-based DS alloy during high temperature oxidation

by

Masaru SEKIHARA, Mechanical Engineering Research Laboratory, Hitachi Ltd.
1-1, Saiwai-cho 3-chome, Hitachi-shi, Ibaraki-ken, Japan
Shigeo SAKURAI, Mechanical Engineering Research Laboratory, Hitachi Ltd.

Abstract

The influence of the direction of solidification of Ni-based directionally solidified super alloy on the growth rate of cracks was investigated on the basis of long-term creep-fatigue experiments. The relationship between the direction of solidification and the growth rate of a crack was evaluated in terms of the formation and growth of the oxidized zone on the crack's surface. The growth rate of a crack is only slightly affected by the direction of solidification while, it seems, the size of the oxidized zone has a strong effect on the crack's growth rate. The effect of the size of the oxidized zone on the crack's growth rate was evaluated by using a J-integral based on the finite element method and an energy release rate based on a virtual crack extension. The size of the oxidized zone was found to have a stronger effect on the energy release rate than on the J-integral.

1. Introduction

•To improve the efficiency of industrial gas turbines, attempts are being made to raise turbine-inlet temperature. This is leading to the adoption of Ni-based single- crystal super alloy (SC-alloy) and directionally solidified super alloy (DS- alloy) materials for the turbine blades.

•The grain boundaries in anisotropic super alloys present a smaller cross-section in the direction perpendicular to the principal direction of stress than the grain boundaries of the materials conventionally used for turbine blades. This can provide improved resistance to creep.

Thermal stress can be decreased by eliminating stiffness in the longitudinal direction of blades. The allowable temperatures for gas-turbine blades can thus be increased dramatically, and, as a result, higher efficiencies can be achieved.

The reliability of the hot parts of gas turbines in which such anisotropic heat-resisting super alloys are used must then be assured by applying lifetime evaluation technology. The initiation and growth of

cracks in these materials under high temperatures involves many unknowns factors, so it has been difficult to evaluate the reliability of such hot parts in terms of their usable lifetimes.

•In this research the effect of the direction of solidification on the rate of crack growth in DS-alloys has been investigated with the aiming of developing a lifetime evaluation method for Ni-based directionally solidified super alloys.

2. Experiments on and finite element analysis of crack growth

•The material tested was Rene'80H, the chemical composition of which is shown in Table1. In this research, specimens of three types were selected for the evaluation of the influence of the direction of solidification on the material's strength and its behavior in terms of the initiation and growth of cracks: in the vertical type (V-type), the direction of solidification is perpendicular to the load direction; in the horizontal type (H-type), the direction of solidification runs across the load direction; in the diagonal type (D-type), the direction of solidification is diagonal to the load direction. A notched-compact (CT) specimen was used, and is shown in Fig. 1. The radius of the notch at its root was 0.2 mm. Load-controlled creep-fatigue experiments were conducted in air at a stress ratio of 0.1, at 600°C and 860°C. Two loading conditions were applied, a triangular waveform at 0.1Hz and a trapezoidal waveform with 60-min tension-hold period and 10 seconds of relaxation. Maximum load was selected such that the initial stress intensity factor $\bullet K$ was equal to about $25\,\text{MPa}\sqrt{m}$.

•The J-integral value and energy release rate G_1 was calculated by using a finite element model (FEM). The model of the CT specimen is shown in Fig. 2. The oxidized layer on a crack's surface was included in the model for finite element analysis, and the thickness of this layer, d, was selected as the parameter for the representation.

The observed growth of an oxidized layer shown in Fig. 3 was used to choose the oxidized layer's thickness d. The exposure experiment was conducted in air at 860°C, using V-type•H-type and D-type specimens, with an 0.4-mm wide slit to model a crack.

3. Results and discussion

3.1 Rate of crack growth and J-integral

•The initiation and growth of cracks in CT specimens is illustrated by Fig. 4. Crack length here is the distance of growth from the notch's root. At 600°C, a dependence of the growth rate on the direction of solidification can clearly be seen. The growth rate of the crack in the V-type specimen was slowest and fastest in the D-type specimen. The result for the H-type specimen was intermediate. At 860°C the effect of the direction of solidification on the growth rate of the crack displayed the same general trend as the result at 600°C, but the effect was relatively weaker. The effect of the direction of solidification on the crack's rate of growth also tended to

lessen as the crack grew.

•The value of the J-integral for a crack of a certain length was calculated as the average value of a 10-path integral obtained by FEM. The calculated relationships between the value of this J-integral and the growth rate of crack are shown in Fig. 5.

•The growth rate of the crack in the V-type specimen shows a slight scattering at 600°C. On the other hand, the effect of the direction of solidification was not visible at 860°C. The behavior of the material in terms of crack growth under a 0.1 Hz cyclic frequency can thus be evaluated by using the J-integral as obtained by FEM with consideration of the anisotropy.

•The initiation and growth of cracks at 860°C under loading according to a triangular waveform at 0.1 Hz and a trapezoidal waveform with 60-min tension-hold period and 10 seconds of relaxation are shown in Fig. 6.

With the triangular waveform at 0.1 Hz, the crack started most slowly in the V-type specimen and most quickly in the D-type specimen. The effect of the direction of solidification tended to diminish with the number of cycles.

•With the trapezoidal waveform, the effect of the direction of solidification was practically invisible. This result was almost the same for the pre-cracked specimen.

Table 1. Chemical composition of tested materials (mass %)

Cr	Co	Mo	Ti	W	Al	Hf	Zr	C	Ni
13.9	9.2	4.0	4.72	4.0	3.03	0.75	0.01	0.15	Bal.

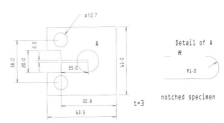

Fig. 1 The specimen's geometry and dimensions (mm)

Fig. 2 Finite-element model

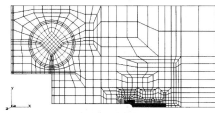

Fig. 3 Oxidized zone at a crack's tip

• The relationship between the growth rate of the crack and the number of cycles at 860°C under loading that followed the same two patterns is shown in Fig. 7. Here, growth is slowest in the V-type and fastest in the D-type specimen, but the effect of the direction of solidification tended to decrease as the number of cycles increased.

• Furthermore, under the trapezoidal waveform pattern of loading, the effect of the direction of solidification on the growth rate of the crack was practically invisible. The growth rate decreased at first, but began to increase again after 5000 cycles.

• It appears first that the direction of solidification had less effect on the crack's growth at 860°C because of the formation of an oxidized layer on the crack's surfaces. The next section is a description of an exposure experiment on specimens with a slit as a model of a crack. The experiment was conducted at 860°C and the behavior of the oxidized layer in terms of growth was observed.

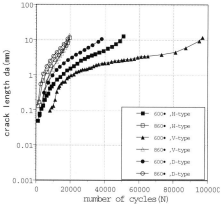

Fig. 4 Growth of cracks (0.1 Hz)

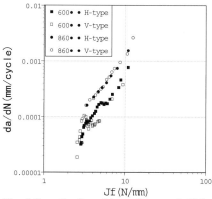

Fig. 5 Growth of crack's vs. J-integral (0.1 Hz)

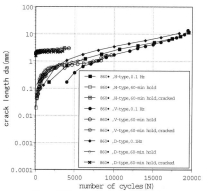

Fig. 6 Crack growth behavior
(0.1 Hz, 60-min hold)

482

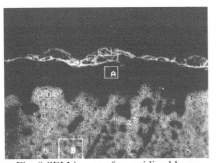

Fig. 7 Crack growth rate vs. number of cycles

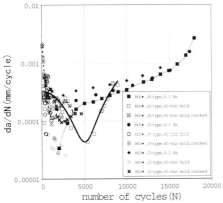

Fig. 8 SEM image of an oxidized layer

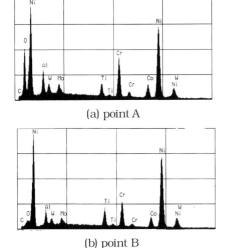

(a) point A

(b) point B

Fig. 9 Results of component analysis

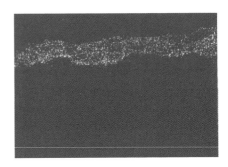

Fig. 10 Distribution of oxygen

3.2•Observation of oxidized layer

•SEM-observed results for a V-type specimen after 100 hours of exposure to air at 860°C are shown in Fig. 8. A dendritic structure and •'phase were observed around point B. The results of component analysis for points A and B of Fig. 8 are shown in Fig. 9. The proportions of O, Al and Cr present at point A are relatively high, so a composite oxidized layer (Al•Cr) O_2 appears to have been formed.

•Fig. 10 shows the observed distributions of oxygen. Oxygen was again shown to be densely distributed around point A and the formation of an oxidized layer was thus confirmed.

•Next, the relationship between the exposure period and the thickness of the oxidized layer was obtained by investigating the rate of increase of the oxidized layer's thickness. The appearance of the oxidized zones is shown in Fig. 11. Three areas were selected for observation, that is, two parallel areas (A,C) and part of the slit's root area (B). An image of a V-type specimen after exposure to air at 860°C for 400 hours is shown in Fig. 12.

•There was little difference between the thickness of oxidized layers in areas A, B and C. It was thus confirmed that the oxidized layer grew at almost the same rate in each area. V-type and H-type specimens showed the same results. No effect of the direction of solidification on the growth of the oxidized layer was discernible.

•Fig. 13 shows the relationship between the

thickness of the oxidized layer and exposure time as obtained by observation of areas A, B and C of V-type, H-type and D-type specimens. The curves shown in Fig. 13 can be approximated as follows,

$$• \, d = a\sqrt{t} \qquad •(1)$$

where d is the thickness of the oxidized layer, t is the exposure time and a is a coefficient.

•The growth of the oxidized layer seems to be according to equation (1), that is, a parabolic relation. With d in •m and t in hours, the coefficient a as obtained from the experimental results was close to 1.

3•3•Energy release rate and thickness of the oxidized layer

•The effect of the growth of the oxidized layer on the parameters of fracture mechanics was also investigated. In addition to the J-integral value obtained by FEM, an energy release rate G_1 based on the extension of a virtual crack was adopted as a parameter of fracture mechanics, because it seemed to be important for a sensitive evaluation of the change in the material's characteristic at the tip of the crack. G_1 is determined by,

$$G_1 = \lim_{\Delta c \to 0} \frac{1}{2t\Delta c} F_a \left(v_c - v_d \right) \quad ••(2)$$

where t is the thickness of the CT specimen, •c is the distance between nodes as shown in Fig. 14, F_a is the reaction force at point a, and v_c and v_d are the displacements in the x_2 direction at points c and d.

•Fig. 15 shows the relations of the J-integral value and the energy release rate G_1 with the thickness of the oxidized layer. The maximum thickness d was selected as 30•m, and this value was selected so that the exposure time would correspond to about 1000 hours according to the results shown in Fig. 13.

•As shown in Fig. 15, the J-integral's value was almost completely unaffected as the thickness of the oxidized layer grew. On the other hand the energy release rate G_1 tended to increase with the thickness of the oxidized layer.

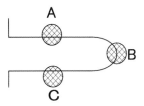

Fig. 11 The areas observed

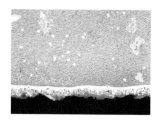

(a) area A   25μm

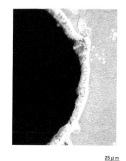

(b) area B   25μm

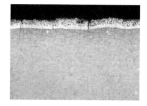

(c) area C   25μm

Fig. 12 The oxidized layer in the three areas

484

On the crack's surface and at its tip, the compressive strain can be expected to grow with the volumetric expansion that occurs as the oxidized layer grows. The energy release rate G_1 was calculated with the local compressive strain (0.5 and 1.0%) in the oxidized layer as a parameter.

When this compressive strain was introduced, the energy release rate G_1 tended to decrease as the thickness of the oxidized layer increased, as shown in Fig. 15. Lagoudas et al reported the same results in terms of the energy release rate G_1 tending to increase with the thickness of the oxidized layer. They also reported, however, that the energy release rate G_1 tends to decrease gradually with the increase in the compressive strain generated by the volumetric expansion in the oxidized layer[4].

•The initial decrease in the crack growth rate shown in Fig. 7 under loading according to the trapezoidal waveform is considered to have been because of the compressive strain field generated by volumetric expansion.

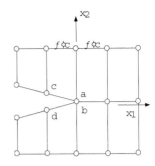

Fig. 14 FEM of the crack's tip

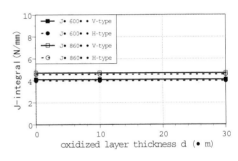

(a)   J-integral

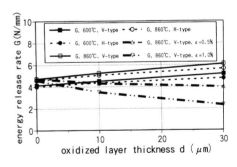

(b)   Energy release rate

Fig. 15 J-integral and energy release rate vs. thickness of the oxidized layer at the crack's tip (da=1 mm)

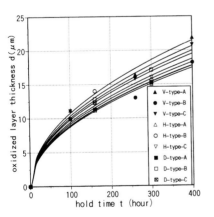

Fig. 13 Thickness of the oxidized layer vs. hold time

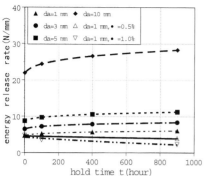

Fig. 16 Energy release rate vs. thickness of the oxidized layer at crack's tip (860°C,V-type)

•The integral's path was selected to cross both the oxidized layer and base metal, so the effect of the oxidized layer at the crack's tip was not apparent. It is thus considered that this is why we were not able to find an effect of the thickness of the oxidized layer on J-integral's value.

•As mentioned above, energy release rate G_1 seems to be adequate to represent the growth of the oxidized layer as a parameter of fracture mechanics.

•Fig. 16 shows the relationship between the exposure time and energy release rate as obtained by using the growth rate of the oxidized layer shown in Fig. 13. Energy release rate G_1 increases with the growth of the crack. Conversely, the energy release rate G_1 tends to decrease when the compressive strain field generated by volumetric expansion is introduced.

•From these facts, energy release rate G_1 is considered to have a tendency to have a single period of decrease in the first part of growth of the oxidized layer. Next, the energy release rate G_1 tends to increase as the crack grows because the effect of the oxidized layer in terms of the crack's growth begins to decrease.

•The reason for the initial one-off decrease in the crack's growth rate seen in Fig. 7 under the trapezoidal waveform load condition can be explained by considering the effect of the oxidation of the crack's surface.

4. Conclusions

•For crack growth, temperature dependence, anisotropy in terms of the direction of solidification, and the effect of the oxidized layer were investigated. Our conclusions are summarized below.

(1) At 600°C and 860°C the crack grows most slowly in V-type and most quickly in D-Type, with H-Type located between them.
(2) The effect of the direction of solidification on crack growth decreases as the crack grows.
(3) The crack's growth in the experimental results without a hold can be evaluated by using a J-integral and considering the effect of anisotropy.
(4) The rate of crack growth in the experimental results for a 60-min hold seemed to be lower.
(5) The rate of growth of the oxidized layer on the crack's surface indicates a parabolic relation with time.
(6) It seems to be possible to evaluate the effect of the growth of the oxidized layer on the crack's growth rate by adopting the energy release rate as derived by the virtual crack extension method as the means for analysis of the crack•

5. References

(1) N.Isobe and S.Sakurai, Transactions of the JSME•Vol. 641•No. 66•2000
(2) T.Kobayashi, M.Shibata, N.Tada and R.Ohtani, Proceedings of the 37th Symposium on Strength of Materials at High temperatures, 1999
(3) H.Ishikawa, H.Kaneko and T.Sakon, Proceedings of the 37th Symposium on Strength of Materials at High temperatures, 1999
(4) D.C.Lagoudas, P.Entchev and R. Triharjanto, Computational Methods in Appllied Engineering, No. 183, p. 35, 2000

Influence of Nickel on the Creep Properties of Ir-15Nb Two-phase Refractory Superalloys At 1650 °C Under 137 MPa

Y. F. Gu, Y. Yamabe-Mitarai, S. Nakazawa, and H. Harada

National Research Institute for Metals

1-2-1 Sengen, Tsukuba, Ibaraki 305-0041, JAPAN

Abstract

The creep behaviors of the Ni added $Ir_{85}Nb_{15}$ alloys were investigated by compression tests in vacuum at 1650 °C under 137 MPa. X-ray diffraction, transmission microscopy and scanning electron microscopy were conducted to characterize the microstructure and lattice misfit change by the addition of Ni in $Ir_{85}Nb_{15}$ two-phase alloy. The results reveal that Ni addition can improved the creep resistance of $Ir_{85}Nb_{15}$ alloy at ultra-high temperature. 1 atom per cent Ni addition can raise creep resistance of $Ir_{85}Nb_{15}$ two-phase alloy in about two orders of magnitude. Basing on the results, the relationship among the Ni addition, the lattice misfit, microstructure development, and the creep resistance in Ni added $Ir_{85}Nb_{15}$ two-phase alloys is discussed in this paper.

1. Introduction

Some Ir-based two-phase alloys, which have a coherent fcc/L1$_2$ (γ/γ') two-phase structure and called refractory superalloys, are superior in high-temperature strength and oxidation resistance, and have good potential as structural materials for use at temperatures up to 1800°C [1-6]. However, binary two-phase Ir-based refractory superalloys normally fracture in an intergranular mode and subject to low crack tolerance at room temperature. High density and high cost are also the big barriers to their practical applications. We tried to replace Ir with some third elements, such as Ni, W, Ta and Mo, in one of these two-phase superalloys, $Ir_{85}Nb_{15}$, to reduce its density and improve the mechanical properties [7]. We found that adding Ni was an effective way to change its fracture mode and to improve its strength and compression ductility [8]. Meanwhile, A decrease of about 7% in density was noted for every 10 at% Ir replaced with Ni in $Ir_{85}Nb_{15}$ alloy. Our results also suggested that the optimum content for Ni to replace Ir in $Ir_{85}Nb_{15}$ two-phase alloy is about 10 at% in terms of compression strength, fracture behavior, and density.

Since two-phase Ir-hased refractory superalloys are tried to be used as structural materials for ultra-high temperature applications, the high temperature properties, such as creep resistance, and stability of the microstructure at ultra-high temperature are important factors for their practical applications. It is believed that alloying additions have the potential of influencing on many factors of γ/γ' two-phase alloys, such as diffusion rates, γ/γ' lattice mismatch, γ' volume fraction, and stacking fault energy, which would significantly affect the mechanical properties and microstructral stability of the alloys at elevated temperatures. However, the detailed effects and mechanisms responsible for Ni addition in $Ir_{85}Nb_{15}$ two-phase alloy have not been determined.

The purpose of present investigation was to examine the effects of adding Ni on the microstructural stability and creep behavior of two-phase $Ir_{85}Nb_{15}$ alloy at 1650°C under 137 MPa, with emphasis on the relationship between the microstructural features, including lattice parameter mismatch between the γ and γ', and the creep behaviors. The results from a previous study are combined with the presented investigation to provide a consistent view of the influence of Ni on the mechanical properties of $Ir_{85}Nb_{15}$ alloy.

2. Experimental Procedures

$Ir_{85-X}Nb_{15}Ni_X$ with X=1, 5, 10 were prepared using raw materials of 99.99 wt% purity iridium, 99.98 wt% purity niobium, and 99.5 wt% purity nickel. The raw materials were mixed and prepared as 50g button ingots by arc melting under argon in a vacuum furnace. To ensure homogeneity, the buttons were re-melted over 10 times.

Specimens 6 mm in height and 3 mm in diameter were prepared by electron-discharge machine (EDM) from the ingots. The specimens were annealed at 1800°C for 72 hours with furnace cooling under vacuum of about 3×10^{-4} Pa. For the microstructure investigated, samples were cut by the EDM from the heat-treated and deformed samples. The samples were prepared using conventional metallographic method and were examined by a scanning electron microscope (SEM) and a transmission electron microscope (TEM). The polished specimens were electrolytically etched in a 5% HCl ethyl-alcohol solution. Accurate lattice parameters for the phases in the tested alloys were identified at ambient and elevated temperatures by an High-resolution X-ray diffraction technique described in detailed elsewhere [9].

Constant-load Compression creep tests were performed at 1650°C under a stress of 137 MPa in a vacuum by an Instron 8560 testing machine. Graphite disks with 10 mm in diameter and 8 mm in height were used as spacers at both ends of the samples to prevent the welding between the samples and tungsten loading rods. The samples were heated to the test temperature for 2 hours and soaked at the test temperature for 15 minutes before testing. Creep strain was determined by recorders from both an CCD camera and a load-strain displacement measurement.

3. Results

3.1 Initial Microstructures

Figure 1 shows typical micrographs of the tested alloys after annealed at 1800 °C for 72 hours. Grain size for all tested alloys was almost same and was about 250 μm. The microstructure of all the tested alloys were characterised by a dentritic structure (Figures 1a to 1d). The γ/γ' two-phase structure, where the γ' phase size was about 0.3 μm, could be seen among the dendritic and interdendritic areas (Figs 1b and 1d, which are enlarged views of the arrows indicating the areas in Figs 1a and 1c, respectively). Larger sized γ and γ' phases, where the phase size varied from 1 to 10 μm, were also observed alone some grain-boundaries in the tested alloys.

The TEM investigation showed that the shape of the γ' phase was different for tested samples with different Ni level, as shown in Figure 2. The γ' precipitate was spherical one with lattice misfit dislocation networks around γ/γ' interfaces for $Ir_{84}Nb_{15}Ni_1$ and $Ir_{80}Nb_{15}Ni_5$ alloys (Figs 2a and 2b). However, the γ' precipitates in the $Ir_{75}Nb_{15}Ni_{10}$ alloy became irregular and some of the r' precipitates coalesce to form γ'-rods (Figure 2c).

3.2 γ-γ' Lattice Misfit

We determined the phase constitute, lattice misfit difference, and volume fraction of the γ' phase change that may be induced by the Ni addition in binary $Ir_{85}Nb_{15}$ alloy. X-ray diffraction patterns of $Ir_{85-X}Nb_{15}Ni_X$ alloys with x = 0, 1, 5, and 10, after annealed at 1800 °C for 72 hours are shown in Figure 3. The peaks in the patterns revealed that these alloys only consisted of γ and γ' phases. Lattice parameters of the γ matrix and the γ' precipitate for $Ir_{85-x}Nb_{15}Ni_x$ alloys were calculated with the fundamental peak 220 from fcc and $L1_2$ structure and superlattice peak 110 from $L1_2$ structure. The lattice misfits between the γ matrix and the γ' precipitate were represented by the parameter $\nabla\alpha$, given by

$$\nabla\alpha = \frac{2(\alpha_{\gamma'} - \alpha_{\gamma})}{(\alpha_{\gamma'} + \alpha_{\gamma})} \tag{1}$$

where $\alpha_{\gamma'}$ is the lattice parameter of the γ' precipitate and α_{γ} is the lattice parameter of the γ phase. The calculation showed that adding Ni changed the lattice misfit from the positive for the two-phase $Ir_{85}Nb_{15}$ alloy (0.44) to the negative for the 1 at% Ni added $Ir_{85}Nb_{15}$ alloy (-0.28). The lattice misfit for the Ni added $Ir_{85}Nb_{15}$ alloys increased with Ni content increasing when Ni addition was below 10 atom percent (at%).

Volume fraction V of the γ' phase in tested alloy was calculated by the equation (2)

$$V = \frac{I_{\gamma'}}{I_{\gamma'} + I_{\gamma}} \tag{2}$$

where I_{γ} and $I_{\gamma'}$ are relative X-ray intensity of the γ and γ' phase. In this calculation, we assumed that the γ and γ' have similar X-ray diffraction characters such as structure, absorption factors. The result of calculation indicated that the volume fractions of the γ' phase were 41% for the $Ir_{85}Nb_{15}$ alloy and 39% for the $Ir_{75}Nb_{15}Ni_{10}$ alloy.

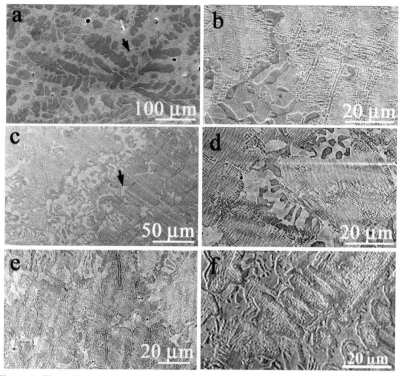

Figure 1. The micrographes of the tested alloys after annealed at 1800 °C for 72 hours:
(a) $Ir_{85}Nb_{15}$; (b) enlarged view of the arrow indicated the area in (a); (c) $Ir_{84}Nb_{15}Ni_1$;
(d) enlarged view of the arrow indicated the area in (c); (e)$Ir_{80}Nb_{15}Ni_5$; (f) $Ir_{75}Nb_{15}Ni_{10}$

490

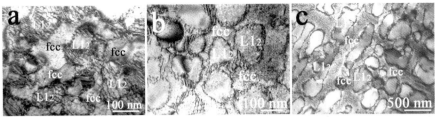

Figure 2. Typical micrographes of the tested alloys illustrating the γ' shape after annealed at 1800 °C for 72 hours in (a) $Ir_{84}Nb_{15}Ni_1$; (b)$Ir_{80}Nb_{15}Ni_5$; (c) $Ir_{75}Nb_{15}Ni_{10}$.

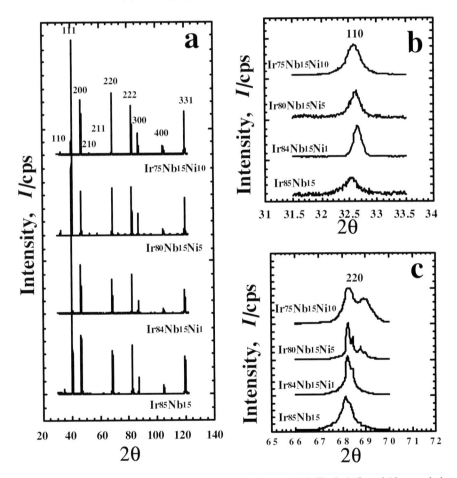

Figure 3. X-ray diffraction patterns for (a) Ir85-XNb15NiX with X= 0, 1, 5, and 10 annealed at 1800 oC for 72 hours; (b) 110 profile, and (c) 220 profile to indicate the lattice misfits of the fcc and L12 phases.

3.3 Creep Properties

Creep tests were conducted in compression at 1650 °C under 137 MPa. Creep curves for the $Ir_{85-X}Nb_{15}Ni_X$ alloys with X=1, 5, 10 are shown in Figure 4. Here we showed the creep curve for the binary $Ir_{83}Nb_{17}$ alloy under same testing condition as comparison. For the binary $Ir_{83}Nb_{17}$ alloy and the Ni added $Ir_{75}Nb_{15}Ni_{10}$ alloy, experiments exhibited a visible primary creep and stable minimum strain rates over varying amounts of strains, followed by a tertiary stage with a rapid increase in the strain rate. However, for the $Ir_{84}Nb_{15}Ni_1$ and $Ir_{80}Nb_{15}Ni_5$ alloys, experiments exhibited stable strain rates over a very long time (298 hours for $Ir_{84}Nb_{15}Ni_1$ and $Ir_{84}Nb_{15}Ni_1$ alloy). Neither clear primary creep nor tertiary creep was observed. When the creep test was stopped after 298 hours, the creep strain of the $Ir_{84}Nb_{15}Ni_1$ alloy was about 1.2%. The steady-state creep strain rate for the $Ir_{84}Nb_{15}Ni_1$ alloy was $1.9×10^{-9}$ s^{-1}, about two orders lower compared with the binary $Ir_{83}Nb_{17}$ alloy ($8×10^{-7}$ s^{-1}). But the steady-state creep strain rate for the $Ir_{75}Nb_{15}Ni_{10}$ alloy was $7.8×10^{-7}s^{-1}$, almost same with that of the binary $Ir_{83}Nb_{17}$ alloy.

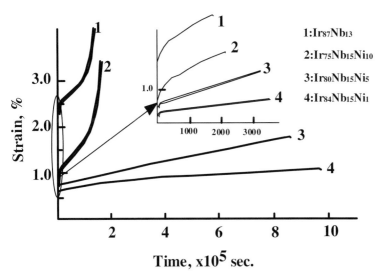

Figure 5. Creep curves for the binary Ir83Nb17 alloy and the Ir85-XNb15NiX alloys with X=1, 5 10 at 1650 °C under 137 MPa.

3.4 Microstructures after Creep Testing

Figure 5 shows the microstructures of the tested alloys after crept at 1650 °C under 137 MPa. Compared with as-heat treatment samples (Fig.1), no great coarsening of the γ' precipitates occurred in $Ir_{84}Nb_{15}Ni_1$ and $Ir_{80}Nb_{15}Ni_5$ alloys. For the $Ir_{75}Nb_{15}Ni_{10}$ alloy, the coarsening of the γ' phases occurred (as the arrows indicated in Fig.5c).

TEM images of the crept specimens show that the γ' precipitates maintained their spherical shapes for the $Ir_{84}Nb_{15}Ni_1$ alloy after creep test (Fig.6a). The shape of the γ' precipitates in the $Ir_{80}Nb_{15}Ni_5$ alloy became irregular and some of the γ' precipitates coalesce to form γ'-rods. For the $Ir_{75}Nb_{15}Ni_{10}$ alloy, most of the γ' precipitates coalesce to form γ'-rods.

492

Meanwhile, the movement of the dislocations during creep tests sheared the precipitates in the tested alloys (as the arrows indicated in Fig.6).

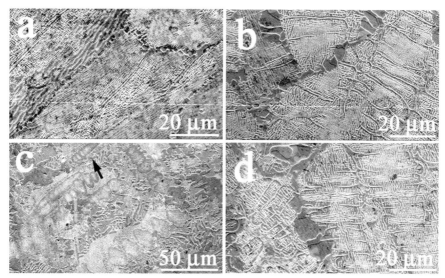

Figure 5. The micrographes of the tested alloys after crept at 1650 °C (a)Ir$_{84}$Nb$_{15}$Ni$_1$; (b) Ir$_{80}$Nb$_{15}$Ni$_5$; (c) Ir$_{75}$Nb$_{15}$Ni$_{10}$; (d) enlarged view of the arrow indicated the area in (c).

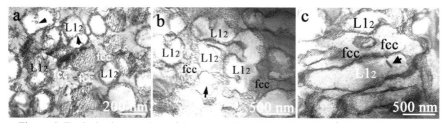

Figure 6. Typical micrographes of the tested alloys Illustrating the γ' shape after crept at 1650 °C in (a) Ir$_{84}$Nb$_{15}$Ni$_1$; (b)Ir$_{80}$Nb$_{15}$Ni$_5$; (c) Ir$_{75}$Nb$_{15}$Ni$_{10}$.

4. Discussion

The compression creep tests of the Ni added Ir$_{75}$Nb$_{15}$ alloys showed a great effect of Ni addition on the creep resistance of the alloys. Adding 1 at% Ni could decrease the creep rate of Ni added Ir$_{85}$Nb$_{15}$ alloys in about two orders of magnitude compared with that of Ni free Ir$_{83}$Nb$_{17}$ alloy. However, the creep rate for the alloy containing 10 at% Ni is almost same with that of the Ni free alloy. Although the detailed creep mechanisms for Ni added Ir$_{85}$Nb$_{15}$ alloys can not be identified now, we can still tell some characters of Ni addition on creep behavior of the alloys basing on available datum.

It is well known that high resistance to creep deformation for two-phase nickel based superalloys at high temperature is due to the γ/γ' microstructure, which means the γ/γ' boundaries are obstacles to dislocation motion. In general, the loss of creep resistance of the polycrystalline two-phase alloys due to high temperature creep is caused by the coarsening of γ' or by mechanical damage such as initiation and propagation of crack at grain boundary [10,11]. Our previous results showed that adding Ni to the $Ir_{85}Nb_{15}$ alloy would improve the grain boundary strength of the alloy. In addition, we couldn't find any mechanical damage at grain boundaries during the creep tests for the Ni added $Ir_{85}Nb_{15}$ alloys in our tests. So we believe that the great effect of Ni addition on the creep behavior of the Ni added alloys comes from effects of Ni addition on the γ/γ' interface and the coarsening process of the γ'.

Firstly, the Ni addition changes the lattice misfit from positive 0.44 for binary $Ir_{85}Nb_{15}$ alloy to negative 0.24 for $Ir_{84}Nb_{15}Ni_1$ alloy. This may be caused by a distribution difference of Ni content in the γ and γ' phases. Since the atom size of Ni is about 10% smaller than that of Ir and Nb, replacing Ir or Nb with Ni in the γ and γ' phases would reduce their lattice parameters and thus caused the lattice misfit change. The lattice misfit may change at high temperature possible toward more negatives, but we couldn't confirm it in this report. It is believed that negative lattice misfit may help to induce dislocation networks at the γ/γ' interfaces in Ni-based two-phased alloys, which believed to be effective to prevent dislocation motion during creep tests. From our tests, we can find many network of misfit dislocations formed at the γ/γ' interfaces (Fig.2) in $Ir_{84}Nb_{15}Ni_1$ and $Ir_{80}Nb_{15}Ni_5$ alloy, which believed to be benefit to their creep resistance.

Secondly, the evolution of the γ/γ' microstructure in the Ni added $Ir_{85}Nb_{15}$ refractory superalloys is classically considered to be a diffusion process at such high temperature, which depends strongly on the misfit between the γ and γ' phase as well as on the applied stress [12]. The γ' particles in the tested alloy with larger lattice misfit seems easy to coarse during creep tests and then lost their resistance to the dislocation movement during creep tests, as that in $Ir_{74}Nb_{15}Ni_{10}$ alloy.

Thirdly, since the compositions of γ and γ' are different, the coarsening of the γ' particles must be accompanied by diffusion of alloying elements between its faces normal and parallel to the applied compression, and rate of coarsening is presumably controlled by the rate of this diffusion. We couldn't determinate exactly diffusion datum for Ni, Nb, and Ir atoms in Ir-Nb-Ni alloy system at 1650°C, but it is reasonable certain that the diffusion of Ni atom is the quickest one in the alloys. More Ni addition would increase the rate of coarsening and then make the alloys easy to lose creep resistance. Therefore, the steady state creep rate for the Ni added $Ir_{85}Nb_{15}$ alloys depends on the balance between adding Ni on the lattice misfit changing and γ' coarsening. We suggest that the optimum content for Ni adding in the $Ir_{85}Nb_{15}$ alloy is below 5 at% to retain Ni added $Ir_{85}Nb_{15}$ alloy with high creep resistance.

Meanwhile, in steady-state creep, the interaction between dislocation and γ' precipitate is the γ' particle shearing by diffusive slip of dislocation (Fig.6), but we couldn't identified the character of the dislocation structure in this paper. More work has been done to clear the dislocations structure and creep mechanisms of the Ni added $Ir_{85}Nb_{15}$ alloy at 1650°C and higher temperatures.

5. Conclusions

The high temperature creep resistance of Ni added $Ir_{85}Nb_{15}$ alloy were investigated in this paper. The results can be summarized in the following:

(1) Lattice misfit between fcc matrix and $L1_2$ precipitate changed from positive for the binary $Ir_{85}Nb_{15}$ alloy to the negative for the Ni added $Ir_{85}Nb_{15}$ alloys

(2) The steady-state creep rate for the $Ir_{84}Nb_{15}Ni_1$ alloy was 1.9×10^{-9} s^{-1}, which was about two orders lower compared with that of the binary $Ir_{83}Nb_{17}$ (8×10^{-7} s^{-1}). The steady-state creep rate for the $Ir_{75}Nb_{15}Ni_{10}$ alloy was 7.8×10^{-7} s^{-1}.

494

6. Acknowledgments

We thank Mr. S. Nishikawa and Mr. T. Maruko of Furuya Metal Co. Ltd. for preparing the alloys.

7. References

[1] Y. Yamabe, Y. Koizumi, H. Murakami, Y. Ro, T. Maruko, and H. Harada, Scripta Mat. 35 (1996) 11.

[2] Y. Yamabe-Mitarai, Y. Koizumi, H. Murakami, Y. Ro, T. Maruko, and H. Harada, Scripta Mat. 36 (1997) 393.

[3] Y. Yamabe-Mitarai, Y. Koizumi, H. Murakami, Y. Ro, T. Maruko, and H. Harada, Mat. Res. Soc. Symp. Proc., 460 (1997) 701.

[4] Y. Yamabe-Mitarai, Y. Ro, T. Maruko, T. Yokokawa, and H. Harada, Structural Intermetallics. (1997) 805.

[5] Y. Yamabe-Mitarai, Y. Ro, T. Maruko, and H. Harada, Met. Trans. 29 (1998) 537.

[6] Y. Yamabe-Mitarai, Y. Ro, T. Maruko, and H. Harada. Intermetallics. 7, (1999) 49.

[7] Y. F. Gu, Y. Yamabe-Mitarai, and H. Harada, *Iridium,* edited by E. K. Ohriner, R. D. Lanam, P. Panfilov, and H. Harada; TMS 2000, p73

[8] Y. F. Gu, Y. Yamabe-Mitarai, Y. Ro, T. Yokokawa, and H. Harada, Met. Trans. 30 (1999) 2629.

[9] S. Yoshitake, T. Yokokawa, K. Ohno, H. Harada, and M. Yamazaki, in Materials for Advanced Power Engineering, Part I, D. Coutsouradis et al eds., Kluwer Academic Publishers, Dordrecht, The Netherlands, 1994, pp875-882.

[10] D. A. Woodford, J. Eng. Mater. Technol., 101(1979),311-316

[11]N. Shin-ya, and S. R. Keown, Met. Sci., 13(1979), 89-93

[12] A. Pineau, Acta Metall., 24(1976), p559

ENGINEERING APPLICATIONS I

A model to reconcile anomalous creep and fractography behaviour of reformer plant materials

Paul R McCarthy[$], Sandra Estanislau*, Margarida Pinto* and Manuel A. Real Gomes**

[$] = ERA Technology Ltd, Cleeve Road, Leatherhead, Surrey, KT22 7SA, UK
* = Integridade-Servicos de Manutenção e Integridade Estrutural, Lda, Rua Francisco Antonio Silva, 2780-055 OEIRAS, Portugal
** = Instituto de Soldadura e Qualidade, Taguspark, Apartado 119, EC OEIRAS 2781-951, Portugal

Abstract

This paper presents the results of investigations into the creep behaviour of a high Cr – Ni steel used in the petro-chemical industry. A programme of creep tests was established within a collaborative project, with the testing distributed between the two participating mechanical testing laboratories. Upon initial examination the test data and metallography of the ruptured testpieces revealed inconsistencies between the two data sets, in terms of both creep behaviour and fractographic appearance. A detailed review was then initiated in order to identify the cause of this mis-match. Aspects examined included not only assessments of the instrumentation capabilities, calibration traceablility and operational procedures used by each laboratory, but also a thorough review of all of the metallographic evidence generated by the project from ruptured testpieces. Detailed consideration of this information led to the development of a model able to reconcile the two patterns of behaviour. This model, and details of its subsequent validation, are presented herein and conclusions are drawn as to the most appropriate testing practice for these materials.

1. Introduction

The work presented in this paper has been undertaken within the European Union BRITE Euram Project 3890 D-LOOP "Design and Life Optimisation of Fired Heater Tubes for Enhanced Process Performance". The participating organisations are DSM Engineering-Stamicarbon, ERA Technology Ltd, Kinetics Technology International, Integridade-Servicos de Manutenção e Integridade Estrutural, Lda and Paralloy Ltd.

One of the tasks in this programme of work was the generation of creep data on microalloyed high Cr-Ni alloys with a range of levels of carburisation. This testing work was divided between two of the participating organisations, each undertaking part of the test matrix. An initial assessment of the data revealed inconsistencies between the two laboratories. Tests at a common set of test conditions were examined and it was found that the data from Laboratory A exhibited shorter rupture life, lower rupture ductility and higher creep rates than the corresponding test performed in Laboratory B. This was a cause for concern; a stress rupture test by a third laboratory in the consortium gave an intermediate rupture life and ductility (Figure 1).

It was decided that a thorough review be made of the test equipment and procedures used by the two laboratories. In addition the testpiece fractography and the microstructural characteristics of the ruptured testpieces would be assessed in order to explain the observed inconsistencies.

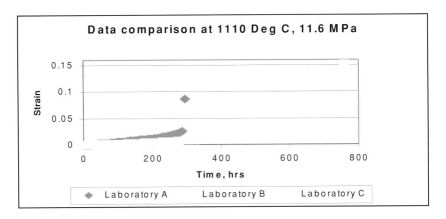

Figure 1. The different creep responses from the three laboratories

2. Investigations

2.1 Test Equipment and Procedures

The testing equipment and procedures used in the two laboratories were examined and compared, the findings being summarised in the following table.

	Laboratory A	Laboratory B
Machine Type	Mayes TC20 lever type	Denison T47 lever Type
Capacity	20kN	10kN
Force Verification	EN ISO 7500-2 : 1999	EN ISO 7500-2 : 1999
Universal Joints	2 sets	2 sets
Thermocouples	Noble metal (Type S)	Noble Metal (TypeR)
Temperature verification	Traceable to National Stds	Traceable to National Stds
Testpiece Production	By Laboratory A	By Laboratory A
Carburisation Process	In common batches	In common batches
Test Atmosphere	Air	Air
Extensometry	Attached to ridges on the testpiece	Attached to ridges on the testpiece
Extensometer Verification	EN 10002-4 : 1990	EN 10002-4 : 1990

Extensometer design	Plate limbs	Rod limbs
Transducers	Capacitance type	LVDT inductive type
Insulation Material	Kaowool ceramic fibre	Kaowool ceramic fibre
Testing Procedure	BS 3500 Part 1	BS 3500 Part 1

Table 1. Testing characteristics of Laboratories A and B

The test equipment at both establishments was thoroughly scrutinised, looking for any differences in design, construction and commissioning which could explain the observed behaviour. Calibration traceability in both laboratories was demonstrated to be to the relevent European standards and national reference standards of measurement. The test techniques adopted by both laboratories were in accordance with the relevant standard and the work performed was technically mutually consistent. The quality assurance records for the identification, manufacture and carburisation of the testpieces were reviewed and found to be self consistent. The only significant differences found were related to the physical structure of the extensometry.

On the basis of these exhaustive examinations it was concluded that the observed behaviours were a realistic representation of physical processes occurring within the testpieces; the investigations then turned to testpiece fractography and data analysis.

2.2 Fractography and Data Analysis

Examination of the testpieces and post-test fractography records revealed two categories of behaviour, roughly aligned with the testing laboratory. These differences manifested themselves as a greater preponderance for visible surface cracks for the tests performed at Laboratory A, as shown in Figure 2, and a more ductile failure for those tested at Laboratory B.

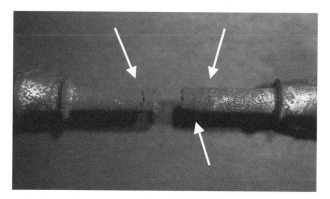

Figure. 2 – Major cracks on the testpiece gaugelength

In parallel with the fractographic examinations studies were made of the creep curves, in particular the strain rate responses. This revealed a pattern of behaviour which also had a bias between the two laboratories. Examples of the strain – time graphs from the two laboratories, plus the corresponding strain rate responses, are set out below.

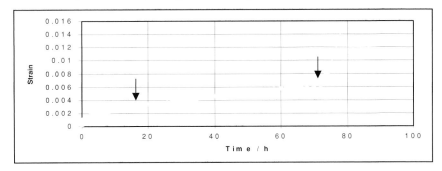

Figure 3 : Creep response, Laboratory A

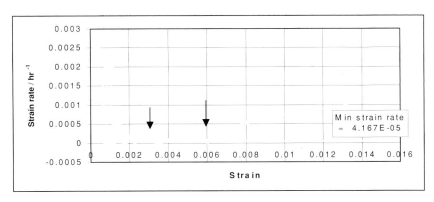

Figure 4 : Sherby Plot, Laboratory A

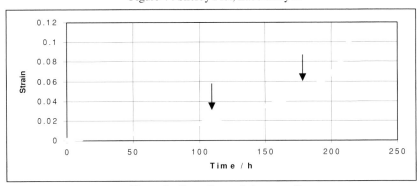

Figure 5 : Creep Data : Laboratory B

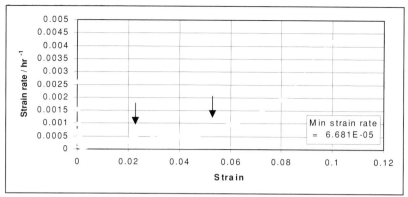

Figure 6 : Sherby Plot : Laboratory B

Present on many of the creep curves were minor deflections which, when examined as Sherby plots, reveal "strain burst" events, indicated by arrows on figures 3 to 6 inclusive. It was considered that these strain bursts and the observed cracks on the testpieces were related and an analysis was undertaken to establish the extent of this correlation. This data was collated and is presented as Figure 7, where the clear relationship between strain bursts and major cracks is shown.

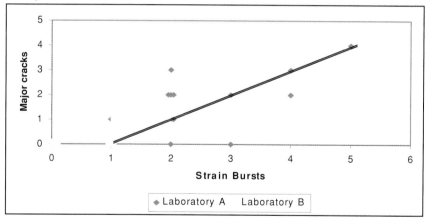

Figure 7 : The relationship between strain bursts and major cracks on the testpiece

2.3 Metallographic examination

Selected testpieces were subjected to both scanning electron and optical microscopy. Differences in oxide morphology were observed, correlating with testing location, and failure was either by cavitation or by the fracture dominated processes. SEM observations revealed differences in the depth of the oxidised layer between laboratories; some corresponding spalling and chromium depletion was also detected. It was noted that the tests from Laboratory A exhibited significant surface decarburisation whereas those performed at Laboratory B revealed it to a lesser extent.

3. Discussion

A review of the available information revealed the following:-

- The test techniques used by both laboratories were compatible and consistent
- The strain bursts could be correlated to the observed major cracks seen on the testpieces
- Differences in oxide morphology correlated to the test laboratory
- Shorter test lives correlated with the presence of major cracks on the testpieces

Consideration of these findings led to a conclusion that the main difference between the results generated by the two laboratories was related to oxygen access and hence in turn to the test furnace insulation practices. The failure morphologies (cavitation and lesser decarburisation; cracking and greater decarburisation) supported this view, a mechanism was therefore required to explain the observed behaviour. Attention turned to the physical properties of the base material and the oxides created during testing. In particular it was noted that the oxide formed at the surface of the testpiece would have different mechanical properties to the base material, in particular being less ductile. This characteristic, coupled with the observations above, led to a mechanism which could explain the shorter test lives of testpieces where failure was dominated by cracking processes.

Step 1 – The test commences, subjected to temperature and load. A coherent surface oxide layer forms, using the oxygen present inside the test furnace (Figure 8.1)

Step 2 – The strain in the testpiece reaches a level where the strain tolerance of the oxide is exceeded. This results in the creation of a crack in the oxide which acts as a starting point for crack initiation in the base material. This initiation process is driven by the stress increase which occurs in the base material, caused by offloading from the now cracked oxide. This mainfests itself as the strain burst observed on the creep curves.

Step 3 – The creep deformation of the testpiece continues. The strain tolerance mismatch between the base material and the surface oxide results in spalling, producing further crack initiation sites; existing cracks continue to grow.

Step 4 – The testpiece starts necking, with a dominant crack developing.

Step 5 – The dominant crack rapidly develops, causing a significant increase in the observed creep deformation rate.

Step 6 - The testpiece fails as a consequence of crack growth and testpiece necking.

Figure 8 : The proposed mechanism

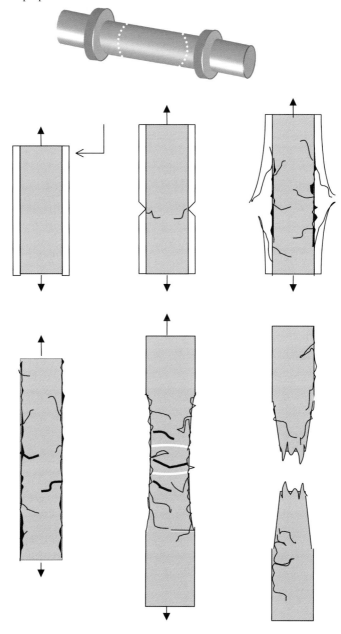

504

This mechanism was consistent with the observed creep and fractographic data; however there was still a need to identify the difference in testing practice which would lead to the different oxidation rates implicit in the theory. Close examination of the extensometry, and the associated furnace insulation practice, led to the discovery of a means of generating the different oxygen levels around the testpieces. Figure 9 presents a schematic arrangement of the two extensometry systems, viewed from the base of the test machines. That used by Laboratory B uses four rods to transmit the testpiece elongation from the testpiece inside the test furnace to the transducers outside the furnace, with ceramic fibre insulation sealing the furnace bore. The system used by Laboratory A is similar in principle, however the testpiece elongation is transmitted by two pairs of extensometer limbs which are of a plate form and, critically, have a small gap between them.

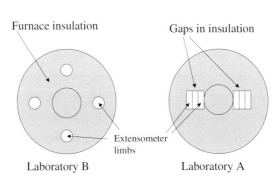

Figure 9 : Extensometry limb schematics for the two laboratories

During testing these two gaps allow a more frequent replenishment of the oxygen inside the test furnace; in Laboratory B this occurs more slowly by diffusion through the insulating ceramic fibre. The consequences of this difference in test practice are higher oxidation rates, greater oxide cracking, enhanced creep rates and hence premature failure when compared with the tests performed by Laboratory B.

The theory fitted the observed facts, all that remained was validation by performing a test at Laboratory A with the modified furnace insulation practice. A repeat test, under identical conditions to those shown in Figure 1, was undertaken. The resulting data are presented as Figure 10, where the good correlation of the two creep curves can be seen.

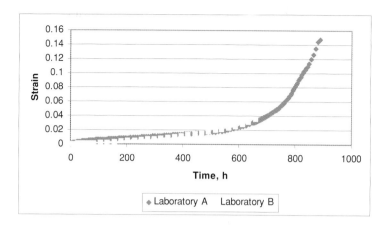

Figure 10 : Comparison of test data from the two laboratories, new insulation practice

4. Conclusions

Following these investigations it was concluded that : -

- Both laboratories were undertaking the tests in a manner conforming to the relevant international standards, with traceability to the appropriate national standards of measurement.
- Different operational practices, consequent upon different extensometry designs, led to the observed differences in creep behaviour.
- Developments in furnace insulation practice at Laboratory A eliminated the differences seen in the original test results. The subsequent creep response, rupture ductility and rupture lives generated by the two laboratories were found to be mutually consistent.
- A knowledge of both the creep and oxidation behaviour of these materials is necessary for accurate modelling and assessment.

5. Acknowledgements

The authors would like to acknowledge the permission of the directors of ERA Technology Ltd, Integridade-Servicos de Manutenção e Integridade Estrutural, Lda and Instituto de Soldadura e Qualidade to publish this paper. The participants of the BRITE EURAM project D-LOOP (DSM Engineering-Stamicarbon, ERA Technology Ltd, Kinetics Technology International, Integridade-Servicos de Manutenção e Integridade Estrutural, Lda and Paralloy Ltd.) are acknowledged, as is the BRITE EURAM office of the European Commission for provision of partial funding for the project.

The help and assistance of colleagues at the partipating organisations is acknowledged, in particular discussions with John Brear, Ainsley Fairman and Andy Tomkings of ERA Technology Ltd.

CREEP PROPERTIES OF 15Cr-15Ni AUSTENITIC STAINLESS STEEL AND THE INFLUENCE OF TITANIUM

S.Latha, M.D.Mathew, K.Bhanu Sankara Rao and S.L.Mannan

Materials Development Group
Indira Gandhi Centre for Atomic Research
Kalpakkam, 603 102, INDIA

ABSTRACT

Titanium modified 15Cr-15Ni austenitic stainless steel, also called alloy D9, is the current choice of material for fuel cladding and wrapper applications in fast breeder reactors. Creep rupture properties of this material have been studied at 923 K and 973 K at different stress levels using tubular specimens in 20% cold–worked condition, and compared with the creep properties of 20% cold worked type 316 stainless steel (316 SS) cladding tubes. Steady state deformation was found to follow power law creep. At 923 K, the value of stress exponent n was 12 whereas at 973 K, a two slope behaviour was noticed with n=2 and n=13. Alloy D9 exhibited almost a factor of six increase in rupture life as compared to 316 SS. The influence of titanium on the creep rupture properties of alloy D9 was studied by varying the titanium content in the range of 0.2 to 0.4 wt% at a fixed carbon content of 0.05 wt%. Rupture strength decreased with increase in titanium content; alloy D9 containing Ti/C= 4 had highest rupture strength. The differences in creep properties have been explained on the basis of the differences in matrix precipitation and intergranular damage during creep.

1.0. INTRODUCTION

Austenitic stainless steels, primarily type 316 stainless steel and its modifications are being used as fuel clad and wrapper material in fast breeder reactors (FBRs). These materials have good elevated temperature mechanical properties including creep, and compatibility with liquid sodium coolant, besides resistance to irradiation induced void swelling and helium embrittlement [1]. Void swelling results in increase in length and diameter of the fuel clad and wrapper tubes and causes the fuel sub-assemblies subjected to gradients in neutron flux and temperature to bow and interact with their neighbours and core restraint structures giving rise to difficulties in fuel handling operations [2]. In order to improve the burn up of FBR fuels, it is necessary to have clad and wrapper material of higher resistance to void swelling. Studies have shown that void swelling reduces with increase in nickel content, and swelling increases with increase in the chromium content [3]. Even minor elements like titanium, silicon and phosphorus are effective in increasing the swelling resistance of austenitic stainless steels [4]. Cold work, grain size refinement and dispersion of fine second phase particles are also known to be effective in lowering void swelling [5,6]. Titanium and niobium additions to austenitic steels improve their resistance to helium embrittlement [7]. Compositional adjustments in terms of decreasing chromium, increasing nickel and addition of titanium has led to the development of a 15Cr-15Ni-2.2Mo-Ti modified austenitic stainless steel designated as Alloy D9 (ASTM designation UNS 38660). This material is now favoured for fuel clad and hexagonal wrapper for fast reactor applications [8]. As cold-work plays an important role in reducing irradiation induced void swelling of stainless steels, alloy D9 would be used with an optimum cold work of 20% in FBRs.

As cladding tubes operate at relatively high temperatures (upto 973 K), their thermal creep behaviour is very important. This paper deals with the thermal creep behaviour of 20% cold-worked alloy D9 cladding tubes at 923 and 973 K The creep properties of alloy D9 are compared with the data obtained on 316 SS concurrently. Effects of Ti/C ratio (keeping carbon level at 0.05%) on creep properties of alloy D9 are also explored.

2.0 EXPERIMENTAL

20% cold worked alloy D9 tubes were procured from M/s.Valinox France. Type 316 stainless steel tubes in 20% cold worked condition were obtained from M/s. Fine tubes, UK. Alloy D9 bars with different titanium levels (Ti/C=4,6 and 8) were processed at MIDHANI, Hyderabad, India. The bars were initially solution treated at 1343 K for 30 minutes and then 20% cold work was introduced by rotary swaging. The chemical compositions of alloy D9 bars, D9 tubes and 316 SS tubes are given in table 1. Creep tests were performed at 973 K at stress levels of 125, 150, 175, 200, 225 and 250 MPa, and at 923 K at stress levels of 175, 200, 225 and 250 MPa. Alloy D9 tubes had a nominal dimension of 6.60 mm outside diameter and 0.45 mm wall thickness. The outside diameter of 316 stainless steel cladding tubes was 5.10 mm and the wall thickness was 0.37 mm. Tubular creep specimens were made by welding tubes of 100 mm length to mandrels (end plugs) at both the ends. The effective gauge length of the specimen was 50 mm. The specimens were checked for weld integrity. Creep tests on Alloy D9 bars were conducted on standard size samples having 10 mm diameter and 50 mm gauge length. Constant load creep tests were carried out by using SATEC single lever creep machines. The test temperature was controlled within ±2K during the tests. Creep elongation was measured using linear variable differential transducers. Scanning and transmission electron microscopic investigations were carried out to examine the fracture and precipitation behaviour respectively.

Table 1

Chemical composition of alloy D9 tube, alloy D9 bar and 316 SS tube.

Element	AlloyD9 cladding	Alloy D9 bar			316SS cladding
		Ti/C=4	Ti/C=6	Ti/C=8	
C	0.04	0.05	0.05	0.051	0.04
Si	0.61	0. 50	0. 51	0. 52	0.33
Mn	1.78	1.51	1.51	1.50	1.88
S	<0.002	0.003	0.003	0.003	0.009
P	0.007	0.011	0.011	0.012	0.009
Ni	15.15	15.1	15.1	15.27	13.59
Cr	14.04	15.04	15.05	15.12	17.10
Mo	2.25	2.26	2.25	2.23	2.38
Ti	0.26	0.21	0.32	0.42	-
Al	0.006	-	-	-	0.005
B	0.0015	0.001	0.001	0.001	-
N	0.0037	0.006	0.006	0.0059	0.044
Fe	balance	balance	balance	balance	balance

3.0. RESULTS AND DISCUSSION

3.1. Comparison of creep properties of alloy D9 and 316 SS

The creep curves for alloy D9 and 316 SS tubes generally exhibited distinct primary, secondary and tertiary stages at all the test conditions. Figure 1 shows the variation of steady state creep rate with applied stress (log-log plot). Alloy D9 crept at a much lower rate compared to 316 SS at all the test conditions.

Table 2
Creep stress exponent n for alloy D9 and 316 SS tubes.

Material	Temperature, K	Stress range, MPa	Stress exponent,n
Alloy D9	973	125-200	2
Alloy D9	973	200-250	13
316 SS	973	100-150	6
316 SS	973	150-225	11
Alloy D9	923	175-250	12
316 SS	923	200-250	12

A power law relation, $\varepsilon_s = A\sigma^n$, (where ε_s is the steady state creep rate, σ is the applied stress, n is the stress exponent, and A is an empirical constant) was found to be obeyed. At 973 K, a two-slope behaviour was observed. At 923 K, there was no change in the slope over the stress conditions for which results are available so far. The n values are summarised in table 2. In the case of 316 SS, the stress exponent decreased from n=11 in the stress range of 150 to 225 MPa, to n=6 in the stress range 100-150 MPa at 973 K. On the other hand, alloy D9 showed a value of n=13 in the high stress range (>200 MPa) and n=2 in the low stress range (125-200 MPa). At 923 K, alloy D9 and 316 SS yielded the same stress exponents n=12. The two-slope behaviour at 973 K suggests that two different creep mechanisms may be operating at this temperature in different stress regimes.

Creep deformation mechanisms are identified in terms of the stress exponent n and the activation energy Q. Generally, in pure metals and solid solutions, values of n=3 to 5 have been

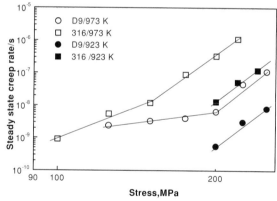

Figure 1
Variation of steady state creep rate with applied stress

3

510

associated with dislocation creep, while diffusion creep is characterised by n=1, and grain boundary sliding controlled creep is identified with value of n=2. In precipitation hardened and dispersion strengthened materials, much higher values of n (as high as 30) have been reported. In

austenitic stainless steels, precipitation usually takes place during creep and values of n intermediate between those of solid solution alloys and precipitation hardened/ dispersion strengthened materials have been generally reported [9,10].

Elevated temperature exposure of austenitic stainless steels during creep leads to precipitation of carbides on grain boundaries and on matrix dislocations. The stress dependence of creep rate in such cases is determined primarily by the stress dependence of the mobile dislocation link density,

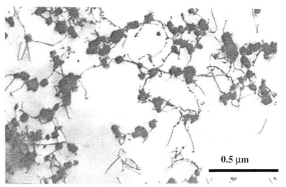

Figure 2

Precipitation of carbides on matrix dislocations in type 316 SS during creep deformation at 873 K.

which is a function of the dislocation link lengths. Typical transmission electron micrographs showing precipitation of $M_{23}C_6$ type carbides on matrix dislocations due to creep deformation in type 316 SS is shown in Fig.2. Precipitation of fine carbides on the dislocations leads to small interparticle spacing and hence short link lengths. The smaller the link length, higher is the link density and thus higher is the creep stress exponent [9,10]. In the present investigation, the high stress exponent n in the range of 6-13 obtained at 923 K and high stress region at 973 K suggest

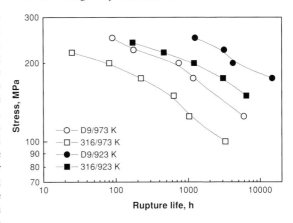

Figure 3
Variation of rupture life with applied stress.

dislocation creep mechanism. In the high stress range at 973 K, steady state creep rate of D9 was more than an order of magnitude lower than that of 316 SS. At lower stress levels, this difference appear to narrow down. At conditions under which n=2 has obtained, the actual mechanism of deformation needs to be established through additional tests.

The variation of rupture life with applied stress for these materials is shown in Fig.3. Creep strength of alloy D9 is greater than 316 SS by a factor of six at 973 K and by a factor of four at 923 K. The improvement in strength was found to be related to the prolonged secondary creep stage in D9 material. Austenitic stainless steels derive their strength due to solid solution strengthening by elements like Mo, Cr, etc. and also from carbide precipitation in the matrix. In the case of 316 SS, fine $M_{23}C_6$ type of carbides are known to form at 873 K. But at the test temperature of 973 K and higher, coarsening of carbides take place enabling recovery and thus decreasing the efficiency of precipitation strengthening. On the other hand, in alloy D9, carbon is partitioned between titanium and chromium, and since the amount of carbon available to form secondary TiC and $M_{23}C_6$ type carbides is less, the carbide particles will be finer in size in D9 than in 316 SS even at 973 K, thereby giving rise to higher creep rupture strength.

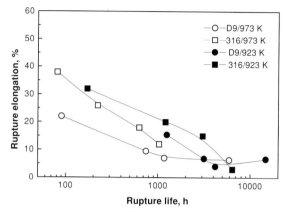

Figure 4
Variation of rupture ductility with rupture life.

The variation of rupture elongation with rupture life is shown in Fig.4. Rupture ductility decreased with increase in rupture life at both the temperatures and for both the alloys. Rupture ductility was rather low (<10%) except at short durations. Microstructural studies indicated the occurrence of numeroud grain boundary triple cracks in alloy D9 is shown in Fig.5.

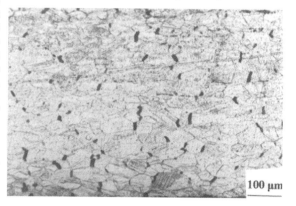

Figure 5
Wedge-type creep cracks in alloy D9 after testing at 923 K; stress = 250 MPa.

3.2. Influence of titanium on creep properties

The influence of titanium on the creep rupture properties of alloy D9 was investigated at 973 K. Alloy D9 was produced with different amounts of titanium (at a constant carbon level of 0.05 wt%) such that the chemical composition corresponds to Ti/C = 4, 6 and 8. It was found that the highest rupture strength was given by the heat containing Ti/C=4 and the lowest rupture strength was given by the heat containing Ti/C=6. Figure 6 shows the stress rupture curves at 973 K. The influence of Ti/C ratio on the creep rupture properties of alloy D9 was scarecely examined.

Fujiwara et al [11] carried out systematic studies on the influence of titanium on rupture strength of 15Cr-15Ni-2.5Mo SS at seven Ti contents ranging from 0 to 1 wt%. The corresponding Ti/C ratios were 0 to 13. Tests were carried out at 973 K in solution annealed and 20% cold worked conditions. It was found that rupture strength increased with increasing Ti/C ratio, reached a peak corresponding to Ti/C = 4 and then rupture strength decreased with further increase in Ti/C to 13. Ti/C=4 corresponds to the stoichiometric ratio for titanium carbide. The highest volume of fine secondary TiC after creep testing has been reported in the heat containing Ti/C=4. Extraction of precipitates followed by X-ray analysis showed presence of TiC and $M_{23}C_6$ precipitates in our samples after creep testing [12]. Similar precipitation behaviour has also been reported by Kesternich and Meertens [13] and Grot and Spruell [14] also. Peak rupture strength observed at Ti/C=4 in our studies is in agreement with the results of Fujiwara et al [11].

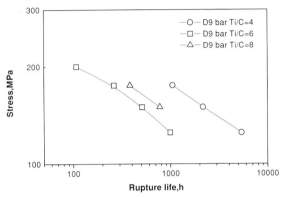

Figure 6
Variation of creep strength with Ti/C ratio.

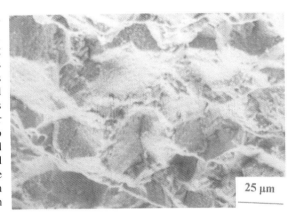

Figure 7
Intergranular creep fracture in alloy D9
(Ti/C=6) after creep testing at 973 K.

Rupture ductility was generally lower than 10%, and the lowest ductility was obtained for the alloy having ratio of Ti/C=4. Scanning electron microscopic investigations showed that failure was intergranular at all test conditions. Figure 7 shows the fracture surface of alloy D9 bar having a Ti/C ratio of six creep tested at 200 MPa. Intergranular failure resulted from extensive grain boundary cracks that formed during creep as a result of matrix strengthening achieved by fine TiC and $M_{23}C_6$ precipitates.

4. CONCLUSIONS

1. The steady state creep rates of alloy D9 cladding tubes were lower than those of 316 SS cladding tubes generally by an order of magnitude. Power law relationship was obeyed between steady state creep rate and applied stress. A two slope behaviour was seen at 973 K, whereas at 923 K there was no change in slope over the stress range investigated.
2. The rupture lives of alloy D9 cladding tubes were higher than that of 316 stainless steel cladding tubes. The improvement in rupture life of alloy D9 is attributed mainly to matrix precipitation of titanium carbides.
3. Rupture elongation of alloy D9 tubes was significantly less than that of 316 stainless steel tubes.
4. Alloy D9 bars with Ti/C ratio of 4 exhibited the higher rupture strength.

ACKNOWLEDGEMENTS

The authors wish to thank Mr. S.B.Bhoje, Director, Indira Gandhi Centre for Atomic Research (IGCAR), and Dr. Baldev Raj, Director, Metallurgy and Materials Group, IGCAR for their constant encouragement and support during the course of this work.

REFERENCES

1. Harries D R, Nucl . Energy, **17**, 4 1978, 301.
2. Sundaram CV, Rodriguez P, and Mannan SL, J. Inst. Engg. (India) **67**,1986, 1.
3. Johnston W G, Lauritzen T Rosolowski J H ,and Turkalo A M ,   Proc.Radiation Damage in Metals, ASM, (eds.) Perterson J E, and Harkness S D, ASM Metals Park, Ohio ,1976, .227.
4. Leitnaker J M, Bloom E E, and Steigler O O, J .Nucl Mater **49** ,1973, 57.
5. Steigler J O, and Bloom E E, J .Nucl.Mater **41** 1971, 341.
6. Masiaz P J, and Roche T K, J Nucl Mater **103 & 104** 1981, 797.
7. Bloom E E, Steigler J O, Rowcliffe A F, and Leitnaker T M, Scripta Met ., **10** 1976 , 303.
8. Hamilton M L, Jhonson G D, Puigh R T, Garner F A, Maziaz P T and Yang W J S, and Abraham N Rep.No. PNL-SA-15303 Pacific Northwest Lab,USA 1988.
9. Morris D G, and Harries D R, Metal Sci **12** , 1978, 525.
10. Mannan S L, and Rodriguez P, Metal Sci, **17** 1983, 63.
11. Fujiwara M, Uchida H, and Ohta S, J Mat Sci Letters , **13** 1994, 908.
12. Latha S, Mathew M D, Bhanu Sankara Rao K and Mannan S L, Trans Ind Inst Met., **49**, (1996) 587.
13. W.Kesternich and D.Meertens, Acta Metallurgica, **34**, No.6 ,1986, 1082.
14. Arnold S. Grot and Joseph E. Spruiell, Metallurigical Transactions A, **6A**, 1975, 2023.

High-temperature Strength of Unidirectionally Solidified Fe-Cr-C Eutectic Alloys

Shoji GOTO, Chungming LIU*, Setsuo ASO and Yoshinari KOMATSU

Department of Materials Science and Engineering,
Faculty of Engineering and Resource Science, Akita University,
Tegata Gakuencho, Akita, 010-8502, JAPAN
*Department of Electronic Material Engineering, Chin Min College,
110 Shefu Rd. Toufen. Miaoli, 351 TAIWAN

Keywords: chromium white cast iron, high temperature strength, M_7C_3 carbide, eutectic alloy, compressive strength, mechanical property, unidirectional solidification, roll material

Abstract

High chromium white cast iron having large amounts of carbides up to 40 vol. % is known to have an extremely high strength and acceptable level of toughness. However, relationship between the microstructure and the mechanical properties has been still unknown though the iron will be widely used for high-strength structural applications at high temperatures. Unidirectional solidification of Fe-25mass% Cr-3.5mass% C fully-eutectic high chromium white cast iron was conducted under a solidification rate range from 3.0×10^{-3} mm/s to 1.3×10^{-1} mm/s to clarify the deformation behavior of the iron at elevated temperatures. The iron specimens were processed in the parallel and vertical directions to the solidification direction, and a compression test was done under a strain rate range from 2.5×10^{-4} to 1.7×10^{-2} at various temperatures up to 1073 K. The relation between the microstructure and high temperature strength was studied. The strength depended on the morphology of eutectic cells and the orientation of carbides. The maximum compressive strength of the parallel direction to the solidification direction was larger than that of the vertical direction within the whole temperature range. The deformation of the composed γ austenite phase occurs by dislocation slip mechanism, while that of the M_7C_3 carbide phase by twinning mechanism. These results suggest that the strength of the parallel direction is mainly supported by the elongated carbide phases, while the strength of vertical direction is supported by the iron matrix and strongly depended on the deformation behavior of the iron matrix. The rule of mixtures was used for analysis of the experimental results described above.

1. Introduction

The continued improvement in efficiency of high-temperature structures depends on improved materials and on designs that utilize these materials more effectively. White cast irons are, because of their high hardness, high strength, net shape castability, corrosion resistance and relatively low cost, attractive candidates for many applications requiring high

516

wear resistance. Some high chromium white cast irons have been used for a material of steel hot working mill roll [1]. The roll fabricated from the high chromium white cast irons was developed and used practically during the last few years. It is said that the mill roll of high chromium white cast iron has extremely high performance and long service life. It can safely be said that the improvement of quality in this roll material achieved both the high speed and high load rolling possible. As for the study on research and development of hot working mill roll material, the approach from various viewpoints is necessary to give superior wear resistance, surface roughing resistance and mechanical properties, which satisfy the purpose of service to mill rolls.

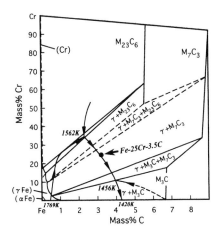

Fig. 1 Fe-Cr-C Phase diagram at 1143K [4]. Solid mark shows the chemical composition of the specimen used for this experiment.

The wear resistance of the roll is closely related to high compressive strength and high hardness at elevated temperatures. However, there is a very few study on the strength of high chromium white cast iron at elevated temperatures.

It is said that the properties mentioned above can be influenced metallographically by matrix structure and the type, morphology, quantity, distribution of carbides which are embedded in the matrix. Furthermore, crystallized grains of primary and eutectic phases are known to directionally grow from the roll surface toward the core region when the roll was made by the CPC (Continuous Pouring for Cladding) process [2]. In order to raise the mill roll performance, therefore the alloy design to control these items and the evaluation for the manufacturing process should be carried out sufficiently. Especially, a study on the relation between the microstructure and strength is important. Recently the multi-component white iron, which consists of hard special carbides and heat-treated tough and hard matrix, has been researched and developed particularly for the hot working rolls.

By the way, the microstructure of directionally solidified high chromium white cast iron containing hard carbide phases is similar to that of fiber reinforced composite materials. The strength of the materials seems to depend on the carbide spaces, their orientation and volume fraction which are denoted by the Cr/C ratio. **Figure 1** shows the austenite liquidus surface phase diagram and constitutional phase diagram at 1143 K for the Fe-Cr-C system[3][4].When only the C quantity is varied with a fixed Cr quantity, it can result in hypoeutectic, fully-eutectic and hypereutectic alloys. But because the Cr/C value is different, the composition contained in carbide phase and the iron matrix is different in each alloy. On the other hand, if the Cr and C quantities are varied but with a fixed Cr/C value, it can also result in hypoeutectic, fully-eutectic and hypereutectic alloys. But the compositions of the carbide phase and the iron matrix phase of each alloy become similar in this case. Therefore, it is necessary to relate the strengthening mechanism of the high chromium white cast iron with their particular microstructures for developing a mill roll of extremely high performance and long service life. Goto et al. studied on the unidirectionally solidified hypoeutectic, fully-eutectic and hypereutectic high chromium white cast irons to clarify their strength and deformation behavior at elevated temperatures [5]. They revealed that the fully-eutectic iron shows the highest strength within the whole temperature range and also suggested that the strength strongly depends on the microstructure.

In this study, a systematic research on the unidirectional solidification of the fully-eutectic white cast iron was conducted to clarify the strengthening mechanism of the iron with particular emphasis on high temperature strength.

2. Experimental procedures
2.1 Specimen preparation

As mentioned on Fig. 1 microstructure of the alloy strongly depends on the chemical composition. A composition of Fe-25mass%Cr-3.5mass% was selected to make an alloy specimen with fully-eutectic structure under a fixed Cr/C ratio of 7 on the eutectic line in Fig. 1. The alloy was melted using a high frequency induction furnace, then poured into a metal mold at 1723 K. The alloy was remelted in an alumina tube (15 mm in diameter and 100 mm in length) using Bridgman type electric furnace with an argon atmosphere. The melt was solidified unidirectionally by moving the furnace vertically upward under a speed range from 3.0×10^{-3} m/s to 1.3×10^{-1} m/s. The directionally solidified samples were machined in the directions, both of parallel and vertical to the solidification direction, into rectangular specimens of $3.0 \times 3.0 \times 10.0$ mm for the compression test.

2.2 Mechanical tests

Shimazu Autograph IS-10T was used for the compression test under argon atmosphere. The stress-strain curve was measured under a strain rate range from 2.5×10^{-4} to 1.7×10^{-2} at various temperatures up to 1073 K. In addition to the entire stress-strain curve, the yield stress ($\sigma_{0.2}$) and their dependence on the microstructure were investigated. A micro-Vickers hardness test was also conducted for analyzing a temperature dependence of the hardness in γ-austenite phase, Fe_7C_3 phase, and the eutectic phase of the alloy specimens.

2.3 Structure observation

To reveal the morphology and distribution of carbides in the specimens and also fracture surface after compression test, an Optical microscope and EPMA (JEOL-JXA 733) were used. Deformation mechanism of the specimens was studied by using a TEM (JEOL-2000 CX) operated at 200 kV.

3. Results and discussion
3.1 Microstructure

Figure 2 shows microstructure of the longitudinal and transverse sections in each directionally solidified specimen. All the specimens show fully-eutectic microstructure composed of the γ austenite phase (dark phase) and Fe_7C_3 carbide phase (white phase). The microstructure grew in the solidification direction (longitudinal direction) and show a fibrous

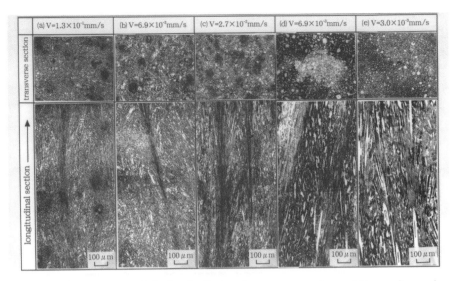

Fig. 2 Microstructures of the longitudinal and transverse sections of the specimens, in which the values of V show the unidirectional solidification rate.

structure. Therefore, the structure of the specimens is similar to that of fiber reinforced metal matrix composite. The microstructure also shows an eutectic cell structure which is characteristic in high chromium white cast iron. The cell structure changes to a finer one as the directional solidification rate increases. The cell diameters for solidification rate (V) of 1.3×10^{-1}, 6.9×10^{-2}, 2.7×10^{-2}, 6.9×10^{-3} and 3.0×10^{-3} mm/s were 158, 227, 299, 417, and 597 μm, respectively.

3.2 Stress-strain curve

Results of the compression test are summarized in **Fig. 3** and **Fig. 4**. Figure 3 shows typical examples of the true stress-true strain curves in the specimen parallel to the solidification direction. The yield stress ($\sigma_{0.2}$) and maximum compressive stress(σ_B) decrease as the test temperature increases in each specimen. The form of the stress-strain curve is different for each solidification rate. The stress-strain curves of almost all the specimens show work hardening occurring at an early stage in the plastic deformation from room temperature up to 773 K. The amount of plastic strain to fracture decreases as the solidification rate decreases. When the temperature is below 773 K, the specimens of V=3.0×10^{-3}mm/s show brittle fracture. The specimen of V=2.7×10^{-2} mm/s shows the highest strength of 2800 MPa at room temperature. Even at 973 K the strength reaches up to 800 MPa. This is an amazing fact because such a high strength material is not easily found in the practical materials for structures. Therefore, it can be said that the high chromium white cast iron has an extremely high strength and acceptable level of toughness at elevated temperatures.

Figure 4 shows typical examples of the true stress-true strain curves in the specimens vertical to the solidification direction. It is clear that the stress level for the vertical direction

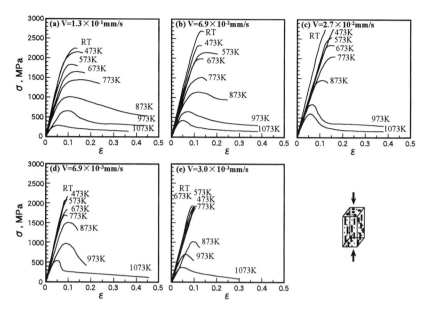

Fig. 3   True stress-true strain curves of the specimens parallel to the solidification direction at various temperatures under a strain rate of $2.5 \times 10^{-4} \text{s}^{-1}$.

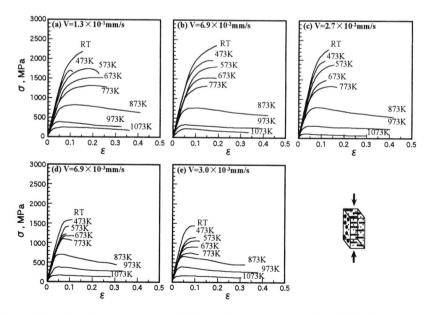

Fig. 4   True stress-true strain curves of the specimens vertical to the solidification direction at various temperatures under a strain rate of $2.5 \times 10^{-4} \text{s}^{-1}$.

520

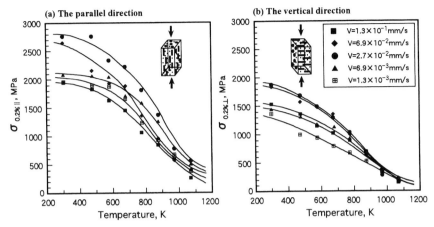

Fig. 5   Temperature dependence of yield stress of the specimens tested under a strain rate
of $2.5 \times 10^{-4} s^{-1}$.

is lower than that for the parallel direction in each specimen. Even so, the strength for V=6.9x10^{-2} mm/s shows 2300 MPa at room temperature. And also, it is found that the plastic strain to fracture is much larger for the vertical direction than for the parallel direction. The form of the stress-strain curves in all the specimens shows work hardening for whole temperatures.

Figure 5 shows the temperature dependence of yield stress of the specimens tested under a strain rate of 2.5x10^{-4} s^{-1}. The strength for V=2.7x10^{-2}

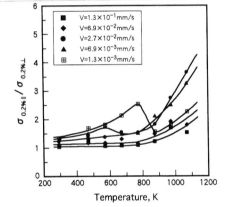

Fig. 6   Anisotropy of the yield strength of the specimens.

mm/s is the highest among the specimens for both the directions.

3.3 Anisotropy of the strength

In a mill roll produced by the CPC process [2], an anisotropy of the strength in the crystallized grains is especially important because these grains tend to grow from the roll surface toward the core region. Then, the anisotropy of strength in the specimens was analyzed from the data of Figs. 3 and 4.

Figure 6 shows the relations between the yield stresses of the parallel direction, ($\sigma_{0.2\% \parallel}$), and the vertical direction, ($\sigma_{0.2\% \perp}$). The ratio of $\sigma_{0.2\% \parallel}$ and $\sigma_{0.2\% \perp}$ increases with the rise in temperature. This phenomenon can be explained as follows. As reported in the previous papers [6], at 0.2 % plastic strain of the iron composite, the strain in the γ austenite phase is mostly plastic, whereas the strain in the M_7C_3 carbide phase is mostly elastic. From this large strain anisotropy, a strong constraining to deformation may occur in each eutectic colony in an unidirectionally solidified specimen of parallel direction. This constraining effect is thought to be the reason why the parallel direction is stronger. On the other hand, for the specimens of the vertical direction, there are eutectic colony boundaries which are shown in Fig. 2 and inclined to the compressive stress axis, and the shear deformation occurs preferentially along the colony boundaries, because there exists a soft layer along the colony boundary, where the composite structure is destroyed. This is thought as being the reason why the vertical direction is much weaker than the parallel one. Therefore, it is concluded that in the unidirectionally solidified materials with eutectic colonies, aligning the colony boundaries parallel to the loading direction is very important for the practical use.

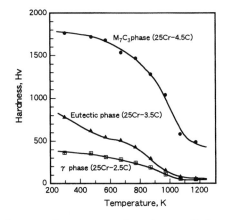

Fig. 7 Temperature dependence of the hardness in the γ -austenite, eutectic matrix and primary M_7C_3 phases of the specimens.

3.4 Strengthening mechanism

According to the rule of mixtures, the strength of composite materials directly depends on the volume fraction of component phases, but not on the microstructure. However, the strength of the specimens in this experiment strongly depended on the microstructure as shown in Figs. 3 and 4, though the volume fraction of M_7C_3 carbide in the alloy is a constant of 35.9 vol. % [7]. To reveal the validity of the rule of mixtures in this alloy, micro hardness test was conducted for the composed phases of γ primary crystal (in Fe-25Cr-2.5C), eutectic structure (in Fe-25Cr-3.5C), and the M_7C_3 primary crystal (in Fe-25Cr-4.5C). The results are shown in **Fig. 7**. The measured hardness of all the phases decreases with increasing temperature. The hardness of the eutectic structure could be explained by the rule of mixtures at lower temperature, while it did not at higher temperatures in which the hardness of the eutectic phase was identical with that of γ austenite phase. To reveal the deformation mechanism, the microstructure of both component phases in the fully-eutectic specimen was observed by TEM.

Figure 8 shows the microstructures of the M_7C_3 carbide and γ austenite phases in the specimen after compression test at room temperature. It is clear that the HCP phase of M_7C_3 carbide shows twinning deformation, while that the FCC phase of γ austenite shows a deformation due to dislocation slip. On the other hand, **Fig. 9** shows the microstructures of both the phases in the specimen after compression test at 1073 K. The microstructure of M_7C_3 carbide is characterized by twinning deformation even at high temperature. However, the microstructure of γ austenite phase shows many precipitates of secondary carbides and recrystallized grains. This means that γ austenite phase deforms due to dislocation slip together with precipitation of the carbide, recovery and recrystallization during compression

522

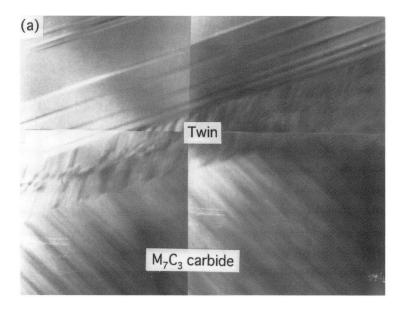

Twin

M₇C₃ carbide

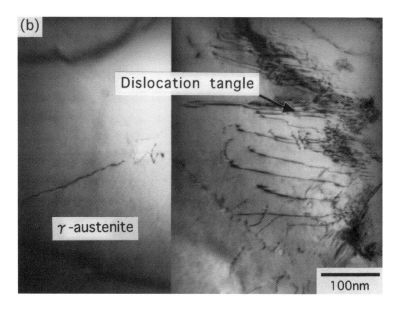

Dislocation tangle

γ-austenite

100nm

Fig. 8 TEM micrographs of (a) M_7C_3 carbide phase and (b) γ austenite phase in the specimen deformed up to a strain of 12.6% under a strain rate of $2.5 \times 10^{-4}s^{-1}$ at room temperature.

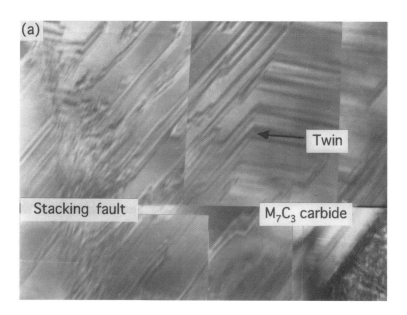

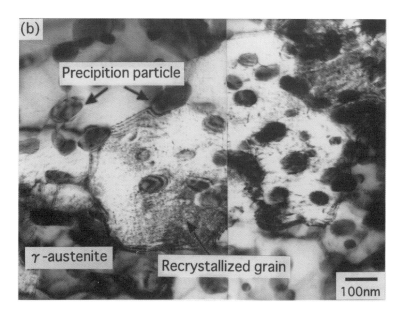

Fig. 9 TEM micrographs of (a) M_7C_3 carbide phase and (b) γ austenite phase in the specimen deformed up to a strain of 24.8% under a strain rate of $2.5 \times 10^{-4} s^{-1}$ at 1073K.

524

and will only be made available with the agreement of NPL, with the exception of the example data set and creep curves for Nimonic 80A used to illustrate this paper.

In addition to this paper, a 'Powerpoint' slide-show has been prepared to enable the existence of the data base to be publicised which contains further details of the contents of the Database; a copy of the presentation is available on a CD-Rom.

The data indexing exercise revealed that a greater amount of data had been generated than previously been suspected, and it is hoped that the information will prove of long lasting value to the creep testing community for many years to come. The custodians of other creep data that have been accumulated over the years in the UK at public expense, eg the NEL data, the DERA Pyestock / Farnborough data and the CEGB data from CERL at Leatherhead, and the data measured at various Universities, are all encouraged to find means of creating and preserving similar user friendly databases.

5.ACKNOWLEDGEMENTS

The DTI CARAD (*Civil Aviation Research and Demonstration*) Directorate is acknowledged for funding the preparation of the index and database.
Mr Jon Ralph is thanked for help in the initial preparation of the Index. Mrs Sandra McCarthy is acknowledged for data entry into the database. Dr T.B. Gibbons and Mr P McCarthy are thanked for constructive discussion through out the work.

6. REFERENCES
[1] Tapsell H.J. & Bradley, J. (1925) *Mechanical tests at high temperatures on a non-ferous alloy of nickel and chromium.* Engineering, 120, 614-615, 648-649, 746-747

[2] Tapsell H.J. (1931) *Creep of Metals.* Oxford University Press, London.

[3] Loveday,M.S. (1982). '*High temperature and time-dependent mechanical testing: an historical introduction.*' Chapter 1 , pp1-12. in *'Measurement of high temperature mechanical properties of materials.'* Ed M.S.Loveday, M.F.Day & B.F.Dyson. Pub. HMSO, London.

[4] Loveday, M.S., Prince,D & Ralph, J.,(2000)'*An Introduction to the NPL Creep Database.*' NPL Measurement Note CMMT(MN) 062. Pub: National Physical Laboratory.

[5] Osgerby,S & Loveday,M.S., (1992)'*Creep Laboratory Manual'*, NPL Report DMM(A) 37.

[6] Loveday, M.S., (1996) '*Creep Testing; Reference Materials and Uncertainty of Measurement'* pp277-293 in '*The Donald McLean Symposium- Structural Material : Engineering Applications Through Scientific Insight.*' Ed: E.D.Hondros & M. McLean, Pub : Inst. Materials, London.

[7] Bullough, C.K, Calvano,F. & Merckling, G. 1999 '*Guidance for the exchange and collation of creep rupture, creep strain-time and stress relaxation data for assessment purposes.*' ECCC-WG1 Recommendations , Vol. 4. Brite-Euram Thematic Network 'Weld-Creep', BET2-0509.Available from ERA Technology, Leatherhead, Surrey, UK.

[8] Over, H.H., 2000   '*Thermo-mechanical fatigue data handling within Alloy-DB'.* Paper 6, issued in abstract form. HTMTC (ESIS TC11) Seminar '*The Practicalities of Thermo-mechanical Fatigue Testing in the New Millennium'*, Darmstadt, Germany. Ed : P. McCarthy. Also see http://matdb.jrc.nl/vanalles/overview

Materials data management

TSP Austin and HH Over
EC-JRC Institute for Advanced Materials, The Netherlands

1. Abstract

Efficient materials data management significantly assists the R&D and design processes, guards against duplication, and addresses the issue of traceability (and hence accountability). At the EC-JRC Institute for Advanced Materials, the Alloys-DB and CMC-DB systems have been developed to meet alloys and advanced ceramics test data management requirements, respectively. Both systems store data from mechanical and thermo-physical tests performed according to recognised standards. Since on-line systems offer considerable advantage in respect of R&D activity, the EC-JRC-IAM database development focus is currently upon implementation of Web-based facilities and the realisation of protocols required to facilitate on-line data exchange. An on-line version of the Alloys-DB system is thus already available to the materials community.

In order to realise true inter-operability in a particular discipline an exchange protocol must be established and, in recent years, many technical disciplines have profited from the use of protocols developed using XML. For the materials community there are a number of advantages to adopting a materials specific equivalent. At the R&D level such a protocol would allow seamless interfacing (i.e. test facility to database, database to design tool, etc.). Web-based activities would enable improved project co-ordination, allow for development of distributed systems, and provide the opportunity to pursue on-line marketing activities. A pre-requisite for the development of a viable protocol, however, is the participation of as many interested parties as possible from the industrial and scientific sectors.

2. Introduction

The issues that are important for materials data management include data conservation, efficient data handling, quality of data and data security.

Modern scientific and technical (S&T) communities inevitably rely upon computerised systems for efficient data handling. Professional database systems consist of a database management system (DBMS) which stores information and an application interface that interacts with the DBMS, allowing data entry and data retrieval. Modern database systems are invariably relational, using joining keys that facilitate rapid selection upon particular criteria. The exploitation of a relational structure is what distinguishes a data management system from a data storage system. For low volume data consisting of a simple structure, electronic spreadsheets are sufficient. The relation of the data to technical reports, similar data sets, etc is known by the individual who uses that information. Any attempt to manage large volume data of a complex structure, which can be accessed and changed by a number of people, requires a dedicated data management system.

The Materials Data Management (MDM) Sector at the Institute for Advanced Materials (IAM) specialises in the development of database applications for handling materials test data. These systems include Alloys-DB (for engineering alloys mechanical and physical properties), Cor-DB (for elevated temperature corrosion) and CMC-DB (for advanced ceramics). These systems rely upon a common application interface that has been designed to guide the user through data entry and data retrieval procedures. Upwards of 20 large European industrial and scientific sector organisations[1] use these systems (the most sophisticated of which is Alloys-DB) for managing materials property data.

Irrespective of data type, structuring data and establishing meaningful relations requires a considerable effort on the part of scientific and industrial experts. Before a relational database can be developed, the way data is structured and the extent of the data required must be established. This information, this data about data, is called metadata[2] and without such ancillary information experimental results lose their contextual meaning. In addition to helping establish data pedigree (i.e. quality), metadata also facilitate comparison of different data sets and helps adjust them to the technological advances that occur with time.

Taking materials property data as an example, a quasi-standard data structure consists of five primary entities: source, material, specimen, test condition and test result information. Source information includes reference to laboratory and facilities, standards, personnel, etc. Material information covers microstructure, composition, supplier, form, etc. Figure 1 shows these five entities together with a sixth 'joining' entity that accommodates welded specimens.

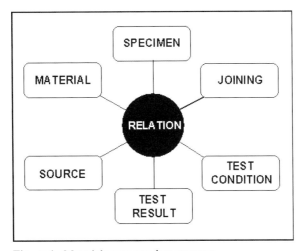

Figure 1. Materials property data structure.

3 Data and competency conservation issues

Data conservation is an aspect of materials research that does not receive a level of attention commensurate with the financial and personnel investment associated with materials testing. Materials related initiatives such as modelling, component and plant design, etc., often suffer because a lack of information prevents a review of available materials.

Competence preservation is just as important an issue as that of data conservation. There is, for example, a real danger that European nuclear technology competence will decline to an untenable level. A recent report from Germany's Federal Ministry of Economics and Technology states that 'maintaining Germany's competence in the field of nuclear safety is a necessary requirement for the next decades'[3]. Since information (i.e. models, codes, procedures, etc.) derives from data, effective data management systems will inevitably contribute to improved competence preservation.

Technical reports alone cannot provide the degree of detailed curve and metadata information required for materials modelling and component simulation studies. Pedigree information[2] is required to determine data quality, a complete description of test conditions is necessary, specimen details are required and curve data is needed to derive materials relations.

Storage of complete records must be considered integral to the S&T process if the investment is to be properly managed. This requires a change in S&T culture, since projects often focus upon a single objective (the design of a particular component, the validation of a certain model, etc.). Whilst S&T activities involve a considerable personnel and capital investment, the data are often not stored in an accessible format. Beyond the immediate relevance to achieving the objectives for which data were first generated, the circumstances that may require access to quality data include validation and revision of models, new technologies (e.g. data mining and neural networks), the data is relevant to other engineering sectors, etc. To persuade the S&T community of the benefits of efficient data management, an effort is required which demonstrates the advantages and includes an element of training.

MDM Sector initiatives are currently oriented to addressing these issues.

4 Dedicated data management solutions

IAM is one of eight institutes of the European Commission Joint Research Centre (JRC). The mission of the JRC is 'to provide customer-driven scientific and technical support for the conception, development, implementation and monitoring of EU policies. As a service of the European Commission, the JRC functions as a reference centre of science and technology for the Union. Close to the policy-making process, it serves the common interest of the Member States, while being independent of special interests, whether private or national'.

The MDM Sector provides services to European industrial and scientific sectors in the context of the stated JRC mission. Efforts have concentrated upon developing well-designed systems. Many years' experience developing systems for various data types (including engineering alloys, corrosion and advanced ceramics), together with close consultation with industrial and research sector clients, has resulted in effective software solutions. Figure 2 shows the application interface of the Alloys-DB system.

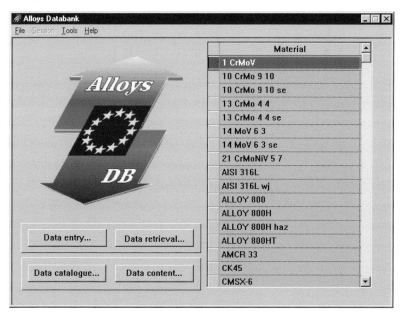

Figure 2. Alloys-DB PC version application interface.

The Alloys-DB system consists of various components for handling materials property data and is built according to the quasi-standard (five-entity) metadata structure for materials property data shown in Figure 1. The interface of the data entry component (see Figure 3) clearly illustrates the reliance upon source, material, specimen, test condition and test result entities.

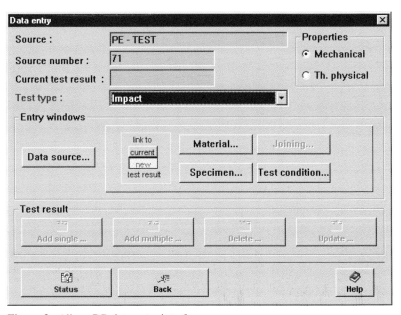

Figure 3. Alloys-DB data entry interface.

5 Distributed data management

A proliferation of isolated database systems and the associated issue of data conservation are problems common to all scientific disciplines. The case for materials property data is especially acute, where a myriad of heterogeneous database systems exist and large amounts of data, 'stored' in technical reports, remain practically inaccessible. Encoding materials property metadata using the eXtensible Markup Language (XML) offers a solution to systems inter-operability and a means to recover and conserve data. The key point of such an exercise is not to convert any existing system, but rather to facilitate communications between heterogeneous systems without any change to their internal structure.

Although the PC and client/server versions of the various systems for managing data that have been developed are well suited to the needs of research institutes and company departments, it is recognised that a different approach to data management is required for collaborative efforts. Amongst EC project partners probably the single most divisive issue at present is the requirement to invest in data management software. This acts as an obstacle to data conservation and is an issue that must be resolved. The use of isolated PC based systems further aggravates the problem of data collection and dissemination.

Since the European Commission funds project collaborations it is reasonable that a JRC institute addresses the issue. The obvious answer to addressing collaborations that involve partners spread a cross a large geographic area is to exploit the Internet and develop integrated on-line networks. The use of existing infrastructures (i.e. the Internet) and software (i.e. browsers) will also free organisations of the burdens associated with database software, namely cost of software, maintenance of the software and training of personnel.

MDM Sector initiatives are currently oriented to developing a 'distributed database framework' for the nuclear energy sector. This first requires that existing systems are Web-enabled and subsequently that the various systems are integrated. On-line data retrieval services are already available. Using any standard browser (Netscape, Internet Explorer, etc.) a client can access the Alloys-DB database via the MDM website, see Figure 4.

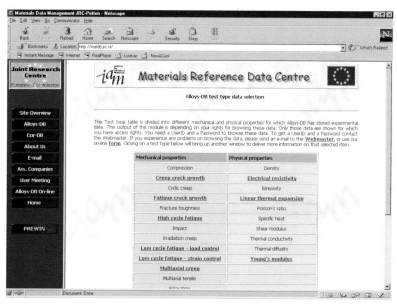

Figure 4. On-line Alloys-DB data retrieval interface.

As a consequence of handling commercially sensitive information, MDM Sector efforts to develop on-line systems place special emphasis upon security. Usernames and passwords are issued to authorised personnel according to strictly defined conditions.

On-line data entry is a considerably more complex issue. A direct connection to the DBMS is not possible since data must first be validated before being uploaded. Figure 5 shows an HTML interface that duplicates the Alloys-DB PC application. The user-guidance is programmed in Java and contains quality procedures to check the input of mandatory information and mismatches. Data entered using the HTML interface will be written to a server-side file. A parser (a program which converts information from one format to another) will next transform the data file into a format compatible with Alloys-DB. Finally, a database transfer application is required that loads the input into Alloys-DB according to the system's existing relational structure.

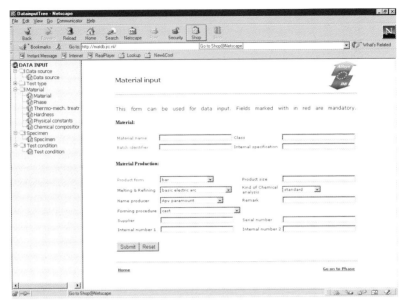

Figure 5. On-line Alloys-DB data entry interface.

The development of Web-enabled systems represents the first stage towards realising systems inter-operability. An integrated on-line framework (of the type that has already been deployed by the JPL OODT task to cater for planetary mission data[4]) requires standard exchange protocols to enable systems connectivity.

6 XML for materials property data

The eXtensible Markup Language (XML) provides S&T communities the possibility to develop standards for transferring information. XML is a generic markup language and is designed to provide S&T communities with the means for developing data specific exchange protocols. 'XML acts like the ASCII of the Internet, allowing interaction between hosts regardless of their operating systems, database managers, or data formats'[5]. XML has been developed by the worldwide web consortium (W3C), an independent body that is not subject to any potential proprietary constraints. XML is a technology available to communities that wish to realise connectivity for the purposes of disseminating (currently isolated) technical data.

Although S&T communities have proved reluctant to adopt Web-based technologies to assist conventional activities[6], the focus of a recent CODATA conference offered some insight into the probable impact that XML will have upon S&T. This 4-day conference[7], consisting of four continuous parallel sessions, focussed almost entirely upon Web-based initiatives. The possibility to use XML to define theme specific transfer protocols has provided a solution to the ubiquitous problem of systems inter-operability. Rather than attempt to align the vast range of electronic systems to adopt a standard internal structure, XML provides a means whereby a diverse range of systems can communicate using a common 'language'.

Both NASA and NIST already have coherent policies for Web-based data management. A combination of on-line access to data at the NASA Space Telescope Science Institute and a commitment to placing data in the public domain has led to over 200 publications citing use of Hubble Deep Field data. The NASA Jet Propulsion Laboratory includes a group responsible for the development of distributed frameworks for integrating science data[4]. The NASA 'Global Exchange' programme manages a facility whereby experts from any discipline can develop and register a metadata structure on-line[8]. Notably, US agencies lead in the use of Web technologies to complement existing S&T activities.

The IT environment is one characterised by rapid change, this is particularly true of Internet related initiatives. The various S&T sectors that have come to rely upon IT to complement conventional activities now realise that contingency plans are required to accommodate obsolescence of technologies if IT manpower and capital investments are to be conserved. In the context of MDM Sector activities, the particular technologies that are of immediate interest include Web and database technologies. The use of XML schemas to encode metadata counters the possible obsolescence of Web technologies and addresses the issue of distributed data management through database inter-operability. Records of these structures, preferably appended to the relevant international testing standards, will find application irrespective of the particular method used to encode the metadata.

Besides providing an opportunity to further exploit the existing framework of communication networks and software, XML offers a means whereby metadata can be encoded and remain accessible for the future. It is the development of metadata structures that will define discipline specific forms of XML. Several specific forms of XML are already available, including CML for on-line transfer of chemical data, MathML for mathematics and BioML for biomolecular data. Development of an equivalent XML protocol for engineering materials data is not the responsibility of IT specialists, but that of the materials community. As detailed in the Nature lead article[6] 'Using XML, CORBA and other languages and protocols, it is already possible for websites and online databases to communicate, interpret and understand one another's content. It is up to individual communities to organise themselves to take advantage of this opportunity'. The effort required is that of defining a data structure that accommodates the requirements of the technical community - industrial and scientific alike.

Efforts to establish a markup for materials property data have already begun at NRIM[9] and NIST. The NIST initiative has resulted in MatML[10]. To achieve a common format for materials property data, contributions to the project are encouraged from across the materials community. To date, NIST efforts have focussed upon defining a basic structure, aligning MatML with conventional XML syntax and integration with related markups. In respect of integration with existing markups, an important activity for all scientific disciplines is under way at the Lawrence Berkeley National Laboratories, where efforts to develop a markup for scientific units is in progress. MDM Sector initiatives include direct contributions to the international initiatives to develop exchange protocols for materials property data. Involvement at an early stage ensures that European scientific and industrial materials interests are represented.

MDM Sector efforts clearly demonstrate the feasibility of Web-enabling and cross-linking existing database systems. The single most important issue for establishing a framework of connected systems is the use of standard protocols. If, through consultation with the European materials community and participation in the international initiatives described, a standard exchange protocol can be achieved the benefits will be considerable. Test facilities could be configured to 'pipe-line' data direct to storage facilities, databases could be interfaced seamlessly with design tools and on-line data services could be offered.

7 Conclusions

A considerable opportunity exists to better manage and exploit the large volumes of data that are generated by the materials community. This requires a more responsible attitude towards data conservation and willingness to exchange data. Before this is possible, however, standard data exchange formats must be developed that will facilitate systems inter-operability and the development of integrated on-line frameworks.

The MDM Sector has deployed an on-line data retrieval system and is developing an on-line data entry program. Such efforts will free the S&T community of the burdens commonly associated with database software and remove the obstacles to data conservation. If successful, MDM Sector involvement in efforts to establish a standard format for exchanging materials property data will provide the materials community with a technology that will remove barriers to inter-operability and enable a more efficient and innovative approaches to handling materials property data.

Bibliography

1. 'Alloys-DB Materials Data Management', Leaflet 2000.

2. Romeu, JL and Grethlein, CE 'Statistical Analysis of Material Property Data'. AMPTIAC, January 2000.

3. 'Nuclear Reactor Safety and Repository Research in Germany', Report of the Working Group Convened by The Federal Ministry of Economics and Technology (BMWi), January 2000.

4. Crichton, D et al 'A Distributed Component Framework for Science Data Product Interoperability'. 17th International CODATA Conference, Baveno (Italy), October 2000. Available from http://oodt.jpl.nasa.gov/doc/papers.

5. Cohen, D 'Construct Your E-commerce Business Tier the Easy Way With XML, ASP, and Scripting'. Microsoft Systems Journal, February 2000.

6. 'Challenges of the Grid'. Nature, July 2000.

7. 'Data and Information for the coming Knowledge Millenium'. 17th International CODATA Conference, Baveno (Italy), October 2000.

8. See the 'Global Exchange Master Directory' at http://globalchange.nasa.gov.

9. Halada, K and Yoshizu, H 'Possibilities and Problems of Generic Data-Sharing Systems of Materials' databases by Use of XML'. 17th International CODATA Conference, Baveno (Italy), October 2000.

10. Begley, EF and Sturrock, CP 'MatML: XML for Materials Property Data'. Advanced Materials & Processes, Volume 158, Number 5, November 2000.

Development of 'The NPL Creep Database'.

Malcolm S Loveday & David Prince
Materials Centre,
National Physical Laboratory, Teddington,
TW11 0LW, UK.

ABSTRACT

Details of an index for the high temperature creep test data accumulated at NPL over approximately the last fifty years, known as '*The NPL Creep Database*' are presented. Details from 3,061 tests are listed, covering more than 66 different materials representing over 5.5 million hours (~ 630 years) of test data.
In addition an example is given of details of part of the creep test database for Nimonic 80A. Information is also presented of the development of the database to enable it to become readily available over the internet.

1. INTRODUCTION

Reliable creep data is essential for design and remnant life prediction of many high technology and safety critical components mainly used in the power generation and aerospace industrial sectors. 'Creep' is the term used to describe slow time-dependent deformation exhibited by materials subjected to a stress, usually at elevated temperature. Such data is measured in specialised testing machines and tests may last anything from a few hours to a few years depending upon testing conditions. Typical creep curves are shown in Figure 1

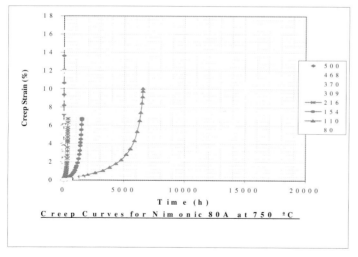

Figure 1: Sample of NPL Creep Data for Nimonic 80A.
Grain size = 42 microns, Stress in MPa (Key in R-H box)

Reliable databases are extremely scarce and in general only provide a summary of the data rather than the full creep curves. With the increasing demand to extend the safe operating life of power stations beyond their 30 year design life, sophisticated models have been developed, usually using Finite -Element (FE) analysis, which require reliable input data and, increasingly, the full strain-time creep curves. Thus there is a need for long-term reliable data on industrial relevant materials for predictive purposes, and extensive data sets for validation of models. A list of the shortest and longest test lifetimes for selected alloys in the NPL Creep Database is given in Table 1, together with the number of test curves.

Material	Shortest Test, Hours	Longest Test, Hours	Number Of Tests
NON FERROUS			
NICKEL BASE ALLOYS			
Pure Nickel Bar Specimens	30	7752	19
Experimental Nickel Base Superalloy	104	7353	8
NiCr-2.5%Ti-1.5%Al	11	1450	28
NiCr-2.5%Nb	0.45	1841	20
NiCr-Al-Nb	6	2421	23
NiCr Alloys	6.5	5696	46
Nimonic 75 Creep Reference Material	51	6962	40
Nimonic 75	7	8950	21
Nimonic 80A	0.55	25,649	454
Nimonic 80A DS	9.5	12434	16
Nimonic 90	7	8079	67
Nimonic 90 (Torsion)	41	26,135	27
Nimonic 100	2	6212	74
Nimonic 101	0.05	820	36
Nimonic 105	4	10,028	96
Nimocast 739	8	11,705	16
MAR M 002	1.5	7429	234
DS MAR M 002	25	2602	14
MA6000	9	7897	6
MAR M 246	31	13282	49
IN597	1	12762	162
IN738	53	18786	106
IN738 DS	384	1550	6
IN939	1	28163	74
FERROUS			
Iron	0.5	61628	163
AISI304 Steel ASTM (Creep Reference Material)	66	233	2
½CrMolyVan	21	22250	17
2¼Cr 1Moly	2	44110	27
Pipe & Rotor Steels CrMoV	2	33544	369
C-Mn-Nb Steel	5.5	50736	33
Stainless Steel	70	8930	35

Table 1. NPL Creep Database: Summary of Shortest and Longest Test Lifetime for Selected Materials

The rapid development of aircraft engines during the First World War as well as the post-war developments in chemical engineering required high temperature material's property data to enable more reliable design of components and prediction of plant life. Details of the creep testing machines developed in the War years and used in Engineering Division of NPL were subsequently published by Tapsell and Bradley in 1925 [1]. They gave a description of an under-slung single lever machine and many machines based on their design are still in use at many laboratories today. Creep strain was measured using mirror and rhomb extensometers which were viewed manually through telescopes and Tapsell (1931) in his book *'Creep of Metals'* [2] gave an interesting review of the apparatus then in use throughout the world for studying the long term creep strength and fracture behaviour of metals, as was noted by Loveday,[3]. In 1939 a special purpose built Creep Laboratory was opened at NPL which had controlled air conditioning giving ambient temperature control of ~ ± 1° C, thus providing the necessary stability for high precision strain measurement with extensometers. It is thought that the building accommodated over 100 machines prior to the transfer of Engineering Division, NPL, to the National Engineering Laboratory at East Kilbride, near Glasgow, in 1954. Creep work continued at NPL under the auspices of the Metallurgy Division, partially using a row of 2 ton Tapsell type machines which were finally removed in the late 1980's, but mainly on automatic-levelling over-slung Mayes 2 ton and 3 ton single lever machines.

Over the years many well known scientists have undertaken research in the creep field at NPL, notable Donald McLean, B.E.Hopkins, L.M.T. Hopkins, H.R.Tipler, B.F.Dyson, T.B.Gibbons, Malcolm McLean and many more; a bibliography of the resulting published papers has now been complied and is available on request.

The creep data currently held at NPL, accumulated over the last fifty years or so, is held in a mixed format of printed paper records, hand written tables and graphs, together with computer data files and plotted graphs. Unfortunately until now there has not been a comprehensive index to this wealth of data, although research workers have been able to readily retrieve the data by searching through a variety of cupboards or filing cabinets. Much of the data has been published in summary form, but, in general, the complete strain-time curves have not reached the public domain. Preliminary details of this database project have been given elsewhere [4]. The majority of the data was accumulated complying with BS 3500 Part 3, which has recently been superseded by BS EN 10291 (2000) and following procedures outlined in the NPL Creep Laboratory Manual [5] . In general terms the Uncertainty of Measurements associated with the creep data is ~ ± 22% with a 95% confidence level as discussed elsewhere [6].

2. OVERVIEW OF INDEX

An Index for the NPL Creep Database has been compiled into a Microsoft Excel spread sheet, which is a very common PC compatible format. The summary index is shown in Table 2, and for each material listed there is a separate spread sheet giving full details of the data. An extract from part of the data sheet for Nimonic 80A is shown as an example in Table 3 for a series of tests undertaken at 750°C. The summary Index shown in Table 2 has been configured so that as more tests are added to the individual data sheets, the index automatically updates the listed total cumulated testing time and number of tests given at the bottom of the Index.

The Summary Index is publicly available indicating the range of a) materials, b) the stress and temperature ranges covered, c) the total number of individual tests, and d) the total cumulated number of hours. It can be seen from the pie-chart shown in Figure 2, that the majority of the data is for Ferrous and Nickel base alloys, with smaller amounts of information relating to Aluminium, Copper and other materials.

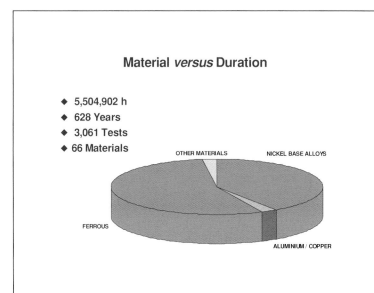

Figure 2. Distribution of database: testing time for each material class.

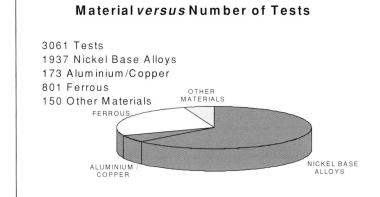

Figure 3. Schematic illustration of distribution of number of tests for different types of material in the database.

Details of the distribution based on total number of tests for the materials in the database is shown in Figure 3.

In addition, as an example, the stresses and temperature ranges for the Ferrous materials listed in Table 1 are shown as bar charts in Figures 4 & 5. The distribution of number of tests for the various ferrous materials is shown in Figure 6

Material	Total Testing Time in Hours	Total No. of Tests	T min °C	T max °C	σ_{Min} MPa	σ_{Max} MPa	File Prefix
NON FERROUS							
NICKE BASE ALLOYS							
Pure Nickel Bar Specimens	25,064	19	600	800	19	646	A
High Purity Nickel	22,917	35	550	800	29	211	B
Experimental Nickel Base Superalloy	16,545	8	700	800	450	750	C
NiCr-2.5%Ti-1.5%Al	8,123	28	750	750	123	216	D
NiCr-2.5%Nb	12,554	20	500	650	92	601	E
NiCr-Al-Nb	12,217	23	750	750	185	277	F
NiCr Alloys	58,883	46	750	750	123	216	G
Nimonic 75 Creep Reference Material	50,640	40	600	700	75	250	H
Nimonic 75	50,904	21	600	750	45	231	H
Nimonic 80A	372,392	454	575	900	30	1100	I
Nimonic 80A DS	20,884	16	600	800	185	308	J
Nimonic 90	80,676	67	750	900	30	450	K
Nimonic 90 (Hollow Specimens Tested in Air)	15,293	15	750	850	200	375	L
Nimonic 90 (Torsion)	71,827	27	850	850	2	230	M
Nimonic 90 (Compression - Prestrain - Stress Drop)	3,946	10	850	850	50	230	N
Nimonic 100	46,815	74	800	1000	7.7	308	O
Nimonic 101	4,310	36	800	850	200	500	P
Nimonic 105	93,075	96	700	900	60	600	Q
Nimocast 739	46,327	16	850	850	150	300	S
MAR M 002	96,703	234	600	1000	80	800	T
DS MAR M 002	8,121	14	900	900	200	400	U
Nickel Alumina	19,669	38	547	800	6.875	92.4	V
NiCr4A1	93	9	750	750	139	447	W
MA6000	12,900	6	1050	1100	100	145	X
MAR M 246	50,453	49	500	900	250	500	Y
IN597	163,729	162	100	850	75	800	Z
IN738	369,834	106	750	950	50	450	AA
IN738 DS	4,549	6	850	850	250	250	AB
IN738 Turbine blades	8,853	9	850	850	250	250	AC
IN738 LC	289,865	76	788	900	105	450	AD
IN738 LC DS	32,127	31	850	850	200	450	AE
IN738 LC, Sulzer Overaged	1,286	6	850	850	250	250	AF
IN738 LS	3,914	9	750	950	100	600	AG
IN939	137,764	74	850	850	100	350	AH
IN939 (Off Axis) 4 Extensometer	30,624	15	850	850	200	200	AI
INCONEL X750	14,860	32	700	700	300	400	AJ
CMSX-4	3,790	4	850	850	490	490	AK
SRR99 (Off Axis)	4,615	7	850	850	320	450	AL
ALUMINIUM							
Aluminium	11,474	24	200	550	4	280	AM
Fibre Reinforced Aluminium	19,062	33	200	400	2	300	AN
Aluminium 4043 +SiCp	865	8	200	200	75	200	AO
Aluminium Alloy	1,149	11	150	180	100	165	AP
2014+ SiCp	29,311	17	176	180	50	150	AQ
2014 Monolithic Alloy	10,348	11	23	180	100	445	AR
2618 Monolithic	5,332	11	23	200	125	200	AS
2618+ 8%SiCp	3,834	5	200	200	100	200	AT
2618+ 12%SiCp	3,034	5	200	200	125	175	AU
2618+ 15%SiCp	4,756	8	180	220	125	200	AV
8090+ 12%SiCp	3,422	8	200	200	70	200	AW
OTHER NON FERROUS							
Copper	35,618	31	250	400	20	280	AX
FERROUS							
Iron	898,928	163	150	750	4	508	AY
AISI304 Steel ASTM (Creep Reference Material)	299	2	732.2	732.2	93.1	93.1	AZ
½CrMolyVan	91,901	17	550	550	16	349	BA
2¼Cr 1Moly	227,663	27	550	800	30.8	200.2	BB
Notch Tests, 2¼Cr 1Moly,1CrMoVTiB	9,756	12	550	600	92	566	BC
2¼Cr 1Moly Welded Joints	207,885	33	565	593	46.2	146.3	BD
Pipe & Rotor Steels CrMoV	949,803	369	550	700	69	416	BE
C-Mn-Nb Steel	196,061	33	450	550	30	300	BF
Stainless Steel	32,360	35	400	800	30	481	BG
Incoloy 800	6,903	17	550	800	90	340	BH
Steel (Cyclic Creep)	-	3	370	420	400	500	BI
Weld Metal	378,942	90	550	800	77	462	BJ
OTHER MATERIALS							
OTHER MATERIALS							
RBSN Batch 5	1,234	6	1,375	1,375	25	100	BL
Syalon 201	5,863	20	1300	1300	50	100	BM
Commercial in Confidence	52,827	61	470	1,350	1	648	BN

Total Testing Time in Hours	5504651	3,061	Tests

Total Testing Time in Years	628

Table 2. Summary Index of NPL Creep Data Base.

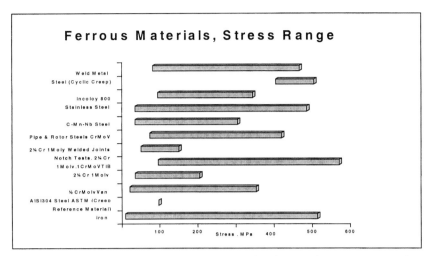

Figure 4. NPL Creep Database : bar-chart showing distribution of test stress ranges for Ferrous materials.

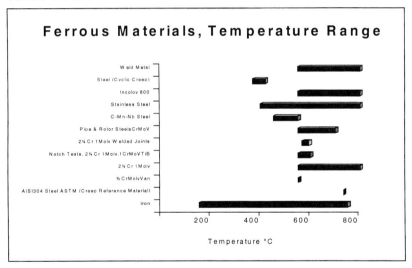

Figure 5. NPL Creep Database : bar-chart showing distribution of test temperature ranges for Ferrous materials.

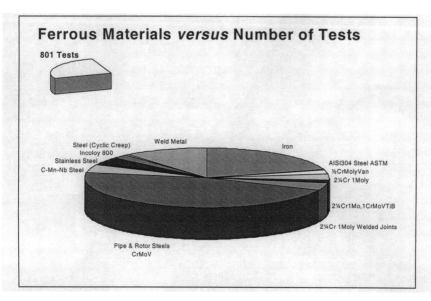

Figure 6.Distribution of number of tests in the database for Ferrous materials

The fields in the data spreadsheet, have been chosen so as to indicate **a) the essential information to specify the material** ie composition, heat treatment, grain size, as-received or re-worked and heat treated, prestrained before creep testing, etc, **b) the testpiece identification number**, usually based on a three letter Code registered in the NPL Materials Index, **c) the type of test**, ie rupture, creep with strain measurement, constant load or constant stress, stress state: uniaxial, bi-axial (torsion), multi-axial (Bridgman notches); cyclic loading, constant strain-rate testing, off-axis loading, (on master data base only, hidden in example given in Table 2), **d) the testing conditions,** eg stress, temperature **e) the recorded material properties**, ie room temperature and high temperature modulus, the test lifetime, the minimum creep rate, the elongation at fracture **f) additional information,** i.e. testing machine, staff undertaking the tests, the year the test was started, etc. An example of the spreadsheet format is given in Table 2.

NPL Creep Database
Nimonic 80A

Test Piece Number	Test Temperature °C	σ MPa	Fracture Time h	F/SD	ε min x10 $^6h^{-1}$	$\varepsilon_f\%$ Measured	$\varepsilon_f\%$ from Graph	Z (reduction in area at fracture) RA%	E_{RT} GPa	E_T GPa	M/C No.	Test Number	Year of Test	Grain Size
LCP49	750	216	373	F	23	7.2			171		34		1980	42µm
LCP54	750	154	1495	F	5	7.62		9.64	179		34	1884	1980	42µm
LCP55	750	468	2.17	F	21757	16.269			166	123	37	1894	1980	42µm
LCP56	750	370	20	F	1101	8.37			154	120	39	1901	1980	42µm
LCP65	750	80	15248	F	1.3	13					38	2238	1982	42µm
LCP66	750	110	6537	F	2.03	10.1	10	5.22	148	114	14	2806	1988	42µm
LCP72	750	500	2.15	F	50337	16.02	13.63	8.76	190	122	34	3001	1990	42µm

* Composition: Ni..Remainder, Cr ..18.0-21.0, Ti..1.8-2.7,Al..1.0-1.8, C..0.1max, Si..1.0max,
Cu..0.2max, Fe..3.0max, Mn..1.0max, Co..2.0max, B..0.008Max, Zr..0.15max,S..0.015

Table 2. Example of format of full data sheet for selected testpieces of Nimonic 80A

540

3. EXAMPLE CREEP DATABASE

From the detailed data sheets it is a relatively easy task to plot the data as shown in Figures 7 - 9, where the data for Nimonic 80A, given in Table 2, has been presented graphically.

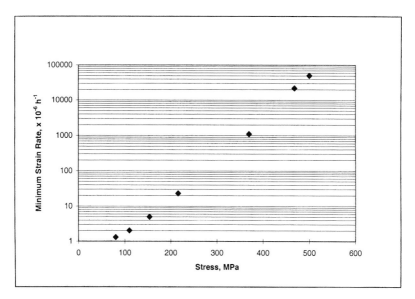

Figure 7. Nimonic 80A: Graphs showing Minimum Strain Rate as a function of Stress.

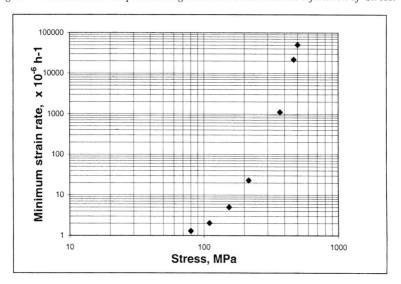

Figure 8.Nimonic 80A Minimum Creep Rate as a function of Stress.
Temperature 750°C, Grain size: 42 micron

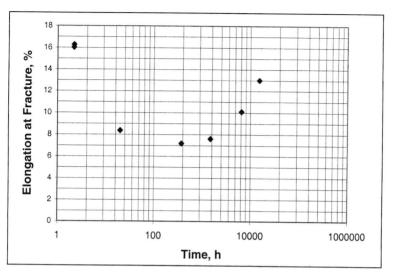

Figure 8. Nimonic 80A Elongation at Fracture versus Time.
Temperature 750°, Grain size: 42 micron

Such graphs can be used for interpolation or the determination of material property parameters used in constitutive laws, eg the stress index, n, for the Norton Creep law,

$\dot{\varepsilon} = A\sigma^n$, where $\dot{\varepsilon}$ is the minimum creep rate, σ is the applied creep stress and A is a constant. The value of n may be determined from the slope of the graph of minimum creep rate plotted against stress on log-log scales. The extracted data set was for tests all undertaken at 750°C, additional data is available for the same material at a single stress over a range of temperatures from which it would be possible to determine the creep activation energy, Q.

For the example data set for Nimonic 80A given in Table 3, the full creep data has been extracted and also listed in a Microsoft Excel spread sheet. A series of curves have been plotted to demonstrate the type of information that would be available for the various materials listed in Table 2. Such data can be obtained in electronic format so that it could be downloaded into FE packages for detailed analysis.

Following consultation with representatives for industry various materials have been selected for the full creep curves to be entered into the database. To date approximately 400 complete creep curves representing ~ 15 materials have been installed following the format recommended by the European Collaborative Creep Committee (EC3) [7]. Arrangements are now in hand to access the NPL data via the Materials Database established by the EC Joint Research Centre, Petten, details of which are given elsewhere [8], and which may be viewed on the web site www.http://matdb.jrc.nl/vanalles/overview.html.

5. CONCLUDING REMARKS

The first phase of indexing the NPL Creep Data Base has been completed and the index is now available in a user friendly spreadsheet format. All the physical files containing the hard copy of the raw data are now coded and can readily retrieved. Access to the full Index and spread sheets containing the details of the measured creep properties have commercial value and will only be made available with the agreement of NPL, with the exception of the example data set and creep curves for Nimonic 80A used to illustrate this paper.

In addition to this paper, a 'Powerpoint' slide-show has been prepared to enable the existence of the data base to be publicised which contains further details of the contents of the Database; a copy of the presentation is available on a CD-Rom.

The data indexing exercise revealed that a greater amount of data had been generated than previously been suspected, and it is hoped that the information will prove of long lasting value to the creep testing community for many years to come. The custodians of other creep data that have been accumulated over the years in the UK at public expense, eg the NEL data, the DERA Pyestock / Farnborough data and the CEGB data from CERL at Leatherhead, and the data measured at various Universities, are all encouraged to find means of creating and preserving similar user friendly databases.

5.ACKNOWLEDGEMENTS

The DTI CARAD (*Civil Aviation Research and Demonstration*) Directorate is acknowledged for funding the preparation of the index and database.
Mr Jon Ralph is thanked for help in the initial preparation of the Index. Mrs Sandra McCarthy is acknowledged for data entry into the database. Dr T.B. Gibbons and Mr P McCarthy are thanked for constructive discussion through out the work.

6. REFERENCES

[1] Tapsell H.J. & Bradley, J. (1925) *Mechanical tests at high temperatures on a non-ferous alloy of nickel and chromium.* Engineering, 120, 614-615, 648-649, 746-747

[2] Tapsell H.J. (1931) *Creep of Metals.* Oxford University Press, London.

[3] Loveday,M.S. (1982). '*High temperature and time-dependent mechanical testing: an historical introduction.*' Chapter 1 , pp1-12. in *'Measurement of high temperature mechanical properties of materials.'* Ed M.S.Loveday, M.F.Day & B.F.Dyson. Pub. HMSO, London.

[4] Loveday, M.S., Prince,D & Ralph, J.,(2000)'*An Introduction to the NPL Creep Database.*' NPL Measurement Note CMMT(MN) 062. Pub: National Physical Laboratory.

[5] Osgerby,S & Loveday,M.S., (1992)'*Creep Laboratory Manual*', NPL Report DMM(A) 37.

[6] Loveday, M.S., (1996) '*Creep Testing; Reference Materials and Uncertainty of Measurement*' pp277-293 in '*The Donald McLean Symposium- Structural Material : Engineering Applications Through Scientific Insight.*' Ed: E.D.Hondros & M. McLean, Pub : Inst. Materials, London.

[7] Bullough, C.K, Calvano,F. & Merckling, G. 1999 '*Guidance for the exchange and collation of creep rupture, creep strain-time and stress relaxation data for assessment purposes.*' ECCC-WG1 Recommendations , Vol. 4. Brite-Euram Thematic Network 'Weld-Creep', BET2-0509.Available from ERA Technology, Leatherhead, Surrey, UK.

[8] Over, H.H., 2000 '*Thermo-mechanical fatigue data handling within Alloy-DB*'. Paper 6, issued in abstract form. HTMTC (ESIS TC11) Seminar '*The Practicalities of Thermo-mechanical Fatigue Testing in the New Millennium*', Darmstadt, Germany. Ed : P. McCarthy. Also see http://matdb.jrc.nl/vanalles/overview

LOW ALLOY STEELS I

Creep Fatigue Problems in the Power Generation Industry

R. Viswanathan PhD, FASM, FASME
Electric Power Research Institute, California, USA
H. Bernstein PhD, FASME
Southwest Research Institute, San Antonio, Texas, USA

Abstract

Creep-fatigue damage induced by thermal stresses is of major concern with respect to the integrity of many high temperature components. The concern has been exacerbated in recent years due to cyclic operation of units originally designed for base load service. Much of the past research has been aimed primarily at crack initiation phenomena and, although useful from a design point of view, it is not always relevant to plant operations who in many instances can run components containing cracks. In terms of both crack initiation and crack growth prediction, variations in material, temperature environment, stress state, etc. have made it impossible to apply a single damage rule for all cases. The need for component-specific life prediction using appropriate material property data generated under conditions relevant to the service and using the proper failure criterion, has become very apparent. In the face of this need, thermomechanical fatigue (TMF) testing, creep-fatigue crack growth testing, and bench marking against field experience is essential. This paper will assess the current state of the art with respect to creep-fatigue life prediction especially with a view to provide a plant user's perspective to the research community, and to present a case study on TMF life prediction of combustion turbine blades.

1.0 Introduction

Many utility power companies in the U.S. are aggressively preparing themselves to face the inevitable transition from being regulated monopoly companies to de-regulated free market competitors. Reducing the cost of power production is paramount for staying competitive in the new scenario. Reducing capital costs by deferring replacement of expensive components and reducing operating and maintenance (O&M) costs by optimizing operation, maintenance and inspection procedures will both be key strategic objectives for utilities. This poses a significant challenge to the technical community since two apparently opposing needs will need to be reconciled. On the one hand, the need for improved plant efficiency and availability will dictate more severe duty cycles such as increased cold starts, fast starts, load cycling and load following, all of which result in more severe creep-fatigue damage and warrant increased attention to the components. On the other hand, the need to reduce O&M costs may result in fewer, shorter and lower quality maintenance and inspection outages; thus, placing the components at greater risk of failure. The challenge to the technical community, therefore, is to develop tools and techniques that will permit <u>more rapid</u>, <u>cost-effective</u> and <u>accurate</u> assessment of condition of critical components, both off-line and on-line. In addition to assessing the current condition, these tools must also be capable of evaluating the impact of alternative strategies for operation, inspection and maintenance. It is crucial therefore that the creep-fatigue research community be more intimately familiar with the specific needs of the industry.

Table 1 is a sample list of fossil plant components in which creep-fatigue has been a dominant failure mode. The list is by no means complete, since many components not included here may also become subject to creep-fatigue if more severe cycling conditions were imposed upon them. The purpose of the table is to make several key points as follows:

(1) Creep fatigue damage is generally the result of thermal stresses induced by constraint to thermal expansion during transient conditions. The constraint may be integral such as in the case of heavy section components (e.g., rotors, headers, drums, casings) where thermal gradients arise between the surface and the interior or vice versa. Internal constraint may also arise from internal cooling of components subject to rapid surface heating such as in combustion turbine blades. The constraint may be external such as in the case of joining of thick sections to thin sections or of materials of different coefficients of thermal expansion (dissimilar metal welds). Since stresses are always thermally induced, and since crack initiation occurs in less than 10^3 cycles, this form of creep fatigue damage is also referred variously as thermal fatigue, thermomechanical fatigue and low-cycle fatigue.[*] These terms will be used interchangeably in this paper.

(2) This form of creep fatigue damage may involve large plastic strains achieved locally at stress concentrations such as rotor grooves, header bore holes, etc. It may also involve primarily elastic strains combined with stress relaxation, as occurs for combustion turbine blades.

(3) Table 1 also shows that the industry view of what constitutes failure is different for stationary components such as headers and casings and rotating components such as blades/rotors. In the former case cracks are tolerated and crack initiation is believed to occur early (10-20% life) in life-component retirement is therefore based on economics of repeated repairs or growth of a crack to a critical allowable size. Hence, excessive concern with refining the damage rules is unwarranted in such cases. In rotating components, such as CT blades and rotor grooves, crack initiation defines failure since upon crack initiation other failure modes such as high cycle fatigue may intervene and cause rapid failure. In these cases, more refined prediction of damage evolution and crack initiation would be useful.

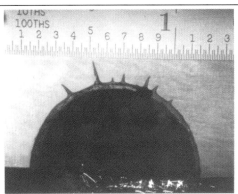

Figure 1. Thermal fatigue cracking at boreholes of a boiler header.

An example of thermal fatigue cracking in a superheater outlet header is shown in Figure 1. Cracks initiate in the tube bore holes and are oriented parallel to the axis of the tube bore hole. Linking up of cracks between holes on the inside surface of the header leads to propagation to form cross ligament cracks. Presence of ligament cracking has been observed in a very large number of superheater headers in the U.S.

The origin of thermal stresses arising from stop-start transients in heavy section components can be illustrated by a brief discussion of the rotor surface groove cracking[1]. Figure 2 illustrates a typical but simple cycle for a rotor in which a major load

[*] Damage to low temperature headers and drums is purely from thermal fatigue with no creep damage component.

increase occurs, followed by steady operation at the high load and then by a major load decrease. The load variation with time is shown in Figure 2(a). The patterns of temperature variation at the surface, middle, and bore of the rotor are shown in Figure 2(b). The rotor surface stress varies with time, as shown in Figure 2(c). The surface first tries to expand but is held in check by the bulk of the rotor, resulting in compressive stresses at the surface. If the load increase is sufficiently severe, compressive yielding occurs so that a residual tensile stress results when the loading cycle is completed. During steady operation, the residual tensile stress relaxes to a degree that depends on the temperature and time of operation at the steady load. When the load is decreased, the rotor surface goes into tension. This tensile stress is superimposed on the residual tensile stress. If tensile yielding occurs during a load decrease, a residual compressive stress results. This stress will not relax appreciably, however, because the temperature has reached a low value by now. The surface thermal strain variation with time is shown in figure 2(d).

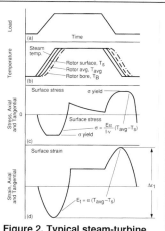

Figure 2. Typical steam-turbine load-change cycle, showing variations in temperature, stress, and strain with time.

During repetition of this simple cycle, at least three damage mechanisms can be operative: (1) fatigue due to the repeated cycles imposed by the strain range; (2) creep damage during stress relaxation at high temperature; and (3) creep damage under the steady operating loads. The extent of damage depends on the strain range, the frequency of cycling, and the time and temperature under steady loading conditions. The crack initiation life under these conditions is predicted using one of several damage summation rules. If crack growth is permitted, available crack growth rate data is utilized to calculate if the initial crack size will reach the critical crack size prior to the next inspection.

2.0 Crack Initiation Life Prediction by Calculation

As discussed earlier, crack initiation life prediction is crucial in the case of components where crack initiation is life limiting, e.g., combustion turbine blades. In addition, calculation based approaches are also useful for screening purposes for prioritization of NDE, knowing where to look, and in those instances where access to the component is limited. In the calculational procedure, information is at first gathered regarding the temperature and cycling history from the plant records. By use of standard material properties and damage rules, the fractional life expended up to a given point in time can be estimated. Unfortunately, historical operating data are usually not available in sufficient detail. Uncertainties in material behavior and deficiencies in the damage rules further compound the problem leading to difficulties in accessing the accuracy of the predictions.

The principal damage rules/methods that are applied to express creep-fatigue damage are: Linear damage summation method [2], the frequency modified Coffin method [3], the strain range partitioning method [4], the Ostergren damage function [5], ductility exhaustion [6], Bicego's energy criterion [7], ductility normalized inelastic strain range [8], damage rate method [9], and the general damage function [10].

The damage rules vary essentially in terms of the fatigue parameter related to fatigue life, i.e., stress range, strain range (plastic vs total) or a combination of stress and strain ranges. A vast amount of literature exists comparing the effectiveness of these damage rules in predicting LCF life expenditures and has been reviewed elsewhere [11,12].

The most common approach in the U.S. is based on linear superposition of fatigue and creep damage. This approach combines the damage summations of Milner for fatigue and of Robinson for creep as follows:

$$\sum \frac{N}{N_f} + \sum \frac{t}{t_r} = D \tag{1}$$

where D is the cumulative damage, as defined by the bilinear damage curve of Figure 3, N/N_f is the cyclic portion of the life fraction, in which N is the number of cycles at a given strain range and N_f the pure fatigue life at that strain range. The time-dependent creep-life fraction is t/t_r, where t is the time at a given stress and t_r is the time to rupture at that stress. The stress-relaxation period is divided into time blocks during which an average, constant value of stress prevails, and for each time block t/t_r is computed and summed. The creep damage occurring under constant stress subsequent stress relaxation is also included in the second term using the same life fraction approach. In the alternative approach, the stress relaxation damage is included in the fatigue term by using fatigue curves that incorporate the effect of hold times, instead of using the pure fatigue curves and value of D is assumed to be unity at failure. The above two procedures are based on the "inelastic" and "elastic" routes described in the ASME Code Case N47 [13].

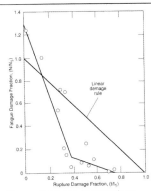

Figure 3. Creep-rupture/low-cycle-fatigue damage interaction curve for 1Cr-Mo-V rotor steel at 540°C (1000°F).

The linear damage rule is purely phenomenological, having no mechanical basis. its applicability is, therefore, dependent upon the available data base for a given material and how well that data base represents the application to be predicted. Contrary experience, it also assumes that tensile and compressive hold periods are equally damaging. The strain softening behavior encountered in many steels and the effect of prior plasticity on subsequent creep are not taken into account. Use of virgin-material rupture life to compute creep-life fractions is, therefore inaccurate. In spite of these limitations, the damage-summation method is very popular because it is easy to use and requires only standard fatigue and stress-rupture curves.

All of the creep-fatigue damage rules, including the linear environment and testing variables, all of which can be difficult to quantify. Thus, the ability to accurately predict life beyond the range of the experimental data base can be quite difficult. The following review of these uncertainties will serve to highlight the complexity of life prediction.

2.1 Scatter in Material Properties

A key source of uncertainty in life prediction is the scatter in the fatigue and creep rupture properties. The scatter can arise due to heat treatment, chemistry, cleanliness, porosity, and other fabrication variables. The greatest amount of scatter is often from vendor to vendor, even when they use nominally the same material.

2.2 Effect of Rupture Ductility

There is ample evidence to show that rupture ductility has a major influence on LCF life [12,14]. Because this effect is believed to be caused by the influence of rupture ductility on the creep fracture component, endurance in continuous cycle and high frequency or short hold time test will be relatively unaffected. Only in long hold times, where the strain rates reach values sufficiently below a critical strain rate, creep cavitation phenomena become operative and the effect of rupture ductility on LCF life becomes pronounced. This behavior is illustrated from the work of Kadoya et al. for a rotor steel in Figure 4 [14]. In Figure 4 the hold-time effects on the LCF lives of two rotors differing primarily in rupture ductility are compared. With decreasing ductility, the hold-time effect becomes increasingly pronounced. This factor is often not taken into account in conducting the appropriate tests and leads to inaccuracies in life prediction using damage rules.

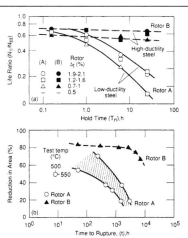

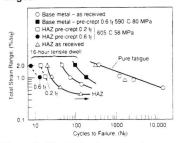

Figure 4a, b. Effect of rupture ductility on hold time effects during creep-fatigue.

2.3 Effect of Prior Degradation

The nature and extent of prior degradation seems to exert an influence on subsequent LCF behavior. This factor is hardly considered in most laboratory studies, which use virgin material tests to predict fatigue life consumption. Borden et al. [15] investigated the creep-fatigue behavior of samples which had been previously testing in creep to life fractions of 0.2 and 0.6. Their results (Figure 5) show that in simulated heat-affected-zone material where prior creep had resulted in cavitation damage, the prior creep greatly affected decreased cyclic life. On the other hand, in the base material of the 1Cr-1/2Mo steel, where prior creep damage consisted merely of softening, subsequent low-cycle fatigue behavior was actually improved in comparison with the non-precrept samples. Softening can lead to increased ductility and hence to improved low-cycle fatigue life. The nature of creep damage thus plays a key role in subsequent LCF damage. In contrast, Kimura, Fujiyama and Muramatsu [16] report decreasing fatigue life due to prior softening and have quantified the effect of hardness on fatigue life of CrMoV rotor steels.

Figure 5. Effect of prior creep damage (to 0.2, 0.5 life fractions) and tensile hold time on the cyclic fatigue life of 1Cr-0.5Mo steel at 535°C.

2.4 Effects of Environment

The possible roles of environment in affecting fatigue life are too numerous to describe. The more important ones include formation of oxide notches, grain-boundary embrittlement, increase of net section stress, corrosion-product wedging, and shielding. The first four are fairly obvious and can cause reductions in fatigue life. The last effect, shielding, is described as a net reduction in the effective strain range or stress-intensity range due to corrosion products which in some instances can result in decreased crack growth and increased life. Where detrimental effects occur, it is to be expected that longer hold times or lower frequencies during testing will display the environmental effects more prominently.

Detrimental effects resulting from oxidation in air also have been reported for Udimet 500, type 304 stainless steel, type 316 stainless steel, Hastelloy, and wrought IN-738 LC. Reduction in fatigue life due to hot corrosion of Udimet 710 and 720 is well documented [11].

While detrimental effects due to air oxidation and hot corrosion have been reported in a number of alloy systems, the effects of environment are not always straightforward. Harrod and Manjoine [17] claim to have found no difference in fatigue behavior of type 304 stainless steel tested in air and in vacuum at 650°C (1200°F). Gell and Leverant [18] showed air to be actually beneficial to the fatigue life of MAR-M 200 at 910°C (1670°F) in comparison with vacuum, presumably because of the oxide shielding effect. Taking credit for beneficial effects of shielding in actual applications can, however, be dangerous, because one cannot ensure that the oxide products will remain intact during service as they do in laboratory tests. The shielding effect also has been found to be beneficial and to lead to increasing threshold stress intensity for crack propagation with decreasing test frequency for a 1Cr-Mo-V steel at 550°C (1020°F) [19]. Once the threshold K value was exceeded, however, the air environment caused accelerated crack growth. In cast Inconel 738, an air environment has been found to be beneficial because of crack branching along oxidized dendritic boundaries, thus resulting in reduced crack growth [20]. The effects of environment are thus found to be complex and to vary with the material, environment, temperature, and test frequency.

2.5 Effect of LCF Cycle Type

The need for choosing the relevant type of strain cycle has not always been realized in the course of many laboratory investigations and has resulted in inaccurate LCF life predictions. This point can be illustrated using the results of Thomas and Dawson [21]. They compared cyclic lives under two types of strain cycles: the laboratory-type cycle, in which the hold time is normally imposed at the maximum strain in the tensile cycle; and a type II cycle in which the hold time is imposed at the zero strain. In the laboratory-type cycle, the maximum tensile stress occurs at the start of the hold period, while in the type II cycle the tensile stress at the start of the hold period depends on the extent of the yielding during the previous compressive part of the cycle. In addition, the hold period is followed by further tensile strain, rather than by a strain reversal.

The two types of cycles and the corresponding fatigue data are illustrated in Figure 6. the type II cycle is more akin to the actual strain cycles expected on the surface of a HP rotor. The data show that for laboratory cycles, at 550°C (1020°F), increasing hold time progressively decreases fatigue life at strain ranges above about 0.4%. At lower strain ranges, all the fatigue curves converge at the same values as those of the 0.5-h hold-time curve. At 500°C (930°F), similar trends are apparent, although the actual convergence of the various hold-time fatigue curves is indicated at lower strains

For the type II service cycle, the effect of a 0.5-h hold period on endurance at both 550°C and 500°C is negligible. The 16-h hold period has a very small effect at low strain ranges; at high strain ranges, the effect increases but is still less than for laboratory cycles.

These data clearly point out the pitfalls of using unrealistic laboratory tests in life prediction of components and the need for simulating the component strain cycles in the laboratory tests in order to generate the appropriate data. The laboratory test produces a conservative life prediction by overpredicting the damage caused by the hold period. While adequate for the design purpose of a safe component, this procedure is inadequate for the plant operator who is trying to obtain as much life as possible out of his equipment.

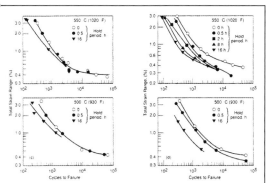

Figure 6. Types of LCF cycles employed, and corresponding endurance data, for a 1Cr-Mo-V rotor steel.

For nickel-base superalloys, it is well known that fatigue cycles with compressive hold times are far more damaging than cycles with tensile hold times. This behavior is thought to be due to the tensile mean stress that develops during the compressive hold time test.

2.6 Non-Relevance of Isothermal LCF Tests to TMF

In most instances of fatigue, the temperature varies along with the strain, giving rise to what is known as thermomechanical fatigue. Two simple waveforms in TMF testing are shown and compared with the LCF test in Figure 7. If maximum temperature corresponds to peak compression, as in the center diagram in the Figure, it is known as out-of-phase cycle (OP). if the maximum tensile stress occurs at the peak temperature, it is known as an in-phase (IP) cycle. Depending further on when the hold time is superimposed, various cycle shapes are possible.

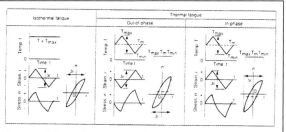

Figure 7. Schematic diagrams showing waveforms of temperature, strain, and stress in thermal and isothermal fatigue tests.

In the past, thermal fatigue traditionally has been treated as being synonymous with isothermal low-cycle fatigue at the maximum temperature of the thermal cycles. Consequently, life-prediction techniques have evolved from the iso-thermal low-cycle-fatigue literature. More recently, advances in finite-element analysis and in servohydraulic test systems have made it possible to analyze complex thermal cycles and to conduct thermomechanical fatigue (TMF) tests under controlled conditions that simulate these complex cycles. The assumed equivalence of isothermal LCF tests and TMF tests has been brought into question as a result of a number of studies.

High tensile strains at high temperature (IP) would favor creep, whereas high tensile strains at low temperature (OP) would favor cracking of oxide and hence accelerated environmentally induced damage during subsequent high-temperature exposure. Hence, Kuwabara et al. Rationalized that in case of materials where damage is driven by creep, IP cycles would be more damaging than OP cycles, for a given strain strange [22]. In other materials, where the environmental contribution is significant, OP cycles may be more damaging than IP or isothermal LCF cycles. In addition to environmental effects, differences also arise between cycles in terms of the relaxed mean stresses. The relative severity or the different cycles can also change with material ductility, maximum temperature and hold time. Consequently, a simple classification of material behavior is not possible.

A case in point is the ligament cracking encountered in CrMo steel header pipes illustrated in Figure 1. The cracking mode has been identified as thermal fatigue. A computer code, Boiler Life Evaluation and Simulation System (BLESS) developed recently, incorporates two alternate approaches for predicting crack initiation; one involving an inelastic linear damage summation method, and a second approach involving repeated cracking of oxide scale and oxide notching [23]. For a variety of cycle histories, the Code predicts crack initiation occurring in short times by the oxide cracking mechanism. The creep-fatigue damage summation approach, on the other hand, is inconsistent with the early initiation of cracks observed in headers. Metallography of cracked headers has shown numerous oxide spikes, see Figure 8 indicating oxide cracking to be the crack initiation mechanism. This example clearly illustrates the need for suing appropriate thermomechanical fatigue data simulative of actual component cycles in predicting crack initiation life of components.

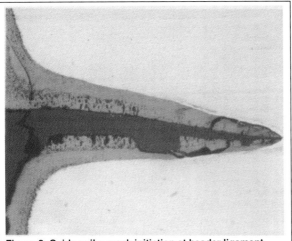

Figure 8. Oxide spike crack initiation at header ligament.

Another example of the critical need for TMF data is in the case of protective coatings. In the case of coated components such as combustion turbine blades, cracking of the coating leads to loss of environmental protection from the coating, and, eventually, to cracking of the base metal. The integrity of the coating depends upon both the ductility of the coating and the strain-time history of the blade, as shown in Figure 9. The strain-to-cracking of the coating is a strong function of temperature, often given by a "DBTT" curve. The strain-temperature cycle in the engine must lie below the DBTT curve. For the example shown in Figure 9 coating A will not crack under normal duty, but will crack during an emergency shutdown. However, coating B will not crack under either condition. In view of the DBTT behavior exhibited by coatings, it becomes even more critical in evaluating coated components that thermo-mechanical fatigue tests be performed. Furthermore, simple TMF cycles, in which the maximum tensile strain is made to coincide with either the peak temperature (in-phase, IP) or with the lowest temperature (out-of-phase, OP) in the cycle, will lead to unrealistic results. If one were to compare the performance of two coatings A and B by conducting an isothermal LCF test or an IP type thermomechanical fatigue strength, one would conclude that both coatings perform well, since both coatings have high strain capability at high temperature. If the same comparison were made under more relevant TMF conditions, coating B would be chosen over coating A. Hence, TMF cycles simulative of actual blade cycles must be performed to evaluate the effect of the coating. Such test data on coated components is extremely scarce in the open literature, and is limited even in proprietary data bases.

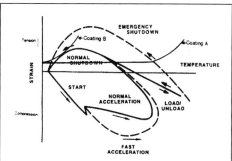

Figure 9. Typical thermomechanical cycle for a first-stage blade, showing leading-edge strain and temperature variations for normal start-up and shut-down.

This type of TMF data is of importance to plant operators, because they often have a choice of coatings to apply during the refurbishment of their blades. To assist plant operators in these choices, EPRI has developed an expert system computer program called Advisor on Blade Coatings, ABC.

2.7 Summary Evaluation of Life Prediction Methods

Detailed review of literature shows that there are divergent opinions regarding which damage approach provides the best basis for life prediction. It is quite clear that a number of variables, such as test temperature, strain range, frequency, time and type of hold, waveform, ductility of the material, and damage characteristics, affect the fatigue life. The conclusions drawn in any investigation may therefore apply only to the envelope of material and test conditions used in that study. The validity of any damage approach has to be examined with reference to the material and service conditions relevant to a specific application. Broad generalizations based on laboratory tests, which often may have no relevance to actual component conditions, do not appear to be productive. Thus, one should use a tailored, case-specific approach for any given situation.

One of the major problems in evaluating the applicability of different life-prediction methods is that in many cases it is necessary to use all the available data in deriving the life-prediction method, and thus it is not possible to examine the accuracy with which a given method describes data not used for the development of the method, or outside of the range of conditions, or lives, considered. There also is a scarcity of instances in which service experience has been compared with specific life-prediction methods. In general, the available methods are utilized only to predict the lives of samples tested under laboratory conditions. Validation against component test data in the laboratory and in-service monitoring of actual equipment would lead to more confidence in the use of the various results.

Results from most studies how that even the best of the available methods can predict life only to within a factor of 2 to 3. Some of the cited reasons for these inaccuracies have already been discussed. Some additional reasons are: failure of the methods to model changing stress-relaxation and creep characteristics caused by strain softening or hardening, use of monotonic creep data instead of cyclic creep data, and lack of sufficiently extended-duration test data. All of the damage rules available today are at least partly, if not totally, phenomenological in nature. They all involve empirical constants that are material-dependent and difficult to evaluate theoretically. Extrapolation of the rules to materials and conditions outside the envelope covered by the specific investigation may result in unsuccessful life predictions and usually result in predictions whose accuracy is difficult to evaluate. One form of extrapolation that is especially difficult to evaluate is the need to use short time data for long time service. For components whose service time is from 3 to 30 years and which may operate continuously for hundreds to thousands of hours, use of test data—which lasts as much as 4 months (1/3 of a year) and has hold times as much as 16 hours—requires extrapolation of one to two orders of magnitude. The statistical methods chosen to make these extrapolations significantly affect the allowable lives of the components. Furthermore, in performing these accelerated tests, either the strain, the temperature, the frequency, or some combination must be increased over actual service conditions in order to produce failure in a reasonable amount of time. Thus, there is always the danger that the physical mechanism of failure in the laboratory test is different from that during service, or that important aspects of the service conditions are not considered during the laboratory tests. For application to service components, the stress-strain variation for each type of transient and its time dependence must be known with accuracy. The importance of using relevant TMF data cannot be overemphasized. This realization has led to several recent studies in life prediction of combustion turbine components using TMF based algorithms.

3.0 TMF Life Prediction—An EPRI Case Study

The following is a case study on TMF life prediction for first stage combustion turbine blades made of INCO 738 LC alloy, both in the coated and uncoated condition as reported by Bernstein et al. [24].

The first step in the project was to generate strain range versus number of cycles to failure data. Specimens were made from hollow bars which had been case, hot isostatically pressed and heat treated to produce a microstructure similar to that of actual blades. They were then tested at the General Electric Company using servohydraulic test machines under strain control. The test program comprised three groups of experiments: isothermal low cycle fatigue (LCF), linear out of phase thermomechanical fatigue (LOP) and "bucket" thermomechanical fatigue simulative of the leading edge of GE Frame 7E turbines, as shown in Figure 9.

As shown in Figure 9, compressive peaks occur after ignition, during acceleration, and at steady state. The tensile peak occurs during shutdown. The ignition and acceleration compressive peaks occur at low and intermediate temperatures while the steady-state (hold time) compressive peak occurs at maximum temperature. The shutdown peak occurs at intermediate temperature. Thus, the linear-out-of-phase TMF cycle best simulates the blade TMF cycle among the simplified cycles that have been studied.

It can be seen from Figure 9 that the magnitude of the strain cycle depends on how the machine is started and shut down. The normal mode of operation is a normal startup, normal acceleration, normal loading to base load, fired shutdown path. The fast start option increases the strain range somewhat. This option is for emergency demands for electricity and is therefore seldom used. The full load trip (or emergency shutdown) increases the strain range significantly and is a major reason for premature TMF cracking in service. Full load trips occur for a variety of reasons, such as excessive vibration, temperature limits being exceeded, by sudden disconnection from the electrical grid, etc.

3.1 Test Details

Isothermal low-cycle fatigue (LCF) tests were conducted at 760, 871, and 982°C (1400, 1600, and 1800°F). The tests were run with compressive holds of 0, 0.33, 2, and 15 min. Total strain ranges varied from 0.2 to 0.8%, and the cycling frequency was 0.33 Hz. Strain amplitude ratio $A = \varepsilon_{amp}/\varepsilon_{mean}$ was held constant at -1. All but three specimens were coated.

Thermo-mechanical fatigue (TMF) tests were on linear out-of-phase (LOP) cycles in which the maximum strain and minimum temperature occurred simultaneously. This cycle is representative of industrial gas turbine buckets in which the highest temperatures occur while the surface is in compression. The minimum temperature was held constant at 427°C (800°F) and the maximum temperature was either 871, 916, or 982°C (1600, 1680, or 1800°F). Compressive hold times of 0, 2, and 15 min were applied with both coated and uncoated specimens. The strain range varied from 0.4% to 0.8% with $A = \infty$ in most cases. Four tests were conducted at $A = -1$ to investigate the significance of mean strain effects.

The "bucket" thermal-mechanical fatigue tests followed a strain-temperature history representative of the leading edge of the first stage bucket in service. This bucket history was developed by GE from their component analysis and was very similar to that shown in Figure 9. Tests were carried out at maximum temperatures of 916 and 960°C (1680 and 1760°F), which are in the range of application. The hold time was held constant as a 2-min compressive dwell (at maximum temperature), and the strain amplitude ratio was maintained at $A = -2.3$. The only free parameters in these tests were the strain range and the presence or absence of coatings.

Life prediction modeling focused on parametric analysis of the test data using a strain-based approach. To accomplish this task, total strain ranges versus fatigue life were plotted holding all but one parameter constant. This facilitated identification of the most important factors to be included in a life prediction model. The total strain range was used because the cyclic inelastic strains are negligible in the range of application, and inelastic strain was reported as zero for a significant fraction of the test data. Furthermore, the prediction of small inelastic strains is difficult with potentially large errors and scatter, which will lead to similar, if not larger, errors and scatter in the life predictions. The failure life N_f corresponded to 50% load drop.

3.2 Effect of TMF Test Variables

The five parameters investigated, apart from strain range, were the compressive hold time, the test temperature, the presence or absence of a coating, the strain A-ratio, and the cycle type—isothermal, LOP and bucket. The effects of these parameters are considered on the basis of the total strain range only, because of the difficulty in accurately determining the inelastic strain range.

556

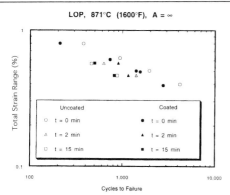

Figure 10. LOP test results at T_{max} = 871°C (1600°F) and A = ∞, showing the effect of compressive hold time and coatings.

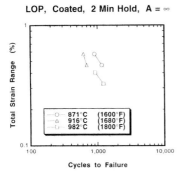

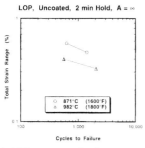

Figure 11. LOP tests results for a compressive hold time of 2 min. and A = ∞, showing the effect of temperature: (a) coated specimens and (b) uncoated specimens.

The effect of compressive hold time upon fatigue life in the LOP tests in shown in Figure 10 for both coated and uncoated specimens. With the exception of one test, there was a clear decrease in life as the hold time increased. Similar decreases in life with hold time were observed for the LOP tests at 982°C (1800°F) and the isothermal tests at 871°C (1600°F). This hold time behavior is similar to that of most nickel-base superalloys.

The effect of temperature upon the fatigue life in the LOP tests with a 2-min hold is shown in Figure 11a for coated specimens and Figure 11b for uncoated specimens. The life at 982°C (1800°F) was consistently shorter than at 871°C (1600°F) for both the coated and uncoated specimens. Similar behavior was seen for the uncoated LOP tests with no hold time. The isothermal tests of coated specimens with $A = -1$ and a 2-min hold time showed decreases in life as the temperature was increased from 760 to 871 to 982°C (1400 to 1600 to 1800°F). For the tests in Figure 11a, the lives at 916°C (1680°F) appeared to be similar to those at 982°C (1800°F). There also was no difference in the fatigue life between the 916 and 960°C (1680 and 1760°F) bucket tests for both coated and uncoated specimens. Thus, increasing temperature decreases fatigue life when considered on a total strain range basis. However, for temperatures that are within about 45°C (80°F) of each other, limited test results may show little difference due to inherent material scatter.

The effect of coatings upon the fatigue life is more difficult to interpret because the inner surface of the hollow specimen was not coated. Furthermore, the tensile properties of the coated specimens showed somewhat greater ductility and lower strength than the uncoated specimens. The reason for this behavior is not known. As shown in Figure 10 for LOP tests at an A-ratio of infinity, the coating reduced the life for zero hold time tests at both 871 and 982°C (1600 and 1800°F). For 2- and 15-min holds, the coated

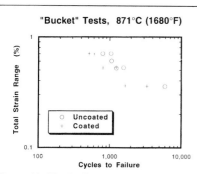

Figure 12. "Bucket" test results at 916°C (1680°F) showing the effect of coatings.

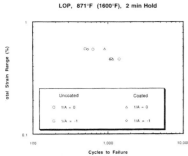

Figure 13. LOP test results at Tmax = 871°C (1600°F) and a compressive hold time of 2 min showing the effect of strain A-ratio for both coated and uncoated specimens.

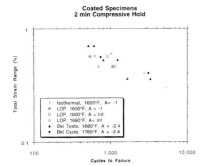

Figure 14. Test results on coated specimens with a 2-min compressive hold time showing the effect of different cycle types.

and uncoated specimens had similar lives. For the bucket specimens, which had a 2-min hold time, the coating caused a significant reduction in the fatigue life of the specimens, as shown in Figure 12 for the 916°C (1680°F) data. The effect of coatings on the bucket tests at 960°C (1760°F) was similar. For the TMF tests, the general conclusion is that the coating acts to reduce the life.

Because the strain A-ratio varied in the test program and in the application, the effect of strain A-ratio on the fatigue life was considered. As shown in Figure 13 for the LOP tests with a 2-min hold at 871°C (1600°F), there was a small but consistent decrease in life for A = -1 compared to A = ∞. Additional data spanning a larger range in A were not available.

The effect of cycle type is somewhat difficult to distinguish in the data because each cycle type was run at a different combination of temperature and A-ratio. The data for coated specimens with a 2-min compressive hold are shown in Figure 14 for the isothermal, LOP, and bucket cycles. It can be seen that the isothermal tests gave the longest life, and the LOP tests gave the shortest life, with the bucket cycles generally being intermediate between the two.

3.3 TMF Life Model

These observations led to the development of a model to predict fatigue life. The model was optimized to those conditions of interest to the fatigue of industrial gas turbine blades: a few well-defined cycles of the LOP type; a relatively narrow range of temperature, strain range and A-ratio; and the presence of coatings. The more difficult problem of a general description of thermal-mechanical fatigue for LIP, LOP, and isothermal cycles, was outside the scope of the modeling effort, as was the description of the effect of temperature upon life. The goal of the effort was to model the fatigue life based upon the independent variables without complicating the analysis by the use of constitutive equations

558

to determine either inelastic strains or mean stresses. (As stated earlier, such analysis, which is in many respects more difficult and less certain than life analysis, introduces additional error and uncertainty in the results.) The independent variables used in the model are those determined from the elastic finite-element analysis.

The model developed for bucket fatigue is based upon strain range, strain amplitude ratio, and dwell time. This combination of parameters is intuitively reasonable, since the strain amplitude ratio can account for mean strain effects, and the dwell time can account for both environmental effects and mean stress effects caused by steady-state stress relaxation, both of which can effect the fatigue life. The general form of the model is

$$N_f = C_0 (\Delta \varepsilon)^{C_1} (t_h)^{C_2} \exp\left(\frac{C_3}{A}\right) \qquad (2)$$

where N_f is the fatigue life, A is the strain A-ratio, $\Delta \varepsilon$ is the total strain range (expressed in percent strain), and t_h is the dwell time (in minutes). Two equations for fatigue life were developed, one for coated and one for uncoated specimens. The empirical constants C_1 in Equation 2 were determined from multiple regression of the LOP data at 871°C (1600°F), and are given in Table below. As shown in Figure 15, the model adequately described the behavior of IN-738LC for the LOP test conditions at 871°C (1600°F). This model was then used to predict the results of the bucket cycle tests, which had *not* been utilized to generate the model constants. Figure 15 shows that the model did indeed adequately predict the bucket cycle lives for the coated specimens. Similar results were obtained for the uncoated specimens.

Table 2
Constants in Equation 2

Type	C_0	C_1	C_2	C_3
Coated	125	-3.40	-0.217	0.247
Uncoated	171	-3.45	-0.381	0.388

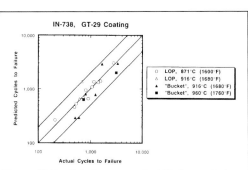

Figure 15. Predicted versus actual life for coated specimens. The linear out-of-phase tests at 871°C (1600°F) were correlated, and the other tests were predicted.

3.4 Benchmarking TMF Model Against Field Experience

The TMF model presented above was used to predict occurrence of cracks at the leading edge of several first stage buckets of General Electric Frame 7E machines which had been operating in base load and intermediate duty cycles. The engines had run at turbine inlet temperatures of 1085 to 1104°C (1985 to 2020°F) mostly on natural gas. The engine history obtained from the field consisted of number of starts, fast starts, trips and fired hours. The total numbers of fired hours was divided by the number of starts to determine the average hold time. The strains and temperatures at the leading edge of the buckets for each cycle type based on finite element analysis was obtained from General Electric. A linear damage rule (Equation 1 but without t/t_r) was used to sum up total damage for different cycle types and a total damage ratio, D, of unity was considered as 100% life consumption, corresponding to crack initiation. The predictions from calculations were then compared with the actual metallographic visual and fluorescent dye penetrant inspection results.

The predictions of the field data are shown in Figure 16 as the percentage of life consumed versus the number of starts. The predictions account for all cycle types and dwell times. Below 100% life, cracks are not expected to be present. Above 100% life, cracks are expected in the leading edge of the blade. As the percentage of life consumed becomes greater than 100%, the extent of cracking is expected to be greater, but the model does not make a quantitative prediction of the crack length. Around 100% life, cracks are expected to be present. Shown in Figure 16 are the field data and the nature of the cracking observed. Cracks are defined for the purposes of this work as engineering size cracks of the order of 0.2 to 0.4 mm (10 to 50 mils) in depth. As can be seen in Figure 16, the agreement of the field data with the predictions was remarkably good. All points below the 100% line were not predicted to have any cracking, and no cracking was found. All points in which cracks were just beginning to form were predicted in 77 and 93% life consumed.[*]

4.0 NDE Methods for Damage Evaluation During Low-Cycle Fatigue

Reasonable progress has been made in quantifying low cycle fatigue damage in rotor steel by X-ray and hardness based techniques. Based on extensive fatigue tests interrupted at various life fractions, Goto found that the X-ray line width from diffraction studies decreases with increasing fatigue damage. This width normalized with respect to the X-ray line width for the steel in the undamaged condition, i.e. Hw/Hwo decreases with the number of cycles at various strain ranges as shown in

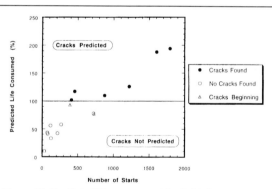

Figure 16. Predicted life consumed for field data and presence of cracks. At 100% life consumed, cracks are predicted to initiate.

Figure 17a. Based on the Hw/Hwo values at failure (N_c[†]), the "failure line" was defined as shown in figure 17b. The figure also illustrates the method for estimating the fatigue life fraction consumed, i.e. N/N_c, for field components operating under unknown strain range levels. Based merely on a knowledge of the number of cycles (e.g. start ups) and measured value of Hw/Hwo at the damage location, a point A is defined in the diagram. From this point, a line which has the same slope as the lines determined from laboratory experiments (Figure 17A) is drawn; the point at which this line meets the failure line defines the

---

[*] Cracks beginning to form refers to metallographic evidence where the coating is cracked and a crack has formed in a grain boundary of the base metal.
[†] Number of cycles at which surface cracks of 1 mm length were observed.

560

number of cycles to failure N_c. N/N_c, i.e. the fatigue life fraction can therefore be readily calculated. This method has been successfully applied by Goto et al. for estimating not only the fatigue damage but also the depth of damage at periphery grooves of numerous service rotors [25].

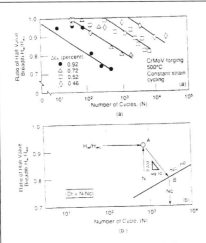

Figure 17. Method for estimating fatigue life fraction using fatigue life expended vs x-ray line width for CrMoV steel.

Goto also found that LCF damage in CrMoV rotor steels results in strain induced softening which can be measured in terms of changes in hardness [25]. The ratio of hardness in the undamaged condition to that in the damaged condition, Hv/Hv, was found to follow a trend similar to that shown in Figure 17. A methodology similar to the X-ray technique described above could be utilized to calculate the fractional fatigue life consumed.

However, similar NDE methods do not exist for other components or materials. In general, the development, and proper application, of these methods is non-trivial. For the case of coated combustion turbine blades, no such method exists.

5.0 Crack Growth Life Prediction

As described in Section 1, life prediction of heavy section components such as headers, drums, casings etc. involves crack growth analysis to determine if initially detected cracks of size a_i, will reach the critical size a_c, between inspections. The early phenomenology for describing crack growth under creep fatigue condition was proposed by Saxena, Williams and Shih [26]. In this method the crack growth, Δa, per cycle is expressed as the sum of three terms

$$\Delta a = \frac{da}{dN} = \left(\frac{da}{dN}\right)_o + C_2(K)^{2m}(t_h)^{1-m} + C_3(C^*)^m(t_h) \tag{3}$$

where t_h is the hold time, K is the elastic stress intensity (function of flaw geometry, a and stress), C^* is the crack tip driving force under creep conditions (function of stress, a strain rate and geometry). C_2, C_3 and m are experimentally determined constants. $(da/dN)_o$ is the amount of crack growth due solely to fatigue. Integration of crack size between the limit a_i to a_c, then yields the number of cycles N_f to cause failure.

The first term in Equation 3 is a pure-fatigue contribution reflecting no effect of hold time and corresponding to crack-growth behavior at short hold times and high frequencies. The second terms shows a nonlinear power-law dependence of crack-growth rate on hold time and pertains to intermediate hold times and frequencies where creep-fatigue interaction is present. The third term, containing C^*, shows a linear dependence of crack growth on hold time and corresponds to purely creep-dominated crack growth occurring at long hold times and low frequencies. Corresponding to these three terms, the plot of da/dN vs t should have three regions as has been experimentally confirmed in terms of frequency dependence of da/dN, shown in Figure 18 [27] for the iron-base alloy A-286. This figure shows that the crack-growth rate is a continuous function of frequency, even thought the actual mechanism varies from transgranular fatigue at one end of the spectrum to intergranular creep at the other end of the spectrum. Hence, no discontinuous changes occur in the crack-growth rates. At high frequencies, fracture is fatigue-dominated in air and in vacuum. At intermediate frequencies where pure creep processes begin to dominate, the date for air and for vacuum start to converge. In the intermediate region where the creep-fatigue interaction will be most pronounced, the effects due to environment also are most obvious. The presence of environment promotes the onset of intergranular fracture mechanisms at higher frequency levels.

Most recently, Saxena and coworkers have proposed a new model in which the second and the third term in Equation 3 are combined into a single term as follows [28].

$$\Delta a = \left(\frac{da}{dN_{Total}}\right) = C\Delta K^n + b\, C_{t(avge)}^m t_h \tag{4}$$

$$\text{where } C_{t(avge)} = \frac{1}{t_h}\int_0^{t_h} C_t dt \tag{5}$$

The $C_{t(avge)}$ was observed to correlate very well with the average crack growth rate during the hold time, expressed as

$$\left(\frac{da}{dt}\right)_{avge} = \frac{1}{t_h}\left(\frac{da}{dN}\right)_{hold} = \frac{1}{t_h}\left[\left(\frac{da}{dN}\right)_{Total} - \left(\frac{da}{dN}\right)_o\right] \tag{6}$$

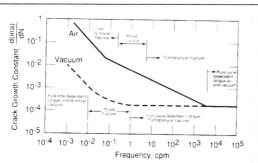

Figure 18. Comparison of crack-growth behavior of Fe-Ni alloy A-286 in air and vacuum at 595°C (1100°F).

The creep-fatigue crack growth rate data with various hold times ranging from 10 seconds to 24 hours for a 1.25Cr 0.5Mo header steel are shown in Figure 19 as a function of ΔK. The regression line representing the cycle-dependent crack growth rate is also shown as a dotted line on the same plot. As expected, the data for various hold times are scattered, because ΔK is not suitable for characterizing the time-dependent crack growth rate during hold period. For a fixed ΔK value, da/dN increased with hold time due to the increasing contribution of time-dependent crack growth.

Figure 19b shows all the creep-fatigue crack growth data from the same study (Figure 19a) plotted in terms of the $(C_t)_{avg}$ parameter. The da/dt vs C_t data obtained from creep crack growth experiments (CCG) are also included in the figure for comparison. It is clear from the figure that all the CFCG (Creep Fatigue Crack Growth) and CCG conditions can be expressed as a single trend if $(da/dt)_{avg}$ is characterized in terms of $(C_t)_{avg}$ or da/dt is characterized in terms of C_t. This is an important

562

conclusion for applications because it considerably simplifies the life prediction procedures. That is CCG and CFCG data can be used interchangeably for life prediction. This observation is material specific and may not be the case for other materials and conditions.

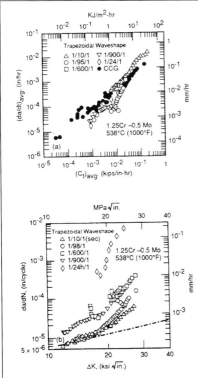

Figure 19. Comparison of creep-fatigue crack growth rates with (a) fatigue crack growth rate data plotted using ΔK; (b) creep crack growth data using $(C_t)_{avg}$.

6.0 Determination of Critical Crack Size

In heavy section components where crack growth is permitted, the ultimate failure is deemed to occur when the crack reaches a critical size. If final failure is envisaged to occur under steady operating conditions, the a_c corresponds to some critical dimension in the component, e.g., wall thickness, ligament, width, etc. Alternately a_c could also represent that crack size at which further crack growth rapidly accelerates. If failure is envisaged to occur at low temperatures under transient conditions, then a_c is derived from fracture toughness, K_{IC}, at the appropriate temperature and stress. Several non-destructive methods for estimating the in-service toughness of steel components have been described elsewhere [29,30].

7.0 Summary and Conclusions

Most common instances of creep-fatigue damage in fossil power plant components are caused by thermal stresses due to internally or externally imposed constraints to thermal expansion. In stationary components, cracks can often be tolerated thus allowing scope for inspection and crack growth based approaches. Refinement of creep-fatigue damage rules for crack initiation in such cases is not necessary and the development of improved crack growth methodologies would be more useful. However, in rotating components, crack initiation may result in rapid failure by the superimposition of other failure mechanisms such as high cycle fatigue. In such cases improved life prediction models based on damage concepts and crack initiation would be most desirable. Most of the current models based on damage concepts and crack initiation would be mos desirable. Most of the current models suffer from inadequate accounting for several material, environmental and testing variables. Most importantly, they fail to simulate the relevant strain-temperature cycles occurring in service leading to gross errors in prediction. In a recent case study, where all the important parameters such as cycle type, strain range, temperature range, strain ratio and hold time were taken into proper account in constructing a thermomechanical fatigue life prediction model, the cracking susceptibility of a number of in-service combustion turbine blades has been successfully predicted.

8.0 Acknowledgments

The authors thank Dr. Barry Dooley for his review of the manuscript and providing useful suggestions.

References

[1] Timo, D.P., 1984, Design Philosophy and Thermal Stress Considerations of Large Fossil Steam Turbines, General Electric Fossil Steam Turbine Seminar, Schenectady, New York, Pub. 84T16.

[2] Taira, S.S., 1962, *Creep in Structures*, Academic Press, p. 96-124.

[3] Coffin, L.F., 1969, Prediction Parameters and Their Application to High Temperature Low Cycle Fatigue, In *Proc. of The Second International Conference on Fracture*, Brighton, London, Chapman Hall, p. 643-654.

[4] Manson, S.S., 1973, in *Fatigue At Elevated Temperatures*, ASTM STP 520, American Soc. For Testing and Material, Philadelphia, p. 744-782.

[5] Ostergren, W.J., 1976, A Damage Function and Associated Failure Equations for Predicting Hold Time and Frequency Effects in Elevated Temperature Low Cycle Fatigue, *ASTM J. Test. Eval.*, Vol. 4, No. 5, p. 327-339.

[6] Priest, R.H., Beauchamp, D.J. and Ellison, E.G., 1983, Damage During Creep Fatigue, in *Advances in Life Prediction Methods*, ASME Conference, Albany, p. 115-122.

[7] Bicego, V., Fossati, C. and Ragazzoni, S., 1982, An Energy Based Criterion for Low Cycle Fatigue Damage Evaluations, in *Material Behavior at Elevated Temperatures and Component Analysis*, Y. Yamada, R.L. Roche and F.L. Cho., Eds., Book No. H00217 PVP Vol. 60, American Society of Mechanical Engineers, New York.

[8] Yamaguchi, K., Nishijima, S. and Kanazawa, K., April 1986, Prediction and Evaluation of Long Term Creep Life, in *International Conference on Creep*, Tokyo, p. 47-52.

[9] Majumdar, S. and Maiya, P.S., 1976, A Damage Equation for Creep-Fatigue Interaction, in ASME-MPC *Symposium on Creep-Fatigue Interaction*, MPC-3, Metal Properties Council, New York, p. 323.

[10] Cincotta, G.A. et al., February 1988, *Gas Turbine Life Management System*, Final Report by General Electric Company, EPRI Project RP2421-2.

[11] Viswanathan, R., 1989, *Damage Mechanisms and Life Assessment of High Temperature Components*, ASM International, Metals Park, Ohio.

[12] Miller, D.A., Priest, R.H. and Ellison, E.G., 1984, Review of Material Response and Life Prediction Techniques Under Fatigue-Creep Loading Conditions, *High Temperature Mat. Proc.*; Vol. 6, No. 3 and 4, p. 115-194.

[13] 1974, ASME Boiler and Pressure Vessel Code, Section III, Code Case N-47.

[14] Kadoya, et al., 1985, Creep-Fatigue Life Prediction of Turbine Rotors, in *Life Assessment and Improvement of Turbo-generator Rotors*, R. Viswanathan, Ed., Pergamon Press, New York, p. 1.101-1.114.

[15] Borden, M.P., Ellis, F.V., Miller, D.A. and Gladwin, D., 1987, *Remaining Life of Boiler Pressure Parts—Creep Fatigue Effects*, Report CS-54588, Vol. 5, Electric Power Research Institute, Palo Alto, CA.

[16] Kimura, K., Fijiyama, K. and Muramatsu, M., 1988, Creep and Fatigue Life Prediction Based on the Nondestructive Assessment of Material Degradation of Steam Turbine Rotors, in *High Temperature Creep-Fatigue*, R. Ohtani, M. Ohnani and T. Inoue, Eds., Elsevier Applied Science and The Society of Materials Sciences, Japan, p. 247-271.

[17] Harrod, D.L. and Manjoine, M.J., 1976, in ASME MPC Symposium on Creep-Fatigue Interaction, MPC-3, Metal Properties Council, New York, p. 87.

[18] Gell, M. and Leverant, G.R., 1973, in *Fatigue at Elevated Temperatures*, STP 520, American Society for Testing and Materials, Philadelphia, p. 37.

[19] Haigh, J.R., Skelton, R.P. and Richards, C.E., 1976, *J. Mater. Sci. Engg.*, Vol. 26, p. 167.

[20] Scarlin, R.B., 1977, in Proceedings of the *Fourth International Conference on Fracture*, ICF4, Vol. 2, p. 849.

[21] Thomas, G. and Dawson, R.A.T., 1980, The Effect of Dwell Period and Cycle Type on High Strain Fatigue Properties of a 1Cr-1MoV Rotor Forging at 500°C-550°C, *International Conference on Engineering Aspects of Creep*, Sheffield, p. 167-175, I. Mech. E., London.

[22] Kuwabara, K., Nitta, A. and Kitamura, T., 1985, in *Advances in Life Prediction*, D.A. Woodford and R. Whitehead, Ed.; ASME, New York, p. 131-141.

[23] *Boiler Life Evaluation and Simulation System, BLESS Code and User Manual*, 1991, Electric Power Research Institute, Palo Alto, CA, Report TR-103377, Vol. 4.

[24] Bernstein, H.L., Grant, T.S. McClung, R.C. and Allen, J.M., 1993, Prediction of Thermal-Mechanical Fatigue Life for Gas Turbine Blades in Electric Power Generation, in ASTM STP 1186, p. 212, Philadelphia.

[25] Goto, T., March 1994, *Steam Turbine Rotor Life Assessment, Vol. 4—Fatigue Life Assessment* Report TR-103619, Vol. 4, Electric Power Research Institute, Palo Alto, CA.

[26] Saxena, A., Williams, R.S., and Shih, T.T., 1981, A Model for Representing and Predicting the Influence of Hold Times on Fatigue Crack Growth Behavior at Elevated Temperature, in *Fracture Mechanics: Thirteenth Conference*, STP 743, American Society for Testing and Materials, Philadelphia, p. 86-99.

[27] Solomon, H.D. and Coffin, L.F., 1973, Fatigue at Elevated Temperatures, STP 520, American Society for Testing and Materials, Philadelphia, p. 112.

[28] Yoon, K.B., Saxena, A. and Liaw, P.K., 1993, Characterization of Creep-Fatigue Crack Growth Behavior under Trapezoidal Waveshape Using C_t Parameter, *International J. of Fracture*, Vol. 59, pp. 95-102.

[29] Viswanathan, R. and Gehl, S., April 1991, Trans ASME, *J. of Engg. Mat. & Tech.*, Vol. 113, p. 263.

[30] Foulds, J. and Viswanathan, R., October 1994, Trans ASME, *J. of Engg. Mat & Tech.*, Vol. 116, p. 457.

Table 1
Fossil Power Plant Components Involving Creep-Fatigue As a Common Failure Cause

Component	Location	Inspection Method	Action at Inspection	Definition of Failure	Act
High Temperature Headers	Bore-hole and ligament at ID.	VT, FPT, (after oxide removal), ET, UT	Set next inspection interval by crack growth analysis, mitigate crack growth, monitor.	Leak or critical thru-wall depth/length of crack.	Rej hea
Economizer Inlet Header and Lower Headers	Tube bore hole and ligament at ID: crotch corner of inlet tee at ID.	VT, PT, ET, UT	Set next inspection interval by crack growth analysis. Mitigate crack growth, monitor.	Leak or critical thru-wall depth-length of crack.	Rej hea inle
Drums	At downcomers and saturation tubes at ID.	VT, PT, ET, UT	Set next inspection interval by crack growth analysis, mitigate crack growth, monitor.	Leak or critical thru-wall depth/length of crack.	Rej if e
Supercritical Unit Waterwall Tubes	Tube OD	VT after grit blast, UT survey for wall loss.	Take corrective actions if wall loss and/or cracking acceptable and continue operation.	Unacceptable wall loss.	Rej
Superheater/Reheater Tubes	Dissimilar metal welds.	VT, UT, RT	Replace if cracked.	Crack detection.	Rej
Steam Turbine HP/IP Rotor	Surface grooves Center bore	VT, PT, MT, ET VT, MT, ET, UT	"Skin peeling" if cracks found; enlarge radius. Size cracks and perform crack growth analysis. Overbore, bottle bore to remove cracks, if cracks approach critical size.	When "peeling" is no longer possible due to diameter loss. When boring is not viable or economical.	We rep Rej
HP Inner Casing Steam Chests and Valve Casings	Stud hole ligament, nozzle ports bridges, shell fit flange ligaments, section transitions	VT, MT, PT, Replication	Grind out large cracks and repair.	Repair is uneconomical or toughness is so degraded that critical crack size is very small.	Rej
Combustion Turbine Vanes, Combustor Cans, Transition Pieces	Stress concentrations, hot spots, trailing edges of airfoils.	VT, PT, FPT	Remove cracks and perform weld repair.	Repair is uneconomical. Extent of damage is too widespread to permit repair.	Rej
Combustion Turbine Blades	Leading and trailing edges, concave and convex surfaces, airfoil to platform fillet, shank, tip.	VT, FPT, ET, UT	Recoat if coating is too poor to make next inspection. Weld repair in low stress regions.	Unrepairable cracks in base metal. Severe loss of toughness. Creep damage, either calculated or observed. Severe environmental attack of base metal. Insufficient remaining wall thickness.	Rej

VT - Visual Exam; UT - Ultrasonic Test; ET - Eddy Current Test; PT - Dye Penetrant Test; FPT - Fluorescent Dye Penetrant; MT - Magnetic Particle Test; RT - Radiography.

R5 CREEP ASSESSMENT OF WELDED
TRUNNION & LARGE BORE BRANCH COMPONENTS

R D Patel, S Al Laham and P J Budden
British Energy Generation Ltd, Barnett Way, Barnwood, Gloucester GL4 3RS
e-mail: rajesh.patel@british-energy.com
Fax: 01452 653025

ABSTRACT

Creep life assessments of two ex-service welded $^1/_2$CMV pressure vessels tested under steady loading are presented. The assessments are performed using the R5 procedure in conjunction with finite element (FE) analyses. In particular, the reference stress is obtained by performing elastic-perfectly plastic FE analyses to determine the limit load. The first component is a trunnion tested under internal pressure and system loading. Using mean materials data, the total creep damage at the end of the test is estimated to be less than 0.5. Using lower bound data, the creep damage is estimated to be less than 0.8. Rupture of the trunnion is then predicted not to have occurred. The post-test experimental results showed that there were no signs of the component as a whole being close to failure, consistent with the predictions. The second component is a large bore branch tested under internal pressure plus out-of-plane bending. A conservative estimate of rupture life is obtained and failure is predicted to occur in the weld Type IV position on the vessel flank. Post-test examination indicated that a crack had initiated at the flank position and had grown to nearly penetrate the vessel wall at the inner surface. The crack had grown mainly through the Type IV zone as predicted.

1   INTRODUCTION

The R5 procedure [1] is routinely used within British Energy for life assessments of high temperature plant. Full-scale vessel tests have been carried out to validate the procedure. This paper describes the assessments of two recent vessel tests, both ex-service welded ferritic $^1/_2$CMV steel components tested under steady loading. As is normal in $^1/_2$CMV components, the weld metal used is $2^1/_4$Cr1Mo. The assessments used R5 in conjunction with FE analyses to determine the reference stress. Similar assessments of other welded ferritic pressure vessel tests have been reported by Budden [2].

The first component assessed is a welded trunnion removed from service after about 92000h at 527°C (Table I) following detection of Type IV cracking in a $2^1/_4$Cr1Mo weld on one arm of the trunnion. Examination of the second trunnion arm revealed the presence of extensive Type IV creep cavitation at the main pipe side of the weld. The main pipe section containing the two trunnion arms was therefore removed from plant and replaced with a new section of pipe. The removed pipe section was fabricated into a test vessel, but including only the cavitated trunnion arm (Fig. 1), and was tested under near-service loading at accelerated temperature (Table I). The second component (Fig. 2) is a welded large bore branch vessel removed from service after approximately 106000h at 535°C and again tested at higher temperature but at the service loads (Table I).

2   REFERENCE STRESS APPROACH TO RUPTURE LIFE CALCULATION

R5 [1] defines the limit load reference stress, σ_{ref}, for a homogeneous component as

$$\sigma_{ref} = \frac{P_W \sigma_y}{P_L} \tag{1}$$

where P_L is the limit load for yield stress σ_y. The reference stress is in proportion to the working load, P_W. The ratio σ_y/P_L is independent of yield stress for a homogeneous component. For combined loads, the limit load is evaluated assuming proportional loading. Then the rupture reference stress for ductile materials is calculated in R5 from

$$\sigma_{ref}^R = \{1 + 0.13[\chi - 1]\}\sigma_{ref} \tag{2}$$

where σ_{ref} follows from equation (1) and χ, the stress concentration factor, is given by

$$\chi = \frac{\overline{\sigma}_{E,max}}{\sigma_{ref}} \tag{3}$$

In equation (3), $\overline{\sigma}_{E,max}$ is the maximum elastically-calculated value of equivalent stress for the same set of loadings as are used to obtain σ_{ref}. Equation (2) applies to rupture of initially uncracked bodies and is limited to moderate stress concentrations with $\chi \leq 4$. Modifications to equation (1) to account for the effect of welds are discussed in Sections 3 and 6. The effects of crack growth are not addressed.

3   RUPTURE DATA

The R5 procedure for a welded feature estimates the time to rupture using rupture data for the various metallurgical zones of the weldment. For the purpose of assessment, $^1/_2$CMV weldments made using $2^1/_4$Cr1Mo filler metal are characterised as consisting of four distinct metallurgical regions: (i) parent $^1/_2$CMV steel; (ii) $2^1/_4$Cr1Mo weld metal; (iii) Type IV zone material; and (iv) coarse grained high temperature heat affected zone (HAZ) material. The high-temperature HAZ is characterised as fully coarse grained for conservatism. For each region, the predicted failure life is obtained based on a calculated σ_{ref}^R for that zone and the corresponding rupture data. The overall failure time of the weldment is then the minimum calculated value for the various weld zones.

3.1 $^1/_2$CMV steel

The mean rupture life, t_r (h), at stress σ (MPa) and temperature T (°C) is given [3] from

$$P(\sigma) = \frac{\log(t_r) - F}{(T - G)^H} = a + b(\log \sigma) + c(\log \sigma)^2 + d(\log \sigma)^3 + e(\log \sigma)^4 \tag{4}$$

where the coefficients a, b, c, d, e, F, G and H are material dependent and are tabulated in Budden [2]. The equation is valid in the stress range $61\text{MPa} \leq \sigma \leq 410\text{MPa}$ and temperature range $450°C \leq T \leq 600°C$. The lower bound may be obtained from –20% of the mean rupture stress. For stresses below the range of validity of equation (4), extrapolation is used on the data.

3.2 $2^1/_4$Cr1Mo weld

Equation (4) again applies but with coefficients derived for the weld metal [2,3]. The validated range of data is $20\text{MPa} \leq \sigma \leq 309\text{MPa}$ and $450°C \leq T \leq 600°C$. The lower bound may be obtained from –20% of the mean rupture stress.

3.3   Type IV zone

The relationship

$$P(\sigma) = \frac{\log(t) - 20.01}{(T + 104)} = -2.081817 \times 10^{-2} - 3.82507799 \times 10^{-5}\sigma + 6.78012145 \times 10^{-8}\sigma^2 \quad (5)$$

is used for rupture of the Type IV zone [4]. The data cover the stress range 35MPa to 154MPa and the temperature range 550°C to 640°C. Equation (5) is considered valid provided the calculated rupture life is less than 5×10^5 h. In the absence of further data, the equations can additionally be used to calculate mean rupture lives for Type IV zone material at temperatures down to 500°C, provided that the predicted life is less than 5×10^5 h. Upper and lower bound rupture lives are obtained by adjusting the mean by ± 20% on stress. It should be noted that, because of the limited extent of the data, estimates of lower bound creep lives are restricted to less than 3×10^5 h.

3.4   Coarse grained HAZ

Few data are available for the HAZ material. The mean parent rupture data (Section 3.1) are used to provide a lower bound (i.e. conservative) estimate for coarse grained HAZ behaviour.

4   FINITE ELEMENT MODELS

The service and test conditions of the $\frac{1}{2}$CMV trunnion and large bore branch vessels (Figs 1,2) are listed in Table I. The trunnion vessel has a mean radius to thickness ratio for the main pipe, $R_m/T = 9.4$. Only one quarter of the vessel is analysed because of symmetry, with the appropriate boundary conditions applied on the planes of symmetry. The mesh (Fig. 3) consists of 1365 three-dimensional ABAQUS [5] C3D20 or C3D20R elements and 7385 nodes. ABAQUS C3D20R elements are used for the elastic-plastic analyses and C3D20 for elastic analysis. The elastic material properties used in the FE analyses are obtained from BS806 [6] and are listed in Table II. For the elastic-perfectly plastic analyses, the material is arbitrarily assumed to have a yield stress $\sigma_y = 100$MPa. The pressure internal to the main pipe was applied as a pressure load to the element faces, and the system loading as pressure load to two element faces of the trunnion arm to avoid excessive local deformation in the mesh. The trunnion arm itself does not contain any pressure loading.

The large bore branch (Fig. 2) has a mean radius to thickness ratio of the main pipe, $R_m/T = 7.7$. The mean radius to thickness ratio of the branch, $r_m/t = 9.7$. The wall thickness ratio $t/T = 0.5$. Half of the vessel is analysed due to symmetry, with the appropriate boundary conditions to represent the test conditions. The mesh (Fig. 4) consists of 1960 three dimensional ABAQUS C3D20 elements and 14114 nodes. Internal pressure is included in the finite element model as a pressure load on all internal element faces. The pressure loads on the end caps are represented by applying equivalent pressure loading to the element faces at the member ends. The out-of-plane bending load was applied in the test via a pneumatic jack at the end of the branch extension. For the FE analysis, an equivalent out-of-plane bending moment and shear force were applied at the end of the branch. The total bending moment at the branch intersection is 30.4kNm.

The effects of non-linear geometry have not been considered for either geometry. Small strain, small displacement FE analyses have been performed. For the calculation of limit loads for reference stress assessment, this agrees with R6 [7] recommendations.

5   FINITE ELEMENT ANALYSES

Elastic FE analyses of the trunnion and large bore branch, for both test and service conditions, were performed to calculate the maximum elastic value of equivalent stress in equation (3). This was then used in the calculation of the rupture reference stress from equation (2). In the elastic-perfectly plastic FE limit load analyses, for both test and service conditions, the internal pressure and system loads were increased in proportion using the RIKS algorithm within ABAQUS [5].

6   RESULTS AND DISCUSSION

6.1   Trunnion

6.1.1   Stress analysis

Under the test condition, a maximum Von Mises stress of 69.2MPa was obtained from the elastic analysis. The reference stress results for the test conditions are given in Table III. The working load, P_W, plastic collapse limit load, P_L, reference stress, σ_{ref}, equivalent elastic stress, $\overline{\sigma}_{E,max}$, stress concentration factor, χ and rupture reference stress, σ_{ref}^R, are tabulated. Note that, as a sensitivity study on the effect of mesh refinement on the computed peak elastic stresses, three maximum equivalent stresses are considered in Table III, namely 75MPa, 85MPa and 95MPa. This covers uncertainty in the details of the weld profile. For the service condition, a maximum Von Mises stress of 70.4MPa was obtained from the elastic analysis. However, as in the test case condition, elastic equivalent stresses of 75MPa, 85MPa and 95MPa were used as a sensitivity analysis. The reference stress results for service conditions are also given in Table III.

Results for the creep damage due to the test and service conditions are given in Table IV. Damage is calculated, using Robinson's rule, by linearly summing the ratios $t/t_f(\sigma_{ref}^R)$ from the service and test conditions. Here, t is the corresponding time at load and t_f is the creep rupture time at the associated calculated value of rupture reference stress, σ_{ref}^R. Following R5, the homogeneous rupture reference stress is multiplied by a factor, k, which quantifies stress redistribution within a weldment. For stress states where the maximum principal stress is essentially parallel to the fusion line and overall stress redistribution occurs, k may differ from unity and varies between zones [1]. This is judged from the FE analyses to be the case for the trunnion vessel, where the stresses near the weld are controlled by the pressure load. For all of the cases examined, the total damage does not exceed 0.8, so that the trunnion is predicted to have not failed at the end of the test. The lives quoted in Table IV correspond to the Type IV material zone, which gives the lowest rupture time and thus limits the predicted life of the vessel. For the other zones, the weld stress redistribution factor, following the advice in R5 [1], is taken as k = 0.7 for the weld and k = 1.4 for the coarse-grained HAZ. For Type IV material, k=1 is assumed for conservatism.

6.1.2 Experimental results and comparison with R5 estimates

Monitoring using potential drop techniques indicated that creep crack initiation occurred at about 12576h and that the crack extended marginally in the circumferential direction during the subsequent 2855h of testing. There was no evidence of through-thickness crack growth from any of the monitoring techniques applied to the vessel throughout the test. After a total of 15431h on test there were no signs of the component as a whole being close to the end of its life. The test was then terminated. This experimental observation is consistent with the life estimate of Section 6.1.1.

6.2 Large bore branch

6.2.1 Stress analysis

The finite element analysis of the large bore branch showed that the tension flank location is the most highly loaded section, with the highest loads at the branch fusion toe. The load against side displacement curves predicted by ABAQUS elastic-perfectly plastic analyses are shown in Figure 5. The effects of using different yield properties for the parent and weld metals in the finite element analysis, with an under-matched weld, were investigated. The results showed that such effects on the limit load prediction are negligible; the collapse mechanism being a localised buckling of the main vessel wall under the compression side of the branch. Hence the results of the model with homogeneous material property have been adopted. The limiting lateral force, or the out-of-plane force, predicted by ABAQUS is approximately 112kN. For a material yield stress of 175.7MPa, and an applied out-of-plane force of 35kN, a reference stress of 54.9MPa follows. The maximum elastically calculated value of the equivalent stress is 102MPa. Thus χ=1.86, giving a rupture reference stress, σ_{ref}^R = 61.08MPa (see Table III). The maximum principal stress is shown in the FE analyses to be effectively transverse to the fusion boundary so that the stress redistribution factor k=1 for each weld zone. The unfactored homogenous rupture reference stress is then adopted for each weld zone.

6.2.2 Experimental results and comparison with R5 estimates

Both the branch and the main pipe sides of the weld were inspected before the test by magnetic particle inspection. Two small surface indications were reported, one of which was non-crack-like, the other indication was about 4mm long and located at the branch intersection with the weld boundary. Testing of the branch was completed without failure after a total of 11726h running time. Although the NDT indicated a through-wall defect at the end of the test, the vessel had not lost pressure through leakage. Destructive metallography subsequently confirmed that a crack had grown from the outer surface tension flank position, on the branch side of the weld and mainly through the Type IV zone, such as to very nearly penetrate the vessel wall. A comparison between the test time and the rupture life based on the calculated rupture reference stress of 61.08MPa, is given in Table V. The table includes estimates of rupture life for the different weld zones. In order to calculate the total creep damage (usage factor), the service time at 535°C was also considered. It is clear from Table V that the maximum total usage factor, for the Type IV zone, is 1.14, predicting failure of the vessel prior to the end of the test in the Type IV zone. Hence the use of mean Type IV material data together with the results of the FE analysis is conservative since the vessel had not failed at the end of the test.

572

It should be noted that the life assessments of both vessels used descriptions of creep rupture data for the generic vessel materials, no cast specific data being available. The use of lower bound data maximises the estimated creep usage factors and hence the conservatism in the assessments.

7   CONCLUSIONS

Life assessments of two ex-service welded $^{1}/_{2}$CMV pressure vessels under steady creep loading have been performed as validation of the R5 assessment procedure. The following conclusions are drawn.

1. For the trunnion, the total damage due to service and test conditions was predicted to be less than 0.5 using mean rupture data.
2. For the trunnion, using lower bound rupture data, the total damage was estimated to be less than 0.8. This suggests that creep rupture of the trunnion will not have occurred during the test.
3. Post-test examination of the trunnion suggested that there were no signs of the component as a whole being close to failure, consistent with conclusions 1 and 2 above.
4. For the large bore branch, the use of mean material properties predicted the vessel to fail within the test duration in the Type IV zone at the flank position of the branch weld.
5. Post-test examination of the large bore branch vessel indicated that a crack had formed at the tensile flank position and had grown, predominantly through the Type IV zone, to near the inside surface. This agrees well with the R5 prediction.
6. The analyses support the use of R5 for the assessment of welded pressure vessels.

8   ACKNOWLEDGEMENTS

The large bore branch test was performed within the LICON project, part-funded by the CEC under its Brite-Euram programme. The permission of LICON to publish the results is gratefully acknowledged. This paper is published by permission of British Energy Generation Ltd.

9   REFERENCES

[1] Ainsworth, R.A. (editor), 1998, R5: Assessment Procedure for the High Temperature Response of Structures, Issue 2, Revision 2, British Energy Generation Limited.

[2] Budden, P.J., 1998, Analysis of the Type IV Creep Failures of Three Welded Ferritic Pressure Vessels, Int. J. Pres. Ves. and Piping, Vol.75, pp. 509-519.

[3] BS PD6525 Part 1, 1990, Elevated Temperature Properties for Steels for Pressure Purposes, British Standards Institution, London.

[4] Sanham, E.J., and Phillips, J.L., 1991, Derivation of a Parametric Equation to Describe Type IV Creep Rupture Lives in $^{1}/_{2}$CrMoV/$2^{1}/_{4}$Cr1Mo Steel Weldments, Nuclear Electric Memorandum TD/SIP/MEM/1239/90.

[5] ABAQUS Version 5.7, Hibbitt, Karlsson & Sorensen Inc.

[6] BS 806:1993, 1993, Design and Construction of Ferrous Piping Installations for and in Connection with Land Boilers, British Standards Institution, London.

[7] Ainsworth, R.A. (editor), 2000, Assessment of the Integrity of Structures Containing Defects, R6 Revision 3, Amendment 11, British Energy Generation Ltd.

Table I: Service and test conditions

	Service conditions	Test conditions
Trunnion		
Temperature (°C)	527	585
Trunnion hanger load (kN)	46.03	50.20
Internal pipe pressure (MPa)	3.60	3.50
Service and test times (h)	91658	15431
Large bore branch		
Temperature (°C)	535	585
Branch bending moment (kNm)	30.4	30.4
Internal pipe pressure (MPa)	3.59	3.59
Service and test times (h)	106000	11726

Table II: Elastic properties

Trunnion	
Young's modulus @ 527°C (MPa)	172300
Young's modulus @ 585°C (MPa)	166500
Poisson's ratio	0.29
Large bore branch	
Young's modulus @ 585°C (MPa)	170,000
Poisson's ratio	0.3

Table III: Reference stress results

	P_W (MPa)	P_L (MPa)	σ_{ref} (MPa)	$\overline{\sigma}_{E,max}$ (MPa)	χ	σ_{ref}^R (MPa)
Trunnion						
Test conditions						
CASE i	3.50	11.935	29.33	75	2.56	35.3
CASE ii	3.50	11.935	29.33	85	2.90	36.6
CASE iii	3.50	11.935	29.33	95	3.24	37.9
Service conditions						
CASE i	3.60	12.1	29.85	75	2.51	35.7
CASE ii	3.60	12.1	29.85	85	2.85	37.0
CASE iii	3.60	12.1	29.85	95	3.18	38.3
Large bore branch						
Test conditions						
CASE I	3.39	11.49	54.9	102	1.86	61.08

Table IV: Trunnion damage calculation results (Type IV zone) for various load cases

	Test time (h)	Estimated test failure time (h)	Test damage	Service time (h)	Estimated service failure time (h)	Service damage	Total damage
mean data							
CASE i	15,431	62363	0.247	91658	500000	0.183	**0.431**
CASE ii	15,431	58214	0.265	91658	500000	0.183	**0.448**
CASE iii	15,431	54361	0.284	91658	500000	0.183	**0.467**
lower bound data							
CASE i	15,431	43112	0.358	91658	300000	0.306	**0.663**
CASE ii	15,431	39789	0.388	91658	300000	0.306	**0.693**
CASE iii	15,431	36742	0.420	91658	300000	0.306	**0.726**

Table V: Large bore branch damage calculation results

Material zone	Rupture reference stress from FE (MPa)	Rupture life based on FE (h)	Test time (h)	Service time (h)	Service rupture life based on the rupture reference stress from FE (h)	Total damage
0.5CMV (parent)		61233			250000*	**0.62**
2.25Cr1Mo (weld)		21943			250000*	**0.96**
Type IV	61.08	17014	11726	106000	237624	**1.14**
HAZ		205096			250000*	**0.48**

*Numbers capped by the limits of the data at the service temperature of 535°C.

Figure 1   General arrangement of trunnion

Figure 2   General arrangement of large bore branch.

576

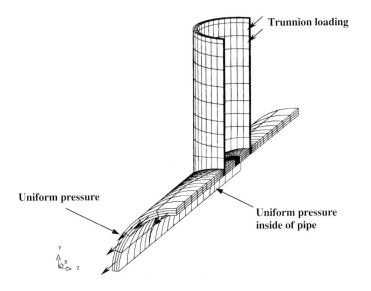

Figure 3   FE mesh of Trunnion.

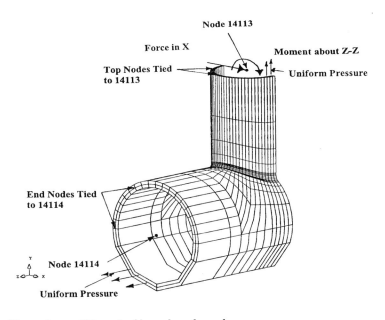

Figure 4   FE mesh of large bore branch.

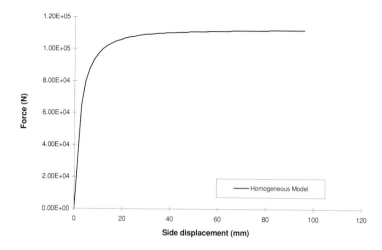

Figure 5   **Load side-displacement curve for limit load assessment of the large bore branch (elastic-perfectly plastic material properties).**

Development of Creep Life Assessment for Boiler Parts of Fossil Power Plants

Junichi KUSUMOTO, Yasufumi GOTOH
Research Laboratory, Kyushu Electric Power Co., Inc., Fukuoka, 815 Japan

Toshimi KOBAYASHI
Kansai Division, Sumitomo Metal Technology, Inc., Amagasaki, 660 Japan

Abstract

The present investigation relates to the creep life assessment method of 2.25Cr-1Mo steel tubes and pipes using hardness measurement, and to the newly developed method for hardness measurement on site.

Authors have continued investigation on the assessment method of remaining life using the hardness measurement of weld metal. This is because the hardness change of weld metal during plant service is larger than that of base metal and variation in measured hardness values is smaller than that in the weld heat affected zone.

Investigation on relation of time-temperature, chemical composition, stress, and post-weld heat treatment to hardness of weld metal revealed that the relation between hardness change of weld metal and stress depends upon chemical composition. And, it was considered that influence of post-weld heat treatment is practically negligible.

In this investigation, development of a new method for hardness measurement and instrument was conducted in order to practice hardness measurement on-sites more accurately.

The remaining life assessment method by hardness measurement in the present investigation was applied to evaluation of the material of headers drawn from an actual plant. The result of the remaining life assessment was in good agreement with that of the evaluation by the microstructure observation.

1. INTRODUCTION

One of the non-destructive methods in a creep remaining life assessment for plant equipment used in the high temperature creep range is the evaluation method by hardness. This method is based upon the fact that hardness of plant equipment decreases and softens as a result of prolonged use at a high temperature. Also, this method has features of simplicity, rapidity and low cost compared with other evaluation methods, such as the microstructural method. However, since the degree of softening of plant equipment depends on chemical composition, stress, and post-weld heat treatment as well as time-temperature, it is necessary to understand effects of these parameters in order to conduct an accurate remaining life assessment.

Although such effects of these parameters seem to have been investigated separately, it should be reasonable to consider that each parameter mutually affects hardness.

While conducting investigations on the remaining life assessment method by hardness measurement of 2.25Cr-1Mo steel tubes and pipes, the authors pointed out that a final layer of weld metal in a butt-welded joint is the most adequate as a location for hardness measurement.

In the present investigation, effects of parameters such as chemical composition, stress, and post-weld heat treatment on hardness change will be considered comprehensibly to obtain some quantitative expressions.

Although hardness measurement with accuracy is necessary to the remaining life assessment by hardness, since hardness measurement on-site in actual plants used to be conducted by portable hardness measuring equipment with an operator manually holding it, it was not necessarily accurate enough to claim accuracy with confidence. Therefore, in the present investigation, a method which makes on-site measurement more accurate will be considered, too.

2. EXPERIMENTAL PROCEDURES

In this investigation, the degree of deterioration on 2.25Cr-1Mo steel tubes and pipes by aging and creep was studied. Specimens for an aging test and internal pressure creep test were taken from a butt-welded portion.

Plate-like specimens for aging tests and creep tests were machined, after applying TIG without filler metal to a block taken from a pipe. Internal pressure creep test to 1200 hour and normal creep test to 600hour with plate-like specimens were performed on three stress levels. The creep tests were interrupted at optional times. Hardness measurement of the interrupted creep test specimens was conducted.

The chemical composition of weld metal is shown in Table 1, and test conditions in the aging test and the interrupted creep test are shown in Table 2. Test conditions related to time-temperature were expressed by Larson-Miller parameters (PLM).

Table 1　Chemical Composition of Samples (mass%)

Sample	C	Si	Mn	P	S	Cr	Mo
W1	0.06	0.65	0.64	0.014	0.009	2.23	1.00
W2	0.07	0.55	0.60	0.007	0.008	2.10	0.95
W3	0.06	0.40	0.74	0.007	0.006	2.22	1.04
W4	0.07	0.35	0.60	0.008	0.004	2.22	1.03
W5	0.09	0.10	0.46	0.005	0.001	2.30	0.95
W6	0.08	0.48	0.92	0.008	0.007	2.28	1.11
W7	0.09	0.26	0.41	0.019	0.005	2.10	0.98

Sample W1, W2: Final layer of weld metal in welded joint of pipe (ϕ350×t60mm).
W3~W5: Final layer of weld metal in welded joint of tube (ϕ50.8×t9.5mm).
W6: Final layer of weld metal in welded joint of tube (ϕ58.8×t13.6mm).
W7: Final layer of weld metal in TIG without filler to a block taken from pipe (ϕ335×t25mm).

Table 2　Condition of aging and creep aging

Test type	Sample	Specimen No.	Aging or PWHT				Creep		
			Temp (K)	Time (Hr)	PLM* (C=20)	Stress (MPa)	Temp (K)	Time (Hr)	PLM* (C=20)
Aging	W1~W7	--	823 ~973	100~ 10000	18000 ~23000	--	--	--	--
Internal pressure creep	W6	S1	993	5.2	20500	58.8	922	~1200	20000 ~22000
		S2	993	5.2	20500	39.2	948		
		S3	988	0.3	19200		948		
		S4	993	5.2	20500	19.6	964		
Normal creep	W7	SW11	--	--	--	58.8	923	~600	20000 ~22000
		SW12	963	19.2	20500		923		
		SW21	--	--	--	39.2	923,973		
		SW22	993	17	21000		923,973		
		SW31	--	--	--	19.6	923,973		
		SW32	993	17	21000		973		

*: PLM = $T (\log t + 20)$, T (K), t (hr)

In order to consider the effect of post-weld heat treatment (hereafter called as PWHT) on hardness change, PWHT was applied to some of the creep test specimens before the test.

To avoid decarburization or carburization during the aging test, either a closed vessel or a Cr-plated specimen was used. For the interrupted creep test, Cr-plated specimens were used. Hardness measurement was conducted on weld-metal of a final layer.

The remaining life assessment method by hardness measurement was applied to the aged headers drawn from the actual plant, and results were compared with those of microstructure observation.

3. RESULTS AND DISCUSSION

3.1 Effect of chemical composition

With respect to samples W1~W7 in Table 1, the relation between hardness of weld metal of a final layer after aging tests and PLM is shown in Fig.1. As shown in Fig.1, softening behavior of any of them could be approximated by a straight line. Fig.1 shows individual data measured that was obtained from only one sample (W4) since individual data fluctuation in every sample was similar.

From Fig 1,it is known that the degree of softening, that is, inclination of a softening straight line is different from sample to sample, and it was recognized that a sample with less content of Si and Mn which has a larger inclination tends to soften more easily. As-welded Vickers hardness of every sample was in the range of $Hv285$~305, and almost the same. If data of each sample is extrapolated within this range, it was revealed that every sample converges on approximately the same PLM ($PLM_W=16500$).

Inclination differs chiefly because chemical composition differs, as was reported in hitherto published papers[1]. It is considered that approximately the same as-welded hardness is due to small difference in carbon content which influences as-welded hardness of material.

Using the softening line of the samples shown in Fig.1, the relation between inclination K and chemical composition was determined by regression analysis. C, Si, Mn, Cr, and Mo were elements subjected to analysis, and it was assumed that each element affects the inclination of a line independently and linearly, for the sake of convenience. Results of the analysis are

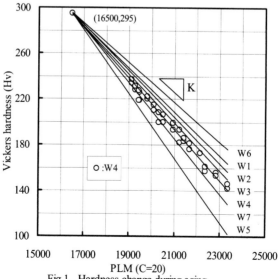

Fig.1 Hardness change during aging

582

shown in Equation (1),

$$K = -0.034949 + 0.028946 \ \phi_C + 0.013329 \ \phi_{Si} + 0.0063712 \ \phi_{Mn}$$
$$- 0.0081887 \ \phi_{Cr} + 0.018556 \ \phi_{Mo} \qquad \dots\dots\dots\dots (1)$$

where ϕ_C, ϕ_{Si}, ϕ_{Mn}, ϕ_{Cr} and ϕ_{Mo} are contents of C, Si, Mn, Cr and Mo respectively.

The relationship between an aging softening line and chemical composition can be expressed by Equation (2), by using Equation (1).

$$H_0 - H_w = K \ (PLM - PLM_W) \qquad \dots\dots\dots\dots (2)$$

where H_0, H_w is hardness, PLM = T (log t + 20), T is temperature (K), t is time (hr).

Since all of Si, Mn, Cr, and Mo are elements which retard softening, it is thought that coefficients of every element should be essentially positive in Equation (1), however, the coefficient of Cr was negative. This is a problem with the regression analysis, but Equation (1) is used as it is for the time being.

3.2 Effect of stress
Results of hardness measurement on interrupted creep test specimens are shown in Fig.2. Fig.2 also displays the simple age softening lines of samples W6 and W7 show in Fig.1 as well as as-PWHT data.
From Fig.2, it is known that softening behavior during creep on the same stress can be approximated by the same straight lines. Also, it is known that the greater the stress, the greater the inclination K_1 of a stress-affected softening line. In addition, if the stress is the same, any of the samples has the same inclination K_1. That is, it is suggested that inclination K_1 is influenced by stress, but has nothing to do with chemical composition of a sample. Evidently from Fig.2, it is known that softening behavior during creep advances along two different softening lines. To be specific, softening behavior during creep advances along the age softening line with inclination K up to a certain value of PLM (hereafter, PLM_C), and

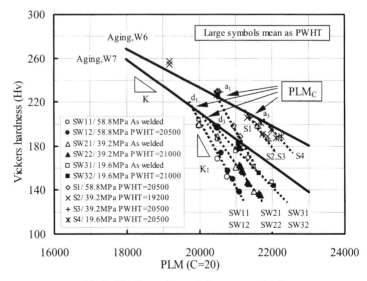

Fig.2   Hardness change during creep and aging

beyond PLM_C softening progresses along softening line with inclination K_1 stress-affected.

For instance, in test No.S4 of sample W6, PLM_C is approximately 21450. If $PLM < PLM_C$, hardness changes with the behavior of an age softening line, while if $PLM > PLM_C$, it does along with a softening line with larger inclination. And, in test No.SW11~SW32 of sample W7, it is known that the softening behavior during creep of as-welded specimens and PWHT applied specimens advance along the approximately same softening line.

PLM_C depends on chemical composition of a sample and stress levels. Generally, the higher the stress or the larger the inclination of age softening straight line, the smaller the PLM_C value of a sample. Thus, acceleration of softening due to stress begins at the early creep stage.

Fig.3 shows results of creep elongation measurement on interrupted creep test specimens. Although the creep elongation was recognized when PLM value was over a certain value, this PLM value was approximately equal to that of PLM_C in Fig.2. The degree of increase in elongation depends on stress, but has nothing to do with chemical composition of a sample. This is more evidence of the relationship among inclination K, stress and chemical composition.

Figs.2 and 3 suggest that effects of stress and chemical composition of a sample on PLM_C and inclination K can be understood as those of stress and chemical composition on creep elongation.

Fig.4 displays typical microstructure of weld metal of specimens for aging tests and interrupted creep tests (test No.S1, 58.8Mpa) of sample W6. Though PLM values of all specimens are the same, the former is $Hv206$ and the latter is $Hv167$. No difference in precipitation was observed between these specimens. This agrees to the past report[2] that the condition of precipitates is not affected by stress, but affected chiefly by PLM.

In general, hardness of metals has something to do with precipitates and dislocation structure, and stress-accelerated softening in the present investigation is considered to be due to change in dislocation structure.

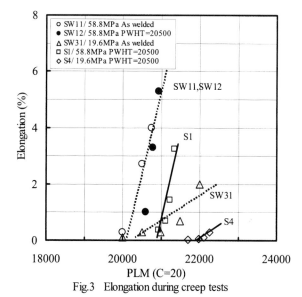

Fig.3   Elongation during creep tests

584

Aged Specimen	Creep Specimen No.S1 (58.8MPa)

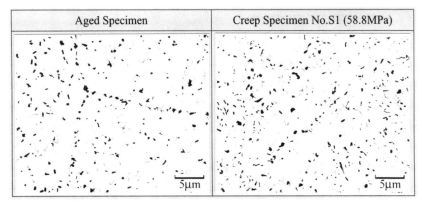

Fig.4   Extraction replica TEM micrograph of welded metals of aged specimen and creep specimen with same PLM=21400.

3.3 Effect of PWHT

In this study, three conditions 19200, 20500 and 21000 of PLM values for PHWT were considered, as shown in Table 2. As mentioned also in Section 3.2, Fig.2 indicates that the softening behavior of interrupted creep test specimens with PWHT is almost similar to that of as-welded interrupted creep test specimens in any stress condition.

According to the author's past investigation, PWHT was applied to the welded joint of 2.25Cr-1Mo steel header of an actual plant around 993K with PLM of about 20500. There was no example of PLM of over 21000. PLM values of actual aged plants that need remaining life assessment mostly exceed 21000.

From the above-mentioned results, it is considered that the effect of PWHT on remaining life assessment of aged actual plants is negligible.

3.4 Evaluation of effects of chemical composition and stress

From the test data hitherto obtained, effects of chemical composition and stress on softening were evaluated. Results of consideration in Section 3.2 suggest that the age softening line and the softening line affected by creep stress must be evaluated. Meanwhile, as a result of Section 3.3, further evaluation of the effect of PWHT was omitted.

Gotoh reported on an evaluation of the effect of stress[3]. His method followed the minimum-commitment station-function method by Manson[4] and Leeuwen[5]. This method is based on the statistical process and needs so much measurement data that it is considered hard to apply it to the evaluation in this study. Evaluation was performed, as shown in schematic diagram of Fig.5.

The softening line in the case where only simple aging without stress is taken into account can be evaluated as a function of only chemical composition by Equations (1) and (2) which were discussed in Section 3.1.

Meanwhile, it is necessary to calculate PLM_C and K_1 for evaluation of the softening line affected by stress. As is suggested by discussion in Section 3.2, it is understood that inclination K_1 can be evaluated as a function of only stress and that PLM_C is affected by stress and chemical composition.

A regression calculation was used to evaluate the relationship between inclination K and stress. As a result, inclination K_1 can be determined from Equation (3), which is the function of only stress.

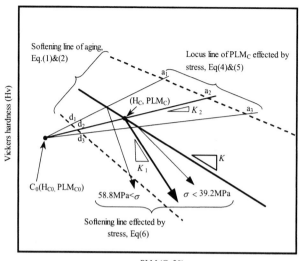

PLM (C=20)

Fig.5   Evaluation of effect of stress and composition (Schematic diagram)

$$K_1 = f_1(\sigma) \qquad \dots\dots\dots\dots (3)$$
where σ is stress.

On evaluating PLM_C, it was assumed that PLM_C of different chemical compositions under the same stress falls on a single line.

It was assumed that the locus of PLM_C is located on a straight-line a_3d_3 when stress is below 39.2MPa and on straight-line a_1d_1 when the stress is 58.8MPa. In addition, it was also assumed that in cases where stress is other than what was mentioned above, the locus of a straight line a_2d_2 passes a cross point C_0 of a straight lines a_1d_1 and a_3d_3.

From inclinations of lines a_1d_1 and a_3d_3 and stresses corresponding to each line, the relationship between inclination K_2 of line a_2d_2 which expresses the locus of PLM_C and stress σ was evaluated. As a result, inclination K_2 was determined by Equation (4), which is the function of only stress.

$$K_2 = f_2(\sigma) \qquad \dots\dots\dots\dots (4)$$
where σ is stress.

By using of Equation (4), hardness H_{C0} and PLM_{C0} at point C_0, the relationship between stress σ and line a_2d_2 which expresses the locus of PLM_C is expressed by Equation (5).

$$H - H_{C0} = K_2 (PLM - PLM_{C0}) \qquad \dots\dots\dots\dots (5)$$
where H, H_{C0} is hardness and K_2 is the inclination of line connecting PLM_C and Point C_0.

H_C for which effect of composition and stress is taken into account can be evaluated by Equations (2) and (5). Using hardness H_C corresponding to PLM_C and K_1 in Equation (3), stress-affected softening line is expressed by Equation (6).

$$H - H_C = K_1 (PLM - PLM_C) \qquad \dots\dots\dots\dots (6)$$

586

3.5 Creep remaining life assessment method by hardness

In this Section, the assessment method will be considered for remaining life of materials.

As was discussed in Sections 3.1~3.4, on the basis of stress and contents of chemical composition, PLM, a parameter of time-temperature, can be evaluated by Equations (1)~(6). If time for operation of plant equipment is known, temperature calculation of a section becomes possible. This temperature becomes the average temperature for use. Chemical contents can be determined by analysis of a chipped sample, etc., and hoop stress by the internal pressure of a pipe is used as stress.

The remaining life assessment of creep can be made by using stress and assessed average temperature to obtain creep rupture data of new materials.

3.6 Consideration of equipment for on-site hardness measurement

Considering that the indentator should be pressed onto the measured surface perpendicularly with stability as a decisive factor of accurate hardness measurements, methods were discussed for assuring verticality and stability.

According to the method newly developed by the authors, a laser beam of about 1mm in diameter is used. As is shown in Fig.6, perpendicularity to the measured surface is assured by adjusting an angle so that incident ray can coincide with reflected ray. In addition, a fixed device with a magnet makes it possible to measure the surface in any position stably. The body of a portable hardness tester is an ultrasonic type.

As a result, the dispersion of hardness measurements and the accuracy of the average value are roughly the same as in the case where fixed laboratory hardness testers are used.

3.7 Evaluation in application of the assessment method of creep remaining life by hardness

The assessment method of creep remaining life by hardness measurement was applied to headers drawn from an actual plant. Fig.7 shows the use of newly developed supporting equipment while hardness of circumferential welded joint is being measured. Fig.8 displays the shape of indentation after hardness was measured. As is clear from the shape shown in Fig.8, it is known that the indentator is correctly pushed into the surface even if it is a curved surface of header's circumferential welded joint.

Table 3 shows the result of remaining life assessment by hardness measurement. The

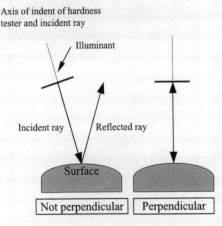

Fig.6   Perpendicularity to surface

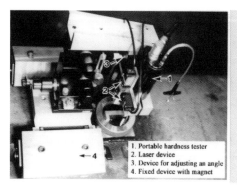

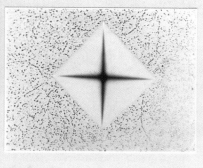

×400

Fig. 7   Hardness measurements on header   Fig. 8   A shape of indentation

Table 3   Sample of pipe which were examined remaining life

	Hardness evaluation method	Micro-structural evaluation method
Operation duration (hr)	202,000	
Portion	Super heater	
Material	2.25Cr-1Mo	
Life consumption (%)	33	17

result of microstructural assessment is also included in Table 3. As is proved by this Table, it was made clear that the assessment result of weld metal by hardness measurement agrees comparatively well to that by micro-structural assessment.

By the way, the author's group confirmed that the accuracy of structural assessment is in the range of factor 2 of that of the creep rupture test[6].

4. CONCLUSIONS

1) Change in weld-metal hardness of 2.25Cr-1Mo steel tubes was affected by parameter of chemical composition and stress as well as time-temperature.

2) It was revealed that chemical composition and stress not independently, but mutually influence each other. Materials in the chemical composition system with a large inclination of age-softening line were found to be accelerated in softening by stress from the early stage of creep. With increased stress, inclination of a softening straight line affected by its influence, became large.

3) As the state of precipitates did not change with stress, softening by stress was considered to be accelerated due to change in the dislocation structure. In addition, the effect of chemical composition and stress on softening corresponds to creep elongation.

4) For the remaining life assessment of aged actual plant equipment, the effect of PWHT is thought to be negligible.

5) The author has proposed a method for remaining life assessment by hardness measurement to consider comprehensively the effects of chemical composition and stress on softening. In addition, our newly developed on-site hardness tester was used for header materials of an actual plant and good results were obtained.

REFERENCES
(1) Jaffe, L.D. and Gordon, E.: Temperability of steels, Transaction of the ASM, 1957, 49, pp.359-371.

588

(2) Dobrzanski, J. and Hernas, A.: Relationship between microstructure and residual life-time of low alloy CR-Mo steels, C494/079 © IMechE, 1996, pp.451-461.

(3) Gotoh, T.: Study on Detection of deterioration and damage due to high temperature service using nondestructive material property tests, Mitsubishi Juko Giho, 1984, 21, pp.389-398 in Japanese.

(4) Manson, S.S.: ASM publication, 1968, No.D8-100.

(5) van Leeuwen, H.P. and Schra, L.: Parametric analysis of the effects of prolonged thermal exposure on material strength, Int. conf. on creep and fatigue in elevated temperature applications C134/73, 1973, 134.1-134.9

(6) Hajime, W. et al: Development of creep life assessment technique for boiler parts of fossil-fired power plants, First international conference on microstructures and mechanical properties of aging materials, AMS, Nov, 1992, pp.243-249

An analytic approach for interpreting creep curves from small punch test

Li Ying Zhi

KEMA Nederland BV

Postbox 9035 6800ET Arnhem, the Netherlands

1   INTRODUCTION

Creep damage is one of the failure mechanism of components in power plants and processing plants. The current conditions of components are needed to ensure safe operation and to extend inspection intervals. Thus it is necessary to develop techniques to assess the current creep damage of components in service. The "Small Punch Test (SPT)" is one of the techniques, which can directly assess the current material properties of components on site.

The pioneer work on small punch test in the creep region has been done by European researchers. In 1994, a COPERNICUS project entitled Small Punch Test Method Assessment for the Determination of the Residual Creep Life of Service Exposure Components (Contract: ERB CIPA CT 94 0103) has been set up and was carried out in laboratories in different European countries. Small punch test method has been explored into the creep region to provide material properties for residual life assessment purpose.

The history of small punch test at creep region is only six years and lot of work should be carried out further. For instance, it is found that the temperature oscillation gives a strong influence on the test results. Therefore, a better temperature control system should be developed for the creep test. For interpretation of creep test results, present work is based either on the equivalent stresses concept (developed by CESI) or empirical equation, such as modified form of Dorn equation (developed by Parker etc.). Encouraged by the EPRI analytical approach to determine fracture toughness, K_{IC}, a similar analytical approach is proposed by the author. Examples are given by interpretation of the observed creep curves (deflection vs. time) from small punch test published by ENEL and the COPERNICUS project.

2   STATE OF ART FOR DETERMINATION OF CREEP PROPERTIES

A European activity was started in 1994 in the frame of a COPERNICUS program to validate the small punch test methodology to assess residual life of components at high temperature. The technique is an alternative with respect to the use of the traditional cylindrical specimens to obtain isostress creep curve. Some achievements in miniature creep test aspect are quoted here.

2.1   CESI approach: equivalent stress concept

CESI is involved for years in a miniaturisation creep methodology and developed a complete creep small punch test procedure for creep that involved particular test facility and correlation law comparable with the common isostress methodology for residual life evaluation. Six small punch tests were made for an aggregate test time of about 2500 hours (Figure 1). Obviously before doing SP tests it is necessary to answer the question: "What is the load value to use in the small punch test to obtain the same time to rupture as in the 1-D tests?"

Two different relations between the SP load and 1-D stress were proposed. In the first of them the SPT load that gives the same creep life as a standard cylindrical specimen under uniaxial stress is achieved from the balance of forces (Tettamanti, Crudeli, 1998):

$$\frac{F_0}{\sigma} = [2\pi h_0 (R + \frac{h_0}{2}) \sin\theta] \sin\theta \qquad (2.1)$$

where:
F_0: applied load on the SPT specimen
h0: thickness of the specimen (typically = 0.5 ± 0.005 mm)
R: radius of the ceramic sphere head, applied at the top of the punch
σ: uniaxial stress applied on standard cylindrical creep specimens
θ: angle between the upright axis and the last contact point between specimen and sphere during the deflection process. θ could be empirically obtained from the comparison among uniaxial and SP tests. From the tests on steel X20 in the COPERNICUS project, $F0/\sigma = 1.876$ and consequently $\theta = \pi/4.6$.

In the second relation the SP load is derived from the deflection process of the disk (applicable on large deflections).

$$\frac{F_0}{\sigma} = K [\pi h_0 d (\frac{D}{d} - \chi)] \qquad (2.2)$$

where:
D: hole diameter of lower die, (typically 4 mm),
d: 2R, diameter of the ceramic sphere head
K: blocking coefficient, K=0.5 for not blocked condition of specimen (not pressed on the edge)
$\qquad\qquad$ K>0.6~0.8, if pressed
χ: friction coefficient, 0.6~0.7, depending on temperature and the applied load
With this second relationship a value of $F0/\sigma = 1.95$~2.06 is obtained, depending on tests characteristics in the not blocked condition of the specimen.

In the first approximation, $F0/\sigma = 2$ is suggested by CESI.

2.2   COPERNICUS & SWANSEA approach: empirical equation

Two alloyed steels were used in the COPERNICUS project. One of them was the low-alloyed steel 14MoV63, whereas the other was the high-alloyed steel X20CrMoV 121. A range of experiments has been performed studying the creep and fracture behaviour. Initially, uniaxial tests were carried out to obtain material with a predetermined level of creep damage. Then, miniature disc tests were undertaken under accelerated conditions. Deflection of the disc plotted against time for the steel 14MoV63 at a temperature of 600^0C for loads of 425, 450 and 475 N is shown in Figure 2.

Parker etc. published their research work in 1998. Disc testing was carried out at 550 ^{0}C and 585 ^{0}C for loads from 250 to 600 N. The rupture lives observed were in the range 20 to 4600 hours. After the initial deformation noted on loading the displacement-time behaviour exhibited a "primary" region of decreasing rate. The deformation rate was approximately constant for a significant part of the curve before the deformation rate increased leading to failure. Tests under the same conditions indicated very good reproducibility.

For a given test temperature, the rupture life increased markedly as the applied load decreased. Moreover, the data at 550°C and 585°C exhibited similar trends with results at the lower temperature being about 6 times the lives noted at 585°C. The results obtained could therefore be described using a relationship of similar form to the uniaxial tests.

$$\frac{1}{t_f} = B(Load)^m \exp(-\frac{Q_c}{RT})$$
(2.3)

where B is a constant and m is the load exponent of rupture life. For the current rotor steel the data could be represented by an m value of 6.5 and Q_c of 330 KJ/mol. As would be expected for tests performed using the same geometry, the applied load in the disc tests is proportional to stress in a conventional uniaxial creep experiment. Moreover, the values of Qc were similar for disc and uniaxial tests indicating that the rate determining damage processes were the same in both cases.

2.3   Summary

In comparison of determination of fracture toughness, determination of creep properties is still in an developing stage. Up to now, only limited work has been carried out on determination of creep properties using small punch test. The difficulties are both in test facility and in interpretation of results.

First of all, for a small punch test at creep region, an environment chamber is needed. In order to protect the specimen from oxidation, the environment chamber should be filled with inertial gas like nitrogen. The temperature oscillation gives strong influence on the test results. According the participants of the COPERNICUS project, temperature change in 1 °C, the deflection will change in 2%. The loading oscillation gives the same influence, but load control is easier than temperature control.

Secondly, most work on creep behaviour using small punch test is based on empirical equations. This work aims to find the relation between rupture time and applied loading by empirical equation. Another way is based on the equivalent stress concept provided by CESI, which aims to find the equivalent loading value in small punch test to obtain the same time to rupture as 1-D test. However, these methods only provide limited information of creep properties. For life time assessment purpose, complete creep properties, including Norton creep law, rupture time vs. stress and Dobes-Milicka relation, are needed. It is the driving force for this investigation.

3   AN ANALYTIC MODEL FOR DETERMINE OF CREEP PROPERTIES

3.1   Analytical procedure

Encouraging by the Manahan method to derive the stress-strain constitutive behaviour from small punch load-deflection curve (Manahan, 1982), a similar analytical procedure can be introduced to interpret deflection vs. time curve from small punch test. The analytical procedures used represent a critical aspect of the small punch test interpretation. The procedures consist of the following three steps:

1. Selection of creep model and identification of material parameters
2. Determination of material parameters by using the SPT creep curves (deflection vs. time) from small punch test at a certain temperature with several loading levels
3. By using the material parameters obtained, determination of creep properties at the test temperature. These creep properties include Norton's creep law ($\dot{\varepsilon} = A\sigma^n$), rupture time vs. stress and Dobes-Milicka relation at the test temperature.

The obtained material parameters are functions of temperatures. If one set of creep curves at different temperatures is available, a least square method can be used to determine the relations between the material parameters vs. temperatures. The form of Dorn equation (see paragraph 3.4) can be used to describe the temperature dependency.

3.2 Kachanov model

There are several general accepted creep models to describe isostress creep properties. In this report, only the Kachanov model is addressed.

According to Kachanov model, a creep curve can be governed by the following coupled equations:

$$\dot{\varepsilon} = At^{-m}\left(\frac{\sigma}{1-\omega}\right)^n \qquad (3.1)$$

$$\dot{\omega} = B.t^{-m}\frac{\sigma^v}{(1-\omega)^\eta} \qquad (3.2)$$

where ω denotes the damage factor, $\omega = 0$ for untouched and $\omega=1$ for failure. The integration of the coupled equations gives a relation between creep strain rate and time.

$$\dot{\varepsilon} = A\sigma^n t^{-m}[1-(t/t_R)^{1-m}]^{-\frac{n}{\eta+1}} \qquad (3.3)$$

Integral the equation above, we have

$$\varepsilon = \varepsilon_R\{1-[1-(\frac{t}{t_R})^{1-m}]^{1-\frac{n}{\eta+1}}\} \qquad (3.4)$$

or

$$t = t_R[1-(1-\varepsilon/\varepsilon_R)^{\frac{\eta+1}{\eta+1-n}}]^{\frac{1}{1-m}} \qquad (3.5)$$

where

$$t_R = [\frac{1-m}{B(\eta+1)\sigma^v}]^{\frac{1}{1-m}} \qquad (3.6)$$

$$\varepsilon_R = \frac{A\sigma^n t_R^{1-m}}{(1-m)(1-\dfrac{n}{\eta+1})} \tag{3.7}$$

Here A, B, m, n,ν and η are six Kachanov parameters. With these six parameters known, all creep properties can be derived. A simplified form of Kachanov model was introduced by Cane (Cane, 1982). In the simplified form, the primary creep is neglected, i.e. m = 0. In addition, if the rate of damage accumulation is strain controlled, as appears to be the case in ferritic steels, i.e. $\omega = \omega(\dot{\varepsilon}, \omega)$, then we expect ν = n. In this way, the number of Kachanov parameters can be reduced to four. In order to give a general approach, the simplified model will be not addressed in this report.

3.3 Determination of Kachanov parameters

A similar procedure to Manahan method to determine the stress-strain constitutive curve from small punch test load-deflection curve can be introduced to interpret creep curves from small punch test. The proposed approach consists simply of matching the observed creep curve (deflection vs. time) from small punch test (up to the point of time to rupture) to a curve from a database of curves corresponding to a range of creep properties. Such a database may be generated via small punch creep test on materials with known creep properties, or via a series of finite element creep analysis of the small punch test.

The matching scheme utilises an optimisation procedure that determines the material parameters that provide the closest match to the observed deflection vs. time curve from small punch test. The closest match is determined through successive searching and interpolating between the database curves by attempting to minimise the sum of the squares of the residuals (difference between observed curve, and matching curve, deflection). Specifically, the procedure attempts to minimise the function, F(x):

$$F(x) = \sum_{i=1}^{n_p} w_i[f(t_i, x) - f_0(t_i)]^2 \tag{3.8}$$

where x represents the vector of material parameters to be estimated; In the Kachanov model, the material parameters are A, B, m, n,ν and η; f is the calculated deflection at each time point in the data base, t_i; f_0 is the observed (measured) deflection at each time point t_i.

And n_p and w_i are the number of discrete points and associated weights used for the procedure. In the procedure, w_i = 1.0 as a default value.

An important element of the "matching" procedure is the database of curves spanning the range containing the measured SPT creep curve (deflection vs. time) from small punch test. Increasing the number of curves within a range generally results in improved accuracy of the creep property prediction. Further, the procedure requires that the user have a database of curves spanning the experimental curve. With current computational capability, such a database may be generated. Finally the database is generated by finite element calculation which curves are corresponding to the combinations of the material parameters A, B, m, n,ν and η.

3.4 Determination of creep properties

It is important to remember that the obtained material parameters are temperature dependent. Creep properties determined by the obtained parameters are only valid at the test temperature.

With the Kachanov parameters known, the creep properties of material can be determined. The creep strain rate can be determined by equation (3.3), the rupture time can be determined by equation (3.6) and the rupture strain can be determined by equation (3.7).

If the primary creep stage can be neglected, i.e., m= 0, equations (3.3), (3.6) and (3.7) can be rewritten as well known expressions of creep properties.

The creep strain rate can be determined by parameters A and n in the form of Norton creep law

$$\dot{\varepsilon}_{min} = A\sigma^n \tag{3.9}$$

The rupture time vs. stress can be derived by parameter B, ν and η as follows,

$$t_R = 1/(B(\eta+1)\sigma^\nu) \tag{3.10}$$

The rupture strain can be determined by parameters A, n and η and notice eqn (3.9), the rupture strain can be expressed in the form of Dobes-Milicka relation

$$\varepsilon_R = \lambda \dot{\varepsilon}_{min} t_R \tag{3.11}$$

where

$$\lambda = (\eta+1)/(\eta+1-n)$$

The obtained material parameters are derived at test temperature. If a creep curve at another temperature is available, a set of parameters is derived at another temperature level. A least square method can be used to determine the relations between the material parameters vs. temperatures. The form of Dorn equation can be used to describe the temperature dependency. For example, the Norton's creep law can be expressed in the form of

$$\dot{\varepsilon}_{min} = A \ \sigma^n \ \exp(-Q \ /RT) \tag{3.12}$$

4 FIRST APPROXATION

The matching procedure for determining creep properties sounds a good idea, however, it is difficult to be realised. Because Kachanov model has six parameters with which thousands of combination can be made. In addition, among the six Kachanov parameters, only A, m, n and η give influence on creep strain rate while B, m, v and η give influence on damage progress. If one wants to use the matching procedure to determine all parameters, both creep law and damage law should be included in finite element calculation, i.e. so-called coupled calculation.

Therefore, an efficient and simple method is needed and put forward here. The method is firstly to find approximations for the Kachanov parameters and then improve them by the matching procedure. This can be done by converting creep curves (deflection vs. time) from small punch test into isostress creep curves (strain vs. time), from which Kachanov parameters can be derived. The approximate solution can be improved by the matching procedure mentioned above.

4.1   Converting deflection curve to isostress creep curve

Actually, the idea of converting creep curves (deflection vs. time) from small punch test to isostress creep curve (strain vs. time) has been put forward by CESI (Tettamanti and Crudeli, 1998). The method can be used here and quoted as follows.

During the deflection the disk shows an extension of the fibres on the tensile side, in first approximation, the disk bends with spherical radius with fold angles from 0 to 180^0. According to this consideration, the extension of the arc, that insists on the same chord (equal to the hole matrix, 4 mm), can be calculated using a simple geometric diagram shown in Figure 3. In this way, creep rate of the fibres on the tensile side of disk can be derived from the creep curve (deflection vs. time) and assumed as the isostress creep rate.

The creep curve (deflection vs. time) published by CESI and shown in Figure 1 can be converted to isostress creep curve (strain vs. time) as shown in Figure 4.

4.2   Determination of Kachanov parameters from isostress creep curves

The method to derive Kachanov parameters from isostress creep curves has been put forward by Prof. Hayhurst and quoted here (Dunne, Othman, Hall and Hayhurst, 1989).

Equation (3.4) can be normalised to give

$$\varepsilon/\varepsilon_R = 1-[1-(\frac{t}{t_R})^{1-m}]^{1-\frac{n}{\eta+1}} \tag{4.1}$$

or $v = 1-[1-\tau^{1-m}]^\Delta$ \hfill (4.2)

where V is the normalised strain, τ the normalised time, and Δ is the material constant group [1-n/(η+1)]. Non-linear least square method can be used for fitting equation (4.2) to isostress creep data to obtain the two material constants m and Δ for a given temperature. In this way, m_i and $(n/(\eta+1))_i$ can be determined for each creep curve with stress level σ_i.

Returning to the original equation (3.4) and (3.7), it can be seen that this equation contains three constant groups; $1-m$, $[1-n/(\eta+1)]$ and $A\sigma^n$. Since the rupture time and rupture strain are known, equation (3.7) may be rearranged to give the third constant group as

$$A\sigma^n = \varepsilon_R (1-m)[1-n/(\eta+1)]/t_R^{1-m} \qquad (4.3)$$

From equation (4.3), the third constant group $(A\sigma^n)_i$ is determined for each stress level σ_i. The constant m and the group $n/(\eta+1)$ are assumed to be independent of stress level, however, it is found that they are not perfectly constant and hence the average values over all stress levels are determined as follows

$$m = \frac{1}{N}\sum_{i=1}^{N} m_i \qquad (4.4)$$

$$n/(\eta+1) = \frac{1}{N}\sum_{i=1}^{N}(n/(\eta+1))_i \qquad (4.5)$$

where N is the number of stress levels or tests carried out.

The group $A\sigma^n$ is dependent upon stress but A and n may be found by putting

$$f(\sigma) = A\sigma^n \qquad (4.6)$$

and taking logarithms

$$\log[f(\sigma)] = \log A + n\log\sigma \qquad (4.7)$$

A straight line may be fitted to the equation (4.7) by means of least square technique to yield best fit values for A and n. Having determined n, η may be determined using equation (4.5).

The material constants B and ν are determined by equation (3.7) which can be rewritten

$$B\sigma^\nu = \frac{1-m}{(\eta+1)t_R^{1-m}} \qquad (4.8)$$

Taking logarithms and B and ν may also be derived by using least square technique.

As the first approximation, the stress level σ_i is assumed to be $F0_i/2$ as suggested by CESI.

5   IMPROVEMENT OF SOLUTION

Actually, the first approximate solution mentioned in paragraph 4 has given a good approximation for the material parameters. The inaccuracy comes from the assumption of stress ($\sigma = F0/2$) and the converting deflection curve to isostress creep curve. To improve the approximate solution a matching procedure mentioned in paragraph 3 is needed and the first approximate values of parameters can be used as the initial values in the matching procedure.

In order to carry out the matching procedure, a series of deflection curves covering a range of material parameters should be built up. These deflection curves are calculated by using finite element package ANSYS 5.6 at KEMA. As the initial values of material parameters are known, the range of material parameters can be determined accordingly.

5.1   Finite element model

Non-linear analyses due to large displacement, plasticity and creep of material are considered in the calculation. In addition, contact elements with sliding property between the disk, upper and lower dies are introduced. The finite elements are eight-node axi-symmetric quadrilateral, and in total 1372 of nodes and 1338 of elements are used in the model.

The model disk uses geometry with thickness 0.5 mm and diameter 8 mm. The diameter of the punch head is 2.5 mm and the hole of the die is equal to 4 mm (see Figure 5). These dimensions are used in the European project and slight different to the dimensions used in EPRI report. A sliding interface (contact elements) is used to account for the changing contact between the punch head and specimen. The friction factor used in calculation is 0.6.

As large displacement is considered, the so-called "element co-ordinate" is used for results output. For example, creep stress distribution at 900 hours is shown in figure 6.

5.2   User defined creep law

In ANSYS package, only primary and secondary creep is included in the standard creep laws. Therefore, a user defined FORTRAN subroutine is needed, in which the creep law is specified by the user. The basic formulas have been described from eqn. (3.3) to (3.7).

The rupture time t_R is known from the test record and can be directly used as input data in stead of parameters B and v. The number of input data is five: A, m, n, η and t_R, among them the rupture time t_R is known. In this way, the number of parameters to be determined reduces to four and that makes the matching procedure easier.

As creep stress redistribution, stresses will change continuously during creep progress. The relation between creep strain rate and time will be changed accordingly. In the calculation, strain- hardening law is used to calculate the accumulated creep strain, i.e., the creep rate is calculated using the accumulated strain rather than using the real time. The calculation steps for creep rate are described as follows:

- use the rupture time t_R from the test record in stead of parameters B and v
- use eqn. (3.7) to calculate the rupture strain ε_R
- use eqn. (3.5) and the accumulated strain ε to calculate a fictitious time t
- use eqn. (3.3) and the fictitious time to calculate the creep rate

The details of program will not be described in the report.

5.3   Build up data base

As mentioned above, a series of deflection curves covering a range of material parameters should be built up. These deflection curves are calculated by using finite element package and stored in a database. However, it is impossible to build up a general database as six Kachanov parameters cover a broad range and thousands combination could be made. It is only feasible if the initial values of parameters are known. In addition, the rupture time t_R is introduced from test record and the number of parameters to be determined reduces to four which makes the matching procedure easier.

For each material with specified material parameters, A, n , m and η, finite element calculation provides a creep curve (deflection vs. time). The range of material parameters can be determined by the initial values of parameters accordingly. For example, the parameter scatter bands can be specified as 90%, 100% and 110%. The number of combinations for four parameters A, n, m and η is 81 and for six parameters will be 729!

The calculated creep curves (deflection vs. time) are used for matching the observed (measured) curve to determine the material parameters. The matching equation is shown as eqn. (3.8). Later on, the parameter B and ν can be improved by using the matching resulted m, η and repeat the regression procedure in eqn (4.8).

6   EXAMPLES

EXAMPLE 1: CESI tests (Tettamanti & Crudeli, 1998)

Six Small Punch tests were made for an aggregate test time of about 2500 hours. All the creep Small Punch tests are at the same load value 74 N but accelerated on temperature to obtain creep life from few hours to 1000 hours. Material is Steel ASTM A335 P12. Tests are carried out in argon atmosphere .

The disk specimen uses geometry with thickness 0.5 mm and diameter 8 mm. The diameter of the punch head is 2.5 mm and the hole of the die is equal to 4 mm. The measured creep curves (deflection vs. time) is shown in Figure 1 and the converted isostress creep curves is shown in Figure 4.

In order to find the approximate Kachanov parameters, the procedure described in paragraph 4.2 is followed. Normalising the converted isostress curves by using eqn. (4.1) and (4.2), parameters m and Δ are derived by using least square technique. Taking average over six curves, we have:

m = 0.27833 and Δ = 0.475

The second step is to calculate constant group $A\sigma^n$ by eqn. (4.3) and determine parameters A and n by using eqn. (4.7). Unfortunately, as all tests carried out at the same stress σ, the least square fitting of eqn. (4.7) can not be carried out.

EXAMPLE 2: COPERNICUS project (Ule etc., 1997)

Two alloyed steels were used in the COPERNICUS project. One of them was the low-alloyed steel 14MoV63, whereas the other was the high-alloyed steel X20CrMoV 121. Miniature disc tests were undertaken under accelerated conditions. The disk specimens have the same geometry as CESI tests in example 1. Deflection of the disc plotted against time for the steel 14MoV63 at a temperature of 600°C for loads of 425, 450 and 475 N is shown in Figure 2.

From the presented diagrams it can be clearly seen that the large initial displacement of the punch (amounting in some cases to more than 1 mm in less than one minute) is the consequence of the initial plastic deformation of the disc while the load is acting on the disc only over small contact area, which causes very high stress. During converting deflection curves to isostress curves, deflection due to plastic deformation should be excluded and only deflection due to creep left. As no original data available, the deflection due to plastic deformation is assumed to be 1 mm and subtracted from the total deflection. Using the geometric relations in Figure 3, the deflection curves can be converted to isostress creep curves and shown in Figure 10a.

Again, the procedure mentioned in paragraph 4.2 is followed to determine Kachanov parameters A, B, m, n,ν and η. Normalising the converted isostress curves by using eqn. (4.1) and (4.2), parameters m and Δ are derived by using least square technique. An example of curve fitting to determine parameters m and Δ is shown in Figure 10b. The fitting results are summarised in table 1.

Table 1   Fitting results for constant group m and Δ

Curve No.	1	2	3	4	5	6	7	8	9	Average
Load (N)	475	475	475	450	450	450	425	425	425	
Temp. (°C)	600	600	600	600	600	600	600	600	600	
m	0.64	0.64	0.60	0.60	0.56	0.56	0.58	0.58	0.57	0.588
Δ	0.38	0.40	0.30	0.38	0.40	0.38	0.30	0.36	0.32	0.358

The second step is to calculate constant group $A\sigma^n$ by eqn. (4.3) and determine parameters A and n by using eqn. (4.7). According to CESI suggestion, the stress σ is taken as F0/2. The fitting results are summarised in table 2.

Table 2   Fitting results for parameter A and n

Curve No.	1	2	3	4	5	6	7	8	9	Fitting results
Stress (MPa)	237.5	237.5	237.5	225	225	225	212.5	212.5	212.5	
Rupture	525	525	552	775	775	840	1520	1440	1325	

time (h)										
Rupture strain	0.181	0.136	0.144	0.144	0.141	0.136	0.156	0.116	0.106	
$A\sigma^n$ (*10^{-3})	2.019	1.512	1.576	1.370	1.342	1.246	1.123	0.856	0.807	
A										$1.69085*10^{-16}$
N										5.474898

With the parameter n known, the parameter η can be determined from the definition of Δ:

$$\eta = \frac{n}{1-\Delta} - 1$$

and η = 7.5249.

With parameters m = 0.5878 and η = 7.5249, the constant group $B\sigma^\nu$ can be calculated by eqn. (4.8). The parameters B and ν can be determined by least square fitting. The fitting results are summarised in table 3 and the six Kachanov parameters are listed in table 4.

Table 3   Fitting results for parameter B and ν

Curve No.	1	2	3	4	5	6	7	8	9	Fitting results
Stress (MPa)	237.5	237.5	237.5	225	225	225	212.5	212.5	212.5	
Rupture time (h)	525	525	552	775	775	840	1520	1440	1325	
$B\sigma^\nu$ (*10^{-3})	3.657	3.657	3.582	3.115	3.115	3.013	2.359	2.413	2.497	
B										$7.96326*10^{-12}$
ν										3.6469

Table 4   Estimated Kachanov parameters

A	B	n	m	ν	η
$1.69085*10^{-16}$	$7.96326*10^{-12}$	5.4749	0.5878	3.6469	7.5249

With Kachanov parameters known, all the creep properties of testing material can be derived. As parameter m ≠ 0, the primary creep can not be neglected. The parameter n = 5.4749 is the exponent of secondary creep stage, i.e. the exponent of Norton law. The slope of relation of rupture vs. stress can be derived as $\nu/(1-m)$ = 8.847. The Dobes-Milicka $\lambda = (\eta+1)/(\eta+1-n)$ turns out to be 2.795.

LOW ALLOY STEELS II

Creep behaviour of a low alloy ferritic steel weldment

S Fujibayashi* and T Endo**

* Idemitsu Engineering Co., Ltd., 37-24 Shinden-cho, Chiba, Japan
** Yokohama National University, 157 Tokiwadai, Yokohama, Japan

1. Abstract

Nowadays the preferential creep damage accumulation at the intercritical HAZ(ICZ), termed Type IV damage, has been a great concern for a power generating and refining industry. In this paper, creep behaviour of the service exposed 1.25Cr-0.5Mo steel weldment has been examined. The influence of multiaxial stress state on the susceptibility to Type IV cracking is assessed experimentally using plain and notched creep specimens. The transition of fracture mode, from a ductile parent material failure to a brittle failure at ICZ, is observed with the increase in constraint. The weldment is composed of two kinds of parent materials, a flange and pipe. The failures with a brittle manner have taken place at the ICZ on the flange side though the flange parent shows higher creep strength. The reason for higher grain boundary damage at the flange HAZ shall be discussed.

2. Introduction

Three circumferential butt welds between a pipe and flange were removed from inlet piping to a catalytic reformer in a refining plant for this work. These welds have got service-induced grain boundary damage with various degrees. Designing and operating conditions for the weldments are as follows.

Design temperature:550^{0}C, Operating temperature: 500^{0}C
Design pressure: 2.65MPa, Operating pressure : 1.2~2.5MPa
Internal fluid: Naphtha +H_2
External diameter: 660mm, Thickness : 22mm
Operating duration: 23 years (184,000 hours)

3. Damage distribution of the service-exposed welds

Though no defect was found by Wet Fluorecent Magnetic Test (WFMT) when three welds were sampled, the grain boundary damage was detected by replication tests. A large difference in damage level was observed among three casts which had been closely located during operation. Most severely damaged weld has got Transverse Weld Metal Cracking (TWMC), Type III cracking, Type IV damage and isolated cavitation at the parent material which was located adjacent to HAZ on the flange side. However, the damage of welds for the counter flange was much less significant, only isolated cavitation at the coarse grained HAZ (CGZ) on the flange side was observed. From the feature of the weldment and high hardness value at the weld metal (240 Hv) of the most heavily damaged cast, it was concluded that this weldment was repaired at the fabrication with unsuitable PWHT. Higher residual stress and brittle microstructure associated with lower temperature for PWHT should be attributed to TWMC and multidirectional cracking at CGZ. No damage was detected on the pipe side in these three casts.

In this paper, a moderately damaged cast containing micro-cracking at CGZ and isolated cavitation at ICZ has been examined. The direction of the localized micro-cracks at CGZ were perpendicular to the weld at the external surface and allied cavities at the grain boundaries parallel to the weld were also observed at the cross-section near the external surface. Chemical compositions of a flange, pipe and weld metal are shown in Table.1.

Table.1 Chemical compositions of materials (wt%)

	Flange	Weld Metal	Pipe
C	0.12	0.066	0.10
Si	0.43	0.43	0.65
Mn	0.51	0.61	0.47
Cr	1.31	1.37	1.21
Mo	0.54	0.57	0.45
Ti	<0.001	0.009	0.002
V	0.005	0.008	0.002
Nb	<0.002	<0.002	<0.002
S	0.015	0.009	0.004
P	0.015	0.009	0.012
Sn	0.023	0.012	0.012
As	0.029	0.019	0.012
Sb	0.003	0.01	0.001
Cu	0.11	0.092	<0.05
C.E.F	0.192	0.180	0.056

A parent material on the flange side, which is associated with the service induced creep damage at HAZ, contains higher tramp elements such as phosphorus, tin , arsenic and antimony than a pipe parent which shows no creep damage. Takamatsu et al.[1] correlated the impurity contents using the following factor with creep ductility at CGZ for 1.25Cr-0.5Mo steel. This factor, termed Creep Embrittlement Factor (C.E.F), was originally proposed by King to assess the susceptibility to stress relief cracking for 0.5Cr-0.5Mo-0.25V steel.

C.E.F=P+3.57Sn+8.16Sb+2.43As

Though threshold value to assess the likeliness of premature cracking at CGZ has not been derived, values of C.E.F for the cracked reactors made from 1.25Cr-0.5Mo steel found in relevant papers were higher than 0.15.

The flange parent produced by forging shows a higher level of bainitic transformation product. On the other hand, the microstructure of the pipe parent is composed of ferrite-pearlite.

Though a detailed welding procedure at the fabrication is not available, it can be presumed from the feature of the welds and the common welding practice employed then that the following procedure should have been applied.

Geometry for a welding groove was V and Tungsten Inert Gas (TIG) welding was applied to a root run and the remaining was filled with Manual Metal Arc (MMA) welding. Presumably, Post Welding Heat Treatment (PWHT) should have been performed at 700^{0}C for two hours.

The hardness profile of the weldment measured in the middle of the wall thickness is shown in Fig.1. No significant difference between both parent materials can be seen.

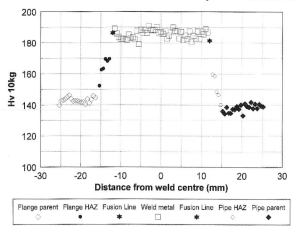

Fig.1 Cross-weld hardness distribution at the position 11mm from the outer surface

4. Experimental Procedure

To examine the cross-weld creep behaviour, two types of specimens were prepared. One was a 14mm×14mm square specimen to observe the grain boundary damage at interruptions during creep tests. The other was a circular specimen with helical vee notches applied to all the microstructures constituting the weldment. The geometry of a spiral notched specimen referred to the relevant paper[*(2)] is shown in Fig. 2. The length of a screwed part is 40mm in 50mm gauge length.

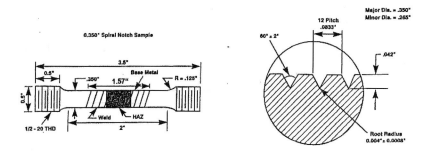

Fig.2 The geometry of a spiral notched specimen (Sizes are denoted in inch.)

The area ratio of a weld metal in the gauge length for a square and spiral notched specimen is approximately 0.19 and 0.34 respectively. Creep and creep rupture tests were conducted in air with a constant load technique. The damage development due to creep for a cross-weld specimen was observed using a replication technique. In order to quantify the damage at ICZ, Cavity Density (C.D.) counted by an optical microscope with 500 times magnification was employed.

5. Experimental Results

5.1. Creep Rupture Properties of service-exposed materials

Results of creep rupture tests for homogeneous specimens, square and spiral notched cross-weld specimens are shown using Manson-Haferd Parameter in Fig.3. Most of rupture data for parent materials, a weld metal and the cross-weld lie above the lower boundary of virgin materials (mean strength -3•which is based upon NRIM 21B for Normalized and Tempered 1.25Cr-0.5Mo steel plate) at the stresses of 30 to 125MPa. Despite similar values of hardness of both parent materials (around 140Hv), the creep lives of a flange material are approximately three to four times longer than those of a pipe. In the case of 2.25Cr-1Mo steel, a relatively good correlation between hardness and creep strength has been found, but it might not be applicable to 1.25Cr-0.5Mo steel. A flange material shows higher creep strength than mean value of NRIM data at stresses lower than 100MPa, to which the actual high temperature components are likely to be subjected.

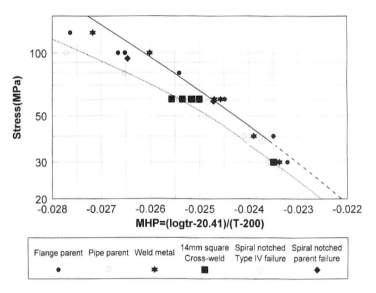

Fig.3 Manson-Haferd Parameter-Stress correlation for service exposed 1.25Cr-0.5Mo steel

As previously noted, the weldment has got service-induced Type IV damage, which is categorized as isolated cavitation, on the flange side. Cavity densities before tests approximately ranged from 200 to 300 (n/mm^2). Pre-existing Type IV damage at the flange ICZ has not caused Type IV failure for a square specimen. Most failures with significant grain deformation have taken place at the pipe parent despite damage development at ICZ on the flange side. Two specimens failed at the HAZ on the flange side. One failed with the mixed mode, with Type III at the external surface, Type IV in the middle and the base metal at the bottom. The other one was broken predominantly with Type III manner. However, in this specimen Type IV damage to the level of micro-cracking was observed in the middle of wall thickness after the failure. The ICZ damage at the external surface was significantly lighter.

5.2. Creep Rupture Tests of Vee Notched Specimens

To increase the triaxiality to all the microstructures constituting the weldment, creep rupture tests using the helical notched specimens were conducted. By applying notches, ultimate failures with the Type IV manner were promoted despite relatively short duration of tests. The crack penetration preferentially through the ICZ was observed in these specimens. In Fig.4, the cross-section failed with the Type IV mode is shown.

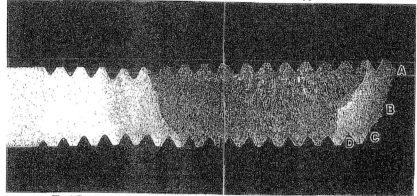

Fig.4 Cross-section of a spiral notched specimen @ 610°C and 60MPa (×2)

Time to rupture for all the specimens tested at the nominal stress of 60MPa is shown in Fig. 5. As for homogeneous specimens, life prediction by extrapolating the results gained at higher temperatures can be achieved with reasonable accuracy. In the case of cross-weld specimens, the failed position of a cross-weld specimen depends upon the test temperature (or test duration). The transition of failure position could result in nonlinear correlation between logarithmic rupture life and temperature. This tendency is more pronounced in notched specimens. At the temperature of 650°C, the final failure of notched specimen is caused by transgranular rupture at a pipe parent associated with significant grain deformation. In this case, time to failure is extended from the lower boundary to the mean value in NRIM 21B due to notch strengthening effect of the ductile pipe parent. It can be considered that the equivalent stress is operative for a pipe parent in the current testing condition. Therefore the decrease in the equivalent stress associated with notches should contribute to the longer rupture life for a pipe parent.

At lower temperatures, failure location changes from the pipe parent to the ICZ on the flange side. Time to rupture at 630°C is approximately the same as that of mean value of virgin materials and it decreases to the lower boundary at the temperature of 610°C.

The reduction of rupture life for those failed at ICZ suggests that decrease in ductility under the multiaxial stress state plays an important role for the Type IV failure.

The stress and strain distribution of notched specimens is now being examined using

Finite Element Analysis.

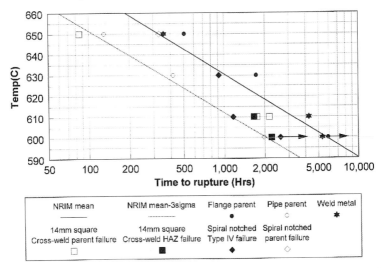

Fig.5 Iso-stress creep rupture behaviour of homogeneous and cross-weld specimens at 60MPa

5.3. Damage development with life consumption

A replication technique has been widely utilized to assess the condition of high temperature components in actual plants. Walker et al. [3] examined the development of grain boundary damage associated with creep using 0.5Cr-0.5Mo-0.25V steel weldments. In their work, fairly good correlation between life consumption and cavity density was found.

The change in Cavity Density (C.D.) with life consumption in this work, measured during and after the creep-rupture tests, is shown in Fig.6. C.D. for a square specimen was measured at the highest position found in the cross-section replica taken at each interruption and that for a spiral notched specimen was done by sectioning a specimen. The large difference in the value and increase rate of C.D. between the flange ICZ and the pipe ICZ can be seen. Cavitation in a square specimen is not generated until the final stage of creep rupture on the pipe side. In contrast, cavity density on the flange side increases with the life fraction consumed. It can be said that ICZ generated on the flange side is more susceptible to cavitation.

As most of ultimate failures of square specimens have taken place at the pipe parent to date, correlation between life consumption and C.D. measured on the flange ICZ of square specimens is somewhat misleading. (C.D. on the flange side at the failure should be higher if the final fracture happened at the flange ICZ.) In the case of helical notched specimens, C.D. for the failed specimen on the flange side becomes higher, approximately ranging from 2,500 to 4,000. Increase in C.D. at failures on the pipe side was also observed in spiral notched specimens. A small number of C.D. measured in the notched specimen sectioned at the half of its life suggests the accelerated increase in C.D. at the final stage of creep life.

The cracking at the flange CGZ was observed in most specimens when the half of a

rupture life was consumed. But mostly Type III crack did not penetrate the wall.
And no creep damage has been detected at a weld metal.

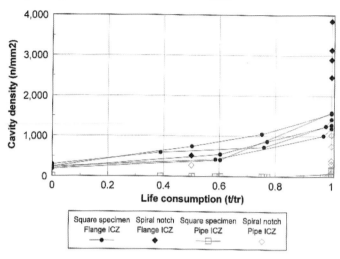

Fig.6 Life fraction-Cavity Density correlation

6. Discussion

The main objective of this work is to establish the procedure for the remnant life
prediction of welds, especially against Type IV damage. The first hurdle is how to
generate Type IV damage experimentally. To obtain the results within the reasonable
time scale, the way of acceleration must be chosen. As the rupture test at high stress
(125MPa) and low temperature (550°C) resulted in ductile failure at the pipe parent and
the iso-stress rupture tests are more convenient to correlate the experimental results with
the remnant life, creep tests with accelerated temperatures have been employed in this
work. However, Type IV failure has not been generated by plain specimens to date.
The correlation between the temperature/rupture life and failure mode is shown in Fig.7.
According to the works on the cross-weld behaviour of low alloy steels by Brear et
al. [4], the transition to Type IV mode takes place with increase in rupture life. And the
final failure mode for 2.25Cr-1Mo steel welds is exclusively determined by the time to
rupture.

On the other hand, Parker et al. [5] succeeded in generating Type IV cracking of 1.25Cr-
0.5Mo steel with short term creep tests at relatively high stresses (135-160MPa) and
low temperature (580°C) where clear decrease in creep ductility was found. The
authors pointed out the importance of ductility to promote the Type IV cracking.

In the current work, Type IV failures were generated in short term creep tests by
introducing notches. This pheomenon can be interpreted as the reduction of ductility at
ICZ under the multiaxial stress state. The tendency of the significant creep damage in
the middle of the wall thickness might be attributed to the same reason.

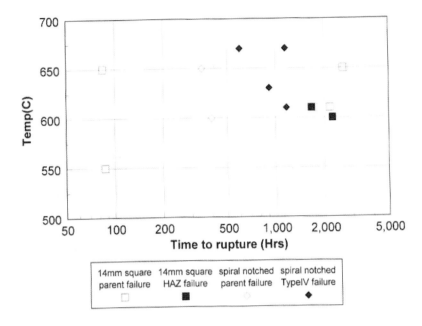

Fig.7 Failure mode versus rupture life and testing temperature

As described above, significant difference in the life consumption-cavity density correlation between both ICZs is observed. Despite higher creep strength of the flange parent, the grain boundaries at the flange HAZ suffer more damage than those of the weaker pipe.

The reason for higher susceptibility to cavitation on the flange side has not been confirmed. But from the higher contents of tramp elements, higher susceptibility to cavitation might be attributed to the grain boundary segregation of the impurities as observed in Type III damage. The similar tendency on the welds between a forged Tee and a casting valve made from 1.25Cr-0.5Mo steel for a power generating plant was observed in the work by Westwood et al.[6]. In this case, ICZ on the dirtier cast side was preferentially damaged.

The role of impurities for Type IV failures shall be examined in further work.

7. Conclusion

1) The ultimate failure position dose not necessarily depend upon the creep strength of parent materials when loaded with cross-weld manner.
   In spite of higher creep strength of the parent material for a forged flange, ICZ on the flange side suffer more damage than that generated in a weaker pipe.

2) Despite short testing duration, Type IV failure for 1.25Cr-0.5Mo steel weldment was generated due to the notch weakening effect of ICZ.

3) Cavity density could be a potential tool for the condition monitoring for welds, especially those located at the stress concentrated area such as a nozzle intersection. However, the susceptibility to cavitation could be cast dependent.

References

(1) K. Takamatsu, Y. Otoguro, K. Shinozuka and K. Hashimoto, Journal of Iron and Steel Institute of Japan, 1979, vol. 17, pp129-138

(2) J.E. McLaughlin, G. G. Karcher and P. Barnes , Service Experience and Reliability Improvement: Nuclear, Fossil and Petrochemical Plants, 1994, ASME PVP-vol. 288, pp351-366

(3) N. S. Walker, D. J. Smith and S.T. Kimmins, International Conference on Creep and Fatigue, 1996, pp341-350

(4) J. M. Brear, A Fairman, C. J. Middleton and L. Polding, International Conference on Creep and Fracture of Engineering Materials and Structures, 1999, pp35-42

(5) J. D. Parker and G. C. Stratford, MATERIALS AT HIGH TEMPERATURES, 1995, vol. 13, pp37-45

(6) H. J. Westwood, M. A. Clark and D. Sidey, International Conference on Creep and Fracture of Engineering Materials and Structures, 1990, pp621-634

A Study of Weld Repairs in a CrMoV Service-Exposed Pipe Weld

W. Sun[1], T. H. Hyde[1], A. A. Becker[1] and J. A. Williams[2]

[1] School of Mechanical, Materials, Manufacturing Engineering and Management
University of Nottingham, Nottingham NG7 2RD, UK

[2] Independent Consultant, East Leake, Leicester LE12 6LJ, UK

Abstract This paper summarises the work on a FE creep failure assessment, for a series of circumferential welds and weld repairs cases, in a service-exposed CrMoV main steam pipe. The material properties used for different material zones are related to a 1/2Cr1/2Mo1/4V: 2 1/4Cr1Mo pipe weld, at 640° C. Failure lives and failure positions predicted using both continuum damage modelling and steady-state analyses are presented. The effects of the differences in the relative properties of the constituents of the welds, the weld geometry or repair profiles as well as the system loading, on the failure life, failure position and the remaining lives of these welds were estimated. The results presented are useful for assessing the performance of service-aged and repaired welds in power plants pipework systems and could be of practical significance for the development of weld repair strategies.

1. Introduction

For the main steam piping systems in power plants, operating at elevated temperatures, weld repairs are often performed, when localised creep cavitation and cracking are found in or in the vicinity of the weldments. Experience from the replacement of the repaired welds has shown that creep damage is often found in or close to the repair earlier than would have been expected [1,2]. Extensive effort has been made to optimise weld repair techniques and to improve the high temperature performance of weld repairs [3,4].

Schematic diagrams of typical weld and weld repair geometries are shown in Fig. 1. A typical weld in a component consists of parent material (PM), heat-affected zone (HAZ) and weld metal (WM), Fig. 1(a). The material models for the new and aged welds are shown in Figs. 1(b) and 1(c). Figs. 1(d) to 1(f) show two possible weld repair profiles used to remove a typical HAZ crack running across the wall thickness. In one case, Fig. 1(d), the whole of the original weld and surrounding parent material is removed (full repair). In the other case, Figs. 1(e) and 1(f), only part of the original weld is removed (partial repair). The partial weld replacement technique has been generally used in the past but the full weld repair technique is now used in the great majority of cases in the UK [5].

The difficulty in weld assessment is primarily due to the variability in the material property distributions in the localised regions of welds and in the variable dimensions and shapes of welds, particularly for repaired welds. In addition, the effect of system loading on the failure behaviour may be significant. For instance, within the low temperature HAZ region of a weld in a main steam pipeline, type IV cracking may occur [5,6], which is directly influenced by the local stress state.

Finite element (FE) creep and damage analyses have been used to assess the creep performance of weld repairs (e.g. [2]), using the material properties generated from creep testing. Creep damage constitutive equations of the form [7]:

$$\dot{\varepsilon}_{ij}^c = \frac{3}{2} A \left[\frac{\sigma_{eq}}{1-\omega} \right]^n \frac{S_{ij}}{\sigma_{eq}} \qquad \text{and} \qquad \dot{\omega} = \frac{M \sigma_r^{\chi}}{(1+\phi)(1-\omega)^{\phi}} \qquad (1)$$

can be used in conjunction with the FE method to study the welds [8], where $\sigma_r = \alpha\sigma_1 + (1 - \alpha)\sigma_{eq}$ is defined as a rupture stress, related to the equivalent stress, σ_{eq}, and the maximum principal stresses, σ_1. ω is the damage variable $(0 < \omega < 1)$, A, n M, ϕ and χ are material constants and α is a tri-axial stress state parameter $(0 < \alpha < 1)$.

Since precise material behaviour models, especially for the HAZ material, and also the FE damage codes required, are not widely available, steady-state FE analyses, using Norton's creep law, i.e.

$$\dot{\varepsilon}_{ij}^{c} = \frac{3}{2} A\sigma_{eq}^{n-1} S_{ij} \tag{2}$$

are widely used in weld studies, particularly in parametric analyses. It has been shown that the creep failure times can be estimated using the steady-state creep rupture stress, σ_r, and the appropriate creep rupture material properties [9], i.e.

$$t_f = 1 / M(\sigma_r)^{\chi} \tag{3}$$

The present investigation is a summary of the work on the assessment of weld repairs in main steam CrMoV pipes, based on a comprehensive approach involving some aspects of metallurgical study, creep testing, material property generation and FE modelling etc [10-12]. The results presented focus on the failure assessment of a series of new, service-aged, fully repaired and partially repaired, internally pressurised, pipe welds, using the FE method with simplified axisymmetric models. The material properties used were related to a 1/2Cr1/2Mo1/4V: 2 1/4 Cr1Mo weldment at 640° C. Results presented allow direct assessment of the effects of weld repair profiles, relative material properties, and system loading on the failure life and position for these welds.

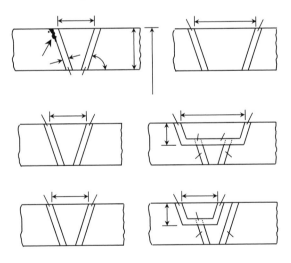

2.   Material properties

In all cases, the welds were assumed to consist of three material zones, i.e. the PM, HAZ and WM, each of which has constant creep properties. The material data used in the FE analyses for different material zones, Fig. 1(a), were obtained from creep tests at $640°$ C, in the stress range of 40 to 70 MPa, on different constituents of the new, service-aged and fully repaired 1/2Cr1/2Mo1/4V: 2 1/4Cr1Mo pipe weld in a main steam pipeline. Details of material descriptions, procedures of creep testing and material property generation and the material constants obtained have been reported [10]. It was found that in this practical stress range, all the HAZ materials are weaker than the parent materials and all weld metals are stronger than the parent materials with respect to the minimum strain rate and rupture strength, in each weld situation. Also, due to the nature of creep testing, the HAZ properties obtained have been assumed to be representative of the type IV properties instead of the average HAZ property [10]. Since the material properties for zones 8 and 9, Figs. 1(e) and 1(f), are unavailable, a number of different assumptions were made for these zones in the FE modelling.

3.   Weld dimensions and FE analyses

The pipe dimensions used were typical of ferritic 1/2Cr1/2Mo1/4V main steam pipe-lines in UK power plants, with an outer diameter, D, of 355.6 mm and a wall thickness, T, of 63.5 mm. The weld dimensions used have been given in Fig. 1(a) and are typical of those expected in repair situations.

A simplified axisymmetric pipe weld model was used in the FE modelling for all weld situations. The pipe welds were subjected to an internal pressure, p_i, and a uniform tensile axial stress, σ_{ax}, at the end of the pipe to represent system loading. In steady-state analyses, the failure lives were predicted using equation (3), using the peak rupture stresses for each material zone. The failure life of each weld is taken to be the minimum of the failure time obtained for all of the material zones, with the failure position being the position where the maximum peak rupture stress occurs in the zone in which failure is predicted [9]. In damage modelling, creep failure was assumed to have occurred when the failure damage, i.e., $\omega \rightarrow 1$, was achieved through a significant part of the wall thickness. Detailed descriptions on the FE modelling can be found in [11,12]. The effects of the residual stresses caused by welding and local stress singularities at the free surfaces of dissimilar material interfaces were not taken into account in the FE analyses; the repaired welds were post-weld heat treated.

4.   Failure assessment of weld repairs

For all weld cases, results presented were obtained from FE calculations with an internal pressure of $p_i = 16.55$ MPa. To investigate the effect of axial (system) loading, the mean axial stresses applied, σ_{ax}, cover a range of $\sigma_{ax}/\sigma_{mh} = 0.306$ (closed-end) to 1.0, which is the most extreme case allowed within some design codes [13], where σ_{mh} is the mean diameter hoop stress.

4.1   Typical behaviour under the closed-end condition

Detailed damage analyses indicate that for all welds, the high damage levels mainly occur in the HAZ (new HAZ of repaired welds) at all times. The failure damage zone ($\omega = 0.99$) first occurred near the outer surface of the pipe in the HAZ and then grew inwards. Figs. 2(a) and 2(b) show typical damage variations, with the normalised distance along the HAZ close to the

616

type IV region, at different times. An example of the damage variations across the HAZ for the fully repaired weld is shown in Fig. 3. It can be seen that the damage variations across the HAZ are small in the first half of the life time. However, with increasing time, high damage moves to the HAZ/PM boundary, and when t → t_f, the failure damage quickly approaches the whole width of the HAZ.

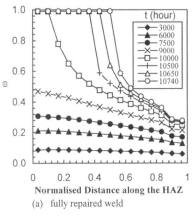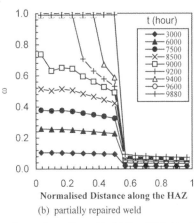

(a) fully repaired weld

(b) partially repaired weld

Fig. 2 Damage variations in and along the HAZ, starting from outer surface, near the type IV region, at different times, obtained for the fully and partially repaired welds, with σ_{ax}/σ_{mh} = 0.306 (p_i = 16.55 MPa).

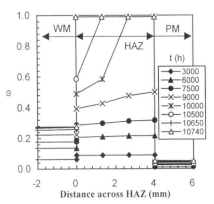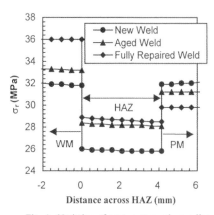

Fig. 3 Variations of damage across the HAZ, along a line parallel to and 5.1mm from the outer surface of the fully repaired weld, at different times, with σ_{ax}/σ_{mh} = 0.306 (p_i = 16.55 MPa).

Fig. 4 Variation of rupture stress along a line across the HAZ, parallel and 3.8mm to the outer surface of the pipe, for new, aged and fully repaired welds, with p_i = 16.55 MPa, under closed-end condition (σ_{ax}/σ_{mh} = 0.306).

Steady-state results showed consistent behaviour. In all cases, the peak rupture stresses in the HAZ gave the shortest failure lives, occurring at the positions close to the outer surface and at the HAZ/WM interfaces. However, it was found that the rupture stresses near the peak rupture stresses positions vary insignificantly across the HAZ [11]. Examples of the rupture stress variations across the HAZ, in the new, aged and fully repaired welds are shown in Fig. 4. It is clear that the failure lives predicted by the rupture stresses at the HAZ/WM interface or at the HAZ/PM interface will not be significantly different, although there is a tendency for the minimum life to occur adjacent to the HAZ/weld metal interface. The failure lives and failure positions predicted are summarised in Table 1, where the results for the partial repairs are based on the assumptions that the properties of zones 8 and 9, Figs. 1(e) and 1(f), are the same as those zone 7, i.e. the new HAZ in aged PM. Results obtained showed that in general, the failure behaviour of the partially repaired welds was similar to that of the fully repaired weld.

Table 1 Failure lives and positions predicted for the CrMoV welds (closed-end condition)

| Welds | | Damage | | Steady-State |
	t_f	Position	t_f (hrs)	Position
New weld	21,018	HAZ, near OD	14,795	HAZ, near OD
Aged weld	15,640	HAZ, near OD	8,942	HAZ, near OD
Full repair	10,803	HAZ, near OD	7,077	HAZ, near OD
Partial repair-i	9,892	HAZ, near OD	6,467	HAZ, near OD
Partial repair-ii	----	----	6,298*	HAZ (7), near OD*

* obtained using properties for zone 8 the same as those for zone 7

4.2   Effect of HAZ properties on the failure position of partial repair-ii

In the partial repair case-ii, the failure life and position presented in Table 1 are based on the assumptions that the properties of zone 8, Fig. 1(f), are the same as those of zone 7. For the purpose of investigating (qualitatively) the effect of the material properties of zone 8, on the failure behaviour of the partial repaired weld, steady-state calculations were also conducted using some other assumed properties for zone 8, by simply changing the A or M values only, and keeping all the other properties for zone 8 the same as those for zone 7. The results obtained are presented in Table 2. It was found that when zone 8 is "stronger" than zone 7, i.e. either having a higher failure ductility (with same rupture strength) or a higher rupture strength (with same failure ductility) than zone 7, the failure life of and failure position in the component is still controlled by zone 7. However, when zone 8 is "weaker", the failure will occur within zone 8, which is within the weld metal with a shorter calculated life, see Table 2.

Table 2 Effect of properties for zone 8 on the failure position of partial repair-ii (steady-state analyses, closed-end condition, with assumed material properties n, χ and α for zone (8) equal to those for zone (7))

$A^{(zone\ 8)}$	$M^{(zone\ 8)}$	t_f (h)	Failure Position	zone (8) compared with zone (7)	
$= A^{(zone\ 7)}$	$= M^{(zone\ 7)}$	6,298	HAZ (7 or 8), near OD	same as zone (7)	
$= A^{(zone\ 7)}$	$< M^{(zone\ 7)}$	6,298	HAZ (7), near OD	higher in rupture strength	"stronger"
$= A^{(zone\ 7)}$	$= 1.2 \times M^{(zone\ 7)}$	5,249	HAZ (8), near OD	lower in rupture strength	"weaker"
$= A^{(zone\ 7)}$	$= 1.5 \times M^{(zone\ 7)}$	3,149	HAZ (8), near OD	lower in rupture strength	"weaker"
$> A^{(zone\ 7)}$	$= M^{(zone\ 7)}$	6,298	HAZ (7), near OD	higher in failure ductility	"stronger"
$= 0.5 \times A^{(zone\ 7)}$	$= M^{(zone\ 7)}$	4,900	HAZ (8), near OD	lower in failure ductility	"weaker"
$= 0.2 \times A^{(zone\ 7)}$	$= M^{(zone\ 7)}$	3,475	HAZ (8), near OD	lower in failure ductility	"weaker"

4.3   Effect of the weld width on the failure life

The main geometry difference between a fully repaired weld and the aged/new welds is the weld width, w. The results obtained, under the closed-end condition, using the new/aged weld geometry, i.e. w = 46 mm, and the full repair geometry, i.e. w = 80 mm, with the new, aged and fully repaired weld properties are summarised in Table 3.

It can be seen that, in the range of w = 46 to 80 mm, the differences in failure lives for the new, aged and fully repaired welds are small (less than 10%). This indicates that, for a full repair situation, choosing different w values which are in this range should not result in a significant difference in the failure lives of the welds, although slightly shorter lives are obtained with the narrower weld width. The failure positions were found to be the same for the two geometries for all the weld cases examined.

Table 3 Failure lives (hrs) of the new, aged and full repaired welds with two w (mm) values (closed-end condition)

Property	Life (hrs) [Damage]			Life (hrs) [Steady-State]		
	w = 46	w = 80	$[t_f^{80}/t_f^{46}]$	w = 46	w = 80	$[t_f^{80}/t_f^{46}]$
New weld	21,018	21520	1.0239	14,795	14,990	1.0132
Aged weld	15,640	16,788	1.0734	8,942	9,357	1.0464
Full repair weld	10,402	10,803	1.0386	6,525	7,077	1.0846

4.4 Effect of the end loading on the failure life and position

The effect of the system loading was simply characterised by using higher axial loading, σ_{ax}, covering a range of $\sigma_{ax}/\sigma_{mh} = 0.306$ (closed-end) to 1.0. The failure lives obtained for different σ_{ax}/σ_{mh} are presented in Table 4. It can be seen that the differences in failure lives of the new, aged and repaired welds reduce significantly with increasing axial load. Detailed analyses showed that in all cases, the failures still occur in the HAZ, and near the outer surface. However, when σ_{ax}/σ_{mh} is high, failure is more likely to occur at the HAZ/PM boundary [12]. An example of the damage variations, across the HAZ, with time for the fully repaired weld, with $\sigma_{ax}/\sigma_{mh} = 1.0$, is shown in Fig. 5 and the corresponding steady-state rupture stress across the HAZ are shown in Fig. 6. It can be seen that, compared with the results for closed-end case for $\sigma_{ax}/\sigma_{mh} = 0.306$, shown in Figs. 3 and 4, the high damage values or the peak rupture stresses now move to the type IV position.

Table 4 Failure lives (hrs) for different end loading

σ_{ax}/σ_{mh}	New weld		Aged weld		Full repair		Partial repair-i	
	Damage	S-S	Damage	S-S	Damage	S-S	Damage	S-S
0.306	21,018	14,795	15,640	8,942	10,803	7,077	9,892	6,467
0.5	16,266	12,815	12,962	8,182	9,305	6,525	8520	6,082
0.75	8,274	6,484	7,197	5,415	6,206	5,185	5,632	4,530
1.0	4,186	3,265	3,966	2,895	3,751	2,971	3,307	2,678

4.5 Failure initiation time and remaining life

By obtaining the failure initiation time, t_o, the remaining life, $t_f - t_o$, can be estimated, where t_o is the time for the failure damage, i.e. $\omega = 0.99$, being reached at the first point in the HAZ. The results obtained for the new, aged, fully and partially repaired (case-i) are presented in Table 5. It can be seen that under the closed-end conditions, the ratios of the remaining times to the total creep lives (failure lives), for the new, aged and repaired welds, are in a similar range, i.e. about 7-10%. This suggests that under the closed-end conditions, cracks will start to initiate after about 90% of the total lives have been used up. Different results were obtained for $\sigma_{ax}/\sigma_{mh} = 1$. The ratio of the remaining time to the total creep life is about 20% for the fully repaired weld and is about 10% for the partially repaired weld. The variations of the length of the failure damage zone, a, in and along the HAZ, near the parent material, with time, under closed-end condition, are shown in Fig. 7.

Table 5   Failure life, t_f, initiation time, t_o, and the remaining life, $t_f - t_o$, estimated from results of damage analyses, for $\sigma_{ax}/\sigma_{mh} = 0.306$ (closed-end) and $\sigma_{ax}/\sigma_{mh} = 1$.

σ_{ax}/σ_{mh}	$t\,(h)$	*New*	*Aged*	*Full*	*Partial-i*
	t_f	21,018	15,640	10,803	9,892
0.306	t_o	19,000	14,500	10,000	9,150
(closed-end)	$(t_f - t_o)/t_f$ %	9.60	7.28	7.43	7.50
	t_f	----	----	3,751	3,307
1.0	t_o	----	----	3,000	3,000
	$(t_f - t_o)/t_f$ %	----	----	20.02	9.28

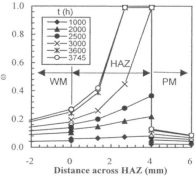

Fig. 5   Variations of damage across the HAZ, along a line parallel to and 5.1mm from the outer surface of the fully repaired weld, at different times, with $\sigma_{ax}/\sigma_{mh} = 1$ ($p_i = 16.55$ MPa).

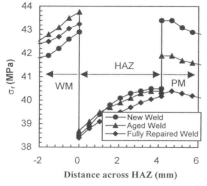

Fig. 6   Variation of rupture stress along a line across the HAZ, parallel and 3.8mm to the outer surface of the pipe, for new, aged and fully repaired welds, with $\sigma_{ax}/\sigma_{mh} = 1$ ($p_i = 16.55$ MPa).

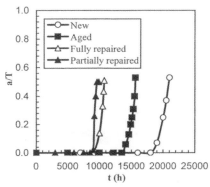

Fig. 7   Variations of a/T with creep time, obtained for the new, aged, fully and partially repaired welds, with $\sigma_{ax}/\sigma_{mh} = 0.306$ ($p_i = 16.55$ MPa).

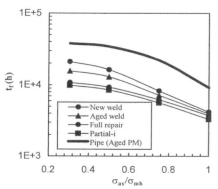

Fig. 8   Failure life versus sax/smh, predicted by damage analyses, for the new, aged and repaired welds, with $p_i = 16.55$ MPa.

5. Discussion and concluding remarks

The ratios of the predicted lives for the aged, the full repair and the partial repair welds are about 1 : 0.7 : 0.65, based on damage analyses, under the closed-end condition. These predictions are directly influenced by the balance of material properties within the welds. It is interesting to note that, for the input data used, the life of the fully repaired weld is similar to that of the aged weld. The difference between the predictions of these two welds is mainly due to the material property difference and not the geometry factors.

The lives predicted for the fully and partially repaired welds when zones 7 and 8 are considered to have identical creep properties are similar, with the fully repaired weld showing a slightly longer life. In general, identical properties were used in the analyses for both of these weld models and therefore, the major difference between these welds is the geometry. For the partial repair cases, the results obtained have shown that the stress and damage distributions in the failure dominating areas are similar to those of the fully repaired weld [11,12], which give failure lives of the order of 0.9 of that of the fully repaired weld. The implication of this result could have practical significance for the development of weld repair strategies but further work is necessary to establish whether this result is of general applicability.

The excavation width, for the full repair procedures, is an important operational parameter to ensure the full removal of the damaged material and to minimise preparation and welding costs. The current results indicate that, choosing a different w value, within the range 46–80 mm, will not make a major difference (<10%) in life as long as all existing damaged material is removed. This is consistent with the results of more general studies of the effect of weld width [14] and provides an indication for practical weld repair procedures.

In all weld cases, failure was predicted to occur within the HAZ regions, and generally to initiate at positions near the outer surfaces. This could be the result of the relatively weaker rupture strength of the HAZ structures in each weld situation, which are, as described before, more closely related to the type IV properties. Under the closed-end condition, the damage and stress distributions were found to vary insignificantly across the HAZ. However, the variations are more significant when additional axial loading is applied. In this case, the high damage or rupture stresses within the HAZ move to a position adjacent to the HAZ/parent material boundary. This suggests that for the welds investigated, excessive axial loading will result in a high probability of type IV cracking occurring, which is consistent with laboratory and plant experience. The life predictions with σ_{ax}/σ_{mh} obtained from damage analyses are presented in Fig. 8, where the results obtained for the plain pipe (aged PM) [15] are shown for comparison. This illustrates the main features of life reduction due to the presence of welds, an additional life reduction due to increased axial stress and the more limited life reduction due to geometry factors. An additional interesting feature is that when $\sigma_{ax}/\sigma_{mh} \rightarrow 1$, where the failure is expected to occur by type IV cracking, the life is relatively independent of geometry and weld metal. The predicted failure positions are summarised in Table 6. For the partial repair case–ii, the results obtained indicate that failure may occur in either of the two HAZs, generated by repair welding, dependent on the specific properties of the HAZ region in the existing weld metal. If the properties of the HAZ within the existing weld metal are "weaker" than those of the HAZ in the aged parent material, failure can occur in the HAZ in weld metal. Such failure has been experienced in a practical seam weld in a 2 ¼Cr1Mo pipe [16].

Table 6 Summary of the failure positions of the CrMoV pipe welds (all occur near OD)

Components	$\sigma_{ax}/\sigma_{mh} = 0.306$	$\sigma_{ax}/\sigma_{mh} = 1$	Comments
Pipe	----	----	----
New, aged and full repair	HAZ, HAZ/WM	HAZ, HAZ/PM	all cases of w = 40–80 mm
Partial repair-i	HAZ, HAZ/WM	HAZ, HAZ/PM	----
Partial repair-ii	HAZ (7 or 8), HAZ/WM**	HAZ (7 or 8), HAZ/PM	zone 8 and zone 7 the same
Partial repair-ii	HAZ (7), HAZ/WM	HAZ (7), HAZ/PM	zone 8 "stronger"* than zone 7
Partial repair-ii	HAZ (8), HAZ/WM	HAZ (8), HAZ/PM	zone 8 "weaker" than zone 7

* higher in uniaxial rupture strength or failure ductility
** HAZ (7 and 8) etc defines the specific HAZ considered; see Fig. 1

The results presented showed that conservative life predictions were obtained from steady-state analyses, compared to the corresponding damage predictions, underestimating the failure life by about 30-40% under the closed-end condition and by about 20-30% when $\sigma_{ax}/\sigma_{mh} \to 1$ (see Table 4). However, the failure positions predicted by continuum damage and steady-state analyses were consistent.

Although all the creep data used in the analyses are applicable to a temperature of 640° C, the constitutive laws obtained were found to be of the similar form as would be expected at lower creep temperatures. In addition, previous work has shown that the creep data at 640° C are consistent with those available for a lower temperature such as 565° C [10]. The work presented in this paper has illustrated some general effects, caused by the difference in weld dimensions, including repair profiles, prescribed material properties balance and system loading etc., on the failure behaviour of these CrMoV welds.

Acknowledgements

The authors wish to acknowledge EPSRC, Nuclear Electric, PowerGen and National Power, UK, for their financial support through an EPSRC/ESR21 grant.

References

[1]   Viswanathan, R., Weld repair of aged piping - a literature review, *2nd Int. EPRI Conf. on Welding and Repair Technology of Power Plants*, Florida, May 1996.

[2]   Samuelson, L. A., Segle, P. and Storesund, J., Life assessment of repaired welds in high temperature applications, *Welding & Repair Technology for Fossil Power Plants, EPRI Int. Conf.*, Virginia, 1994.

[3]   Meyer, H. J., Mayer, K. H. and Streatz, A. J. A., Upgrading and retrofitting of earlier steam power plants up to and beyond the year 2000, *FWP Journal*, Vol. 29, 1989, pp. 15-25.

[4]   Gandy, D. W., Findlan, S. J. and Viswanathan, R., A comparison of conventional post weld heat treatment and temperbead weld repair techniques for service-aged 2-1/4Cr-1Mo girth weldments, *2nd Int. EPRI Conf. on Welding and Repair Technology of Power Plants*, Florida, May 1996.

[5]   Coleman, M. C. and Miller, D. A., Inspection, assessment and repair strategies for type IV cracking in power plant components, *AWS/EPRI Conf. on Maintenance and Repair Welding in Power Plants V*, Florida, 1994, pp. 5-15.

[6]   Williams, J. A., Methodology for high temperature failure analysis, *Proc. of the European Symposium, Behaviour of Joints in High Temperature Materials*, Ed. T. G. Gooch, R. Hurst and M. Merz, 1982, Petten (N.H), Netherlands.

[7]   Hayhurst, D. R., The role of creep damage in structural mechanics, *Engineering Approaches to High Temperature*, Ed. B. Wilshire and D. R. J. Owen, 1983, pp. 85-176, Pineridge Press, Swansea.

[8]   Hall, F. R. and Hayhurst, D. R., Continuum damage mechanics modelling of high temperature deformation and failure in a pipe weldment, *Proc. R. Soc. London*, A443, 1991, pp. 383-403.

[9]   Hyde, T. H., Sun, W., Becker, A. A., Assessment of the use of finite element creep steady-state stresses for predicting the creep life of welded pipes, *Proc. 4th Int. Conf. on Computational Structures Technology*, 1998, Ed. B. H. V. Topping, pp. 247-251, Civil-Comp Press, Edinburgh.

[10]   Hyde, T. H., Sun, W. and Williams, J. A., Creep behaviour of parent, weld and HAZ materials of new, service-aged and repaired 1/2Cr1/2Mo1/4V: 2 1/4Cr1Mo pipe welds at 640° C, *Material at High Temperatures*, 16 (3), 1999, pp. 117-129.

[11]   Hyde, T. H., Sun W. and Becker, A. A., Life assessment of weld repairs in 1/2Cr1/2Mo1/4V:2 1/4Cr1Mo main steam pipes using the finite element method, *J. Strain Analysis*, Vol. 35, 2000, pp. 359-372.

[12]   Sun, W., Hyde, T. H, Becker, A. A. and J. A. Williams, Comparison of the creep and damage failure prediction of the new, service-aged and repaired thick-walled circumferential CrMoV pipe welds using material properties at 640° C, *Int. J. Pres. Ves. & Piping*, Vol. 77, 2000, pp. 389-398.

[13]   Specification for unfired fusion welded pressure vessels BS 5500, BSI, London, 1997.

[14]   Hyde, T. H., Williams, J. A. and Sun, W., Assessment of creep behaviour of narrow gap welds, *Int. J. Pres. Ves. & Piping*, Vol. 76, 1999, pp. 515-525.

[15]   Hyde, T. H., Sun, W. and Williams, J. A., Prediction of creep failure life of internally pressurised thick walled CrMoV pipes, *Int. J. Pres. Ves. & Piping*, Vol. 76, 2000, pp. 925-933.

[16]   Hickey, J. J., Bernard, P. J., Bissell, A. M. and Jirinec, M. J., Investigation and repair of a failed seam welded reheater outlet header, *2nd Int. EPRI Conf. on Welding and Repair Technology of Power Plants*, Florida, May 1996.

Potential correlation between the creep performance of uniaxial welded specimens and heavy section ferritic welds

by

J.A.Williams*, T.H.Hyde and W. Sun

School of Mechanical, Materials, Manufacturing Engineering and Management,
University of Nottingham, Nottingham, NG7 2RD
*Independent Consultant, East Leake, Loughborough, Leics. UK

Abstract

Welds are complex, both in structure and properties, although the geometry and the extent of the different microstructural regions, within the welds, can be controlled during fabrication. Any property differences directly influence the stress redistribution and damage accumulation that occur under creep loading.

Different failure modes occur during service, dependent on the form of the loading. The cross weld specimen geometry is generally used to provide guidance on weld performance under pressure and additional loading conditions. However, such a specimen design may not be applicable to all expected practical weld loading situations.

This paper considers the calculated weld performance of a circumferential pipe weld subjected to different loading, ranging from internal pressure to a simple end load case without internal pressure. Based on an initial comparison between the calculated life of a cross weld and a pipe weld specimen subjected only to axial loading, the other loadings are considered. The internally pressurised pipe weld is then modelled using a three bar structure and the results obtained are compared with some published results from large scale component tests. Conclusions are drawn from these data.

Notation.

A, m, M, n, χ, ϕ, α	Constants in constitutive equations
t, t_f	Time and failure time
T, D	Wall thickness, outer diameter
P	Internal pressure
ε, $\dot{\varepsilon}^c$	Creep strain, strain rate
σ_{ij}, σ, σ_θ, σ_{eq}, σ_1, σ_r,	Deviatoric stresses; hoop, equivalent, maximum principal and rupture stresses, respectively
σ_{mdh}	Mean diameter hoop stress
ω $\dot{\omega}$	Damage and damage rate

1. Introduction

High temperature component performance is generally limited by the creep performance of welds that are used in the fabrication, [1,2]. Thus, an understanding of the creep performance of welds is essential so that maximum safe economic use of plant can be achieved.

Welds are metallurgically complex due to the local effects of heat from the welding process and the subsequent effect of this heat on the local metallurgical structures, [3,4]. These structures can have different mechanical and creep properties and these differences directly influence the stress redistribution and damage accumulation that occurs by creep within the weld. The form and distribution of such structures within the heat affected zone, HAZ, of a weld are better understood and can be defined, [3].

The creep performance of welds has been extensively studied over many years using a variety of techniques. Firstly, axially loaded cross weld specimens, taken from a full cross section of the weld, have been tested under creep conditions. This simple specimen design, although easy to test, behaves in a complex manner when loaded under creep and its interpretation is not straightforward. Analytical, [5,6], numerical, reviewed by Hyde and Tang, [7], and other studies, for example, [8,9] have improved our understanding of their behaviour.

A second evaluation uses finite element, FE, numerical studies, using steady state or more complex one or two parameter continuum damage constitutive laws, to evaluate the expected full life of a realistically loaded weld, [10,11,12,13].

Stationary state creep solutions, using a Norton's law, i.e.

$$\dot{\varepsilon}_{ij}^{c} = \frac{3}{2} A' \sigma_{eq}^{n'-1} S_{ij} \qquad (1)$$

can be used to obtain stress distributions in a relatively short time compared to that required for damage modelling. For multi-material weld models where the individual HAZ, weld metal, WM, and parent metal, PM, properties can be represented by different A and n values, the creep failure times can be estimated from the stationary state rupture stress, σ^{ss}_{r}, at the most critical position and the appropriate creep rupture material properties by:-

$$t_{f} = \left[\frac{1+m}{M(\sigma_{r})^{\chi}} \right]^{1/(1+m)} \qquad (2)$$

The more advanced continuum damage constitutive laws are typified for a one parameter model by:-

$$\dot{\varepsilon}_{ij}^{c} = \frac{3}{2} A \left[\frac{\sigma_{eq}}{1-\omega} \right]^{n} \frac{S_{ij}}{\sigma_{eq}} t^{m} \qquad (3)$$

$$\dot{\omega} = \frac{M\sigma_{r}^{\chi}}{(1+\phi)(1-\omega)^{\phi}} t^{m} \qquad (4)$$

where ω is the damage parameter which varies from 0, (no damage) to 1, (failure). The representative rupture stress, σ_{r}, is given by:

$$\sigma_{r} = \alpha\sigma_{1} + (1 - \alpha)\sigma_{eq} \qquad (5)$$

A third evaluation involves the testing of full size welded components, [14,15,16].

Although these studies have improved our understanding of the creep behaviour of welds, there has been less effort on the correlation of these different approaches and to support simple methods for creep life evaluation of welds.

The paper examines this more general problem. It briefly summarises the expected modes of failure in heavy section ferritic welds. Then, the calculated creep behaviour of a selected range of weld/specimen geometries using FE models is described. These range from cross weld to heavy section welded pipe models subjected to an axial load and include simple three bar structures. Finally, it considers more limited data for the intermediate cases of combined internal pressure and axial loads. In all cases, identical weld geometry and constitutive laws are used, allowing an initial evaluation of small tests on cross sections of welds.

2. Failure modes in heavy section ferritic welds under creep conditions

The current study considers a typical weld, fabricated in CrMoV main steam pipe and welded with a 2Cr1Mo weld metal. This combination has been studied and cracking/failure statistics are available for high temperature plant, [1,2,17]. Table 1 summarises the cracking forms.

In practice, the design stresses for a pipe, under pressure loading, are characterised by the mean diameter hoop stress, defined as:-

$$\sigma_{mdh} = p \, (D - T)/2T \qquad (6)$$

Most design codes limit any additional loading to a maximum axial stress equal to the mean diameter hoop stress, or a hoop to axial stress ratio of 1.0. The allowed stress can be factored for the presence of welds dependent on the degree of inspection and welding process and in some cases, weld factors based on weld metal strength.

Crack form	Orientation *	Contributing * material factor	Main driving stresses	Time range, h For detection
A. Stress relief	Circ.	CGB HAZ Trace elements	Residual stress	0 – 5,000
B. Transverse weld metal	Axial	Coarse grain WM Trace elements	Residual stress Operating stress	0 – 30,000
C. "Type 4"	Circ.	HAZ Type 4 structure	Additional axial loading	60 – 70,000
D. "Type 3A"	Circ.	C depletion near weld interface	Additional axial loading	60 – 80,000
E. End of life failure	Early axial Final axial + circ.	WM/HAZ structure	Pressure and additional loading	~ 200,000

*Circ. = circumferential; CGB = low ductility coarse grained Bainite, HAZ = heat affected zone; WM = weld metal.

Table 1. Typical crack forms found in CrMoV:2CrMo:CrMoV welds.

Post weld heat treatment, PWHT, which tempers the material and reduces the residual stress, is required for ferritic welds greater than around 11 mm thick. This paper considers the crack types C to E as these lead to the premature medium and normal long term creep failures.

Any simplified laboratory test should meet specific criteria. Firstly, the test must both generate and fail by the expected long term cracking modes. Secondly, the off loaded hoop stress redistribution must be incorporated for the pure pressure loaded components. Finally, as creep damage will form normal to the maximum principal stress, the predominant crack form and direction must be correct.

Currently, these criteria can be met by cross weld tests where the failure is due to a high axial additional loading. It cannot be assumed that such tests characterise the failure of heavy section welds under only internal pressure loading.

3. Analysis and experimental models.

Various weld/specimen geometries, Figure 1, are considered with emphasis on the two extremes of high axial stresses and pure internal pressure, see Table 2. The circumferential weld in a plain pipe, Figure 1, is typical of a conventional fusion weld. The pipe has an outer diameter of 356mm, a wall thickness of 63.5mm and is fabricated with a 2.25Cr1Mo weld metal. It is subjected to PWHT, at 700°C and any residual stress component is considered negligible. The ABAQUS finite element programmes were used to obtain the local stress and strain distributions under steady state conditions, using the conventional Norton's law. The life estimate is then obtained from the representative rupture stresses and the uniaxial data for constituent parts of the weld as noted earlier. This has been shown to give a conservative estimate of the creep life relative to the life estimates made using the continuum damage constitutive laws.

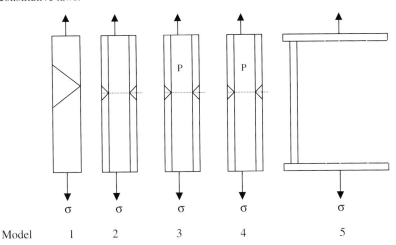

Figure 1. Models used to represent weld under creep conditions.

Specimen/weld design	Loading
1.Cross weld specimen	Uniaxial normal to weld interface
2. Full size pipe weld	Axial, normal to weld interface
3. Full size pipe weld	Internal pressure and pressure induced end loading only
4. Full size pipe weld	Internal pressure + range of different axial stresses up to hoop/axial=1
5. Three bar structure	Uniaxial stress parallel to weld interface

Table 2. Summary of the considered weld models.

The creep relationships used for FE analysis, are obtained from laboratory tests. These were generated at accelerated temperatures but the constitutive laws obtained are similar in form to those at operating temperatures.

In addition to the FE analysis used for models 1 to 4, a simplified three bar model was used for comparison with the pressurised pipe weld using steady state creep solutions, [12]. Their model, Figure 1e, identified the properties of the three bars as the parent, (1), the HAZ, (2), and the weld metal, (3). As an example, the stress in the HAZ, bar 2, is given by equation (7) as shown below:-

$$\left(\frac{A_1}{A}\right)\left(\frac{\dot{\varepsilon}_{o2}}{\dot{\varepsilon}_{o1}}\right)^{1/n_1}\left(\frac{\sigma_2}{\sigma_{nom}}\right)^{n_2/n_1}+\left(\frac{A_2}{A}\right)\left(\frac{\sigma_2}{\sigma_{nom}}\right)+\left(\frac{A_3}{A}\right)\left(\frac{\dot{\varepsilon}_{o2}}{\dot{\varepsilon}_{o3}}\right)^{1/n_3}\left(\frac{\sigma_2}{\sigma_{nom}}\right)^{n_2/n_3}-1=0 \qquad (7)$$

where the bar lengths are equal. A_1/A to A_3/A are the area fractions, for bars 1, 2 and 3. $\dot{\varepsilon}_{o2}/\dot{\varepsilon}_{o1}, \dot{\varepsilon}_{o2}/\dot{\varepsilon}_{o3}$ are normalising strain rates calculated at the nominal stress levels in the relevant materials. Similar forms exist for stresses in bar 1, (PM), and bar 3, (WM).

Two materials data sets were available for the CrMoV:2CrMo:CrMoV pipe welds. The first was generated from a service weld, operated for ~ 180,000 h at 565° C with the constitutive law constants generated at 640°C, [18]. The second set was generated from a virgin weld, subjected to full PWHT, [10,19].

3.1 Models where the main loading is in the axial direction

Models 1 and 2 are considered first. Model 1 is a conventional cross weld geometry and involves a waisted specimen design where the HAZ is set at the gauge length centre. Model 2 is a full size circumferential pipe weld of identical form and structure to the cross weld but subjected only to an end load equivalent to axial stresses of 40, 54 and 70 MPa. Unless otherwise stated, the constitutive laws used are those for the service exposed material quoted, [18].

The steady state representative rupture stress values, equation (5), and the life estimates, equation (2), are given in Table 3 with the predicted failure position in bold print.

Nominal stress MPa	Rupture stress, MPa, (Life) h PM	Rupture stress. MPa, (Life), h. HAZ	Rupture stress. MPa, (Life), h. WM	Geometry
25	25.3 (134955)	**23.8 (15740)**	26.7 (148739)	Model 2
40	40.6 (8823)	**38.15 (3478)**	45 (11831)	Model 2
54	54.9 (1548)	**51.73 (1312)**	62.2 (2462)	Model 2
70	**71.3 (343)**	67.4 (563)	82 (644)	Model 2
40	43.7 (5772)	**42.4 (2480)**	46.5 (10091)	Model 1
54	59.2 (1002)	**57.5 (936)**	65.4 (1930)	Model 1
70	**77 (220)**	74.8 (403)	87.8 (463)	Model 1

Table 3. Steady state rupture stresses for PM, HAZ and WM using Model 1 and model 2 geometries.

The maximum rupture stress, when applied to the uniaxial rupture data for each constituent part, gives the estimated rupture life of that part. Failure is defined to occur in the region with the lowest rupture life. The bracketed values are the lives calculated using the corresponding rupture stress and relevant uniaxial data. The data show that:-
- If the nominal stress is less than ~63MPa, failure occurs in the HAZ whereas parent material failure occurs at stresses greater than 63MPa.
- The rupture stress in the cross weld specimen is around 5 – 10 % higher than in the full sized weld subjected only to an end load with the largest effect in the HAZ.
- With one exception, the rupture stresses are greater than the nominal stress, for these weld properties. The exception is the HAZ in the Model 2 pipe weld.

These data suggest the rupture stress value, defined from the cross weld specimen is conservative when applied to an axially loaded welded pipe. One reason may be that the rupture stress, as defined by analysis of the cross weld specimen, is a function of the specimen size. It has been suggested that the rupture stress reduces as the specimen size increases, [20], and they quoted stress differences of the order of 10% which are compatible with the current data. A further comparison for the failure position, Figures 2a and 2b, shows the local distributions of estimated life for Models 1 and 2, respectively, near the outer surface. These clearly illustrate that the failure points tend to be close to the parent material side of the HAZ and that, at 70 MPa, the failure is within the parent material.

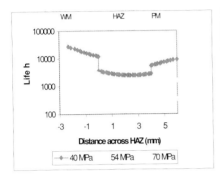

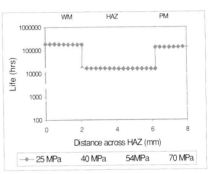

a. Along the centre line of the CrMoV waisted cross weld specimen, for a range of nominal stress.

b. Along a line parallel to and 3.8 mm from the outer surface of the cross weld pipe, for a range of nominal stress.

Figure 2. The variation of failure lives, predicted by the steady state rupture stresses across the HAZ for Models 1 and 2.

These data suggest that the failure life of an unpressurised circumferential weld in the pipe, subjected to an end load could be generally conservatively estimated from small, uniaxially loaded cross weld samples subjected to the same nominal stress levels. The comparison of life, failure position and damage orientation is consistent and meets the criteria for development of a simplified laboratory test and specimen for high axial loading only.

3.2 Models where the main loading is in the hoop direction

Model 3 is a circumferential weld in a heavy section pipe, subjected to internal pressure only and the axial stress is relatively small. As the maximum principal stress is in the hoop direction and experimentally confirmed by short axial cracks in the weld metal/HAZ regions, cracking mode E in Table 1, the cross weld specimen does not satisfy the criteria for generating the correct failure mode. For this loading, an alternative geometry is considered; namely the three bar structure, Model 5.

3.2.1 Consideration of the three bar structure

The first question is the definition of a specimen size which is representative of the real weld. A range of sizes were considered, taking a nominal HAZ width of 3 mm, and a symmetrical arrangement of bar sizes for the parent and weld metal, giving total specimen widths of 13, 25, 63 and 123 mm. These sizes cover the range from small samples to those where the parent metal has a width approximately equal, for main steam geometries, to the pipe thickness. To evaluate general trends, it was assumed that n_1, n_2 and n_3, the Norton's Law indices, were equal. Values of n of 2, 4 and 6 were considered and the trends are approximately the same. However, the data for n = 4 are given in Figure 3. These show the effect of specimen width on the calculated stress levels in each of the bars/ weld constituents for different uniaxial creep strain rate ratios for the HAZ/PM and PM/WM and illustrate:-.

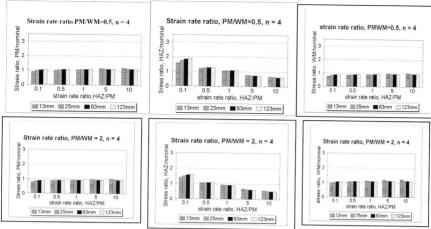

Figure 3. The effect of specimen size on the stresses generated in a three bar structure as a function of the material properties.

- For values of PM/WM creep strain rate ratio of 0.5 and 2, compatible with a conventional ferritic weld, the effect of specimen size on stress in individual bars is relatively small. Differences are generally found where the HAZ/PM strain rate ratios are 0.1 or less. Subsequently, a specimen width of 63 mm was used.
- The stresses in a three bar array showed off loading; namely the load is transferred from the weaker to the stronger materials, modelling stress redistribution in the pressurised weld case.
- The maximum principal stress will be parallel to the HAZ and thus any damage will occur normal to this, typical of the final failure mode, E, Table 1.

Thus, the model is compatible with the pressurised pipe weld and specimen sizes can be defined where the stresses are relatively independent of specimen size. Note however, that this is a steady state analysis using the representative rupture stress criterion and should be an underestimate of weld life. The model can be used to obtain a creep life as follows.

- The stress levels are calculated in each bar of the structure and converted into time using the appropriate rupture life equations. A life is obtained for the first bar to fail, defining an initiation time and cracking position for the weld.
- The life fractions used up on the remaining two bars are calculated and a new nominal stress based defined.
- As before, the new bar stresses and the new lives, in the two bar structure, are calculated. Taking account of the remaining life fraction for each bar allows a time to be estimated for the second bar to fail. This process can be continued to obtain a total three bar life. The normal Robinson's life fraction rule is used with a sum equal to unity.

The failure of the first bar does not signify complete weld failure which is given by failure of all three bars, similar to the real case. The failure of the first bar is strictly comparable to a time for formation of short axial cracks in the HAZ/Weld metal. The combined three bar life should correspond to some value less than the life obtained from a full damage analysis. It should be noted that the stresses in the subsequent two bar structure formed will be potentially more size dependent, particularly if the first failure occurs in the weld or parent metal, due to their larger areas.

3.2.2 Consideration of the pressurised circumferential pipe weld, (Model 3)

The main comparison is that generated from a full FE analysis of a pressurised circumferential weld in a plain pipe, using data taken from pipe sections removed from power plant, [18]. Both steady state and continuum damage analyses are available at a range of internal pressures from 40 to 70 MPa. Other cases considered, [10,19], involve a pressure vessel test at 565°C.

3.2.3 Comparison with the numerical results on service exposed material at 640° C

The three bar structure is used to represent the pressurised pipe weld. The limited effect of specimen size was initially confirmed for this range of material properties where n_1, n_2 and n_3 are not equal. The stresses for these particular creep data, with n not constant, are size dependent in the smaller sizes but relatively constant for specimen widths >= 63 mm. For all subsequent estimates, a width of 63 mm was used. The HAZ width is 3 mm and representative of practical cases. The parent and weld metal sections are taken to be equal in width, with the width of 63 mm made up of 30:3:30 in mm for the parent, HAZ and weld widths respectively.

The results are summarised in Table 4 which shows the calculated failure times of the failure dominant zone of each stage, for a 63 mm wide specimen, only showing the minimum values and the main conclusions are :-
- The failure sequence at mean diameter hoop stresses of 38.065, 40, and 54 MPa is initially at the HAZ followed by the parent and then the weld metal. At 70 MPa, the sequence is slightly changed with the parent failing first followed rapidly by the WM and the HAZ.
- Although there is no major size effect for this particular three bar model, if the first failure is in the parent or weld metal sections, the new nominal stress in the remaining structure will be sharply increased once the two bar model is used. At this stage, specimen size could be important.
- Comparison with published FE analysis, Table 5, illustrates the reasonable agreement with model 5. The failure site is similar in all cases, even reproducing the change at 70 MPa. In addition, the three bar life to the first bar failure is compatible with the steady state failure stage as defined by the finite element analysis, [11]. The damage analysis, by allowing for damage distributions etc., should be a better comparison with the total three bar model, and this is confirmed.

Nom. Stress (Size)	3 bar model parent Life, h	3 bar model HAZ Life, h	3 bar model WM Life, h	2 bar model PM Life, h	2 bar model HAZ Life, h	2 bar model WM Life, h	1 bar model PM Life, h	1 bar model HAZ Life, h	1 bar model WM Life, h
38.065 (63)	-	6735	-	8042	-	-	-	-	124
40 (63)	-	5747	-	5674	-	-	-	-	25
54 (63)	-	2379	-	177	-	-	-	-	98
70 (63)	658	-	-	-	-	10	-	~0	-

Table 4. Details of the individual bar failure times for the service exposed weld data at 640 C.

Nominal stress MPa	FE steady state Life, h	FE damage Life, h	3 bar model life, First bar	3/2/1 bar model Life, h
38.065 (63)	8942	15640	6735 (HAZ)	14901

Table 5. Comparison of the three bar data with the FE results for a plain pressurised pipe weld.

A second comparison is taken from a full size component test, [19]. The required constitutive laws are taken from Hall and Hayhurst, [10], based on the original creep data of Cane, [21]. The original vessel data was generated from a circumferentially welded vessel tested at 565° C and a mean diameter hoop stress of 110 MPa. The results from Coleman and reported in [10], showed short axial cracking in the low ductility regions of the HAZ and weld metal before 20,000h and the failure occurred in ~ 46,000 h, by massive transverse and circumferential cracking.

The three bar results suggested that:-
- First failure occurs in the weld metal as short axial cracks after 9590h
- Similar damage in the HAZ within a very short time
- Total failure through the parent in 9661h.

The correct failure sequence is identified and the time for crack initiation would be around 10,000 h which is compatible with Coleman's data. The final life was not however, defined with any accuracy but is conservative.

3.3. Models where there is combined loading due to pressure and additional axial loading, Model 4

Very few experimental data exist for welds subjected to combined pressure and axial loading. For comparison purposes, the steady state and continuum damage calculations for the service exposed CrMoV weld, [22], are taken, Table 6. The constitutive laws and geometry are identical to those presented in Section 3.1. The calculations suggest that the life is markedly reduced on increasing the axial stress, cases 1-4. There is also an influence of internal pressure as illustrated by comparison of cases 4 and 5. Life was reduced by ~ 30% with the addition of internal pressure to the end loaded pipe.

For a fixed stress of 38.065 MPa, the life ratio for a pipe weld axially loaded in tension, case 5, and loaded in the hoop direction by pressure to the same level, case 1, is 0.457, again showing the major effect of the load direction.

Case No.	Axial stress MPa	Axial: hoop stress ratio	Steady state Life, h.	Damage analysis, Life, h	Mean diameter hoop stress, MPa
1	11.65	0.306	8942	15640	38.065
2	19.03	0.5	8182	12962	38.065
3	28.55	0.75	5415	7197	38.065
4	38.065	1.0	2895	3966	38.065
5	38.065	Infinity	4088	-	

Table 6. Calculated effects of end load and internal pressure on the creep life of a circumferential weld,[22].

There are insufficient data available to unambiguously define the characterising stress at this time and further work is necessary.

632

4. Concluding remarks

The weld model, which is typical of the geometry used for a main steam pipe weld, has been subjected to a number of different loading conditions. In addition, two laboratory type test specimens, the cross weld and the 3 bar structure or parallel weld, have been examined under identical loading conditions. As only limited experimental data are available, the major comparison is with life calculations using steady state or continuum damage FE analysis.

The results clearly show that the cross weld specimen underestimates the life of an axially loaded circumferential weld in a pipe and thus is conservative. However, the presence of an imposed internal pressure in the pipe to produce a hoop : axial stress ratio of 1.0 will reduce the life further and the conservatism.

For circumferential welds in pipes subjected to internal pressure only, the three bar structure or wide plate test with the stress parallel to the weld, produces off loading from the weaker to the stronger materials and produces a reasonable estimate of the expected life of the weld, based on a calculated weld life. In addition, the time and position of first cracking can be estimated.

The application of such a specimen design to a full size vessel test derived the correct failure position and gave a reasonable estimate of the crack initiation time but was an underestimate of the total life. Such specimen designs are useful, to provide simple estimates of weld life, the effect of changing weld variables and properties and defining the first cracking point. However, in general, they will underestimate the total life and their testing will be diffiicult.

This general approach can provide useful information on weld performance and future work should continue to consider the use of such designs.

5. Acknowledgements.

The authors wish to acknowledge EPSRC, National Power, Brtitish Energy and Powergen for financial support through an EPSRC/ESR21 grant.

6. References

[1]. Brett, S.J., 1994, Cracking experience in steam pipework in National Power, VGB. Conf., Materials and weld technology in power plants", Essen, Germany.
[2]. Cheetham, D., Fidler, R., Jagger, M. and Williams, J.A., 1977, Relationships between laboratory data and service experience in cracking of CrMoV welds, Conf. "Residual stresses and their effects on welded construction", London, TWI, UK.
[3]. Alberry, P.J. and Jones, W.K.C., 1982, A computer model for the prediction of HAZ microstructures in multipass welds, Metals Technol., 9, 10, 419.
[4]. Easterling, K., 1983, "An introduction to the physical metallurgy of welding", Butterworth, London.
[5]. Craine, R.E. and Hawkes, T.D., 1993, J. Strain Analysis, 28, 303.
[6]. Craine, R.E. and Newman, M.G., 1996, J. Strain Analysis, 31, 117.
[7]. Hyde, T.H. and Tang, A., 1998, Creep analysis ands life assessment using cross weld specimens, Int. Mats. Rev., 43, 6, 221.
[8]. Williams, J.A.,1982, A simplified approach to the effect of specimen size on the creep rupture of cross weld samples, J.Eng. Mat. & Technol., ASME, 104, 36.

[9]. Parker, J.D., 1998, The creep fracture behaviour of thick section multipass welds, Conf. "Integrity of High Temperature Welds", Nottingham, UK, IOM Communications /Professional Eng. Pub., UK.

[10]. Hall, H.R. and Hayhurst, D.R., 1991, Continuum damage mechanical modelling of high temperature deformation and failure in a pipe weld, Proc. Roy. Soc. London, A, 433, 383.

[11]. Sun, W, 1996, Creep of service aged welds, PhD Thesis, University of Nottingham, UK.

[12]. Hyde, H.H., Tang, A. and Sun, W., 1998, Analytical and computational stress analysis of welded components under creep conditions, Int. Conf. "Integrity of High Temperature Welds", Notingham, UK, Prof. Eng. Pub. Ltd., UK.

[13]. Perrin, I.J and Hayhurst, D.R., 1999, Continuum damage mechanics analyses of Type 4 creep failure in ferritic steel cross weld specimens, Int. J. Press. Vess & Piping, 76, 599-617.

[14]. Browne, B.J., Cane, B.J., Parker, J.D. and Walters, D., 1981, Creep failure analysis of butt welded tubes, Conf. "Creep and Fracture of Engineering Materials and Structures", Swansea, U.K., Pineridge Press, Swansea.

[15]. Williams, J.A.,1987, The effect of loading cycle form on the local creep strains near the interface of austenitic-ferritic welds fabricated with a Type 316 consumable, Conf. "Creep and Fracture of Engineering materials and Structures", Swansea, U.K., Pineridge Press.

[16]. Van Wortel, H., Heerings, J.H. and Arav, F., 1995, Integrity of creep damaged components in power generation plant after welding without PWHT, Conf. BALTICA III, Helsinki, Finland, VTT.

[17]. Price, A.T., 1982, CEGB experience with small diameter DMWs in coal fired boilers, Conf. "Joining Dissimilar Metals", Pittsburgh, USA, AWS/EPRI, Palo Alto, Ca, USA.

[18]. Hyde,T.H., Sun,W., Becker,A.A. and Williams, J.A., 1997, Creep continuum damage constitutive equations for the base, weld and HAZ materials of a service aged CrMoV weld at 640 C., J. Strain Anal., 32, 4, 273-285.

[19]. Rowley, T.R. and Coleman, M.C.,1971, Collaborative programme on the correlation of test data for high temperature design of welded steam pipes, CEGB Report RD/M/N.710.

[20]. Storesund, J. and Tu, S-T., 1995, Geometrical effect on creep in cross weld specimens., Int. J. Press. Vess. & Piping, 62, 179.

[21]. Cane, B.J.,1981, Collaborative programme on the correlation of test data for high temperature design of welded steam pipes, CEGB Report No., RD/L/2101N81, CEGB, UK.

[22]. Hyde, T.H., Sun, W., and Becker, A.A., 2000, Effects of end loading on the failure behaviour of CrMoV welds in main steam pipe lines, Conf. "Computer aided assessment and control, (Damage and Fracture Mechanics 2000), Montreal, Canada.

Extending the Life of Steam Turbine Rotors by Weld Repair

K.C.Mitchell, Innogy

The weld repair of steam turbine rotors by OEM's and utilities has been performed over a number of years, in order to extend the life of a rotor and improve the efficiency, in some instances. These repair techniques have been adopted, since significant cost savings are achievable by avoiding rotor replacement costs and reducing outage times. For Innogy, the capability to perform such weld repairs at their own workshops would lead to significant cost savings and increased flexibility in making run/replace decisions. This paper will describe in detail the work performed by Innogy to underwrite these weld repairs.

To develop this rotor weld repair capability for HP, IP and LP rotor steels, there are a number of key issues which need to be understood by the repairer. Before considering any potential welding procedures, there is a need to characterise the rotor steel in terms of composition, microstructure and mechanical properties.

For an actual repair job, the location and extent/cause of damage would need to be identified, along with service history of the rotor. Once this is complete, then a matching welding consumable is selected or developed depending on the composition/mechanical properties of the rotor steel and the repair location on the rotor. Welding would only commence, if a fully approved weld procedure had been developed using the selected consumable on a matching rotor steel to the repair rotor. It is important that the weld procedure states the range of desired welding parameters along with the weld bead deposition sequence required to ensure a uniform weld deposit and an acceptable heat input level. The trial weldments should be characterised fully in terms of microstructure and mechanical properties for the parent, HAZ and weld metal regions. At this stage, additional testing could be performed depending on the service conditions, the rotor would experience. eg. creep testing for high temperature operation, stress-corrosion cracking tests for low temperature operation.

If weld repair is still considered a viable option after these tests, then a quality plan for the job would be prepared which would include the weld repair procedure, NDT inspection requirements and the post-weld heat treatment procedure for the rotor repair. This paper will evaluate the relationship between microstructure and properties of welds on steam turbine rotor steels, after applying appropriate welding parameters and weld bead deposition sequence.

1. Introduction

Turbine rotors are among the most critical and highly stressed components in steam power plants. Although relatively few instances of catastrophic rotor bursts have occurred, they have resulted in lengthy forced outages and severe economic penalties to the affected utilities. To forestall the possibility of a catastrophic burst, utilities will retire the rotors affected, generally to the original equipment manufacturer's (OEM's) recommendations. The criteria and methodology for determining which rotors should be retired are proprietary and vary among manufacturer's. If utilities could extend the life of these rotors by 10-20 years, then substantial savings would be made. The principal method for extending their life is weld repair and over the last 20 years, there has been a substantial increase in the number of repairs combined with the complexity of repair adopted.

The work described in this paper was designed to evaluate the relationship between microstructure and properties of welds on ex-service steam turbine rotor steels, after applying appropriate welding parameters and weld bead deposition sequence to minimise heat input and produce acceptable microstructures. Innogy were particularly interested in developing weld procedures with matching filler wires, utilising the submerged arc (SA) welding technique to ensure maximum productivity benefits could be gained without sacrificing weld quality.

The inherent risks associated with high rotational velocities and containment of high energy steam, result in a need for weld repairs to steam turbine rotors being carefully engineered and executed (1). Often an essential part of that engineering process is verifying that the weld metal and the HAZ are adequate for the intended service. A mechanical and metallurgical test programme in series with, or parallel to, the repair process is often required. A significant amount of time and costs are included in the verification of a weld repair to ensure it sees safe and reliable service. Both non-destructive and destructive testing forms part of this evaluation/verification process for each repair.

While non-destructive examination expectations are defined on the basis of original requirement and industry standards, metallurgical and mechanical property requirements are not as clearly specified. The default position taken by OEM's, owners and insurer's is to require the weld repair to meet the original blade or rotor material specifications. Although this is a safe position, it may not be the best choice in all cases. If the rotor has failed and the failure was assisted by rotor steel property deficiencies, then returning it to original condition might not be good enough. To address this scenario, the repair must have better properties. Each repair must be reviewed considering the "cause" of the previous failure, and some thought given to enhancing the component's resistance to failure.

A generally accepted proof for a repair is to make a weld coupon or weld mock-up that is subjected to a destructive testing programme. The challenge is to generate a mock-up that is representative in size and shape to the actual repair and to conduct a test programme that addresses the properties needed to mitigate the expected failure modes. In addition, the coupon or mock-up must avoid size effects in the mechanical test programme but cannot be so large as to be prohibitively expensive. Tests performed should also reflect the nature of the failure and operational environment of the rotor.

2. Experimental

The materials selected for this work came from ex-service steam turbine rotors. Since the project deals with the repair of these components, it was imperative to perform welding trials and subsequent weldment testing on rotor steels which have typical chemical compositions and properties to rotors still in-service.

The materials involved in the project are listed below along with their area of application:-

1. 3NiCrMoV - Low Pressure (LP) rotor steel.

2. 1CrMoV - High/Intermediate Pressure (HP/IP) rotor steel.

The 3NiCrMoV steel samples were taken from a disc on an LP rotor by arc air gouging off around the periphery of a disc. This rotor was retired from service due to stress corrosion cracking problems in the keyway of a disc. This rotor had seen approximately 130,000 hours service at the time of retirement. Steel of this composition is used extensively for monobloc LP rotors.

The 1CrMoV steel samples were taken off the periphery of a disc on an IP rotor by arc air gouging. This steel has been used for a number of years on monobloc HP/IP rotors due to it's good creep resistance at elevated temperatures. This rotor was retired from service following the failure of a thrust bearing resulting in heavy rubbing of some discs. The rotor had seen approximately 65,000 hours service at the time of failure and 119 starts.

The chemical composition of the rotor steels is shown in Table 1.

Alloy Content	3NiCrMoV	1CrMoV
%C	0.32	0.25
%Si	0.29	0.24
%S	0.007	0.005
%P	0.016	0.013
%Mn	0.46	0.74
%Ni	3.01	0.60
%Cr	0.73	1.36
%Mo	0.28	0.89
%V	0.11	0.32

Table 1 – Chemical Analysis of Rotor Steels

Each rotor steel was fully characterised by standard metallographic techniques to determine the microstructure, grain size and hardness of the steel, before welding of test samples commenced.

For the rotor steels, a series of weld pads were manufactured using the SAW process. Each weld pad was designed to ensure the appropriate number of mechanical test specimens could be machined from them. If more than one weld pad was produced per rotor steel, the same batch of wire consumable/flux would be used to ensure consistency of weld deposit for testing. Typically, the weld metal pads were 70mm thick. The submerged arc (SA) consumables used for all the weld pads were produced via the tubular cored wire route. These metal cored wires (3.2mm diameter) were used in conjunction with basic agglomerated fluxes (basicity index of 2.3 and 3.1 (2)). A total of four consumable/flux combinations were tested (two for each rotor steel). Two of the consumables were commercially available (2½Ni½CrMo and 1CrMoV (low C)), while two consumables were manufactured to Innogy specification (3NiCrMoV and 1CrMoV (high C)). The four consumable/flux combinations used are detailed, in terms of chemical composition in Table 2. Consumables were selected for use depending on their chemical composition and mechanical properties, which would hopefully match those of the rotor steels. The commercially available wires were used inconjunction with the higher basicity flux.

Alloy (%)	3NiCrMoV	2½Ni½CrMo	1CrMoV (low C)	1CrMoV (high C)
C	0.1	0.038	0.072	0.11
Si	0.31	0.16	0.10	0.42
S	0.008	0.007	0.006	0.01
P	0.012	0.18	0.023	0.016
Mn	0.62	1.26	0.68	0.92
Ni	2.83	2.52	0.47	0.03
Cr	1.11	0.60	1.34	1.70
Mo	0.56	0.39	1.13	1.38
V	0.08	0.006	0.23	0.35
Fe	balance	balance	Balance	balance

Table 2 - Chemical Analysis of Weld Metals

The resulting weld pads were sectioned to characterise the weldment microstructure fully and allow for standard mechanical property tests to be performed (ie. tensile, charpy, hardness and bend tests). For the 1CrMoV weld metals, additional pads were deposited, in order to creep test the weldments.

3. Results

Microstructural characterisation of the LP rotor weldments revealed the parent material exhibited a fine grained microstructure which is a mixture of upper (white areas) and lower bainite (dark areas), see Figure 1. The typical grain size within this fine matrix is 14μm (ASTM 9). The microstructure is mixed bainitic in nature and typical for these 3NiCrMoV LP rotor steels. The macrospecimens of the weldments revealed weld bead placement had been acceptable with a minimum of 50% overlap, combined with a relatively low heat input for SAW. This was evident from the fine grained HAZ microstructure, see Figure 2. In the inner HAZ, there are larger grains present but generally it is fine grained in structure (<50μm).

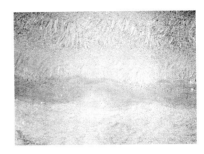

Figure 1 - 3NiCrMoV Parent Material, x100   **Figure 2 - 2½Ni½CrMo HAZ, x5**

The 2½Ni½CrMo and 3NiCrMoV weld metals possess a columnar structure with large areas of tempered/refined microstructures and were bainitic in nature. The determination of grain size was not possible with these microstructures.

The 1CrMoV HP/IP rotor steel possessed a bainitic microstructure which was quite lath-like in structure, see Figure 3. The typical grain size for this steel was 190μm (ASTM 1.5), which was large for this type of steel. Accurate determination of grain size was dificult due to the problems of identifying prior austenite grain boundaries by etching and the presence of sub-grains within the prior austenite grains. This type of microstructure is typical of upper bainite.

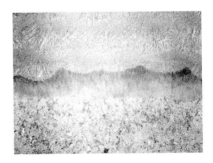

Figure 3 - 1CrMoV Parent Material, x100   **Figure 4 - 1CrMoV (low C) Weld, x5**

Macrospecimens were prepared through both of the weldments and they revealed that weld bead placement had been acceptable for both and heat inputs had been relatively low for SAW (width 3-4mm for HAZ), see Figure 4. The inner HAZ microstructure was generally fine grained (typically 7μm) with some larger grains present (maximum size - 91μm). Both weld metals were columnar in structure, with areas of tempered/refined microstructures present. By optical microscopy, there was no clear difference between the weld metals. Both possess a bainitic microstructure. The determination of grain size was not really possible for these weld metals.

The mechanical property test results for the LP rotor weld metals and parent material are shown in Table 3.

Test Location	Proof Stress (MPa)	Ultimate Tensile Strength (MPa)	Elongation (%)	Reduction Of Area (%)	Charpy FATT (°C)	Hardness 30Kg Load (Hv)
2|Ni|CrMo Weld Metal	474	741	24	66	-40	220
3NiCrMoV Weld Metal	708	766	19	61	1	255
3NiCrMoV Rotor Steel	690	875	16	50	80	260

Table 3 - Mechanical Property Test Results for LP Rotor Welds

The mechanical property test results for the HP/IP rotor weld metals and parent material are shown in Table 4.

Test Location	Proof Stress (MPa)	Ultimate Tensile Strength (MPa)	Elongation (%)	Reduction Of Area (%)	Charpy FATT (°C)	Hardness 30Kg Load (Hv)
1CrMoV (low C) Weld Metal	520	648	24	77	12	220
1CrMoV (high C) Weld Metal	644	779	18	70	19	260
1CrMoV Rotor Steel	628	773	21.5	68	45	230

Table 4 - Mechanical Property Test Results for HP/IP Rotor Welds

A series of uniaxial creep tests were performed on the 1CrMoV weldments at a temperature of 575°C and over a range of stresses between 125 and 240MPa. The creep test matrix for this work was designed on lower bound properties of 1CrMoV rotor steel forgings (3). The results of the weld metal creep tests are shown in Figure 5, where they are compared against weld metals used by Westinghouse for repair of 1CrMoV rotors (4).

642

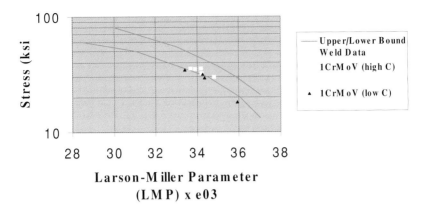

Figure 5 - 1CrMoV Weld Metal Creep Tests

4. Discussion

The 3NiCrMoV and 1CrMoV rotor steels used in this work were representative, in terms of microstructure and mechanical property values, of other rotor steels of a similar vintage (early 1960's).

Macrosections taken through the weld pads, clearly showed that the weld beads were uniform and that the HAZ was narrow with a relatively fine microstructure, compared with the parent rotor steel. Also there were no welding defects such as slag or lack of fusion. These factors demonstrate that excellent control of welding was achieved during pad deposition. Microstructural characterisation of the weldments on the rotor steels revealed a relatively fine grained HAZ microstructure, which clearly showed the success of limiting weld heat input to minimise peak HAZ grain size, whilst adopting a 50% weld bead overlap to refine the grain size. Thus, the refinement technique developed for MMA welding appears to be satisfactory for SAW, when the process is carefully controlled to give weld bead uniformity. The weld metals have a columnar grain structure, which is bainitic in nature, but a large degree of tempering/refinement is evident between grains.

The tensile strength properties of the LP weld metals undermatched those of the parent rotor steels, but compared favourably with weld metals used in LP rotor repairs reported in open literature. The 3NiCrMoV weld metal was substantially stronger than the 2½Ni½CrMo weld metal, while still undermatching the 3NiCrMoV rotor steel. The weld metals had significantly greater ductility than the parent rotor steels. The fracture properties of the weld metals were substantially better than those of the LP rotor steels. The FATT values of the weld metals were significantly lower than the parent rotor steels.

The tensile strength properties from the 1CrMoV (low C) weld metal were significantly worse than those of the parent steel, while the 1CrMoV (high C) weld metal matched the parent tensile properties. The ductility of the 1CrMoV (low C) weld metal was higher than those of the parent or the 1CrMoV (high C) weld metal. The higher strength weld metal matched the ductility of the parent. The fracture properties of both weld metals were significantly better than those of the parent rotor steel. The FATT values determined are similar to undermatching or better than matching SA consumables used for rotor repairs, which have been reported in the open literature (5).

Uniaxial creep testing revealed that the parent rotor steel was within the scatterband for 1CrMoV forgings. The 1CrMoV (low C) weldment was significantly weaker than the parent. The 1CrMoV (high C) weld metal matched the parent creep properties, which compares favourably with data produced by other researchers on weld metals (4).

From the discussion presented above, it is clear that the LP and HP/IP rotor steels selected were representative of contemporary UK manufactured rotor forgings, that were never intended for welding repair. The approach adopted by Innogy has clearly demonstrated that these steels can be repaired successfully by welding and that weldments with acceptable microstructures and mechanical properties can be produced.

5. Practical Experience

Since the completion of the development work on SA welding of LP and HP/IP rotors, Innogy have performed full-scale welding trials on a scrap LP rotor, see Figure 6, where journal, disc and shaft repairs were performed.

Figure 6 - LP Rotor Welding Trials

To date Innogy have performed two SAW repairs on main boiler feed pump (MBFP) turbine rotors, which are 3NiCrMoV monobloc rotor forging. One repair was performed on a bent rotor shaft and involved weld build-up of the journal and interstage gland regions. The other repair involved extensive disc weld build-ups. The weld metal deposition sequence followed previous weld pad development work with a minimum weld bead overlap of 50% on preceding weld bead and weld heat input levels maintained below 2KJ/mm. After welding, the rotors were stress relieved in a "hot" box, while still under rotation, see Figure 7. These repairs were successful and saved the station the expense of new MBFP rotors and the time delay asociated with replacement.

644

Figure 7 - MBFP Rotor Weld Repair

6. Summary

This programme of work has developed a number of rotor weld repair procedures and underwritten the use of these procedures, in terms of acceptable weldment microstructures and mechanical/creep properties, for Innogy.

7. Acknowledgements

I would like to acknowledge my colleagues in Innogy who helped me with the work, in particular Mr D.H.Dawson and Mr M.Blackburn.

8. References

1. "Verification of Fitness for Service of Weld Repaired Steam Turbine Rotors"; K.Fuentes, R.Munson; EPRI 2nd Int. Conf. on Welding and Repair Technology for Power Plants, Daytona Beach, Florida, May 1996.

2. "Specification for Electrode Wires and Fluxes for the Submerged Arc Welding of Carbon Steel and Medium-Tensile Steel"; BS 4165: 1984.

3. "CrMoV Turbine Rotor Forgings: High Temperature Properties"; ERA Project 2021 - Creep of Steel; ERA 84-0142, Nov 1984.

4. "Refurbishment and Upgrading of Steam Turbine Rotor Blade Attachments by Welding"; T.L.Driver and R.E.Clark; EPRI Conf. on Steam and Combustion Turbine Blading, Jan 1992.

5. "Metallurgical Aspects in Welding CrMoV Turbine Rotor Steels; Part 1 - Evaluation of Base Material & HAZ"; G.S.Kim, J.E.Indacochea and T.D.Spry; Mat. Science and Tech, Jan 1991, Vol 7.

"Cold" Weld Repairs using Flux Cored Arc Welding (FCAW)

K.C.Mitchell, Innogy
Engineering, Swindon

Abstract

The development and application of cored wire welding techniques within the power industry, over the last couple of years has been significant and lead to substantial savings, in terms of component replacement costs and outage times. Until recently, the use of cored wires was restricted for many repair welding situations in power plant and only acceptable for welding of structural steelwork. This was due to the previously poor experiences with Metal Inert Gas (MIG) welding in the power industry for pressure parts and the rejection of Flux Cored Arc Welding (FCAW) techniques by the Central Electricity Generating Board (CEGB). This rejection was based on lack of weld metal composition consistency, handleability of wires and poor creep crack growth rates in the weld metal.

The application of "cold" weld repair techniques in the power industry has been well documented. This type of repair is only considered, when a conventional repair (involving post-weld heat treatment) is impracticable or the penalties of time and cost for conventional repair are sufficiently high. A typical "cold weld" repair in the UK has involved low alloy ferritic steel ($\frac{1}{2}$Cr$\frac{1}{2}$Mo$\frac{1}{4}$V, 2$\frac{1}{4}$Cr1Mo) components welded with nickel based MMA consumables. The use of these consumables can pose problems with the volumetric inspection of welds and residual welding stresses present after completion. An alternative "cold weld" repair technique utilises a matching ferritic consumable to the component for repair. This technique will allow full volumetric inspection of the component combined with a higher rate of residual stress decay, while the component is in-service. However, the ferritic "cold weld" has a higher risk of hydrogen cracking, higher initial residual welding stress, poorer fracture toughness and is dependent on controlled weld deposition rates. The last factor is very important to ensure the weldment does not suffer from reheat cracking. CrMoV components are particularly susceptible to this problem.

Innogy have tried to develop the ferritic "cold weld" repair technique further by adopting the use of FCAW equipment and consumables, whilst maintaining the basic deposition sequence and heat input ratios between the initial layers. The use of FCAW instead of MMA should allow for more rapid deposition, reduced risk from hydrogen cracking and a reduction in the time required for welder training to achieve controlled weld deposition rates. The object of the work was to develop ferritic "cold weld" repair procedures using the FCAW technique on mild steel (<0.25%C) and $\frac{1}{2}$Cr$\frac{1}{2}$Mo$\frac{1}{4}$V steels.

This paper will describe in detail the development work on "cold" welding performed to date on these steels and provide an example of where the technique has been used on an in-service power plant component.

"Cold" Weld Repairs using Flux Cored Arc Welding (FCAW)

K.C.Mitchell, Innogy
Engineering, Swindon

1. Introduction

Innogy have been involved with the development and application of Flux Cored Arc Welding (FCAW) within the power industry for 10 years. Initial interest in the technique was due to type IV cracking/damage found on $\frac{1}{2}Cr\frac{1}{2}Mo\frac{1}{4}V$ steam pipelines, where a faster repair technique would reduce outage times and costs significantly. The FCAW process, shown in Figure 1 (1), uses a continuous tubular flux cored wire with gas shielding.

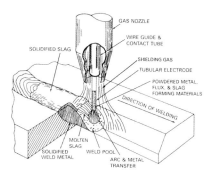

Figure 1 - Flux Cored Arc Welding (FCAW) Process

The flux cored wire consists of a metal sheath with a core of various powdered materials. The benefits of FCAW are achieved by combining three principal features; the productivity of continuous wire welding, the metallurgical benefits derived from a flux and a slag which supports/shapes the weld bead.

An extensive development programme on $2\frac{1}{4}Cr1Mo$ flux cored welding consumables has been performed in the UK (2) and this has led to the adoption of FCAW for type IV repairs on conventional power plant. All of the in-situ repairs on high integrity components have utilised standard MIG welding equipment, as shown in Figure 2 (3).

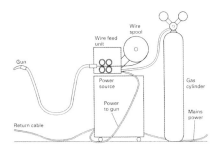

Figure 2 - Gas Shielded FCAW Equipment

The equipment typically comprises a transformer-rectifier, operating as a direct current (DC) constant voltage type power supply, portable wire feeder unit and appropriate welding gun with associated cabling. Utilising this FCAW repair technique, a typical main steam butt weld could be completed within 10 hours as opposed to 22 hours for the normal Manual Metal Arc (MMA) weld repair, see Figure 3.

Figure 3 - Flux Cored Arc Weld Type IV Repair

The application of "cold" weld repair techniques in the power industry has been well documented (4). This type of repair is only considered, when a conventional repair (involving post-weld heat treatment) is impracticable or the penalties of time and cost for conventional repair are sufficiently high. A typical "cold weld" repair in the UK has involved low alloy ferritic steel (½Cr½Mo¼V, 2¼Cr1Mo) components welded with nickel based MMA consumables, see Figure 4.

Figure 4 - Typical "Cold" Weld Repair to Header Stub

The use of these consumables can pose problems with the volumetric inspection of welds and residual welding stresses present after completion. An alternative "cold weld" repair technique utilises a matching ferritic consumable to the component for repair. This technique will allow full volumetric inspection of the component combined with a higher rate of residual stress decay, while the component is in-service. However, the ferritic "cold weld" has a higher risk of hydrogen cracking, higher initial residual welding stress, poorer fracture toughness and is dependent on controlled weld deposition rates. The last factor is very important to ensure the weldment does not suffer from reheat cracking. CrMoV components are particularly susceptible to this problem.

The deposition technique adopted for "cold" weld repairs is similar to the one used for prevention of reheat cracking on CrMoV components, see Figure 5.

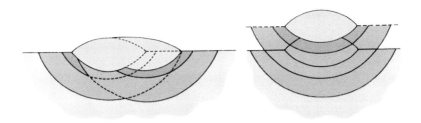

Figure 5 - Grain Refinement by Succeeding Weld Beads

Generally this technique involves depositing the first layer of weld metal with 2.5mm electrodes as single stringer weld beads with 50% overlap on the preceding weld bead in that layer. The second layer is deposited with 4.0mm electrodes, as single stringer weld beads again with 50% overlap. The rest of the excavation is then filled normally with 4.0mm electrodes adopting a maximum weave width of 3 times the core wire diameter. The ferritic "cold" weld repair is further complicated by ensuring the heat input from the second layer must be at least double that of the first layer. This is to ensure grain refinement is achieved in the HAZ of the weldment. Extensive welder training is required to ensure welders achieve these heat input ratios consistently.

Over the last two years, Innogy have tried to develop the ferritic "cold" weld repair technique further by adopting the use of FCAW equipment and consumables, whilst maintaining the basic deposition sequence and heat input ratios between the initial layers. The use of FCAW instead of MMA should allow for more rapid deposition, reduced risk from hydrogen cracking and a reduction in the time required for welder training to achieve controlled weld deposition rates.

2. Weld Procedure Development

The object of this work was to develop ferritic "cold" weld repair procedures using the FCAW technique on mild steel and ½Cr½Mo¼V steels, by simulating a typical repair to an eroded header component.

To achieve the two layer refinement required using the FCAW technique needed the development of 1.0mm flux cored wires in mild steel and 2¼Cr1Mo steel. These wires can produce heat inputs equivalent to that of a 2.5mm electrode, hence replicate the initial weld layer to be deposited. The second weld layer is deposited using a 1.2mm wire which can be equivalent to a 4.0mm electrode, in terms of heat input. For the initial layer, the weld bead deposited was 6-8mm wide with 50% overlap on preceding weld bead. The second layer was 10-12mm wide with 50% overlap on preceding weld bead.

For welding trial purposes, sections of mild steel (555mm OD, 37mm WT) and ex-service ½Cr½Mo¼V (367mm OD, 64mm WT) pipe were excavated to a suitable profile, see Figure 6, by arc air gouging followed by grinding.

80mm

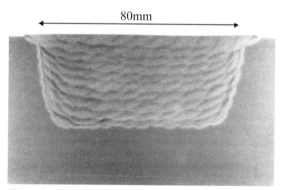

Figure 6 - Cross-Section through CrMoV Pipe Excavation

Before welding commenced, a preheat of 100°C for the mild steel and 250°C for the ½Cr½Mo¼V pipe was applied. After the initial two layers were deposited adopting the two layer refinement technique, see Figure 7, then the excavation was filled normally with the 1.2mm flux cored wire.

Figure 7 - 1ˢᵗ Layer Deposit on Mild Steel Excavation

On completion of welding, the preheat was maintained for 4 hours to allow for any hydrogen present to diffuse away.

The welds were non-destructively tested (NDT) by magnetic particle inspection (MPI) and ultrasonics (UT) followed by microstructural analysis and standard mechanical property testing (tensile, charpy and hardness) to determine whether the two layer refinement technique had been a success or not.

3. Results and Discussion

For both weldments, macrostructural examination revealed no evidence of any defects such as entrapped slag or porosity, which agreed with the NDT inspections. Weld bead shape and size was uniform throughout and the HAZ width was typically 3mm. Microstructural examination of the weldments revealed a high degree of grain refinement in the HAZ, producing a relatively fine grain size of less than 50 micron, see Figure 8 and 9.

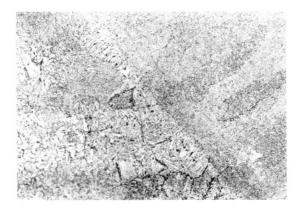

Figure 8 - Mild Steel Fusion Boundary, x900

Figure 9 - 2¼Cr1Mo Fusion Boundary, x900

The mechanical properties of the mild steel weldment is shown below in Table 1.

	Proof Stress (0.1%) MPa	Ultimate Tensile Strength (UTS) MPa	Charpy Impact Toughness (Room Temp) J	Hardness (Hv) 30Kg Load
Parent	261 - 348	519 - 528	132 - 150	142 - 170
HAZ	321 - 333	524 - 536	233 - 236	176 - 242
Weld	366 - 370	509 - 524	200 - 219	167 - 189

Table 1 - Mechanical Properties of Mild Steel Weldment

The values obtained during mechanical testing of the mild steel weldment are satisfactory, with no high hardness values or low charpy impact values being recorded in the weldment.

The mechanical properties of the ½Cr½Mo¼Vweldment are shown below in Table 2 :-

	Proof Stress (0.1%) MPa	Ultimate Tensile Strength (UTS) MPa	Charpy Impact Toughness (Room Temp) J	Hardness (Hv) 30Kg Load
Parent	260 - 271	439 - 450	16 - 18	138 - 146
HAZ	260 - 266	463 - 470	38 - 65	200 - 311
Weld	621 - 700	871 - 908	16 - 20	290 - 350

Table 2 - Mechanical Properties of ½Cr½Mo¼VWeldment

The mechanical properties determined from the ½Cr½Mo¼Vweldment reveal the poor charpy impact values of the ex-service pipe material, which is below the HAZ charpy values. The 2...Cr1Mo weld metal possesses high tensile properties with a corresponding high hardness and poor charpy impact values.

4. Further Developments and Practical Experience of Flux Cored "Cold" Weld Repairs

Following on from our initial welding trials on mild steel/CrMoV pipework and subsequent microstructural evaluation and mechanical testing of these weldments reported in the last section, Innogy decided to pursue the development of flux cored "cold" weld repairs further.

4.1 Further Developments on CrMoV Pipework

The practicality of applying the two layer refinement technique with the FCAW process was shown to be feasible, following the microstructural examination of the 2¼Cr1Mo weldment. However, the high hardness noted in the weld metal could lead to potential reheat cracking problems in the weldment. In an attempt to reduce these hardness values, further 1.0/1.2mm flux cored wires were produced with 2CrMoL (<0.05%C) composition.

A main steam pipe butt weld was produced in ex-service CrMoV material utilising the 2CrMoL flux cored wires and adopting the two layer refinement technique, in order to produce an approved weld procedure to BSEN 288, Pt3 and ASME IX standards. The results of the mechanical testing are shown in Table 3.

	Proof Stress (0.2%) MPa	Ultimate Tensile Strength (UTS) MPa	Charpy Impact Toughness (Room Temp) J	Hardness (Hv) 30Kg Load
Parent		498 - 518 (*)	4 - 5	148 - 164
HAZ			4 - 28	195 - 285
Weld	747	844	12 - 32	249 - 306

Table 3 - Mechanical Properties of 2CrMoL Weld
(*) - Cross-weld tensile tests failing in parent.

The remaining sections of the flux cored weld were creep tested at 600°C and varying stress levels alongside 2CrMoL MMA weldments. The results of these tests are shown in Figure 10.

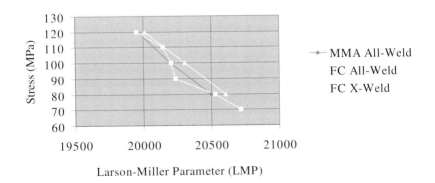

Figure 10 - Stress Rupture Data for 2CrMoL Weld Metals

The stress rupture lives of the MMA and flux cored weld metals are very similar, but the ductilities of the flux cored samples are lower than the MMA ones. Innogy are currently funding further work to underwrite these repairs to enable their use on site.

4.2 Practical Experience on Flux Cored "Cold" Weld Repairs

While still developing the flux cored "cold" weld repair technique for CrMoV and 2¼Cr1Mo steels, Innogy have applied this technique to thick section mild steel components, where post weld heat treatment was impracticable. One particular example of this type of repair is described in detail below.

4.2.1 Repair of HP Feedheater Waterbox

A series of thermal fatigue cracks developed around the tubeplate drain hole as it goes through the wall of an HP feedheater waterbox. These cracks were extensive upto 170mm depth in 200mm thick wall, and would have potentially led to the waterbox being scrapped. A conventional weld repair would have been difficult due to the size of the component and the close proximity of the tubeplate. The excavation was prepared in the waterbox by arc air gouging followed by preheating to 100°C and subsequent welding using 1.0/1.2mm mild steel flux cored consumables, see Figure 11 and 12.

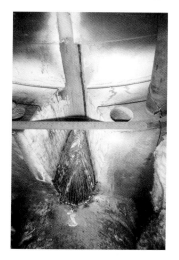 

Figure 11 - Weld Excavation on Waterbox   Figure 12 - Completed Repair on Waterbox

After welding, the repair was soaked at 250°C/4h before cooling to ambient for NDT inspection. The repair was successful and saved Innogy the replacement costs of £300,000 and time involved in delivery of new waterbox/tubeplate. The total cost of the repair was £30,000 and took three weeks to complete, which clearly shows the cost benefit to Innogy of attempting such a repair.

5. Summary

The development of flux cored "cold" weld repair techniques looks promising and although further work is needed to underwrite the use of this technique on high temperature/ high pressure systems, it could prove to be an effective repair technique where PWHT is impracticable and time consuming.

6. Acknowledgements

I would like to acknowledge my colleagues in Innogy who have helped me with this work, in particular Dr S.J.Brett and Mr D.H.Dawson.

7. References

1. AWS Welding Handbook, Volume 2, p158-190.
2. Mitchell, K.C., Allen, D.J. and Coleman, M.C., 1996, "Development of Flux Cored Arc Welding for High Temperature Applications", EPRI, TR-107719.
3. Widgery, D.J., 1994, "Tubular Wire Welding", Abington Publishing, Cambridge, UK.
4. Mitchell, K.C., 1994, "Cold Weld Repair Applications in the UK", EPRI Conf, Welding & Repair Technology for Fossil Fired Plants, Virginia, USA.

AUSTENITIC STEELS I

Microstructural Evolution of Type 304 Austenitic Heat Resistant Steel after Long-Term Creep Exposure at the Elevated Temperatures

M. Yamazaki, K. Kimura、 H. Hongo, T. Watanabe and H. Irie
National Research Institute for Metals
T. Matsuo
Tokyo Institute of Technology

Abstract

Microstructural evolution of Type 304 austenitic heat resistant steel after long-term creep exposure up to about 100,000h over a temperature range from 723 to 823K, has been examined. In the specimens creep ruptured for about 100,000h in the lower temperatures of 723 and 773K, the precipitations of $M_{23}C_6$, G-phase and alpha-Fe were detected at grain boundaries. In the specimen creep ruptured for about 100,000h at 823K, the fine $M_{23}C_6$ particles were confirmed within grain and a sigma-phase were detected at grain boundary combined with G-phase and alpha-Fe. G-phase was divided into two types based on the difference in the chemical compositions, which were Ni-Mn-Si type and Cr-Ni-Si type. Ni-Mn-Si type G-phase was observed in the specimens ruptured at 723 and 773K, and the other one of Cr-Ni-Si type was found in the specimens ruptured at 773 and 823K. Long-term creep done at low temperature gives the loss of creep ductility. A precipitation of alpha-Fe is correlated with the depletion of Ni due to the precipitation of G-phase. Consequently, the decrease in creep rupture ductility in the long-term duration would be caused by embrittlement due to a lot of precipitations such as $M_{23}C_6$, G-phase and alpha-Fe at grain boundary.

1. Introduction

Austenitic heat resistant steels are used for high temperature structural components in the field of nuclear application. The design temperature range of structural components for construction of nuclear power plant is settled from 700 to 1090K against 304 steel, which was specified in the ASME Boiler & Pressure Vessel Code Section III Division 1-Subsection NH Class 1 components in elevated temperature service [1]. Therefore, long-term exposure in the temperature range of 700 to 1090K should be considered for the design of structural components in nuclear reactor [2,3]. Furthermore, according to the length of the creep life, creep features turns to transient type from accelerating type, so the creep rate-time curve on a logarithmic scale is available, because transient creep region is expanded. By using the creep rate-time curve on a logarithmic scale, the mechanism of creep strengthening would be effectively discussed [4,5]. While Time-Temperature-Precipitation (TTP) diagrams for 304 reported by a few researcher are available for discussion of microstructural evolution during creep [6,7]. A TTP diagram of 304 steel in the Metallograpic Atlas of Long-Term Crept

Materials No.M-1 is prepared from the head portion of crept specimen as shown in Fig.1 [6]. This figure is very nice indicator to understand the microstructural change in 304 steel, but information at 723 and 773K are not indicated. And it is emphasized that in the TTP diagram, the precipitations of $M_{23}C_6$ and TCP phases would be promoted in the specimen subjected to creep deformation [8-13]. The first information of precipitation in the specimen subjected to long-term creep at 723K is very available for the design of high temperature component in nuclear power plant.

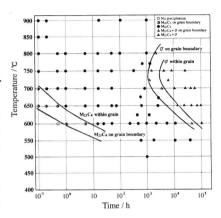

Fig.1 Time-temperature-precipitation diagram of 304 steel [6].

In this study, the creep tests at 723, 773 and 823K were done and the change in the shape of the creep rate-time curves with temperature was discussed in conjunction with changes in microstructure during creep.

2. Material and experimental method

The material was sampled from a commercial plate [14] which was produced as follows; an ingot of 16 ton which were produced from 30 ton melts in an electric furnace was hot-rolled to plate in a thickness of 25mm and solution treated for 0.5h at 1373K. The chemical composition and the average austenite grain diameter are shown in Table 1.

The specimens for creep test with 50mm in gauge length and 10mm in diameter were sampled along a rolling direction of the plate. Tensile creep tests were carried out under the constant load condition at 723K-294MPa, 773K-216MPa and 823K-157MPa. Microstructure of the specimens subjected to creep was examined by scanning electron microscope (SEM) and transmission electron microscope (TEM). Precipitates at grain boundary were identified by electron diffraction analysis and chemical composition analysis by EDX.

Table 1 Chemical composition and austenite grain diameter of 304 steel

Material	Chemical composition								mass %	A.G.D
	C	Si	Mn	P	S	Ni	Cr	Mo	N	μ m
304 steel	0.07	0.59	1.05	0.026	0.005	9.21	18.67	0.1	0.025	85

Solution treatment: 1373K for 0.5h $\rightarrow$ water quench

3. Results and Discussion

3.1 Temperature dependence of instantaneous strain and creep rate-time curve

The creep rate-time curves at 723K-294MPa, 773K-216MPa and 823K-157MPa of the steel are shown in Fig.2. Time to rupture of the steel was 116,363.2h, 90,106.2h and 105,351.9h at 723K-294MPa, 773K-216MPa and 823K-157MPa, respectively. The creep rate decreases up to about 5,000h at 723K-294MPa, and increases gradually after showing minimum creep rate. In the creep rate-time curves at 773K-216MPa and 823K-157MPa, the slope of the curves increase at about 3,000h and the curves indicate significant decrease in creep rate after 3,000h. Such decreases in creep rate continue up to about 30,000h and 50,000h at 773K and 823K, respectively.

The relationships between creep rate and strain at 723K-294MPa, 773K-216MPa and 823K-157MPa of the steel are shown in Fig.3. The strain is a sum of instantaneous strain at the loading and creep strain. The temperatures of 723, 773 and 823K are regarded as low temperature for a high temperature creep. Since stress conditions of 294MPa at 723K, 216MPa at 773K and 157MPa at 823K are characterized as relatively high, relatively large instantaneous strain of 0.074, 0.029 and 0.005 were observed at 723K-294MPa, 773K-216MPa and 823K-157MPa, respectively. The magnitudes of creep strain after subtracting instantaneous strain from rupture elongation are estimated as 0.08, 0.14 and 0.115 for 723K-294MPa, 773K-216MPa and 823K-157MPa, respectively.

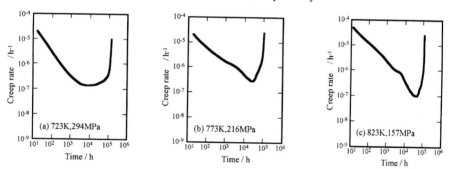

Fig.2 Creep rate – time curves at 723, 773 and 823K.

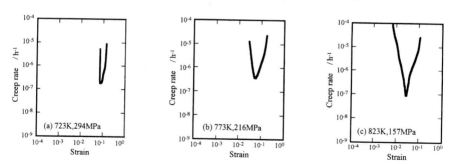

Fig.3 Creep rate – strain curves at 723,773 and 823K.

3.2 Changes in microstructure near the prior grain boundary with temperature

Scanning electron micrographs of the specimens ruptured for 116,363.2h at 723K-294MPa, 90,106.2h at 773K-216MPa and 105,351.9h at 823K-157MPa are shown in Fig.4. All these photographs show the microstructure near the grain boundary. In the specimens creep ruptured at 723K and 773K, a series of granular particles precipitate on the grain boundary. And the area near the granular particles on the grain boundary is heavily etched. In the specimen creep ruptured at 823K, the size of the precipitates on the grain boundary is smaller than those in the specimens creep ruptured at 723 and 773K, irrespective to higher temperature.

Transmission electron micrograph of the specimen creep ruptured for 116,363.2h at 723K-294MPa is shown in Fig.5. Precipitations of alpha-Fe and Ni-Mn-Si type G-phase which will be mentioned later in detail (Figs.8, 11). These phases precipitate in contact with $M_{23}C_6$ carbide particle on the grain boundary. The dislocation density near the grain boundary particles is very high.

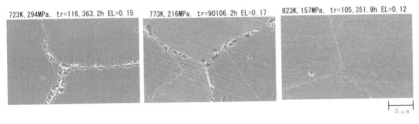

Fig.4 SEM micrographs of crept specimen at 723,773 and 823K.

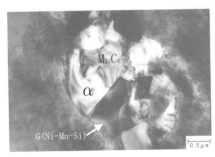

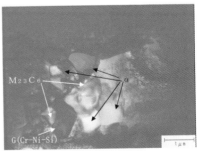

Fig.5 TEM micrograph of particles of precipitate at 723K and 116363.2h.

Fig.6 TEM micrograph of particles of precipitate at 773K and 90106.2h.

Transmission electron micrograph of the specimen creep ruptured for 90,106.2h at 773K-216MPa is shown in Fig.6. Alpha-Fe and Cr-Ni-Si type G-phase precipitate in contact with $M_{23}C_6$ carbide particle on the grain boundary. Although the microstructural feature in the specimen crept at 773K is very similar to that crept at 723K, there is difference in type of G-phase, almost of the G-phase at 723K is Ni-Mn-Si type and Cr-Ni-Si type at 773K.

Fig.7 TEM micrograph of particles of precipitate at 823K and 10531.9h.

Transmission electron micrograph of the specimen creep ruptured for 10,531.9h at 823K-157MPa is shown in Fig.7. Precipitation of alpha-Fe was detected in contact with sigma-phase on the grain boundary, however precipitate of alpha-Fe is smaller number than that in the specimen crept at 723 and 773K. Considerable amount of dislocations are still remained near the grain boundary in the specimen crept at 823K.

Comparing the microstructural features along grain boundary in the specimens creep ruptured at 723,773 and 823K with those examined by scanning electron microscopy as described in Fig.4, the region observed in transmission microscopy is corresponded to the heavily etched region. In these region, alpha-Fe and Ni-Mn-Si type G-phase would be etched deeply, because of their low chromium contents.

Experimental results on the identification of the two types G-phases, $M_{23}C_6$ and alpha-Fe by the electron diffraction and EDX analysis are shown in Figs.8, 9, 10 and 11, respectively. Stoichiometric relationship of G-phase is $A_6B_{16}C_7$ and C is silicon. Two types of G-phases, Ni-Mn-Si and Cr-Ni-Si types, were detected in the 304 steel creep ruptured for about 100,000h at 723 and 773K. Crystal structure of face centered cubic (fcc) of G-phase is the same as that of $M_{23}C_6$ carbide as shown in Fig.10. Therefore, two types G-phases and $M_{23}C_6$ carbide are distinguished by the difference in chemical composition.

Precipitation of alpha-Fe shown in Fig.11 can be distinguished from delta-Fe that exist in the as solution treated condition by the higher chromium content in delta-Fe and a lot of precipitation of $M_{23}C_6$ within delta-Fe. Precipitation sequence in the type 304 steel during long-term creep at the low temperature is proposed as follows;

(1) Precipitation of $M_{23}C_6$ carbide decreases chromium content of its surrounding.

(2) Low chromium content promotes the precipitation of G-phase.

(3) Precipitation of G-phase decreases nickel content of its surrounding.

(4) Low nickel content decreases stability of austenite and results in precipitation of alpha-Fe.

664

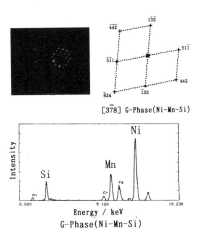

[378] G-Phase(Ni-Mn-Si)

G-Phase(Ni-Mn-Si)

Fig.8 Electron diffraction analysis of G phase of 304 steel at 723K.

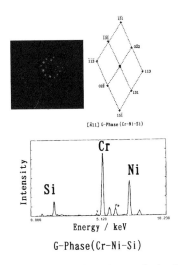

[411] G-Phase(Cr-Ni-Si)

G-Phase(Cr-Ni-Si)

Fig.9 Electron diffraction analysis of G phase of 304 steel at 773K.

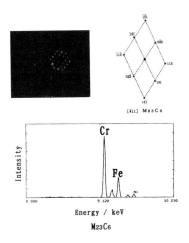

[411] M₂₃C₆

M23C6

Fig.10 Electron diffraction analysis of $M_{23}C_6$ carbide of 304 steel at 723K.

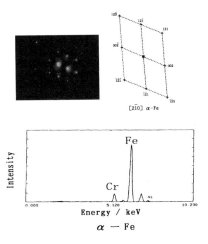

[210] α-Fe

α − Fe

Fig.11 Electron diffraction analysis of alpha-Fe of 304 steel at 773K.

3.3 Changes in precipitates within grain with temperature

Transmission electron micrographs of the specimens creep ruptured for 116,363.2h at 723K-294MPa, 90,106.2h at 773K-216MPa and 823K-157MPa are shown in Figs.12, 13 and 14, respectively. In these specimens, dislocation density within grain is higher in the specimen crept at 723 than that of the other specimens. Precipitate of $M_{23}C_6$ on dislocation is most frequent in the specimen creep ruptured at 723K, and very fine $M_{23}C_6$ particle can be observed.

The creep condition in this study is regarded as relatively lower temperature and higher stress condition. A lot of dislocations introduced by instantaneous strain strongly influence on the precipitation behaviour of $M_{23}C_6$, because dislocation act as a precipitation site of $M_{23}C_6$.

Time-Temperature-Precipitation diagram of the type 304 steel has been rearranged by the present results and shown in Fig.15. By adding our data the following new evidences appear. ① Sigma-phase is only detected in the specimen crept at 823K, and nice correlation with previously reported data is confirmed. ② The appearance of G phases is confirmed in the temperature range of 723 to 823K. ③ $M_{23}C_6$ precipitation on dislocation is confirmed in the specimens crept for 100,000h even at the temperature of 723K.

Fig.12 Dislocation substructure of crept specimen at 723K and 116363.2h.

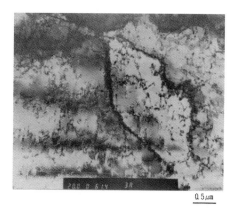

Fig.13 Dislocation substructure of crept specimen at 773K and 90106.2h.

Fig.14 Dislocation substructure of crept specimen at 823K and 10531.9h.

666

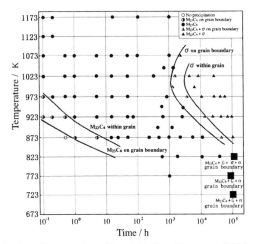

Fig.15 Time-Temperature-Precipitation diagram of 304 steel

4. Conclusions

The shape of creep rate-time curve of the type 304 austenitic heat resistant steel in the long-term condition at 723, 773 and 823K has been discussed in conjunction with changes in microstructure during long-term creep and the following results are obtained.

1) Large instantaneous strain has been observed at the loading of the stress and the magnitude of creep strain has been evaluated to be small. Creep rate-time curves at 773 and 823K indicate inflections and large decreases in creep rate after such inflection in the transient creep stage.

2) Precipitations of $M_{23}C_6$ carbide, G-phase and alpha-Fe on grain boundaries and fine $M_{23}C_6$ carbide on dislocations have been detected after long-term creep deformation of about 100,000h at 723, 773 and 823K. Moreover, precipitation of sigma-phase on grain boundary has been also observed in the specimen creep ruptured at 823K.

3) Two types of G-phase whose stoichiometric relationship is $A_6B_{16}C_7$, have been identified as Ni-Mn-Si type and Cr-Ni-Si one. Crystal structure of G-phase is a face centered cubic which is the same as that of $M_{23}C_6$ carbide.

4) Precipitation sequence on grain boundary has been proposed as follows; (1) precipitation of $M_{23}C_6$ carbide, (2) precipitation of G-phase and (3) precipitation of alpha-Fe. Precipitation of such phases on the grain boundary has been supposed to be a reason of low rupture ductility.

5) It has been concluded that changes in microstructure with a sequence of precipitation should be taken into account for the evaluation of long-term creep strength property of the type 304 austenitic heat resistant steel even at relatively low temperature of 723 to 823K.

References

[1] ASME Boiler & pressure vessel code section III division 1-subsection NH, Class 1 components in elevated temperature service, ASME (1998)

[2] NRIM CREEP DATA SHEET No.32A, Data sheets on the elevated temperature properties for base metals, weld metals and welded joints of 18Cr-8Ni stainless steel plates, National Research Institute for Metals, (1995).

[3] T.NAKAZAWA and H.ABO, The effect of some factors on the creep behavior of type 304 stainless steel, Transactions ISIJ, 18(1978), p602-p610

[4] M.YAMAZAKI,HONGO,T.WATANABE, K.KIMURA,T.TANABE,H.IRIE and T.MATSUO, Correlation Between Creep Rate – Time curve and Microstrucural Parameter in SUS304, CAMP-ISIJ Vol.11(1998),1169 (in Japanese).

[5] M.YAMAZAKI, HONGO,T.WATANABE, K.KIMURA,T.TANABE,H.IRIE and T.MATSUO, 100 000h Creep Behavior and Microstructure of Ruptured Specimen for SUS304, CAMP-ISIJ Vol.11(1998),1170 (in Japanese).

[6] NRIM CREEP DATA SHEET No.M-1, Metallographic Atlas of Long – Term Crept Materials, micrographs and microstructural characteristics of crept specimens of 18Cr-8Ni stainless steel for boiler and heat exchanger seamless tubes (SUS 304H TB), National Research Institute for Metals (1999)

[7] George F. Vander Voort, Atlas of Time – Temperature Diagrams for Irons and Steels, ASM international (1991)

[8] NRIM CREEP DATA SHEET No.4B, Data sheets on the elevated temperature properties for of 18Cr-8Ni stainless steel boiler tube, National Research Institute for Metals, (1986).

[9] F.R.BECKITT and B.R.CLARK, The shape and mechanism of formation of $M_{23}C_6$ carbide in austenite, Acta Metallurgica, 15(1967), p113-p129

[10] F.G.WILSON, The morphology of grain-and twin-boundary carbides in austenitic steels, Journal of The Iron and Steel Institute, (1971), p126-p130

[11] V.A.BISS and V.K. SIKKA, Metallographic study of type 304 stainless steel long-term creep-rupture specimen, Metallurgical transaction 12A(1981), p1360-p1364

[12] V.Biss,D.L.Sponseller and M.Semchyshen,Metallographic examination of type 304 stainless steel creep rupture specimens, Journal of materials,7(1972), p88-p94

[13] D.J.WIlSON, The influence of simulated service exposure on the rupture strengths of grade11, grade22, and type 304 steels, Journal of Engineering Materials and Technology, (1974), p10-p21

[14] SUS304-HP, Hot rolled stainless steel plates, sheets and strip, JIS G 4304 (1991).

Long-term studies on the creep behaviour of the structural material 316 L (N) in the low-stress range at 550 and 600°C

M. Schirra

Forschungszentrum Karlsruhe

Institut für Materialforschung I

Postfach 3640, D-76021 Karlsruhe

Abstract

Experiments were performed at 550°C in the stress range of 250-100 MPa. A further series of tests were run at 600°C in the range of 170-60 MPa. Continuous recording of the very small creep rate necessitated a special double-extensometer with an enlarged 200 mm gauge length in combination with a recorder of high resolution. When plotting the determined or estimated values for the minimum creep rate in the $\dot{\varepsilon}_{pmin}$ versus σ diagram, it was found that the experimentally determined creep rates in the low-stress range were much higher than the values that would have been obtained by extrapolation of the data observed at higher stress levels. The stress exponent decreased from 12.8-10.5 (550-600°C) to $n \approx 6.8$ at low stress levels. The technically relevant time-strain limits t_{ef} also exhibited significantly changed dependencies in the upper and lower stress ranges.

1. Introduction

The austenitic 17/12/2-CrNiMo steel 316 L (N) - DIN 1.4909 is used in both conventional and nuclear power plants and envisaged for use as structural material in the new ITER. A number of experimental investigations were carried out to determine the creep and creep-rupture strength behaviour of this steel in the normal stress and temperature range [1-5]. In the design-relevant low-stress range (<150 MPa) at 550 and 600°C, however, data allowing statements to be made with regard to the stress dependence of the minimum creep rate or the technically relevant creep strain limits are completely lacking. Therefore, evaluation of creep data was considered to be absolutely necessary for setting up reliable constitutive equations for the material behaviour. This report will deal with experiments which started in 1991 and meanwhile extend over a duration of 80,000 h (status: January 2001).

2. Material and test program

Experiments were performed at 550°C in the stress range of 250 - 100 MPa. A further series of tests were run at 600°C in the range of 170 - 60 MPa. These tests partly complement and partly overlap with previous experiments performed on this steel. [4]

Table 1:
Chemical composition (wt%) of 316 L(N) heat: 11477 (Creusot), plate 40 mm

C	Si	Mn	P	S	Cr	Ni	Mo	Cu	N	Al	B
0.020	0.32	1.80	0.020	0.0006	17.34	12.50	2.40	0.12	0.08	0.018	0.0014

solution annealed: 1100°C/water   hardness:132-151 HV30   grain-size ASTM: 4

Table 2:
Stress range (MPa) of creep rupture tests performed at 550°C and 600°C

MPa: 380 _ _ _ _ 550° _ _ 240

250 - 210-180-150-135-120-100 MPa

MPa :300 _ _ _ _ 600° _ _ 150

170 - 120 - 100 - 80 - 70 - 60 MPa

phase I [4]   phase II

creep-rupture-tests   creep-tests
specimen: ⌀5x30mm(doxLo)   ⌀ 8x200mm (doxLo)

3.   Experimental

Specimens with a screw head of 8 mm in diameter and 200 mm in gauge length were selected. This length ensured that the measurement of the creep behaviour was much more accurate than for a standard specimen. The test temperature was controlled by means of three Pt/Rh-Pt thermocouples and kept constant at ± 2°C by means of three PID control units. For creep determination the specimen was equipped with a double-coil extensometer, by the ε-t recorder of which the creep behaviour was registered continuously. The resolution was ≤ 0.001 mm (1:1250). Loading took place via a lever arm (1:15) using plates (Fig. 1).

4.   Test results

At first, the state of knowledge concerning the creep rupture strength and creep behaviour of the steel shall be summarised. At the Institute for Materials Research (IMF III), various heats were investigated in the temperature range of 500 to 750°C [4]. For the design-relevant temperatures of 550 and 600°C, the creep rupture strength ranges covered by the experiments and the lower 1% time-strain limit curve are represented in Fig. 2a. The values for the minimum creep rate are plotted in Fig. 2b as a function of the test stress (phase I). The current experiments (phase II) are shown on the left. They demonstrate how the experimentally covered stress range is extended to 100 MPa at 550°C and to 60 MPa at 600°C and illustrate superposition of the results from phase I. For the design of the levers it is of particular interest to have the 1% time-strain limit curve up to ≥ 10^5/h and the creep rates of ≤ 10^{-8}/h confirmed experimentally by these long-term creep tests.

The experimental creep rupture data published by Japanese authors [2] are located within the hatched areas and do not provide for the knowledge being extended towards lower stresses.

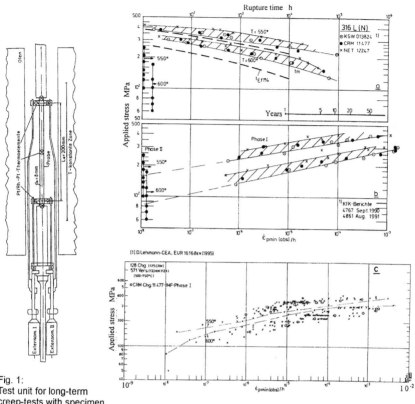

Fig. 1:
Test unit for long-term creep-tests with specimen ∅ 8x200 mm(doxLo)

Fig. 2: Experimental status after phase I

The evaluation of European creep rupture data (including IMF data) by R. Lehmann [1] provides for a small extension of the stress range down to smaller values as far as the 1% time-strain limit values and the minimum creep rate are concerned. However, these values ($\dot{\varepsilon} < 10^{-6}$/h) were obtained for various heats and do not allow any clear statements to be made with regard to the stress dependence due to their scattering (Fig. 2c).

4.1 Creep behaviour up to a test duration of about 80,000 h

As an example, the creep behaviour measured so far is represented in Fig. 3 on double-logarithmic ε-t scales. This type of representation gives a complete survey for each test temperature and allows a good determination of the time-strain limits in the first decades. Furthermore, superposition of the previous test data by the 250 MPa and 170 MPa tests can be noticed. Both tests were completed without fracture at 2% and 6% creep strain, respectively. The initial elongation (ε_{in}) existing at the start of the tests is given for each experiment. Due to the low yield strength typical of austenites, considerable deformations ($\geq 3\%$) occur in the

672

experiments performed in the usual stress range (> 150 MPa). Their effect corresponds to that of previous cold deformation. According to definition, the statements made with regard to the creep behaviour of these tests do not refer to the state of solution annealed, but to solution annealed + deformed. The tests performed in the σ range of < 150 MPa allow first statements to be made with regard to the creep behaviour of the design-relevant solution annealed state. At the test temperature of 550°C, the curves obtained at a high ε_{in} (250 - 150 MPa = > 1%) differed considerably from those obtained at a small ε_{in}. Therefore, another experiment was started later on at 135 MPa and has now reached a total duration of about 65,000 h.

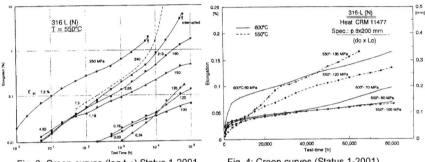

Fig. 3: Creep curves (log t, ε) Status 1-2001   Fig. 4: Creep curves (Status 1-2001)

To obtain an impression of the actual creep stage and the creep rate, a linear ε-t representation (as recorded continuously, but with a higher t/ε resolution) is required. The creep behaviour is represented in Fig. 4. It is obvious that some of the experiments running at lower stresses have not yet advanced beyond the primary creep stage.

This statement becomes even more clear when determining the linear creep rate at regular intervals and plotting it versus the experimental duration. As an example, the creep rates in 1000 h and 2000 h intervals are shown in Fig. 5 for the experiments at 500°C. As far as the tests at 250 - 135 MPa and creep rates of up to 10^{-7}/h are concerned, the values are described well by a mean curve without large scattering. This curve allows to determine the time fraction of primary creep (= decreasing creep rate), the minimum creep rate, and, if applicable, a quasi-stationary creep range with a transition to tertiary creeping (= increasing creep rate). Below 10^{-7}/h, periodic fluctuations can be observed frequently. They are partly caused by measurement technology, because differences of some μm/time period lead to stronger deviations in these decades. Other reasons are the microstructural changes with time and creep hardening, which, to a small extent, cause a constant alternation of relaxation and deformation hardening. In the tests performed at 600°C, the wave-like behaviour of the curves can be noted more clearly. This can also be explained by the reasons mentioned above.

For this range of very small creep rates, averaging has to be done over larger time intervals in order to obtain a better statement with regard to the minimum creep rate. In Figs. 6 and 7, the respective creep rate is represented in intervals of 5,000 h and 10,000 h. From these curves, the minimum creep velocity can be derived or estimated. If these values are plotted as a function of the test stress, Fig. 8 is obtained as the experimental extension of Fig. 2b. It allows the following preliminary statements:

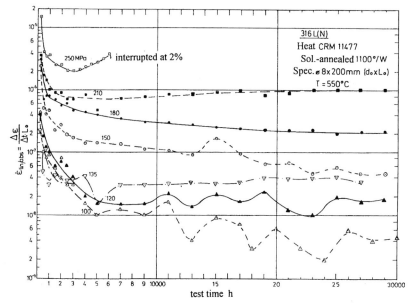

Fig. 5: Creep rate vs. test time

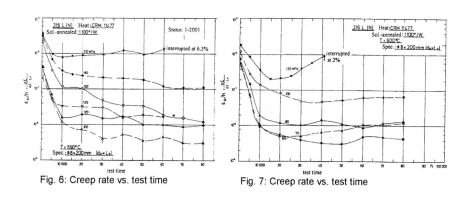

Fig. 6: Creep rate vs. test time

Fig. 7: Creep rate vs. test time

Experiments in the low stress range at 550 and 600°C yield far higher values for the minimum creep velocity than expected from the estimation based on the results of the experiments performed in the higher stress range (hatched area). This significantly changed stress dependence of $\dot{\varepsilon}_{pmin}$ results in smaller values of the stress exponent n (according to Norton), which are in the range of n ≤ 6.8 at test temperatures of 550°C and 600°C.

674

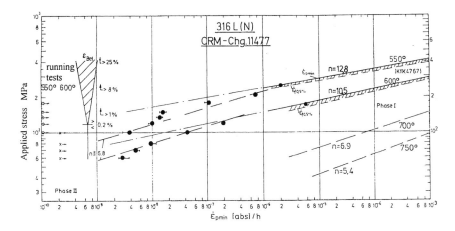

Fig. 8: Minimum creep rate vs. applied stress (Status 1-2001)

In connection with Fig. 3, it was pointed out already that test stresses > 150 MPa result in a plastic deformation, as the yield point at elevated temperature is exceeded. This plastic deformation amounts to σ = 400 MPa $\approx$ 25%. In Fig. 8 on the left, this is represented as a hatched wedge. Since this deformation takes place below the recrystallisation temperature, it acts like a previous cold deformation [7]. It is known from past investigations of several austenitic steels that the optimum degree of deformation amounts to 8 - 15% and has been applied specifically for improving the creep rupture strength and creep behaviour [8].

It is also obvious from Fig. 8 that practically all test stresses applied in phase I by far exceed 200 MPa. Hence, the creep behaviour data refer to the state of solution annealed + deformed ($\geq$ 5%). In contrast to this, the experiments in phase II mainly cover the stress range of < 150 MPa. In this range, only an elastic or a maximum of 1% plastic deformation occurs at the beginning of the experiment. Consequently, the creep behaviour refers to the solution annealed state as delivered. For this reason, the stress dependence, expressed by the n value, approaches the values obtained from experiments at 700 - 750°C in the stress range of < 150 MPa [4].

4.2 Time-strain limits

In normal long-term creep rupture tests, strain limits starting at 0.1% are usually determined. For measurement reasons, significant scattering of the 0.1% and 0.2% time-strain limits has to be expected. When using the overlong sample of 200 mm gauge length, strain limits starting at 0.01% may be determined already. However, also scatterings are shifted to values smaller by an order of magnitude. Starting at 0.1% creep strain, creep behaviour is in agreement with stress, as obvious from Fig. 3.

For all experiments of phase II, time-strain limits of up to 0.05% at least and sometimes of up to 2% are available. All technologically relevant time-strain limits that have been determined from continuous strain recording are given in the diagram according to DIN 50118 (Fig. 9). In each partial diagram the curves obtained in the upper stress range for 0.1 - 5% strain and the rupture curve for the experiments of phase I are plotted for both test temperatures of 550 and 600°C. The corresponding values of the stress overlapping experiments at 250 and 170 MPa performed under phase II are found to be in good agreement.

It can be noticed, however, that at a test temperature of 550°C and a decreasing test stress (up to 150 MPa) the various time-strain limits are reached much earlier than expected from the respective curve in the upper stress range, i.e. the range, where deformation hardening does no longer occur due to the reduced initial elongation of < 1%. At test stresses of ≤ 120 MPa, a considerably changed time dependence of the determined time-strain limits is found. In the 600°C experiments, it even seems to behave asymptotically. Also this representation clearly distinguishes between results obtained at test stresses above and below the yield point at elevated temperature.

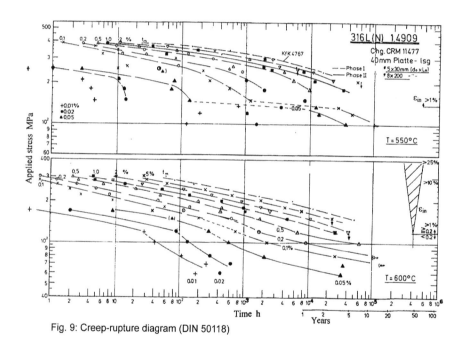

Fig. 9: Creep-rupture diagram (DIN 50118)

4.3 Creep ranges

When representing the linear time-strain diagram (DIN 50118), the time until rupture is divided into a first, second, and third creep and time range. The first (primary) range is characterised by a decreasing creep velocity until the minimum creep velocity $\dot{\varepsilon}_{pmin}$ is reached. The second (secondary or stationary) range is characterised by a linear curve fraction, while the third range (tertiary range) shows an increasing creep velocity until rupture. If the stationary range is not clearly defined or if the transition to the tertiary creep range cannot be determined clearly (as is partly the case with the steel used here), a line parallel to the second creep range shifted upwards by 0.2% and its point of intersection with the third creep range are specified to be the end of secondary creeping (0.2% offset method).

676

In Fig. 10, the thus obtained time fractions of all experiments of phases I and II are plotted as a function of the test stress and a test temperature of 550 and 600°C, respectively. Based on the current state of the investigations, both partial diagrams allow the conclusion to be drawn that only primary creeping takes place at 550°C until 180 MPa and 25,000 h and at 600°C until 100 MPa and 35,000 h. It still remains to be found out whether the curve may simply be continued towards lower stresses.

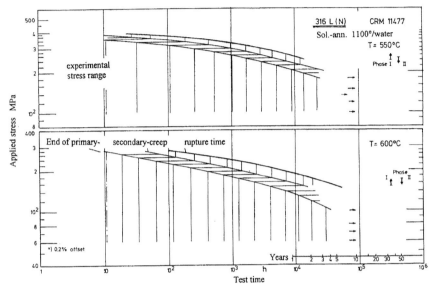

Fig. 10: Part of the primary, secondary and tertiary creep stage

5.   Summary

The EFR structural material 316 L(N) - DIN 1.4909 (Creusot-Marrell heat 11477) was subjected to creep tests in the range of low stresses at 550°C (250 - 100 MPa) and 600°C (170 - 60 MPa), respectively. This range was not covered experimentally before. The tests were supposed to yield data with regard to the technologically relevant time-strain limits (0.01% - 1%) in this design-relevant stress range. They were to demonstrate whether the dependence of the creep behaviour determined for the upper stress range also applied to the lower stress range.

Up to date, the experiments have advanced to about 80,000 h. Evaluation as performed so far indicates that far higher creep rates are measured in the stress range of < 150 MPa than expected from the extrapolation of the results obtained in the stress range that has been covered experimentally so far. According to Norton's creep law, this results in a much smaller value of the stress exponent n.

When transferring the data of the minimum creep velocity shown in Fig. 8 to the representation of a CEA evaluation based on the European data [1] in Fig. 11, it can be noticed that only the FZK-IMF experiments of phases I + II using the CRM heat (using the same material) allow a clear and well founded statement to be made with regard to the stress dependence of the minimum creep velocity over six decades. This statement may be confirmed by individual values (x) obtained for the CEA heats.

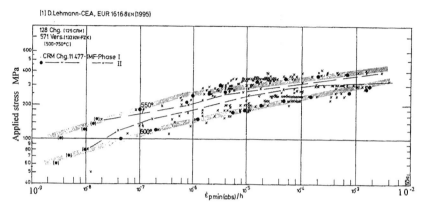

Fig. 11: Minimum creep rate vs. applied stress (Status 1-2001)

6. References

[1] Lehmann, D.;
Evaluation of the stress to rupture and creep properties of type 316 L (N) steel for design use.
Final report EUR 16168 EN, 1995

[2] NRIM Creep Data Sheet No. 42 - 1996,
National Research Institute for Metals, 2-2-54, Nakameguro, Tokyo 153, Japan

[3] Mathew, M.D. et al.;
Influence of carbon and nitrogen on the creep properties of type 316 stainless steel at 873 K.
Mat. Sc. Eng. A 148, 253-260, 1991

[4] Schirra, M., Heger, S.; 1991. Zeitstandsfestigkeits- und Kriechversuche am EFR-Strukturwerkstoff 316 L (N), DIN 1.4909, KfK-Bericht 4767, September 1990. KfK-Bericht 4861, August 1991

[5] Tavassoli, A.A., Touboul F.;
Austenitic stainless steels, status of the properties database and design rule development.
Journal of Nuclear Materials 233-237, 1996, Part A, pp. 52 - 61

[6] Nakazawa, T. et al., 1999. Long-term creep-rupture properties and precipitation in type 316 stainless steels. The Minerals, Metals and Material Society, Proc. Conf. San Diego, 1.-4.3.99, pp. 181-189

678

[7]   Schirra, M.;
        Das Zeitstandsfestigkeits- und Kriechverhalten des SNR-300 Strukturwerkstoffes
        X6CrNi 1811 (1.4948),
        KfK-Bericht 4273, February 1988

[8]   Böhm, H., Schirra, M.;
        Vortrag VDEh-Düsseldorf, 1972; Archiv f.d. Eisenhüttenwesen 44/73, Nr. 10; KfK-
        Bericht 1892, Oct. 1973

Appendix A. Nomenclatura

ITER	=	International Thermonuclear Experimental Reactor
FZK/IMF I	=	Forschungszentrum Karlsuhe, Institut für Materialforschung I
σ	=	applied stress [MPa]
T	=	test temperature [°C]
Lo	=	gauge length [mm]
ε	=	elongation [mm or %]
t	=	test time [h]
$\dot{\varepsilon}$	=	creep-rate [%/h]
$\dot{\varepsilon}_{pmin(abs)}$	=	minimum creep rate [mm/mm/h] (secondary or stationary creep rate)
$\dot{\varepsilon}_{lin}$	=	creep rate by 5000 or 10000 h-steps $\dfrac{\Delta\varepsilon}{\Delta_t \cdot Lo}$
n	=	stress exponent (by Norton $\dot{\varepsilon} = k \cdot \sigma^n$)
t_{ε_f}	=	time to reach a strain limit [h]

THE MULTIAXIAL CREEP DUCTILITY OF AN EX-SERVICE TYPE 316H STAINLESS STEEL

M W Spindler[1], R Hales[2] and R P Skelton[3]

[1]British Energy, Barnett Way, Barnwood, Gloucester, GL4 3RS, UK.
e-mail mike.spindler@british-energy.com, Fax +44 (0)1452 653025.
[2]Consultant, Woodstock, Upton Lane, Upton St. Leonards, Gloucester GL4 8EQ, UK.
e-mail royden_hales@totalise.co.uk, Fax +44(0)1452 618886.
[3]Dep. of Mech. Eng., Imperial College, Exhibition Road, London, SW7 2BX, UK.
e-mail r.p.skelton@ic.ac.uk, Fax +44(0)20 7594 7017.

ABSTRACT

Calculations of creep damage under conditions of strain control are often carried out using either a time fraction approach or a ductility exhaustion approach. For the time fraction approach there are a number of models which can be used to predict the effect of state of stress on the creep rupture duration of stainless steels. However, the published models for the effect of state of stress on creep ductility have not been compared with data for stainless steels. British Energy has developed both empirical and mechanistic models for the effect of multiaxial states of stress on creep ductility, which are intended for use in calculations of creep damage using the ductility exhaustion approach. This paper outlines the results of a series of notched bar creep rupture tests on an ex-service Type 316H stainless steel and compares the data with the predictions of the multiaxial creep ductility models. The interpretation of the results of the notched bar tests is complicated by the variations in stress and strain through the section of the specimen and with time. Notwithstanding this, it is shown that the triaxial states of stress present in the notch lead to reductions in the creep ductility of up to an order of magnitude.

1. INTRODUCTION

The assessment of engineering components subjected to multiaxial loading is usually performed by reducing the combined stresses or strains to a single "equivalent" value. The most widely used relationship is the von Mises equivalent stress. Although this relationship was originally devised to predict the effect of a multiaxial state of stress on yielding, it has subsequently been used for many other applications without full justification. A number of authors have proposed alternative relationships for calculating equivalent stresses to describe isochronous creep rupture [1] to [3]. In particular, Huddleston [3] developed a model from data on stainless steels. However, the published models for the effect of states of stress on the creep ductility [4,5,6] have not been compared with data for stainless steels.

More recently, British Energy has developed relationships to take account of biaxial states of stress on ductility. In the first instance Hales [6] derived equations based on simple models of cavity growth which were further modified to incorporate an empirical description of cavity nucleation. Later, Spindler [7] derived an empirical equation from biaxial creep data on Type 316 and Type 304 stainless steels. The empirical equation is of the form

$$\frac{\bar{\varepsilon}_f}{\varepsilon_{fu}} = \exp\left[p \left(1 - \frac{\sigma_1}{\bar{\sigma}} \right) \right] \exp\left[q \left(\frac{1}{2} - \frac{3\sigma_p}{2\bar{\sigma}} \right) \right] \tag{1}$$

where $\bar{\varepsilon}_f$ and ε_{fu} are the von Mises equivalent and uniaxial strains to failure, respectively, and

σ_1, $\overline{\sigma}$ and σ_p are the maximum principal, von Mises equivalent and hydrostatic stresses, respectively. The values of p and q determined in [7] are p=2.38 and q=1.04 (for materials where ε_{fu} decreases with decreasing strain rate), and p=0.15 and q=1.25 (for materials where ε_{fu} is constant). These values were obtained from published biaxial creep tests on Type 304 steel at 593°C and various Type 316 steels at 593 and 600°C. However, the extent of the reduction in ductility, which is predicted under triaxial states of stress, is sensitive to the values of the coefficients which are used. Consequently, a series of tests has been conducted to measure the effect of triaxial states of stress on creep ductility and hence validate the general applicability of equation (1).

The present paper describes a series of creep rupture tests on notched bars designed to determine the effect of triaxial state of stress on the rupture ductility of an ex-service Type 316H stainless steel. The results of the notched bar tests are also compared with the results of uniaxial tests on the same material [8]. The analysis of notched bar tests is complicated by the fact that the state of stress varies across the section of the notch and the stress distribution changes with time. This problem is often addressed by considering the magnitude and state of stress at the 'skeletal point'. At this point the stresses vary only slightly with time and are taken to be representative of the whole section. The current analysis has also used the "skeletal point" to characterise the effect of triaxial stresses on ductility. The basis of the analysis is the general empirical equations due to Spindler [7]. Finally, the models due to Hales [6] have been extended to a general multiaxial case (see Appendix A) and have also been used to assess the results.

2   EXPERIMENTAL

Double notched bar creep specimens were made from ex-service Type 316H superheater headers, which had been exposed to temperatures in the range 490 to 520°C for approximately 58,000 hours. The chemical composition of the material is shown in Table 1. The dimensions of the specimens and the notches are shown in Figure 1. The specimens conform to the code of practice for notched bar creep rupture testing [9]. Notch acuities (a/R) of 1.5, 2.41, 5 and 15 were chosen to give a wide range of triaxial states of stress (see Section 4). All tests were performed at 550°C and the stresses are shown in Table 2. For the tests with durations less than 1000 hours the diameter of one of the notches was monitored during loading, and subsequent creep, using a diametral extensometer. After failure specimens were measured on a profile projector.

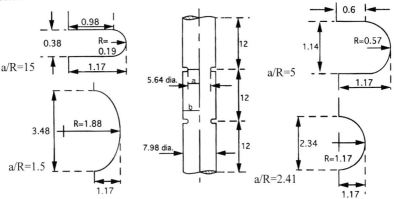

FIGURE 1 – Dimensions of Notched Bar Creep Specimens in mm.

C	Si	Mn	S	P	Ni	Cr	Mo	Co	B
0.06	0.4	1.98	0.014	0.021	11.83	17.17	2.19	0.1	0.005

TABLE 1 – Chemical Composition of the Type 316H in Weight Percent.

3   RESULTS

The results of the tests are reported in Table 2 and are shown graphically in Figures 2 and 3. The surface hoop creep strains, in a notched bar, have been calculated from

$$\varepsilon_h = \ln(a_f/a_0) \tag{2}$$

where a_0 and a_f are the net section radii after initial loading and after failure, respectively. The instrumented notch was not necessarily the notch that failed and it has been assumed that the loading strains will be the same for both notches. For tests without a diametral extensometer a_0 was estimated from the results of the instrumented tests.

a/R	Net Stress (MPa)	Time (hours)	ε_h (abs.)	$\overline{\varepsilon}_{skf}$ (abs.)	a/R	Net Stress (MPa)	Time (hours)	ε_h (abs.)	$\overline{\varepsilon}_{skf}$ (abs.)
1.5	357	372	-0.0368	0.0518	5	405	92	-0.0211	0.0264
2.41	382	175	-0.0285	0.0344	5	374	267	-0.0165	0.0207
2.41	382	248	-0.0317	0.0383	5	340	1940	-0.0243	0.0304
2.41	340	2241	-0.0376	0.0454	5	320	5718	-0.0096	0.0120
2.41	320	5673	-0.0347	0.0419	5	300	>27732	-	-
2.41	300	15923	-0.0368	0.0444	15	429	47	-0.0132	0.0176
5	500	15	-0.0270	0.0338	15	397	108	-0.0118	0.0157
5	438	55	-0.0130	0.0163	-	-	-	-	-

TABLE 2 - Results of Notched bar Creep Rupture Tests on ex-service Type 316H at 550°C.

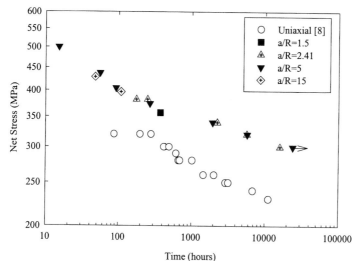

FIGURE 2 –Notched Rupture Behaviour of an ex-Service Type 316H at 550°C.

682

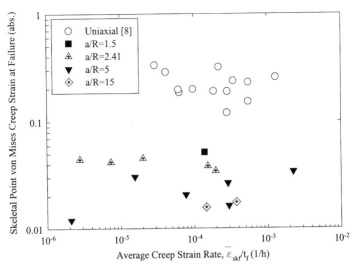

FIGURE 3 – Notched Rupture Ductility of an ex-Service Type 316H at 550°C.

4   INTERPRETATION OF NOTCH CREEP RUPTURE DUCTILITY

In order to compare the notched bar data with equation (1) and with the equations in Appendix A it is necessary to normalise the von Mises creep strain at failure by the uniaxial creep strain at failure. The uniaxial creep strain at failure has been determined from tests on the same heat of material [8], which have been averaged to give 20.2%. This value is the true creep strain at failure of uniaxial tests and has been calculated from equation (3) (below) using measurements of the final diameter at the neck minus the plastic loading strain, which has also been converted into true strain. This definition of the uniaxial creep strain at failure has been used because it is consistent with the definitions of notch creep strain at failure, in that both are calculated from measurements of specimen diameter at the failure location and exclude any loading strain.

It has been proposed [11] that the equivalent strain at failure is given by:

$$\overline{\varepsilon}_f = 2.\ln(a_0/a_f) \tag{3}$$

The factor of 2 is only strictly applicable when the hoop and radial strains are equal and reduces with increasing notch acuity. Since the present work has investigated the effect of a range of notch acuities this approach was considered inappropriate. However, it is used as a measure of uniaxial ductility for comparison with the notched bar data. In the present work the strain at failure of the notched bars is taken to be the von Mises strain at the skeletal point. The values of this strain are deduced from the measured surface hoop strain, equation (2), and the ratio between von Mises strain at the skeletal point and surface hoop strain, $\overline{\varepsilon}_{sk}/\varepsilon_h$, as determined from finite element analysis, FE.

Non-linear FE analyses of the stress and strain distributions in the notches have been conducted using the BERSAFE stress analysis programme. An axisymmetric mesh of half the specimen was modelled. Boundary conditions of zero axial displacement on the plane of symmetry together with an axial load were applied to the ends of the specimen. Initial loading strains were derived using tensile data obtained from a tensile test at 550°C. The subsequent creep behaviour was calculated using the RCC-MR creep equations for Type 316 [10]. The FE analysis described the initial distribution of stresses across the notch and their redistribution as a function of time up to 300 hours (Fig. 4). It should be noted that steady state conditions were reached after typically 1 hour. The skeletal point was defined as the point at which the von Mises stress was approximately constant with time and the FE results have been used to determine values for both $\sigma_p/\overline{\sigma}$ and $\sigma_1/\overline{\sigma}$ at the skeletal point (Table 3). These ratios are used to represent the degree of triaxiality and both increase with increasing a/R. In addition, the distributions of creep strain across the notch are predicted (Fig. 5) and were used to determine $\overline{\epsilon}_{sk}/\epsilon_h$ (Table 3). The values of $\overline{\epsilon}_{sk}/\epsilon_h$ have been used with equation (2) to calculate $\overline{\epsilon}_{sk}$ for the notched tests (Table 2). It should be noted that tertiary creep was not modelled and that heat-specific creep properties were not used in the FE. Thus, the FE analysis could only predict the distribution of creep strain and not the magnitude at failure.

a/R	$\overline{\epsilon}_{sk}/\epsilon_h$	$\overline{\sigma}/\sigma_{net}$	$\sigma_p/\overline{\sigma}$	$\sigma_1/\overline{\sigma}$
1.5	-1.407	0.705	0.745	1.408
2.41	-1.207	0.646	0.848	1.511
5	-1.253	0.611	1.089	1.745
15	-1.334	0.581	1.526	2.174

TABLE 3 – Results of FE Analyses Used to Calculate the von Mises Creep Strains and the States of Stress at the Skeletal Point.

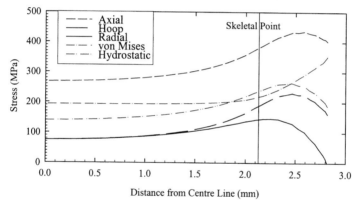

FIGURE 4 – Predicted Stresses after 300hours, a/R=5 and Net Stress of 374MPa.

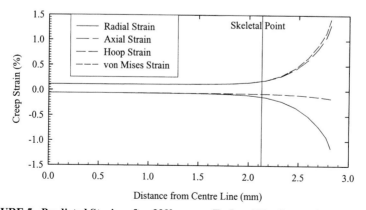

FIGURE 5 - Predicted Strains after 300hours, a/R=5 and Net Stress of 374MPa.

684

5   DISCUSSION

The observed reduction in ductility with triaxiality and the predictions of equation (1) for different values of p and q are shown in Figure 6. It should be noted that $\sigma_p/\bar\sigma$ has been used in this paper to represent the degree of triaxiality. In general the results lie between the two extremes of the predictions, with p=0.15 and q=1.25 giving an upper bound and p=2.38 and q=1.04 giving a lower bound to the data. These values for p and q were derived from biaxial data on Type 316 at 600°C and biaxial data on Type 304 at 593°C respectively [7]. Clearly the ex-service Type 316H tested here at 550°C shows a greater effect of state of stress than Type 316 tested at 600°C, although a lesser effect than Type 304 tested at 593°C. New values of p=1.18 and q=1, which fit the data more closely, have been found by using non-linear regression to determine a new value for p. Since both $\sigma_p/\bar\sigma$ and $\sigma_1/\bar\sigma$ increase with increasing a/R (Table 3) it is not possible to separate the effects of the two stress ratios from notched bar tests alone and it was necessary to assume that q=1.

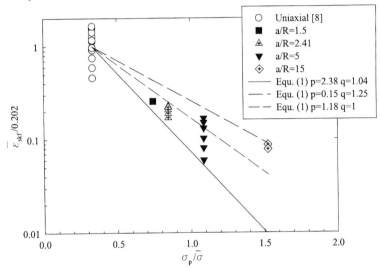

FIGURE 6 – The Effect of Triaxiality on Creep Ductility Compared with Equ. (1)

The mechanistic model for constrained cavity growth (Appendix A) has also been compared with the results of the notched bar tests (Fig. 7). When constant nucleation is assumed the constrained cavity growth model (equation (A10)) underestimates the observed effect of triaxiality. If, however, the nucleation of cavities is assumed to be dependent on σ_1^{-m} (equation (A6)) then larger reductions in ductility with increasing triaxiality are predicted. Indeed non-linear regression has been used to determine a value of m (2.76), which gives a good prediction for the effect of triaxiality. It should be noted that the diffusion controlled cavity growth model has not been used here because it predicts that a plot of $\log(\varepsilon_{fu})$ versus $\log(\dot\varepsilon)$ would have a slope of (n-1-m)/n which is clearly not the case for this material (see Fig. 3).

Notwithstanding the fact that both equations (1) and (A11) have been used to represent the effect of state of stress on ductility observed in these notched tests, there are a number of factors which mean that the effects can not be unambiguously determined from the skeletal point. In practice, there is no unique point at which the von Mises stress remains constant with time. Therefore, there is an uncertainty in both the position of the skeletal point and in the value of the von Mises strain there. Since the von Mises strain can vary rapidly close to the skeletal point (see Fig. 5) this uncertainty can be significant.

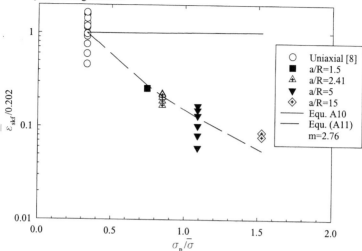

FIGURE 7 - The Effect of Triaxiality on Creep Ductility Compared with the Constrained Cavity Growth Models.

Measuring the hoop strain from the unfailed notch caused an additional problem with the present work. After failure of the specimen, the unfailed notches contained intergranular cracking to depths between 0.06 and 0.65mm from the surface so that the strain of the unfailed notch relates to crack initiation. In contrast, the uniaxial ductility relates to total failure of the specimen. Since the cracks in the notch specimens were always surface breaking whereas the skeletal point was located at depths of between 0.35 and 1.27mm below the surface (depending on a/R), clearly the skeletal point strains are not representative of the failure of the notched specimens. Indeed this observation is predicted from the creep damage, D_c, calculated using ductility exhaustion, which for the multiaxial case where the ductility is independent of strain rate is given by

$$D_c = \bar{\varepsilon}_c \Big/ \bar{\varepsilon}_f \qquad (4)$$

where $\bar{\varepsilon}_c$ is the accumulated von Mises creep strain. The distribution of creep damage for a notch with a/R=5 has been predicted from the FE results, where $\bar{\varepsilon}_f$ is given by equation (1) with p=1.18 and q=1 (Fig. 8). It should be noted that, since the FE analysis was only used to predict the distribution of creep strain and not the magnitude at failure, the creep damage has been factored to give unity at the surface. It can be seen from Figure 8 that ductility exhaustion predicts that, as observed in the tests, that failure will occur at the surface earlier than it does at the skeletal point. Nevertheless, as failure did not occur at the skeletal point in the unfailed notch it can be concluded that the von Mises strain at the skeletal point should represent a conservative underestimate of the creep strain at failure.

Because of the above problems it is not possible to conclude which of the models gives the better prediction of the effect of state of stress on the creep ductility of this ex-service Type 316H. British Energy is currently conducting a number of biaxial creep tests on the same

686

material with the aim of elucidating this point.

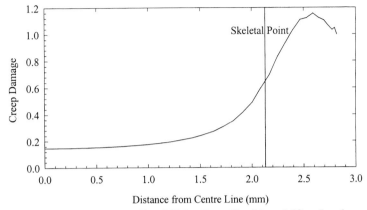

Distance from Centre Line (mm)

FIGURE 8 – Creep Damage Predicted Using Equ. (1) with p=1.18 and q=1.

6   CONCLUSIONS

1. The results show that the use of the empirical multiaxial ductility model, with p=2.38 and q=1.04, gives a conservative prediction of the effect of stress state on creep ductility. Better agreement is achieved with p=1.18 and q=1.
2. The constrained cavity growth with constant nucleation does not predict the observed effect of state of stress on creep ductility. However the inclusion of a cavity nucleation term, which is dependent on the maximum principal stress gives good agreement when m=2.76.
3. The interpretation of the results of the notched bar tests is complicated by the large variations in stress and strain through the section of the specimen and with time. These variations mean that the effect of state of stress on creep ductility can not be unambiguously determined from the strain at the skeletal point.

7   ACKNOWLEDGEMENT

This paper is published by permission of British Energy Generation Ltd.

8   REFERENCES

[1]   Hayhurst D. R., Dimmer P. R. and Morrison C. J., 1984, Development of Continuum Damage in Creep Rupture of Notched Bars, Phil. Trans. R. Soc., Vol. A311, pp. 103-129.

[2]   Cane B. J., 1980, Creep Cavitation and Rupture in 2¼Cr1Mo Steel Under Uniaxial and Multiaxial Stresses, 3rd Int. Conf. on Mechanical Behaviour of Materials, Aug. 1979, Cambridge; Proceedings Vol. 2, pp. 173-182, eds. K J Miller & R F Smith, Pergamon Press, Oxford.

[3]   Huddleston R. L., 1985, An Improved Multiaxial Creep Rupture Strength Criterion, ASME J. of Press. Vess. Tech., Vol. 107, pp. 421-429.

[4]   Cocks A. C. F. and Ashby M. F., 1980, Intergranular Fracture During Power-Law Creep Under Multiaxial Stresses, Metal Sci., Vol. 14, pp. 395-402.

[5]   Rice J. R. and Tracey D. M., 1969, On Ductile Enlargement of Voids in Triaxial Stress Fields, J. Mech. Phys. Solids, Vol. 17, pp. 201-217.

[6]   Hales R., 1994, The Role of Cavity Growth Mechanisms in Determining Creep-Rupture under Multiaxial Stresses, Fatigue. Fract. Engng. Mater. Struct., Vol. 17, No. 5, pp 579-291.

[7]   Spindler M. W., 1994, The Multiaxial Creep of Austenitic Stainless Steels, Nuclear Electric Report TIGM/REP/0014/94.

[8]   Spindler M.W., 2000, Creep Behaviour of ex-Service Type 316H, Unpublished Work.

[9]   Webster G. A., Cane B. J., Dyson B. F. and Loveday M. S., 1991, A Code of Practice for Notched Bar Creep Rupture Testing: Procedures and Interpretation of Data for Design, Pub. NPL, UK, ISBN 0-946754-13-6.

[10]   RCC-MR,1985, Section 1, Sub-section Z, Technical Appendix A3, AFCEN, Paris.

[11]   Contesti E., Cailletaud G. and Levaillant C., 1987, Creep Damage in 17-12 SPH Stainless Steels Notched Specimens: Metallographical Study and Numerical Modelling, ASME J. of Press. Vess. Tech., Vol. 109, pp. 228-235.

APPENDIX A - MECHANISTIC MODELS FOR CREEP CAVITY GROWTH AND THE PREDICTION OF MULTIAXIAL CREEP DUCTILITY

There are three generally recognised mechanisms of cavity growth in deforming metals; plastic hole growth, diffusion controlled growth and constrained cavity growth. The effects of biaxial states of stress have been derived for all three mechanisms [6]. The effects of triaxial state of stress are derived here.

Steady state creep deformation is assumed throughout and a simple deformation law of the form

$$\varepsilon = A\sigma^n t \tag{A1}$$

is used where ε is the creep strain, σ is the stress, t time, and A and n are constants.

DIFFUSION-CONTROLLED-CAVITY-GROWTH

It has been shown that, under diffusion control, the time to failure, t_f, is given by

$$t_f = \lambda / (2B_2 \, \sigma_1) \tag{A2}$$

where λ is the cavity spacing, B_2 is a temperature dependent constant and σ_1 is the maximum principal stress. This gives the uniaxial strain to failure, ε_{fu}, as

$$\varepsilon_{fu} = \lambda A\sigma^{n-1} / (2B_2) = \lambda A / (2B_2) \, (\dot{\varepsilon}/A)^{(n-1)/n} \tag{A3}$$

where $\dot{\varepsilon}$ is the creep strain rate. Under multiaxial loading the von Mises creep strain to failure, $\overline{\varepsilon}_f$, is given from equations (A1) and (A2) as

$$\overline{\varepsilon}_f = \lambda A\overline{\sigma}^n / (2B_2 \, \sigma_1) = \lambda A / (2B_2) \, (\dot{\overline{\varepsilon}}/A)^{(n-1)/n} (\overline{\sigma}/\sigma_1) \tag{A4}$$

where $\overline{\sigma}$ is the von Mises equivalent stress and S_1 is the maximum deviatoric stress. Thus, at the same equivalent strain rate, the effect of state of stress can be expressed as a factor of the uniaxial ductility

$$\overline{\varepsilon}_f / \varepsilon_{fu} = \overline{\sigma}/\sigma_1 \tag{A5}$$

A simple empirical model of cavity nucleation is introduced such that the cavity spacing is inversely proportional to the maximum principal stress according to

$$\lambda = C\sigma_1^{-m} \tag{A6}$$

where C and m are constants. Proceeding as before, the uniaxial creep ductility is given by

$$\varepsilon_{fu} = AC/(2B_2) \, (\dot{\varepsilon}/A)^{(n-1-m)/n} \tag{A7}$$

and the corresponding value under multiaxial loading is

$$\overline{\varepsilon}_f / \varepsilon_{fu} = (\overline{\sigma}/\sigma_1)^{m+1} \tag{A8}$$

CONSTRAINED-CAVITY-GROWTH

Under constrained cavity growth the ductility is controlled by the cavity spacing and for constant nucleation the maximum principal strain at failure is given simply by

$$\varepsilon_{f1} = B_1 \lambda \tag{A9}$$

where B_1 is a constant. Under multiaxial loading the maximum principal strain at failure is constant and hence the equivalent strain at failure is given by

$$\bar{\varepsilon}_f / \varepsilon_{fu} = 2\bar{\sigma}/(3S_1) \tag{A10}$$

If nucleation based on the dependency given by equation (A6) is assumed, it can be shown that the effect of multiaxial stress on ductility is given by

$$\bar{\varepsilon}_f / \varepsilon_{fu} = 2\sigma_1 / (3S_1)(\bar{\sigma}/\sigma_1)^{m+1} \tag{A11}$$

ENGINEERING APPLICATIONS II

Initiation and Growth of Creep Voids in Austenitic Steel SUS310S Under Multi-axial Stress Conditions

L.-B. Niu[*], A. Katsuta, M. Nakamura, H. Takaku and M. Kobayashi

Faculty of Engineering, Shinshu University
4-17-1 Wakasato, Nagano, 380-8553 Japan
*E-mail: niulibn@gipwc.shinshu-u.ac.jp

Abstract:
　　High temperature materials in actual uses are normally subjected to multi-axial stress states. However, creep fracture mechanisms in multi-axial stress states have not been made clear. In this study, using ductile austenitic steel SUS310S, tensile creep, torsional creep and tensile-torsional creep rupture tests and interrupted tests with tubular specimens are conducted at 973K. Creep fracture modes and creep voids formed in specimens are examined in detail by observation on the scanning electron micrographs. The initiation and growth behaviors of creep voids under multi-axial stress conditions are discussed. It is suggested that the von Mises equivalent stress may be a dominant component for the initiation of creep voids, and also that the mean stress component may promote their growth strongly.

Key Words: Multi-axial Stress State, Austenitic Steel, Creep Void, Initiation and Growth.

1. Introduction

　　Creep fracture properties such as the initiation and growth behaviors of creep voids, creep rupture modes and lives are influenced strongly by material characteristics, temperature, applied stress states, and so on. Many investigations have been made upon the creep fracture properties in a uniaxial tension stress state, using smooth or notched specimens [1, 2]. Some investigations have also been made on the creep rupture properties of materials loaded in torsion, combined tension and torsion, internal pressure or bending stress states, and so forth [3 - 5]. It has been known that the creep rupture properties of materials under multi-axial stress conditions are influenced by the multi-axial stress components, such as the maximum principal stress σ_1, the von Mises equivalent stress σ_e and the mean stress σ_m [6, 7]. However, the initiation and growth behaviors of creep voids under multi-axial stress conditions have been not made clear, even though they are closely related with the creep rupture life.

　　In this study, the initiation and growth behaviors of creep voids under multi-axial stress conditions are investigated, using ductile austenitic steel SUS310S. For this purpose, tensile creep, torsional creep and tensile-torsional creep rupture tests and interrupt tests with tubular specimens are conducted at 973K. Creep fracture modes and creep voids in interrupted specimens and ruptured specimens are examined in detail by observation on the scanning electron micrographs.

Table 1 Chemical composition of SUS310S (wt%)

C	Si	Mn	P	S	Ni	Cr	Fe
0.05	0.42	1.19	0.03	0.025	19.36	24.32	Bal.

Table 2 Mechanical properties of SUS310S

Temp.	σ_B (MPa)	$\sigma_{0.2}$ (MPa)	ε_r (%)	ψ (%)
R.T.	608.0	255.0	54.0	73.0
973K	325.4	122.5	67.6	47.3

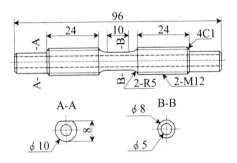

Fig. 1 Geometry of a tubular specimen.

2. Experimental Procedures

2.1 Experimental Material

The material used in the present work is commercial austenitic steel, SUS310S, with high ductility and high metallographic stability. Table 1 shows the chemical composition of the steel. Specimen blanks with a diameter of 16mm were solution treated at 1373K for 1hr, then air cooled to room temperature. The mechanical properties of the steel measured after heat treatment at room temperature and 973K are listed in Table 2.

2.2 Creep rupture tests and creep interrupt tests

Tubular creep specimens were machined to 8mm external diameter with 1.5mm wall thickness by a gauge length of 10mm. Figure 1 shows the geometry of a tubular specimen. In this study, creep rupture tests and creep interrupt tests in uniaxial tension, torsion and tension-torsion combined stress states were conducted using the tubular creep specimens. In the tension-torsion combined stress states, the ratios of applied tensile stress σ to applied shear stress τ on the external surface are 1:1 and 1:2, respectively. These tests were performed in air at 973K using a tensile-torsional creep test machine. The temperature was kept constant within ± 1K and was controlled by means of a thermoelectric couple placed on the external surface of each specimen. Tensile creep elongation and torsional creep rotation angle were measured continuously by a dial gauge and a rotary encoder device, respectively.

To investigate the effect of multi-axial stress states on the initiation and growth behaviors of creep voids in specimens, 3 cases of the creep interrupt tests mentioned above were mainly carried out: (1) creep tests were interrupted at the initial and final stages of steady state creep of specimens loaded as the maximum principal stress σ_1 was a same value; (2) creep tests were interrupted at the initial and final stages of steady state creep of specimens loaded as the von Mises equivalent stress σ_e was the same value; and (3) creep tests of specimens loaded under a same maximum principal stress σ_1 were interrupted when effective strain reached a same value.

2.3 Observation of creep fracture modes and measurement of creep voids

For creep-ruptured specimens, fracture surfaces and their longitudinal sections were observed with a scanning electron microscope (SEM). The fracture modes and the creep voids formed near the fracture surfaces were examined in detail. Also, for interrupted specimens the longitudinal sections were observed with SEM micrographs, and the creep voids formed in specimens were examined in detail. *Fraction of creep voids on grain boundary lines* [8], i.e., a fraction of total length of creep void areas on the grain boundary lines, was measured on the SEM micrographs of longitudinal sections. And, *mean diameter of creep voids* was also calculated. In fact, each void area observed on the longitudinal sections was just only one section of a 3D-shape void. So, the mean diameter of the 3D-shape voids was calculated approximately from these void areas using Fullman's method [9, 10]. About 100 creep voids were measured for the calculation.

3. Results

3.1 Creep rupture properties

The effective strain ε_e was calculated with tensile elongation and torsional rotation angle measured in testing. The typical creep curves of the steel, showing the relation of effective strain and time, are plotted in Figure

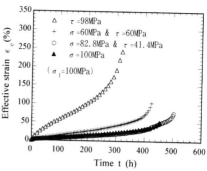

Fig. 2 Creep curves of specimens undergoing a same maximum principal stress of 100MPa.

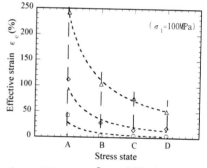

A: $\tau =98$MPa   B: $\sigma =60$MPa & $\tau =60$MPa
C: $\sigma =82.8$MPa & $\tau =41.4$MPa   D: $\sigma =100$MPa

- -○- - Initial stage of sready state creep
- -◆- - Final stage of sready state creep
- -△- - Rupture strain

Fig. 3 Creep rupture strains as well as the strains at initial and final stages of steady state creep.

2. At each stress state, only one curve is shown as an example in this figure. From these curves, it is found that the torsional creep specimen had larger creep deformations than the others, even though all of them were undergone at nearly the same maximum principal stress of 100MPa. Figure 3 shows the rupture effective strains as well as the effective strains at the initial and final stages of steady state creep of these specimens. At each stage, the effective strain showed the largest value in the torsional creep specimen and decreased from torsional stress state to tensile one.

Figures 4 are SEM micrographs showing the fracture surfaces and their longitudinal sections. It can be found that all of the present specimens exhibited a ductile creep fracture mode, and many creep voids formed in these specimens. Therefore, these creep ruptures are assumed to be caused by the initiation and growth of creep voids and the large creep deformations.

694

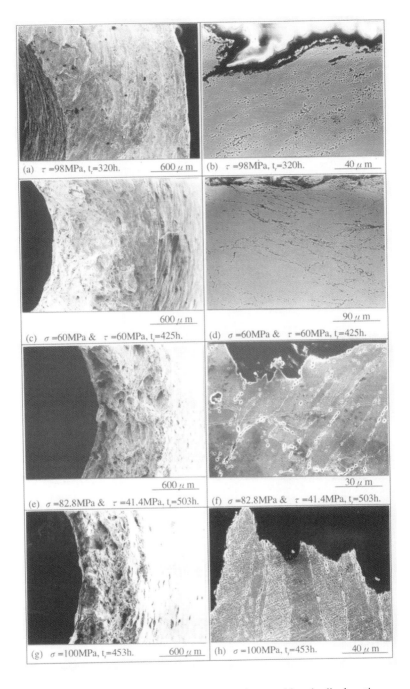

(a) τ =98MPa, t,=320h.　　600 μ m

(b) τ =98MPa, t,=320h.　　40 μ m

(c) σ =60MPa & τ =60MPa, t,=425h.　　600 μ m

(d) σ =60MPa & τ =60MPa, t,=425h.　　90 μ m

(e) σ =82.8MPa & τ =41.4MPa, t,=503h.　　600 μ m

(f) σ =82.8MPa & τ =41.4MPa, t,=503h.　　30 μ m

(g) σ =100MPa, t,=453h.　　600 μ m

(h) σ =100MPa, t,=453h.　　40 μ m

Fig.4 SEM micrographs showing fracture surfaces and longitudinal sections
of specimens loaded at a same maximum principal stress of 100MPa.

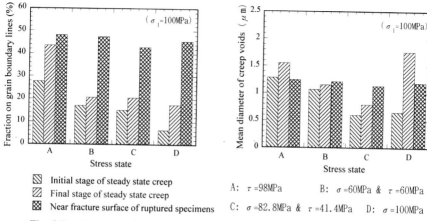

Fig. 5 Fractions on grain boundary lines and mean diameters of creep voids formed in specimens loaded under a same maximum principal stress of 100MPa.

3.2 Initiation and growth behavior of creep voids

At first, creep voids formed in specimens interrupted at the initial and final stages of steady state creep and loaded under a same maximum principal stress of 100MPa were investigated. The fractions of creep voids on grain boundary lines and their mean diameters were measured on SEM micrographs of the longitudinal sections. For comparison, they were also measured on the SEM micrographs of longitudinal sections of specimens ruptured under a same maximum principal stress of 100MPa as mentioned above. The results are plotted in Figure 5, in which the abscissas show torsional stress state (σ=0 & τ=98MPa), 2 tension-torsion

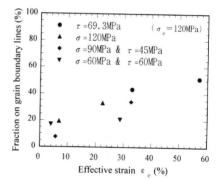

Fig. 6 Relation between fraction of creep voids on grain boundary lines and effective strain.

combined stress states (σ=60MPa & τ=60MPa, σ=82.8MPa & τ=41.4MPa) and uniaxial tensile one (τ=0 & σ=100MPa). From this figure, it can be seen that from torsional stress state to tensile one, i.e., with decreasing of applied shear stress τ and increasing of applied tensile stress σ, at initial stage of steady state creep the fraction of creep voids on grain boundary lines and the mean diameter decreased. However, in each stress state both the fraction of creep voids on grain boundary lines and the mean diameter grew from the initial stage of steady state creep to rupture. Furthermore, they reached almost the same levels to rupture. It needs a little explanation that in the measurements of creep voids, larger cracks observed in interrupted specimens and which should be considered to lead final rupture, were taken into account. While in ruptured specimens only voids near fracture surfaces were calculated. So, the mean diameters in the torsional ruptured specimen and the tensile one exhibited smaller than those in specimens interrupted at the final stages of steady state creep.

Secondly, creep voids formed in specimens interrupted at the initial and final stages of

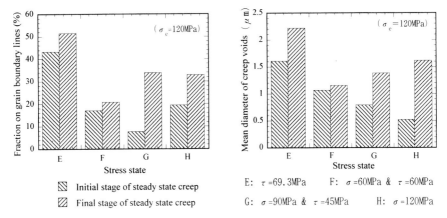

Fig. 7 Fractions on grain boundary lines and mean diameters of creep voids formed in specimens loaded under a same von Mises equivalent stress of 120MPa.

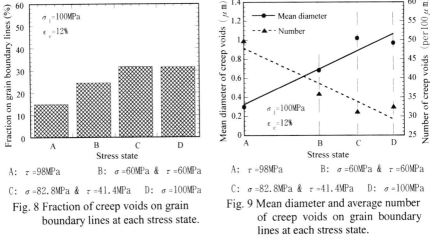

Fig. 8 Fraction of creep voids on grain boundary lines at each stress state.

Fig. 9 Mean diameter and average number of creep voids on grain boundary lines at each stress state.

steady state creep and loaded under a same von Mises equivalent stress of 120MPa were examined. The relation of the fraction of creep voids with the effective strain at the initial and final stages of steady state creep is plotted in Figure 6. The fraction of creep voids on grain boundary lines can be found to become larger with the increasing of effective strain. Figure 7 shows the fractions on grain boundary lines and the mean diameters of creep voids at each stress state. In this figure, the abscissas show torsional stress state (σ =0 & τ =69.3MPa), 2 tension-torsion combined stress states (σ =60MPa & τ =60MPa, σ =90MPa & τ =45MPa) and uniaxial tensile one (τ =0 & σ =120MPa). At initial stages of steady state creep, the fraction of creep voids on grain boundary lines and their mean diameter are found to decrease with the decreasing of applied shear stress τ and the increasing of applied tensile stress σ. However, at final stages of steady state creep they show a growing tendency conversely.

From above experimental results, fraction of creep voids on grain boundary lines was found to become larger with increasing effective strain. So creep voids observed in specimens, which were loaded under a same maximum principal stress σ_1 of 100MPa and interrupted when the effective strain reached a same value of 12%, were investigated. Figures 8 and 9 show the fraction of creep voids on grain boundary lines and their mean diameter at each stress state, respectively. From torsional stress state to tensile one, both of them were shown to become larger. Using these results, the average number of creep voids on per 100 μ m of grain boundary lines was also calculated (=*Fraction on grain boundary lines* $\times$ *100 μ m* $\div$ *Mean diameter*). For contrasting with the mean diameter, the result was plotted in Figure 9 too. From this figure, it was found that the average number of creep voids decreases from torsional stress state to tensile one, showing a converse tendency with the mean diameter.

4. Discussion

It has been reported that for metals pre-cavitated or cracking continuously during creep, the maximum principal stress σ_1 will determine the rupture life under a multi-axial stress state [3, 11, 12]. For the present ductile austenitic steel SUS310S, the maximum principal stress σ_1 has been found to correlate the rupture life under different stress states [13]. So, the initiation and growth behavior of creep voids is considered to influence the rupture life directly for this steel. In Figure 3, the rupture strain as well as the strains at the initial and final stages of steady state creep were found to decreased from torsional stress state to tensile one, even though all of the specimens were undergone a same maximum principal stress and thereby they should have almost same rupture lives. However, the fraction of creep voids on grain boundary lines and their mean diameter measured in the specimen ruptured under each stress state reached almost the same level with others, as shown in Figure 5. Therefore, it is assumed that these specimens ruptured as creep voids initiated and grew to almost a same extent although in different courses in creep testing. So, the initiation and growth behavior of creep voids under each stress state is discussed as below.

In the periods of primary creep, fractions on grain boundary lines and diameters of creep voids formed in specimens loaded under both a same von Mises equivalent stress and a same maximum principal stress were found to grow down from torsional stress state to tensile one. The reason may be that the creep deformation became smaller as the applied shear stress decreased. On the other hand, in specimens loaded under a same maximum principal stress and interrupted specially when the effective strain reached a same value, they increased from torsional stress state to tensile one as shown in Figures 8 and 9, respectively. Therefore, the initiation and growth of creep voids in the present stress states should be investigated separately. Generally, it has been assumed that the von Mises equivalent stress control creep rate and govern the nucleation of creep voids [12, 14], while mean stress component promote the growth of creep voids or cracks [15, 16]. For the specimens loaded under a same maximum principal stress and interrupted especially when the effective strain reached a same value, in torsional stress state the von Mises equivalent stress was larger than in other stress states and it became smaller from the torsional stress state to tensile one. However, the mean stress was 0 in torsional stress state and became larger conversely with increasing applied tensile stress. Therefore, it may be suggested that in specimens loaded under a same maximum principal stress, creep voids nucleate easily in torsional stress state and grow easily in tensile one. For this reason, the average number of creep voids on per 100 μ m of grain boundary line, as shown in Figure 9, decreased from torsional stress state to tensile one, while their mean diameter exhibited a converse tendency.

Consequently, in torsional creep stress state with a larger von Mises equivalent stress and having a larger creep deformation, creep voids nucleated easily and grew to a certain extend in the period of primary creep. But because the mean stress was 0, they grew little to rupture, even though they initiated continuously and parts of them connected to large cracks in the period of tertiary creep and resulted in the final rupture. On the other hand, from torsional stress state to tensile one, the von Mises equivalent stress became smaller while the mean stress was raised, so the creep voids became easy to grow meanwhile the initiation of them occurred till rupture.

5. Conclusions

The results obtained are summarized as follows:

(1) For the austenitic steel SUS310S with high ductility, initiation and growth behavior of creep voids influences the creep rupture life directly

(2) With decreasing applied shear stress and increasing applied tensile stress, fractions on grain boundary lines and diameters of creep voids in specimens loaded under a same von Mises equivalent stress or a same maximum principal stress were found to become smaller in the periods of primary creep. However, alone with further testing to ruptures they grew fast conversely.

(3) In torsional creep stress state, it was found that creep voids formed easily to a certain size in period of primary creep, but they grew little till the rupture, even though they initiated continuously and parts of them connected to large cracks in the period of tertiary creep and resulted in the final rupture. While from torsional stress state to tensile one, creep voids were found to become easy to grow.

(4) Under a multi-axial stress state, it is further suggested that the von Mises equivalent stress component and the mean stress component promote the initiation and the growth of creep voids, respectively.

References

1) For example, H. R. Voorhees, J. W. Freeman and J. A. Herzog: *Trans. ASME.*, **D84**(1962), 207.
2) M. Kobayashi, Y. Maeda, K. Sugimoto and K. Nakamura: *Trans. JSME.*, **56A**-531(1990), 2241.
3) J. Henderson: Trans. ASME., **101**(1979), 356.
4) L. –B. Niu, A. Futamura, K. Sugimoto and M. Kobayashi: *Trans. JSME.*, **61A**-589(1995), 2037.
5) W. Sawert and H. R. Voorhees: *Trans. ASME.*, **D84**(1962), 228.
6) A. Yousefiani, F. A. Mohamed and J. C. Earthman: *Metal. and Mater. Trans.*, **31A**(2000), 2807.
7) L. –B. Niu, M. Nakamura, A. Futamura and M. Kobayashi: *ISIJ Inter.*, 40(2000), 511.
8) N. Tada, T. Kitamura and R. Ohtani: *J. Soc. Mat. Sci. of Japan*, **45**(1996), 110.
9) R. L. Fullman: *Trans. Metall. Soc. AIME*, **197**(1953), 447.
10) J. Kyono, N. Shinya, H. Kushima and R. Horiuchi: *Tetsu-to-Hagane*, **79**(1993), 604.
11) S. E. Stanzl, A. S. Argon and E. K. Tschegg: *Acta Metall.* 31(1983), 833.
12) M. M. Abo El Ata and I. Finnie: Creep in Structures 1970, ed. by J. Hult, Springer-Verlag, (1972), 80.
13) L. –B. Niu, M. Nakamura and M. Kobayashi: *Proceedings of the 1999 Annual Meeting of JSME/MMD*, Kyoto, (1999),139.
14) F. C. Monkman and N. J. Grant: *Proc. Am. Soc. Test. Metals*, **56**(1956), 295.
15) D. Hull and D. E. Rimmer: *Philos. Mag.*, **4**(1959), 673.
16) R. T. Ratcliffe and G. W. Greenwood: *Philos. Mag.*, **12**(1965), 59.

Creep of 2.25Cr-1Mo tubular test pieces under hydrogen attack conditions

by

P. Castello[o], A. Baker[*], G. Manna[o], P. Manolatos[o], R.Hurst[o] and J.D. Parker[§]

[o] *IAM/JRC of the EC – Petten – The Netherlands*
[*] *IAM/JRC of the EC – Petten – The Netherlands; now at British Energy PLC. Barnett Way, Barnwood, Gloucester GL4 3RS, UK.*
[§] *Dept. of Materials Engineering, University of Wales, Swansea, SA2 8PP, UK*

ABSTRACT

A comparative study was carried out on the creep behaviour of several tubular test pieces of 2.25-1Mo steel, one bearing a circumferential weldment at mid-length. The samples were internally pressurized with hydrogen at 180-260 bars and exposed at temperatures in the range 570-600°C; most of the tests were interrupted just prior to failure to avoid damaging the experimental setup. For comparison, creep testing of tubes pressurized with argon was performed. In general, a decrease of the creep properties of 2.25Cr-1Mo in hydrogen was observed due to the formation of methane-filled voids at grain boundaries in the steel, i.e. hydrogen attack. A peculiar distribution of damage through the wall of the tube was observed, which can be attributed to the boundary conditions of stress and hydrogen concentration inherent in this type of testing. For the welded tube, the point of maximum deformation was located in the weld metal, although cracks formed preferentially in the base material, immediately adjacent to the weldment. Microstructural examination revealed extended hydrogen damage both in the base metal, where voids coalesced into cracks, and in the Heat-Affected Zone where, on the contrary, little link-up of cavities was observed. In contrast with these regions, the weld metal showed only moderate evidence of hydrogen attack.

1. Introduction

The study of creep under multiaxial stress states combined with hydrogen attack on low-alloy ferritic steels and weldments is a subject of valuable interest in view of applications such as hydrocrackers and ammonia reactors. These are typically made of low-alloy ferritic steel grades like 1Cr-0.5Mo or 2.25Cr-1Mo, and they operate with hydrogen pressures varying from 10.3 MPa to 34.5 MPa in the temperature range 399°C to 482°C [1]. Under such conditions, hydrogen can react with the carbon in the steel according to the reaction:

$$2H_2 + C_{(steel)} = CH_4 \qquad (1)$$

Reaction 1 can take place at the steel surface, where it is sustained by hydrogen in the gas phase and by an outward flux of carbon diffusing from the steel, or it can occur internally in the material, due to the high diffusivity of hydrogen. The first case is normally regarded as not yielding serious consequences from the engineering viewpoint, because the thickness of the decarburized layer is quite small; on the contrary, methane molecules which are formed internally accumulate in microvoids along the grain boundaries, and these, on coalescence, can lead to crack formation and premature failure of the material [2, 3]. Hydrogen attack resistance can be achieved by alloying the steel with elements capable of forming very stable carbides, such as, typically, molybdenum and vanadium. The addition of these elements lower the activity of carbon in the ferrite matrix, thus making it much less readily available for reaction 1.

Several factors have to be taken into account for pressure vessel design, based on appropriate codes and standards[4,5]. The considerations pertinent specifically to hydrogen attack are summarised in the so-called Nelson Curves[6]. These indicate, for different types of steels, the limits of temperature and hydrogen pressure below which the damage incubation period is so extended that it goes beyond the operational life of the plant. However, the Nelson curves do not explicitly account for the possibility of the synergistic effect of creep stress and hydrogen attack in causing failures. This effect has been demonstrated elsewhere[7], and it will become quite important in view of the growing use of V-modified steel grades for hydrogen service[8]. Because of their improved creep strength, these steels will in fact lead to a reduction of reactor wall thickness, with a consequent increase of stress.

Problems in design arise also from the fact that the material data which are available are very frequently generated by using uniaxial tests, which are relatively simple to perform. However, typical industrial components operate under multiaxial states of stress as a result of the internal pressure, temperature gradients and system stresses. This complication is generally overcome by using very general effective stress concepts, based on the principle that the life of a component with a multiaxial effective stress corresponds to the rupture time at the same uniaxial stress[9, 10]. Multiaxial stress states are thus commonly described via the Mean Diameter Hoop stress:

$$\sigma_{MHD} = \frac{P(D_o - t)}{2t} \tag{2}$$

where P is the pressure, D_o is the outer diameter of the tube and t the tube wall thickness. One possible alternative is that of the skeletal point Von Mises stress under steady state creep:

$$\sigma_{SP} = \frac{P\sqrt{3}}{K^2 - 1} \left(\frac{K^2 - 1}{n(K^{2/n} - 1)} \right)^{n/n-1} \tag{3}$$

where n = Norton exponent and $K = D_o/D_i$, with D_i = inner diameter of the tube. Especially the design based on Eq. 2 has proved quite conservative, but the particularly severe conditions to be encountered in hot hydrogen service suggest deeper investigation on this subject.

Finally, the problem gets further complicated when the presence of weldments has to be accounted for. First, the creep properties of weldments are usually different from those of base material[11]; secondly, it is known that weldments are normally more sensitive to hydrogen attack than the base materials[3]. This is considered to be due to the detrimental effect of welding residual stresses, and to the fact that carbides, which dissolved during the welding process, are not sufficiently stabilized by the post-weld-heat-treatment. In particular, the heat-affected zones are known to be quite vulnerable to this form of degradation.

The present work describes the results of multiaxial testing of 2.25Cr-1Mo steel tubes, non-welded and welded, internally pressurized with hydrogen. The results have been compared with data coming from reference tube testing carried out with argon and/or uniaxial testing in air and hydrogen. With respect to uniaxial creep testing, the procedure described in this paper adds two parameters, which are encountered in industrial situations, namely the multiaxial stress state and a gradient of hydrogen concentration across the tube thickness.

2. Experimental

2.1 Materials
Two different heats of 2.25Cr-1Mo steel, grade P22[12] were selected for this study, and they were used respectively for testing on non-welded and welded tubes. The chemical compositions of both heats are given in Table 1, together with that of the flux-and-wire combination used for submerged arc welding heat 2.

Table 1: Composition of the test steels and of the flux-and-wire combination used for weldment of heat 2.

	C	Si	Mn	P	S	Cr	Mo	Ni	Al	Cu	Sn	V
Heat 1	0.10	0.21	0.51	0.006	0.01	2.18	0.92	0.20	0.005	0.15	0.008	0.007
Heat 2	0.14	0.18	0.59	0.004	0.003	2.25	1.05	0.14	0.01	0.07	0.006	-
Cons.	0.10	.11	.75	.009	0.002	2.37	1.01	0.11	-	-	0.002	-

Heat 1 was received as a 6 m section of a hot finished seamless steel tube, normalised and tempered; optical microscopy revealed that the structure consisted mostly of bainite with scattered ferrite grains. Because the material strength was found to be excessively high and the ductility too low for the purposes of this work, the as-received material was further submitted to an in-house tempering treatment, in order to reduce its strength. This consisted of 4 hours at 750°C in Ar, followed by air cool[13]. Preliminary creep testing in air under 100 MPa at 600°C showed that this treatment greatly increased the creep ductility and reduced the creep rupture life by a factor of about 10. Heat 2 was water quenched from 940°C and tempered at 650°C. The post-weld heat treatment consisted of 30 h at 690°C. The material was received in the form of a square-base ingot, approximately 80x80x400 mm, bearing the weldment at mid-length. Also in this case, the structure of the base material (BM) was bainitic, with some fine carbides precipitated mainly along the grain boundaries at the edges of bainite plates. The Heat-Affected-Zone (HAZ) was also mostly bainitic, with some precipitates of proeutectoid ferrite. Finally, the weld metal (WM) was a mixture of ferrite and a lath arrangement of upper bainite. The Vickers hardness profile for the transition from the BM to the WM was determined, and a zone of low hardness (~ 185 HV) was identified between the BM (~ 198 HV) and the WM (~ 210 HV). Round bars approximately 60mm in diameter and 400 mm length were cut by means of spark erosion; tubular samples were machined out of the bars into the final shape shown in Fig.1, in conformity with the code of practice suggested by the High Temperature Mechanical Testing Committee[14].

2.2 Test conditions
Two test temperatures were selected, namely 570 and 600°C for welded and non-welded samples, respectively. In total, five tests were carried out in hydrogen and three in argon. In the case of the welded tube, the choice of a lower test temperature was aimed to achieve longer test times without excessively decreasing either the hydrogen partial pressure or the stress level. Table 2 summarizes the experimental conditions selected, including data on the geometry of samples tested and the corresponding values of reference stress.
Detailed specifications on the experimental setup have been given elsewhere[15]. In short, it consisted of the tubular specimen itself, pressurized with hydrogen or argon supplied by means of ¼" pipe mounted on caps which were electron beam welded at the ends of the test piece. The sample was mounted coaxial with another tube supporting a capacitive hoop-strain

multiple extensometer, which allowed deformation to be monitored at three different locations, namely top, bottom and centre of the tube gauge length. The system so constructed was placed in a furnace, which in turn was covered with an aluminium bell housing, filled with inert gas for safety purposes. The entire rig was situated in a reinforced concrete cell designed to stop debris in the unlikely event of a gas explosion. The experiments were initially brought to failure, thus allowing the tubes to burst. However, this damaged the hoop strain measurement devices and so the practice was stopped. Future tests were interrupted when the rapid increase of the strain rate suggested that the tubes were close to failure.

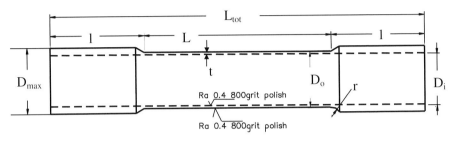

Figure 1: Schematic layout of the tubular test pieces

3. Results

3.1 Creep testing
Table 3 presents the results of all the experiments, by giving the time to 1, 2 and 5% hoop deformation, the total experiment times and the associated final strains, as well as all the corresponding values of the minimum creep rate.

All data are given with reference to deformation at the centre of the tube. At 600°C, deformation occurred rather uniformly along the gauge until 80-90% of total test time; after that, bulging began to form normally between the top and the middle gauges, with maximum strain occurring near the mid-length of the test piece. By plotting the strain rate data as a function of the mean diameter hoop stress, "n" values of 5.7 and 6.6 were found for the Norton Equation applied to the non-welded tubes in argon and hydrogen, respectively. The relevant creep curves are shown in Fig. 2; hydrogen caused a reduction of failure times and an increase in the creep rate with respect to argon. As it is shown in particular by the experiments carried out to failure at 240 bars, ductility was also reduced by approximately 30% with respect to the argon case. In the case of the test performed on the welded tube at 570°C, an evident bulging of the sample progressively developed during the test in proximity to one of the fusion lines, i.e. close to the tube mid-length. The test time was almost one order of magnitude longer than at 600°C under a comparable level of mean diameter hoop stress. Much lower values of the minimum creep rate and ductility were observed.

3.2 Microstructural examination.
Metallographic examination of the tubes pressurised with argon revealed ductile behaviour. No cavitation was seen in any of the tubes, with the exception of the burst tube which showed small amounts of localised ductile voiding next to the fracture surface. The burst failure occurred mainly as a result of local thinning followed by tearing of the remaining ligament. The thinning appeared to have started from the outside of the tube wall and moved inwards.

Table 2: Test conditions.

Gas used	Material tested	Sample Geometry* [mm]		Pressure [bar]	Mean Diameter Hoop stress (MPa)	Skeletal Point Von Mises stress (MPa)
H₂, Ar	Heat 1	T=	4.4	260	113.5	98.0
		Dmax =	47.2			
		D₀ =	42.8	240	104.7	90.5
		Dᵢ =	34.0			
		(D₀+Dᵢ)/2=	38.4	220	96.0	82.9
		L =	170.0			
		l =	67.5	195**	85.1	73.5
		r =	8.8			
H₂	*Heat 2 weldment*	t =	3.8	180	100	86.5
		Dmax =	50.4			
		D₀ =	46.6			
		Dᵢ =	38.9			
		(D₀+Dᵢ)/2=	42.8			
		L =	167.8			
		l =	73.5			
		r =	7.7			

* See Fig. 1; ** Hydrogen test only

Table 3: Creep results

Gas	Temp. [°C]	Pressure [Bar]	Time to 1% strain [h]	Time to 2% strain [h]	Time to 5% strain [h]	Test time [h]	Minimum Strain-rate [%/h]	Failure Hoop strain [%]
Ar	600	260	40	98	171	190	1.7x10⁻²	8.0
Ar	600	240	41	144	258	307*	1.2x10⁻²	35.6*
Ar	600	220	94	248	477	574	6.5x10⁻³	10.7
H₂	600	260	44	81	109	123	1.7x10⁻²	9.4
H₂	600	240	28	88	187	226*	1.6x10⁻²	25.4*
H₂	600	220	43	161	286	342	8.4x10⁻³	12
H₂	600	195	69	277	651	715	3.5x10⁻³	9.5
H₂	570	180	1163	2280	-	3230	8x10⁻⁴	4.3

* Test carried out to failure

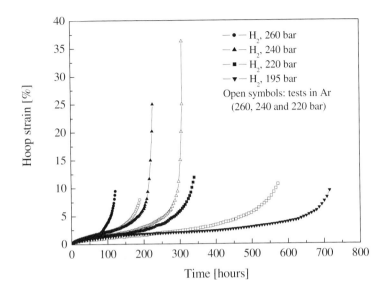

Figure 2: Creep behaviour of heat 1 at 600°C, for different H_2 and Ar pressures

Non-welded, hydrogen pressurised tubes showed increasing evidence of hydrogen attack with increasing testing times. In particular, the tube tested with 260 bar hydrogen showed practically no evidence of hydrogen-induced cavitation or surface decarburisation. In all the other cases both decarburisation of the hydrogen-exposed surface and internal cavitation were observed. The cavities were preferentially aligned at those grain boundaries approximately perpendicular to the hoop stress direction, a feature which became increasingly evident with longer test times. Gradients of cavity density were observed through the wall thickness, with cavitation being more severe close to the inner wall, and along the tube length. Close to the inner surface of the tube, the coalescence of cavities resulted in the formation of open and/or subsurface microcracks, propagating from the interior to the exterior of the tube.

With reference to the welded tube, different cavity morphologies were observed in the BM, the HAZ and the WM, respectively. In agreement with the bulging of the sample at centre, cracks were detected by non-destructive analysis in the BM adjacent to the HAZ (Fig. 3).

Microstructural analysis revealed that these were related to void coalescence (Fig. 4a). Although the

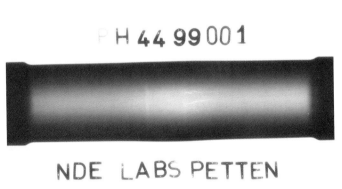

Figure 3: Cracking of welded tube in proximity the fusion line.

adjacent HAZ also showed extended cavitation, the tendency of the voids to link-up into cracks was less evident, and the cavity distribution was rather random, as is shown in Fig. 4b. Finally, the WM appeared as the least damaged zone, with little cavitation and no evident sign of cracking.

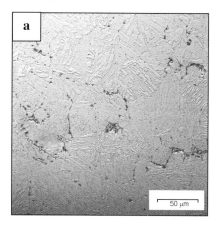 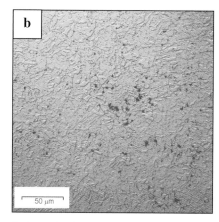

Figure 4: Cross sections of the welded tube in proximity to the fusion line.
a: Base Metal;   b: Heat-Affected Zone

4. Discussion

The use of the effective reference stress concept derives from the principle that a uniaxial specimen exposed to the same reference stress should fail in the same time. In Fig. 5a and 5b respectively the tube rupture lives and the minimum strain rates at 600°C have been correlated with both the mean diameter hoop stress, σ_{MDH}, and the steady state skeletal point Von Mises stress, σ_{sp}, and a comparison has been made with the results of uniaxial testing of in air and hydrogen[7]. With respect to Fig. 5a, it has to be noticed that because many experiments were interrupted before tube bursting, the exact time to failure is often not known. However, the rupture plot shows that the results of multiaxial tests fit reasonably well with a power law, which confirms that the experiments were indeed interrupted very close to failure, and that the relevant times can be regarded as failure times within a small margin of error. σ_{MDH} and σ_{sp} give the best correlation for the rupture lives and for the minimum strain rates, respectively. The one or the other can consequently be used either for lifetime estimation or for strain behaviour prediction, which is in agreement with literature findings [9, 16].

Unlike samples pressurised with argon, hydrogen-tested tubes underwent a serious reduction of lifetime and ductility. This was due to the combined effects of mechanical stress and methane formation in the steel, the latter being related to the hydrogen concentration in the material through equation 1. Because the hoop stress and the hydrogen concentration reached their respective maxima at the tube mid-length and close to the inner surface, cavities formed preferentially in that area, grew until coalescence under the influence of the applied stress and caused crack initiation immediately inside the decarburised zone. The crack became open to the inner surface after tearing of the thin, ductile decarburised layer and, more important, they propagated towards the exterior of the tube, ultimately leading to failure. The crack propagation mechanism also probably consisted of a combination of stress and hydrogen attack mutually assisting each other. In fact, the stress concentration at the crack tip is known to favor the nucleation and growth of new cavities, which then link up with the crack, thereby accelerating its propagation[17].

Because no uniaxial testing was carried out at 570°C, the behaviour of the welded tube can not be compared directly with uniaxial data. Discussion can be based on the results of a recently concluded EC-funded research project, in which the creep properties of this weldment were investigated as a function of temperature using uniaxial testing[18]. Based on those results, the rupture life of the same weldment in the same environmental conditions and under an equivalent superimposed uniaxial load is expected to be almost three times lower than that observed in the present case. This result may be a consequence of the geometry of the weldment. Similar to that observed by Eggeler et al.[19], the hard WM may in fact have acted as a constraining ring, thereby limiting the deformation of the test piece and determining a lifetime extension. As a consequence, the hydrogen exposure was prolonged, which seems a reasonable cause for low ductility in relation with cavitation-induced embrittlement. However, creep damage did not concentrate in the HAZ, which is somewhat surprising in view of the relatively low hardness combined with a concentration of tensile stress which is expected on the inner side of the tube in proximity to a rigid circumferential constraint[19]. Cavities in the HAZ show less tendency to link-up into cracks than they do in the BM, and it appears from Fig. 3 that the failure was more likely to occur in the latter, although the crack paths partially cross the fusion lines in a few cases. However, the relatively large size of the voids in the HAZ compared with the BM corroborates the idea that their growth may have been assisted by a fairly high stress concentration in that area[7]. Further testing on 2.25Cr-1Mo weldments is needed in order to clarify these points. In particular, stress-free exposures and uniaxial creep testing must be carried out in order to assess the effects of creep and hydrogen attack separately; secondly, tubes bearing the weldment in a different position have to be tested in order to investigate the influence of the geometry of weldments.

5. Summary and conclusions

Tubular samples made out of two different heats of 2.25Cr-1Mo steel for petro-refinery applications were tested under multiaxial creep conditions in hydrogen and argon at 570 and 600°C. The following points were investigated:

  i. the influence of the internal pressure on the rupture life and the hoop strain behaviour,
  ii. the effect of using hydrogen as a pressurising gas in comparison with argon, and the combined effects of hydrogen attack and superimposed multiaxial creep load,
  iii. the presence of a weldment in the tube, in a circumferential position at mid length.

Comparison of creep results with uniaxial data showed that predictions of the rupture lives and creep strain rates can be based on the mean diameter hoop stress and the skeletal point Von Mises stress respectively. In comparison with argon, hydrogen caused a reduction in both lifetime and ductility, with loss of ductility due to the nucleation and growth of methane filled grain boundary cavities. The two main factors controlling the process were the hydrogen concentration in the steel and the hoop stress. The combination of these two factors yielded the maximum damage close to the inner side of the tube wall, at mid tube gauge-length. Extended cavitation developed at this location, with void coalescence along grain boundaries perpendicular to the hoop stress direction. This caused the cracking of the material starting from the inside of the tube, in contrast with the failure mechanism in argon, which was due to ductile thinning starting from the outside. The presence of the weldment resulted in a strong concentration of cavities in the HAZ, whose sensitivity to hydrogen attack was thus confirmed. However, link-up of voids and consequent cracking occured mostly in the base material adjacent to the HAZ; this was possibly due to the hardness of the weld metal, which may have acted as a constraining ring during the creep test, thereby yielding a concentration of tensile stresses in the adjacent regions

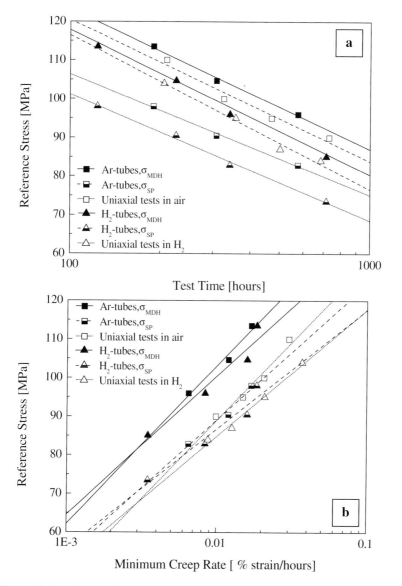

Figure 5: Correlation of experimental data with calculated values of the Mean Diameter Hoop stress, σ_{MDH}, and the Skeletal Point Von Mises stress, σ_{SP}, in comparison with uniaxial data[7]. **a**: rupture plots;   **b**: minimum creep rates.

708

6. Aknowledgements
The authors would like to thank their colleagues from IAM (Petten) Mr. J. Morrisey and Mr. F. Harskamp for their support in carrying out the experiments as well as Mr. P.D. Frampton, Mrs. E. Conceicão and Mr. L. Metten for their help in structural analyses.

7. References

[1] Antalffy, L.P. and Chaku, P.N., 1994, Metallurgical, design and fabrication aspects of modern hydroprocessing reactors, Proc. of 2nd Int. Conf. on Interact. of Steels with H_2 in Petr. Ind. Pres.Ves. and Pipeline Serv., Oct. 19-21, Vienna, Austria, MPC, pp. 619-657.

[2] Shewmon, P., 1985, Hydrogen attack of pressure vessel steels, Mat. Sci. Tech. 1, pp. 2-11

[3] Prescott, G.R., 1994, history and basis for prediction of hydrogen attack of C-1/2Mo steel: a state-of-the-art review, see Ref. 1, pp. 301-329

[4] Canonico, D.A., 1994, ASME boiler and pressure vessel code steels for petroleum industry service, see Ref. 1, pp. 49-53

[5] Tahara, T., Ishiguro, T., Iga, H. and Motoo, A., 1994, Effect of metallurgical factors on hydrogen attack resistance in C-0.5Mo steel, see Ref. 1, pp. 366-377.

[6] RP Publ. 941 (1997) Steels for Hydrogen Service at Elevated Temperatures and Pressures in Petroleum Refineries and Petrochemical Plants, 5th Ed. American Petroleum Institute

[7] Baker A.,Manolatos P., Castello, P. and Hurst, R.C., 2000, Role of stress on hydrogen attack, Proc. EUROMAT 2000, "Adv. in Mech. Behaviour, Plasticity and Damage", Eds. D. Miannay, P. Costa, D. Francois, A. Pineau, Elsevier Sci. Ltd., Vol. 2, pp. 1231-1236.

[8] Berzolla, A., Bertoni A., Boquet, P., and Hauck, G., 1994, Application of advanced Cr-Mo pressure vessel steel for high temperature service, see Ref. 1, pp. 541-553.

[9] Hurst R.C. and Rantala J.H., 2000, Influence of multiaxial stress on creep and creep rupture, ASM Handbook, Vol. 8, "Mechanical Testing and Evaluation", pp. 405-411

[10] Hyde, T.H., Sun, W., Williams J., 1999, Prediction of creep failure of internally pressurised thick walled CrMoV pipes, Int. J. of Pressure Vessels and Piping 76, pp. 925-933

[11] Hyde, T.H., Sun, W. and Williams J.A., 1999, Creep behaviour of parent, weld and HAZ materials of new, service-aged and repaired 1/Cr1/2Mo1/4V: 2 ¼ Cr1Mo pipe welds at 640°C, Materials at High Temperatures 16 (3), pp. 117-129

[12] ASTM Standard A335/A 335M –95 A, 1996, Standard specification for seamless ferritic alloy-steel pipe for high-temperature service, Annual Book of ASTM Standards, 01.01.

[13] Baker, A., 2000, Combined creep and hydrogen attack of petro-refinery steel, PhD Thesis, Swansea University, UK

[14] Internal Pressure Testing Working Party HTMTC/IP/6/89, 1989, "A code of Practice for internal pressure testing of tubular components at elevated temperatures".

[15] Baker, A., Morrisey, J., Manolatos, P., Hurst R.C., 1999, Multiaxial creep testing of internally pressurised tubular steel components under HA conditions, Proc. Conf. CHIFI-99, V.Bicego, A.Nitta, J.W.H. Price and R.Viswanathan (Eds.), Eng. Mat. Adv. Serv. Ltd., pp. 157-167

[16] Church J.M., 1992, Creep and creep crack growth behaviour of ferritic tubular components under multiaxial stress states, PhD. Thesis, University of Sheffield, UK

[17] Chen, L.-C. and Shewmon P., 1995, Stress-assisted hydrogen attack cracking in 2.25Cr-1Mo steels at elevated temperatures, Met. and Mat. Trans. 26 A, pp. 2317-2327

[18] Bocquet, P., Castello P., De Araujo C.L., Ohm R., Van Wortel H., Schlögl S., Prediction of pressure vessel integrity in creep hydrogen service (Predich), Proc. European Symposium on Pressure Equipment-ESOPE 2001-Paris 23-25 Oct. 2001, *To be published*

[19] Eggeler, G., 1993, Analysis of creep in a welded P91 pressure vessel, COST 501/II/WP5 B Final Coordinator's Report.

CREEP BEHAVIOUR OF CrMo(V) STEEL GRADES AND THEIR WELDS IN HOT HYDROGEN ENVIRONMENT

Andrew GINGELL, Pierre BOCQUET & Pascal BALLADON
USINOR INDUSTEEL-France

ABSTRACT

Steels used for pressure vessels for petrochemical applications may be submitted to high pressures at temperature where creep needs to be considered for the design. In addition, for hydrotreatment reactors, the hydrogen-rich environment may produce a phenomenon, named hot hydrogen attack, which results from a reaction between hot hydrogen and the carbides of the steel which may be destabilised. Newly developed steel grades (Vanadium modified versions of Cr Mo steels with 2¼Cr, 3Cr and 9Cr) offer a combination of improved creep and hydrogen attack resistance.

1. INTRODUCTION

In a research of the Brite-Euram programme, PREDICH [1], an European consortium conducted by Creusot-Loire Industrie (the previous name of Usinor Industeel) investigated the different aspects of the problem of creep-hydrogen interactions with the following objectives:

- demonstrate the capability of the manufacturers in Europe to control the production of these new materials and their welds with the required properties, even for very large thickness (200 mm and more).

- evaluate the best way for the control of the most critical properties such as temper embrittlement and creep resistance of base and weld metals, through the selection of appropriate welding products and parameters and the use of optimised Post Weld Heat Treatment (PWHT).

- demonstrate that the new steels and their welds offer a lower sensitivity to hydrogen damage, even when operating in more severe conditions of hydrogen environment than the reference material (2¼Cr1Mo steel grade).

- for a best prediction of the residual life of the vessels, develop a numerical model taking into consideration the microstructure stability (for the resistance to hot hydrogen attack), the creep damage in air and the interaction between creep and hydrogen damages.

After developing the steels and the welding consumables, the work was mainly focused on tests at high temperature and under high hydrogen pressure. High temperature exposure in autoclave, without stress, for test duration up to 4000 hours, as well as creep tests in air and in hydrogen environment for up to 10000 hours were performed.

This paper presents the main experimental results from this project, notably concerning the hydrogen effect on creep strength. The correlation with hot hydrogen attack is also made.

2. MATERIALS

The reference material was the 2¼Cr1Mo steel grade (ASTM A387 Gr22 or 10CrMo9-10 of the EN 10028-2). V-modified 2¼, 3 and 9 Cr1Mo steels as per ASME Code Cases 2098, 1961 and 1973 were compared to the reference. All grades were supplied in the form of thick plate (200mm to 300mm).

The following comments can be made on the chemistries of the plates supplied by Usinor Industeel (Table 1). Due to the high thickness of the plates, Carbon content was aimed near the upper limit of the required range, whereas Silicon content was aimed towards the lower limit. The low Sulphur content ranging between 16 to 40 ppm reflects the normal level of purity of modern steels. For all the materials, J factor values are lower than the requirement of $J \leq 100$, which may be found in very severe specifications. Extra low P content is the major key for the reduction of the sensitivity to temper embrittlement [2].

Grade Steel Plate	Standard 2¼ Cr Steel (1) 200 mm		V mod. 2¼ Cr Steel (2) 200 mm		V mod. 3 Cr Steel (4) 200 mm		V mod. 9 Cr Steel (6) 300 mm	
	ASTM gr22	Heat	case 2098	Heat	case 1961	Heat	Gr 91	Heat
C	.05-.15	.138	.11-.15	.118	.1-.15	.13	.08-.12	.105
Si	< .50	.177	< .10	.026	< .20	.057	.20-.50	.24
Mn	.30-.60	.593	.30-.60	.433	.30-.60	.524	.30-.60	.44
P	< .035	.004	< .015	.004	< .020	.006	< .020	.008
S	< .035	.003	< .010	.002	< .020	.004	< .010	.0016
Cr	2.0-2.50	2.25	2.0-2.50	2.36	2.75-3.25	3.02	8-9.5	8.41
Mo	.90-1.10	1.051	.90-1.10	1.017	.90-1.10	.972	.85-1.05	.95
Ni		.14	< .25	.073	< .25	.196	< .40	.08
Cu		.069	< .20	.062	< .25	.048		.035
V			.25-.35	.276	.20-.30	.237	.18-.25	.25
N							.03-.07	.033
Ti			< .030	< .003	.015-.035	.021		
Ca			< .015	< .0005				
Nb			< .07	< .005			.06-.10	.08
B			< .0020	.0005	.001-.003	.0019		
J		77		41		58		82

$J = (Mn\% + Si\%)(P\% + Sn\%)10^4$

Table 1 Chemical Analysis Of Base Materials

Test plates were welded using procedures identical to those used in fabricating heavy wall reactors. The process used was SAW single wire, typically used in the fabrication of nozzle-shell or nozzle-head joints, known to be the most critical joints for creep damage. After NDT examination, the weldments were characterised mechanically and selected for hydrogen attack and creep in hydrogen testing.

3. CHARACTERISATION OF PLATES AND WELDMENTS

3.1. Mechanical Properties of Plate Material
3.1.1. Tensile properties at room temperature
The materials were characterised conventionally at quarter-thickness, with specimens transverse to the rolling direction. An important parameter acting on mechanical properties is the tempering effect of the tempering heat treatment performed by the steel manufacturer and that of the Post Weld Heat Treatment (PWHT) performed after the welding of the reactors.
For hydro-treating reactors, these materials are commonly heat treated after welding at elevated temperature (required T > 675°C by ASME) to improve the hydrogen resistance and the toughness of welds and Heat Affected Zones (HAZ). Consequently, the tempering heat treatment performed by the steelmaker has to be at a significantly lower temperature to avoid excessive softening of the base material. In such conditions, the base material delivered to the manufacturer may be very hard, and subsequent cold forming may be difficult.

The graph of Figure 1 gives the evolution of tensile strength (UTS) at room temperature as a function of TP[†] for the 4 categories of materials: standard 2¼Cr1Mo, V-modified 2¼Cr1Mo, 3Cr1Mo and V-modified 9Cr1Mo and the results plotted on this figure show that all materials meet the required values.

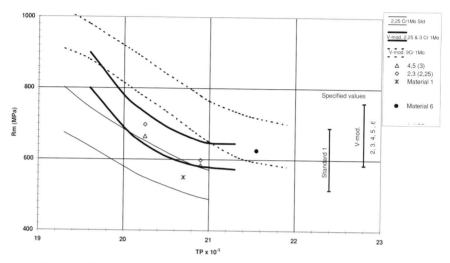

Figure 1. Effect of Tempering Parameter on tensile strength of CrMo(V) steels

Considering the risk of reducing the base metal tensile properties to below the minimum required value with too high a PWHT condition, the upper limit for TP for these high thickness materials is about:

- 20 600 for standard 2¼Cr-1Mo
- 21 000 for V-modified 2¼Cr-1Mo and 3Cr-1Mo
- 21 800 for V-modified 9Cr-1Mo

3.1.2. Impact properties - Resistance to temper embrittlement
For standard material, it is well known that toughness may be deteriorated by two mechanisms of embrittlement :

1) overtempering may produce the upper nose temper embrittlement (irreversible).
2) slow cooling or ageing at intermediate temperature may introduce the lower nose temper embrittlement (reversible) by segregation of impurities (mainly P, Sn) to the grain boundaries.

The first type of embrittlement appears when the tempering parameter (TP) is above about 20 500 to 20 800 for standard CrMo and above about 21 000 for V-modified CrMo. So, the toughness properties for this material are optimal for these TP values.
To avoid the lower nose temper embrittlement at operating temperature, it is necessary to produce high purity materials with extra-low J- factor, as mentioned above. Table 2 summarises

---

[†] Larson-Miller Tempering Parameter: TP = T·(20 + log (t))

the results of Charpy-V transition temperatures (at 54J or 40ft.lb) for the base materials before and after step-cooling. The insignificant differences on TT 54 J or FATT before and after step cooling confirm that very pure materials are not sensitive to the temper embrittlement.

Material	P + Sn [%]	J factor	TT 54 J / FATT before S.C. [°C]	TT 54 J / FATT after S.C. [°C]	Δ SC. [°C]
(1) Standard 2¼Cr-1Mo	.010	77	-95 / -95	-90 / -80	+5 / +15
(2) V-modified 2¼Cr-1Mo	.009	41	-75 / -70	-85 / -75	-10 / -5
(4) V-modified 3Cr-1Mo	.010	58	-110 / -90	-120 / -100	-10 / -10
(6) V-modified 9Cr-1Mo	.012	82	-15 / 0	-30 / -10	-15 / -10

Table 2. Effect Of Step Cooling On Toughness Of Base Materials

3.2. Characterisation of Weldments

3.2.1. Cracking Sensitivity

The sensitivity to cold cracking was evaluated by implant tests for high stress level conditions (500 MPa). The addition of vanadium to the steel had no significant effect on the sensitivity to cold cracking, in agreement with previous work [3], and the precautions used for welding the standard 2¼Cr-1Mo material (minimum preheating temperature of 200°C) may also be used for welding V-modified 2¼Cr and 3Cr-1Mo materials.

For V-modified 9Cr-1Mo, an extra low hydrogen content of the weld metal (3-4cm^3/100 g) is necessary to use a pre-heating temperature of 200°C. For a higher H content (11-14cm^3/100g) this temperature will be increased up to 250-275°C as shown in Table 3.

Welding Heat input (kJ/cm)	Weld metal hydrogen content (cm^3 / 100 g)	Maximum stress (MPa) for 200°C preheating	Minimum preheat temperature (°C) for stress applied 450 MPa
10	4.0-4,3	475	200
20	3.3-4.2	500	200
10	14.4-14.7	250	225
20	10.8-11.7	300	275

Table 3. Conditions And Results Of Implant Tests On V Modified 9 Cr 1 Mo

The sensitivity to re-heat cracking has been evaluated by the measurement of the HAZ ductility at the PWHT temperature. Only 2¼Cr1Mo¼ V material shows a significant decrease compared to standard 2¼Cr1Mo steel. The favourable effect of increasing Cr in V and V/Nb steels was clearly observed: ductility increases when Cr increases from 2¼ to 3 to 9 % (Figure 2).

As a result of these tests, it appears that the sensitivity to reheat cracking is quite low for materials with Cr ≥ 3% and acceptable for the 2¼Cr1MoV grade (striction at 700°C > 20%).

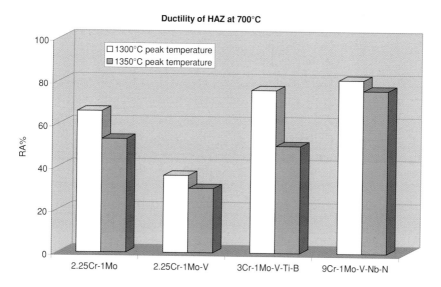

Figure 2. Reheat cracking sensitivity evaluation of simulated HAZ at PWHT temperature.

3.2.2. Mechanical Testing Of Weldments

In order to consider the effect of an eventual repetition of PWHT (for instance in the case of repairing after first PWHT), a characterisation after minimum expected PWHT and maximum expected PWHT is required. For this project, it has been decided to characterise the weldments in minimum PWHT conditions (duration of 10 hours) and in maximum PWHT conditions (duration of 30 hours). It should be noticed that 2¼Cr1Mo Standard has been tested only in maximum PWHT conditions.

All welded joints must demonstrate to guarantee at least the minimum mechanical properties required for base materials. The results of mechanical testing of weldments selected for the hydrogen attack and creep in hydrogen programmes are reported in Table 4.

Steel grade		2.25Cr1Mo	2.25Cr1MoV		3Cr1MoV		Grade 91
Weld N°		1A	2B	2D	4B	4D	6A1
PWHT	T(°C)	690	705	705	705	705	760
	t (h)	30	10	30	10	30	4
YS at 525°C in MPa on crossweld specimens							
Required		>210	>317				>285
Actual		333	343	306	354	324	336
Vickers hardness (HV30)							
BM(Average)		184	198	182	181	171	206
HAZ(Max.)		198	211	194	217	205	232
WM(Max.)		187	235	214	190	190	216
Charpy V properties (TT54J in°C)							
HAZ PWHT		-114	-112	-106	-118	-109	-97
HAZ SC		-78	-126	-102	-127	-107	-72
WM PWHT		-65	-77	-90	-78	-86	-9
WM SC		-69	-76	-80	-92	-91	-4

Table 4 – Results of mechanical tests on welds selected for the creep programme

The tensile properties of the weldments are in accordance with the code, with the exception of high temperature yield strength for steel 2, 2¼Cr1MoV with PWHT 705°C / 30h, which is lower than expected. This discrepancy is probably due to a combination of relatively low carbon content and slight differences in the actual PWHT temperature. The Charpy-V impact properties are highly acceptable for materials for heavy wall reactors.

4. HIGH TEMPERATURE BEHAVIOUR IN HYDROGEN ENVIRONMENT

4.1. Hydrogen Attack Sensitivity

The evaluation of hydrogen attack sensitivity of the materials cannot be performed at the service temperature (450 to 480°C) because the process of carbide destabilisation by hydrogen needs a very long time at that temperature [4]. It is admitted that accelerated damage may be observed after exposure for about 1000 hours at 600°C and classification of the materials is commonly based on this type of test.

Previous studies [3] have shown that the temperature range of exposure to H_2 must be adapted to the sensitivity of materials. Figure 3 shows the results of tensile tests performed after similar exposure conditions at 600°C, showing significant damage on the standard material, but not on the V modified materials. However it is clear that tests conditions are more severe than service conditions because the test temperature is highly increased.

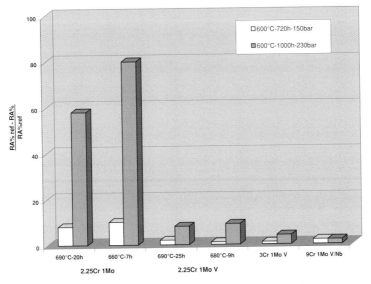

Figure 3 - Ductility loss due to hydrogen attack
(variation in reduction in area from uncharged to hydrogen charged specimen)

4.2. Creep Testing in Air and in Hydrogen

Creep experiments have been carried out in hydrogen environment in order to establish the combined effect of creep and hydrogen attack. The objective was to determine the creep strength and creep ductility in air and in hydrogen of the base materials and of selected weldments. The results are compared to the relevant codes.

Iso-stress creep tests in air reveal the life-time at design stress and temperature, whereas the same experiments in 180 bar H_2-pressure establish the combined effect of creep and hydrogen attack. Iso-stress creep testing uses the service stress but temperatures higher than the operating temperature. The data are plotted as log (rupture time) vs. temperature and a straight line fit is extrapolated to the operating temperature. In case the service stress is not known, testing at different stress levels is performed.

Table 5 identifies the tested materials, base materials and weldments and their PWHT, as well as the stress levels used in the experiments. The creep experiments in hydrogen were based on the results in air. The temperatures were intended to be identical to those selected for the creep experiments in air. The hydrogen pressure was 180 bar, similar to the autoclave testing. The creep curves of the base materials were recorded using continuous strain measurement.

Steel	Stress Level [MPa]	Base metal		Weldment*	
		Air	H_2	Air	H_2
1 standard 2¼Cr-1Mo	100			30h 690°C	30h 690°C
2 V-modified 2¼Cr-1Mo	120				
	150			10h & 30h 705°C	30h 705°C
	190				
4 V-modified 3Cr-1Mo	120				
	150			10h & 30h 705°C	10h 705°C
	190				
6 V-modified 9Cr-1Mo	127				
	166			4h 760°C	4h 760°C
	201				

* Also indicated is PWHT (duration and temperature)

Table 5. Stress levels and weldment parameters used in creep experiments

4.3. Results & Observations of Creep Testing

4.3.1. Standard 2¼Cr-1Mo

This material was included in the test programme as the reference material. There is ample information available on this material to make a comparison between published values and the experimental results obtained in this project. This comparison can validate the experimental results on all the steels tested in the project.

The creep experiments on steel 1 were performed at one stress level, 150MPa (Fig.4). The experiments show a clear effect of hydrogen on the creep behaviour of the material. The graph of the rupture time vs. temperature for the results in hydrogen has a different slope, indicating a larger effect of hydrogen at lower temperatures (down to 530°C). The reduction caused by hydrogen dominates the reduction caused by welding at lower temperatures. All the tested weldments ruptured in the FG-HAZ.

716

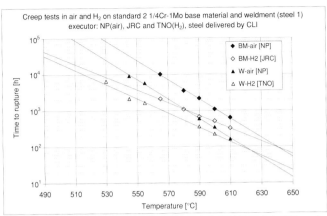

Figure 4. Effect of hydrogen on rupture time of steel 1 (2¼Cr1Mo) BM and weldments at 100MPa.

4.3.2. V-modified 2¼Cr-1Mo

The creep experiments in air on steel 2 were performed at three stress levels (Fig.5). The addition of 0.25%V to 2¼Cr-1Mo steel has a profound effect on the creep strength of the steel. The rupture times for steel 2 at 150MPa are comparable to those of steel 1 at 100MPa (Fig.5).

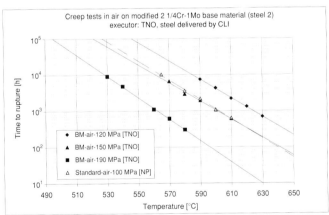

Figure 5. Effect of stress on rupture time of steel 2 (2¼Cr1Mo¼V) BM in air.

The reduction of the rupture time as a result of welding is considerable (Figure 6). The slope of the rupture times in hydrogen deviates less from the graph in air than in the case of steel 1. The effect on the ductility is also smaller in steel 2. All the weldments ruptured in the FG-HAZ.

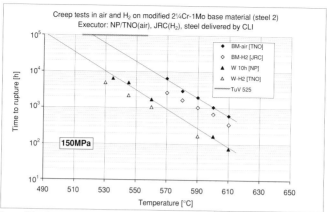

Figure 6. Effect of hydrogen on rupture time of steel 2 (2¼Cr1Mo¼V) BM and weldments at 150MPa.

4.3.3. V-modified 3Cr-1Mo

The creep experiments in air on steel 4 were also performed at three stress levels (Fig.7). The creep strength is reduced compared to the V-modified 2¼Cr1Mo grade, the rupture times for steel 4 at 120MPa being similar to those for the standard material at 100MPa.

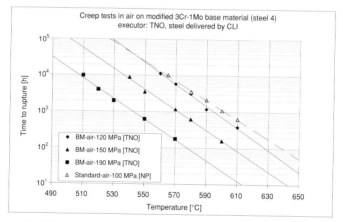

Figure 7. Effect of stress on rupture time of steel 4 (3Cr1Mo¼V) BM in air.

Welding reduces the rupture time dramatically, as does the longer PWHT (Figure 8). The weldments rupture in the BM. The reduction of area is similar to the base material experiments, but the rupture time is 5 to 10 times lower. The experiments in hydrogen show results that are largely identical to the related experiments in air. There is virtually no effect of the hydrogen atmosphere on the rupture-time or ductility of the specimens. The same large effect of welding on the rupture time is evident with the rupture occurring in the BM. The extrapolated values are roughly the same as those obtained for the experiments in air.

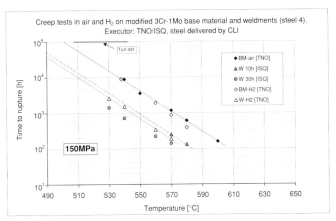

Figure 8. Effect of hydrogen on rupture time of steel 4 (3Cr1Mo¼V) BM and weldments at 150MPa.

4.3.4. V-modified 9Cr-1Mo

The experiments in air on steel 6 base material were again performed at three stress levels, based on the relevant code for P91 (Figure 9). The rupture time graphs are steeper than the graphs for the other materials. This grade shows significantly higher creep strength than the standard material.

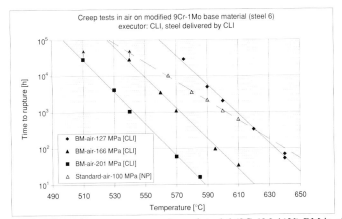

Figure 9. Effect of stress on rupture time of steel 6 (9Cr1Mo¼V) BM in air.

The weldments rupture in the FGHAZ with little change in the reduction of area as compared to the base material. However, the rupture time is much shorter for the weldments. The experiments on steel 6 in hydrogen show no change in rupture times and ductility compared to the tests in air. The material seems to be unsusceptible to hydrogen damage (Fig.10).

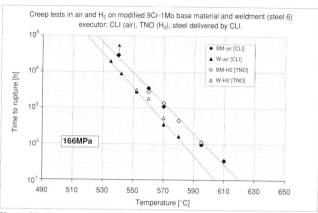

Figure 10. Effect of hydrogen on rupture time of steel 6 (9Cr1Mo¼V) BM and weldments at 166MPa.

4.4. Discussion on creep results

The hydrogen damage susceptibility of standard 2¼Cr-1Mo steel has been confirmed by the autoclave experiments. It was concluded (from tensile tests) that a significant hydrogen attack is only noted for standard steel i.e. steel 1 and that the ductility of all V-modified material remains unchanged even for the longest (autoclave) exposure conditions.

This ductility loss under hydrogen was clearly present in the creep experiments on steel 1. It manifested itself in a reduced creep strength of steel 1 in hydrogen atmosphere. The weldments of standard 2¼Cr-1Mo steel ruptured in the FG-HAZ. This is in accordance with the common knowledge about 2¼Cr-1Mo steel weldments.

The creep data of steel 2 show a limited susceptibility for hydrogen damage, but the effect is much smaller than in steel 1, even though these data were determined at the higher stress level of 150MPa. Based on these results we can conclude that the addition of ¼ %V increases the creep strength of the 2¼Cr-1Mo steel in air and in hydrogen, and that the most susceptible region in the weldment remains the FG-HAZ.

The effect of hydrogen exposure on 3Cr steels (steel 4) is small, reflecting the excellent hydrogen attack resistance of this grade. However the weldments fail in the BM zone at times much shorter than the base material specimens, with or without hydrogen.

Steel 6 shows no effect of hydrogen damage and a stronger temperature dependence of the creep strength. The weldments pose the same problem as in steel 4. The location of rupture is in the FGHAZ zone, but the rupture time is much shorter than the base material specimens.

A general remark for all weldments, but particularly shown in Figure 8, is the marked deterioration in creep strength with increasing PWHT time. This has been observed previously for short test times [5], and confirms the necessity to limit TP in order to retain sufficient creep strength.

The creep results can also be used for the calculation of the life-time at design temperature and stress for the steels and weldments. These values are necessary to adapt the construction codes so that the steels and welding consumables can be used to their full potential.

The values for the temperature of $1 \cdot 10^5$h creep strength at test stress, have been plotted in the code graphs of the relevant steels after extrapolation of the creep testing results [6]. From this we can observe the following:

720

Steel 1 has a high enough creep strength in the welded and base material condition to comply with the DIN 17176 code (Fig. 11). This is to be expected, since the base material meets the requirements of the code in terms of chemical composition. The service in hydrogen reduces the creep strength considerably to a point far outside the margins set by the code. This too is expected since the material is well-known to be susceptible to hydrogen attack.

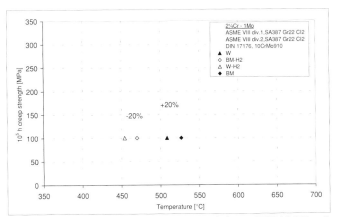

Figure 11. Extrapolation results for steel 1 relative to DIN 17176 code

Steel 2 in the BM condition in air has a creep strength in accordance with the code (Fig.12). This is due to the strengthening effect of the vanadium addition. The effect of hydrogen is large enough to lower the creep strength below the code margin. The weldment and the hydrogen data for steel 2 do not agree with the code margins.

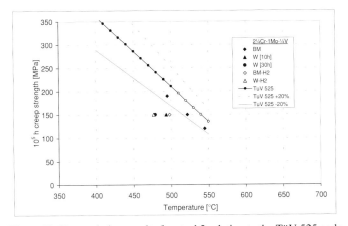

Figure 12. Extrapolation results for steel 2 relative to the TüV 525 code

Steel 4 has creep strengths bordering on the lower code margin (Fig.13), for the BM condition in both air and hydrogen. The weldments have much lower creep strengths in air and hydrogen. The V-modified 3Cr-1Mo steels have a lower creep strength than the V-modified 2¼Cr-1Mo steels. The extrapolated values border on the code margin of TüV 491.

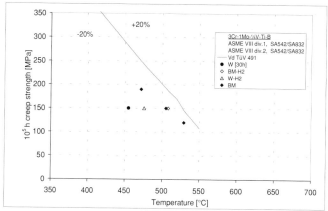

Figure 13. Extrapolation results for steel 4 relative to TüV 491 code

Steel 6 shows compliance with code, be it on the lower margin of the code (Fig. 14). The effect of hydrogen service on the creep strength is nil. V-modified 9Cr-1Mo steel has creep strength values in accordance with the relevant code TüV 511/2.

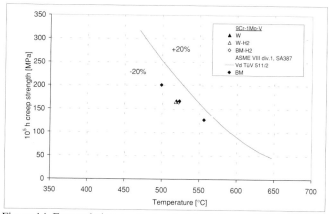

Figure 14. Extrapolation results for steel 6 relative to TüV 511/2 code

5. CONCLUSIONS

The results from this programme of hot hydrogen testing enable the following conclusions to be drawn :

(1) Addition of V provides excellent resistance to hot hydrogen attack.
(2) V addition improves creep strength of base metal in air.
(3) The creep strength of weldments is sensitive to the PWHT conditions. Minimum PWHT conditions are optimal.

(4) Hydrogen has a marked effect on the creep strength for the standard material. This effect is reduced for the V-modified 2¼Cr1Mo material and is virtually non-existent for 3Cr1Mo¼V and 9Cr1Mo¼V grades.

(5) Hydrogen shifts creep values outside of the code limits for standard and V-modified 2¼Cr1Mo material.

Further work is required to investigate the effect of hydrogen at temperatures closer to the actual service temperature. In parallel, further work is necessary in order to improve the creep resistance of weldments for V modified material.

ACKNOWLEDGEMENTS

This work has been carried out as part of a European Commission-funded project, PREDICH, in the BRITE-EURAM programme, with the participation of Shell Global Solutions, TNO, Delft University of Technology, Joint Research Centre IAM, Petrogal, Dillinger Hutte & Nuovo Pignone.

REFERENCES

[1] Bocquet, P. et al., 2000, Prediction of Pressure Vessel Integrity in Creep Hydrogen Service (PREDICH), Final Report, BRITE/EURAM project N° BE-1835, European Commission.

[2] Bocquet, P., Coudreuse, L. & Cheviet, A., 1995, The New 2¼Cr and 3Cr1Mo¼V Steels for Pressure Vessels Operating in Hydrogen Environment, AFIAP, Paris.

[3] Cheviet, A., Bocquet, P. & Coudert, E., 1998, Elements de Choix des Aciers CrMo(V) Evolués pour Appareils à Pression Utilisés en Pétrochimie, AFIAP, Paris.

[4] Schlögl, S.M., Van Der Giessen, E. & Van Leeuwen, Y., On Methane Generation and Decarburisation in Low Alloy Cr-Mo Steels During Hydrogen Attack, Metall. Mater. Trans. A, Vol. 31, pp. 125-137.

[5] Bocquet, P., 1998, Technical News on CrMo Materials – Mechanical Properties, Cresuot-Loire Industrie Seminar, Gilly, Oct. 1998

[6] Balladon, P., 1998, Les Méthodes d'Extrapolation en Fluage et Leurs Applications, AFIAP, Paris.

CRACK GROWTH

THE EFFECT OF CYCLIC LOADING ON THE EARLY STAGES OF CREEP CRACK
GROWTH

K. Nikbin, M.R.Chorlton, and G.A.Webster

Department of Mechanical Engineering,
Imperial College of Science, Technology and Medicine,
London SW7 2BX.

Summary

Crack growth experiments have been carried out on two low alloy steels under combined
creep and fatigue loading over a frequency range of 0.2E-5 Hz to 1 Hz. The presence of a
superimposed cyclic component eliminated the transient regime of cracking at frequencies
above 0.0001 Hz. At about 0.0001 Hz approximately the first millimeter of crack growth was
by transgranular fatigue fracture. This crack extension corresponded to between one third and
one half of the specimen lifetime. At 0.001 Hz, in this region, cracking was dominated by
both intergranular fracture and the striation mechanism. At and below this frequency steady
state creep crack growth dominated the remaining lifetime and could be correlated by the
creep fracture mechanics parameter $C*$ to within a factor of two or three. At frequencies
above and including 0.01 Hz fracture was almost entirely by fatigue crack growth, which
could be correlated in terms of elastic stress intensity factor range ΔK.

1.0 Introduction

Under constant load at elevated temperature, failure of a cracked component may occur by
steady state creep crack growth. This corresponds to the situation whereby there is a steady
state distribution of creep damage ahead of the crack tip [1]. Some structural integrity
assessment procedures, such as BS PD 7910 [2] and R5 [3], assume that up to this point there
is an incubation period during which there is an accumulation of damage and no cracking
occurs. An equivalent approach which has been applied successfully to compact tension
specimens is to allow for a transient regime of cracking during the build up of damage at the
crack prior to the onset of steady state growth [4].

Components subjected to cyclic loading are found in aero engines, electric power plant and
chemical process plant. Under such conditions at elevated temperatures, cracks may initiate
and propagate by a combination of both creep and fatigue mechanisms. Start-up and shutdown will cause a cyclic stress history whereas the steady state conditions will determine the
extent of creep deformation and stress redistribution. It has been found that crack growth
under both creep and fatigue processes can be accounted for simply by a linear cumulative
damage approach, adding the cracking contributions due to each process [5].

The cumulative damage approach describes creep-fatigue crack growth most accurately under
steady state conditions. The factors affecting the period prior to this stage are undetermined.
For conservatism it may be assumed that where a cyclic component is present initiation
occurs immediately and growth occurs at the steady state creep-fatigue rate [2,3]. However,
if a transient stage is still to be considered then lifetime predictions based simply on linear
cumulative damage throughout may prove to be over-conservative. The practical relevance

of this investigation was, therefore, to determine the influence of cyclic loading on the early stages of creep-fatigue crack growth.

The study was therefore carried out on two low alloy ferritic steels of relevance to the electric power generation industries - a 1Cr1Mo1/4V rotor steel (hereafter referred to as 1CMV) and a 1/2Cr1/2Mo1/4V (hereafter referred to as 1/2CMV) pipe steel. The static creep crack growth properties of these steels had also been well characterized in a separate study [6]. A series of superimposed cyclic loading experiments was carried out at a range of frequencies from 2E-5Hz to 1 Hz on compact tension specimens. Experimental results were then interpreted using a fracture mechanics approach as described in the following section.

2.0 Theory

2.1 Steady State Creep Crack Growth Model

Material deformation is assumed to occur in the power law creep regime. The uniaxial creep rate $\dot{\varepsilon}_{cr}$ is then given by

$$\dot{\varepsilon}_{cr} = \dot{\varepsilon}_o \left(\frac{\sigma}{\sigma_o} \right)^n \tag{1}$$

where $\dot{\varepsilon}_o$, σ_o and n are material constants given in Table 1.

Material/Temp	$\dot{\varepsilon}_o$ 1/h	σ_o, MPa	n	R,A
1/2CMV - 565 °C	1	450	10.7	0.64
1CMV - 550 °C	1	715	9.5	0.59

Table 1: Uniaxial creep material properties

Creep crack growth is based on a ductility exhaustion model, which postulates a fracture process zone, length r_c (usually taken as the average grain size [7]), ahead of the crack tip. Upon the application of a constant load the initial elastic and or plastic stresses redistribute to a steady state creep stress state, described by eqn. 1, accompanied by material degradation in the process zone. This period of creep damage development and stress redistribution is referred to as the incubation (or transient) period. Eventually a steady state distribution of stress anddamage is accumulated in the process zone. Experience has shown that the steady state creep crack growth rate, $\dot{a}_{ss}$, can then be successfully correlated by the creep fracture mechanics parameter C^*

$$\dot{a}_{ss} = D C^{*\phi} \tag{2}$$

where D and ϕ are related to the uniaxial creep properties by a ductility exhaustion model where

$$D = (n+1)\frac{\dot{\varepsilon}_o}{\varepsilon_f^*} \left[\frac{1}{I_n \sigma_o \dot{\varepsilon}_o} \right]^{n/(n+1)} r_c^{1/(n+1)} \tag{3}$$

and

$$\phi = n/(n+1) \qquad (4)$$

Nikbin et al. [7] showed that for a wide range of materials eqn. 2 approximates to

$$\dot{a}_{ss} = \frac{3C^{*\phi}}{\varepsilon_f^*} \qquad (5)$$

with $\dot{a}$ is in mm/h and C^* is in MJ/m²h In eqns. (3) and (5) I_n is a non-dimensional factor and ε_f^* is the creep ductility appropriate to the state of stress at the crack tip.

2.2 Cumulative Creep-Fatigue Damage Law

For the case of a superimposed cyclic component the fatigue contribution is described by a power law of the form:

$$\frac{da}{dN} = C\Delta K^m \qquad (6)$$

where da/dN is in mm/cycle, ΔK is the stress intensity factor range in MPa√m, and C and m are material constants. Where both creep and fatigue crack growth mechanisms exist it is assumed that the total crack growth rate can be partitioned into a cycle-dependent component, arising from the rise and fall of the load waveform, and a time-independent part corresponding to the hold period of a loading cycle [9,10]. Hence, if crack propagation is expressed in terms of number of cycles, N,

$$\left(\frac{da}{dN}\right)_{tot} = C\Delta K^m + \frac{DC^{*\phi}}{3600f} \qquad (7)$$

or, in time-dependent form:

$$\left(\frac{da}{dt}\right)_{tot} = C\Delta K^m 3600f + DC^{*\phi} \qquad (8)$$

where f is the frequency in Hz and growth rate is /h. Hence, creep-fatigue crack growth rate is assumed to be simply the linear superposition of the creep and fatigue damage components.

3.0 Experimental Procedure and Results

Static load creep crack growth tests and cyclically loaded creep-fatigue crack growth tests were performed using compact tension (CT) specimens. Specimens were heated in split three-zone electrical resistance furnaces to a temperature of 565 °C for the 1/2CMV and 550 °C for the 1CMV. Fluctuations in temperature were to within no more than ±1 °C in order to keep within ASTM validity limits [11]. Temperature was measured using Type K chrome-aluminel thermocouples welded to the specimen body.

Cyclic waveforms were generated using servo-hydraulic or pneumatic ram loading machines. All cyclic tests were load controlled; hence, these tests could simulate the failure mechanism experienced at rotor bores subjected to rotational stress for example. All waveforms were trapezoidal except for the lowest frequency test (6 hour dwell (2.3e-5Hz) at maximum load) where the hold time at minimum load was zero and the highest frequency tests conducted at 0.1 and 1 Hz where a sinusoidal waveform was employed. All cyclic tests reported here were performed at an $R(P_{min}/P_{max})$ ratio of 0.1 in order that all creep crack growth could be attributed to the hold period at maximum load, to simplify interpretation of the experimental results. Furthermore, C^* was always calculated at the maximum load of the cycle. For all tests the static load of the pure creep crack growth test was taken as the maximum load of the cyclic loading waveform; thus the superimposed cyclic component was cycled 'down' from the creep load.

All CT specimens were fatigue pre-cracked at room temperature at high frequency (80Hz). The purpose of the fatigue pre-crack was to simulate an ideal plane crack with essentially zero tip radius. This condition will then agree with the assumptions made in fracture mechanics analyses. After pre-cracking CT specimens were side-grooved to a depth 14 % of the thickness on both sides. Side-grooving is intended to promote straight cracks under high constraint (plane strain) conditions.

The effect of frequency on crack growth is plotted in figs. 1-2 for the 1/2CMV CT and 1CMV specimens respectively. All CT specimens were loaded as accurately as possible to the same initial stress intensity factor, K_o, in order for valid comparisons to be made against the static load 'benchmark' tests. As can be seen from the legend in the figures, not all tests were loaded to the same level. The initial crack length at the start of the test could only be determined accurately after a specimen has been broken open. The frequency range for the tests on 1/2CMV was 2.3E-5 Hz to 0.01 Hz and for the 1CMV the range was 1e-4 Hz to 1 Hz.

As shown in fig. 1 the lifetime of the test cycled at 0.0001 Hz was twice that of the static load test. Furthermore, the initial load of this cyclic test was 17 % greater than the initial load of the static load test, and therefore would have been longer had the initial loading conditions been the same. Extension of the lifetime of the 0.0001 Hz test is believed to be due mainly to the fact that there was virtually no crack growth during the hold time at the minimum load. Hence, an approximate doubling of the predicted lifetime based upon a pure creep crack growth test might be expected. In addition, it is likely that creep stress relaxation during the hold time at minimum load would reduce the rate of damage accumulation immediately ahead of the crack tip.

It is believed that reduced damage formation and creep stress relaxation during the hold times at minimum load would have the effect of reducing the overall crack growth rate and consequently extend the expected lifetime of a slow creep-fatigue crack growth test which involved long hold times at low load. An indication of the sole effect of the fatigue component on crack growth is given by the slow cyclic test with a zero hold-time at minimum load and 6 hour hold at maximum load. Zero hold-time at minimum load would mean insufficient time for substantial creep stress relaxation to occur. Thus, any reduction in the lifetime of the specimen would be a result of the fatigue component alone.

Using eqns. 7 - 8 the expected lifetime for the 6-hour hold-time specimen was approximately 1000 hours. The actual lifetime, therefore, of 800 hours was a 20 % reduction due to the cyclic component. However, the result of Fig.1 does not indicate whether, for this case, the 20 % reduction in lifetime was during the early stages of cracking or if it was spread over the entire lifetime of the specimen. The results of an examination of the fracture surface indicated that for the slow frequencies the early stages of crack growth were dominated by fatigue mechanisms. The fractographic results also indicated that creep crack growth did not commence until significant creep damage had accumulated ahead of the crack tip.

Fig. 2 shows the crack growth data for the 1CMV CT specimens cycled between 0.0001 Hz and 1 Hz. For these tests the variation of initial stress intensity factors, that is, loading conditions, was always less than 4 %. As can be seen, the test cycled at 0.0001 Hz had an extended lifetime, but in this case less than twice the expected (static load) lifetime. This was probably due to a faster accumulation of creep damage, enhanced by the cyclic component. This might also account for the abrupt change in crack growth rate during the last 10 % of life. The results at 0.0001 Hz imply that a finite time, analogous to an incubation period, was still required for a steady state distribution of creep damage to accumulate ahead of the crack tip.

At 0.001 Hz the results were markedly different for the two steels. The results showed an experimental scatter of ±20% and the mean values are presented here. For the 1/2CMV CT specimen the expected lifetime was 350 hours (due to the high load) and the actual lifetime was 170 hours. In this case the hold-time at minimum load apparently had a negligible effect upon reducing the overall crack growth rate, and the crack growth rate enhancement was most probably the consequence of the fatigue component. For the 1CMV CT specimen the expected lifetime was 925 hours and the actual lifetime was approximately 1100 hours. This result suggests that there was a balance between the reduction in crack growth rate due to the effect of the holds at minimum load and the increase in crack growth rate due to fatigue. The result for the 1CMV specimen implies that the cracking mechanism was the same throughout the entirety of the specimen lifetime as opposed to fatigue domination during the early stages progressing to a creep cracking mechanism during the later stages of life.

At 0.01 Hz the reduction in lifetimes for both steels was significant - approximately a factor of 5 for the 1/2CMV steel and a factor of 10 for 1CMV steel. The crack growth rate increase can be attributed to fatigue. Crack growth rate is plotted as a function of ΔK for each material in figs 3 - 4. These show that no correlation was achieved for the ½ CMV steel. This indicates that pure fatigue did not control fracture and that a mixed fracture mode was most likely.

For the 1CMV steel crack growth is correlated successfully in terms ΔK of for frequencies above and including 0.001 Hz. The results of fig.4 suggest that 0.001 Hz is a threshold above which crack growth is fatigue controlled.

The crack growth rate data are plotted in terms of the creep fracture mechanics parameter C^* for each material in figs. 5 - 6. In addition, data from the static load tests are included for comparison. Fig. 5 shows the case for the 1/2CMV steel. From fig. 5 it can be seen that only the 2.3E-5 Hz test correlated with the steady state creep crack growth rate of the static load tests. This is probably due to the long hold time at maximum load during which creep crack growth could occur. In addition a zero hold time at minimum load meant there could be no reduction in crack growth rate.

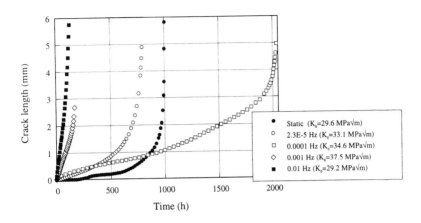

Figure 1: Creep-fatigue crack growth for 1/2CMV CT specimens.

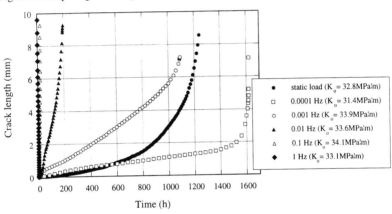

Figure 2: Creep-fatigue crack growth for 1CMV CT specimens.

This is sensible when one considers that the lifetime of this test was twice that of the static load test. The steady state crack growth rate at 0.001 Hz lies near the static load data. At 0.01 Hz the crack growth rate is almost an order of magnitude (factor of 10) higher than the static load crack growth rate at low C^* magnitudes. Reflecting the contribution of the fatigue controlled component of cracking behaviour.

For the 1CMV steel only the crack growth data up to and including 0.01 Hz were plotted as a function of C^* since crack growth above this frequency is in the pure fatigue regime, as shown in fig. 4. Crack growth rate at 0.0001 Hz is approximately 3 times slower than the static load creep crack growth rate. At 0.01 Hz the crack growth rate was approximately a factor of two slower than the static data.

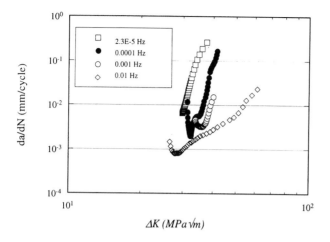

Figure 3: Crack growth as a function of ΔK *(MPa√m)* for 1/2CMV CT specimens

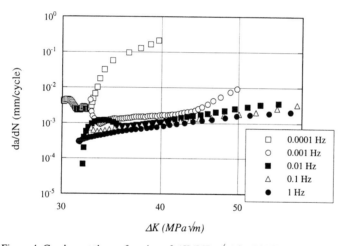

Figure 4: Crack growth as a function of ΔK *(MPa√m)* for 1CMV CT specimens.

To within a factor of three the steady state creep crack growth rates of the creep/fatigue crack tests can be correlated using C^*. The apparent correlation of the crack growth rate at 0.01 Hz with the static load data should be treated with caution since at 0.01 Hz crack growth rate had been correlated using ΔK as shown in fig. 4, and the results of the fracture surface examination indicated that fracture was predominantly transgranular.

732

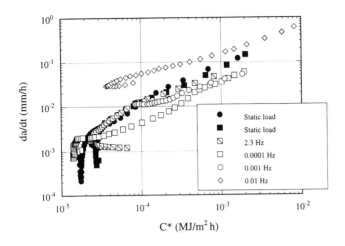

Figure 5: Crack growth rate as a function of C^* for 1/2CMV, CT specimens.

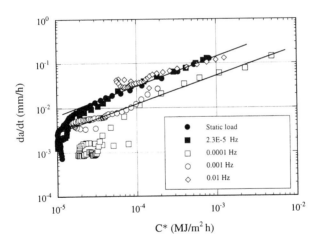

Figure 6: Crack growth rate as a function of C^* for 1CMV, CT specimens.

3. Prediction of Creep-Fatigue Crack Growth Rate

When cycling occurs at low frequencies deformation becomes increasingly time dependent and creep mechanisms may interact or become dominant, depending on the period of the cyclic excursion (usually referred to as the dwell or hold time). It has been shown that by converting da/dN in the low frequency time-dependent region to da/dt and plotting against C^* (calculated at the maximum load of the cycle) that these cyclic results correlate well with the static data. This suggests that low frequency cyclic data can be predicted satisfactorily from creep crack propagation rates [10]. This result also implies that a similar stress distribution to that formed ahead of a crack in static creep tests is developed at the maximum load in cyclic tests irrespective of the dwell times on and off load.

Creep crack growth rates are characterized by C^* using eqn. 2. Pure fatigue crack growth rate is assumed to be correlated by eqn 6. The major assumption in the prediction of creep-fatigue crack growth is the partitioning of the total crack growth rate into a cycle-dependent part, corresponding to the loading and unloading portions of the load cycle, and a time-independent part corresponding to the hold period [3,10]. The model assumes linear superposition of these two parts. Two forms are available as proposed in eqns. 7 and 8.

The strength of the cumulative damage law lies in the fact that creep-fatigue crack growth, for any frequency, can be predicted from a knowledge of only the pure creep cracking and pure fatigue crack growth behaviour of a particular material. It should be noted that only the steady state crack growth rates are predicted using the cumulative damage law. As material data-bases for the high temperature response of materials expand to include behaviour of cracked components this method is becoming increasingly viable for industrial applications and is already included in the R5 procedure for assessment of cracked components at elevated temperature [3]. The model assumes that, for a cracked specimen or component subjected to creep and fatigue loading, only knowledge of C^* and K are required, where these can be estimated from reference stress methods.

For this investigation, the pure fatigue stage II crack growth rate was evaluated at 1 Hz and R=0.1. Since the highest frequency test for the 1/2CMV was 0.01 Hz, which did not represent pure fatigue crack growth, the fatigue crack growth law for the 1CMV was used. Using the pure creep and fatigue crack laws fitted to the experimental data, eqns. 2,8 were used to predict the cumulated creep-fatigue creep crack growth in terms of either da/dt or da/dN. The results are shown in figs. 7-10.

For the 1/2CMV (figs. 7-8) specimens the model predicted the crack growth rates for the slow frequencies (0.001 and 0.0001 Hz) to be faster than the experimental rate to within a factor of two, as indicated by the scatter band. This was expected since the creep crack growth rate for the static test was a factor of two faster than the cracking rate for the 0.0001 Hz cyclic test, as shown in fig. 5 At 0.01 Hz the crack growth rates both in terms of time- and cycle-dependency were predicted below the experimental values to within a factor of two. The reason for this is possibly due to the modelling of the 1/2CMV pure fatigue crack growth rate based on the 1CMV material behaviour.

For the 1CMV steel, the crack growth rates for all three frequencies were predicted to be faster than the experimental results. The greatest difference was for the slowest frequency, 0.0001 Hz. This latter result was not unexpected since from fig. 6 it can be seen that the steady state creep crack growth rate at 0.0001 Hz was approximately a factor of two to three times slower than that of the static load creep crack growth rates. The main reason for the conservatism of the predictions is due to the fact that the static load steady state crack growth rate was always faster than the steady state crack growth rates for the cyclic tests.

734

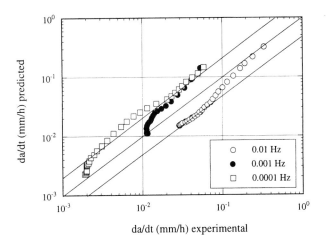

Figure 7: Prediction of 1/2CMV creep-fatigue crack growth rate in terms of *da/dt*.

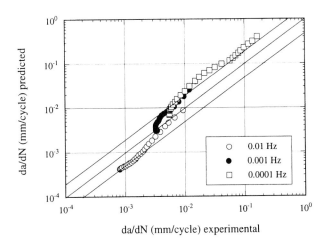

Figure 8: Prediction of 1/2CMV creep-fatigue crack growth rate in terms of *da/dN*.

4. Discussion

In figs. 5-6 it can be seen that, for the static load tests, a vertical 'tail' exists prior to the onset of steady state creep crack growth. This is known to be due to transient creep crack growth. For most static load tests the tail is vertical since the uncracked ligament is in a state of extensive creep deformation, that is, the stress state is stationary and C^* is therefore fairly constant. It might be expected, therefore, that the existence of a tail is evidence of a build up of damage at a crack tip creep crack growth. However, for the case of creep-fatigue this was not always the case as the fractographic examinations have shown that the cyclic component eliminates the transient cracking regime even at low frequencies. The experimental results for both steels cycled at 0.0001 Hz were very similar. The cyclic component caused the crack to initiate soon after loading - approximately 4 % into the life for the 1CMV specimen and almost immediately for the 1/2CMV specimen.

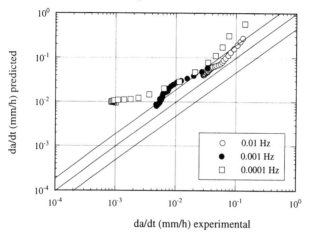

Figure 9: Prediction of 1CMV creep-fatigue crack growth rate in terms of *da/dt*.

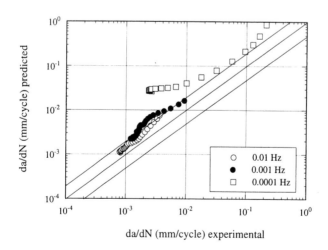

Figure 10: Prediction of 1CMV creep-fatigue crack growth rate in terms of *da/dN*.

The cyclic component eliminated the transient regime of intergranular creep cracking and effectively replaced it with transgranular crack growth. This was identified by examination of the fracture surfaces using scanning electron microscopy: the transgranular crack extended approximately 1 mm for both steels. This amount of initial cracking usually corresponds with the period of build up of damage under constant loading conditions.

A finite period was still required therefore to accumulate a steady state distribution of damage ahead of the crack tip. Once this had been achieved the cracking mechanism was dominated by steady state creep crack growth and the cyclic component had a minor role in crack propagation. For the 1/2CMV specimen cycled at 0.0001 Hz steady state creep growth commenced around 1300 hours, or after 65 % of the lifetime.

At 0.0001 Hz the overall crack growth rate was also slower than that of the static load creep crack growth tests. This was attributed to a very low crack growth rate during the hold time at minimum load. In addition it was believed that stress relaxation during the dwells at minimum load would reduce the overall creep deformation rate. Combined, these factors reduced the overall crack growth rate due to creep. The reduction in steady state creep crack growth rate at this frequency was about a factor of two for the 1/2CMV CT specimen and about a factor of three for the 1CMV CT specimen. As a result there was appreciable scatter in the correlation of steady state creep crack growth rates with the static data using C^*.

At 0.001 Hz the fracture surfaces of both steels were macroscopically intergranular but at high magnification fatigue striations were visible. The conclusion is that fracture was by both creep and fatigue mechanisms but the crack path was entirely intergranular. For the 1/2CMV material, test duration was half the predicted duration. This implies that the fatigue mechanism was the main cause for the increase in crack growth rate during the early stages despite the absence of transgranular fracture. For this steel, at about 50 % through the lifetime the crack growth rate reached steady state and could be correlated by C^* to within a factor of two.

For the 1CMV specimen transgranular fracture was seen after initiation but areas of intergranular fracture soon appeared. The duration of this test was close to the predicted duration. There was neither a substantial increase in crack growth rate due to fatigue nor a decrease due to the dwells at low load. The conclusion is that at 0.001 Hz a balance exists between fatigue and creep mechanisms for the 1CMV steel. Steady state creep crack growth commenced after the first 25 % of life.

At 0.01 Hz fracture was almost entirely transgranular for the majority of the test. For the 1CMV specimen, crack growth rate as a function of ΔK correlated with the high frequency data indicating that this frequency was in the pure fatigue range. However, it was still possible to find a correlation between crack growth rate and C^* indicating that crack growth was creep controlled for the last 65 % of lifetime. Furthermore these data lay on the static load data. Correlation of crack growth rates was therefore achieved using both ΔK and C^*. However, since no intergranular growth, characteristic of creep cracking, was found on the fracture surfaces of the specimens cycled at this frequency except during the last stages of lifetime, the use of C^* should be treated with caution.

5. Conclusions

The presence of a superimposed cyclic component eliminates the transient regime of cracking at frequencies at and above 0.0001 Hz. At 0.0001 Hz approximately the first millimeter of crack growth was by transgranular fatigue fracture. This crack extension corresponded to between one third and one half of the specimen lifetime. At 0.001 Hz between the first quarter and first half of lifetime was dominated by both intergranular fracture and the striation mechanism. At both frequencies steady state creep crack growth dominated the remaining lifetime and was correlated by the creep fracture mechanics parameter C^* to within a factor of two or three. At frequencies above and including 0.01 Hz fracture was almost entirely by fatigue crack growth and could be correlated with high frequency data using the mode I elastic stress intensity factor range ΔK.

6. References

[1] Webster, G.A., Int. J. Pressure Vessels and Piping, Vol. 50, pp 133 - 145, 1992.

[2] British Standards PD 7910 :1999, Guide to methods of assessing the acceptability of flaws in fusion welded structures, BSI, London, 1999.

[3] Ainsworth, R. A, editor R5:' assessment procedure for the high temperature response of structures', Nuclear Electric procedure R5 Issue2, 1996.

[4] Austin, T.S.P. and Webster, G.A., 'Prediction of Creep Crack Growth Periods', Fatigue Fract. Engng. Mater. Struct., Vol. 15, No. 11, 1992, pp. 1081 - 1090.

[5] Winstone, M.R., Nikbin, K.M and Webster, G.A., 'Modes of Failure under Creep-fatigue loading for a Nickel Based Superalloy', J. Mat. Sci., 20, pp. 2471 - 2476, 1985.

[6] Chorlton, M.R., Nikbin, K. and Webster, G.A., 'Characterization of the Early Stages of Creep Crack Growth'. Lifetime Management and evaluation of structures and components', Eds. J. H. Edwards et.al., EMAS, Proc. 'Eng. Struct. Integ. Assessment, Cambridge, pp47-58 , 1998.

[7] Nikbin, K.M., Smith, D.J., and Webster, G.A., 'Prediction of Creep Crack Growth from Uniaxial Creep Data'. Proceedings of the Royal Society of London, A396, pp. 183 - 197, 1984.

[8] Riedel, H. and Rice, J.R., 'Tensile Cracks in Creeping Solids'. ASTM STP 700, pp. 112 - 130, 19080

[9] Nikbin, K.M. and Webster, G.A., 'Prediction of Crack Growth under Creep-Fatigue Loading Conditions'. ASTM STP 942, pp. 281 - 292, 1988.

[10] Dimopulos, V, Nikbin, K.M., and Webster, G.A., 'Influence of Cyclic to Mean Load Ratio on Creep/Fatigue Crack Growth'. Metallurgical Transactions A, 19A, pp. 873 - 880, 1988.

[11] Standard Test Method for Measurement of Creep Crack Growth Rates in Metals', ASTM E 1457 - 92.

[12] Webster, G.A., 'Methods of Estimating C*'. Materials at High temperatures, 10, No. 2, pp. 74 - 78, 1992.

Investigation on the cracking behaviour of 316L stainless steel circumferentially notched tubular components under conditions of creep interacting with thermal fatigue

L. Gandossi [a], N.G. Taylor [a], R.C. Hurst [a], B.J. Hulm [b], J.D. Parker [b]

[a]*Institute for Advanced Materials, SIC Unit, P.O. Box 2, 1755 ZG Petten, The Netherlands*
[b]*Department of Materials Engineering, The University of Wales Swansea, Singleton Park, Swansea SA2 8PP, Wales, UK*

This paper is concerned with the investigation of the cracking behaviour of austenitic stainless steel 316L under the interaction of creep and thermal fatigue. The test piece consists of a hollow cylinder with a full circumferential notch machined onto the external surface to act as crack starter. A primary load can be axially applied in order to introduce creep stresses in the specimen section. Thermal stresses are cyclically induced by heating and cooling the external surface by means of induction heating. Under the applied loading, a crack initiates and propagates across the test piece section. The potential drop method is used for continuos in-situ crack growth monitoring. A series of test with different loads and duration of the hold time at high temperature were carried out. The experimental crack growth was correlated in term of fracture mechanics parameters evaluated by means of finite element techniques. This route proved to yield a good prediction of the crack growth rate and has potential to be applied to actual industrial components.

Key words: *Thermal fatigue, Creep, Crack growth, 316L steel*

1 Introduction

In order to aid design and support remaining life assessment of high temperature plant components, analytical techniques, which purport to predict the behaviour of actual components from laboratory based uniaxial materials data, require validation. The performance of this validation procedure on operating plant components is largely impractical since there is rarely a satisfactory record of the components primary and thermal loading history, to say nothing of the potential risk of component failure. By contrast, the use of idealised component geometries such as tubes, tested in the laboratory environment, permits the accurate and controlled variation of the primary and thermal stress systems, offering the potential for a better understanding of the influence of these variables upon the component's mechanical behaviour and usable life.

The present work describes the application of the component validation test philosophy to the problem of crack growth for simulated thermal fatigue loading. Two objectives are highlighted for the experiments: the first, to demonstrate that significant crack growth could be achieved on laboratory scale components exposed to conditions simulative of plant operating conditions; the second, to establish a benchmark for the validation of the predicted crack growth behaviour from analytical methods based on conventional specimen data. The experimental work to date has largely focused on 316L stainless steel which is widely used in heat exchanger structures for a broad range of elevated temperature engineering applications.

2 Test methodology

2.1 SIMULATION OF CREEP INTERACTING WITH THERMAL FATIGUE

A unique experimental facility has been developed at the JRC Petten for the investigation of the cracking behaviour of structural materials subjected to conditions of creep and thermal fatigue acting together (CTF). This rig consists of an improved, more flexible version of an existing apparatus [1-4] that has been used in the past to investigate crack growth in the austenitic stainless steel 316L under prevalent conditions of thermal fatigue (TF).

The TF rig generated severe temperature gradients through the specimen cross-section by repeatedly heating and cooling its external surface whilst a forced convection water flow continuously maintained constant the temperature of the inner surface. Under the applied thermal loading, a crack could quickly initiate and grow from artificial defects introduced to act as stress concentrators. The hold time cycles provided a first step towards investigating the interaction of creep and fatigue, however the relaxation of the thermal stresses limited the extent of interaction obtained. This lead to the modification of the rig to allow axial tensile stressing of the test-piece.

Figure 1 shows a schematic of the new TFC test system. The specimen is a thick-walled pipe-like component, whose typical dimensions are an inner radius of 10 mm, an outer radius of 24 mm and a length of 220 mm. Induction heating is utilised to cyclically heat the outer surface of the test-piece whilst water flowing in its bore provides cooling. The water flow can be interrupted at any time and the water can be removed from the specimen bore by blowing compressed air. A primary (creep) load may be axially superimposed by means of a dead-weight creep machine, an ESH Testing Limited apparatus. It is able to introduce in the test piece a maximum nominal load of 280 kN. Such a high load is required by the very thick cross section of the specimen (1500 mm^2), which in turn is necessary in order to reproduce the required thermal gradients during testing. Artificial defects (typically full circumferential 1 mm-deep notches) were machined onto the specimen to act as crack starters. The crack growth from the starter notch was monitored using a potential drop method [5].

3 Analysis route

The validation of thermal fatigue crack growth assessment methods relies typically on the representation of the component's TFCG behaviour in a da/dN vs. ΔK format to facilitate direct comparison with conventional isothermal fatigue crack growth data, typically generated on CT type specimens. This requires the determination of the stress intensity factor K for the defects within the component as a function of crack depth. The choice of the appropriate parameter to correlate the experimental data produced by the TF and TFC facilities was by no means straightforward, since the material was subjected to a wide variety of conditions ranging from extensive reverse plasticity, ratchetting, primary and secondary creep.

Following the definition of the stress/strain response, the complex problem of thermal loading resolves itself into one where the FE solutions for stress and strain are converted into fracture mechanics terms and then related to the observed crack growth rates. Some work involving crack growth in stainless steel has been conducted under thermal shock loading conditions, where the use of fracture mechanics has been carried out for turbine blade applications [6, 7], valve chests [8], heat exchangers [4] and fusion reactor first wall [9] components. For these applications, it was assumed that the crack growth rates were controlled by an elastically calculated stress intensity factor. In view of these considerations, it was decided to attempt an approach involving a similarly defined parameter.

The application of an elastically calculated parameter may not at first seem reasonable for high temperature fatigue and creep behaviour and some approaches [10] adopt the J-integral approach, an obvious choice for conditions where non-linear material behaviour such as plasticity occurs. Also creep crack growth data are usually correlated using different crack-tip parameters, designed to characterise the stress field at the crack tip (e.g. the C* parameter). Both these approaches have been investigated, however for reasons of space the findings can only be discussed elsewhere [11].

4 Results and discussion

4.1 THERMAL FATIGUE (TF) TESTING

In devising the test programme two factors, duration of the thermal up-shock and duration of hold time, were identified as the principal parameters, although the latter was to assume the predominance, in view of the interest in creep damage on fatigue cracking. As previously reported [12] five tests (codes A-E) were carried out in on the TF rig using an 80-600°C cycle. The experimental conditions are summarised in Table 1. Fig. 2 plots the crack growth rate versus crack depth curves. The introduction of a progressively longer hold time had a clear effect on the peak crack growth rate, although the mode of crack growth for the whole set of tests was found to be transgranular. Using Test B as reference, a 90-sec dwell (Test C) yielded a peak rate 9-12% higher, a 10-min dwell (Test D) yielded a peak rate 16-20% higher and a 30-min dwell (Test E) yielded a peak rate 38-45% higher. The rate peak was achieved at very different numbers of cycles (700-800 for test A, 1400-1600 for the others), but always for a crack depth in the restricted range 2.6-3.2 mm. The ratio a_{max}/a_{min} was always in the range 1.4-1.6, attesting the fact that the development of the crack front from the circumferential starter notch was not perfectly circular.

4.2 CREEP/THERMAL FATIGUE (TFC) TESTING

The first test carried out on the TFC facility (test F) was a control to reproduce the TF rig results. Test G was characterised by a 14-sec up-shock and a dwell at 600°C of 90 seconds with no load applied. Tests H and I were characterised by a 14-sec up-shock and a dwell at 600°C of 30 minutes with loads respectively of 20 and 100 kN. In Test J the duration of the up-shock was increased to 60 seconds. Test K was characterised by a 14-sec up-shock and a long hold-time at 600°C (2.5 hours) with a 100 kN primary load, Table 1.

The crack growth vs. cycles for each TFC test (except for F) is plotted in Fig. 3, and the crack growth rate vs. crack depth in Fig. 4. The following results were highlighted.

Crack initiation: Very few cycles (<100) were required to achieve crack initiation. The remaining part of the test was spent for growth. Longer hold-times appeared to reduce the number of cycles to initiation.

Crack growth: In some TFC tests the crack growth presented a pattern similar to the TF, with a rate first increasing and subsequently decreasing, Fig. 3 (Tests G and H). In the remaining TFC tests the crack growth rate presented the highest value at the end of each test. In test K a local peak was achieved, the rate then dropped for a limited period before the increase was resumed. No crack arrest occurred in the range of crack depths investigated (up to roughly 44% of the specimen thickness). The ratio a_{max}/a_{min} was always in the range 1.24-1.29, lower than in TF testing. This proved that a better axisymmetry of the boundary conditions could be achieved in the TFC facility.

Accumulation of cyclic inelastic deformation: Due to the severity of the applied loading, ratchetting deformation occurred in some of the TFC tests. The overall variation of the specimen length was irrelevant during Tests F, G and H. Tests K, J and I were characterised by a significant accumulation of cyclic deformation. This deformation followed a very similar pattern in the latter tests: (a) It was very high in the first 50 cycles, with the specimen elongating of approximately 1.5 mm. The rate in this stage was independent of test conditions; (b) It decreased due to hardening effects and approximately constant between 100 and 300 cycles. The rate was dependent on test conditions during this phase; (c) It showed progressively increasing values, related to a reducing uncracked ligament and increasing stress over the section.

Metallography: The cross section of each specimen under investigation was polished with series of diamond pastes (10-5-1-0.5 μm) to a surface roughness of 1 to 0.5 μm. The polished surfaces were etched in a solution of 10% oxalic acid, with a potential drop 10 volts for 5 to 20 seconds as required. Transgranular crack growth was predominant, but some intergranular crack growth was also found in tests H and K. No creep damage (in form of voids at the grain boundaries) was found. A typical crack cross-section obtained in Test H is reported in Fig. 5.

4.3   CRACK DRIVING FORCE ANALYSIS

Fig. 6 summarises the results obtained for TF cycling (tests A-E) using the Weight Function approach. The pattern shown by the stress intensity factor range calculated with the weight function method was similar for all the tests considered. The peak was achieved for a depth of approximately 2 mm from the crack mouth for tests A and B. The presence of increasingly long hold-times seems to "push" back the peak. This is clearly an effect of the creep strains, which are present exclusively at the external surface in TF cycling. The duration of the up-shock had a major influence over the magnitude of ΔK (test A and B). The peak in test A was 90% higher than in test B (141 against 74 MPa√m). The duration of the hold-time had a minor (but significant) influence over the predicted peak value: 74 MPa√m for test B, 84 MPa√m for test C, 91 MPa√m for test D, 99 MPa√m for test E. ΔK tended to the same values moving towards deeper cracks for all the tests with equal up-shock duration.

The results obtained for TFC cycling (tests G-K) are summarised in Fig. 7. The pattern shown in TFC cycling by the stress intensity factor range was similar to the TF distributions. Two major differences could be identified: the magnitude of the range (considerably higher in TFC) and the behaviour of deeper cracks. Fig. 8 shows the comparison between the ΔK distributions calculated for test C (TF) and test G (TFC). Both cycles were characterised by a 14-sec up-shock and a 30-sec hold-time without superimposed load. The predicted maximum value for ΔK was 84 MPa√m in test C and 209 MPa√m in test G, 2.48 times higher. The difference is attributed to the interruption in the bore water flow during dwells.

The presence of a superimposed load modified the pattern of ΔK distribution at depths higher than 6 mm. Whereas in test G a decreasing ΔK was predicted, in tests I, J and K (100kN), the calculations showed that ΔK would increase again after having reached a minimum. In test H this tendency was less, due to the lower load (20 kN). This effect is clearly due to the increasing primary stress applied over the uncracked ligament, a consequence of the tests being carried out under constant-load conditions. The effect of the primary load was also to augment the peak value of ΔK. Considering test G (0 kN), test H (20 kN) and test I (100 kN), this value passed from 209 to 220 to 262 MPa√m. Together, the stress analysis provides a plausible interpretation of the increased crack growth rates observed under the thermal fatigue – creep conditions

The experimental crack growth rates for both the TF and TFC cases were plotted against the calculated ΔK_{eff}, see respectively Fig. 9 and 10. Comparison of the experimental TF and TFC crack growth rates with isothermal fatigue tests on conventional specimens of 316 and 316L stainless steel at different temperature is also illustrated in these pictures. The experimental crack growth rates exhibited a fairly linear dependence on the theoretical equivalent elastic stress-intensity factors both in TF tests and in TFC tests. In Figures 9 and 10 data obtained for cracks shorter than 0.4 mm were not reported for clarity.

5 Summary and conclusion

- The new TFC facility was successfully used to reproduce creep-fatigue crack growth in 316L austenitic stainless steel components, subjected to different temperature cycles and superimposed primary loads.
- The effect of hold time in both TF and TFC cycles was to increase the equivalent creep strain range through the wall thickness, giving higher crack growth rates. In

TFC this effect was not as high as expected, due to stress redistribution that un-loaded a large proportion of the section.

- The effect of load was to increase the net-section stress and therefore the crack growth rates. Calculations revealed that this effect was more biased towards increasing the plastic strain range (and thus the fatigue component of the damage) rather than the creep strain range. Ratchetting (cyclic accumulation of inelastic deformation) occurred for all the tests carried out under a load of 100 kN. Since the tests were carried out under constant load conditions, the effect of the load was also to promote exponentially increasing rates as the crack grew deeper into the component.

- The equivalent elastic stress intensity factor, ΔK, based on the weight function method applied to solutions from thermomechanical finite element analyses, was effective in describing the crack driving force under multiaxial primary and thermal loading conditions.

7 References

(1) Kerr, D.C., An Investigation of fatigue crack-growth in thermally loaded components, Ph.D. Study, University of Glasgow,1993.
(2) Kerr, D.C., Andritsos, F. and Hurst, R.C., Crack growth determination in thermally loaded components, in Proc. SMIRT 11, pp. 265-270, 1991.
(3) O'Donnell, M.P., The effect of cyclic thermal fatigue/creep loading on the crack growth behaviour from notches in cylindrical austenitic stainless steel components, Ph.D. thesis, Trinity College, Dublin, Ireland.
(4) O'Donnell, M.P., Hurst, R.C. and Taylor, D., The effect of cyclic thermal stress on crack propagation in cylindrical stainless steel components, Proceedings ECF11, 1996.
(5) L. Gandossi, S. Summers, N. Taylor, R. Hurst, B. Hulm, J. Parker , The Potential Drop Method for Monitoring Crack Growth in Real Components subjected to combined fatigue and creep conditions – Application of FE techniques for deriving calibration curves, HIDA II Conference, Advances in Defect Assessment in High Temperature Plant, Stuttgart, Germany, 4 – 6 October 2000.
(6) Morwbay, D.F., Woodford , D.A. and Brandt, B.E., in Fatigue at elevated temperatures, ASME STP 520, pp. 416-426, (1973).
(7) Morwbay, D.F. and Woodford, D.A., Inst. For Mech. Engs., Conf. publication 13, pp. 179.1-179.11., (1973).
(8) Skelton, R.P., Cyclic crack growth and closure effects in low alloy ferritic steels during creep fatigue at 550°C., High temperature technology, Vol. 7, No. 3, Aug. (1989).
(9) Merola, M., Thermal fatigue life time of the first wall of a fusion reactor. Numerical predictions and experimental results. Produced by the IAEA Vienna Austria, (1994).
(10) Merola, M., Beghini, M., Bertini, L. and Sevini, F., Numerical analysis of crack growth under thermal fatigue loading. Report EUR 16194 EN (1994).
(11) L. Gandossi, Crack growth behaviour in austenitic stainless steel components under combined thermal fatigue and creep loading, PhD Thesis, University of Wales Swansea, November 2000.
(12) L. Gandossi, N.G. Taylor, R.C. Hurst, Crack growth behaviour in 316L stainless steel components under thermal fatigue loading, Proceeding of the Thermal Stresses '99 – Third International Congress on Thermal Stresses, Krakow University of Technology, June 13-17, 1999, Krakow, Poland, Edited by J. J. Skrzypek and R. B. Hetnarski, pp. 233-236.

Test	Rig	Up-shock (s)	Hold (s)	Water flow	Load (kN)	Cycles	Duration (h)
A	TF	7	0	ALWAYS	0	1990	35.5
B	TF	14	0	ALWAYS	0	2440	50.7
C	TF	14	90	ALWAYS	0	2400	118.7
D	TF	14	600	ALWAYS	0	2730	516.8
E	TF	14	1800	ALWAYS	0	1610	845
F	TFC	14	1800	ALWAYS	0	1637	859
G	TFC	14	90	TRANSIENTS	0	913	48
H	TFC	14	1800	TRANSIENTS	20	954	504.6
I	TFC	14	1800	TRANSIENTS	100	533	282
J	TFC	60	1800	TRANSIENTS	100	790	428
K	TFC	14	9000	TRANSIENTS	100	400	1011

Table 1   Summary of TF and TFC tests.

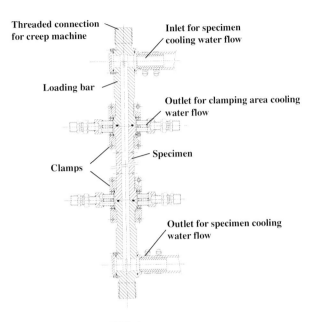

Fig. 1 The loading train of the TFC rig.

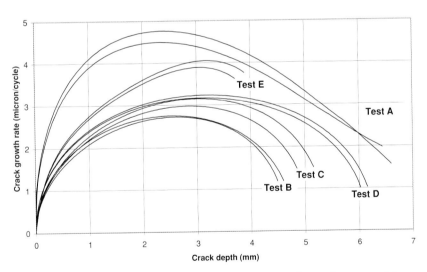

Fig. 2 Comparison between TF tests (A-E): crack growth rate (in μ per cycle) versus crack depth.

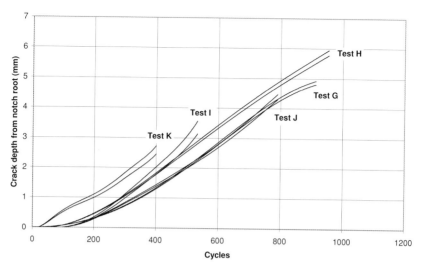

Fig. 3   Comparison between TFC tests (G-K): crack depth versus number of cycles.

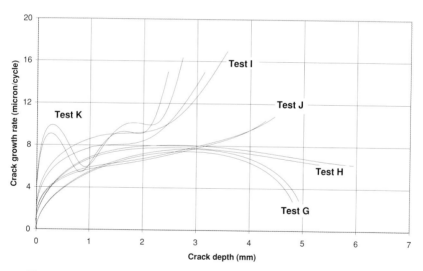

Fig. 4   Comparison between TFC tests(G-K): crack growth rate (in μ per cycle) versus crack depth.

748

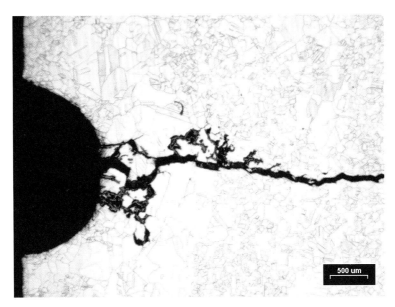

Fig. 5 Typical crack cross-section obtained in Test H. The crack initiated transgranularly, grew for the first 1.5 mm in a mixed mode (transgranular-intergranular) and eventually reverted to transgranular behaviour to reach its final length.

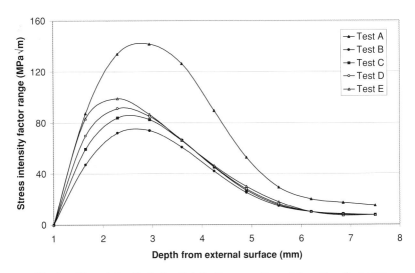

Fig. 6 Stress intensity factor distributions over the notch section. Summary of results obtained for TF cycling using the weight function method.

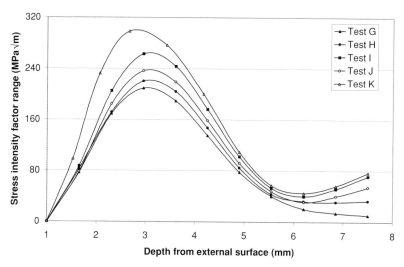

Fig. 7   Stress intensity factor distributions over the notch section. Summary of results obtained for TFC cycling using the weight function method.

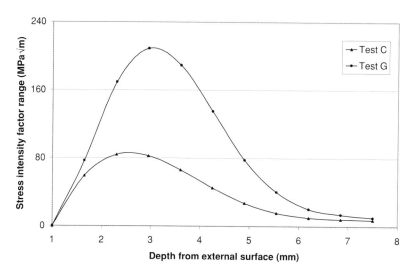

Fig. 8   Stress intensity factor distributions over the notch section. Comparison of the results obtained for Tests C (TF cycling, 90-sec hold-time) and G (TFC cycling, 90-sec hold-time, no load) using the weight function method.

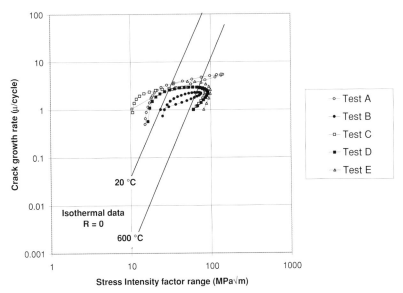

Fig. 9   Dependence of the experimental crack growth rates on ΔK_{eff} in TF cycling. Tests A-E are represented.

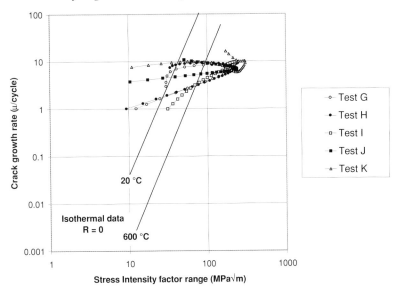

Fig. 10   Dependence of the experimental crack growth rates on ΔK_{eff} in TFC cycling. Tests G-K are represented.

CHARACTERISATION OF CREEP CRACK GROWTH BEHAVIOUR IN TYPE 316H STEEL USING BOTH C* AND CREEP TOUGHNESS PARAMETERS

D W Dean and D N Gladwin

Structural Integrity Branch, British Energy Generation Ltd.,
Barnwood, Gloucester GL4 3RS, United Kingdom
e-mail: david.dean@british-energy.com

Abstract

Experimental creep crack growth data are generally obtained by following standard test methods such as ASTM E1457-98 and subsequently characterised using the C* parameter. These data are then used in assessment procedures, such as R5, together with reference stress estimates of C* in the component, to predict creep crack growth behaviour. However, incubation and the early stages of creep crack growth are not generally well characterised by C*. Alternative methods for assessing incubation and the early stages of creep crack growth are currently being developed within the R5 procedures, including the time dependent failure assessment diagram (TDFAD) approach. This is similar to the approach adopted in the established R6 low temperature defect assessment procedure except that a "creep toughness" is used instead of the conventional fracture toughness and time dependent stress and strain parameters are required. This paper describes the results of a series of creep crack growth tests performed on Type 316H stainless steel at a temperature of 550°C and then examines characterisation of the observed behaviour using both C* and creep toughness parameters. It has been shown that difficulties in interpreting the relationship between creep crack growth rate and C* mean that it is difficult to discern any trends which clearly reflect observed changes in fracture mode with test duration. However, creep toughness values for a given crack extension exhibit a clear reduction with increased test duration in accordance with the observed changes in the creep fracture mode.

1 Introduction

Type 316H steel is used in the steam headers and other high temperature components of British Energy's Advanced Gas Cooled Reactors (AGRs). Defect assessments are performed to demonstrate the long term safe operation of these components, which require creep crack growth behaviour to be characterised using appropriate fracture mechanics parameters. This paper first describes in Section 2 some experimental creep crack growth data generated on Type 316H steel according to the standard ASTM E1457-98 [1].

Section 3 describes the assessment approaches within R5 [2, 3] for predicting creep crack growth behaviour in components. The basic procedure uses reference stress estimates of C* in the component, together with laboratory creep crack growth data, to predict creep crack growth behaviour. However, incubation and the early stages of creep crack growth are not generally well characterised by C* [4]. For low temperature fracture, the simplified R6 procedure [5] has been developed, which uses the concept of a Failure Assessment Diagram (FAD) to avoid detailed calculations of crack tip parameters such as J or C*. In recent years,

FAD approaches have been extended to the creep regime [6, 7, 8 and 9] and the high temperature Time Dependent Failure Assessment Diagram (TDFAD) method has been formally incorporated into R5 [3]. This method is similar to that adopted in the established R6 low temperature defect assessment procedure [5] except that a "creep toughness" is used instead of the conventional fracture toughness and time dependent stress and strain parameters are required.

Results of the creep crack growth tests are interpreted according to the methods given in Section 4 both in terms of C* and the creep toughness required to follow the TDFAD approach. These interpretations are presented in Section 5, where difficulties in applying the ASTM standard [1] are discussed.

2   Experimental Details

2.1   Material

The Type 316H stainless steel material used in the present study was taken from two ex-service headers, denoted 1C2/3 and 2D2/2. The chemical composition of the two headers is very similar as shown in Table 1.

2.2   Test Procedure

The creep crack growth tests were performed on standard 19mm thick compact tension (CT) specimens [10]. Prior to testing, the specimens were fatigue pre-cracked to a crack length to width (a/w) ratio of approximately 0.5 and were then side-grooved by 20% of their thickness using a Charpy V-profile cutter.

Testing was carried out on constant load creep machines at a temperature of 550°C with load line displacements monitored during load-up and at the (constant) test load using capacitance gauges. Crack lengths during the tests were estimated by using the DCPD technique to linearly interpolate between the initial and final crack sizes measured from the specimen fracture surfaces.

2.3   Test Results

Test details are summarised in Table 2. Table 3 gives further information for each test [10], including times for the cracks to extend by 0.2, 0.5 and 1.0 mm together with the associated load line displacements and relevant fracture mechanics parameters which are described below.

3   Assessment Procedures

3.1   Basic R5 Procedure

The basic R5 procedure [3] for assessing the initiation and subsequent growth of defects under creep conditions utilises reference stress methods to estimate the required fracture mechanics parameters. Initiation times are predicted using either critical crack opening displacements or experimentally derived relationships between initiation time and the steady state crack tip parameter, C*. Subsequent creep crack growth is then estimated using either the transient crack tip parameter, C(t), which depends on time, t, or C*, for times less than or greater than the redistribution time respectively, together with experimentally derived relationships between creep crack growth rates and C*.

3.2 TDFAD Approach

The TDFAD is based on the Option 2 FAD in R6 [5] and involves a failure assessment curve relating the two parameters K_r and L_r, defined in equations (1) and (2) below, and a cut-off L_r^{max}. For the simplest case of a single primary load acting alone

$$K_r = K / K_{mat}^c \tag{1}$$

where K is the stress intensity factor and K_{mat}^c is the appropriate creep toughness value, and

$$L_r = \sigma_{ref} / \sigma_{0.2}^c \tag{2}$$

where σ_{ref} is the reference stress and $\sigma_{0.2}^c$ is the stress corresponding to 0.2% inelastic (plastic plus creep) strain from the average isochronous stress-strain curve for the temperature and assessment time of interest.

The failure assessment diagram is then defined by the equations

$$K_r = \left[\frac{E\varepsilon_{ref}}{L_r\sigma_{0.2}^c} + \frac{L_r^3\sigma_{0.2}^c}{2E\varepsilon_{ref}} \right]^{-1/2} \qquad L_r \leq L_r^{max} \tag{3}$$

$$K_r = 0 \qquad L_r > L_r^{max} \tag{4}$$

In equation (3), E is Young's modulus and ε_{ref} is the total strain from the average isochronous stress-strain curve at the reference stress, $\sigma_{ref} = L_r\sigma_{0.2}^c$, for the appropriate time and temperature. Thus, equation (3) enables the TDFAD to be plotted with K_r as a function of L_r, as shown schematically in Figure 1. The cut-off L_r^{max} is defined as

$$L_r^{max} = \sigma_R / \sigma_{0.2}^c \tag{5}$$

where σ_R is the rupture stress for the time and temperature of interest. For consistency with R6 [5] at short times, the value of L_r^{max} should not exceed $\bar{\sigma}/\sigma_{0.2}$ where $\bar{\sigma}$ is the short term flow stress and $\sigma_{0.2}$ is the conventional 0.2% proof stress. As in R6 [5] $\bar{\sigma}$ may be taken as $(\sigma_{0.2} + \sigma_u)/2$ where σ_u is the ultimate tensile strength.

A central feature of the TDFAD approach is the definition of an appropriate creep toughness, K_{mat}^c, which, when used in conjunction with the failure assessment diagram, ensures that crack growth in the assessment period is less than a value Δa.

The point (L_r, K_r), from equations (1) and (2), using the current values of stress intensity factor and reference stress respectively, is plotted on the failure assessment diagram. If the point lies within the failure assessment curve of equation (3) and the cut-off of equation (4), then the crack extension is less than Δa and creep rupture is avoided.

4 Analysis of Experimental Creep Crack Growth Data

4.1 Characterisation Using C*

Following ASTM E1457-98 [1], the parameter C* is evaluated using

$$C^* = \left(\frac{n}{n+1}\right)\frac{P\dot{\Delta}_c}{B_n(w-a)}.\eta \tag{6}$$

where n is the exponent in Norton's law, P is the applied load, B_n is the net specimen thickness, w is the specimen width, a is the current crack length, $\dot{\Delta}_c$ is the creep component of load line displacement rate and

$$\eta = 2 + 0.522(1 - a/w) \tag{7}$$

The creep component of load line displacement rate is given by

$$\dot{\Delta}_c = \dot{\Delta}_T - (\dot{\Delta}_e + \dot{\Delta}_p) \tag{8}$$

where $\dot{\Delta}_T, \dot{\Delta}_e$ and $\dot{\Delta}_p$ are the total, elastic and plastic load line displacement rates respectively,

$$\dot{\Delta}_e = \frac{2K^2 B_n \dot{a}}{PE} \tag{9}$$

and

$$\dot{\Delta}_p = \frac{(m+1)J_p B_n \dot{a}}{P} \tag{10}$$

where K is the stress intensity factor, $\dot{a}$ is the creep crack growth rate, E is Young's modulus, m is the exponent in a Ramberg-Osgood fit to the tensile data and J_p is the plastic component of the J-integral.

4.2 Characterisation Using Creep Toughness

Creep toughness values can be derived directly from experimental load-displacement data from creep crack growth tests by using methods for estimating the J-integral given in low temperature fracture toughness standards such as the ESIS procedure [11]. The ESIS procedure evaluates experimental total J values, J_T, using a relationship based on the total area under the load-displacement curve, U_T. Thus,

$$J_T = \frac{\eta U_T}{B_n(w - a_0)} \qquad (11)$$

where a_0 is the initial crack length. Values of creep toughness, J_{mat}^c and K_{mat}^c, may then be derived from creep crack growth tests as a function of crack growth increment, Δa, where J_{mat}^c is obtained using equation (11), and

$$K_{mat}^c = \sqrt{E'J_{mat}^c} \qquad (12)$$

where $E' = E/(1 - v^2)$ and v is Poisson's ratio.

5   Results and Discussion

5.1   Characterisation Using C*

Creep crack growth data from each of the tests are correlated with C* in Figure 2. A creep exponent, n, of 8.2 has been assumed based on the secondary part of the RCC-MR creep law for Type 316 steel at 550°C [12]. Data obtained prior to $\Delta a = 0.5$mm have been excluded as recommended in ASTM E1457-98 [1] and the remaining data points are all for times in excess of the transition time, t_T. However, problems were encountered in estimating elastic and plastic displacement rates using equations (9) and (10) respectively. In both cases, the analytical estimates were unreliable and sometimes resulted in calculated creep displacement rates which were negative. Further work is clearly required to establish reliable methods for estimating $\dot{\Delta}_e$ and $\dot{\Delta}_p$. C* values presented in this paper are therefore based on total displacement rates $\dot{\Delta}_T$.

In applying the ASTM E1457-98 validity criteria to censor the raw experimental data, information relating to incubation and the early stages of creep crack growth are removed from the characterisation in terms of C*. Figure 3 shows da/dt versus C* data obtained from test 2D2/2 CT2 prior to application of the validity criteria. It can be seen that the 'tail', characterised by reducing values of da/dt and C*, is completely removed by rejecting data obtained prior to $\Delta a = 0.5$mm. If the analytical correction for elastic displacement rate, $\dot{\Delta}_e$, given by equation (9) is applied to the measured total displacement rate, $\dot{\Delta}_T$, the criterion $\dot{\Delta}_c / \dot{\Delta}_T > 0.5$ from [1] is violated for almost all of the data points which satisfy $\Delta a > 0.5$mm. Although the negative values of $\dot{\Delta}_c / \dot{\Delta}_T$ obtained in the latter part of this test indicate that $\dot{\Delta}_e$ is being overestimated analytically, it is likely that realistic values of elastic displacement rate would still result in a significant proportion of the test data failing to meet the criterion $\dot{\Delta}_c / \dot{\Delta}_T > 0.5$. It is also worth noting that although the transition time, t_T, of 470h calculated for this test represents a small proportion of the test duration, the redistribution time, t_{red}, used in the R5 procedures [3] as a measure of the time to establish widespread creep conditions is a factor $(n + 1) \approx 9$ times higher than t_T and is therefore comparable with the test duration. During the redistribution period, the crack tip stress and strain rate fields are characterised by the parameter C(t) which is in excess of C* and this may explain the relatively high values of da/dt obtained from this test (see Figure 2).

Metallographic examination of the sectioned creep crack growth specimens revealed a change in fracture mode with test duration. Whilst initiation and subsequent crack growth was intergranular in all cases, there was a systematic change from a ductile intergranular fracture mode at short test durations (Figure 4) to a brittle intergranular fracture mode as the test duration was increased (Figure 5). In the former case, initiation occurred by the formation of a bifurcation along planes of maximum shear stress (at approximately ±45° to the plane of the initial fatigue pre-crack) and subsequent crack propagation was accompanied by deformation of the grains. However, in the latter case, initiation and subsequent crack growth occurred in a narrow band close to the plane of the initial fatigue pre-crack with little or no evidence of deformation of the grains.

In summary, characterisation of the creep crack growth data for Type 316 steel at 550°C with C* revealed a number of difficulties in applying the validity criteria specified in ASTM E1457-98 [1] and, for the longest term test, the use of C* is questionable as most of the crack growth occurred prior to establishing widespread creep conditions. These problems in interpreting the da/dt versus C* data meant that it was difficult to discern any trends which clearly reflect the observed change in mode from ductile to brittle intergranular fracture with increasing test duration.

5.2   Characterisation Using Creep Toughness

Creep toughness, J^c_{mat}, values have been evaluated from each of the creep crack growth tests as a function of crack growth increment, Δa. The creep crack initiation and growth resistance reduces with increasing test duration; this trend is evident in Figure 6, which shows creep toughness, K^c_{mat}, values for crack growth increments, $\Delta a = 0.2$, 0.5 and 1.0mm as a function of time. The creep toughness values for each crack growth increment exhibit a clear reduction with increasing time, which is consistent with observed changes from a ductile to brittle intergranular creep fracture mode.

In contrast to the test standard ASTM E1457-98 [1], which concentrates on the latter parts of the creep crack growth tests ($\Delta a > 0.5$mm, $t > t_T$ etc.), the creep toughness provides an integrated description of the earlier parts of the tests ($\Delta a \leq 0.2$, 0.5 or 1.0mm here). The former approach can lead to difficulties as the validity criteria in [1] can result in the exclusion of data from a significant proportion of the duration of long term tests. For component applications, where only small crack extensions are often allowed, the latter approach using a creep toughness has attractions.

6   Concluding Remarks

This paper has described the results of a series of creep crack growth tests performed on Type 316H stainless steel at a temperature of 550°C and then examined characterisation of the observed behaviour using both C* and creep toughness parameters. It has been shown that difficulties in interpreting the relationship between creep crack growth rate and C* mean that it is difficult to discern any trends which clearly reflect observed changes in fracture mode with test duration. However, creep toughness values for a given crack extension exhibit a clear reduction with increased test duration which reflects observed changes in the creep fracture mode.

7   Acknowledgements

This paper is published with permission of British Energy Generation Ltd.

8   References

1   E1457-98 Standard Test Method for Measurement of Creep Crack Growth Rates in Metals. Philadelphia : American Society for Testing and Materials, 1998.

2   Ainsworth R A, Chell G G, Coleman M C, Goodall I W, Gooch D J, Haigh J R, Kimmins S T and Neate G J. CEGB Assessment Procedure for Defects in Plant Operating in the Creep Range. Fatigue Fract. Engng. Mater. Struct., 1987;10:115-127.

3   British Energy Generation Ltd. An Assessment Procedure for the High Temperature Response of Structures, R5 Issue 2 Revision 2, 1998.

4   Webster G A and Ainsworth R A: High Temperature Component Life Assessment. London : Chapman & Hall, 1994.

5   British Energy Generation Ltd. Assessment of the Integrity of Structures Containing Defects. R6 Revision 3, Amendment 11, 2000.

6   Ainsworth R A. The Use of a Failure Assessment Diagram for Initiation and Propagation of Defects at High Temperatures. Fatigue. Fract. Engng. Mater. Struct. 1993; 16:1091-1108.

7   Hooton D G, Green D and Ainsworth R A. An R6 Type Approach for the Assessment of Creep Crack Growth Initiation in 316L Stainless Steel Test Specimens. Proc. ASME PVP Conf., Minneapolis, 1994; 287:129-136.

8   Ainsworth R A, Hooton D G and Green D. Further Developments of an R6 Type Approach for the Assessment of Creep Crack Incubation. Proc. ASME PVP Conf., Honolulu, 1995; 315:39-44.

9   Ainsworth R A, Hooton D G and Green D. Failure Assessment Diagrams for High Temperature Defect Assessment. Engng. Fract. Mech.1999; 62: 95-109.

10   Gladwin D N. Creep Crack Growth and Creep Toughness Data for Austenitic Type 316 Steels, British Energy Generation Ltd. Report, E/REP/AGR/0128/00, 2000.

11   European Structural Integrity Society, ESIS Procedure for Determining the Fracture Behaviour of Materials, ESIS P2-92, 1992.

12   RCC-MR, Design and Construction Rules for Mechanical Components of FBR Nuclear Islands, AFCEN, Paris, 1985.

Table 1 Chemical Composition of the Type 316H Stainless Steel from Headers 1C2/3 and 2D2/2

Header	C	Si	Mn	P	S	Cr	Mo	Ni	Al	As
1C2/3	0.05	0.3	1.5	0.02	0.01	17.1	2.4	11.1	<0.005	0.008
2D2/2	0.04	0.29	1.49	0.02	0.014	17.1	2.38	11.0	<0.005	0.008

Header	Co	Cu	Nb	Pb	Sn	Ti	V	W	Sb
1C2/3	0.05	0.09	<0.005	<0.005	0.006	0.01	0.042	0.03	<0.01
2D2/2	0.09	0.09	<0.005	<0.005	0.006	0.013	0.042	0.02	<0.01

Table 2 Summary of Type 316H Creep Crack Growth Data at 550°C

Test ID	Temp. (°C)	Initial Plane Stress Reference Stress (MPa)	Initial Stress Intensity Factor (MPa$\sqrt{m}$)	Duration (h)	Creep Crack Growth (mm)	Creep Load Line Displacement (mm)
1C2/3 CT5	550	322.3	43.0	146	2.25	1.84
1C2/3 CT6	550	302.1	40.1	1081	3.46	1.23
2D2/2 CT2	550	269.5	34.5	4698	2.63	0.17
2D2/2 CT4	550	286.5	38.6	1921	1.09	0.23
2D2/2 CT5	550	289.2	39.6	1589	1.26	0.25

Table 3 Summary of Type 316H Creep Toughness Data at 550°C

Test ID	Load, P (kN)	Specimen Width, w (mm)	Net Thickness, B_n (mm)	Initial Crack Depth, a_0 (mm)	Crack Growth Increment, Δa (mm)	Time (h)	J^c_{mat} (MN/m)	K^c_{mat} (MPa$\sqrt{m}$)	Δ^c_{LL} (mm)
1C2/3 CT5	14.40	38.02	15.2	20.15	0.2	28	0.110	137.9	0.21
					0.5	58	0.135	153.1	0.40
					1.0	100	0.185	178.8	0.76
1C2/3 CT6	13.25	38.1	15.21	20.35	0.2	108	0.045	88.4	0.16
					0.5	394	0.053	95.4	0.22
					1.0	802	0.069	109.3	0.35
2D2/2 CT2	10.17	38.01	15.62	21.62	0.2	1346	0.017	54.5	0.09
					0.5	2123	0.018	56.6	0.10
					1.0	3093	0.021	60.0	0.12
2D2/2 CT4	12.83	37.9	15.59	20.24	0.2	592	0.026	67.7	0.08
					0.5	1334	0.032	73.8	0.12
					1.0	1877	0.043	85.8	0.21
2D2/2 CT5	13.68	37.97	15.5	19.82	0.2	493	0.035	77.0	0.07
					0.5	1048	0.041	83.4	0.12
					1.0	1516	0.051	93.2	0.19

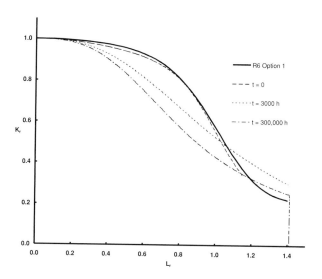

Figure 1   Schematic Failure Assessment Diagram Based on Data from an Austenitic Steel

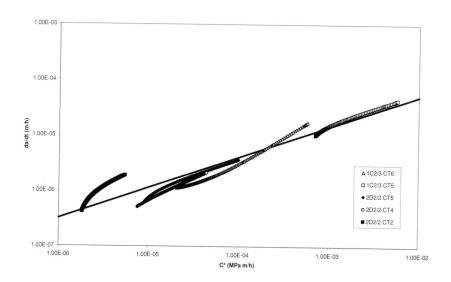

Figure 2   Correlation of Creep Crack Growth Rates with C* for Type 316H Steel at 550°C

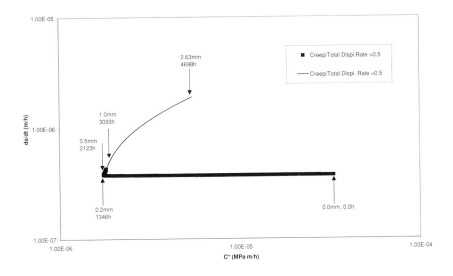

Figure 3   Creep Crack Growth Rates Data for Test 2D2/2 CT2

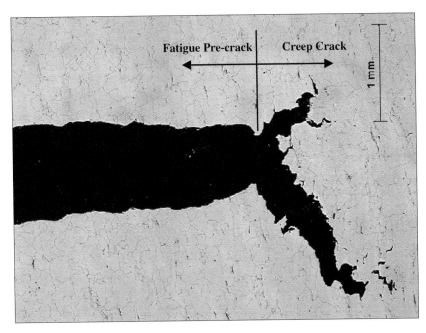

Figure 4   Ductile Intergranular Creep Crack Growth (Specimen 1C2/3 CT5)

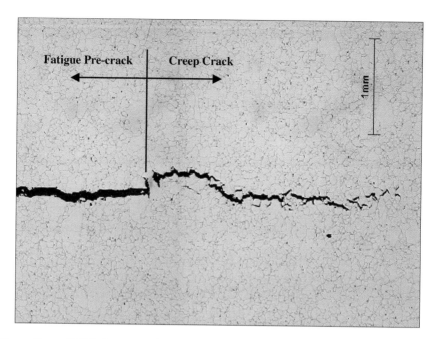

Figure 5   Brittle Intergranular Creep Crack Growth (Specimen 2D2/2 CT2)

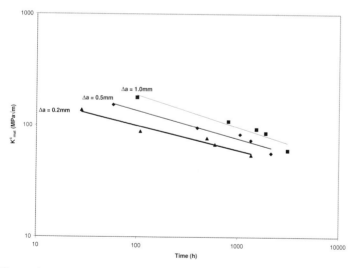

Figure 6   Creep Toughness as a Function of Time for Crack Growth Increments,
Δa = 0.2, 0.5 and 1.0 mm

HIGH TEMPERATURE BEHAVIOUR OF P91-STEEL WELDMENTS

B.Dogan* and B.Petrovski [+]
*GKSS Research Centre, Max-Planck-Str., D-21502 Geesthacht, Germany
[+]IfW-TUD, D-64283 Darmstadt, Germany

ABSTRACT

The creep crack growth (CCG) studies to date concentrated mostly on homogeneous materials. The concepts used for time dependent fracture analysis of homogeneous bodies are commonly applied for CCG in weldments. However, the crack growth in weld joints may involve periods of fast and slow propagation as crack traverses regions of different microstructures, such as in weldments with softer or harder weld metal than the base metal. Hence, the applicability of the crack tip parameters in crack growth of weldments with microstructural variations requires systematic studies to identify the relevant correlation parameter.

The present paper reports on a study on the deformation and CCG behaviour of somewhat brittle high strength martensitic steel Grade P91 steel weldments at 600°C. The weld joints were produced with a spectrum of industrially relevant weldment properties. It included materials with weld metal resistance to creep deformation and crack growth was a) lower, produced by butt welding (BW), and b) higher, produced by electron beam welding (EBW). Therefore, the studied materials cover a spectrum of microstructures and ductility to give representative data that applies to a range of weld types. The high temperature properties were determined on standard round tensile and creep specimens. The crack growth was studied using CT25 (B=12.5, W=25mm) and CT50 (B=12.5, W=25mm) type specimens, under constant load that produced CCG data in the small-scale creep to extensive creep ranges. The starter cracks were spark eroded due to the difficulty of producing fatigue pre-cracks located in the narrow weld and HAZ regions. The emphasis is placed on the crack growth initiation and applicability of test procedures and the crack tip parameters that will lead to provision of guidelines for industrial weldment crack growth assessment and analysis.

Key Words: Crack growth initiation, Creep crack growth, C*(t)-integral, Data analysis, Defect assessment, P91 steel, Butt welding, Electron beam welding, Similar welds, HAZ.

INTRODUCTION

The elevated temperature deformation and fracture behaviour of components containing welds is a major design and safety concern. The economic and environmental drive lead to improved thermal efficiency and increased service life of power generation and petrochemical plants. Therefore, the industrial components containing welds, such as turbines, headers, steam pipes and pressure vessels are subjected to stresses at higher temperatures.

The failures in industrial components operating under creep conditions occur more frequently in welds and associated heat affected zones (HAZ). The composite structure of weldments consisting of a base metal (BM) and a weld metal (WM), possess different mechanical and creep properties. Furthermore, the interface region between BM and WM show considerable microstructure/property gradients that extends from the fusion region of the weld metal through the HAZ of the BM. Hence, defect assessment of weldments of the properties of the constituent microstructural resistance to crack growth.

In high temperature plant components exercising primary (applied stress) and/or secondary requires (residual stress) loading the damage is dominated by creep mechanisms [1]. The weldments with microstructural gradients exercise complex distribution of stress and determination temperature. However, the creep properties from uniaxial testing under isothermal conditions are used for crack growth and defect assessment of components. Hence, the prerequisite for a reliable defect assessment is that the laboratory data are meaningful [2].

The grade P91 steel offers good high temperature properties and oxidation resistance that qualifies it for use in new plants and the upgrade of existing plants. For industrial applications good weldment creep properties are essential along with adequate fracture toughness. Therefore, welding method, procedure and consumables need to be selected such that toughness and creep properties are optimised to avoid problems during pressure testing, start-up and shut-down without loss in creep resistance to avoid problems during high temperature operation. The butt weld (BW) and electron beam weld (EBW) materials were studied in the reported work to address these aspects.

The testing and analysis techniques for welds require consideration of homogeneous microstructure, limited ductility and smaller deformation zone size associated with defect incubation and propagation. In the case of HAZ the change of microstructure in the range of fine to coarse grain size require notch positioning inside this region affect the crack growth behaviour of HAZ material. At present, guidelines are not available for the consideration and treatment of these aspects. Current approaches for predicting creep crack growth in weldments are based entirely on concepts developed for homogeneous materials [3]. The crack growth rate is correlated with $C^*(t)$ in the extensive creep regime according to ASTM E1457 [4], the only available standard for creep crack growth testing of metallic materials. Linear elastic and elastic-plastic analyses of bimaterial interface cracks show that the crack tip fields, and shapes and sizes of plastic zones are different for homogeneous bodies and bimaterials. Also, even though the nominal loading is Mode I, the interface experiences a combined Mode I and Mode II loading. Hence, a rigorous treatment of elastic-plastic fracture and creep crack growth in welded structures is beyond the current capability of fracture mechanics. Consequently, somewhat ad hoc correlation procedures are presently available, although without rigorous justification [3]. Considerable progress has been made in the determination, analytical representation and application of creep crack growth (CCG) property data during the past three decades [5-11]. The recent experimental work [4,5,12] clearly demonstrated the promise of such approaches. However, experimental data [12] and analytical evidence [13,14] has shown that $C^*(t)$ is not suitable for characterising creep crack growth behaviour in materials with low ductility in which the crack tip can advance at a rate comparable to the creep zone expansion rate. This will have direct implementation in characterisation of weldments and directs attention to the applicability of the available standard in weldment testing and characterisation along with crack tip parameters that correlate data in both creep ductile and creep brittle materials. An additional issue is arised in treatment of early crack growth data [1,11]. The crack tip parameters include C_t parameter which is proposed to correlate data in small scale creep to extensive creep regimes and local crack tip opening displacement (CTOD) rate that potentially may also include transition tails in crack growth correlations [6,15,20].

The present paper reports on a systematic work on Grade P91 steel weldments produced by both BW of pipes and EBW plates. The crack growth tests were carried out on compact tension (CT) type specimens at 600°C. The sharp starter slit notches were introduced by electric discharge method (EDM) in test locations in WM and HAZ. The crack initiation and growth data is analysed and correlated with crack tip parameters. The technical and industrial need for guidelines for testing and data analysis of weldments is addressed.

EXPERIMENTAL PROCEDURE

<u>Experimental material</u>: ASTM A387 Grade P91 modified 9Cr1Mo steel is a high strength, high ductility martensitic steel that offers good high temperature properties, oxidation resistance and resistance to hot hydrogen attack [16]. It finds applications in pressure parts in conventional and nuclear power plant and petrochemical reactor vessels and pipework.

BW pipes were produced by circumferential butt welding of pipes of 295mm outer diameter and of 55mm wall thickness, using shielded metal arc welding (SMAW) process. Post weld heat treatment (PWHT) on the welded pipes was performed at 760°C for 2h, followed by air cooling in still air. Metallographic sample sections showed a good quality weld with refined microstructure and little evidence of coarse grains of the weld fusion line that sometimes observed in industrial welds [17].

The EBW plates were produced using BM P91 steel that was received in reaustenitised (1050°C/1/2h/Air Cooling) and tempered (800°C/1h/AC) condition. Single pass full penetration EBWs were produced by Siemens using an electron gun with a capacity of 150kV, 19.5kW beam power, welding speed of 4mm/sec. and without filler material.

The simulated BW HAZ material used for mechanical and creep tests, was produced by thermal cycling in a Gleeble weld simulator machine. The simulated material had the microstructure and hardness of real weld HAZ's. An analytical method [18,19] was used in HAZ simulation for the calculation of the time between 800°C and 500°C during welding. This time is considered critical for the resulting microstructure and hardness of the material.

<u>Specimens</u>: Standard tensile and creep test specimens were machined from the welded material (both BW pipe and EBW plate) and from simulated HAZ material. Crack growth test specimens were machined from the weldments with starter notches introduced by EDM in WM and HAZ (i.e. in the centre of HAZ, WM and TypeIV region). The CCG specimens of type CT25 (W=25mm) and CT50 (W=50mm) of BW pipes and EBW plates, respectively, were side grooved 20% after EDM notching to $a_o/W=0.5$. The notch tip radius of 0.05mm served as a sharp starter crack.

<u>Testing</u>: Mechanical and uniaxial creep rupture tests were carried out at 600°C. Creep crack growth tests were done at the same temperature as mechanical and creep rupture tests, under constant load. The applied constant loads were calculated from predetermined initial K levels for the given test durations. Target test durations were 500h, 2000h and 4000h, with the main concern of acquisition of valid creep crack growth data.

Direct current potential drop (DCPD) method was used for crack size monitoring during testing. The displacement was directly measured on the load line and at the initial crack tip, (i.e. local CTOD, δ_5), measured on the specimen surface, 5mm apart across the main crack plane [5,20] using laser scanner system [21].

<u>Data Analysis</u>

<u>Determination of crack length</u>: The crack length was determined from the DCPD and using the Johnson's formula, for the lead positions on the specimen as described in ASTM E1457 Annex A1.1 [4].

The predicted crack extension, Δa_{pf}, was calculated by subtracting the initial crack length, a_o, from the predicted value of the final crack length, a_{pf}. The initial crack length, a_o, and the final crack size, a_f, are calculated from the measurements made on the fracture surface at nine equally spaced points centered on the specimen mid-thickness line.

Validation of Test: The ASTM E1457 [4] is followed for data validation for further processing, i.e.:

$$0.85 \leq (\Delta\, a_{pf}/(a_f - a_o)) \leq 1.15 \tag{1}$$

However, if the above is not satisfied the difference between the predicted and measured crack growth was noted and the data was further processed.

Determination of crack growth and displacement (load-line and crack tip) rate: The crack length, load-line deflection, local CTOD and time data were processed to determine creep crack growth and displacement rates using secant method [4]. The time and load-line displacement were set to zero at initial crack size a_o. Subsequent data points were chosen consisting of crack length and the corresponding load-line displacement and time such that crack extension between successive data points is 0.005W or less. For small crack growth smaller Δa values were chosen such that minimum nine successive rate data points will be determined for the total crack growth range.

Determination of C*(t)-integral: From the recorded data the magnitude of C*(t)-integral was determined at each point following the test standard ASTM E1457 [4].

Validity Criteria: Only data for which the test time is higher than transition times, t_T, are taken for valid for C*(t) correlation.

$$t_T = \frac{K^2(1-v^2)}{E(n+1)C*t_T} \tag{2}$$

Although the crack growth rate data that meet the displacement rate requirement of $\dot{V}c/\dot{V} \geq 0.8$ are considered valid, the data with $0.5 \leq \dot{V}c/\dot{V} < 0.8$ were plotted separately for further consideration.

The crack growth data obtained prior to the initial 0.2mm and 0.5mm crack extension, that is alternatively defined as the transient creep crack growth, is identified separately and correlated with C*(t), K and local CTOD.

The amount of crack deviation where crack growth path oriented at an angle, θ, to the main crack plane [22] and the amount of accumulated load-line deflection at the end of the test [4] were noted.

RESULTS

Materials: The chemical composition of the experimental materials, and the mechanical and creep properties determined at 600°C are given in Table 1 and Table 2, respectively.

Table:1.Chemical composition of Butt Weld and Electron Beam Weld P91 steels.

Material	C	Mn	Si	P	S	Cr	Ni	Mo	V	Nb	Cu
BW-BM	0.091	0.409	0.369	0.028	0.013	8.44	0.272	0.922	0.24	-	0.04
BW-WM	0.087	0.692	0.285	0.013	0.007	9.39	0.63	0.98	0.267	0.04	0.64
EBW-BM	0.096	0.384	0.244	0.011	0.001	8.4	0.146	0.943	0.203	0.082	0.098

Table: 2. P91 Materials data determined in tensile and creep tests at 600°C.

Material	$\sigma_{0.2}$ MPa	σ_{US} MPa	E GPa	D_1	m	A_1	n
BW-BM	441	463	164	0.0018	27.73	1.57E-45	18.51
BW-WM	362	385	125	0.0015	23.86	5.99E-24	8.55
EBW-BM	294	329	156	0,0019	13,97	1,02E-29	12,40
BW-SIM. HAZ Type IV	320	333	155	0.0016	17.38	7.16E-35	14.35
BW-SIM. HAZ Centre	293	317	139	0.0016	20.74	7.16E-35	14.09

Note: BW-BM data were taken from EC HIDA Project (BE 1702) that used the same experimental material [21].

The yield strength of WM and particularly in HAZ-Mid. section (centre) is lower than that of the BW-BM. Therefore, the welds are undermatched in terms of yield strength. However, note that the creep resistance of material rather than the yield strength determine the crack growth behaviour under creep conditions.

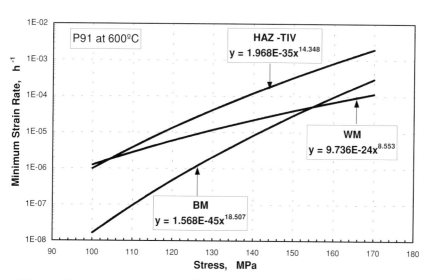

Figure 1: Steady-state creep rate as a function of stress for BW P91 at 600°C.

The creep properties were determined at 600°C as depicted in Table 2. It is seen in the table that the variation in creep properties does not follow the same sequence as the yield strength data. The creep exponent, n, is the lowest for the BW-WM whereas EBW-WM has the lowest m value determined in tensile tests. The creep data of BW material is shown in Figure 1. The change in the creep resistance with applied stress is seen which is an important aspect of creep crack tip behaviour.

Metallography: The CCG test specimens were sectioned in the mid-thickness after testing and prepared for optical microscopy both in as polished and etched conditions. Figures 2a and b show micrographs of the specimen with starter notch in the WM and HAZ in EBW and BW specimens, respectively. The crack growth and secondary crack formation in HAZ (Type IV crack) in EBW and fusion line (FL) crack in BW are seen. The microstructure and damage at the location of starter sharp crack (EDM Slit) and along the crack growth path is examined in etched specimens. Extensive damage at the crack tip is seen in Fig.2b.

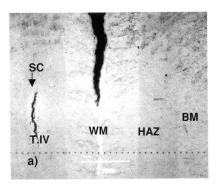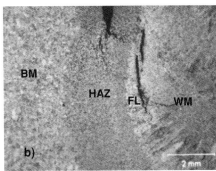

Figure 2: Optical micrographs of tested specimens sectioned and etched to show starter crack tip zone and secondary cracking (SC) a) in Type IV HAZ with microhardness indentations in EBW, and b) at FL with extensive damage in HAZ in BW specimen.

Figure 3 shows fracture surface of a specimen with crack front irregularities and unbroken ligaments seen as unoxidised islands when broken open at room temperature. These examinations direct attention to the problems associated with the use of the DCPD method for crack growth monitoring and crack length determination at high temperatures.

Figure 3: Fracture surface of a tested specimen of BW P91 with starter crack in WM. Unbroken ligaments are seen light on the oxidised fracture surface.

Data Assessment and CCG Correlations: The complete experimental data set was assessed and the CCG rate of both EBW and BW materials was correlated with crack tip parameter C*(t) in Figure 4. A good correlation is seen, particularly at early crack growth following the transition range. The scatter is high in tails and in later stage of crack growth of EBW material. The specimen BW-PH2 had a crack growth initiation at a location on fusion line away from the starter crack tip, therefore invalid. This specimen is included to show the crack growth behaviour at secondary crack that is the case in most of the large size EBW specimens (Fig.2a) examined in the present work.

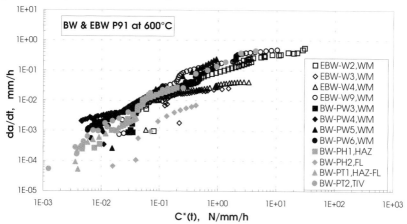

Figure 4: Creep crack growth rate as a function of C*(t) for EBW and BW.
Complete data set without reduction.

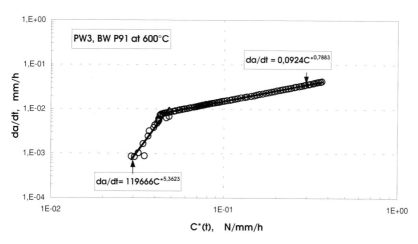

Figure 5: Creep crack growth rate as a function of C*(t) for specimen BW-PW3, starter crack in WM. Complete data set without reduction.

The crack growth initiation and early crack growth that includes the tails in crack growth rate correlations is shown for the specimen BW-PW3 in Figures 5 and 6. As distinct tails and transition to steady state crack growth is noted. The initial crack growth data of all specimens with data $\Delta a<0.5$mm is correlated with $C^*(t)$, K and local CTOD rate in Figures 7a, b and c, respectively. Despite the scatter in initial data the potential of the parameters to correlate early crack growth is seen. The scatter in K correlation is due to test methodology where the specimens were tested with different initial K_0 values.

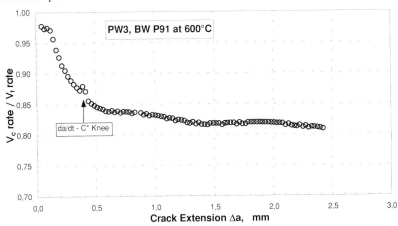

Figure 6: Ratio of load-line deflection rate due to creep (dVc/dt) to total load-line deflection rate (dVt/dt) as a function of crack extension in WM for specimen BW-PW3.

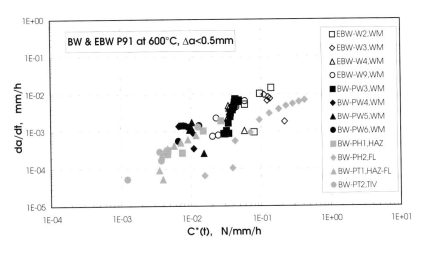

Figure 7a: Creep crack growth rate as a function of a) $C^*(t)$ for EBW and BW P91. Complete initial crack growth data set of $\Delta a<0.5$mm.

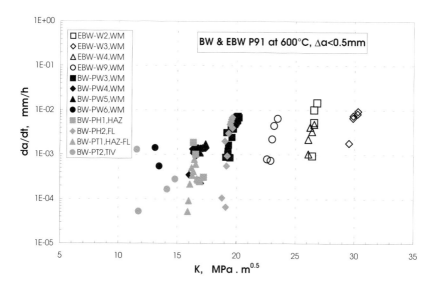

Figure 7b: Creep crack growth rate as a function of K for EBW and BW.
Complete initial crack growth data set of Δa<0.5mm.

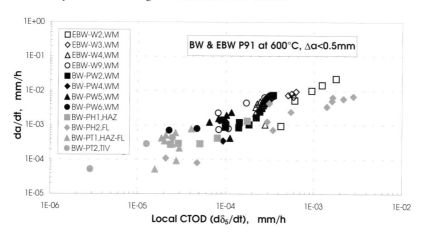

Figure 7c: Creep crack growth rate as a function of local CTOD for EBW and BW P91
Complete initial crack growth data set of Δa<0.5mm.

For structural assessment, the time for crack growth initiation is determined. Therefore, the C*(t) is presented as a function of time needed for crack initiation in specimens defined at Δa=0.2mm and 0.5mm in Figure 8. The scatter in data is high reflecting the different loading, specimen size and weld type used for P91 steel. Linear fit to the data for time needed for 0.2mm and 0.5mm crack growth is shown in the figure. A set of 2.25CrMo data taken from literature [1] is also included in the figure for comparison. It is important to note that P91 may be considered as replacement for 2.25CrMo steel in industrial applications.

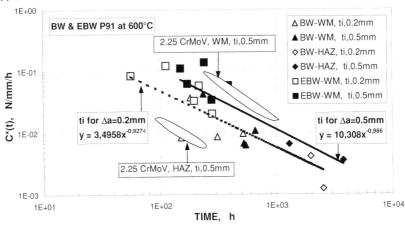

Figure 8: C*(t) as a function of test time for Δa=0.2mm and 0.5mm for EBW and BW P91. The 2.25CrMo data [1] bands are included for comparison.

DISCUSSION

The reliability of the mechanical and creep data of materials is of atmost importance for the elevated temperature fracture mechanical characterisation and assessment of service performance of welded joints with microstructural and property gradients. Therefore, the need for cross-weld creep data for assessment of fracture data is supported by the observations (Figures 2a and b) where crack initiation and growth span a wide range of microstructures. With introduction of advanced joining techniques such as EBW, machining of standard tensile and creep specimens from narrow WM and HAZ may not be feasible. As in the present study the weak base metal properties or available data, e.g. BW data, may be used for fracture assessments.

The minimum strain rate of weldment components vary with stress, such that BW-WM is the weakest, with highest minimum creep rate at 100MPa whereas it is the most creep resistant at 170MPa (Fig.1). In order to elaborate on the strength and deformation behaviour of weldments microhardness measurements were made on tested specimens. Average hardness, HV0.2, was the highest in WM of both materials (EBW:265, BW:251), followed by HAZ (EBW:227, BW:230), and lowest in BM (EBW:205, BW:209). Hence, the crack tip deformation, secondary cracking and crack path with crack deviation is controversial in EBW and BW materials (Fig.2a and b) if judged based on strength data. Therefore, creep resistant or creep weak bond definition of welds rather than relative yield strength values in terms of overmatch-undermatch is recommended in addressing the weld behaviour under creep conditions.

The calibration and correction of the predicted crack lengths with final crack length measured on fracture surface is important. The unbroken ligaments need to be considered fractured if left behind the advancing irregular crack front. Note that the validity criteria as in Eq.(1) need to be modified for the assessment of weldments to account for the difference in crack length measurements due to the crack front irregularities and crack tunneling.

The complete CCG rate data set correlate well with $C^*(t)$ despite different weldments and specimen sizes with initial K_0 values were tested. The scatter in the higher crack growth rate range may be attributed to crack deviation in EBW specimens with increased elastic and plastic component of displacement that correlates better with J-integral [5].

The early crack growth need to be addressed due to the scatter in correlated data and its importance in defect assessment of components. Therefore, each specimen may be examined for tails that includes crack growth initiation and early crack growth. The early crack growth rate correlates also well with $C^*(t)$ where the rate is higher than that of the steady state crack growth rate (Fig.5). That is directly related to the crack tip deformation processes in terms of creep and elastic-plastic components of displacement rates (Figure 6) which agrees with the ductile material behaviour of the numerical work reported [9]. The creep component of displacement reaches a minimum at transition knee and remains almost constant thereafter, at a value of larger than 0.8.

The crack growth behaviour at lower crack growth rates is of industrial interest because a large portion of a component life may be spent in the initiation and early crack growth regime. Due to the importance given to the crack growth initiation, the early crack growth data from all specimens is correlated with $C^*(t)$, K and local CTOD. Note that ASTM Standard E1457 [4] does not address testing weldments and only $C^*(t)$ parameter is recommended. A good correlation of data with $C^*(t)$ is seen if the first two data points (i.e. $\Delta a=0.2mm$) and invalid specimen data BW-PH2 is omitted (Fig.7a). Similarly, good correlation is also seen with local CTOD rate, a directly measured and not a calculated value. The scatter in early crack growth rate correlation with K is attributed to the test method. The predetermined K_0 values were taken for determining the test loads to obtain data with predetermined test durations. However, the good correlation of individual specimens directs attention to the importance of testing methodology.

The material behaviour and crack tip parameter are studied for structural assessment where the time for crack growth initiation is determined. Therefore, the $C^*(t)$ is presented as a function of time for a defined crack initiation of $\Delta a=0.2mm$ and 0.5mm (Figure 8). The data falls in three distinct regions. EBW-WM is in the upper region of $C^*(t)$-time plot, followed by BW-WM and BW-HAZ with decreasing $C^*(t)$ and increasing time. Despite the fact that the data scatter is high, the data for time needed for 0.2mm and 0.5mm crack growth is linear fitted. A set of 2.25CrMo data taken from literature [1] is also included on the figure for comparison. It is important to note that P91 may be considered as replacement for 2.25CrMo steel in industrial applications. Therefore the comparison is helpful for material developers and designers alike. EBW-WM is not inferior to the 2.25CrMo-WM steel. The data of BW-HAZ that is failure critical zone is superior to the 2.25CrMo-HAZ with one order of magnitude in terms of time.

CONCLUSIONS

The reported work is undertaken to address the technical and industrial need for guidelines for CCG testing and data analysis of weldments. Grade P91 steel weldments produced by both BW of pipes and EBW plates were taken for crack growth testing of CT specimens at 600°C.

• Slit notches introduced by EDM in WM and HAZ serve as sharp starter cracks.

• Tests must be terminated prior to final fracture to determine final crack length on the fracture surface.

- The CCG data are assessed based on the metallographic information on the starter notch location (WM, HAZ) and crack growth path.
- Crack growth initiates and grows at a strained and creep weak zone regardless the introduced starter crack tip location.
- The CCG rate data correlates well with the $C^*(t)$ provided the crack initiates at the starter crack and grows on the main crack plane. Out of plane crack growth data is invalid. It must be corrected for the crack tip stress distribution if such data is needed for component assessment.
- Crack growth initiation is defined at $\Delta a = 0.2$mm.
- Transition tails are defined at $\Delta a < 0.5$mm of early crack growth. This data should not be discarded due to its importance for engineering assessments.

ACKNOWLEDMENTS

The authors would like to thank the European Commission for the financial contribution and the partners for their contribution to the EC Project 'SOTA': SMT 2070 - 'Development of Creep Crack Growth Testing and Data Analysis Procedures for Welds'. The partners> ENEL (Italy), ERA Technology (UK), Iberdrola (Spain), JRC-Petten (Netherland), EDP/ISQ (Portugal), SAQ Kontroll/SIMR (Sweden).

REFERENCES

[1] Holdsworth, S., 1998, Proc. Int. Conf. On Integrity of High Temperature Welds, Org. by IOM Comm. and I Mech E, Professional Eng. Publ. Ltd. U.K., pp.155-166.

[2] Parker, J.D., 1998, Proc. Int. Conf. On Integrity of High Temperature Welds, Org. by IOM Comm. and I Mech E, Professional Eng. Publ. Ltd. U.K., pp.143-152.

[3] Saxena, A., 1988, Nonlinear Fracture Mechanics for Engineers, CRC Press LLC, Boca Raton FL, USA, pp.384-389.

[4] ASTM E1457-98, 1998, Standard Test Method for Measurement of Creep Crack Growth Rates in Metals, ASTM 03.01, ASTM, Philadelphia, PA 19103, USA.

[5] Dogan, B., Saxena, A. and Schwalbe, K.-H., 1992, Materials at High Temperatures, Vol.10, No.2, May, pp.138-143.

[6] Saxena, A. and Yokobori, T., Eds., 1999, Engineering Fracture Mechanics, Special Issue on Crack Growth in Creep-Brittle Materials, Vol.62, No.1.

[7] Ainsworth, R.A., 1982, The initiation of creep crack growth, Int J. Solids Structures, Vol.18, No.10, pp.873-881.

[8] Neate, G.J., 1986, Creep crack growth behaviour in 0.5CrMoV steel at 838K, Mat Sci. Eng., Vol.82, Part I: Behaviour at a constant load, pp.59-76. Part II: Behaviour under displacement controlled loading, pp.77-84.

[9] Saxena, A., 1986, Creep crack growth under non steady-state conditions, ASTM STP905, pp.185-201.

[10] Riedel, H., 1987, Fracture at High Temperatures, MRE-Springer Verlag.

[11] Holdsworth, S.R., 1992, Initiation and early growth of creep cracks from pre-existing defects, Materials at High Temperatures, Vol.10, No.2, pp.127-137.

[12] Dogan, B., 2000, Proc.Int.Conf. 'Life Assessment of Hot Section Gas Turbine Components', 5-7 Oct.1999, Edinburgh, U.K., IOM Communications Ltd., Book B731, Eds.: R.Townsend et.al., pp.209-228.

[13] Riedel, H. and Detampel, V., 1988, International Journal of Fracture, Vol.36, pp.275-289.

[14] Hawk, D.E. and Bassani, J.L., 1986, Journal of the Mechanics and Physics of Solids, Vol.34, No.3, pp.191-212.

[15] Saxena, A and Liaw, P.K., 1986, "Remaining Life Assesment of Boiler Pressure Parts-Crack Growth Studies", EPRI CS 4688, Electric Power Research Institute, Palo Alto, CA, USA.

[16] Blondeau, R, Bocquet, P. and Cheviet, A., 1990, "New alloys for pressure vessels and piping", Pro. Conf. Pressure Vessels and Piping, Nashville, Tennessee, USA, ASME Publ., pp.49-53.

[17] Dogan, B. and Petrovski, B., 2000, Int. Conf. HIDA-2, MPA-Stuttgart, 04-06.10.2000, Stuttgart, Germany, to be published in a special issue of Int. J.of Pressure Vessels and Piping.

[18] Stahl-Eisen-Werkstoffblatt 088 1. Ausgabe.

[19] AWS Welding Handbook, Vol.1, p.86.

[20] Saxena, A., Dogan, B. and Schwalbe, K.-H., 1994, Evaluation of the Relationship Between C*, δ_5 and δ_t during Creep Crack Growth, ASTM STP 1207, ASTM, Philadelphia, pp.510-526.

[21] Dogan, B., Martens, H., Blom, K.-H., and Schwalbe, K.-H., 1989, Proc. Int. Conf. LASER 5, IITT Int. Conf., 10-11 April 1989, London, pp.188-194.

[22] ASTM E647-95, Standard Test Method for Measurement of Fatigue Crack Growth Rates, ASTM, Philadelphia, PA 19103, USA.

[23] EC Project SOTA: SMT 2070. Development of Creep Crack Growth Testing and Data Analysis Procedures for Welds.

Creep and fatigue crack growth in P91 weldments

I A Shibli, European Technology Development, Surrey, UK.
N Le Mat Hamata, ERA Technology, UK.

ABSTRACT

Creep and creep-fatigue crack growth tests on the high strength pressure vessel steel P91 (9CrMoVNb) were carried out in the European Commission supported project 'HIDA'. These tests were conducted at 625°C using compact tension (CT) and single edge notch tension (SENT) specimens and large size seam and butt welded pipes with starter notches/defects machined in the HAZ and the base metal constituents. The work has shown that the HAZ region in P91 could be vulnerable to Type IV cracking resulting in large stress reduction factors. This finding is supported by parallel research work carried out elsewhere at lower (service) temperatures.

The use of 9Cr type high strength martensitic steels was first introduced on the basis of their superior base metal design strength. However, the weaker Type IV position does not necessarily support this position with regards to the welded components. Furthermore, adverse effect of low cycle fatigue (with large hold times) on crack growth was observed in feature specimen tests. This can have implications for the so called 'two shifting' operation of power plant which is now becoming more common due to the privatisation and competition in electricity generating industry. The reasons for this and the vulnerability of Type IV position are discussed. Although this work was limited in nature it nevertheless shows the need to evaluate the behaviour of the weldments of this type of steel at service temperatures.

1. INTRODUCTION

A number of steels were tested in the European Commission supported project BE1702 'HIDA'. This paper deals with the steel P91 which is a martensitic steel used as a replacement header and pipe work material in the existing boilers and as high strength steel in the new ultra supercritical boilers. The P91 replacement headers in the existing boilers are used at relatively lower temperatures of 540 to 570°C but in the new ultra-supercritical plant this use could be in the region of 600°C. P91 was introduced as a new high strength material on the basis of its superior base metal creep rupture strength. However, as is well known, most components in boilers fail due to problems associated with welds. So it is now opportune to look at the performance of weldments (i.e. weld metals and associated heat affected zone) in this steel.

Due to competition and privatisation of electricity generating industry world over cycling of power plant to meet customer demand at a short notice is now fast becoming common. This can have serious implications for larger power plants with large headers built for steady load operation, as cycling can induce damage to these components. In terms of cyclic damage, the advantage of P91 over traditionally used ferritic steels such as P22 has been that because of its high strength components can be built in smaller wall thickness and this has been considered to be advantageous from the viewpoint of plant cycling induced fatigue damage, amongst many other benefits. Again, the comparison so far has been based on the base metal performance.

This paper discusses the performance of P91 weldments from the viewpoint of creep and fatigue crack growth as investigated in the project HIDA and supports this with evidence from the work carried out elsewhere on different casts of steel and welds.

2. TEST MATERIAL

All laboratory and feature test specimens were machined from the same cast of material. The pipes were supplied by Mannessman to ASME P91 specification. . The chemical composition and tensile properties are given in Tables 1 and 2 and nominal pipe dimensions are shown in Table 3.

Longitudinal and circumferential welds were made using the Shielded Metal Arc Welding (SMAW) process, except for the root run. The weld was 20 to 30 mm wide.

Table 1: Chemical composition of the test material

Materl	C	Mn	Si	P	S	Cr	Ni	Mo	V	Nb	Ni+Cu
Base	0.091	0.409	0.369	0.028	0.013	8.44	0.272	0.922	0.24		0.04
Weld	0.087	0.692	0.285	0.013	0.007	9.39	0.63	0.98	0.267	0.04	0.64

Table 2: Tensile properties

Source	Temp.	0.2PS (MPa)	UTS (MPa)	A (%)
ASTM (min)	Room Temp.	415	585	20
Test Pipe (Mannessmann)	Room Temp.	505	674	26
		508	671	28

Table 3: Nominal dimensions of the test pipes, including the initiation defect size

	Outer Diameter OD (mm)	285
Specimen Dimensions		
	Inner Diameter ID (mm)	225
	Wall Thickness t (mm)	20
	Overall Length (mm)	600
Defect Dimensions		
Defect 'D' in HAZ	a x 2c (mm)	8x40
Defect 'E' in Base	a x 2c (mm)	8x40
Defect 'F' in Base	a x 2c (mm)	5x25

3. EXPERIMENTAL WORK

All tests were conducted at 625°C.

Uniaxial creep rupture tests were carried out on base metal, weld metal and cross weld (containing base and weld metal and HAZ) specimens. For brevity this paper describes the results of the cross weld specimens only. These specimens were taken out of both the seam and butt welded pipes and are respectively identified as transverse (T) and longitudinal (L) specimens in Table 4. Table 5 gives the rupture location which will be discussed later.

Table 4: Creep rupture results for cross-weld specimens

Orientation	Stress (MPa)	Test duration (h)	Elongation (%)
T	51/80*	9713/10492**	
T	80	1996	2.7
L	100	323	2.8
L	60	3906	

* 51 MPa then 80 MPa (51 MPa was due to an error of loading)
** cumulated duration. The specimen remained 779 hours under 80 MPa.

Table 5: Rupture location in the cross-weld specimens

Total HAZ width* (mm)	HAZ width between the crack and the weld* (mm)	HAZ width between the crack and the unaffected base metal* (mm)
2.0	2.0	≈ 0
2.3	1.9	0.4
1.9	1.5	0.4
2.0	1.5	0.5

* average value computed from 5 measurements.

Creep and fatigue crack growth tests were carried out on Compact Tension (CT) specimens (of two sizes, W=25 and 50 mm) and on Single Edge Notch Tension (SENT) specimens. Tests both under static and cyclic loads (frequency range of 0.001 to 0.1 Hz) were conducted. Details of laboratory testing have been described elsewhere [1]. It suffices to state here that the tests were performed according to ASTM E1457 high temperature CCG testing standard [2].

In the case of feature tests, all pipes contained three elliptical shape initiating defects introduced by electrical discharge machining (edm) [3], the notch root radius being 0.05mm. (EDM was also the technique used for CT and SENT laboratory specimens). Two notches, one machined in the base metal and one in the centre of the HAZ, were of the same size and had been machined to compare the behaviour of the base and weld metal constituents, while the third notch was a shallow notch and had been machined in the base metal to study the metallography of crack initiation, Table 3.

Two tests were conducted on <u>seam welded pipes</u> under internal gas pressure. One pipe was tested under steady load and the second under load cycling conditions. The load cycling component was introduced by cycling the pressure eight times in twenty-four hours giving a frequency of 10^{-4} Hz. Full details of these tests are given in [3] and [4]. The test information together with the results (to be discussed later) is given in Tables 6 and 7.

Table 6: Results of CCG tests on axially-notched pipes with longitudinal seam welds

Pipe	Internal Pressure (Mpa)	Frequency (Hz)	Duration (hrs) Cycles	Crack extension (mm) (for Defect D* in HAZ)	Failure mode
SP-P1	15	0	1430 (hrs)	12.55	Type IV
CP-P2	0 - 11	10^{-4}	5550 (hrs) 1850 cycles	9.06	Type IV

*Initial defect depth $a_0 = 8mm$.

Table 7: Crack extension in the HAZ and base material in the seam welded pipes

Test designation	SP-P1		CP-P2	
Defect	D (HAZ)	E (BM)	D (HAZ)	E (BM)
Crack Extension (mm)	12.55	0.66	9.06	1.2

Two tests were conducted on <u>butt welded pipes</u> under internal gas pressure and four point bending mode. One pipe was tested under steady load and the second under load cycling conditions, Table 8. The cycling component was introduced by cycling the bending moment on the pipes while pressure in the pipes remained steady. The test frequency used was 10^{-3} Hz.

Table 8: Details of tests on butt welded pipes

Specimen No	Test type	Internal pressure (MPa)	External load F, kN	External load Δ, kN	Frequency Hz	Duration h	Crack growth
P91/P1	Steady load	20	96	-	-	3648 h	No initiation
		20	120	-		1440 h*	
P91/P2	Cyclic load	20	-	96	10^{-3}	3648 h 12740 cycles -	3 mm growth
		20	-	120	10^{-3}	1440 h* 4634 cycles*	

* Time/ cycles at increased load.

4. RESULTS

4.1. Laboratory Specimen Tests

In the case of <u>uniaxial creep rupture</u> tests the cracking in the cross-weld specimens was observed to be in the Type IV position, Table 5, and the ductility was very poor, 2.7 %, Table 4. The low ductility can be attributed to the Type IV region, as the ductility for the tests on the base metal and the all-weld metal specimens obtained from the same pipes and at the same temperature was found to be in the region of 8 to 24% respectively [5]. The cross-weld specimens also showed very low creep rupture strength, exhibiting a strength drop of between 30 to 40% [5].

In the case of <u>creep and fatigue crack growth tests</u> the cracking rate was correlated with C* for the CT and the SENT specimens, Figure 1. For cyclic tests only data for low frequencies (up to 0.01 Hz) have been presented in this paper. It is clear that, within the typical range of scatter for cracking data, no specimen size or geometry effect is observed both for base and HAZ material conditions tested under static and low frequency cyclic loading. No effect of low frequency cycling was observed either.

However, the crack growth rate in the HAZ was higher by a factor of about 5 compared with the base metal. The respective best fit lines are shown in Figure 1.

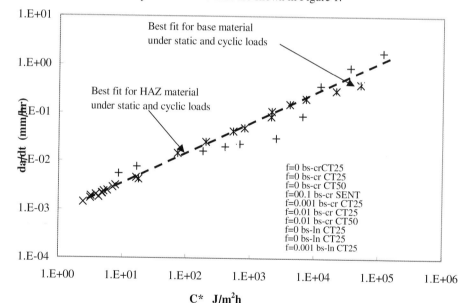

Fig. 1: Creep and fatigue crack growth rates for the laboratory specimen tests

(**Note:** CT25 = CT specimen W=25mm, CT50= CT specimen W=50mm, SENT=single edge notch tension specimen, f=frequency in Hz, bs-cr and bs-lr are base metal specimens with notch orientations parallel respectively to pipe circumferential and longitudinal directions, HAZ-cr and HAZ-ln are specimens with initiation notch in the centre of the HAZ.)

4.2. Feature Test Results

4.2.1. Seam Welded Pipes

In this case crack growth through the HAZ was faster by a factor of 10 [4]. The maximum crack growth in the base and HAZ metal constituents is shown in Table 7. As the crack growth in the base metal was small for both the steady and cyclic load conditions only crack growth in the HAZ metal was analysed in terms of C* correlation.

The analysis for crack growth in the HAZ region is shown in Figure 2. This correlation of the component CCG data was found to fall within HIDA Task 2 scatter band as shown. The HIDA Task 2 scatter band was established within the project HIDA from the published and unpublished laboratory data accessed from organisations in various countries [6].

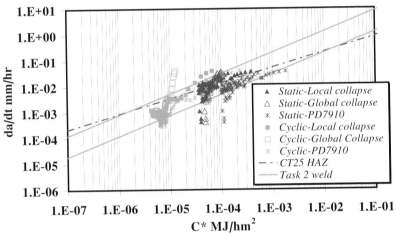

Fig. 2: Creep and fatigue crack growth rates in HAZ of the seam welded pipes (based on the minimum creep rate of the cross-weld specimen tests)

Three reference stress solutions - based on limit load concepts - were selected for the formulation of C*-Integral. Although the BS7910 formulation gave a better correlation with conventional data, the difference was small, implying that it is unimportant exactly how the stress state is characterised at the notch tip. Furthermore, both minimum and average creep rate laws were used to accommodate the strain rate in the C* relationship in the analysis. A better correlation was obtained when using the average creep rate law, confirming the extensive creep cavitation observed for tested pipes.

As can be seen in Figure 2, for a given reference stress solution, the P91 steel exhibits higher crack growth rates under cyclic loading.

In post-test metallography tip of the original notch was observed to be in the centre of the HAZ. However, crack growth was observed to occur along the fine grain HAZ/ Type IV position. This was observed in both the pipes (i.e. steady and cyclic load tests) and a typical case is shown in Figure 3.

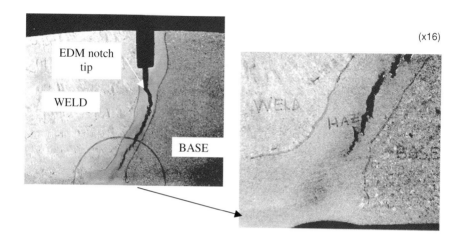

(x16)

Fig. 3. Seam welded pipe test – crack growth through the pipe wall thickness

4.2.2. Butt Welded Pipes

In this case, because of the tests running for longer durations full analysis of the data is being carried out only now and will be reported in future publications. Therefore only preliminary analysis will be reported here.

The comparison of the test results of the two four-point bending pipes, one tested under steady load and the second under cyclic load, showed crack growth of 3mm in the HAZ of the cyclic test i.e. pipe P91/P2. In contrast the steady load pipe P91/P1 did not even show crack initiation during the same testing time and same maximum loads, Table 8. However, differences in wall thickness between both pipes in the area of the starter notches were observed. The wall thicknesses were 29.9 mm for P91/P1 steady load test, and 25.5 mm for P91/P2 cyclic load test. The eccentricity of the pipe P91/P1 was much larger than that of P91/P2.

Although detailed metallography is in progress preliminary examination has revealed the cracking to be in the Type IV region, as was the case with the laboratory specimens and the seam welded pipes.

5. DISCUSSION

a) The feature tests showed the crack growth rate in the HAZ to be higher by a factor of 10, although the prediction from the laboratory specimens was a factor of 5 only. This can perhaps be attributed to the higher constraint situation in the feature tests and thus highlights the importance of conducting feature tests for validation purposes. This work has shown that Type IV position in P91 steel is much more vulnerable to crack growth than the base metal.

b) The finding with regard to the crack growth occurring close to the Type IV position in the laboratory specimen tests and in the tests on pipes, when the starter notch was placed in the centre of HAZ, is a cause for concern. This may imply that in-service situations P91 can be vulnerable to type IV cracking.

Recent work carried out elsewhere [7] on high alloy martensitic steels of 9-12%Cr type testing uniaxial creep rupture specimens has shown that type IV cracking in longer term tests could be a serious problem in these steels. As a result it has been shown that stress reduction factor due to this type of cracking in E911 steel, when tested at 625°C could be 30% for 8000 hour tests (similar to that reported in this paper for P91), which can be extrapolated to 50% for 100000 hours tests [7]. This reduction factor is very high and highlights the need to categorise P91 and other similar steels for their weldment behaviour.

It also throws light on the short sightedness of designing plant on the basis of the base metal strength and behaviour. It can perhaps be argued that 625°C is a high temperature and that P91 at present is only used at up to 600°C. However, work at 600°C has also shown the vulnerability of this steel to Type IV cracking [7a]. This steel is also being used at present as a replacement material in the older power plant at lower temperatures of 540 to 570°C. It can be therefore be argued that in the older plant the use of this steel will be safe. On the other hand, it can also be argued that in the presence of any manufacturing or service induced defects stress at the tip of the defect will relax at a much lower rate (due to lower creep) and therefore the risk of cracking at lower temperatures could be higher. Performance in this region is not known and therefore plans are now underway to investigate the behaviour of the weldments of these steels at service temperature range.

All this is contrary to the belief in some quarters that Type IV cracking is not expected to be a major problem in high strength steels as the strength loss at a test temperature decreases with increasing material strength [8].

c) Another aspect of concern is that the low cycle feature test in seam welded pipes showed higher crack growth rate than the steady load test. This inspite of the fact that the cyclic test sees shorter time at high stress than the steady load test. Similar tests carried out on P22 (2.25Cr1Mo) steel at its typical service temperature of 565°C showed slower crack growth in the cyclic tests [4]. Although tests on butt welded pipes are still being analysed in terms of C*, the preliminary indications, not withstanding the differences in pipe nominal wall thickness and eccentricity, are that the cyclic test is showing higher crack growth rate. As can be seen from Figure 1 this difference was not shown by tests on the standard laboratory specimens, which again shows the significance of conducting validation type tests using feature specimens.

Results from creep and creep fatigue crack growth tests on three materials tested within the project HIDA have been plotted in Figure 4. Although these data are limited they nevertheless for the first time have shown that the creep fatigue interaction is more severe for P91 than for P22 or AISI 316 stainless steel. The HIDA work has shown that the cross weld creep rupture ductility of P91 tests was only about 2.7% while that of cross weld P22 specimens tested at 565°C this was up to 8% [5]. Thus ductility exhaustion in P91 would be a contributory factor to the potential vulnerability of this steel to cracking due to plant cycling.

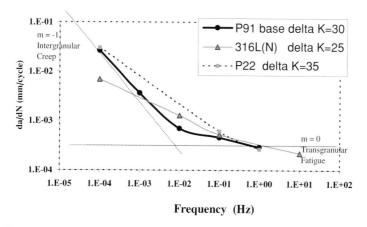

Frequency (Hz)

Fig 4.   Creep-fatigue crack growth curves for some of the high-temperature alloys (including P91)

The worry now is that, for example, in plant cyclic operation, which is now becoming common due to privatisation and is resulting in sever competition in the electricity generating industry, P91 could show Type IV cracking at a reasonably early stage in life. In the past P91 was considered to be advantageous steel for the plant cyclic operation as thinner wall and smaller component size were supposed to reduce thermal gradients and hence the adverse effect of cycling.

Furthermore, recent work of Tabuchi et. al. [9] has shown that $M_{23}C_6$ precipitates and lave phases form faster at the fine grain HAZ region in 9Cr martensitic type steels and this makes the Type IV position in these steels very vulnerable.

d) In terms of the plant experience, not much can be said with certainty at present because of the relatively short service experience of this type of steels. However, there has been one P91 Type IV failure in the UK after only 36000 hours of service at 568 °C ! [10]. Although this was attributed to the design of the endplate and the metallurgical condition of the forging, could it possibly be due to the higher vulnerability of P91 to Type IV cracking? Type IV position of P91 is known to be much softer than the base metal. Is it, therefore, possible that P91 and other 9-12Cr steels will start showing cracking problems at an early stage in life?

There is also an observation that in-plant oxidation of P91 could be high and this also makes this steel vulnerable [11, 12].

In the light of the new information emerging the question that we face is, "Are P91 and other 9Cr type martensitic steels still considered to be the wonder materials for use in high temperature plant?"

6. CONCLUSIONS

Creep and fatigue crack growth work on laboratory and feature test specimens of P91 steel weldments tested at $625^{\circ}C$ has shown the following:

1. P91 weldments are vulnerable to failure in the fine grain HAZ (Type IV). This can result in large stress reduction factors for this steel.
2. In the feature tests both creep and creep-fatigue crack growth rates were 10 times faster in the HAZ compared with the base metal.
3. Creep fatigue interaction has adverse effect on time dependent crack growth in this steel. Limited comparison has shown that this effect can be more severe than that in the traditional P22 (2.25Cr1Mo) steel used in power plant. This has implications for the cyclic operation of power plant now using P91 and other similar high strength martensitic steels.
4. This work has shown the importance of conducting feature tests which under higher constraint can show behaviour more severe than that predicted by testing laboratory size specimens.
5. These findings highlight the need to do comprehensive evaluation of P91 type high strength martensitic steels for their weldment behaviour at service temperatures..

7. ACKNOWLEDGEMENTS

This project HIDA was a joint venture between the following partners: ERA Technology (UK), MPA Stuttgart (Germany), EDF (France), DNV (Sweden), Imperial College (UK), CEA (France), Framatome (France), Petrogal (Portugal), Metsearch (Netherlands), ENEL (Italy), Siempelkamp (Germany). The financial support of the European Commission is gratefully acknowledged.

8. REFERENCES

1. M Tan, N J Cellard, K M Nikbin, G A Webster, 'Comparison of creep crack initiation and growth in four steels tested in the HIDA project', 2nd International HIDA Conference on Advances in Defects Assessment in High Temperature Plant, 4-6 Oct. 2000, MPA, Stuttgart, Germany.

2. ASTM E1457-92, Standard Test Method for Measurement of Creep Crack Growth Rates in Metals, ASTM, Philadelphia, PA 19103, USA.

3. N Le Mat Hamata , I A Shibli, Creep Crack Growth of Seam-welded P22 and P91 Pipes with Artificial Defects. Part I: Experimental Study and Post-Test Metallography, 2nd International HIDA Conference, Advances in Defects Assessment in High Temperature Plant, 4-6 Oct. 2000, MPA, Stuttgart, Germany.

4. N Le Mat Hamata, I A Shibli, Creep Crack Growth of Seam-welded P22 and P91 Pipes with Artificial- Part 2: 2nd International HIDA Conference, Advances in Defects Assessment in High Temperature Plant, 4-6 Oct. 2000, MPA, Stuttgart, Germany.

5. I A Shibli, HIDA Final Report.

6. Al-Abed B. HIDA Project Task 2 – Collation and Review of High Temperature Crack Growth Data from Research Experience. ERA Report 97-0586, Leatherhead, ERA Technology Ltd, July 1997.

7. Allen D J, Fleming A. 'Creep performance of similar and dissimilar E911 steel weldments for advanced high temperature plant', Published in the Proceedings of the 5th Charles Parsons 2000 Conference on 'Advance materials for 21st century turbines and power plant', 3-7 July, Churchill College, Cambridge, UK. Pp 276-290.

7a. I A Shibli, N Le Mat Hamata, 'Creep crack growth in P22 and P91 welds – overview from SOTA and HIDA projects', 2nd International HIDA Conference, Advances in Defects Assessment in High Temperature Plant, 4-6 Oct. 2000, MPA, Stuttgart, Germany.

8. J Hald, 'Integrity of high temperature ferritic steel weldments', paper presented at International conference on 'Integrity of High Temperature Welds', 3-4 November, 1998, Nottingham, UK. Organised by the Institute of Materials, London.

9. Masaaki Tabuchi, Takashi Watanabe, Kiyoshi Kubo, Masakazu Matsui, Junichi Kinugawa and Fujio Abe, 'Creep Crack Growth Behavior in HAZ of Weldments for W containing High Cr Steel', 2nd International HIDA Conference, Advances in Defects Assessment in High Temperature Plant, 4-6 Oct. 2000, MPA, Stuttgart, Germany.

10. Brett S J, Allen D J, Pacey J, Failure of a modified 9Cr header endplate, Proceedings of the International Symposium on case histories on Integrity and failures in Industry, Milan, Italy, 28 Sep. -1 Oct. 1999.

11. B Dooley, EPRI, Palo Ato, California, USA. Private Communication.

12. A Fleming, R V Maskell, L W Buchanan, T Wilson, 'Materials development for supercritical boiler and pipework', Published in Conference Proceedings of Materials Congress, UK, 1998, on session on 'Materials for high temperature power generation and process plant applications', Edited by A Strang.

Fracture analysis of individual wires in wire ropes for petroleum drilling

Cheng Liu[1], Zhenbo Zhao[1], Yunxu Liu[2] and Derek O. Northwood[3]

[1] Mechanical, Automotive & Materials Engineering, University of Windsor,
Ontario, Canada, N9B 3P4
E-mail: lcheng@uwindsor.ca ; zhao3@server.uwindsor.ca

[2] Materials Engineering Department, Jilin Institute of Technology,
Changchun, Jilin Province, China, 130012

[3] Engineering & Applied Science, Ryerson Polytechnic University, Toronto,
Ontario, Canada, M5B 2K3
E-mail: dnorthwo@acs.ryerson.ca

ABSTRACT The in-service fracture of individual wires in wire ropes (steel composition: 0.65wt%C, 0.50wt%Mn, 0.22wt%Si, 0.03wt%S and 0.035wt%P) for petroleum drilling was investigated by means of optical microscopy, SEM and XRD. Microstructural observations show that a white layer is formed on the surface of the individual wires and is composed primarily of abnormal quenched martensite with a hardness of HV1080. Crack initiation and propagation were found to occur in this white layer. Experimental results from a stimulated friction test indicate that the depth of a white layer increases with increase in normal load, sliding rate and friction time. The formation of the white layer includes austenitization of the steel surface during friction at a high sliding rate, and rapid cooling to a martensite structure. The martensite structure in the white layer exhibits a finer grain size and higher dislocation density than that obtained by normal quenching.

1. Introduction

Surface white layers have been observed in various materials under a variety of severe conditions, such as wear, ballistic impact, shear, machining and high velocity shaping [1-7]. A white layer is defined as a surface layer which appears featureless and white under optical microscopical observation following normal etching procedures. Usually, any white layer has a resistance to etching by conventional agents and a high hardness [8].

Because the white layer has a close relationship to fracture, it has become a focus of research in material behavior under dynamic loading since the 1970s [9-11]. Consequently, the formation mechanism of a white layer is still a matter of controversy [12-15]. Some investigators attributed the white-etching behavior in steels to a phase transformation from the parent phase to untempered martensite via reverse transformation to austenite. Some suggested that a white layer could be generated by plastic deformation, mechanical machining or chemical processes, and that the transformation due to the thermomechanical processing included carbide dissolution and dynamic recovery. In fact, white layers have been a focus of attention for the investigation of friction and wear since 1912 when Stead first identified them on the surface of steel wire ropes [16]. However, little work has been carried out on the study of microsructures of white layers on wire ropes.

In this paper, the microstructures of surface white layers on individual wires for petroleum drilling were investigated. The relationship between the depth of the white layers and normal load, sliding speed and friction time was examined using laboratory friction testing. The possible mechanisms for white layer formation are discussed.

2. Experimental Details

A total of 16 fractured individual wires in wire ropes for petroleum drilling were subjected to chemical analysis and microanalysis. The chemical composition is given in Table 1.

Table 1   Chemical composition of steel wires,   wt%

C	Mn	Si	S	P
0.64	0.50	0.22	0.030	0.035

SAE1065 steel samples (1.6mm diameter ×150mm length) with the same chemical composition as the individual wires were used in the stimulated friction test. A friction tester which simulated the conditions in petroleum drilling was used in the experiments (see Fig.1). The friction wheel, 100mm in diameter, was made from SAE1045 steel, which was quenched and tempered at a high temperature. In the test, the sliding speed and normal load can be adjusted by changing the rotation speed of the friction wheel and the weight, respectively.

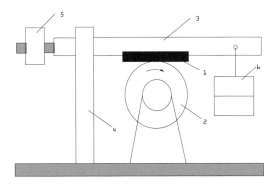

Fig.1   Schematic of friction tester used to study the formation of the white layer
1-sample, 2-friction wheel, 3-sample rack,
4-supporting stand, 5-balanced weight, 6-weight

Longitudinal and transverse sections were examined by optical microscopy and microhardness testing to identify the white layers in term of etching resistance (to 6%HF) and hardness, relative to the bulk material. Scanning electron microscope (SEM) was used to examine cross-sections of the individual in-service wires. The dislocation density, c/a ratio (c and a represent, respectively, the long and short unit cell dimensions of the hexagonal close-packed (HCP lattice)) and grain size were measured by X-ray diffraction analysis (the software used to calculate the dislocation density, c/a ratio and grain size was developed at Jilin University [17]). Tensile tests and bending tests were performed for sections of the wires that were distant from the fracture surface according to the testing standard of American Petroleum Institute (API) [18].

3. Results and discussion

The microstructure of transverse section of a wire rope which is far away from the fractured surface is shown in Fig.2, The structure consists of elongated fine pearlite and ferrite grains from the deep drawing.

Fig.2   Microstructure of transverse section of steel wire rope

The mechanical properties of 16 individual wires for petroleum drilling are shown in Table 2. Their strength and bending numbers are in accord with the API (American Petroleum Institute) Standard [18] for wire ropes.

Table 2   Mechanical properties of individual steel wires

number	diameter, mm	tensile strength, MPa	number of bending cycles to fracture (bend angle = 180°C)
1	1.62	1796	18
2	1.62	1893	19
3	1.59	1864	18
4	1.60	1900	20
5	1.59	1864	18
6	1.62	1741.5	18
7	1.63	1655	20
8	1.61	1699	17
9	1.63	1797	20
10	1.61	1891	20
11	1.63	1725	16
12	1.615	1756	19
13	1.62	1820	14
14	1.60	1860	19
15	1.615	1763	18
16	1.61	1866.5	19
average	1.61	1805.	18.3

A typical longitudinal section of a fractured steel wire exhibited a region of graded microstructure from the fractured surface to the bulk (see Fig.3(a)). A closer examination of areas A, B and C in Fig.3(a), showed a fine needle martensite microstructures (gray and white) in area A in Fig.3(b), 90% martensite (white) and 10% fine pearlite (black) microstructure in area B in Fig.3(c) and 65% martensite (white) and 35% fine pearlite network microstructure (black) in area C in Fig.3(d). The hardness of the three areas and the fractured surface is given in Table3. It is seen from Fig.3 and Table3 that the closer to the friction fractured surface, there is a higher martensite content of the microstructure and a higher hardness.

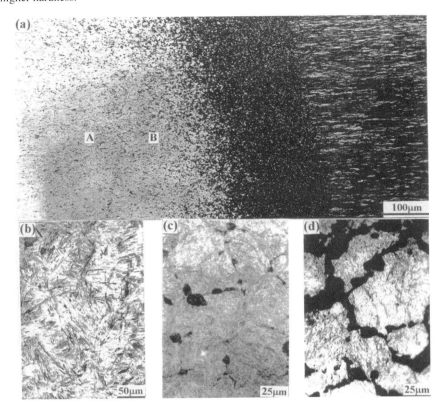

Fig.3 Micrograph of (a) longitudinal section through the surface of steel wire
(b) enlargement of area A;   (c) enlargement of area B;   (d) enlargement of area C

Table 3   Hardness of areas A, B and C and the fractured surface

	Fractured surface	area A	area B	area C
Microhardness	HV1080	HV950	HV890	HV680

SEM micrographs of the longitudinal fracture section are shown in Fig.4. Fig.4(a) shows several score marks resulting from heavy localized friction. Some melted metal drops can be seen in Fig.4(b). As indicated in Table 4, the dislocation density of the martensite in the white layer formed by friction is much higher than that of the "normally" quenched martensite, although the c/a ratio for both is over 1. Measurement of grain sizes by SEM showed that the average grain size of the white layer formed on the steel wire was 100nm.

Based on the above observations, the white layers on the steel wires are considered to be caused by rapid cooling from a temperature approaching the fusion point of the steel caused by friction under conditions of high speeds, heavy loads and short times. It can be seen from Fig.5 that the weights of both the drill rod and the drill pipes are supported by the steel wire ropes, and the weight of drill rod itself is generally over 200 tons. Also, the steel wire ropes must move up and down rapidly (the average speed is 500m/min) driven by the hoist. In this case, the temperature at the friction surface can reach more than 1000°C [19] (see Fig.4(b)) due to the high heat provided by rapid friction. There is local austenitization and this is then transformed into martensite by the rapid cooling since the localized heat can be easily conducted away by the other 125 individual wires when the steel rope moves away from the pulley [20-22]. It can be seen from Fig.3 that the microstructure consists predominately of fine martensite in areas which are close to the friction surface (such as A in Fig.3(b)). The martensite is formed following full austenitization due to the high temperature. Fine pearlite microstructures are obtained in the regions away from the friction surface because of partial austenitization.

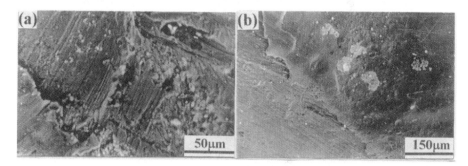

Fig. 4   SEM micrographs of longitudinal section of fractured surface of steel wires, showing (a) ploughing;   (b) melted droplets

Table 4   Dislocation density and c/a ratios of martensites

martensite	friction martensite in white layer	normally quenched martensite
a (nm)	0.2860746	0.2859578
c (nm)	0.2930554	0.2928485
c/a ratio	1.024402	1.024097
dislocation density mm^{-2}	1.21×10^{12}	9.66×10^{11}

The white layer is generally taken to refer to a hard layer formed on the steel surface. For these steel wires, the microhardness of the white layer was as high as HV1080. The "friction martensite", which appears as the white layer with characteristics of a high dislocation density (see Table 4) and a fine grain size (average grain size is 100nm), exhibits properties different from "normal" martensite. It has been shown [23-25] that the high hardness of the white layer with values in excess of those normally attainable for the steel wire composition, may be attributed to phase transformation hardening and the extremely fine grain size hardening, combined with the strengthening interaction with dislocations.

Mechanical behavior, such as fatigue and wear of high strength steel wire, are dependent on the surface roughness and microcrack growth conditions [22, 24, 26-28]. It can be seen from Fig.6 that cracks are initiated at the surface of the white layer and extend into the material from the surface. Thus, white layers are liable to fracture because they have a high hardness and are easily delaminated due to the maximum tensile residual stress which is located just beneath the the hardened layer [29].

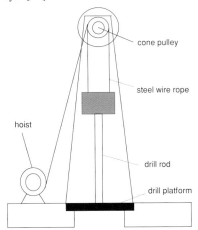

Fig. 5   Schematic diagram showing service conditions for the steel wire ropes

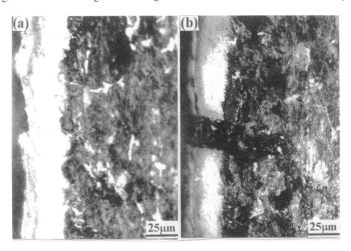

Fig.6   Micrographs showing crack (a) initiating in the white layer and (b) propagating into the SAE1065 steel

The effects of normal load, sliding speed and friction time on the depth of the white layer are shown in Figs.7 to 9.

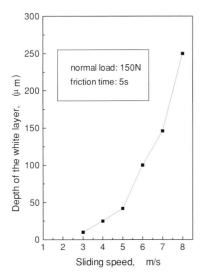

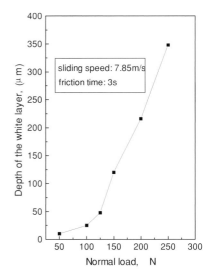

Fig.7　Effect of sliding speed on the depth of
the white layer in SAE1065 steel

Fig.8　Effect of normal load on the depth of
the white layer in SAE1065 steel

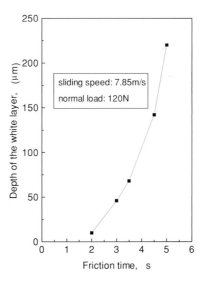

Fig.9　Effect of friction time on the depth of the white layer in SAE1065 steel

Fig.7 shows that the depth of white layer increases with increasing sliding speed for a given normal load and friction time. The same trends are seen for normal load and friction time in Fig.8 and Fig.9, respectively. When the steel wire is used for petroleum drilling, the effects on the formation and depth of white layers are additive, and due to much heavier loads and higher friction speeds than in the stimulated test, white layers are readily produced and lead to fracture.

4. Conclusions

The in-service fracture of individual wires in high strength steel wire ropes for petroleum drilling was caused by the formation of white layers. It has been demonstrated that the formation of a white layer with high microhardness relative to that of the parent material is attributed to a transformation of "friction" martensite due to the high friction heat and then rapid cooling. The "friction" martensite structure obtained in the white layer exhibits a finer grain size and higher dislocation density than that obtained by "normal" quenching. A stimulated friction test showed that the depth of the white layer increases with increase in normal load, sliding rate and friction time. It is proposed that the crack initiation and propagation occurring in the white layer results in fracture. The SAE1065 high strength steel wire ropes readily form white layers under the service conditions in petroleum drilling.

References

[1] M. D. Rogers, 1969, Metallographic Characterization of Transformation Phase on Scuffed Cast-iron Diesel Engine Components, Tribology, Vol.2, pp.123-127.

[2] W. M. Steen and C. Courtney, 1979, Surface Heat Treatment of En8 Steel Using a 2kW Continuous-Wave CO Laser, Metals Tech., Vol.6,pp.456-462.

[3] H. C. Rogers, 1979, Adiabatic Plastic Deformation, Annual Review of Materials Science,Vol.9,pp. 283-311.

[4] A. K. Gangopadhyay and J. J. Moore, 1987, The Effect of Impact on the Grinding Media and Mill Liner in a Large Semiautogeneous Mill, Wear, Vol.114, pp.249-260.

[5] Shirong Ge, 1992, Friction Coefficients Between the Steel Rope and Polymer Lining in Frictional Hoisting, Wear, Vol. 152, pp. 21-29.

[6] G. Baumann, H. J. Fecht and S. Liebelt, 1996, Formation of White-Etching Layers on Rail Treads, Wear , Vol.191, pp. 278-287.

[7] A. Vyas and M.C. Shaw, 2000, Significance of the White Layer in a Hard Turned Steel Chips, Machining Science and Technology, Vol.4, pp. 169-175.

[8] B. J. Griffiths, 1987, Mechanism of White Layers Generation with Reference to Machining and Deformation Processes, J. Tribol., Vol.109, pp. 525-530.

[9] M. A. Meyers and H. R. Pak, 1986, Observation of an Adiabatic Shear Bands in a Titanium by High-Voltage Transmission Electron Microscopy, Acta Metallurgica, Vol. 34, pp.2493-2499.

[10] Y. B. Xu, X. Wang, Z. G. Wang, L. M. Luo and Y. L. Bai, 1990, Formation and Microstructure of Localized Shear Band in a Low Carbon Steel, Scripta Metallurgica et Materialia, Vol. 24, pp. 571-576.

[11] Y. Y. Yang, H. S. Fang, Y. K. Zhang, Z. G. Yang and Z. L. Jiang, 1995, Failure Models Induced by White Layers During Impact Wear, Wear, Vol.185, pp.17-22.

[12] J. F. Archard, 1959, The Temperature of Rubbing Surfaces, Wear, Vol.2, pp. 438-455.

[13] D. M. Turley, E. D. Doyloe and L. E. Samuels, 1974, The Structure of Damaged Layers on Metals, Pro. Int. Conf. Metal. Eng., Tokyo, Part2, pp.142-147.

[14] B. Zhang and W. Shen, 1997, Microstructure of Surface White Layer and Internal White Adiabatic Shear Band, Wear, Vol. 211, pp. 164-168.

[15] Y. K. Chou and C. J. Evans, 1999, White Layers and Thermal Modeling of Hard Turned Surface, International Journal of Machine Tools and Manufacture, Vol. 39, pp.1863-1881.

[16] J. W. Stead, 1912, West. Scot. Iron & Steel Inst., Vol.19, pp.169-204.

[17] Z. Zhao, D. O. Northwood, C. Liu and Y. Liu, 1999, A New Method for Improving the Resistance of High Strength Steel Wires to Room Temperature Creep and Low Cycle Fatigue, Journal of Materials Processing Technology, Vol.89-90, pp.569-573.

[18] American Petroleum Institute (API), 1995, API-9A Steel Wire Rope Standard, API Spec., pp.1-10.

[19] H. S. Fang, 1997, New Development of Mn-B Bainitic Steel, Heat Treat of Metal (Chinese), Vol.4, pp.9-12.

[20] W. J. Tomlinson, L. A. Blunt and S. Spragett, 1991, Effect of Workpiece Speed and Grinding Wheel Condition on the Thickness of White Layers in EN.24 Ground Surface, Journal of Materials Processing Technology, Vol.25, pp105-110.

[21] H. Winter, G.Knauer and J. J. Grabel, 1990, The "White Layer " Phenomenon on Gear Teeth, Tech. Mecc., Vol.21, No.1, pp.144-150.

[22] Y. X. Liu, Z.B. Zhao, C. T. Ji and Q. H. Zhu, 1995, A Study on the Mechanism of Single Wire Fracture in Steel Wire Rope for Oil Drilling, Journal of Steel Wire Products, Vol.21, pp.8-11.

[23] L. Xu and N. F. Kennon, 1992, Formation of White Layer During Laboratory Abrasive Wear Testing of Ferrous Alloys, Materials Forum, Vol.16, pp.43-49.

[24] C. Liu, 1998, A New 20Mn2WNbB Steel Used in Steel Wire Ropes for Petroleum Drilling, Ph.D Thesis, Harbin Institute of Technology, China, pp.39-59.

[25] R. Bulpett, T. S. Eyre and B.Ralph, 1993, The Characterization of White Layers Formed on Digger Teeth, Wear, Vol.162-164, pp.1059-1065.

[26] A. N Sinha and V. Rao, 1995, Premature Failure of SS Tensing Rope-An Investigation, Wire Industry, Vol.62, pp.734-736.

[27] V. A. Dikshit, 1992, Rolling Contact Fatigue Behavior of Pearlitic Rail Steels, Dissertation Abstract International, Vol.53, No.4, pp.276-279.

[28] B. A. Miller, 2000, Failure Analysis of Wire Ropes, Advanced Materials & Processes, Vol.157, pp.43-46.

[29] K. Weiss, V. Rudney, R. Cook, D. Loveless and M. Black, 1999, Induction Tempering of Steel, Advanced Materials & Processes, Vol.156, pp.19-23.